Paleobotany, Paleoecology, and Evolution

Paleobotany, Paleoecology, and Evolution

Editor:

Linda Lait

ArclerPress

www.arclerpress.com

Paleobotany, Paleoecology, and Evolution

Editor: Linda Lait

708 3rd Avenue, 6th Floor
New York
NY 10017
United States of America
www.arclerpress.com

ISBN: 978-1-68094-518-8

Library of Congress Control Number: 2016951373

Arcler Press LLC publishes wide variety of books and eBooks. For more information about Arcler Press and its products, visit our website at www.arclerpress.com

About the Editor

Linda Lait

Linda obtained her MSc from the University of Lethbridge in 2011 and will be completing her PhD at Memorial University of Newfoundland in 2016. Her interests include the evolutionary history of species, particularly those of conservation concern, which she studies with molecular markers. Increasingly complex questions can be answered with the improved technologies that are emerging. She is currently finishing up her PhD in molecular genetics, and will be starting as a Postdoctoral Fellow studying phylogeography and molecular evolution in insects.

Contents

List of Contributors

Franco Biondi
DendroLab, Graduate Program of Ecology, Evolution and Conservation Biology, University of Nevada, Reno, NV, USA

Lynn L Anderson-Carpenter
Department of Biological Sciences, University of Notre Dame

Jason S McLachlan
Department of Biological Sciences, University of Notre Dame

Stephen T Jackson
Department of Botany and Program in Ecology, University of Wyoming

Melanie Kuch
McMaster Ancient DNA Centre, Department of Anthropology and Biology, McMaster University

Candice Y Lumibao
Department of Biological Sciences, University of Notre Dame

Hendrik N Poinar
McMaster Ancient DNA Centre, Department of Anthropology and Biology, McMaster University

Mike Viney
Poudre School District, Ft. Collins, CO 80526, USA
College of Natural Sciences Education and Outreach Center, Colorado State University, Fort Collins, CO 80526, USA

Dagmar Dietrich
Professur Werkstoff- und Oberflächentechnik, Technische Universität Chemnitz, Chemnitz 09107, Germany

George Mustoe
Geology Department, Western Washington University, Bellingham, WA 98225, USA

Paul Link
Department of Geosciences, Idaho State University, Pocatello, ID 832091, USA

Thomas Lampke
Professur Werkstoff- und Oberflächentechnik, Technische Universität Chemnitz, Chemnitz 09107, Germany

Jens Götze
Institut für Mineralogie, Technische Universität Bergakademie Freiberg, Freiberg 09599, Germany

Ronny Rößler
Museum für Naturkunde Chemnitz, Chemnitz 09111, Germany
Institut für Geologie, Technische Universität Bergakademie Freiberg 09599, Germany

Virginia Iglesias
Montana Institute on Ecosystems, Montana State University, Bozeman, MT, USA

Gabriel I. Yospin
Montana Institute on Ecosystems, Montana State University, Bozeman, MT, USA

Cathy Whitlock
Montana Institute on Ecosystems, Montana State University, Bozeman, MT, USA
Department of Earth Sciences, Montana State University, Bozeman, MT, USA

Mark P. Witton, Darren Naish
School of Earth and Environmental Sciences, University of Portsmouth, Portsmouth, United Kingdom

Gabriel S. Ferreira
Laboratorio de Paleontologia de Ribeir ´ ao Preto, FFCLRP, Universidade de S ˜ ao Paulo, ˜ Ribeirao Preto, SP, Brazil ˜

Ascanio D. Rincón
Laboratorio de Paleontolog´ıa, Centro de Ecolog´ıa, Instituto Venezolano de Investigaciones Cient´ıfcas (IVIC), Caracas, Venezuela

Andrés Solórzano
Laboratorio de Paleontolog´ıa, Centro de Ecolog´ıa, Instituto Venezolano de Investigaciones Cient´ıfcas (IVIC), Caracas, Venezuela

Max C. Langer
Laboratorio de Paleontologia de Ribeir ´ ao Preto, FFCLRP, Universidade de S ˜ ao Paulo, ˜ Ribeirao Preto, SP, Brazil ˜

Laura Domingo
Departamento de Geología Sedimentaria y Cambio Medioambiental, Instituto de Geociencias IGEO-CSIC-UCM, Madrid, Spain,
Departamento de Paleontología, Universidad Complutense de Madrid, Madrid, Spain,
Earth and Planetary Sciences Department, University of California Santa Cruz, Santa Cruz, California, United States of America,

Fernando Barroso-Barcenilla
Departamento de Paleontología, Universidad Complutense de Madrid, Madrid, Spain,
Grupo de Investigación IberCreta, Departamento de Geología y Geografía, Universidad de Alcalá de Henares, Alcalá de Henares, Spain,

Oscar Cambra-Moo
Laboratorio de Poblaciones del Pasado, Departamento de Biología, Facultad de Ciencias, Universidad Autónoma de Madrid, Madrid, Spain

Steffen Kiel
Georg-August-Universita¨t Go¨ttingen, Geowissenschaftliches Zentrum, Abteilung Geobiologie, Go¨ttingen, Germany,

Johannes Glodny
Deutsches GeoForschungsZentrum GFZ, Sektion 4.2, Anorganische und Isotopengeochemie, Telegrafenberg, Potsdam, Germany,

Daniel Birgel
Universita¨t Wien, Erdwissenschaftliches Zentrum, Department fu¨r Geodynamik und Sedimentologie, Wien, Austria,

Luc G. Bulot
FRE CNRS 2761, Centre de Se´dimentologie-Pale´ontologie, Universite´ de Provence, Marseille, France,

Kathleen A. Campbell
University of Auckland, Geology Programme, School of Environment Science, Auckland, New Zealand,

Christian Gaillard
Universite´ de Lyon-1, UMR CNRS 5125 Pale´oenvironnements et Pale´obiosphe`re, Villeurbanne, France,

Roberto Graziano
Dipartimento di Scienze della Terra, dell'Ambiente e delle Risorse, Universita` di Napoli Federico II, Largo S. Marcellino, Napoli, Italia,

Andrzej Kaim
Instytut Paleobiologii PAN, Warszawa, Poland,

Iuliana Laza˘ r
University of Bucharest, Faculty of Geology and Geophysics, Department of Geology, Bucharest, Romania,

Michael R. Sandy
University of Dayton, Department of Geology, Dayton, Ohio, United States of America

Jo¨ rn Peckmann
Universita¨t Wien, Erdwissenschaftliches Zentrum, Department fu¨r Geodynamik und Sedimentologie, Wien, Austria,

Antonella Cinzia Marra
Department of Physics and of Earth Sciences, University of Messina, Viale Ferdinando Stagno d'Alcontres 31, Messina 98166, Italy

Robert W. Boessenecker
Department of Geology, University of Otago, Dunedin, Otago, New Zealand,
University of California Museum of Paleontology, Berkeley, California, United States of America,

Morgan Churchill
Department of Geology and Geophysics, University of Wyoming, Laramie, Wyoming, United States of America,
Program in Ecology, University of Wyoming, Laramie, Wyoming, United States of America

Ryosuke Motani
Department of Earth and Planetary Sciences, University of California Davis, Davis, California, United States of America,

Cheng Ji
Department of Geology and Geological Museum, Peking University, Beijing, China,

Taketeru Tomita
Department of Earth and Planetary Sciences, University of California Davis, Davis, California, United States of America,
Hokkaido University Museum, Hakodate, Japan,

Neil Kelley
Department of Earth and Planetary Sciences, University of California Davis, Davis, California, United States of America,

Erin Maxwell
Pala¨ontologisches Institut und Museum, Universita¨t Zu¨rich, Zu¨rich, Switzerland,

Da-yong Jiang
Department of Geology and Geological Museum, Peking University, Beijing, China,

Paul Martin Sander
Steinmann Institute, Division of Palaeontology, University of Bonn, Bonn, Germany

Christopher R. Noto
Department of Biomedical Sciences, Grand Valley State University, Allendale, Michigan, United States of America,

Ari Grossman
Department of Anatomy, Midwestern University, Glendale, Arizona, United States of America

John E McCormack
Department of Ecology and Evolutionary Biology, University of California, Los Angeles
Center for Tropical Research, Institute of the Environment, University of California, Los Angeles

Bonnie S Bowen
Department of Ecology, Evolution, and Organismal Biology, Iowa State University

Thomas B Smith
Department of Ecology and Evolutionary Biology, University of California, Los Angeles
Center for Tropical Research, Institute of the Environment, University of California, Los Angeles

Aldo Piombino
Earth Sciences Department, Florence University, 50125 Florence, Italy

Neil Brocklehurst
Museum fu¨r Naturkunde, Leibniz-Institut fu¨r Evolutions und Biodiversita¨tsforschung, Berlin, Germany,

Paul Upchurch
Department of Earth Sciences, UCL, London, United Kingdom,

Philip D. Mannion
Museum fu¨r Naturkunde, Leibniz-Institut fu¨r Evolutions und Biodiversita¨tsforschung, Berlin, Germany,
Department of Earth Science and Engineering, Imperial College London, London, United Kingdom,

Jingmai O'Connor
Institute of Vertebrate Paleontology and Paleoanthropology, Beijing, China,
Dinosaur Institute, Natural History Museum of LA County, Los Angeles, California, United States of America

S. Ivan Perez
Divisio´n Antropologı´a at Museo de La Plata, CONICET, La Plata, Buenos Aires Province, Argentina,

Marcelo F. Tejedor
Centro Nacional Patago´nico, CONICET, Puerto Madryn, Chubut Province, Argentina

Nelson M. Novo
Centro Nacional Patago´nico, CONICET, Puerto Madryn, Chubut Province, Argentina

Leandro Aristide
Divisio´n Antropologı´a at Museo de La Plata, CONICET, La Plata, Buenos Aires Province, Argentina,

Martı´n D. Ezcurra
School of Geography, Earth and Environmental Sciences, University of Birmingham, Edgbaston, Birmingham, United Kingdom,
GeoBio-Center, Ludwig-MaximilianUniversita¨t Mu¨nchen, Munich, Germany,

Torsten M. Scheyer
Pala¨ontologisches Institut und Museum, Universita¨t Zu¨rich, Zurich, Switzerland

Richard J. Butler
School of Geography, Earth and Environmental Sciences, University of Birmingham, Edgbaston, Birmingham, United Kingdom, GeoBio-Center, Ludwig-MaximilianUniversita¨t Mu¨nchen, Munich, Germany,

Mark A. Elgar
School of BioSciences, University of Melbourne, Melbourne, VIC, Australia

Preface

The fields of paleobotany, paleoecology, and evolution all aim to uncover the environment of the past, and how we arrived at the state we are today. Paleobotany uses the fossils of both terrestrial and aquatic plants to study prehistoric environments and the evolution of organisms. Paleobotany is the scientific study of ancient plants, using plant fossils found in sedimentary rocks. These fossils can be impressions or compressions of the plants left on the rock's surface, or "petrified" objects, such as wood, which preserve the original plant material in rocklike form. Still other specimens are found in calcified lumps called coal balls, so named because they are usually found in or near coal deposits. As most organisms decompose rapidly after death, their preservation in nature is a rare event. Most individuals are not represented in the fossil record, and even many species that must have existed have vanished without a trace. As a branch of botany, paleobotany is of importance primarily because the record of fossil plants helps scientists understand the long process of plant evolution. Especially since the 1940's, fossil evidence has helped to explain the origin of major classes of organisms, such as algae and fungi. Paleoecology is the scientific study of past environments. Paleoecologists are interested in the ecosystem as a whole and derive their understanding of past environments from different lines of evidence, including fossil plants and animals, ancient soils and rocks. This field of study is important for anyone interested in past organisms because it provides the context for understanding the origin, extinction and adaptation of any particular organism. A related discipline, palynology, uses fossilized spores and pollen to address similar questions. Paleoecology is more diverse, looking at both plant and animal fossils to try and discover how species evolved and what the environmental conditions were like. There are a number of techniques that are often applied in paleobiology: morphological analysis has long been a commonly used tool in fossil and evolutionary studies; molecular methods such as the extraction of ancient DNA is becoming increasingly common as techniques are improving; chemical analysis such as the study of stable isotopes can provide information on the movement of a species as well as the environment as it changed through time. It describes and explains the origin and evolution of plants as revealed by the fossil record. Paleobotany, Paleoecology, and Evolution aims to examine the role of paleoecology in constructing evolutionary explanations for taxic-phylogenetic, morphological, and biogeographic patterns. We will advocate greater awareness of the vast amount of accessible ecological data and incorporation of ecological thinking into paleobotanical scenario-building. We begin this book with a discussion of the subdiscipline of paleoecology, and some of the challenges it is faced with in regards to the typical scientific method. We continue with a number of that use paleobotanical and paleoecological methods to study questions of evolution in plants and animals. At the end we discuss the evolution of social behavior, and how this fits with Darwin's original theories. It highlights grand challenges in paleoecology focuses on paleoecology and genetics.

Editor
Linda Lait

INTRODUCTION

Paleoecology (also spelt paleoecology) uses data from fossils and sub fossils to reconstruct the ecosystems of the past. It involves the study of fossil organisms and their associated remains (e.g. shells, teeth, pollen, and seeds), which can be used to interpret their life cycle, living interactions, natural environment, communities, and manner of death and burial. Such interpretations aid the reconstruction of past environments (i.e. paleoenvironment).

The fossil record has been studied to try to clarify the relationship animals have to their environment, in part to help understand the current state of biodiversity. A close link has been found between vertebrate taxonomic and ecological diversity, that is, the diversity of animals and the niches they occupy.

Reconstruction and measurement

Zygospira modesta, atrypid brachiopods, preserved in their original positions on a trepostome bryozoan; Cincinnatian (Upper Ordovician) of southeastern Indiana.

The aim of paleoecology is therefore to build the most detailed model possible of the life environment of previously living organisms found today as fossils. Such reconstruction takes into consideration complex interactions among environmental factors such as temperatures, food supplies, and degree of solar illumination. Often much of this information is lost or distorted by the fossilization process or diagenesis of the enclosing sediments, making interpretation difficult.

The environmental complexity factor is normally tackled through statistical analysis of the available numerical data (quantitative paleontology or paleostatistics), while the study of post-mortem processes is known as the field of taphonomy.

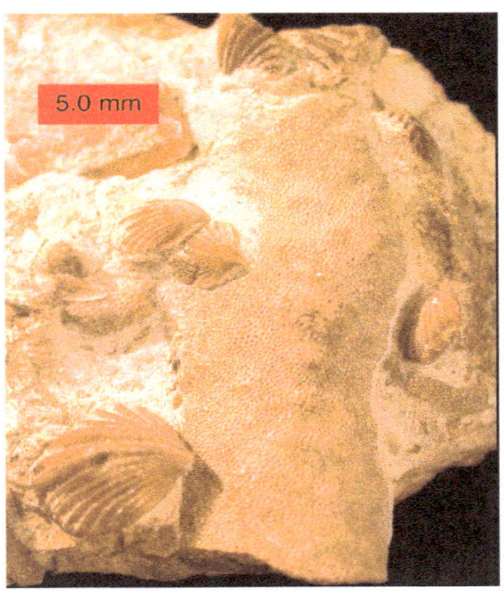

Quaternary

Much of the original paleoecological research has focused on the last two million years (the Quaternary period), because older environments are less well represented in the fossil timeline of evolution. Indeed, many studies concentrate on the Holocene epoch (the last 11,500 years), or the last glacial stage of the Pleistocene epoch (the Wisconsin/Weichsel/Devensian/Würm glaciation of the ice age, from 50,000 to 11,500 years ago). Such studies are useful for understanding the dynamics of ecosystem change and for reconstructing pre-industrialization ecosystems. Many public policydecision-makers have pointed to the importance of using palaeoecological studies as a basis for choices made in conservation ecology.

Understanding ancient environments

Paleobotany is the scientific study of ancient plants, using plant fossils found in sedimentary rocks. These fossils can be impressions or compressions of the plants left on the rock's surface, or "petrified" objects, such as wood, which preserve the original plant material in rocklike form. Still other specimens are found in calcified lumps called coal balls, so named because they are usually found in or near coal deposits.

Paleoecology is the scientific study of past environments. Paleoecologists are interested in the ecosystem as a whole and derive their understanding of past environments from different lines of evidence, including fossil plants and animals, ancient soils and rocks. This field of study is important for anyone interested in past organisms because it provides the context for understanding the origin, extinction and adaptation of any particular organism. The Department of Paleobotany & Paleoecology at the Cleveland Museum of Natural History houses fossil plants from around the world, from the Precambrian (more than 542 million years ago) to the Pleistocene (as recently as about 10,000 years ago), sampling almost the entirety of the evolutionary history of plants during that span. It is also home to one of the world's largest collections of modern African pollen, with more than 35,000 slides. Research in the department focuses on a global view of paleoecology and plant evolution, with emphasis on the interactions between biotic and abiotic elements of an ecosystem, and how these interactions might have impacted the evolution of mammalian lineages within that ecosystem.

Paleobotany and paleoecology are related fields of study into ecological conditions of the past.

Ecology of prehistoric times, extending from about 10,000 to about 3.5 × 109 years ago. Although the principles of paleoecology are the same as those underlying modern ecology, the two fields actually differ greatly. Paleoecology is a historical science that must rely on empirical data from fossils and their enclosing sedimentary rocks to make inferences about past conditions. Experimental approaches and direct measurement of environmental parameters, which are critical components of modern ecology, are generally impossible in paleoecology. Furthermore, distortion and loss of information during fossilization means that fossil assemblages and distributions are rarely congruent with living communities. Hence, the resolution of ancient ecosystems must remain relatively imprecise. The lack of precision is compensated for by the fact that paleoecology deals with processes occurring over vast spans of time that are unavailable to modern ecology. Long-term changes in communities (replacement) may be discerned and related to patterns of environmental change. More significantly, overall patterns of ecological change in the global biosphere may be documented; evolutionary paleoecology focuses on recognition and interpretation of long-term ecological trends that have been critical in shaping evolution. Among the goals of paleoecology are the reconstruction of ancient environments (primarily depositional environments), the inference of modes of life for ancient organisms from fossils, the recognition of recurring groupings of ancient organisms that define relicts of communities (paleocommunities), the reconstruction of the interactions of organisms with their environments and with each other, and the documentation of large-scale and long-term patterns of stasis or change in ecosystems.

Taxonomic uniformitarianism

One of the most useful, but also potentially misused, aspects of paleoecological application is known as taxonomic uniformitarianism. This concept relies on studies of modern organisms to determine limiting environmental factors, such as salinity tolerance, temperature preference, or depth ranges. Fossils of the same or closely related species are then inferred to have had similar environmental preferences, and their occurrence is judged to indicate that particular strata were deposited under a comparable range of environmental conditions. Such an approach is valid for very closely related organisms in relatively recent geologic time. Species and even genera may have relatively uniform environmental ranges through time, but the same cannot always be said of higher taxa such as families. At the level of order or class, only the broadest uniformitarian generalizations apply. For example, it is probably valid to consider fossil nautiloids or echinoderms as indicators of normal marine salinities, as all living representatives of these taxa have very limited abilities for osmotic regulation and therefore are restricted to near-normal salinity. Similarly, the restriction of photosynthetic organisms (such as algae) to the euphotic zone may be useful in determining relative depth. However, the precision and reliability of taxonomic uniformitarianism breaks down in increasingly ancient samples.

Paleoautecology

Paleoautecology, the interpretation of modes of life (broadly, niches) of ancient organisms, involves a multidisciplinary approach. Although ancient modes of life cannot be determined completely, paleoecologists can often assign fossils to generalized guilds in terms of types of feeding, substrate preference, and degree of activity. A thorough understanding of the biology of closest modern analogs is particularly important in any attempt to reconstruct paleoautecology. If the species or a closely related species is extant, then its mode of life, general physiology, and even behavior can be inferred with some confidence through the use of taxonomic uniformitarianism, provided that the biology of living relatives is well understood. "Living fossils," or relict extant taxa, such as Nautilus, sclerosponges, horseshoe crabs, and modern stalked crinoids provide valuable clues for interpreting the paleobiology of extinct organisms.

Paleoecological Methods in Paleoanthropology

Paleoanthropology is a scientific discipline that receives a substantial amount of media attention, especially when new hominin species are discovered. Scientists as well as the public in general are interested in human evolution. While paleoanthropological research is scientific, there are aspects of the discipline that are emotionally charged. The stakes seem high and tempers run short because we are writing the story of ourselves, not just any old mammal. Debates about which species belong to the hominin lineage and which do not will always be passionate. Recent fossil hominin discoveries illustrate this point. Debates continue to rage because unusual morphological characters present in these hominins leave workers with many questions. Why does a bipedal Ardipithecus ramidus also have a divergent big toe? Why are Paranthropus molars so large and why did they need such strong chewing muscles? These questions, among many others, are complex, and workers today are only just beginning to answer these questions. One of the ways to continue shedding light on these topics is to better understand the selective pressures under which these species evolved. Since no species can live outside of ecological pressures, discovering factors that were influencing early hominin evolution can illuminate what caused the human lineage to evolve, as well as why certain morphological characters are seen at certain times in the fossil record. There now is a growing consensus in paleoanthropology that paleoecological studies are integral to understanding hominin evolution. Thus, Early Hominin Paleoecology, edited by Matt Sponheimer, Julia Lee-Thorp, Kaye Reed, and Peter Ungar, is a timely volume. It provides critical details about the current state of the field of paleoecology, particularly how paleoecological methods can be applied to paleoanthropology, and how those methods can help workers understand additional details about early hominin evolution. It is a high-quality book, great for advanced students and professionals interested in current methods in paleoecology. This volume has three sections: paleoclimate and paleoenvironment, hominin adaptations and behavior, and analogues and models. The paleoclimate and paleoenviornment section focuses on paleoecological methods that utilize faunal assemblages, facies analyses, stable isotope analyses, and global climate indicators. One contribution by Kaye Reed, Lillian Spencer, and Amy Rector evaluates a range of methods that can be used to analyze faunal assemblages from paleontological or archaeological sites. They examine the differences between taxonomic and taxon-free methods and highlight the biases inherent in these methods. The hominin adaptation and behavior section focuses on hominin dietary adaptations. Articles in this section examine methods ranging from isotope and trace element analyses, micro wear analyses, and functional morphology. Specifically, one article from Mark Teaford, Peter Ungar, and Frederick Grine evaluates the fundamental principles of dental micro wear analyses and explains how those principles can be applied to paleontological and archaeological samples. The authors also explain the importance of comparing dental microwear patterns of living animals, both in the lab and in the wild, to what is seen in ancient assemblages. Additionally, David Braun discusses some methods that archaeologists use to infer hominin behaviors from the archaeological record. He details the importance of examining artifacts and bones found at archaeological sites and the methods workers can use to infer details about early hominin behavior. Lastly, the analogues and models section focuses on the use and misuse of models in current paleoanthropological research. For example, John Mitani focuses on wild chimpanzee behavior models and how those models can be incorrectly applied. He also highlights how some workers who use wild chimpanzee behavior as a model for early hominine behavior use incorrect or outdated data. Additionally, Jeanne Sept focuses on methods used for paleobotanical reconstructions, where she explains that the structure of many plants has remained stable over the last few million years and that this allows workers to use plant remains as a model for building an environmental reconstruction. All of the contributors to Early Hominin Paleoecology focus on how to correctly use the method they highlight, the biases or assumptions inherent in that method, the ways that methods have been used incorrectly, as well suggesting directions for future research. For this reason, this is an invaluable volume for any person interested in conducting research within the field of paleoecology, whether they are a student or professional. The editors of this volume write in the introduction that paleoecology is a dynamic and increasingly specialized field. Because of this increasing specialization, it is impossible for a single researcher to be an expert in all of these methods. Thus, volumes such as this are essential for those conducting research, as well as anyone who wishes to have a coherent understanding of the state of the discipline at large. The editors of this volume have shown how critical it is for workers to communicate with one another in a way that others outside of their specialization can understand. They also show that it is possible for highly specialized scientists to work together to highlight important details about hominin evolution. Paleoecology has become an integral aspect of paleoanthropological research. As more hominins are added to the growing fossil record, the number of studies conducting paleoecological research will only increase. Thus, as new hominin species are discovered and new data is published, debates about hominin evolutionary history will continue. These new discoveries will be published in many different media outlets, not just scientific journals. Publications in the mainstream media make the debates about hominin evolutionary history accessible to not just scientists, but to the general public as well. It is important that human evolution is shared with the general public, not just with scientists within increasingly specialized disciplines, because human evolution is the story of all people. Because new hominin discoveries cause such intense debates in the field, it is incredibly important that workers today strive to discern more details about the paleoecological pressures that influenced the human lineage. It is only by discovering these paleoecological pressures that unusual morphological characters will begin to make more sense. It is for that reason that this volume, Early Hominin Paleoecology, is an extremely useful and timely volume for advanced students and professionals interested in learning more about hominin paleobiology and frequently used paleoecological methods.

Combining Morphometric and Paleoecological Analyses

The fossil record clearly documents both changes in taxonomic lineages and community composition. However, it is unclear how closely related the processes and patterns of species evolution and community changes may be. The evidence to suggest that the role of biotic interactions shape the evolutionary paths of lineages remains controversial. Among neontologists, however, local ecological interactions between species have been cited as a major influence on speciation and evolution. The geographic mosaic theory of co-evolution has provided a new way of considering the role of local ecology and biotic interactions in allopatric speciation processes and evolutionary patterns. For example, in examining the interactions between a species of moth and a plant in western US states, Thompson documented that there existed mutualistic, commensalistic, or even antagonistic relationships between the two species depending on the presence or absence of other insect species. These differences in biotic interactions can be correlated with different genetic populations and thus, can drive speciation through time. Another implication of these observations and interpretations is that within small geographic areas, "Interactions may show selection mosaics, such that different traits and outcomes are favored by natural selection in different communities". In a paleontological context, the geographic mosaic theory of co-evolution suggests that paleocommunity composition and biotic interactions may vary geographically and temporally even at small scales, and that these differences may result in evolutionary changes. This theoretical framework may, therefore, bridge the fields of evolutionary biology and paleoecology as well as neoecology and population genetics. In this paper, we explore the potential of using quantitative morphometric and paleocommunity composition data to examine whether correlations exist between communitylevel and species-level patterns. Quantitative morphometrics allows for the evaluation of specieslevel microevolutionary changes whereas the analysis of paleocommunity composition provides a first-level proxy for evaluating biotic interactions and ecological fluctuations. If evolutionary changes within lineages as evidenced by morphological change are concurrent with major changes in community structure, it would allow us to seriously reconsider the role of biotic interactions in determining patterns of species evolution in the fossil record. In order to investigate the relationship between ecological and evolutionary processes, one must have appropriate depositional systems and fossil samples which will allow for robust quantitative analyses. One such depositional system is in the Neogene marine sequences of the Cibao Valley of the northern Dominican Republic. This system has become a major focal point of many important paleobiological studies for many good reasons. The fossils have been studied since the mid 1800s with much modern study based on collections made by E. and H. Vokes and the Naturhistorisches Museum of Basel, Switzerland, since the 1970s. The marine sequences exposed in river gorges in the north central highlands are very fossiliferous and are believed to represent fairly continuous deposition in a gradually deepening basin. Many paleontologists around the world have been involved in examining the Dominican Republic Neogene fauna. Experts in taxonomic groups have systematically surveyed the samples collected by the Naturhistorisches Museum of Basel (many monographs have been published in Bulletins of American Paleontology). Given the long history of taxonomic study, the alpha taxonomy and basic identification of fossils found in these rocks is comparatively good even if higher level taxonomic reorganizations have not yet been completed for most of the groups. In addition, paleobiologists have long used this system to examine patterns of evolution, including documenting rates of change and investigating the role of heterochrony in speciation processes.

CHAPTER 1 PALEOECOLOGY GRAND CHALLENGE

Franco Biondi

Keywords: Paleoecology, retrospective studies, proxy interpretations, paleoscience, paleoclimatology

Exactly 50 years ago, in 1964, a landmark article whose two-word title, "Strong inference," revealed little and evoked plenty, appeared in the hallowed pages of *Science* (Platt, 1964). Simply stated, Platt›s view of scientific research in its highest form proceeds through three steps: development of alternative hypotheses; design of crucial experiments to exclude one or more hypotheses; execution of careful studies to obtain clear-cut results. It was a powerful reminder of how problem solving can be performed, written in a style both informal and authoritative. In those days it was not unusual for the most prestigious scientific outlets to publish articles whose language seems to nowadays have been relegated to blog posts. Just seven years earlier, another *Science* paper had appeared with a whimsical title, "Biological clock in the unicorn" (Cole, 1957), addressing the excessive reliance on statistical analysis of cycles and correlations that was plaguing the biologists of the time. Among the many citable passages in either one of those articles, one may dwell on a segment in Platt›s text: "Whether it is hand-waving or number-waving or equation-waving, a theory is not a theory unless it can be disproved. That is, unless it can be falsified by some possible experimental outcome." Literal interpretations of such statements, especially when translated to all possible scientific research, are problematic, and Platt›s impact is best described as "more an inspirational tract than the development of a formal scientific methodology" (Davis, 2006). It is in that vein that I decided to open this Grand Challenge editorial by evoking a paper from half a century earlier. Paleoecology, as any other branch of retrospective science, does not easily lend itself to experimentation. Not because of any fault of its own, but simply because there is no experimental outcome capable of reversing the unidirectionality of time. Elsewhere (Biondi, 2013) I have explored the tension created by the imperative to understand ecosystem legacies and past environmental drivers against the impossibility of distinguishing between true and false when dealing with the past. As it turned out, I was partly rephrasing the meaning of "historicity," i.e., "the necessity of working with complex and unique events in time" (Gould, 1980). In the same vein, but even earlier, Nagel (1952) had discussed in detail the distinction and connections between the "nomothetic" aspects of science, "which seek to establish abstract general laws for indefinitely repeatable processes," and the "idiographic" ones, "which aim to understand the unique and nonrecurrent." These issues cut across the timescales used by Jackson (2001) to distinguish between "real time" (for modern ecology), "Q-time" (for Quaternary ecology), and "deep time" (for paleobiology and paleontology, which is the subject of another section of this journal).

Prescribing, even more than describing, the type of research that this journal section is aiming to attract and publish requires some epistemological considerations. Science is a method for reducing uncertainty (Lindley, 2006): thus, paleoecology could be considered a method for reducing uncertainty about past ecological patterns and processes. A subdivision into *preliminary, exploratory*, and *confirmatory* research has been proposed (Flueck and Brown, 1993), although the last term suggests an inherent bias, as even "validation" and "verification" are not truly possible when dealing with open systems (Oreskes et al., 1994). Choice of words notwithstanding, the most productive approach to science was demonstrated recently by the insightful views on *exploratory*vs. *confirmatory* research made by Wagenmakers et al. (2012) with regard to psychology—another discipline where experimentation is rarely an option. What they define as *confirmatory* is in fact the type of rigorous scientific inquiry that begins with *a priori* definitions of expected outcomes.

Citation: Franco Biondi, Paleoecology grand challenge, http://dx.doi.org/10.3389/fevo.2014.00050.

This practice is diametrically opposed to *a posteriori* analyses and rationalizations that are inevitably more influenced by investigator's bias. Their language is explicit, as in stating that "In fact, the advice to torture the data until they confess is not wrong—just as long as this torture is clearly acknowledged in the research report."

For clarity, I have found useful to categorize studies as (a) *experimental*, which are performed under controlled conditions in a laboratory or in the field with random assignment of treatments to subjects; (b) *observational*, when conditions are not entirely controlled or the assignment of treatments to subjects is not random; (c) *iterative*, whereby no treatments are possible, but a specific question is asked *a priori* and constantly refined by comparing predictions with outcomes (e.g., forecasting, opinion polls); (d) *monitoring*, when questions may not exist or are asked *a posteriori* (e.g., surveys, inventories, proxy reconstructions, data mining, remote sensing, cartography, automated recording). It is better to envisage these categories as part of a continuum rather than as discrete entities, so that at one end of the spectrum are experiments, and at the other end are observations designed to obtain descriptive measures of natural phenomena without including control units or manipulation of any sort. All research activities, including those labeled as "monitoring," can lead to scientific discoveries, as repeatedly shown in the atmospheric sciences (e.g., Keeling, 1960), forestry (e.g., Schreuder and Thomas, 1991), and epidemiology (e.g., Goldstein and Goldstein, 1984).

A large body of literature on what separates an experimental from an observational study can be found in the medical field. A typical example is the link between smoking and disease, which could not be established experimentally. In observational studies "the investigator is restricted to taking selected observations or measurements that seem appropriate for the objectives, either by gathering new data or using those already collected by someone else" (Cochran, 1983; p. 2). Cochran›s book, which was published posthumously and mostly written before 1980, is widely recognized as a classic reference. Estimates from an observational study are likely to be biased, not because of the investigator›s shortcomings but because of potential differences in starting conditions, which make responses not comparable. Such confounding can easily mislead researchers, as shown by Simpson›s paradox, whereby a medical treatment that is judged beneficial for all is also detrimental for both men and women (Lindley, 2006). A prominent role is therefore assigned to minimizing bias using three general methods: (1) refinement of techniques through devices, (2) blocking and matching, and (3) statistical adjustments (Cochran, 1983). Updated monographs on this topic (Rosenbaum, 2002, 2010) have emphasized that without treatments neither experimental nor observational studies can be performed. A list of procedures for improving research quality was provided by Rosenbaum (2010; pp. 5–7), but studies of past ecological changes are rarely able to meet these expectations.

In fact, all retrospective investigations fall in the "monitoring" category mentioned above because of the impossibility of either applying specific treatments, thus preventing experimental or observational investigations, or of using future outcomes to measure model skill, hence excluding an iterative approach. Although not necessarily a hierarchy for discovery potential, the study categories mentioned above are still in descending order for likelihood of raucous debate. As we transition from experiments to, for instance, proxy reconstructions of climate, there is a greater and greater chance that studies addressing the same questions but conducted by different investigators will provide conflicting results. Clever strategies have been proposed to circumvent and ameliorate some of these issues in an earth system science (ESS) modeling framework (e.g., Hughes and Ammann, 2009). On the other hand, any working scientist is well aware of the fact that results based on statistical correlations or simulation models, however complex they may be, are not necessarily harbingers of cause-effect relationships. In Fanelli (2010)›s words, "the nature of hypotheses tested and the logical and methodological rigor employed to test them vary systematically across disciplines and fields, depending on the complexity of the subject matter and possibly other factors (e.g., a field›s level of historical and/or intellectual development)."

The purposely bare-bone taxonomy of scientific studies mentioned above has a biological perspective, but in a geophysics framework one would be confronted with the "forward" and "inverse" problem of linking data with underlying processes (Snieder and Trampert, 1999). A similar concept emerges when using a "state-space" approach, whereby "hidden" or "latent" variables are responsible for generating the observations (Harvey, 1990). Paleo reconstructions, and in general any proxy-record study, can be considered an inverse problem because incomplete information is used to infer the underlying parameters and processes (the "true" model) that generated them. However, it is an *ill-posed* inverse problem, because it has no unique solution, and small errors in the data cause large changes in the inferred explanation (King, 1997). The various assumptions made by the investigator to make sampling decisions, to account for measurement errors, to select theoretical probability distributions, to perform data reduction steps, etc. are needed to approximate a solution, but different approaches and options easily lead to diverging results (Biondi, 2014). The mission statement of the Paleoecology section outlines research topics that are pressing and essential in scope, but the major challenge for the field may be to remain well grounded in what can and cannot be accomplished. When investigating the past, and reporting research results, it is then appropriate to remind ourselves that we cannot prove hypotheses, and rather accept that some of them have not yet been proven false (Kinraide and Denison, 2003).

Another insightful argument was made by Cleland (2002) concerning scientific disciplines that attempt to predict the future and those that focus on postdiction of the past. The former proceed "from causes (test conditions) to effects, with the concomitant worries about ruling out false positives and false negatives," while the latter go "from effects (traces) to causes, with the concomitant worries about ruling out alternative explanations." A "predictive science" is therefore confronted with "the underdetermination of the future by the localized present," whereas a "postdictive science" deals with "the overdetermination of the past by the localized present." As argued by Pigliucci (2005), "the price for our ability to predict the future is that we can do so accurately only under very restrictive conditions," while accurate inferences about a past event can in some cases be derived using more than one type of empirical evidence. The main consequence of this asymmetry is that historical sciences tend to accumulate evidence in support of a hypothesis, rather than trying to disprove it.

Paleoecology cannot be segregated from the rest of ecological science (Rull, 2010), but its philosophical characterization as a postdictive field of study provides a useful frame of reference, particularly when confronted with the emphasis increasingly being placed on all-encompassing simulations models (e.g., Anderson et al., 2006). A more effective approach was laid out in Harte›s (2002) inspirational article on ESS: "There is, in ESS, a growing infatuation with ever more complex models. It's gotten to the point where some models look as inscrutable as nature itself. With numerous adjustable parameters, these models are generally unfalsifiable, so that the opportunity to learn from a wrong prediction is shortcircuited. A "Fermi approach" based on models that capture the essence of the problem, but not all the details, might get us farther. We need to develop simple, mechanistic models. They will, perforce, be caricatures of the Earth system, but they must be falsifiable." It may not be too farfetched to infer that Platt›s approach reverberates in Harte›s words, as well as in the prominence assigned to manipulation for distinguishing ecology from natural history (e.g., Ernest et al., 2009).

Science is a relatively young enterprise, as it emerged in its modern form at the time of Galileo, about 400 years ago. Even younger is the systematic study of uncertainty, which only dates back about a century (Lindley, 2006). These timeframes are miniscule compared to those during which humans have behaved as "political animals," using Aristotle›s definition. To protect the infancy of scientific thought, especially given the nature of "postdictive" research, formal training in how to detect and reduce bias should be an essential component of graduate curricula in paleoecology, with the papers by Wunsch (2010) and Jackson (2012) among the required readings. With regard to this journal section, it is expected to value and attract studies of past ecological changes that foster analytical rigor and constant innovation, to provide an outlet where intellectual creativity and discovery-oriented efforts are wholeheartedly embraced, and to become an arena where vigorous debate on fundamental assumptions is allowed as long as it relies on evidence-based worldviews, thereby keeping in line with the principles that inspired Galileo's revolution.

Conflict of Interest Statement

The author declares that the research was conducted in the absence of any commercial or financial relationships that could be construed as a potential conflict of interest.

ACKNOWLEDGMENTS

The author is grateful for the financial support received from a Charles Bullard Fellowship in Forest Research awarded by Harvard University to visit Harvard Forest in Petersham, Massachusetts. This material is based upon work supported by the US National Science Foundation under grants AGS-EAGER-1256603, P2C2-0823480, and BSC-1230329. The comments made by Michal Kowalewski helped improve an earlier version of the manuscript. The views and conclusions contained in this document are those of the author and should not be interpreted as representing the opinions or policies of the funding agencies.

REFERENCES

1. Anderson, N. J., Bugmann, H. K. M., Dearing, J. A., and Gaillard, M.-J. (2006). Linking palaeoenvironmental data and models to understand the past and to predict the future. *Trends Ecol. Evol.* 21, 696–704. doi: 10.1016/j.tree.2006.09.005
2. Biondi, F. (2013). The fourth dimension of interdisciplinary modeling. *J. Contemp. Water Resour. Educ.* 152, 42–48. doi: 10.1111/j.1936-704X.2013.03166.x
3. Biondi, F. (2014). "Dendrochronology and volcanic eruptions," in *Encyclopedia of Scientific Dating Methods*, eds W. J. Rink and J. W. Thompson (Dordrecht: Springer Science+Business Media), 1–13.
4. Cleland, C. E. (2002). Methodological and epistemic differences between historical science and experimental science. *Philos. Sci.* 69, 474–496. doi: 10.1086/342455
5. Cochran, W. G. (1983). *Planning and Analysis of Observational Studies*. New York, NY: John Wiley and Sons. doi: 10.1002/9780470316542
6. Cole, L. C. (1957). Biological clock in the unicorn. *Science* 125, 874–876. doi: 10.2307/1754178
7. Davis, R. H. (2006). Strong inference: rationale or inspiration? *Perspect. Biol. Med.* 49, 238–249. doi: 10.1353/pbm.2006.0022
8. Ernest, S. K. M., Valone, T. J., and Brown, J. H. (2009). Long-term monitoring and experimental manipulation of a Chihuahuan Desert ecosystem near Portal, Arizona, USA. *Ecology* 90, 1708. doi: 10.1890/08-1222.1
9. Fanelli, D. (2010). "Positive" results increase down the hierarchy of the sciences. *PLoS ONE* 5:e10068. doi: 10.1371/journal.pone.0010068
10. Flueck, J. A., and Brown, T. J. (1993). Criteria and methods for performing and evaluating solar-weather studies. *J. Climate* 6, 373–385. doi: 10.1175/1520-0442(1993)006<0373:camfpa>2.0.co;2
11. Goldstein, M., and Goldstein, I. (1984). "Snow on cholera," in *The Experience of Science: An Interdisciplinary Approach* (New York, NY: Plenum Press), 29–71.
12. Gould, S. J. (1980). The promise of paleobiology as a nomothetic, evolutionary discipline. *Paleobiology* 6, 96–118. doi: 10.2307/2400239
13. Harte, J. (2002). Toward a synthesis of the Newtonian and Darwinian worldviews. *Phys. Today* 55, 29–34. doi: 10.1063/1.1522164
14. Harvey, A. C. (1990). *Forecasting, Structural Time Series Models and the Kalman Filter*. Cambridge: Cambridge University Press. doi: 10.1017/CBO9781107049994
15. Hughes, M. K., and Ammann, C. M. (2009). The future of the past—an earth system framework for high resolution paleoclimatology: editorial essay. *Clim. Change* 94, 247–259. doi: 10.1007/s10584-009-9588-0
16. Jackson, S. T. (2001). Integrating ecological dynamics across timescales: real-time, Q-time, and deep-time. *Palaios* 16, 1–2. doi: 10.1669/0883-1351(2001)016<0001:iedatr>2.0.co;2
17. Jackson, S. T. (2012). Representation of flora and

vegetation in Quaternary fossil assemblages: known and unknown knowns and unknowns. *Quaternary Sci. Rev.* 49, 1–15. doi: 10.1016/j.quascirev.2012.05.020

18. Keeling, C. D. (1960). The concentration and isotopic abundances of carbon dioxide in the atmosphere. *Tellus* 12, 200–203. doi: 10.1111/j.2153-3490.1960.tb01300.x
19. King, G. (1997). *A Solution to the Ecological Inference Problem: Reconstructing Individual Behavior from Aggregate Data.* Princeton, NJ: Princeton University Press.
20. Kinraide, T. B., and Denison, R. F. (2003). Strong inference: the way of science. *Am. Biol. Teach.* 65, 419–424. doi: 10.1662/0002-7685(2003)065[0419:sitwos]2.0.co;2
21. Lindley, D. V. (2006). *Understanding Uncertainty.* Hoboken, NJ: John Wiley and Sons, Inc. doi: 10.1002/0470055480
22. Nagel, E. (1952). Some issues in the logic of historical analysis. *The Sci. Monthly* 74, 162–169. doi: 10.2307/20798
23. Oreskes, N., Shrader-Frechette, K., and Belitz, K. (1994). Verification, validation, and confirmation of numerical models in the earth sciences. *Science* 263, 641–646. doi: 10.1126/science.263.5147.641
24. Pigliucci, M. (2005). Are the historical sciences sciences? *Skeptical Inquirer* 29, 20–21.
25. Platt, J. R. (1964). Strong inference. *Science* 146, 347–353. doi: 10.1126/science.146.3642.347
26. Rosenbaum, P. R. (2002). *Observational Studies.* New York, NY: Springer. doi: 10.1007/978-1-4757-3692-2
27. Rosenbaum, P. R. (2010). *Design of Observational Studies.* New York, NY: Springer Science+Business Media. doi: 10.1007/978-1-4419-1213-8
28. Rull, V. (2010). Ecology and palaeoecology: Two approaches, one objective. *The Open Ecol. J* 3, 1–5. doi: 10.2174/1874213001003020001
29. Schreuder, H. T., and Thomas, C. E. (1991). Establishing cause-effect relationships using forest survey data. *Forest Sci.* 37, 1497-1512.
30. Snieder, R., and Trampert, J. (1999). "Inverse problems in geophysics," in *Wavefield Inversion*, ed A. Wirgin (New York, NY: Springer-Verlag), 119–190.
31. Wagenmakers, E.-J., Wetzels, R., Borsboom, D., Van Der Maas, H. L. J., and Kievit, R. A. (2012). An agenda for purely confirmatory research. *Perspect. Psychol. Sci.* 7, 632–638. doi: 10.1177/1745691612463078
32. Wunsch, C. (2010). Towards understanding the Paleocean. *Quaternary Sci. Rev.* 29, 1960–1967. doi: 10.1016/j.quascirev.2010.05.020

CHAPTER 2

PALEOECOLOGY AND GENETICS

Lynn L Anderson-Carpenter, Jason S McLachlan, Stephen T Jackson, Melanie Kuch, Candice Y Lumibao and Hendrik N Poinar

BACKGROUND

Knowledge of the long term population history of a plant species provides clues to its current distribution [1, 2,3, 4, 5], abundance [2, 6] and genetic structure, [7, 8, 9, 10, 11], and can also help assess its potential response to future environmental change [12, 13]. Ecological and biogeographic history have been inferred primarily from paleoecological studies (fossil pollen and macrofossils in sediments of lakes and peatlands, middens accumulated by packrats and other rodents, tree-rings from living and dead trees) and phylogeographic studies (spatial patterns of molecular and other genetic markers in living populations). These records indicate that terrestrial plant species have undergone population expansion, contraction, isolation, fragmentation, and coalescence of previously isolated populations since the last glacial maximum [2, 3, 4, 14, 15].

Paleoecology and phylogeography each have unique advantages and liabilities in historical inference. Pollen assemblages incorporate pollen grains from a variety of local and distant sources [16], and morphological similarities frequently prevent taxonomic resolution below the genus or even family level [14]. Plant macrofossils offer greater taxonomic and spatial precision [17], but sample sizes may be small. Phylogeographic and population genetic analyses often rely on broad assumptions to infer population histories, but lack sufficient information to connect genetic indices like effective population size (*Ne*) to the census population size (*N*). Uniting the disparate fields of genetics and paleoecology could compensate for their respective weaknesses [18, 19].

Citation: Lynn L Anderson-Carpenter, Jason S McLachlanEmail author, Stephen T Jackson, Melanie Kuch, Candice Y Lumibao and Hendrik N Poinar, paleoecology and genetics, DOI: 10.1186/1471-2148-11-30.

Most studies linking plant paleoecological data to measures of genetic change have relied on indirect links between independent datasets [10, 11, 15, 20, 21, 22]. Quite a few studies take advantage of time series of genetic data provided by ancient DNA (aDNA), usually without explicitly linking them to changes in fossil data [23]. Studies directly linking ancient DNA (aDNA) to fossil material [24, 25, 26, 27, 28] can explicitly characterize the genetic consequences of observed fluctuations in census populations. Until now, these examples apply exclusively to changes in animal populations. Progress for plant species has been limited by technical difficulties in extracting aDNA from the sort of fossils commonly used in paleoecological studies, though recent advances in this area show promise [29, 30]. Here, we introduce a new approach to aDNA extraction and sequencing: the genetic analyses of lake sediments and associated macrofossils.

The successful amplification of aDNA from a variety of materials suggests that identifiable DNA may occur in plant macrofossils and sediments from lakes. However, a wealth of experience from other studies, as well as trial and error from our own work, indicate that the search for aDNA must be approached with caution [31, 32,33]. In waterlogged environments, such as those most common in Holocene fossil studies, hydrolysis of DNA molecules presents a special challenge [30], but there are notable successes in extracting plant DNA from waterlogged material [34, 35, 36]. Generally, DNA degradation, low DNA yields, and the risk of sample contamination by modern DNA have led researchers to develop strict protocols for aDNA extraction and amplification [32, 33, 37, 38, 39].

The purpose of our paper is to develop and evaluate a reliable and repeatable protocol for aDNA extraction, amplification, and sequencing from plant macrofossils and bulk sediment from small lakes typically used in paleoecological studies. We sampled macrofossils of four tree species from sediments spanning the last 8000 years from seven lakes with contrasting bathymetry and water chemistry in the western

Great Lakes region. We also analyzed eight bulk-sediment samples from four of these lakes. We address the following questions: Can aDNA be reliably extracted from macrofossils and bulk sediments? Does preservation differ between sites, species, or ages? How much DNA can be extracted from fossil samples relative to the DNA yield in senescent plant tissue that initially falls into lakes? Do macrofossils contain exogenous DNA? And, to what extent are detailed aDNA protocols to prevent sample contamination necessary?

METHODS

Field and Laboratory Sampling

Our samples come from unconsolidated sediments of seven small lakes in Upper Michigan and adjacent Wisconsin (Figure 1, Table 1). The upper 60-100 cm of sediment from Canyon, Trout, and Ackerman Lakes were obtained using a wedge- type freeze-corer [40, 41]. Upper sediments (topmost 60-80 cm) from the other sites, as well as deeper sediments from Trout and Ackerman Lakes, were obtained using a 7.5-cm diameter modified Livingstone piston corer [42]. Cores were extruded in the field, wrapped in plastic film, and transported to the paleoecological laboratory at University of Wyoming where they were stored at 4°C. Since the cores were collected independently of our studies of aDNA, they were not coated with an exogenous DNA tracer (as per reference [43]). This raises the possibility of contamination by exogenous plant DNA during core processing. We partially compensated for this risk by sampling tissues and bulk sediment from interior portions of the core rather than from the core edge, where exogenous DNA could most easily contaminate the sediments. Cores were sampled in increments 1 cm in thickness, representing time-intervals ranging from 5 to 20 years per sample. All of the cores were rich in macrofossils, owing to small lake size and proximity of coring sites to steep upland slopes [17]. Sampling of materials for aDNA analysis was biased toward the last 3000 years, primarily because Jackson's ongoing paleoecological studies in the region were concentrated on late Holocene dynamics, so more material was available for study. Most of our samples (106/126) came from Ackerman or Tower Lakes, which contained some of the richest and most diverse macrofossil records.

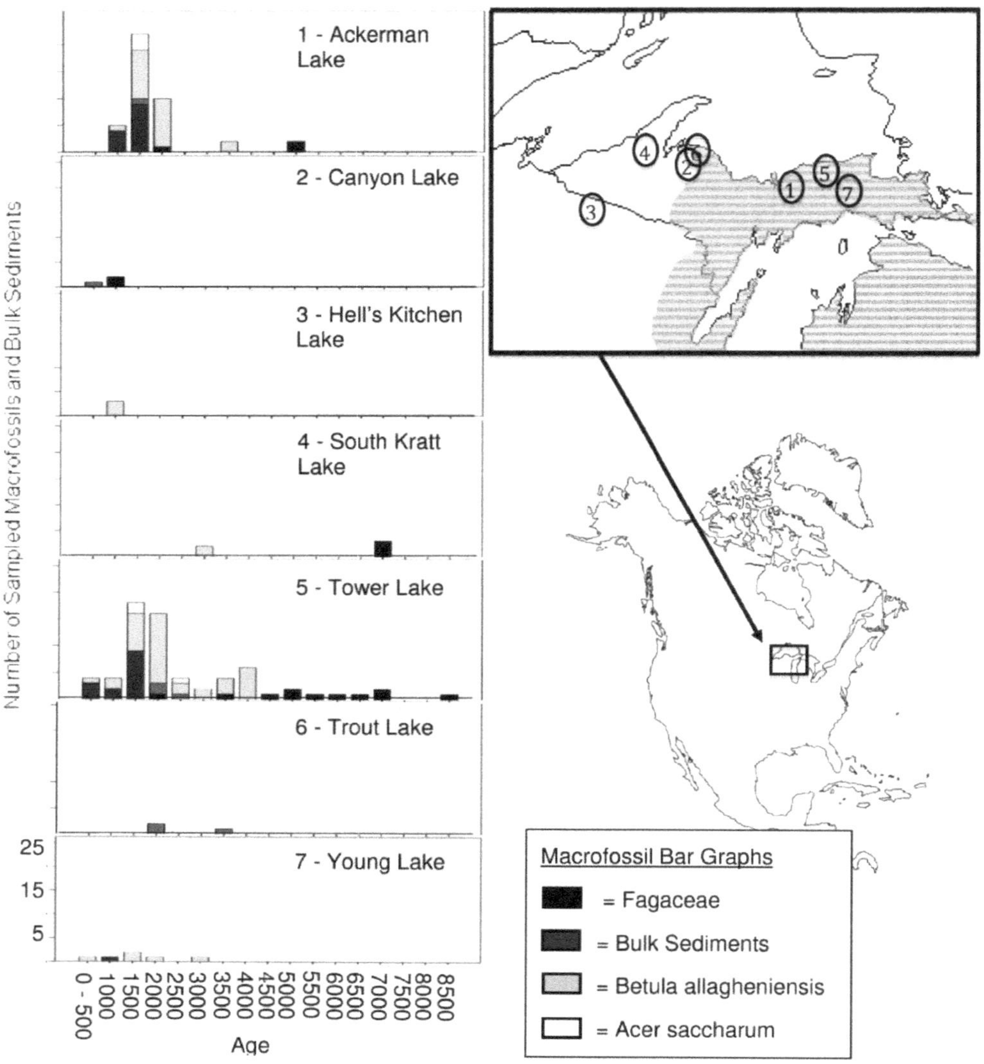

Figure 1 Sample origins. North American map indicating the location of sampled lakes (numbers). The modern range of beech is shaded in gray. The contemporary ranges of all other sampled species are throughout the sampled range. Bar graphs depict the number of macrofossils and bulk sediments taken from each lake's sediments through time.

Table 1 Sample lakes (corresponding to Figures 1 and 6).

Lake Name	Lat/Long Coordinates	Surface Area (ha)	Maximum Water Depth (m)	Water Depth at Coring Site (m)
Ackerman Lake	46.33N/-86.790W	5	14	8
Canyon Lake	46.83N/-87.790W	1	23.7	23.7
Hells Kitchen Lake	46.19N/-89.7W	3	19	10
South Kratt Lake	46.84N/-88.860W	2.5	9	6.9
Tower Lake	46.54N/-86.040W	3	11	8.9
Trout Lake	46.86N/-87.890W	10	18	15
Young Lake	46.44N/-85.710W	2.5	9.5	8.7

All macrofossil and bulk sediment samples were extracted in the paleoecology laboratory at the University of Wyoming. Bulk sediment samples (1 cm^3) were taken from the center of 1-cm slices of the sediment cores. Sediment samples for macrofossil analysis (10-100 cm^3) were prepared by dispersal in distilled water, followed by sieving (710 µm mesh) under a stream of tap water. Sieve residues were rinsed and stored in distilled water and scanned under a stereomicroscope for macrofossil sorting. Macrofossils and bulk-sediment samples were placed in Falcon tubes after separation, and immediately frozen (-20°C) or freeze-dried. All instruments (forceps, sieves, vials, etc.) were sterilized between samples by flaming or bleach treatment. All samples were shipped to the aDNA laboratory at McMaster University in either a frozen or freeze-dried state.

Macrofossil samples included red oak (*Quercus rubra*), American beech (*Fagus grandifolia*), sugar maple (*Acer saccharum*), and yellow birch (*Betula allagheniensis*). Each species was typically represented by distinct tissue types. For example, samaras and catkin-bracts were most common for yellow birch, while beech was most commonly represented by bud scales. Macrofossil composition differed significantly over time and space (Fisher's exact test, $p < 0.0001$). These differences partially result from changing species' distributions over time (e.g., yellow birch and beech immigrated relatively late to the study area, and beech does not occur near our northeastern study sites).

Plant tissues reaching sediments range from living tissue to tissue that has persisted in aerobic conditions for long periods before being washed into the lake. To assess the yield (concentration) of aDNA extracted and amplified from our samples relative to this spectrum of tissues, we quantitated cpDNA in the type of senescent material that might wash or fall into a lake, as well as from analogous fresh tissues, using one 82-bp qPCR-assay. This allowed us to determine whether most DNA degradation/loss takes place after tissue senescence or after deposition in the sediments. Our senescent material consists of tissues from 10 individuals from the macrofossil reference collection at the University of Wyoming, which were all collected from live trees in the field, usually at the time of tissue senescence and/or shedding of organs. These samples were of the same organs as the plant macrofossils, and included seeds, bracts, and/or bud scales from American beech, sugar maple, and yellow birch. We did not have suitable reference samples of red oak, so we substituted bud scales from black oak, Gambel oak, and white oak (*Q. velutina, Q. gambelii, Q. alba*). Senescent tissues were original collected up to several decades ago, so DNA yield in these tissues reflects the original DNA content plus degradation during storage. Analogous fresh material was collected from trees on the University of Notre Dame campus. In the absence of a separate study of the DNA content of plant tissues arriving in sediments, we assume these end members bracket the likely DNA content of the source material reaching sediments.

DNA extraction

DNA extraction protocols were the same for macrofossils, modern samples, and bulk sediments. All aDNA analyses were conducted at McMaster University's ancient DNA facilities, while modern analyses were conducted at facilities at the University of Notre Dame. Published aDNA protocols were used at McMaster including the division of sample preparation, extraction and PCR setup rooms; the use of protective clothing; and techniques to minimize the contamination risk including UV light exposure, bleach sterilization, use of PCR workstations, and filtered pipette tips [32].

Within the designated preparation room, samples were stored at -20°C until use. Immediately after removal from -20°C, samples were submerged into 600 µl CTAB buffer (2% CTAB; 2% PVP; 1.4 M NaCl; 10 mM EDTA, pH 8.0; and 100 mM Tris-HCl, pH 8.0) and chopped into very small pieces using sterile scalpels, after which they were taken to the extraction room for a 2 hour incubation at 65°C. Subsequently 600 µl of Phenol/Chloroform/Isoamyl alcohol (25:24:1) were added to each tube and tubes were vortexed and centrifuged at maximum speed (14,000 × g) for 5 minutes. The aqueous solution was transferred to another tube and 500 µl of $CHCL_3$ was added. Samples were once again vortexed and centrifuged at maximum speed for 5 minutes. The resulting aqueous phase was removed for concentration using microcon cartridges (50 bp cutoff YM-10) (Millipore). Microcon membranes were first primed with 100 µl of 0.1× TE (10 mM Tris-HCl, and 1 mM EDTA, pH 8.0), after which samples were added and centrifuged at maximum speed (14,000 × g) for 10 minutes. Microcon membranes were then washed by adding 300 µl of 0.1× TE, pH 8.0, shaking at room temperature on an Eppendorf thermomixer at 650 rpm for 3 minutes and centrifuging at maximum speed (14,000 × g) for 10 minutes. To elute the DNA we added 100 µl of 0.1× TE pH 8.0 to the cartridges, which we then incubated at room temperature on a thermomixer at 1000 rpm for 5 minutes. Cartridges were then inverted into a new tube and centrifuged to acquire samples.

Negative and blank controls

We processed one negative extraction control for each six fossil samples extracted (total of 28 blank extraction controls) and followed these samples through PCR amplification. Negative extraction controls followed the identical procedure as aDNA extractions, minus the addition of fossil material. We ran 48 additional negative PCR controls, where PCR reagents were mixed without template DNA in the McMaster University cleanroom (prePCR) facilities, and then transported to a separate facility for PCR amplification. We highlighted the effectiveness of our strict protocols against contamination by supplementing these negative PCR controls with the type of negative controls typically used in molecular laboratories focusing on modern tissue. In these "weak" negative PCR controls, we added PCR reagents to sterile water in the same room in which PCR products are analyzed.

In addition to PCR and extraction controls, we processed nine extraction blank controls to test directly for modern DNA contamination along the DNA extraction pipeline. Eight water samples were obtained from sieving and sorting sediment residues from Ackerman Lake (4 samples) and Tower Lake (2 samples) and from tap water in the University of Wyoming's lab (2 samples). We also processed one ambient air sample (obtained from that lab) from a Falcon tube that was opened in his lab to test for airborne DNA particles. DNA extractions of these samples were conducted using the same extraction protocol as for aDNA samples. Positive aDNA controls in this analysis consisted of three macrofossils (two oak leaf fragments and a leaf fragment of yellow birch), which had amplified and quantified reliably in previous PCR reactions.

DNA amplification

For all species, we amplified and sequenced a single 82-bp section of the spacer region between the coding genes atpB and rbcL in the chloroplast genome [44]. Sequence variation in this region within and between species has been characterized by extensive sampling throughout eastern North America (including the Upper Midwest) in our focal species as well as in other common eastern trees [15] and McLachlan unpublished. Polymorphisms in this fragment reliably distinguish three taxonomic groups among the species we investigated: Fagaceae (beech and red oak), sugar maple, and yellow birch, each distinguished by one 2 bp polymorphic site with three individual character-states for each group (See Table 2). We also detected within-group variation among Fagaceae through a transversion mutation. PCR amplification reactions consisted of 1× PCR buffer, 2.5 mM of $MgCl_2$, 400 uM of dNTPs, 300 nM of each primer, 0.05 units of Taq Gold (Applied Biosystems), 1 mg/ml of BSA, 30 mM of Rox Dye as a reference dye, and 0.167 × of SybrGreen. Cycle conditions consisted of: 95°C-7 min initial denaturation, followed by 50 cycles of 95°C-30 sec, 56°C-30 sec, 72°C-30 sec, and a final extension of 72°C for 15 min. Primers used were Forward: 5'-ATTGGGTTGCGCCATACATA-3', Reverse: 5'-GGTTCGTTATTAGACCATGATATTTGAT-3'.

Table 2 Genetic variation within the 82 bp sequenced region in modern tissue of known species and unknown variants in fossil material.

Taxonomic ID [Genbank Number]	Mutations															
Fagaceae [HQ333118] △																
American Beech 1		G	G	C	A	A	T	A	A	G	G	T	T	G	C	G
Chestnut		G	G	C	A	A	T	A	A	G	G	T	T	G	C	G
Red oak		G	G	C	A	A	T	A	A	G	G	T	T	G	C	G
Northern Pin Oak		G	G	C	A	A	T	A	A	G	G	T	T	G	C	G
Pin Oak		G	G	C	A	A	T	A	A	G	G	T	T	G	C	G
Black Oak		G	G	C	A	A	T	A	A	G	G	T	T	G	C	G
American Beech 2 [HQ333119] +		G	G	C	A	A	T	A	**C**	G	G	T	T	G	C	G
Sapindaceae [HQ333120] ×																
Red Maple		G	G	C	A	A	T	A	A	**A**	G	T	T	G	**T**	G
Sugar Maple		G	G	C	A	A	T	A	A	**A**	G	T	T	G	**T**	G
Betulaceae [HQ333117] ○																
Yellow Birch		G	G	C	A	A	T	A	A	G	G	T	T	G	**A**	**A**
Paper Birch		G	G	C	A	A	T	A	A	G	G	T	T	G	**A**	**A**
Black Birch		G	G	C	A	A	T	A	A	G	G	T	T	G	**A**	**A**
UNKNOWN 1 [HQ333121] ▽	*	G	**A**	C	A	A	T	A	A	**C**	G	T	T	G	**T**	G
UNKNOWN 2 [HQ333122] ◇	*	G	G	**T**	A	A	T	A	A	G	G	T	T	G	**A**	**A**
UNKNOWN 3 [HQ333123] □		G	G	C	A	A	**A**	A	A	G	G	**A**	T	G	**A**	G
UNKNOWN 4 [HQ333124] ✳	*	G	G	C	A	A	T	A	A	**A**	G	T	T	G	**A**	**A**

UNKNOWN 5 [HQ333125] ✥		G	G	:	:	:	:	:	:	G	G	T	T	G	C	G
UNKNOWN 6 [HQ333126] ⊾	*	G	G	C	A	A	T	A	A	G	G	T	T	**A**	**A**	**A**
UNKNOWN 7 [HQ333127] ✡	*	**A**	G	C	A	A	T	A	A	G	G	T	T	G	**A**	**A**
UNKNOWN 8 [HQ333128] ⊞	*	G	G	C	A	A	T	A	A	G	**A**	T	T	G	**A**	**A**
UNKNOWN 9 [HQ333129] ⊗		G	G	C	A	A	T	A	A	G	G	T	:	G	**A**	**A**
UNKNOWN 10 [HQ333130] ⍓	*	G	G	C	A	A	T	A	A	**A**	G	T	:	G	**A**	**A**
UNKNOWN 11 [HQ333131] ■	*	G	**A**	C	A	A	T	A	A	G	G	T	T	G	**A**	**A**
UNKNOWN 12 [HQ333132] ◆	*	G	G	C	A	A	T	A	A	**A**	G	T	T	G	C	G

Mutations are in bold. Symbols correspond with Figures 5 and 6. Asterisks indicate mutations that are consistent with potential DNA damage.

Amplification success

Amplification success (indicated by the presence of a band on an agarose gel, or a successful qPCR reaction) could be influenced by several factors, including differential DNA survival over time or in different lakes, primer specificity, PCR inhibition, etc. We fit a binomial generalized linear model (GLM) to amplification success per sample with species identity, lake of origin, tissue type, sample age, and number of PCR attempts as explanatory variables. To more flexibly estimate the relationship of amplification success to sample age, we fit a binomial general additive linear model (GAM) to the success data with age as a covariate. All statistical modeling was conducted using R (http://www.r-project.org).

DNA yield

We used qPCR reactions to test for primer sensitivity and efficiency of the reaction by amplifying a dilution series of known DNA copy numbers (31250, 6250, 1250, 250, 50 and 10 copies/5 μL). High R^2 values (> 0.994) in a regression of the number of DNA copies against PCR cycle at a standardized threshold (Ct) provide evidence that our reaction is efficient and, as the low end of the standard amplifies well, that the reaction is also sensitive. Quantitative PCR is thus used to ensure that our estimates of DNA quantity are reliable, even for smaller DNA quantities. We then quantified DNA yields (in copy numbers) of our ancient samples by comparing amplification plots of the samples to those of the standard dilution series [45]. We used Fisher's exact test to assess the efficiency of our reactions at low copy number: we predicted that samples that amplified with low DNA yield would also amplify in a second quantitation attempt and that samples that had previously failed to amplify would not amplify in this second attempt. We tested whether DNA quantity (calculated as number of DNA copies 5 μl^{-1}) was influenced by the same variables we analyzed for PCR success (sample age, species, lake of origin, and number of attempted PCR reactions) using a GLM with lake and species as categorical variables. We ran a separate GLM to find out whether bulk sediments or macrofossils yielded higher DNA concentrations with sample age, lake of origin and number of attempted PCR reactions as covariates. For all analyses of yield, DNA quantities were log transformed to normalize the data (*ln*DNA).

The PCR reaction can be inhibited by the coeluates of extraneous compounds (e.g., tannins) in the DNA extracts as well as damage to the DNA molecule in the form of covalently bound peptides or crosslinks [43]. Such inhibition can reduce estimates of total amplifiable DNA. We tested for the presence of inhibition by incorporating a mammoth (*Mammuthus*) cytB DNA standard into our qPCR analysis [24]. Mammoth DNA is unlikely to have been present in any of our relatively young lake sediment samples. By adding known quantities of a PCR product derived from mammoth mtDNA into qPCRs containing our macrofossil and bulk sediment extracts, any deviation from the expected mammoth values during qPCR amplification would indicate inhibition from our aDNA extracts. We amplified 250 copies of a mammoth amplicon in the presence of 10 representative extracts, as well as 1:10 dilutions of each extract. In the absence of polymerase inhibition, these reactions should amplify at the same Ct as the reaction with 250 copies of mammoth DNA without added extracts. Polymerase inhibition would be expected to increase the Ct of samples, and we expect that concentrated extracts would amplify at higher Ct than the diluted extract from the same sample in the presence of polymerase inhibition.

To illustrate the risk of contamination in more typical laboratory settings, we supplemented the four true negative controls in this PCR run with an additional four "weak negative controls" (see *negative and blank controls*). PCR amplification reactions consisted of 1× PCR buffer, 2.5 mM of MgCl2, 400 uM of dNTPs, 300 nM of each primer, 0.05 units of Taq Gold (Applied Biosystems), 0.75 mg ml^{-1} of BSA, 30 mM of Rox Dye as a reference dye, and 0.167 × of SybrGreen. Cycle conditions consisted of: 95°C-7 min initial denaturation, followed by 45 cycles of 95°C-30 sec, 60°C-30 sec, 72°C-90 sec, and a final extension of 72°C for 15 min.

Finally, we wanted to identify the point(s) along the taphonomic process where DNA loss occurs. We did this by looking at DNA yields between fossil, modern senescent, and living material. Using the same protocols described above, we extracted DNA and quantified DNA yield from ten living and ten modern senescent tissues. We used a t-test to compare ancient *ln*DNA yields against modern senescent *ln*DNA yields.

Phylogeny

Following the cpDNA phylogeography literature (e.g., Petit et al 2003), we refer to each sequence variant as a "haplotype". We constructed a cladogram of aDNA haplotypes in our samples using TCS (version 1.21)[46]. TCS calculates a pairwise distance matrix between sequence variants and calculates the probabilities for the mutational steps detected using frequency data and parsimony criteria (as defined in [47]). Using a 95% cutoff, a graphical output is generated representing the relationships between the haplotypes.

Multiple haplotypes from single samples

Multiple sequence variants derived from a single PCR reaction could indicate DNA damage, or the different sequences could be derived from multiple species or mixtures of individuals of the same species with different haplotypes. DNA damage is difficult to ascertain with certainty. Here, we consider polymorphic sequences to be more likely the result of DNA damage if they differ from sequences identified in living samples by only one base-pair difference and if that difference is a G to A or a C to T transition [48]. Because we don't know which of these transitions represent DNA damage and which might be true polymorphic variation, we conducted analyses relevant to genetic composition and relevant to the predominance of multiple sequences twice: once including all sequence variation (assuming no DNA damage) and once excluding sequences that differ from known sequences by a single transition (assuming that all these sequences indicate DNA damage). Differences between the results of these paired analyses indicate the extent to which DNA damage might alter the interpretation of our results.

To detect multiple haplotypes within a single PCR product, we cloned the products using a Topo TA cloning kit (Invitrogen) as per manufacturer's instructions and sequenced 4 clones on average. To ensure that the consensus we derived for the remaining sequences were authentic we directly sequenced representative PCR products multiple (at least two) times from two different PCR reactions. We failed to produce a second PCR for eight of the 66 samples which had previously amplified. For these eight samples, we either only have one direct sequence or clones from a single PCR reaction. For cloning, bacterial colonies were grown on LB-agar plates and then lysed using a boil-prep protocol. Samples were then PCR amplified (1× PCR buffer, 2.5 mM $MgCl_2$, 0.2 μM PrimerF, 0.2 μM PrimerR, 0.2 mM dNTPs, 1 U AmpliTaq Gold) under the cycling conditions of 94°C-5 min, (94°C-30 sec, 57°C-30 sec, 72°C-30 sec) 29 times, and a final 72°C extension for 15 minutes. In preparation for BigDye cycle sequencing, all colony-PCR products were purified using Millipore plates (YM-30), and all PCRs subjected to direct sequencing were cleaned using ExoSap (USB Scientific) according to manufacturer›s protocols. BigDye cycle sequencing was conducted according to manufacturers protocols (Version 3.1), and reactions were run on an ABI 3730 for data collection.

We often found more than one cpDNA sequence among colony PCR products from the same reaction and also among separate direct PCR reactions from the same sample. We predicted that we would find multiple haplotypes more commonly in bulk sediments than in macrofossils. We conducted analyses first including all sequences (assuming no DNA damage), and then excluding sequences with polymorphisms associated with DNA damage. We used Fisher›s exact test to determine if the probability of multiple haplotypes was greater in extractions from bulk sediments. We also tested this prediction with a logistic regression on the probability of multiple haplotypes in one sample extract with bulk sediment vs. macrofossil and the number of PCR attempts as explanatory variables. We also predicted that we would see *a higher number* of distinct haplotypes in bulk sediments than in macrofossils. We used a Poisson regression to test whether the number of haplotypes differed between DNA extractions from bulk sediments and macrofossils. To the extent that we can distinguish taxonomic groups using developed molecular markers, we assessed whether the most common (‹dominant›) haplotype found in each individual macrofossil corresponded to that macrofossil›s morphologically based taxonomic assignment. Our aim was to determine whether the dominant haplotype was more likely to represent endogenous DNA (i.e., from tissues of the macrofossil) rather than exogenous DNA (from sediment adhering to the macrofossil or from sample contamination). This analysis requires that we assign each sequence as correctly or incorrectly matching the macrofossil species identity. Unknown sequences, those that either did not correspond to any sequence in our modern data set or had ambiguous base-pair reads at sites that are polymorphic in modern samples, were rare (34/298 sequences). To conservatively bias our test, we coded these sequences as *incorrectly* matching macrofossil ID. This classification only altered the dominant haplotype code (to «incorrect») in one sample (of 57) and did not influence the significance of terms in our model. We modeled the probability that the dominant haplotype from each macrofossil corresponded to the taxonomic identification of that tissue as a binomial response variable. Predictor variables included in a generalized linear regression were number of sequences, number of attempted PCR reactions, DNA quantity, and age of samples. To assess the possible impact of DNA damage on our results, we conducted this test first using all sequences, and then repeated the test without the suspect sequences (WOSS).

Changing genetic composition

As described above, species distributions and abundances have shifted over the last 5000 years at our study sites. For instance, beech macrofossils appear in our dataset only within the last 1500 years. We used pairwise Fisher's exact tests to test if the genetic composition of the fossil material we studied also changed over time. We divided our samples into three time periods with roughly equal sample sizes (0 - 1200 yr BP; 1200 - 1700 yr BP; and > 1700 yr BP) and compared relative haplotype frequencies between the oldest and youngest periods. Again, we conducted this test first under the assumption that no sequence variation was due to DNA damage, then excluding samples with Polymorphisms suggestive of DNA damage (WOSS).

RESULTS

Efficiency and reliability of our approach

Numerous checks and controls illustrate the effectiveness of our protocols for extracting and amplifying low-copy DNA while controlling for contamination by exogenous DNA. The tight match between predicted qPCR threshold cycle number and known DNA quantity in our dilution series ($R^2 > 0.994$) showed that the PCR reaction was sensitive and efficient and that even low numbers of DNA molecules could be amplified. In repeated runs of this dilution series throughout the experiment, however, the most dilute samples (10 copies $(5\ \mu l)^{-1}$) occasionally failed to amplify, presumably due to the stochasticity of PCR reactions with low template. Thus, we were not always able to reliably detect the lowest concentrations of DNA molecules. We found no evidence of PCR inhibition in our extracts. The amplification of known quantities of spiked mammoth DNA were not inhibited by our DNA extracts in twenty independent qPCR reactions (Figure 2). This result imparts confidence in our ability to detect and quantify the small amounts of DNA in our ancient samples.

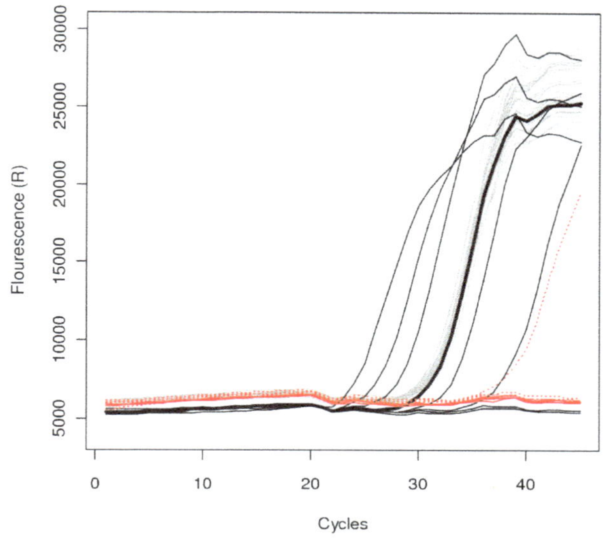

Figure 2 Inhibition quantitation assay. Graph depicts fluorescence values against cycle of appearance along a 45 cycle quantitative PCR reaction. Gray lines are 10 samples from our ancient DNA extracts and 10 1:10 dilutions of those samples, all spiked with a 250 copy standard. Black lines are the dilution series of mammoth DNA from 31,250 copies to 0.4 copies, with the 250 copy standard highlighted in bold. Solid red lines are negative controls and dashed red lines are our "weak" negative controls, used here to illustrate the risks associated with not separating pre- and post-PCR processing.

An array of 85 negative controls indicated that low template samples are unlikely to result from contamination by ambient exogenous DNA. Quantitative PCR did not reveal DNA in the nine extraction blank controls of ambient water and air samples from the macrofossil extraction facilities at the University of Wyoming, despite a sensitivity of 10 to 50 DNA copies (5 μl)$^{-1}$. Of 28 negative extraction controls and 48 negative PCR controls employed in the ancient DNA laboratory at McMaster University, we found one contaminated extraction negative control. The DNA sequence of this contaminant matched the species of the sample that was consistently placed next to it along the extraction process. (All samples are extracted in a particular order of alternating species to track potential contamination). It is likely that the single contamination event resulted from pipetting error, rather than contaminants in the laboratory environment. Of the four "weak" negative PCR controls we used to illustrate the risk associated with not separating low-template DNA preparation from PCR processing facilities, one was contaminated with approximately ten copies of DNA (Figure 2), illustrating the risks of weak controls when working with low-copy DNA template.

PCR amplification success

PCR amplification success rate for samples less than 2000 years old was 64%, but the best fit of our binomial general additive model (Figure 3) indicates a sharp decline in successful amplification at this time. This dropoff in amplification success might be confounded by a coincidental decrease around this time in sampling intensity and in the number of PCR reactions we attempted for each extraction. This leaves open the possibility that older samples might have a relatively higher PCR success rate with more effort. However, we attempted to account for unequal sampling effort across time and unequal number of PCR attempts per sample by simulating samples of equal effort. In ten simulated datasets, we randomly thinned the data to a uniform sampling distribution across time and substituted a Bernoulli draw from the relative probability of success (number of successful PCR reactions/number of PCR attempts) for each individual's success. The thinned data are noisier, as we would expect, but confirm a drop in amplification success around 2000 yr BP independent of sampling effort (Figure 3).

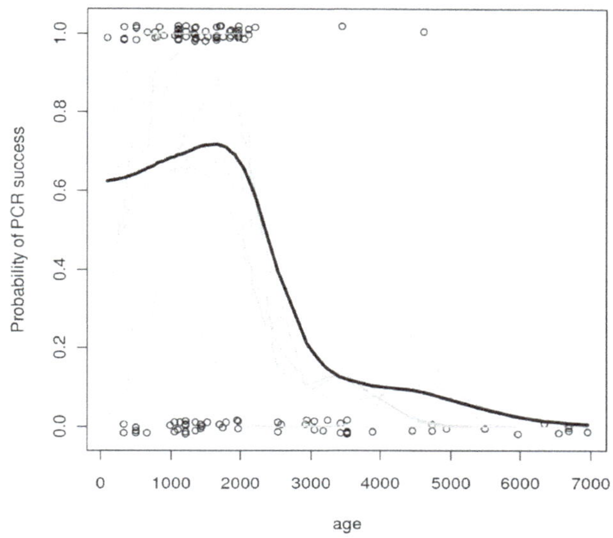

Figure 3 Probability of PCR success. Results of the binomial general additive model (black curve) comparing PCR success (1) to failure (0) relative to sample age. Data are slightly 'jittered' to reduce overlap of points. Gray curves are simulations in which data were thinned to an equal sampling effort.

We used a binomial generalized linear regression to confirm that the probability of successful DNA amplification decreased significantly ($p < 0.001$) with age. However, the confounding decline in the number of attempted PCR reactions was also significant ($p < 0.001$), which means that we are unable to discriminate the effect of age on PCR success from that of sampling effort in this analysis. We therefore accounted for unequal sampling effort across time by simulating samples of equal effort. In ten simulations, where data were thinned to a constant sampling effort across time and number of PCR attempts per sample, age significantly affected amplification success ($p < 0.001$), confirming a decline in PCR success in older samples. The binomial GLM also tested whether PCR amplification success could be explained by lake of origin, species, or whether DNA extractions were from macrofossils or bulk sediments. We found that amplification success was independent of lake of origin. Within the limits imposed by our sampling strategy, it appears that tissue type, species, and sedimentary environment do not strongly influence our ability to detect DNA in samples less than 2000 years old.

DNA yield

Among the samples with successful DNA amplification, DNA yield for the 82-bp cpDNA region was generally low for both macrofossils and bulk sediments (Figure 4). Our median DNA yield across all samples was < 10 copies (5 μl)$^{-1}$, though aDNA yields also reached as high as 76,060 copies of DNA (5 μl)$^{-1}$. We note that DNA quantitation was a good predictor of sequencing reaction success, even for samples with low DNA yield.

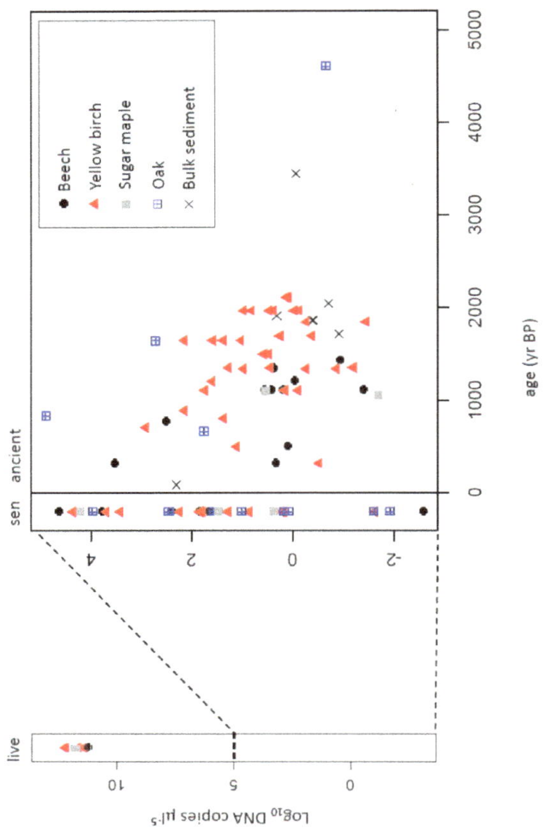

Figure 4 DNA yield. DNA quantity extracted from modern live material (live), senescent material (sen), and macrofossils and bulk sediments (ancient).

Of samples that quantitated with <10 copies (5 μl)$^{-1}$ in a final quantitation assay of all our DNA extracts, 39/41 had previously produced unambiguous sequence data, while samples that did not amplify in this reaction (68 samples) had usually failed to produce sequence data in previous positive PCRs (though 8 had previously produced successful sequences). A Fisher's exact test shows highly significant association ($p < 0.0001$) between the probability of a reaction failing or succeeding (despite low DNA yield) in the final pPCR reaction given results of previous reactions. This result supports our finding (reported in *Efficiency and reliability of our approach*, above) that even exceptionally low DNA concentrations are reliably amplified with the protocols presented here.

A GLM showed that the quantity of DNA extracted decreased significantly on a log scale with increasing age ($p < 0.01$). *ln*DNA yield also varied significantly among species ($p < 0.01$), with oak and yellow birch samples having significantly more amplifiable DNA than other sampled species. Increasing the number of PCR attempts also resulted in significantly higher DNA yield. In a second GLM, we found macrofossils had significantly more DNA than bulk sediments ($p < 0.05$).

DNA yields from ten samples of modern live material analogous in weight and tissue type to our ancient samples were more than six orders of magnitude larger than the highest yield from senescent or ancient material (Figure 4). Plant tissue arriving in sediment, however, originates from both living material and senescent material, including material that might persist in aerobic conditions for years before being washed into sediments. We found the average DNA yield from senescent modern material was significantly greater than average aDNA yield (t test: $p = 0.04$, 3.6031 ± 0.7747 SE, and 1.5115 ± 0.4195 SE, respectively), but the range of yields overlap broadly, with DNA yields from senescent tissue ranging from < 10 copies (5 μl)$^{-1}$ to ~ 43,000 copies (5 μl)$^{-1}$, and aDNA yields ranging from < 10 copies (5 μl)$^{-1}$ to ~ 76,000 copies (5 μl)$^{-1}$ (Figure 4). Thus, while we can say that over 99.9999% of living tissue DNA is lost through taphonomic processes, we can not say what proportion of this loss occurs within the aquatic environment.

DNA sequence data

We successfully sequenced ancient cpDNA across multiple species, lakes, and ages, for both macrofossils and bulk sediment samples. Despite the small size of our cpDNA fragment, we identified 16 haplotypes in sediments, though only four of them matched sequences from known taxa (Table 2, Figure 5). Of all samples that amplified and were subsequently sequenced, 50% of the sequences were consistent with *Betula spp.*, 29% were consistent with Fagaceae, 0.003% were consistent with *Acer saccharum*, 10% were of unknown origin, and 10% had at least one ambiguous base. Seven of twelve unknown haplotypes were differentiated from the haplotype of a known species by a single transition (G to A or C to T), a pattern consistent with DNA damage (see Table 2). These sequences make up less than ten percent (30/310) of all unambiguous sequences. We cannot say with confidence whether these haplotypes represent DNA damage, error in the PCR reaction, or real but unidentified polymorphic variation in plants. We highlight the potential impact of DNA damage on relevant analyses below by repeating statistical tests without suspect sequences (WOSS).

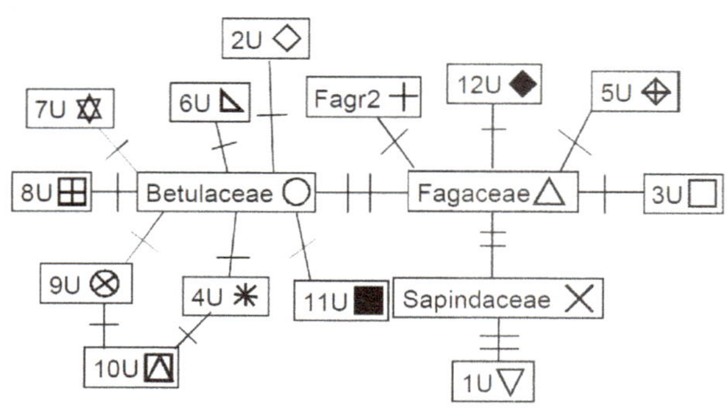

Figure 5 Sequence variation. Cladogram representing the relationship between haplotypes. Each Symbol represents a different haplotype. Each mutational step is represented by a hash mark. Cladogram symbols and identifications correlate with Table 2. Cladogram was constructed using TCS [46].

Multiple haplotypes were obtained from single extractions from both macrofossils and bulk lake sediments. The frequency of multiple haplotypes in DNA extracts from macrofossils (0.45) was not significantly lower than the frequency of multiple haplotypes in bulk sediment samples (0.57) (Test of equal proportions, $p = 0.86$). When haplotypes suggestive of DNA damage are removed (WOSS), the frequency of multiple haplotypes decreases strongly in macrofossils (0.08), but it also decreases in bulk sediments (0.33) and these frequencies are not significantly different ($p = 0.11$). However, we attempted more PCR reactions from macrofossils than from bulk sediments. When we standardized for effort, the probability of sequencing multiple haplotypes from a single macrofossil was still not statistically distinguishable from the probability of sequencing multiple haplotypes from bulk sediments. When this analysis was repeated after removing haplotypes suggestive of DNA damage (WOSS), however, bulk sediments were shown to be more likely to produce multiple haplotypes than macrofossils ($p < 0.05$). This is the only case where excluding haplotypes consistent with patterns of DNA damage (WOSS) changed the significance of a statistical analysis. Neither the number of sequences recovered nor any of our usual explanatory variables (species, age, lake of origin, number of PCR attempts, bulk/macrofossil) significantly explained the number of haplotypes found in our samples in a Poisson regression ($p > 0.05$; WOSS, $p > 0.05$).

In macrofossil samples with at least two sequences, the dominant (most common) haplotype was consistent 80% of the time with the morphological identification of the source macrofossil. This suggests that endogenous DNA in macrofossil tissue is amplified more often than ambient exogenous DNA either because it is present in higher quantities or because it is better preserved. None of the variables we used to predict the probability that the dominant haplotype correctly matched the macrofossil's taxonomic ID (tissue type, sample age, DNA yield, number of attempted PCR reactions, or the number of sequences examined) were significant in our binomial regression. The same non-significant results were found using WOSS data.

In summary, finding multiple haplotypes in DNA extractions from macrofossils and finding evidence of DNA damage both present technical challenges in our results, but most sequences in our dataset did not have these problems. Of the 57 macrofossils with unambiguous sequences, 40 (74%) had only one haplotype. Of these single haplotype samples, 35 (88%) had haplotypes matching the species ID of the macrofossil in question and the remainder had haplotypes matching other species. Of the 14 macrofossils with multiple haplotypes, 9 (64%) contained a mixture of the "correct" and sequences that differed from this "correct" sequence by a single transition. Such samples appear to contain DNA from a single individual that has received a degree of DNA damage over the years. Of the remaining multi-sequence macrofossils, four had a mixture of "correct" sequences and sequences of other species and one had a mixture of sequences from other species.

Changes in genetic composition over time

We found a significant difference between the genetic composition of young (<1200 yr BP) and old (>1700 yr BP) material (Fisher's exact test using all sequences, $p = 0.01$; WOSS, $p = 0.02$); (Figure 6). Younger sediments contained a higher proportion of Fagaceae haplotypes, consistent with the abundant beech budscales found only in younger sediments. This genetic change over time is potentially confounded by the inclusion of locations with only older sequences (Trout Lake) or only younger sequences (Canyon Lake, Hell's Kitchen Lake). We reran the analysis using only samples from the two sites that had relatively abundant sampling in both periods, Ackerman Lake and Tower Lake, which are located 100 km apart in the eastern Upper Peninsula of Michigan (Figure 1).

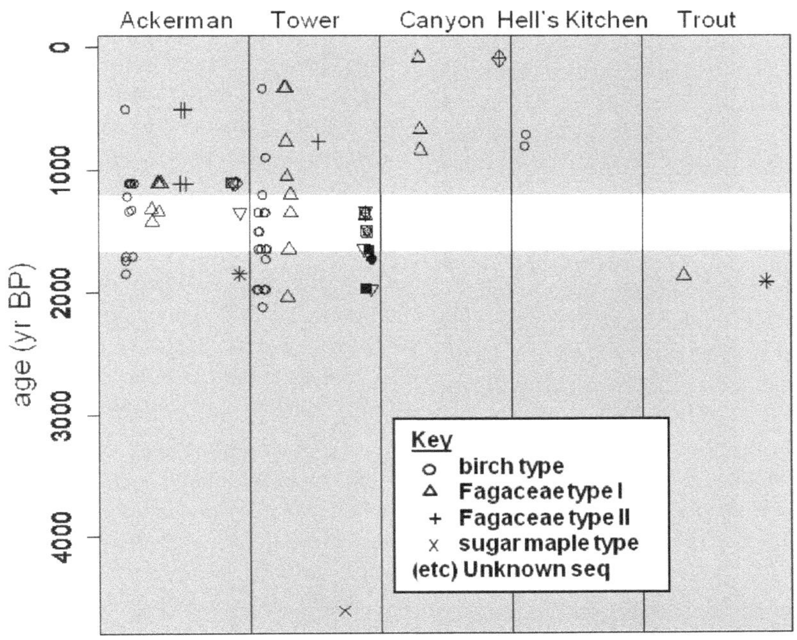

Figure 6 Shifts in genetic composition. Sediments less than 1200 years old (upper grey band) are genetically distinct from sediments over 1700 years old (lower grey band). This trend holds in comparisons limited to Ackerman and Tower Lakes, with samples spanning both time periods.

There is no significant difference between the genetic composition of material in these two locations when samples are pooled across all time periods ($p = 0.61$; WOSS $p = 0.45$), but younger sediments pooled across the two sites contained material with a different genetic composition from older material ($p = 0.02$; WOSS $p = 0.01$). We found no significant differences between the genetic composition of old and young sediments of individual lakes or in pairwise comparisons between individual lakes, but sample sizes for these tests were small, so tests are unlikely to identify significant differences.

DISCUSSION AND CONCLUSIONS

A recent review [49] argued that opportunities for aDNA recovery from sediments in temperate climates are likely to be worse than those in permafrost environments. Indeed, ancient plant DNA as old as 300,000 to 400,000 years old has been recovered from permafrost [43]. However, DNA preservation, degradation and leaching in permafrost settings are likely to be very different from that in lacustrine sedimentary environments and we want to be careful about apples to oranges comparisons. Successful amplification of plant DNA in temperate anaerobic sediments has, until now, been difficult. Holocene aDNA sequences have been recovered from macrofossils [50] and from individual pollen grains [51] in anaerobic peats and lake sediments, but success rates were very low (<10%) in these studies, even for recently deposited material. In our study, success rates are higher than previous analyses and inhibition is lower. On the other hand, the concentration of DNA extracted in our work is often extremely low, highlighting the importance of careful protocols against contamination. Comparisons in results between existing studies are ultimately somewhat weak, however, because each study sampled different tissues from different environments.

Application of aDNA data to the same plant tissues from the same depositional assemblages used in most temeprate paleoecological work has potential for greatly advancing and amplifying understanding of the ecological, evolutionary, and biogeographical consequences of the dramatic environmental changes of the recent past [18, 19, 29] further informing assessments of risks and appropriate actions in the face of rapid environmental change [11, 52, 53]. Our study constitutes a critical first step towards the goal of aDNA application to terrestrial plant paleoecology and biogeography. We document that small amounts of cytoplasmic DNA of multiple species can be extracted and amplified from materials, including a variety of plant macrofossils, preserved in Holocene lake sediments. The vast majority of DNA from living plant tissue appears to be lost before it reaches lake sediments and there is a further slower decline in DNA yield after sediment burial. Under rigorous protocols and safeguards, aDNA from lake sediments can be reliably extracted, amplified, and sequenced with a low threat of contamination by modern DNA. Despite the DNA depletion in sedimentary material, quantitative PCR allowed us to confidently produce sequence data after millennia of its burial in sediments.

Successful extraction and amplification from sedimentary materials younger than 2000 yr BP indicates that further applications to the late Holocene should be straightforward under our protocols. The steep decline in success beyond 2000 yr BP remains enigmatic, and deserves additional study. Age-dependent aDNA degradation processes may be responsible, but there is no obvious mechanism for rapid loss at the 2000-year threshold. A regional environmental signal is suggested by the simultaneous decline in PCR success across lakes. Paleohydrological and paleoecological studies in the region indicate that effective moisture increased substantially during the late Holocene (between 4000 and 2000 yr BP) [54, 55, 56]. Water-levels in all of our study lakes were probably substantially shallower before at least 3000 yr BP, and hence the sediment surface when early to mid-Holocene macrofossils were deposited was better-ventilated and perhaps more conducive to rapid DNA degradation. This hypothesis can be assessed with more-intensive sampling of older sediments, particularly comparing lakes and coring sites of differing water depths.

DNA extractions from most bulk-sediment samples and from some individual macrofossils contained mixed sequences representing multiple individuals and/or DNA damage. The occurrence of aDNA from multiple individuals is unsurprising for bulk sediments, which integrate diverse material from multiple sources (pollen, plant tissues, organic flocculents). Ancient DNA from multiple sources in individual plant macrofossils, particularly representing taxa other than that of the macrofossil itself, may derive from small sediment particles clinging to the macrofossil surface or trapped in small cracks. These sources are difficult to exclude in sample preparation; the rough microtopography of many macrofossil surfaces, as well as the frequent occurrence of tears or fissures, provide opportunity for materials from the surrounding sediment matrix to attach or infiltrate. Complete removal of these by either chemical or mechanical treatment before DNA extraction may be difficult without risking damage to the tissues and the aDNA contained within. We are now experimenting with alternative treatments to minimize extraneous aDNA from sediment matrix while also minimizing damage to the material and aDNA of interest.

Most phylogenetic and population genetic analyses, including those used in other aDNA studies [25], assume that the number of sampled individuals is known. Single macrofossils containing multiple haplotypes can hamper such applications. The aDNA sequences endogenous to individual macrofossils might be probabilistically distinguished from exogenous sequences; endogenous DNA dominated the mixture of sequences in most macrofossils. Additional evidence based on mechanism of DNA damage can be used to statistically identify sequence variation caused by DNA damage [57]. Distinguishing between the genetic identity of individual macrofossils, DNA damage, and the genetic composition of the sediment matrix, even in a probabilistic sense, would allow us to apply standard population genetic analyses to macrofossils (e.g., Beast, Serial SimCoal).

Applications of aDNA to macrofossils and other materials from lake sediments will require careful consideration of taphonomic issues, including the sources of the aDNA. Plant macrofossils in sediments of small lakes generally represent trees growing within 10^0-10^1 meters of the lake margin [17, 58]. Such spatial precision has considerable advantages in application, but it also ensures that only a small number of individual trees are likely to be represented in a macrofossil assemblage. DNA studies of surface sediments, linked with genetic census of trees near the lake margin, can help assess these and other issues of sampling and representation. DNA from pollen, whether analyzed from individual pollen grains [51] or integrated into bulk-sediment samples, represents a broader range of sources. Pollen assemblages in lake sediments comprise a distance-weighted integration of surrounding vegetation; although local individuals will be better represented than those farther away, a substantial amount of the pollen deposited in small lakes derives from plants growing 10^2-10^5 m from the lake margin [16]. Bulk sediments may contain plant aDNA from a variety of sources, including pollen and plant tissues. Comparison of DNA signatures from bulk surface sediments with genetic data from multiple spatial scales may help identify the potential sources. Genetic signals at different spatial scales, for example, might be extracted using a combination of macrofossils and pollen grains as aDNA sources.

The degree of temporal integration or smoothing in aDNA signatures is also of concern. Our lake-sediment samples each represented narrow intervals of deposition (5-20 years), although deposition rates can be far slower in many lake settings [59]. Once entombed, individual plant macrofossils and smaller particles, including pollen grains, are unlikely to move vertically. Although vertical migration or leaching of DNA has been observed in sediments of caves, [60], we view this as highly unlikely in lake sediments. Periodic downward percolation of water can evidently move DNA in porous, granular sediments of caves and perhaps soil profiles. Lake sediments, in contrast, are permanently saturated and gravitational percolation of water does not occur. Once sediments are compacted, vertical advection of pore water is minimal, and multivalent metals and organic compounds (pigments, polycyclic aromatic hydrocarbons (PAHs), organic molecules with more than 15 carbon atoms) are immobilized in the sediment matrix. Large organic molecules such as DNA are likely to adhere to solid-phase sediments (particles, particulate organic matter). Industrially derived organic molecules (PAHs, PCBs) show marked patterns of increase in lake sediments at locally appropriate time-horizons, with no evidence of vertical leaching [61, 62]. Despite the important work left to be done, our study clearly shows significant changes in the genetic composition of plant communities in Holocene lake sediments (Figure 6). Indeed, the significant increase in Fagaceae sequences in recent sediments matches the increased representation of Fagaceous macrofossils in those sediments [56].

DECLARATIONS

Acknowledgements

This research was supported by grants from the Huron Mountain Wildlife Foundation (to STJ, JMcL & HP), the National Science Foundation (DEB-0345012) (to STJ). Field and laboratory assistance was provided by Mark Lesser, Bob Booth, Kelly Reeves, Jennifer Andersen, Tom Minckley, and Mike Urban. We thank Dan Engstrom, Phil Meyers, and John Smol for discussions and insights on DNA mobility within sediments. We thank two anonymous reviewers for improving the manuscript.

Authors' contributions

JSM, STJ, and HNP conceived of the study. LLA-C, JSM, HNP, STJ, MK and CL wrote the manuscript, conducted the molecular studies, and analyzed the data. STJ extracted and processed the macrofossils. HNP and MK developed the protocols and supervised all laboratory activities. All authors read and approved the final manuscript

REFERENCES

1. Woods KS, Davis MB: Paleoecology of range limits: beech in the upper peninsula of Michigan. Ecology. 1989, 70: 681-696. 10.2307/1940219.
2. Jackson ST, Booth RK: The role of late Holocene climate variability in the expansion of yellow birch in the western Great Lakes region. Diversity and Distributions. 2002, 8: 275-284. 10.1046/j.1472-4642.2002.00152.x.
3. Lyford ME, Jackson ST, Gray ST, Eddy RJ: Validating the use of woodrat (Neotoma) middens for documenting natural invasions. Journal of Biogeography. 2004, 31: 333-342. 10.1046/j.0305-0270.2003.00987.x.
4. Gray ST, Betancourt JL, Jackson ST, Eddy RG: Role of multidecadal climate variability in a range extension of pinyon pine. Ecology. 2006, 87: 1124-1130. 10.1890/0012-9658(2006)87[1124:ROMCVI]2.0.CO;2.
5. Grimm E: Fire and other factors controlling Big Woods vegetation of Minnesota in the Mid-Nineteenth Century. Ecological Monographs. 1984, 54: 291-311. 10.2307/1942499.
6. Calcote R: Mid-Holocene climate and the hemlock decline: the range limit of *Tsuga canadensis* in the western Great Lakes region, USA. The Holocene. 2003, 13: 215-224. 10.1191/0959683603hl608rp.
7. Petit RJ, Pineau E, Demesure B, Bacilieri R, Ducousso A: Chloroplast DNA footprints of postglacial recolonization by oaks. Proceedings of the National Academy of Sciences. 1997, 94: 9996-100001. 10.1073/pnas.94.18.9996.
8. Hewitt GM: The genetic legacy of the Quaternary ice ages. Nature. 2000, 405: 907-913. 10.1038/35016000.
9. Davis MB, Shaw RG: Range shifts and adaptive responses to Quaternary climate change. Science. 2001, 292: 673-679. 10.1126/science.292.5517.673.
10. Brewer S, Cheddadi R, de Beaulieu J-L, Reille M: The spread of deciduous *Quercus* throughout Europe since the last glacial period. Forest Ecology and Management. 2002, 156: 27-48. 10.1016/S0378-1127(01)00646-6.
11. Hu FS, Hampe A, Petit RJ: Paleoecology meets genetics: deciphering past vegetational dynamics. Frontiers in Ecology and the Environment. 2009, 7: 371-379. 10.1890/070160.
12. Overpeck JT, Bartlein PJ, Webb T: Potential magnitude of future vegetation change in Eastern North America: comparisons with the past. Science. 1991, 254: 692-695. 10.1126/science.254.5032.692.
13. Williams JW, Jackson ST, Kutzbach JE: Projected distributions of novel and disappearing climates by 2100 AD. Proceedings of the National Academy of Sciences. 2007, 104: 5738-5742. 10.1073/pnas.0606292104.
14. Williams JW, Shuman BN, Webb T, Bartlein PJ, Leduc PL: Late-Quaternary vegetation dynamics in North

America: scaling from taxa to biomes. Ecological Monographs. 2004, 74: 309-334. 10.1890/02-4045.

15. McLachlan JS, Clark JS, Manos PS: Molecular indicators of tree migration capacity under rapid climate change. Ecology. 2005, 86: 2088-2098. 10.1890/04-1036.
16. Jackson ST, Lyford ME: Pollen dispersal models in Quaternary plant ecology: assumptions, parameters, and prescriptions. Botanical Review. 1999, 65: 39-75. 10.1007/BF02856557.
17. Jackson ST, Booth RK: Validation of pollen studies. Encyclopedia of Quaternary Science. Edited by: Elias SA. 2007, Elsevier Scientific Publishing Inc
18. Jackson ST: Late Quaternary biogeography: linking biotic responses to environmental variability across time scales. Frontiers of Biogeography. Edited by: Lomolino M, Heaney L. 2004, Sinauer Associates
19. Jackson ST: Forest genetics in space and time. New Phytologist. 2006, 171: 1-3. 10.1111/j.1469-8137.2006.01781.x.
20. Cwynar LC, MacDonald GM: Geographical variation of lodgepole pine in relation to population history. American Naturalist. 1987, 129: 463-469. 10.1086/284651.
21. Demesure B, Comps B, Petit RJ: Chloroplast phylogeography of the common beech (*Faguls Sylvatica* L), in Europe. Evolution. 1996, 50: 2515-2520. 10.2307/2410719.
22. Magri D, Vendramin GG, Comps B, Dupanloup I, Geburek T, Dušan G, Latalowa M, Litt T, Paule L, Roure JM, Tantau I, Van Der Knapp WO, Petit RJ, De Beaulieu JL: A new scenario for the Quaternary history of European beech populations: paleobotanical evidence and genetic consequences. New Phytologist. 2006, 171: 199-221. 10.1111/j.1469-8137.2006.01740.x.
23. Ramakrishnan U, Hadly EA: Using phylochronology to reveal cryptic population histories: review and synthesis of 29 ancient DNA studies. Molecular Ecology. 2009, 18: 1310-1330. 10.1111/j.1365-294X.2009.04092.x.
24. Debruyne R, Chu G, King C, Bos K, Kuch M, Schwartz C, Szpak P, Gröcke DR, Matheus P, Zazula G, Guthrie D, Froese D, Buigues B, de Marliave C, Flemming C, Poinar D, Fisher D, Southon J, Tikhonov AN, MacPhee RDE, Poinar HN: Out of America: Ancient DNA evidence for a New World origin of Late Quaternary woolly mammoths. Current Biology. 2008, 18: 1320-1326. 10.1016/j.cub.2008.07.061.
25. Hadly EA, Ramakrishnan U, Chan YL, van Tuinen M, O›Keefe K, Spaeth PA, Conroy CJ: Genetic response to climate change: insights from ancient DNA and phylochronology. PLoS Biology. 2004, 2: 0001-0010. 10.1371/journal.pbio.0020290.
26. Leonard JA, Wayne RK, Cooper A: Population genetics of ice age brown bears. Proceedings of the National Academy of Sciences. 2000, 97: 1651-1654. 10.1073/pnas.040453097.
27. Barnes I, Matheus P, Shapiro B, Jensen D, Cooper A: Dynamics of Pleistocene population extinctions in Beringian brown bears. Science. 2002, 295: 2267-2270. 10.1126/science.1067814.
28. Shapiro B, Drummond AJ, Rambaut A, Wilson MC, Matheus PE, Sher AV, Pybus OG, Thomas M, Gilbert P, Barnes I, Binladen J, Willerslev E, Hansen AJ, Baryshnikov GF, Burns JA, Davydov S, Driver JC, Froese DG, Harington CR, Keddie G, Kosintsev P, Kunz ML, Martin LD, Stephenson RO, Storer J, Tedford R, Zimov S, Cooper A: Rise and fall of the Beringian steppe bison. Science. 2004, 306: 1561-1565. 10.1126/science.1101074.
29. Bennett KD, Parducci L: DNA from pollen: principles and potential. The Holocene. 2006, 16: 1031-1034. 10.1177/0959683606069383.
30. Schlumbaum A, Tensen M, Jaenicke-Despres V: Ancient plant DNA in archaeobotany. Vegetation History and Archaeobotany. 2008, 17: 233-244. 10.1007/s00334-007-0125-7.
31. Gugerli F, Parducci L, Petit RJ: Ancient plant DNA: review and prospects. New Phytologist. 2005, 166: 409-418. 10.1111/j.1469-8137.2005.01360.x.
32. Cooper A, Poinar HN: Ancient DNA: Do it right or not at all. Science. 2000, 289: 1139-10.1126/science.289.5482.1139b.
33. Hofreiter M, Jaenicke V, Serre D, Haeseler A, Pääbo S: DNA sequences from multiple amplifications reveal artifacts induced by cytosine deamination in ancient DNA. Nucleic Acids Research. 2001, 29: 4793-4799. 10.1093/nar/29.23.4793.
34. Cappellini E, Gilbert MT, Geuna F, Fiorentino G, Hall A, Thomas-Oates J, Ashton P, Ashford D, Arthur P, Campos P, Kool J, Willerslev E, Collins M: A multidisciplinary study of archaeological grape seeds. Naturwissenschaften. 2010, 97: 205-217. 10.1007/s00114-009-0629-3.
35. Elbaum R, Melamed-Bessudo C, Boaretto E, Galili E, Lev-Yadun S, Levy AA, Weiner S: Ancient olive DNA in pits: preservation, amplification and sequence analysis. Journal of Archaeological Science. 2006, 33: 77-88. 10.1016/j.jas.2005.06.011.
36. Pollmann B, Jacomet S, Schlumbaum A: Morphological and genetic studies of waterlogged *Prunus* species from the Roman viscus Tagetium, Switzerland. Jounal of Archaeological Science. 2005, 32: 1471-1480. 10.1016/j.jas.2005.04.002.
37. Poinar HN, Schwartz C, Shapiro Ji, Qi B, MacPhee RDE, Buigues B, Tikhonov A, Huson DH, Tomsho LP, Auch A, Rampp M, Miller W, Schuster SC: Metagenomics to paleogenomics: large-scale sequencing of mammoth DNA. Science. 2006, 311: 392-394. 10.1126/science.1123360.
38. Pääbo S, Poinar H, Serre D, Jaenicke-Després V, Hebler J, Rohland N, Kuch M, Krause J, Vigilant L, Hofreiter M: Genetic analyses from ancient DNA. Annual Review of Genetics. 2004, 38: 645-679.
39. Willerslev E, Cooper A: Ancient DNA. Proceedings of the Royal Society B-Biological Sciences. 2005, 272: 3-16. 10.1098/rspb.2004.2813.
40. Shapiro J: The core-freezer--A new sampler for lake sediments. Ecology. 1958, 39: 748-10.2307/1931618.

41. Wright HE: Coring of soft lake sediments. Boreas. 1980, 9: 107-114. 10.1111/j.1502-3885.1980.tb01032.x.
42. Wright HE: A spare-rod piston sample for lake sediments. Journal of Sedimentary Petrology. 1967, 37: 975-976.
43. Willerslev E, Hansen AJ, Binladen J, Brand TB, Gilbert MTP, Shapiro BA, Bunce M, Wiuf C, Gilichinsky DA, Cooper A: Diverse plant and animal DNA from Holocene and Pleistocene sedimentary records. Science. 2003, 300: 791-795. 10.1126/science.1084114.
44. Hodges SA, Arnold ML: Columbines: a geographically widespread species flock. Proceedings of the National Academy of Sciences. 1994, 91: 5129-5132. 10.1073/pnas.91.11.5129.
45. King CE, Debruyne R, Kuch M, Schwartz C, Poinar HN: A quantitative approach to detect and overcome PCR inhibition in ancient DNA extracts. Biotechniques. 2009, 47: 941-949. 10.2144/000113244.
46. Clement M, Posada D, Crandall KA: TCS: a computer program to estimate gene genealogies. Molecular Ecology. 2000, 9: 1657-1659. 10.1046/j.1365-294x.2000.01020.x.
47. Templeton AR, Crandall KA, Sing CF: A cladistic analysis of phenotypic associations with haplotypes inferred from restriction endonuclease mapping and DNA sequence data. III. Cladogram estimation. Genetics. 1992, 132: 619-633.
48. Hofreiter M, Serre D, Poinar HN, Kuch M, Pääbo S: Ancient DNA. Nature Reviews Genetics. 2001, 2: 353-359. 10.1038/35072071.
49. Haile J, Froese DG, Macphee RD, Roberts RG, Arnold LJ, Reyes AV, Rasmussen M, Nielsen R, Brook BW, Robinson S, Demuro M, Gilbert MT, Munch K, Austin JJ, Cooper A, Barnes I, Möller P, Willerslev E: Ancient DNA reveals late survival of mammoth and horse in interior Alaska. Proceedings of the National Academy of Sciences. 2009, 106: 22352-22357. 10.1073/pnas.0912510106.
50. Suyama Y, Gunnarsson U, Parducci L: Analysis of short DNA fragments from Holocene peatmoss samples. The Holocene. 2008, 18: 1003-1006. 10.1177/0959683608093540.
51. Parducci L, Suyama Y, Lascoux M, Bennett KD: Ancient DNA from pollen: a genetic record of population history in Scots pine. Molecular Ecology. 2005, 14: 2873-2882. 10.1111/j.1365-294X.2005.02644.x.
52. Jackson ST, Hobbs RJ: Ecological restoration in the light of ecological history. Science. 2009, 325: 567-569. 10.1126/science.1172977.
53. McLachlan JS, Hellmann JJ, Schwartz MW: A framework for assisted migration in an era of climate change. Conservation Biology. 2007, 21: 297-302. 10.1111/j.1523-1739.2007.00676.x.
54. Brugam RB, McKeever K, Kolasa L: A diatom-inferred water depth reconstruction for an Upper Peninsula, Michigan lake. Journal of Paleolimnology. 1998, 20: 267-276. 10.1023/A:1007948616511.
55. Booth RK, Jackson ST, Gray CED: Paleoecology and high-resolution paleohydrology of a kettle peatland in upper Michigan. Quaternary Research. 2004, 61: 1-13. 10.1016/j.yqres.2003.07.013.
56. Reeves K: Holocene vegetation development of hemlock-hardwood forests in the Upper Peninsula of Michigan. M.S. thesis. 2006, University of Wyoming, Department of Botany
57. Mateiu LM, Rannala BH: Bayesian inference of errors in ancient DNA caused by postmortem degradation. Molecular Biology and Evolution. 2008, 25: 1503-1511. 10.1093/molbev/msn095.
58. Dunwiddie PW: Macrofossil and pollen representation of coniferous trees in modern sediments from Washington. Ecology. 67: 58-68. 10.2307/1938503.
59. Webb RS, Webb TI: Rates of sediment accumulation in pollen cores from small lakes and mires of eastern North America. Quaternary Research. 1988, 30: 284-297. 10.1016/0033-5894(88)90004-X.
60. Haile J, Holdaway R, Oliver K, Bunce M, Gilbert MTP, Nielsen R, Munch K, Ho SYW, Shapiro B, Willerslev E: Ancient DNA chronology within sediment deposits: Are paleobiological reconstructions possible and is DNA leaching a factor?. Molecular Biology and Evolution. 2007, 24: 982-989. 10.1093/molbev/msm016.
61. Meyers PA: Application of organic geochemistry to paleolimnological reconstructions: a summary of examples from the Laurentian Great Lakes. Organic Geochemistry. 2003, 34: 261-289. 10.1016/S0146-6380(02)00168-7.
62. Smol JP: Pollution of lakes and rivers: a paleoenvironmental perspective. 2008, Oxford: Blackwell Publishing

CHAPTER 3

MULTI-STAGE SILICIFICATION OF PLIOCENE WOOD: RE-EXAMINATION OF AN 1895 DISCOVERY FROM IDAHO, USA

Mike Viney, Dagmar Dietrich, George Mustoe, Paul Link, Thomas Lampke, Jens Götze and Ronny Rößler

Keywords: Chalcedony, Glenns Ferry Formation, Snake River Plain, Idaho, opal-CT, opalized wood, paleobotany, petrified wood, Quercinium pliocaenicum

INTRODUCTION

This report describes the rediscovery of a historic North American fossil wood locality, its geologic setting and a detailed re-ex amination of newly-collected specimens using a variety of petrographic and geochemical methods.

In 1895, land surveyors working 11 km northwest of Bliss, Idaho, discovered intact branches of a fossil tree protruding a meter above the ground surface [1]. When excavated down to the 8-m level, the trunk's diameter was nearly 2 m. The fossil tree was believed to be standing where it grew, and no other fossil wood was found in the vicinity. Approximately two tons (1800 kg) of the opalized wood were shipped to Philadelphia mineral dealer A.E. Foote by 1897 [1]. All Foote labels corresponding to the Clover Creek material bear the number 212 [2] as a reference to opal's order in the sixth edition of Dana's System of Mineralogy. The Twentieth Report of the United States Geological Survey of 1898 [3] included Foote's description of the Clover Creek fossil find and highlights the amazing preservation of the cellular tissue and apparent woody structure and color. The report indicated that the largest intact branch sections measured 12 inches (30 cm) in diameter.

Clover Creek specimens found their way into museums in the United States, Canada and Europe, commonly displayed in systematic mineral collections as an example of opalization (Figure 1). Clover Creek samples are likewise described in the mineralogical literature as examples of wood preserved by opal [4,5,6]. Even philatelists are confronted with a picture of a Smithsonian Clover Creek sample on two first-day covers of the 1974 stamp issue of the U.S. postal office celebrating U.S. mineral heritage; another serious reason to clearly establish the provenance of Clover Creek opalized wood.

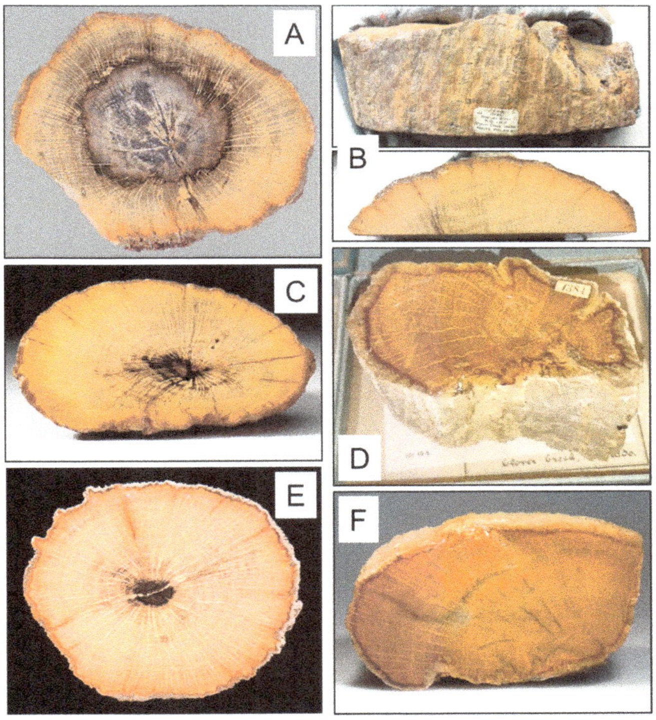

Figure 1. Opalized wood from C+bblover Creek, Idaho. (**A**) Specimen at School of Mines, Golden, Colorado; (**B**) Naturhistorisches Museum der Burgergemeinde Bern, Switzerland; (**C**) Royal Ontario Museum Toronto, Canada; (**D**) Statens Naturhistoriske Museum Kobenhavn, Denmark; (**E**) Natural History Museum (British Museum) London, U.K.; (**F**) Senckenberg Naturhistorische Sammlungen Dresden, Germany (for more information, refer to the Supplementary Material).

Citation: Mike Viney, Dagmar Dietrich, George Mustoe, Paul Link, Thomas Lampke, Jens Götze and Ronny Rößler, Multi-Stage Silicification of Pliocene Wood: Re-Examination of an 1895 Discovery from Idaho, USA, doi:10.3390/geosciences6020021.

Prior to our investigations, scientific study of the Clover Creek fossil wood was limited to a single report in 1908 [7]. The author Julius Schuster received Clover Creek samples from Johann Traugott Sterzel, a German teacher, paleobotanist and first director of the Natural History Collections of Chemnitz, Germany, who obtained them from the German mineral dealer Dr. Friedrich. Krantz. Sterzel assigned the specimens to Quercinium and asked Schuster for a detailed examination of thin sections prepared by the company Voigt and Hochgesang, Goettingen (Figure 2), today part of the Sterzel collection at Museum für Naturkunde Chemnitz, Germany. Schuster established a new species name, Q. pliocaenicum [7]. Subsequently, the German mineral dealer Krantz offered Clover Creek specimens for sale in 1908 [8], adding the new scientific name in 1909 [9].

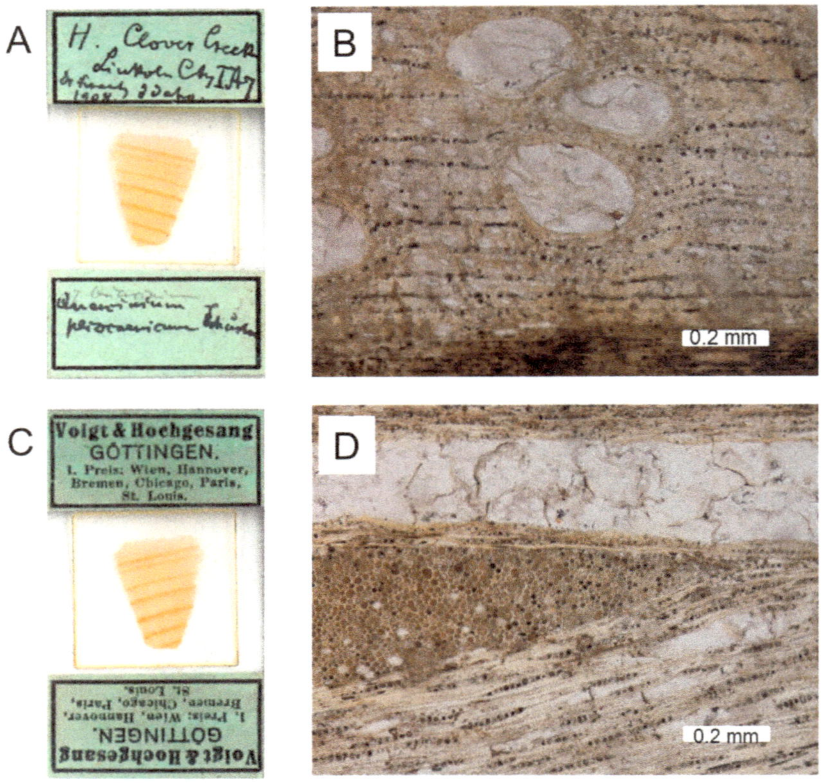

Figure 2. (**A**,) back side of an original thin section (transverse plane) examined by Schuster [7]; (**B**) recently taken OM micrograph showing transverse plane; (**C**) front side of original thin section; (**D**) recent OM micrograph showing an original thin section in the tangential longitudinal plane.

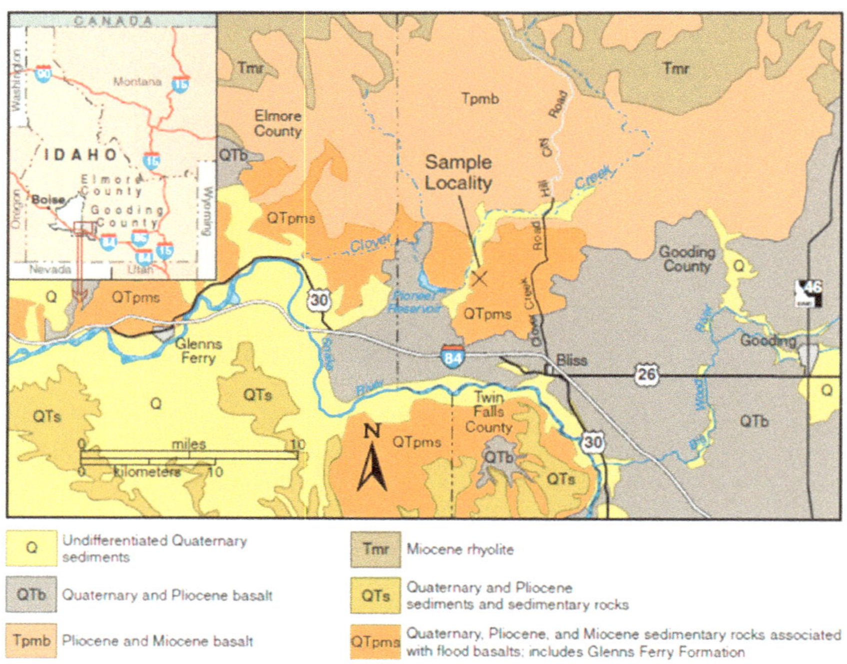

Figure 3. Locality map of the fossil site in Idaho, modified from Lewis et al. [19] (courtesy of William W. Besse).

Quercinium was established as a morpho-genus by Unger [10] to describe petrified wood having a cellular anatomy characteristic of the modern oak family (Fagaceae). In later years, fossilized oak-like wood has been described as Quercoxylon [11,12,13,14], but Quercinium continues to be used [14,15]. We have not attempted to emend the taxonomy of Q. pliocaenicum, but instead focus on geologic and mineralogical aspects.

Clover Creek opalized wood has been referenced in several compendia [16,17,18], but an exact location or story behind the discovery is not given. Modern paleobotany books do not mention the Clover Creek find, even though many specimens are displayed in major museums. Originally, the locality was reported as being in Lincoln County, Idaho, but present county boundaries now place the site in Gooding County (Figure 3, [19]). The 2014 rediscovery of the site is described by Viney et al. [20]. The GPS coordinates are 42°59′46.70″N, 115°00′28.66″W, elevation 1009 m (3300 feet).

The purpose of our study is to provide a comprehensive re-evaluation of this locality, using contemporary knowledge of the geologic setting, and to analyze newly-discovered specimens by contemporary analytical methods not available in 1908. This research is based on examination of the original 1908 thin sections and new thin sections prepared from specimens we collected in 2014. In addition, we extend the taxonomic work of Schuster [7] by providing analyses of the composition and microstructure of Clover Creek Q. pliocaenicum.

GEOLOGY OF THE AREA

The opalized remnants of oak wood are found in flat-lying sandstone and siltstone of the Glenns Ferry Formation, at an elevation 1009 m, on the east side of Clover Creek. The light-gray and light-brown siltstone and sandstone of the fossil site are in the upper member of the Glenns Ferry Formation, stratigraphically above the Clover Creek basalt flow and below the Shoestring Road basalt. The strata are of Pliocene age and lie above the rhyolitic Fossil Gulch ash, with dates of about 3.3 Ma [21,22] (Figure 4).

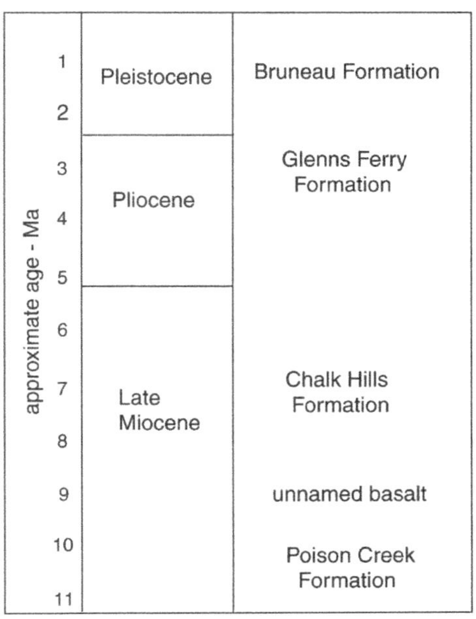

Figure 4. Stratigraphy of the western Snake River Plain, Idaho.

The Pliocene Glenns Ferry Formation of the Idaho Group contains sediment deposited in Lake Idaho, an inland lake that occupied the Snake River Plain from about 6.5 to 1.5 million years ago. The initial spillover of Lake Idaho was about four million years ago, but the age is not tightly constrained [23]. At the Clover Creek fossil site, about 60 m of sediment are exposed below the Shoestring Road Basalt, which forms the upland land surface (Figure 5, stratigraphic column). Exposure is intermittent. The sediment contains poorly- to well-sorted, subangular, medium to granule-sized volcanic lithic sandstone, interbedded with white to gray siltstone that bears small rhizoliths. The volcanic lithic grains are mainly rhyolite and pumice, with subordinate basalt. The sandstone also contains feldspar, quartz and biotite, attesting to the derivation from granitic rocks of the Cretaceous Idaho batholith to the north.

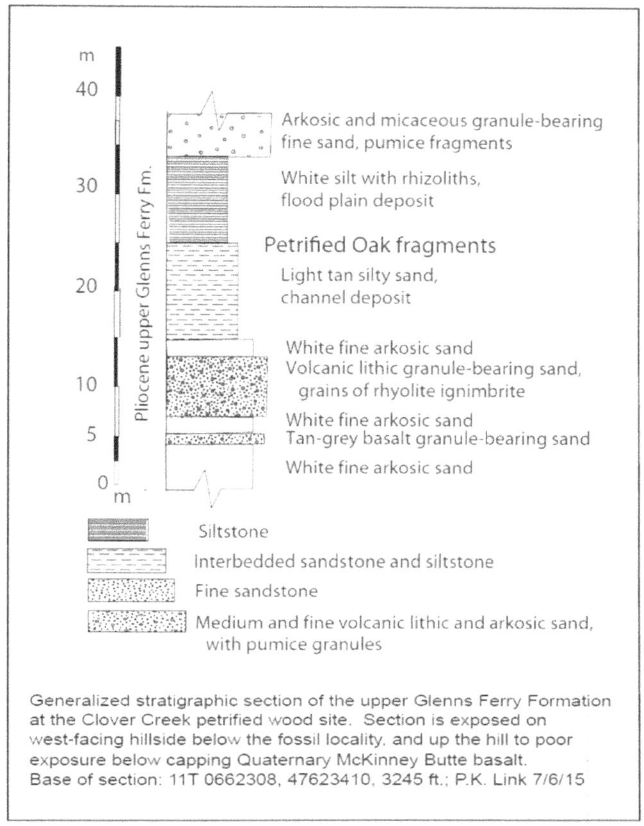

Figure 5. Generalized stratigraphic section of the upper Glenns Ferry Formation at the Clover Creek petrified wood site. Base of the section is in UTM coordinates.

Though exposure is poor, we estimate that about half of the strata are sand and half are siltstone. The sandstone represents fluvial channels, and the siltstone represents the adjacent flood plain. The most abundant fossil wood fragments are in a siltstone interval, suggesting the tree grew in the flood plain. Fragments of well-cemented coarse sandstone are also found with the wood fragments, suggesting that the sand acted as a permeable zone for silica-bearing fluids.

Detrital-zircons from central Idaho suggest that the Glenns Ferry Formation represents the paleo-Wood River, rather than the paleo-Snake River [24]. The Glenns Ferry Formation contains extraordinary vertebrate fossils at Hagerman Fossil Beds National Monument, 20 km to the southeast [25].

MATERIALS AND METHODS

Specimens were cut to reveal transverse, tangential and radial planes of subsequently polished thin-sections. Optical Microscopy (OM) was performed using a Nikon SMZ1500 stereomicroscope (Tokio, Japan), a Nikon Eclipse ME600 petrographic microscope equipped with a Nikon DS-5M-L1 digital camera and a Zeiss Photomicroscope II equipped with a Spot Infinity digital camera. Specimens, thin sections and digital photomicrographs are archived at the Geology Department, Western Washington University, USA, at Museum für Naturkunde Chemnitz and Professur Werkstoff- und Oberflächentechnik, Technische Universität Chemnitz, Germany.

Bulk powdered samples were analyzed by means of an X-ray diffractometer (Bruker AXS D8, Berlin, Germany) using Ni-filtered Cu Kα radiation.

Scanning Electron Microscopy (SEM) was done using a VEGA (Tescan, Brno, Czech Republic) for secondary electron (SE) images, a NEON40EsB (Zeiss, Jena, Germany) combined with an GEMINI (EDAX, Mahwah, Mahwah, NJ, USA) energy-dispersive X-ray detector for backscatter electron (BSE) images and elemental analysis and an Nova NanoSEM (FEI, Hillsboro, Hillsboro, OR, USA) equipped with a cathodoluminescence (CL) detector (GATAN monoCL, Pleasanton, CA, USA). CL spectroscopy and CL imaging (CL-OM) were done on carbon-coated, polished thin sections in the SEM operated at 20 kV and a "hot cathode" CL microscope HC1-LM (lumic, Dortmund, Germany) operated at a 14-kV accelerating voltage and a current of 0.2 mA (current density of about 10 μA/mm^2). Luminescence images were captured "on-line" during CL operations using a Peltier cooled digital video camera (OLYMPUS DP72, Tokio, Japan). CL spectra in the wavelength range 380 nm to 1000 nm were recorded with an Acton Research SP-2356 digital triple-grating spectrograph with a Princeton Spec-10 CCD detector that was attached to the CL microscope by a silica-glass fiber guide. CL spectra were measured under standardized conditions (wavelength calibration by an Hg halogen lamp, spot width 30 μm, measuring time 2 s). Raman spectroscopic measurements were performed with a 785-nm laser for excitation in a confocal Raman microscope (Renishaw inVia Reflex, Wotton-under-Edge, Gloucestershire, UK).

RESULTS

Microscopy

Analytical methods for studying silicification have greatly advanced since the work of Schuster [7]. However, his keen powers of observation are evident in the description of the anatomical characteristics with which he established the new species. Petrographic thin sections made from a sample collected in 2014 compare closely to photomicrographs of Schuster's original thin sections (Figure 6 and Figure 7), confirming that the recently-collected material matches the original 1895 discovery.

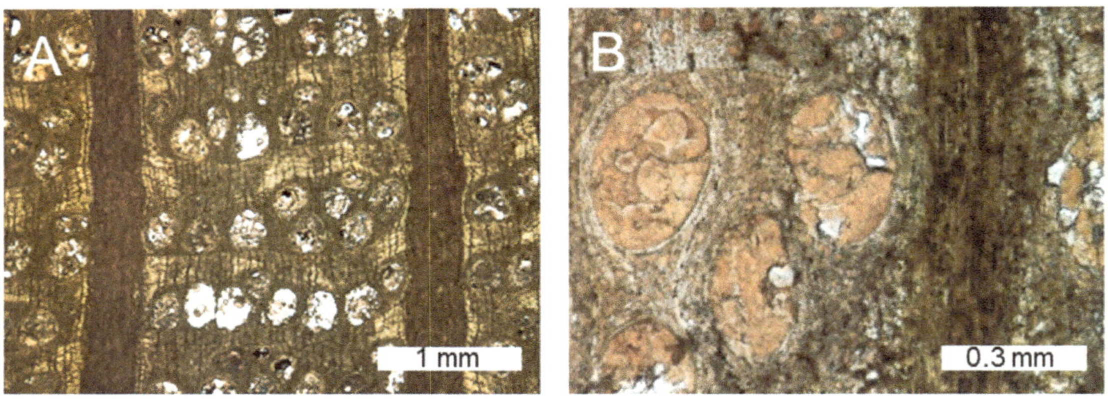

Figure 6. Petrographic thin sections of the 2014 specimen, transverse plane. (**A**,**B**) show anatomical detail matching Schuster's Table III, Figure 1 and Figure 2, respectively [7].

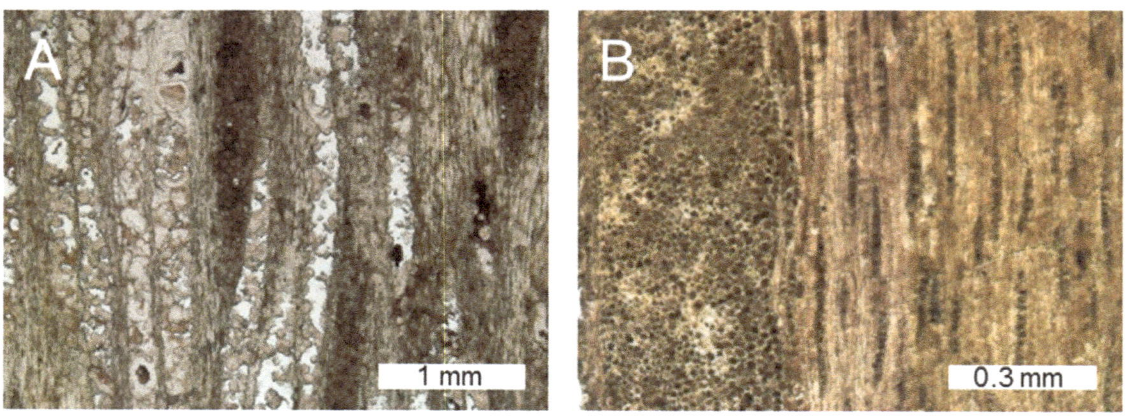

Figure 7. Petrographic thin sections of the 2014 specimen, tangential longitudinal plane. (**A**,**B**) show anatomical detail matching Schuster's Table IV, Figure 1 and Figure 2, respectively [7].

Schuster [7] focused his work on taxonomy, based on the well-preserved cellular detail. However, he also provided a description of petrographic features, observing that polished wood cross-sections were light yellow, with brownish discoloration along the outer margins. Schuster believed that brownish zones are probably caused by the presence of hydrated ferric oxide. In radial sections, the opal was reported to have a blue color.

Schuster's comments on mineralogy are brief, but he made an important observation: the silicification primarily consists of common opal, isotropic under polarized light, but most of the vessels are filled with chalcedony. Schuster [7] believed that the "more or less spherulitic chalcedony" originated from devitrification of the original opal. As discussed later, a more likely explanation is that the chalcedony filling the vessels represents a second stage of silica deposition, not a transformation of opal to chalcedony. The coexistence of opal and chalcedony is evident in thin sections prepared from the 2014 material (Figure 8).

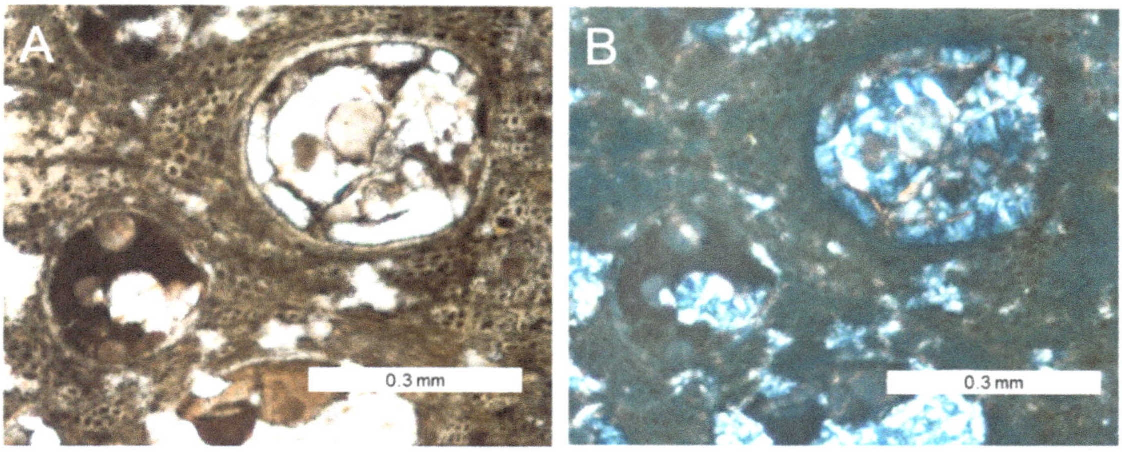

Figure 8. Petrographic thin sections of the transverse plane. (**A**) Vessels surrounded by cellular tissue in plain light illumination and (**B**) in cross-polarized light illumination.

Most of the cellular tissue is mineralized with opal-CT. Opalization of tracheids commonly includes mineralization of spaces between adjacent cells, accounting for the vitreous luster of hand specimens. The transparency of the silicified tissue makes transmitted light microscopy an effective method for revealing the detailed anatomical preservation. Cellular tissue has been mineralized with opal-CT, but vessels remained open during this phase of silicification. Later, chalcedony was precipitated by silica-bearing fluids permeating these vessels. Frequently, vessels were not completely filled with silica. Chalcedony surfaces that face these openings typically have a botryoidal texture. Polarized light microscopy allows for a clear distinction between opal-CT, chalcedony and open spaces. Opal-CT appears dark in cross-polarized light illumination; open spaces appear bright; and chalcedony exhibits birefringence (Figure 8B).

Opalized regions typically appear rather featureless in SEM views because the mineralization of intercellular spaces obscures cellular topography. In regions where intracellular spaces are not completely mineralized, the morphology of individual adjacent cells is visible. In areas of incomplete silicification, the morphologies of individual tracheids and mineral textures are evident (Figure 9). Tracheids are mineralized with tabular microcrystals that are a common form of opal-CT (Figure 9B,C). Mineralization of vessels is very different from that of the surrounding wood. Vessels remain partially unmineralized. Silica in the vessels is commonly present in the form of chalcedony (Figure 10). SE SEM images clearly show the cellular tissue mineralized with opal-CT to form a vitreous groundmass, in contrast to adjacent vessels that have been partially filled with chalcedony. In both transverse and radial views (Figure 10B,C), the semi-hemispherical shape of individual chalcedony masses facing the hollow interior of vessels can be observed.

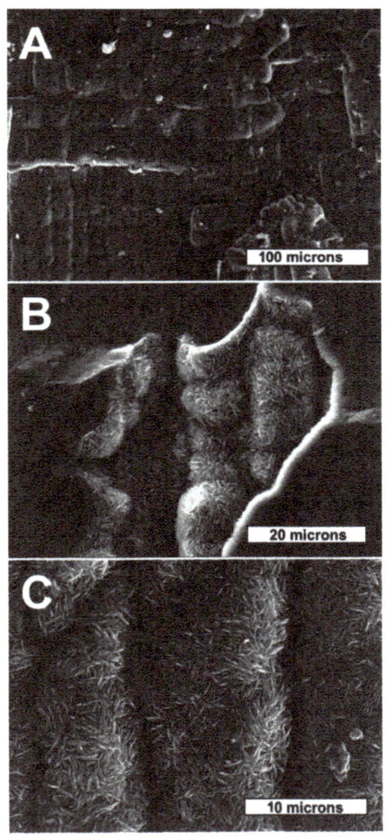

Figure 9. SEM SE images showing the external

morphologies of petrified cells. (**A**) Radial view of tracheids and ray cells; (**B**) silicification of intracellular spaces results in relatively smooth vitreous fracture surfaces, with areas of incomplete silicification revealing the morphologies of individual tracheids; (**C**) individual cell walls mineralized with tabular opal-CT microcrystals.

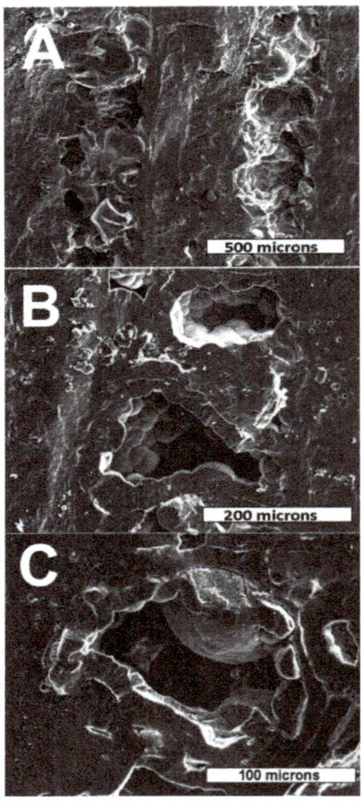

Figure 10. SEM SE images showing vessels partially filled with chalcedony. (**A**) Tangential view of two vessels surrounded by vitreous opal; (**B**) tangential view, fibrous chalcedony forming walls in the vessel and semi-spherical masses (botryoidal texture); (**C**) transverse view of a chalcedony-mineralized vessel.

The coloration of the fossil wood may be at least in part the result of trace amounts of iron. In some regions, silicified cells contain iron-rich microcrystals (Figure 11, Table 1). Atomic percentages calculated from EDS spectra indicate the presence of several iron minerals. Crystal shapes most commonly appear to be variations of the trigonal system, suggesting hematite as a possible parent mineral. Other symmetries are present, however, including orthorhombic forms that could have originated as marcasite. The presence of diverse iron minerals is perhaps the result of alteration related to multiple episodes of mineralization. The internal architecture of these crystals suggests complex paragenesis.

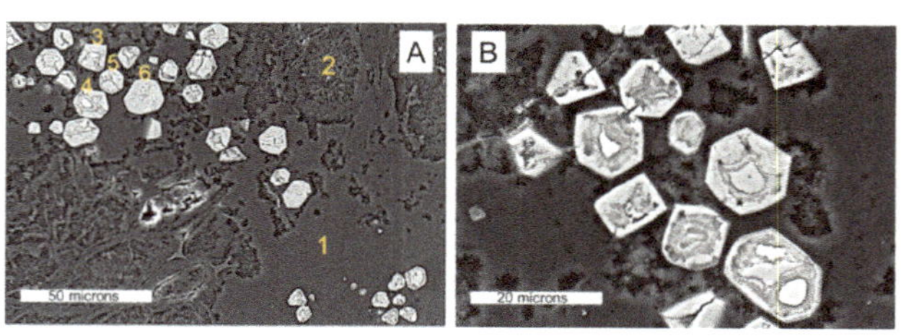

Figure 11. Elemental analyses of opalized wood. (**A**) Positions of the analyses (1–6) shown in Table 1; (**B**) detailed SEM BSE image of iron-rich microcrystals.

Table 1. Elemental analysis by EDS of Clover Creek opalized wood, the part with iron-rich microcrystals.

Location	Element Percent				Inferred Mineralogy
	Si	O	Fe	S	
1	32	68	-	-	opal
2	30	70	-	-	opal, chalcedony?
3	1	23	29	47	marcasite?
4	1	17	50	32	marcasite?
5	3	65	32	-	hematite?
6	4	54	32	1	hematite?

Diffraction and Spectroscopy

X-ray diffraction, Raman spectroscopy and cathodoluminescence microscopy were used to better understand the mineralogy. The X-ray diffraction pattern from a bulk powdered sample (Figure 12) shows the presence of two forms of silica, consistent with the evidence from optical and scanning electron microscopy. Opal-CT peaks are very evident. Modeling this mineral phase as interstratified layers of cristobalite and tridymite [26] results in a cristobalite/tridymite ratio around 0.54 and crystallite sizes around 25 nm. Quartz peaks apparently indicate chalcedony present as vessel filling; no other forms of quartz were observed by microscopy. The size of those crystals is between 60 and 80 nm, which is consistent with the expectation of fibrous chalcedony.

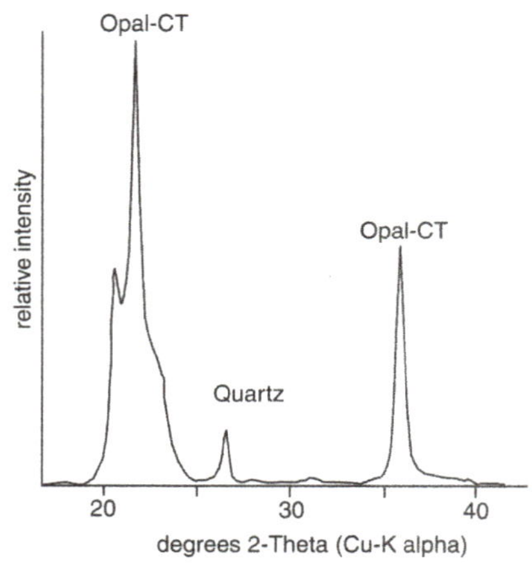

Figure 12. X-ray diffraction pattern for bulk powdered Clover Creek fossil wood.

Raman spectroscopy provides molecular information by inelastic scattering of photons and complements the crystallographic information achieved by elastic scattering of X-rays. Structure-building units, which are composed of a few silicon-oxygen tetrahedra, can be identified by means of vibrational, rotational and other low-frequency modes in the silica system. Different parts of the Clover Creek sample selected by OM indicate different silica polymorphs (Figure 13). Spectra gained from tracheids and medullary rays of the Clover Creek sample (Figure 13, Spectra c to h) exhibit peaks at 220 cm^{-1}, 310 cm^{-1}, 365 cm^{-1} and 415 cm^{-1} with varying intensities. These peaks are attributed to Si–O–Si bending vibrations, suggesting that the structure is close to an opal-CT type consisting of a matrix of intergrown tridymite- and cristobalite-type nano-domains according to [27]. Additionally, a higher portion of tridymite-like structural units can be inferred, which is in accordance with our XRD measurements.

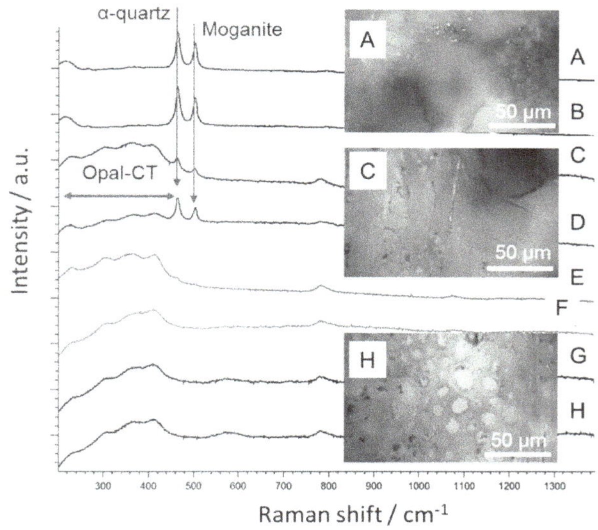

Figure 13. Raman spectra observed at different parts of the xylem. (**A**,**B**) Early wood vessels; (**C**–**E**) medullary rays; and (**F**–**H**) tracheids. Photomicrographs are OM images of a polished specimen.

Raman shifts at 464 cm^{-1} and 502 cm^{-1} are evident at early wood vessels (Figure 13, Spectra A,B). The peak at 464 cm^{-1} is attributed to α-quartz and the peak at 502 cm^{-1} to moganite. First described in 1984, but not formally recognized as a valid mineral species until 2007, moganite is a monoclinic SiO_2 polymorph that commonly occurs as a constituent in chalcedony and other varieties of microcrystalline quartz [28]. The moganite content of chalcedony has been reported to range from 26% to 82%, [29], but because moganite identification is usually based on Raman spectroscopy, the possible presence of this mineral in silicified wood has historically been overlooked. Götze et al.[30] provided a calibration curve for determining the moganite content in nano-crystalline α-quartz (chalcedony). Accordingly, the measured peaks were fitted by Lorentzian distributions for calculating the peak integral ratio. The result (0.5 to 0.7) indicates moganite content between 70 and 78%. Moganite occurs in chalcedony as nano-range lamellae down to single Brazil law twin planes detectable by Raman spectroscopy, but too small for the long-range order-dependent XRD. The spatially-resolved distribution of moganite in a chalcedony-preserved early wood vessel is shown by an intensity map of the moganite-related Raman peak (Figure 14).

CL spectroscopy in combination with microscopy provides information about the distribution of impurities and structural defects. This information can indicate mineralization sequences or hydrothermal mineralization and alteration of silica polymorphs. Four CL peaks were detected, with intensities that varied with respect to sample position and the CL analytical method. CL peaks at ~380 nm, 450 nm and 650 nm were detected by electron beam excitation with the SEM. Peaks at ~550 nm and 650 nm were detected using the optical microscope hot cathode CL system [30,31]. In the latter system, UV radiation cannot be detected because of the absorption effects of the glass optics.

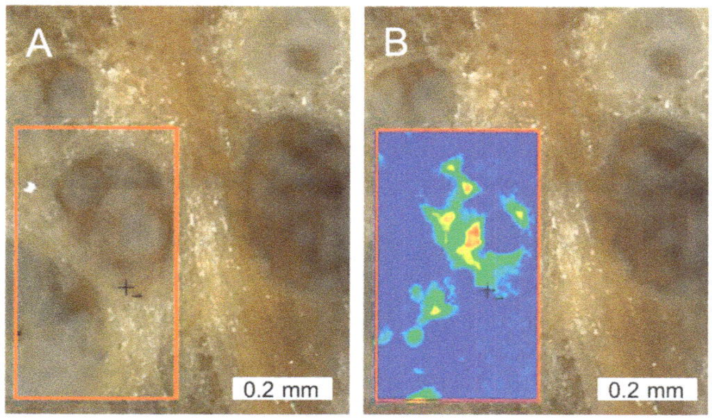

Figure 14. Moganite in a chalcedony-preserved early wood vessel. (**A**) OM image with field of view; (**B**) OM image with overlay of the Raman map generated from the 502-cm^{-1} peak intensity showing high (red) and low (blue) moganite content.

A silicified early wood vessel in the transverse section is shown by SEM BSE imaging, panchromatic SEM-CL imaging and corresponding CL spectra (Figure 15). The peaks at 450 nm, attributed to an oxygen-deficiency center [32], and 650 nm, attributed to a non-bridging oxygen hole center [30], are common in quartz and opal from igneous, volcanic and metamorphic processes. The chalcedony, which filled the early wood vessel, is the source of an additional peak at ~380 nm, which is typical for cryptocrystalline quartz with a hydrothermal origin [33].

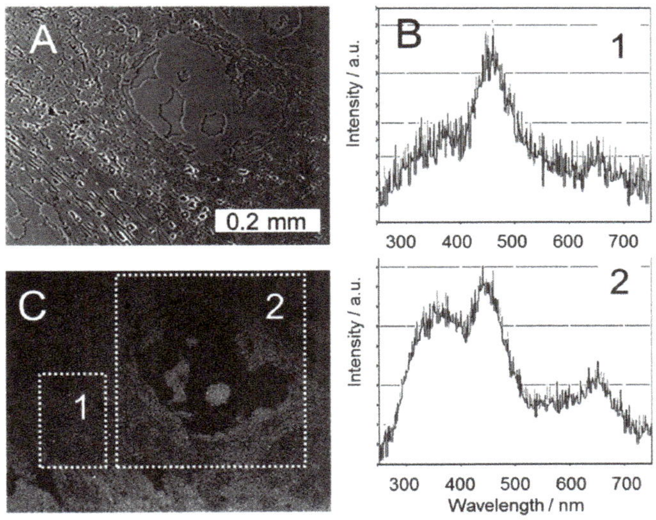

Figure 15. SEM-CL emerging from opalized tissue surrounding an early wood vessel (transverse plane). (**A**) SEM BSE image; (**B**) CL spectra for regions 1 and 2 marked in (**C**) panchromatic SEM-CL image.

The observation of CL peaks was accomplished using an optical CL microscope with a defocused electron beam. OM-CL images (Figure 16) show early wood vessels in the tangential longitudinal section by polarized light microscopy and OM-CL. The polarized light micrograph confirms that the silicified wood consists of a relative dense opal and larger vessels partially lined with chalcedony. The OM-CL images reveal bluish luminescence in the opal matrix and blue-violet luminescence in the vessels. OM-CL spectra are similar for Opal-CT and Chalcedony. The OM-CL spectrum shows only a slight shoulder of the UV band because of the system's detection limits. Cathodoluminescence changed in intensity with increasing radiation dose, this change being more evident in opal as compared to chalcedony (Figure 17). The bluish CL of opal turns to slight yellowish due to the activation of more defects. The CL spectrum of chalcedony is characterized by an increase of the 650 nm band and a decrease of the UV band around 380 nm (not visible in the OM-CL spectra). This is typical behavior for quartz crystallized from low temperature aqueous solutions (e.g., hydrothermal fluids) [33] and indicates a multi-stage process of silicification for the Clover Creek samples.

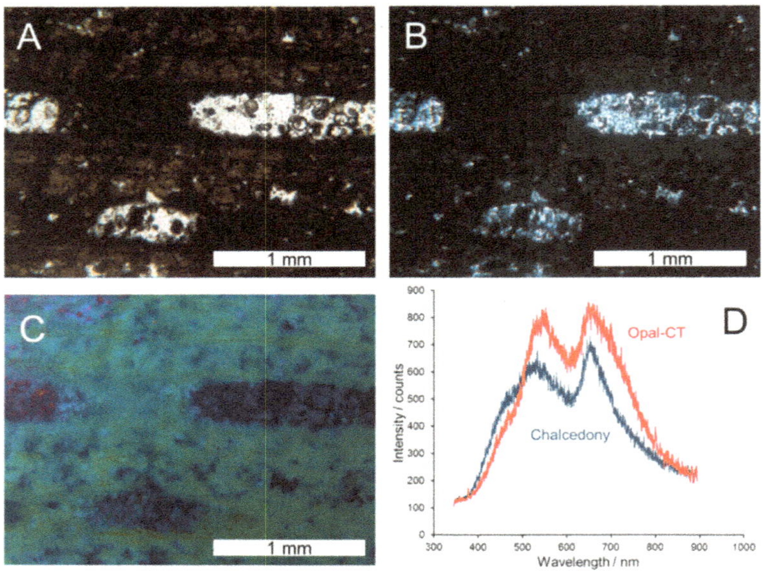

Figure 16. Photomicrographs of chalcedony-mineralized early wood vessels surrounded by opalized tissue (tangential plane). (**A**) Transmitted light OM imge; (**B**) Cross-polarized light OM image; (**C**) panchromatic OM-CL image; (**D**) CL spectra taken form opal/chalcedony rich regions.

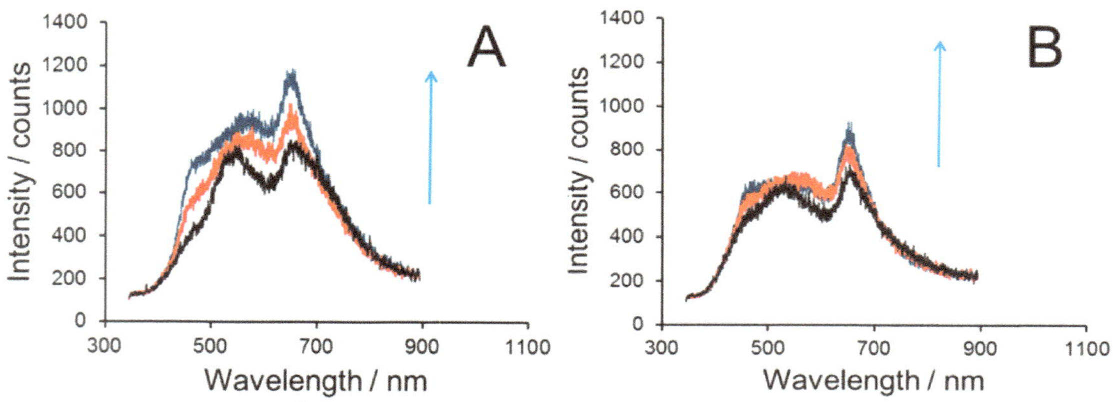

Figure 17. CL spectra taken at the initial state (black), after 60 s (red) and 180 s (blue). (**A**) Opaline matrix; (**B**) vessel region with chalcedony.

DISCUSSION

Taxonomy

The proposed taxonomy for fossil oak wood has not received universal agreement. The first description of fossil oak was by G□ppert [34] who introduced Klödenia quercoides as a name for wood from northern Germany and Silesia. This taxonomy became confused because Klödenia was later accepted as a name for an ostracod genus [35]. Unger [10] proposed Quercinium as the genus to be used for fossil oak wood exhibiting the cellular anatomy characteristic of the modern oak family (Fagaceae). This taxonomy received scant attention until Felix [36,37] published descriptions of known species and added four new ones.

As noted, Schuster [7] named the Clover Creek wood Q. pliocaenicum, and Boeshore and Jump [38] described Q. album from the Miocene Payette Formation of southwestern Idaho. Boeshore and Jump [38] include a table in their paper that listed all Quercinium species recognized in the literature at the time (1938); interestingly, Schuster's Q. pliocaenicum was not included. Other investigators have used other taxonomic strategies. Penhallow [39] described Holocene oak wood from Illinois as Quercus, a practice that has enjoyed continued use, e.g., [40]. Hofman [11] suggested Quercoxylon as a form genus name for fossil oak wood, but did not name a type species. Kräusel [12] designated Q. retzianum as a type species for this new genus.

It is not our intent to resolve these taxonomic complexities. We have retained the originalQuercinium pliocaenicum nomenclature because of historical precedence and because this fossil genus has been applied to most other North American occurrences of fossil oak wood (e.g., [15,41,42]). For a more detailed discussion of these taxonomic issues, see Iamandei et al. [14].

Paleoclimate and Paleoenvironment

At the time of the 1895 discovery and the subsequent published study [7], knowledge of the regional geology was rather scanty. Prior to the advent of radiometric dating, fossils offered a way to estimate ages of sedimentary strata and, by inference, the ages of associated igneous rocks. In recent decades, the geology of the Snake River Plain has received considerable attention, and our knowledge of the ages of rock units has been greatly refined thanks to precise radiometric ages determined for the many volcanic interbeds [21,43,44,45]. Snake River Plain sediments range from late Miocene to early Pleistocene. Quercus leaves are preserved in the late Miocene to early Pliocene Poison Creek and Chalk Hills Formations; Cherry Creek fossil oak wood is probably from the late Pliocene Glenns Ferry Formation.

While Schuster did not offer any thoughts on the paleoclimate associated with Q. pliocaenicum, he did use the presence of narrow growth rings to infer that the tree was deciduous and growing in poor conditions. Today, the Snake River Plain sediments provide one of the best-documented records of Neogene paleoclimate in the western USA. Paleoclimatic inferences have been derived from fossil mammals [46], fish [47,48] and pollen [49,50], as well as isotopic investigations [51]. The paleoflora document the presence of diverse deciduous trees in the central Idaho region during the late Miocene. The Pliocene Glenns Ferry flora records a significant climate change, with the appearance of Artemisia (sagebrush), Sarcobatus (saltbush) and abundant Poaceae (grasses). Pollen from aquatic plants indicates the presence of ponds and lakes. Tree pollen includes Pinus (pine) as the most common conifer, but diverse hardwood pollen includes Quercus. Mustoe and Leopold [50] believed that this floral evolution is the result of the late Miocene or early Pliocene rise of the Cascade Range, resulting in a rain shadow. Glenns Ferry sediments appear to have been deposited in a region of gentle topography, with grasslands, areas of temperate forest and scattered wetlands. The overlying early Pliocene Bruneau Formation shows a marked increase in the abundance ofArtemisia and Sarcobatus and a decline in Poaceae and aquatics, indicating increasing aridity. Mustoe and Leopold [50] used Jaccard similarity coefficients to estimate paleoclimatic conditions for Neogene paleoflora of Washington, Oregon and Idaho. These data suggest that the floral composition of the Glenns Ferry Formation flora resembled modern vegetation from Lewiston, Idaho, ~600 km to the northwest. The Glenns Ferry paleoflora is estimated to have had a mean annual temperature (MAT) of ~12 °C and annual precipitation of 300 to 325 mm. This compares to the modern climate at nearby Hagerman, Idaho, with MAT of 10 °C and annual precipitation of ~250 mm. Pollen studies for Glenns Ferry Formation and Idaho Group strata between 12 and 4 Ma suggest a savannah-like setting, with an annual precipitation of up to ~500 mm [22,49,52].

The ancient oak tree whose remains were preserved at Clover Creek presumably flourished at a time when the climate was somewhat warmer and wetter than at present. One of the most significant climatic differences is seasonality; estimated July precipitation was ~17 mm during Glenns Ferry time, compared to ~5 mm in the modern era. This increased seasonality was an important factor in the appearance of sagebrush steppe flora and explains the decline in habitat for oaks and other broadleaf hardwoods in the central Idaho region.

Silicification Pathway

One of the most significant aspects of the Clover Creek fossil wood is the evidence it provides for the silicification processes. Schuster [7] observed that the wood cells were mineralized with common opal, but that vessels contain chalcedony. Schuster believed that the "more or less spherulitic chalcedony" originated from devitrification of the original opal. During that era, processes of wood silicification were poorly understood, and although Schuster did comment on transformation processes, he did not propose possible mineralization processes.

Pathways for the formation of petrified wood have been discussed as early as 1665 when Robert Hooke used his newly-invented microscope to inspect the cellular structure of petrified wood for the first time [53]. In 1976, Leo and Barghoorn [54] established a key concept for wood silicification that includes two basic assumptions. First, wood acts as a template for silica, even as it slowly degrades and is replaced. Second, silica sol is the silicifying agent that forms hydrogen bonds with hydroxyl groups exposed on cell walls. This concept has been incorporated into every silicification model that has since been proposed. Dietrich et al. provide a review of studies on artificial and natural silicification [55]. More recently, Mustoe [56] studied the evidence of multiple silicification pathways of Neogene fossil woods. Mustoe's model includes early mineralization of cell walls followed by a later silica deposition in lumina, vessels and rot pockets from groundwater that permeated open spaces. He also described the process of open-space filling as analogous to the genesis of geodes and veins, where multiple episodes of hydrothermal precipitation produce opal, chalcedony and quartz as deposits within a single cavity.

Clover Creek fossil wood is a good example of polyphasic and polymineralic petrifaction. Although most of the cellular tissue has been mineralized with opal, the extent of mineralization is not homogeneous. In many regions of the specimens, cell walls, lumen and intracellular spaces are completely mineralized to produce a solid mass of opal. X-ray diffraction and Raman spectroscopic data reveal this material to consist of opal-CT. SEM images show the platy crystalline structure of this opal, but there is no visible evidence of opal lepispheres, the common architecture of opal-A. Instead, opal-CT appears to have formed as a primary mineral. During this initial phase of silica deposition, vessels remained open.

At a later stage of mineralization, many of the vessels became partially filled with chalcedony. The geode-like architecture suggests that the chalcedony was precipitated from hydrothermal fluids during a later episode, unrelated to the initial silicification phases that mineralized both tracheids and ray parenchyma with opal-CT. The observation that vessels are only partially filled with chalcedony is evidence that a silica-bearing hydrothermal fluid was able to permeate the wood even after the tracheids enclosing the vessels were thoroughly opalized. This characteristic suggests that the conductive components of the xylem (tracheids, vessels and fibers) were an important conduit for the entry of silica sol during early stages of petrifaction. This interpretation is consistent with the experimental evidence [54] that suggests that the earliest phase of silica deposition involves precipitation on cell wall surfaces because of the affinity between cellulose and hemicellulose with silicic acid (silica sol). Because some intracellular spaces and some vessels remain open, Clover Creek fossil wood remains potentially susceptible to additional future episodes of mineralization. The initial mineralization of cell walls followed by a later episode of silica deposition in vessels from hydrothermal fluids that permeated these open spaces is consistent with similar observations [56]. The coloration of the fossil is in part the result of hydrated iron oxide. Some silicified cells additionally contain trigonal and orthorhombic pseudomorphs. The internal architecture of these crystals suggests complex paragenesis and may provide additional evidence in support of multiple episodes of mineralization.

CONCLUSIONS

Clover Creek opalized wood is displayed as a mineral variant instead of a fossil in major museums in the U.S., Canada and Europe, even though the mineral dealer Foote [2] described the specimen as "the most perfect and beautiful petrifaction in existence". Interestingly, only a single scientific work in German describing the anatomy and providing the scientific name of Quercinium pliocaenicum for this historic find existed up until this publication. Schuster's paper [7] has recently been translated to English (see the Supplementary Material). Historical research uncovering the provenance of museum specimens, rediscovery of Schuster's original thin section slides, along with rediscovery of the original site and the collection of new specimens allowed for a thorough scientific investigation of the geologic setting and wood mineralization. Clover Creek museum specimens can now be understood in the context of modern analyses of the geologic setting and mineralization.

The mineralization of Clover Creek opalized wood adds to a growing body of evidence suggesting a multistage pathway for wood silicification. X-ray diffraction indicates that the wood is mineralized with opal-CT and chalcedony. OM and SEM reveal that most of the cellular tissue is mineralized with tabular opal-CT microcrystals with no evidence of opal lepisphere precursors, whereas inner vessel walls are lined with botryoidal chalcedony. Raman spectroscopy provides evidence for the intergrowth of chalcedony and moganite. The observation that opal-CT has replaced most of the tissue, with vessels partially filled by chalcedony, each silica polymorph originating as a primary mineral, was substantiated by CL. This is interpreted as evidence for a multiple stage mineralization process. EDS spectra reveal the presence of authigenic iron-rich microcrystals having both trigonal and orthorhombic shapes. The low abundance of these crystals suggest they are not a major cause of the wood's lustrous brown coloration, which is more likely caused by submicroscopic particles. However, the internal architecture of these crystals suggests complex paragenesis, perhaps representing further evidence for multiple episodes of mineralization.

The current study verifies that the fossil tree was recovered from the upper Glenns Ferry Formation along Clover Creek in Idaho. The geologic setting in which the tree was deposited and fossilized reveals that it grew in a flood plain along a fluvial channel. The tree grew at a time when the climate was warmer and wetter; however, the influence of the Cascade rain shadow slowly produced a dryer climate that over time would be less conducive to growing oaks, such as Q. pliocaenicum.

Supplementary Materials

The following are available online at http://www.mdpi.com/2076-3263/6/2/21/s1: pdf file 1: Translation Schuster Q pliocaenicum, pdf file 2: Supplement XRD", pdf file 3: Clover Creek Museum specimens.

ACKNOWLEDGMENTS

The authors would like to thank the following people for answering requests and providing photographs: Katherine Dunnell, Royal Ontario Museum, Canada; Peta Hayes and Mike Rumsey Natural History Museum London, U.K.; Bruce Geller, Beth Simmons, and Sam Allen, Colorado School of Mines Geology Museum, USA.; Klaus Thalheim, Senckenberg Naturhistorische Sammlungen Dresden/Museum für Mineralogie und Geologie, Germany; Achim Reisdorf, Geologisch-Paläontologisches Institut der Universität Basel and Thomas Burri, Naturhistorisches Museum der Burgergemeinde Bern, Switzerland; Zina Fihl, Statens Naturhistoriske Museum Kobenhavn, Denmark; Ursula Müller-Krantz, Dr. F. Krantz Rheinisches Mineralien-Kontor GmbH and Co. KG, Germany. We thank William W. Besse, Associate Cartographer for Rocks and Mineralsmagazine, Philadelphia, USA, for use of his Clover Creek map.

Thomas Mehner, Professur Werkstoff- und Oberflächentechnik, TU Chemnitz, performed the XRD analyses. Daniel Wett, Professur Verbundwerkstoffe, TU Chemnitz, and Massimiliano Rocchia, Thermo Fisher Scientific, Milano, Italy, provided Raman data. Steffen Schulze, Professur Analytik an Festkörperoberflächen, TU Chemnitz, achieved the SEM-CL data. We are much obliged to Alex Brandl, Colorado State University Colorado, USA, for his assistance in translating the paper of Schuster into English (see the Supplementary Material).

Author Contributions

Mike Viney revived the history of the samples. Mike Viney and Paul Link rediscovered the fossil site and investigated the stratigraphy. George Mustoe performed Optical Microscopy (OM) and Scanning Electron Microscopy (SEM) studies. Ronny Rößler provided OM on ancient and newly-prepared thin sections. Jens Götze performed CL measurements. Dagmar Dietrich performed energy dispersive spectroscopy (EDS) analysis. Thomas Lampke contributed analytical tools and supervised Dagmar Dietrich, who was responsible for collecting, analyzing and interpreting the analytic results. Mike Viney, George Mustoe, Paul Link and Dagmar Dietrich wrote the paper.

Conflicts of Interest

The authors declare no conflict of interest.

ABBREVIATIONS

The following abbreviations are used in this manuscript:

XRD	X-Ray Diffraction
SEM	Scanning Electron Microscopy
EDS	Energy Dispersive Spectroscopy
T	Transverse
TL	Tangential Longitudinal
RL	Radial Longitudinal
OM	Optical Microscopy
SE	Secondary Electron
BSE	backscattered electron
CL	cathodoluminescence

REFERENCES

1. Penrose, C.W. A Singular Natural Curiosity on Clover Creek Idaho; The Salt Lake Herald: Salt Lake, UT, USA, 1897; p. 3.
2. Foote, W.M. A complete catalogue of minerals. Available online: http://catalog.hathitrust.org/Record/008423524 (accessed on 3 March 2016).
3. Walcott, C.D. Twentieth Annual Report of the United States Geological Survey. 1898–1899, Part VI—Mineral Resources of the United States 1898; USGS: Reston, VA, USA, 1899.
4. Bauer, M.H. Edelsteinkunde; C.H. Tauchnitz: Leipzig, Germany, 1909.
5. Brauns, R.A. Das Mineralreich; Fritz Lehmann Verlag: Stuttgart, Germany, 1903.
6. Whitlock, H.P. The Story of Minerals; The American Museum of Natural History Handbook Series No. 12; American Museum of Natural History: New York, NY, USA, 1932.
7. Schuster, J. Über ein pliocänes Eichenholz aus Idaho. Neue Jahrb. Mineral. Geol. Pal. **1908**, 2, 49–54.
8. Müller-Krantz, U.; (Dr. F. Krantz Rheinisches Mineralien-Kontor GmbH & Co. KG, Bonn, Germany). Mineralogisches Semester-Verzeichnis Nr.8, Dezember 1908. personal communication, 2014.
9. Müller-Krantz, U.; (Dr. F. Krantz Rheinisches Mineralien-Kontor GmbH & Co. KG, Bonn, Germany). Paläontologisches Semester-Verzeichnis Nr.35, Dezember 1909. personal communication, 2014.
10. Unger, F. Synopsis lignorum fossilium plantarum acramphibryarum. In Genera Plantarum Secundum Ordines Naturales Disposita (Suppl. II), Appendix, Apud Fridericum Beck; Endlicher, S., Ed.; Universitatis Bibliopolam: Wien, Austria, 1842; pp. 100–102.
11. Hofmann, E. Verkieselte Hölzer aus dem Sarmat des Tokaj-Eperjeser-Gebirges. Föld. közlöny**1939**, 12, 260–270.
12. Kräusel, R. Ergebnisse der Forschungsreisen Prof. E. Stromers in den Wüsten *Ägyptens*. IV. Die fossilen Floren *Ägyptens*; Antiquariat Thomas Haker GmbH and Co. KG: Berlin, Deutschland, 1939; Volume 47, p. 140.
13. Gros, J.P. La dénomination des bois fossiles identifiés à des chênes. Bull. Mens. Soc. Linn. Lyon **1988**, 57, 250–260.
14. Iamandei, S.; Iamandei, E.; Bozukov, V.; Tsenov, B. Oligocene woods from Rhodopes, Bulgaria. Acta Palaentol. Rom. **2012**, 9, 15–25.
15. Wheeler, E.F.; Scott, R.A.; Barghoorn, E.S. Fossil dicotyledonous woods from Yellowstone National Park II. J. Arnold Arbor. **1978**, 59, 1–25.
16. Nickles, J.M. Bibliography of North American Geology for 1908 with Subject Index; United States Geological Survey: Reston, VA, USA, 1909; Volume 409.
17. Edwards, W.N. Dicotyledones (Ligna). In Fossilium Catalogus II; Jongmans, W., Ed.; Plantae. Pars 17; W. Junk: Berlin, Greman, 1931; pp. 63–69.
18. LaMotte, R.S. Catalogue of the Cenozoic plants of North America through 1950; Geological Society of America: New York, NY, USA, 1952; Volume 51.
19. Lewis, R.S.; Link, P.K.; Stanford, L.R.; Long, S.L. Geologic Map of Idaho 2012: Scale 1:750,000; Idaho Geologic Survey: Moscow, ID, USA, 2012.
20. Viney, M.; Mills, J.; Link, P. Opalized wood from Clover Creek, Gooding County, Idaho: Museum Mineral Specimens lead to the Rediscovery of an 1897 Fossil Dig. Rocks Miner. **2016**,91, 258–268.
21. Malde, H.E.; Powers, H.A. Geologic map of the Glenns Ferry-Hagerman area, west-central Snake River Plain, Idaho; scale 1:48,000; Map I-696; U.S. Geological Survey Miscellaneous Investigations. U.S. Geological Survey: Reston, VA, USA, 1972.
22. McDonald, H.G.; Link, P.K.; Lee, D.E. An overview of the geology and paleontology of the Pliocene Glenns Ferry Formation, Hagerman Fossil Beds National Monument. Northwest Geol. **1996**, 26, 16–45.
23. Wood, S.H.; Clemens, D.M. Geologic and Tectonic History of the Western Snake River Plain, Idaho and Oregon. Tectonic and Magmatic Evolution of the Snake River Plain Volcanic Province; Bonnichsen, B., White, C.M., McCurry, M., Eds.; Idaho Geologic Survey Bulletin: Moscow, ID, USA, 2002; Volume 30, pp. 69–103.
24. Link, P.K.; McDonald, H.G.; Fanning, C.M.; Godfrey, A.E. Detrital zircon evidence for Pleistocene drainage reversal at Hagerman Fossil Beds National Monument, central Snake River Plain, Idaho. Tectonic and Magmatic Evolution of the Snake River Plain Volcanic Province; Bonnichsen, B., White, C.M., McCurry, M., Eds.; Idaho Geological Survey Bulletin: Moscow, ID, USA, 2002;

Volume 30, pp. 105–119.

25. Baxter, K.; Shallat, T. Secrets of the Magic Valley and Hagerman's Remarkable Horse; Faculty Authored Books, Book 367; Black Canyon Communications: Boise, ID, USA, 2002.
26. Guthrie, G.D., Jr.; Dish, D.L.; Reynolds, R.C., Jr. Modeling the X-ray diffraction pattern of opal-CT. Am. Miner. **1995**, 80, 869–872.
27. Ilieva, A.; Mihailova, B.; Tsintsov, Z.; Petrov, O. Structural state of microcrystalline opals: A Raman spectroscopic study. Am. Miner. **2007**, 92, 1325–1333.
28. Heaney, P.J.; Post, J.E. The widespread distribution of a novel silica polymorph in microcrystalline quartz varieties. Science **1992**, 255, 441–443. [PubMed]
29. Bustillo, M.A.; Pérez-Jiménez, J.L.; Alonzo-Zarza, A.M.; Furio, M. Moganite in the chalcedony variations of continental cherts (Miocene, Madrid basin, Spain). Spectrosc. Lett.**2012**, 45, 109–113.
30. Götze, J.; Nasdala, L.; Kleeberg, R.; Wenzel, M. Occurrence and distribution of—Moganite" in agate/chalcedony: A combined micro-Raman, Rietveld, and cathodoluminescence study.Contrib. Miner. Petrol. **1998**, 133, 96–105.
31. Neuser, R.D.; Bruhn, F.; Götze, J.; Habermann, D.; Richter, D.K. Kathodolumineszenz: Methodik und Anwendung. Z. Geol. Paläontol. Teil I **1995**, 1, 287–306.
32. Götze, J.; Kempe, U. A comparison of Optical Microscope (OM) and Scanning Electron Microscope (SEM) based cathodoluminescence (CL) imaging and spectroscopy applied to geosciences. Miner. Mag. **2008**, 72, 909–924.
33. Götze, J.; Plötze, M.; Habermann, D. Origin, spectral characteristics and practical applications of the cathodoluminescence (CL) of quartz—A review. Miner. Petrol. **2001**, 71, 225–250.
34. Göppert, J.H.R. Über die fossile Flora der Quadersandsteinformation in Schlesien, als ersten Beitrag zur Flora der Tertiärgebilde. Verh. Kais. Leopoldinisch Carol. Akad. Natur. **1842**, 19, 99–134.
35. Jones, T.R. Notes on Palaeozoic bivalved Entomostraca II. Some British and foreign species of Beyrichia. Ann. Mag. Nat. Hist. **1855**, 2, 163–176.
36. Felix, J. Untersuchungen über fossile Hölzer. Z. dt. Geol. Ges. **1883**, 35, 59–91.
37. Felix, J. Die Holzopale Ungarns in palaeophytologischer Hinsicht. Mitth. Jahrb. Ungar. Geol. Anst. **1884**, 7, 1–43.
38. Boeshore, I.; Jump, J.A. A new fossil oak wood from Idaho. Am. J. Bot. **1938**, 25, 307–311.
39. Penhallow, D.P. Two Species From the Post-Glacial of Ilinois. Available online: http://www.biodiversitylibrary.org/bibliography/59738#/summary (accessed on 3 March 2016).
40. Cheng, Y.; Yi, T.; Li, C. Identification of the Early Pliocene fossil woods from Ninghai in Zhejiang Province and its palaeoenvironmental implications. J. Palaeogeogr. **2013**, 15, 105–112.
41. Knowlton, F.H. Fossil flora of the Yellowstone National Park. Geology of the Yellowstone National Park; Geological Survey Monograph: Washington, DC, USA, 1899; Volume 32, pp. 651–882.
42. Platen, P. Untersuchungen Fossiler Hölzer aus dem Westen der Vereinigten Staaten von Nordamerika; Leipzig: Quelle, Meyer, 1908.
43. Malde, H.E. Stratigraphy of the Glenns Ferry from Hammet to Hagerman, Idaho; Geological Survey Bulletin 1331-D; United States Government Printing Office: Washington, DC, USA, 1972; p. 19.
44. Kimmel, P.G. Stratigraphy, age and tectonic setting of the Miocene-Pliocene lacustrine sediments of the western Snake River Plain, Oregon and Idaho. Cenozoic Geology of Idaho; Bonnichsen, B., Breckenridge, R.M., Eds.; Idaho Bureau of Mines and Geology Bulletin. Idaho Geological Survey: Moscow, ID, USA, 1982; Volume 26, pp. 559–578.
45. Jenks, M.D.; Bonnichsen, B.; Godchaux, M.M. Geologic Map of the Grand-View Bruneau Area, Owyhee County, Idaho: Digitally Revised Version, Idaho Geological Survey Technical Report 98–1; unpublished work. 1998.
46. Ruez, D.R., Jr. Middle Pliocene paleoclimate in the Glenns Ferry Formation of Hagerman Fossil Beds National Monument, Idaho: A baseline for evaluating faunal change. J. Idaho Acad. Sci. **2006**, 42, 1–16.
47. Smith, G.R. Fishes of the Pliocene Glenns Ferry Formation southwest Idaho. Univ. Mich. Museum Paleontol. Pap. **1975**, 14, 1–68.
48. Smith, G.R.; Swirydczuk, K.S.; Kimmel, P.G.; Wilkinson, B.H. Fish biostratigraphy of late Miocene to Pleistocene sediments of the western Snake River Plain, Idaho. Cenozoic Geology of Idaho; Bonnichsen, B., Breckenridge, R.M., Eds.; Idaho Bureau of Mines and Geology Bulletin. Idaho Geological Survey: Moscow, ID, USA, 1982; Volume 26, pp. 519–541.
49. Leopold, E.B.; Wright, V.C. Pollen Profiles from the Pliocene-Pleistocene Transition in the Snake River Plane, Idaho. In Late Cenozoic history of the Pacific Northwest: Pacific Division; Smiley, C.J., Ed.; American Association for the Advancement of Sciences: Washington, DC, USA, 1983; pp. 323–348.
50. Mustoe, G.E.; Leopold, E.B. Paleobotanical evidence for the post-Miocene uplift of the Cascade Range. Can. J. Earth Sci. **2014**, 51, 809–824.
51. Smith, G.R.; Patterson, W.P. Mio-Pliocene seasonality of the Snake River plain: Comparison of faunal and oxygen isotopic evidence. Palaeogeogr. Palaeoclimatol. Palaeoecol. **1974**, 107, 291–302.
52. Davis, O.K.; Ellis, B. Early occurrence of sagebrush steppe, Miocene (12 Ma) on the Snake River Plain. Rev Palaeobot. Palynol. **2010**, 160, 172–180.
53. Hooke, R. Observ. 17. Of wood, and other bodies, petrified. In Micrographia or Some Physiological Descriptions of Minute Bodies Made by Magnifying Glasses with Observations and Inquiries Thereupon; Royal Society London: London, UK, 1665; pp. 107–110.
54. Leo, R.P.; Barghoorn, E.S. Silicification of wood. Bot. Mus. Leafl. Harv. Univ. **1976**, 25, 1–47.
55. Dietrich, D.; Viney, M.; Lampke, T. Petrifactions and

wood-templated ceramics: Comparisons between natural and artificial silicification. IAWA J. **2015**, 36, 167–185.

56. Mustoe, G.E. Late Tertiary petrified wood from Nevada, USA: Evidence of multiple silicification pathways. Geosciences **2015**, 5, 286–309.

CHAPTER 4

PALEOECOLOGICAL PROXY DATA AND ECOLOGICAL MODELING

Virginia Iglesias, Gabriel . Yospin1 and Cathy Whitlock

Keywords: fire, charcoal, reconstruction, modeling, paleoecology

INTRODUCTION

Fire has been an integral part of the Earth system for the past 350 million years (Scott and Glasspool, 2006). Today, it is the dominant disturbance agent in most terrestrial ecosystems (Bowman et al., 2009), and its legacy effects–especially in forests–may persist for centuries or millennia. Projected changes in fire are expected to trigger biotic reorganizations with broad consequences for land-surface feedbacks and the global carbon cycle (Flannigan et al., 2009; Parisien and Moritz, 2009). Such scenarios will profoundly impact ecological services ranging from water availability to nutrient cycling and recreation, and have an economic, esthetic, and cultural cost.

The fire regime (i.e., the characteristic frequency, size, and intensity of fire) results from complex interactions between long-term trends in climate and local fuel availability and probability of ignition (Heyerdahl et al., 2008). In most parts of the world, human use and control of fire has led to fire regimes that reflect not only natural processes but also social-economic drivers and their evolution through time (Bowman et al., 2011). Thus, fire regimes can only be understood through the assessment of climate-vegetation-human-fire interactions at a variety of temporal and spatial scales (Whitlock et al., 2010; Moritz et al., 2012).

Paleoecological studies offer information on long-term fire variability and its relationship with a broader array of environmental conditions than can be observed at present. Such information is central to providing a context for modern and future fire activity and helping to identify the causes and consequences of fire at local-to-global scales. Robust assessments of environmental change, however, require the incorporation of quantitative measurements and uncertainty estimation to the typically qualitative nature of paleoecological studies.

An integration between paleoecology and ecological modeling is thus needed to fully understand climate-vegetation-human-fire linkages (Davis, 1994; Anderson et al., 2006; Peng et al., 2011). Recent studies exemplify integrative research that effectively combines both disciplines (Table 1; e.g., Ganopolski et al., 1998; Heiri et al., 2006; Bradshaw, 2008; Brubaker et al., 2009; Colombaroli et al., 2010; Henne et al., 2013; Brücher et al., 2014). These studies demonstrate the power of empirical paleoecological records to describe different components of past fire regimes, and the use of modeling approaches and data-model comparisons to offer broader inference of top-down drivers of past fire activity. In this manuscript, we (1) review the use of sedimentary charcoal as a fire proxy and the methods used in charcoal-based fire history reconstructions; (2) identify existing techniques for paleoecological modeling; and (3) evaluate opportunities for coupling of paleoecological and ecological modeling approaches to better understand the causes and consequences of past, present and future fire activity (Figure 1).

Citation: Virginia Iglesias, Gabriel . Yospin1 and Cathy Whitlock, Paleoecological proxy data and ecological modeling, http://dx.doi.org/10.3389/fpls.2014.00785.

Table 1. Studies utilizing a combination of paleoenvironmental data and ecological modeling mentioned in this paper.

Study	Charcoal analysis	Pollen analysis	Archeology	Other paleodata	Mathematical	Regional model	Global model	Review calling for synthesis
Anderson et al., 2006								X
Boyle et al., 2011		X	X					
Bradshaw, 2008		X				X		
Brubaker et al., 2009	X					X		
Brücher et al., 2014	X						X	
Brunelle et al., 2005	X							
Carcaillet, 1998	X							
Carcaillet and Muller, 2005		X				X		
Colombaroli et al., 2010	X	X				X		
Conedera et al., 2009								X
Davis, 1994								X
Flannigan et al., 2001	X				X			
Froyd and Willis, 2008								X
Ganopolski et al., 1998		X					X	
Gimmi and Bugmann, 2013								X
Heiri et al., 2006		X				X		
Henne et al., 2013		X				X		
Higuera et al., 2007	X				X			
Kaplan et al., 2009								X
Kaplan et al., 2011			X				X	
Keller et al., 2002		X				X		
Kleinen et al., 2011		X					X	
Lorenzen et al., 2011				X				
Peng et al., 2011								X
Perry and Millington, 2008					X	X		
Perry et al., 2012					X			
Power et al., 2008	X							
Scholze et al., 2003							X	
Whitlock and Bartlein, 2003	X	X						
Whitlock et al., 2007	X	X						
Wick and Mohl, 2006		X				X		

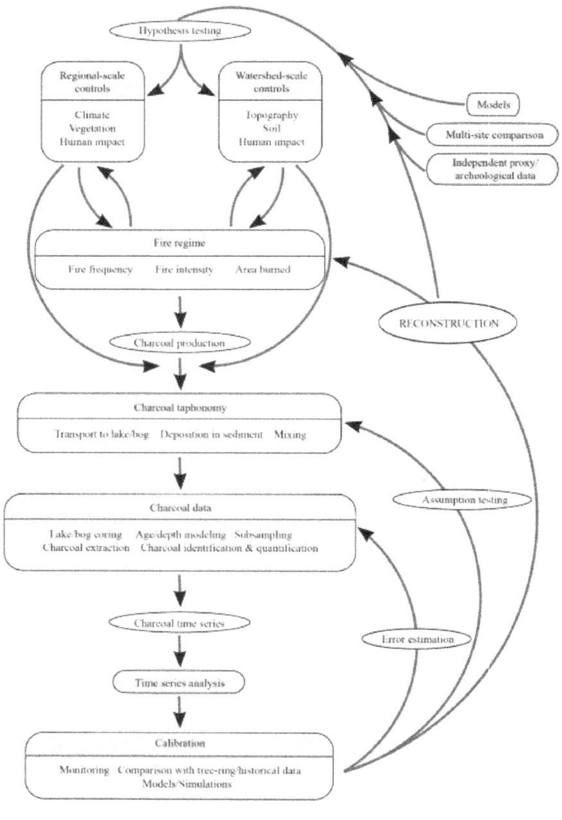

Figure 1. Reconstructing fire histories from lake sediments. Upper-left: charcoal production and taphonomy (i.e., natural sources of variability in the charcoal record). Lower-left: the charcoal record (i.e., analytical sources of variability in the charcoal record). Right: Steps in the reconstruction of past fire activity.

CHARCOAL DATA AND RELATED MODELING TECHNIQUES

Information on past fire activity comes from a variety of sources, including geochemistry, tree-rings, charcoal, archeology, and written documents. We focus on sedimentary charcoal analysis and, in particular, the study of lake- and bog-sediment records, because it has become the most widely used approach for high-resolution reconstructions spanning millennia. Charcoal is formed from the incomplete combustion of fuel under reducing conditions. Charcoal particles are carried aloft during the fire, transported to lakes by airborne and surficial processes, and incorporated into the sediments where they are preserved. Year by year, lake and bog sediments collect and bury charcoal from local and distant fires, along with pollen, macrofossils, and diatoms (Whitlock and Larsen, 2001; Conedera et al., 2009). Sediment cores are thus excellent repositories of paleofire information, and analysis of the charcoal, pollen, lithology, and geochemical components preserved in sediments is widely used to reconstruct watershed history, as well as the changing character of climate-vegetation-fire linkages.

Charcoal analysis is based on the assumption that stratigraphic intervals with abundant charcoal particles represent a primary contribution that is deposited mainly through aerial fallout during or shortly after a fire (Whitlock and Larsen, 2001). Temporal changes in charcoal abundance are inferred from downcore variability in charcoal concentrations. Changes in sedimentation rates through time affect the interpretation of charcoal concentration data, and charcoal accumulation rates (CHAR; charcoal particles * sediment volume^{-1} * time^{-1}) are generally considered a better proxy of past fire activity (Long et al., 1998).

Since the early reviews of Tolonen (1978) and Patterson et al. (1987), charcoal-based studies have become a primary source of information about long-term fire dynamics in most terrestrial ecosystems. Today, well-dated charcoal records are available for most parts of the world. For example, charcoal records from South America suggest that fire is a natural component of arid (Paduano et al., 2003), temperate and cool-temperate (Huber et al., 2004; Whitlock et al., 2007; Markgraf et al., 2013) and tropical wet ecosystems (Cordeiro et al., 2008), and changes in past fire activity have been linked to climate variability in Asia (e.g., Sun et al., 2000), Africa (Mworia-Maitima, 1997), Oceania (Black and Mooney, 2006), Europe (Daniau et al., 2007; Zumbrunnen et al., 2009), and the Americas (Behling, 1997; Moreno, 1997).

The interpretation of sedimentary charcoal data is not straightforward. Stratigraphically ordered sampling (i.e., temporal autocorrelation), non-random data, non-linear trends, uneven sampling over time and heterogeneous variance pose methodological and statistical challenges to the analysis of time series data. Moreover, processes affecting charcoal production, transport, and deposition and issues associated with the creation of an independent radiometric chronology compound uncertainties in fire history reconstructions. Assessment of climate-vegetation-fire dynamics therefore requires clever experimental designs, cautious inspection of the data and refined statistical analyses. Modeling can effectively address all of these requirements because it can fill in the information gaps inherent in paleoecological data and contribute to a mechanistic understanding of charcoal taphonomy, fire behavior, and fire-vegetation dynamics. We highlight several studies that meet these criteria throughout this review.

Ecological models incorporate varying degrees of complexity, resolution, and processes to portray fire-vegetation-climate interactions. Modeling approaches may be conceptual, statistical, or mechanistic. Conceptual models are based on sets of logical rules, and are often used where processes are complex, poorly understood, or both. Statistical models rely on empirical measurements, and often reflect a better understanding of the processes being modeled. Finally, mechanistic models incorporate equations that describe individual processes, implying a high degree of understanding of the system. Examples of these three approaches within the field of small-scale fire modeling include conceptual models like the USDA Forest Service›s standard fuel models (Scott and Burgan, 2005), statistical models such as Rothermel›s (1972) derivation of fire spread rates, and sophisticated and computationally demanding fluid dynamic modeling of fire spread (Dupuy and Morvan, 2005). These degrees of model complexity are evident in larger-scale fire models as well, such as those designed to simulate fire over landscapes. Statistical correlations between fire weather and area burned in the recent past have been used to reconstruct the size and distribution of fires during the Holocene, and to predict future area burned (Flannigan et al., 2001; Bergeron et al., 2006). Statistical modeling has also been used to simulate the production, transport, and deposition of charcoal during and following fires (Higuera et al., 2007), and these insights have helped justify the use of charcoal data as a proxy for past fires.

Models may include greater degrees of complexity by allowing some of the data and conditions that drive the model to update during the course of a simulation based on the ongoing conditions within the model. These "dynamic" (as opposed to "stationary") ecological models utilize interacting modules, each of which simulates different ecosystem properties and processes (e.g., Sitch et al., 2003) to extrapolate from micro- (e.g., carbohydrate production during photosynthesis) to macro-scale dynamics (e.g., global carbon cycling). Such models can increase understanding of complex, emergent phenomena (Perry and Millington, 2008).

Dynamic global vegetation models (DGVMs) are a class of dynamic models used to simulate shifts in potential vegetation at the regional to global scale as a response to shifts in climate. DGVMs simulate biogeochemistry, hydrology, vegetation, and disturbance, by attempting to explicitly represent processes like photosynthesis, plant competition, and mortality. Most DGVMs include several mechanistic modules, while still relying on some statistical and conceptual modules. Simulations develop an "equilibrium" vegetation that establishes initial values for various components, and experiments are then run in a spatially distributed mode across thousands of grid cells, each of which is assumed to have relatively homogeneous conditions. DGVMs allow the conditions from a prior time-step to strongly influence the system response at the current time-step, which makes possible the representation and description of changing ecosystem properties like a disturbance regime. DGVMs with interactive fire components describe biomass burning at regional to global scales by considering climate, ignition, surface fire spread, fire behavior, post-fire mortality, and carbon emissions to estimate area burned and biomass burned over the simulation area (Thonicke et al., 2010; Pfeiffer et al., 2013).

Because paleoecological studies are often used to infer information at the scale of landscapes, there are substantial benefits to research that integrates paleoecology with models that are intended to function at the landscape scale. At this scale, successful models usually include components that are conceptual, statistical, and mechanistic. Modules that are more conceptual within one landscape modeling system may be more mechanistic within another landscape modeling system. For example, seed production is more conceptual within LANDIS-II (Scheller et al., 2007), due to a low level of specificity with regard to plant community composition, whereas it is more mechanistic in FireBGCv2, which tracks more detailed information about individual trees (Keane et al., 2011). The variety of approaches in landscape models can be especially useful in systems where prehistoric human land use and anthropogenic burning are difficult to infer from the paleoecological record. Modeling approaches that range from stationary and largely conceptual (e.g., Perry et al., 2012) to those that are dynamic and largely mechanistic (e.g., Colombaroli et al., 2010) have been useful in evaluating competing hypotheses regarding the role of prehistoric humans in shaping the fire regime and vegetation.

Paleoecological reconstructions often motivate particular modeling experiments, and data, in turn, are used to validate model simulations of past conditions (e.g., COHAMP members, 1998; Bartlein et al., 2011). Data-model comparisons constitute a powerful tool for understanding particular processes (Higuera et al., 2011) as well as exploring explanations for past variations in vegetation and biomass burning (Daniau et al., 2012; Marlon et al., 2013; Molinari et al., 2013). Such comparisons require identifying the scales and precision inherent in the research question, understanding the resolution and uncertainties inherent in the data, and selecting a model in which the degree of integration, number of processes explicitly simulated, and temporal and spatial resolution are appropriate to the question (Bartlein and Hostetler, 2004). Even in cases where the assumptions embedded in charcoal analysis prevent direct quantitative comparisons with modeled output, non-parametric techniques can be employed in model-data comparisons (e.g., Brücher et al., 2014).

Local Fire Activity: Charcoal Source Area

Time series of sedimentary charcoal abundance have become a widely used tool for reconstructing long-term trends in biomass burning. Interpretation, however, rests on understanding the source area of charcoal particles (i.e., the area burned and distance to the site) and the ability to accurately infer other properties of the fire regime, such as frequency, intensity, and fuel type. Constraining the charcoal source area is not a trivial task because charcoal abundance is influenced by vegetation, fire climate, and fire weather as well as site-specific processes related to charcoal deposition and burial. Clark (1988)used particle physics to model the aerial transport and deposition of charcoal particles. In the model, a distance-decay function was used to predict larger transport distances for increased wind speeds and smaller particles. Subsequent calibration studies focused on monitoring the incorporation of charcoal in lake sediments after recent fires partially support Clark's model. In the International Crown Fire Modeling Experiment (ICFME) in the boreal forest of Siberia, for example, Lynch et al. (2004) reported high deposition rates close to the fire and an abrupt decrease in charcoal abundance at farther distances (i.e., >10 km), which matched the Clark (1988) model predictions of non-linear charcoal deposition pattern. Conversely, a broad range of particle sizes were observed within the boundaries of low- and high-intensity fires in the mid- and high-latitudes (Whitlock and Millspaugh, 1996; Pitkänen et al., 1999; Asselin and Payette, 2005), suggesting that particle size alone was not a reliable prediction of transport distance.

Correlation between charcoal records and known fire events (e.g., Gardner and Whitlock, 2001) indicate that charcoal from long distances can comprise a significant proportion of the total amount of particles. Such particles tend to be small (<100 mm), whereas dispersal of large charcoal particles is strongly biased toward short distances (Peters and Higuera, 2007). Much attention has therefore turned to the use of macrocharcoal (particles >125 mm in diameter) as a proxy for local fire history. Charcoal within this relative large size class is expected to have high settling velocities and to be associated with local source areas (<20 km from the lake) (Higuera et al., 2011). Similar conclusions were reached by Oris et al. (2014) after a 3-year study to monitor charcoal dispersion and deposition in boreal lakes. Their findings suggest that both local and regional fires are associated with higher concentrations of charcoal particles of all sizes in the sediments of lakes. Charcoal assemblages resulting from regional fires, however, tend to be characterized by higher proportions of small particles, whereas large particles represent more local events. The analysis of particle size distribution thus emerges as a tool to discriminate charcoal source area.

Analyses of other paleoenvironmental proxies complement this approach and aid in the identification of local fires. A number of studies have noted the correspondence between charcoal variability and changes in the pollen of fire-adapted plants (e.g., Swain, 1973; Patterson and Backman, 1988; Pitkänen et al., 1999) and diatom taxa (Ponader et al., 2002). Local high-severity fires have indirectly been inferred from increases in varve thickness, anomalously high content of aluminum, vanadium, and silt in sediments containing abundant charcoal (Cwynar, 1987), and peaks in carbon, nitrogen (Dunnette et al., 2014), and magnetic susceptibility (Rummery et al., 1979; Rummery, 1983) associated with post-fire erosion, terrestrial nutrient loss, and topsoil oxidation. Other studies, however, find the relationship between erosion and charcoal accumulation less straightforward (Long et al., 1998) and note a complex relationship between fires and sediment transport, which is dependent not only on fire location but also on fire intensity, fuel, soil, and substrate (Meyer et al., 1995).

In spite of advances in linking fire activity and charcoal accumulation in lakes, reconstructing area burned remains a challenge with great potential for model-data integration. Such efforts have been very effective for some aspects of charcoal analysis. For example, Higuera et al. (2007) used a statistical model that incorporated size, location, and frequency of fire as well as charcoal taphonomy and sampling techniques to evaluate assumptions of charcoal analysis. Simulations of charcoal accumulation are consistent with empirical data that suggested that sedimentary charcoal is an excellent proxy of fire activity within a 1–10 km radio of the lake (Higuera et al., 2011; Kelly et al., 2013). These studies helped validate charcoal-based fire reconstructions and identify uncertainties associated with natural (e.g., charcoal production, transport, and deposition) and analytical processes.

Local Fire Activity: Fire Frequency, Severity, and Fuel Type

Variability in the charcoal record comes from a variety of sources, including temporal variations in primary charcoal production, transport, and deposition; bioturbation and addition of secondary (remobilized) charcoal; and differences in sampling, chemical treatment, and charcoal identification and quantification. These sources can obscure the relationship between charcoal accumulation rates (i.e., CHAR) and fire. Consequently, interpretation of the charcoal time series usually involves decomposing the record into low- and high-frequency components (Higuera et al., 2005). The low-frequency component, also known as background charcoal, has been linked to regional charcoal production (Whitlock and Millspaugh, 1996) and long-term changes in charcoal delivery mechanisms before its inclusion in lake sediment (Long et al., 1998). More recent calibration studies in Yellowstone National Park and the boreal forests of Alaska suggest correlations of background CHAR and area burned within a radius of <30 km from the lake (Higuera et al., 2011; Kelly et al., 2013). Conversely, positive anomalies in the high frequency component of the charcoal time series (i.e., charcoal peaks) result from local fire episodes (i.e., one or more fires occurring within a few km of the lake during the deposition time) (Higuera et al., 2010).

Time and frequency domain approaches are used to decompose the charcoal time series into background and peaks (Clark and Royall, 1996), thus accounting for long-term changes in charcoal production and allowing a more sophisticated reconstruction of the fire history. In most studies, background charcoal is defined by fitting non-parametric smoothers to the charcoal time series (Higuera et al., 2007; Higuera, 2009; Blarquez et al., 2013). The positive residuals of the model (i.e., charcoal peaks) are inferred to represent fire episodes as well as natural and statistical noise. A user-defined threshold is frequently applied to the charcoal peaks to separate fire episodes from noise. Although based on poorly understood assumptions regarding charcoal taphonomy (MacDonald et al., 1991; Whitlock and Millspaugh, 1996), fire reconstructions following this method have been shown to correspond well with historical (Mensing et al., 1999; Kelly et al., 2013) and dendrochronological records of local fires (Higuera et al., 2005). However, as noted by Higuera et al. (2010), not all charcoal records are suited for peak detection, and in many cases the signal-to-noise ratio is too low to detect significant charcoal peaks from background trends. As a result, the limitations and assumptions of each fire history reconstruction should be assessed and clearly stated.

Identifying the location (i.e., local vs. regional) and time of individual fire events allows estimation of changes in fire frequency (i.e., number of fire events per unit time) and fire return interval (i.e., years between fire events) over millennia.

Assuming that CHAR time series represent trends in total biomass burned, Ali et al. (2012) proposed the CHAR-to-fire frequency ratio as a semi-quantitative proxy of fire severity. Reconstruction of long-term trends in fire frequency and severity aids in the understanding of the temporal and spatial variability of fire and its impact on landscape heterogeneity.

In addition to fire frequency (fire return intervals) and severity, additional information about past fires can be inferred from the identification of the charcoal particles. Umbanhowar and Mcgrath (1998), for example, used charcoal morphology as an indicator of type of biomass burned and fire intensity. A similar approach allowed the identification of changes in the dominant fuel (i.e., grass or wood) in Patagonia (Whitlock et al., 2006; Iglesias et al., 2012) and tropical Africa (Aleman et al., 2013). Vannière et al. (2003) analyzed the structure and morphology of charred particles from soil samples and noted distinguishing characteristics between airborne and reworked charcoal. Based on this idea, Enache and Cumming (2007) used morphological features in charcoal particles to infer taphonomic processes, and interpreted the presence of fragile particles as an indicator of fires in close proximity to the study site.

Anthracological studies (i.e., analyses of the anatomical structure of charred fragments in soils and archeological sites) have allowed the identification of the botanical origin (i.e., taxon) of large charcoal particles (>250 μm), as well as information on the structural characteristics (i.e., branch vs. trunk) and state (i.e., dead vs. alive) of the fuel before carbonization (e.g., Bégin and Marguerie, 2002; Marguerie and Thibaudeau, 2004; Allué et al., 2005; Asouti and Austin, 2005; Dufraisse, 2006; Carcaillet, 2007; Braadbaart and Poole, 2008; Thery-Parisot et al., 2010). Charcoal-based fire history reconstructions would greatly benefit from the application of techniques used in anthracolology to better characterize changes in fuel composition and fire intensity.

Regional Fire Activity: Climate-vegetation-fire Linkages

Fires are complex ecological processes. Hierarchy theory states that ecological processes are driven by factors acting at different scales, in which higher-level spatial factors constrain the effects of more local ones to various degrees (Turner, 2001). Studies of past biomass burning are grounded in this conceptual understanding of top-down and bottom-up influences on fire occurrence, and the interpretation of past fire activity often focuses on disentangling broad-scale slowing varying drivers, such as climate and vegetation, from local controls related to site characteristics, fuel conditions, and human activity (Whitlock et al., 2010).

The interactions between fire and climate are the most direct. Dendroecological studies, for example, have shown that changes in large-scale atmospheric circulation and seasonal weather patterns synchronize fire activity across broad regions by affecting the length of the fire season and the moisture content of fuels (Swetnam, 1993; Heyerdahl et al., 2001; Kitzberger et al., 2012). Similarly, composites of charcoal time series across continents reveal trends in biomass burning related to millennial scale variations in seasonal cycle of insolation and ocean-land interactions (Carcaillet et al., 2002; Power et al., 2008, 2013; Daniau et al., 2012; Marlon et al., 2012). In both the northern hemisphere (Daniau et al., 2007) and Australasia (Mooney et al., 2011), fire activity matches Dansgaard-Oeschger cycles observed in ice cores from Greenland, indicating long-term associations between regional-to-continental-scale biomass burning and warming. Additionally, a comparison of charcoal records from Indonesia, Papua New Guinea, Central and South America shows that fire is not only dependent on mean atmospheric conditions but also on hemispheric climate teleconnections that define interannual variability (Haberle and Ledru, 2001).

At the regional-scale, comparison of charcoal records between adjacent watersheds has been used to identify the importance of climate in synchronizing fire activity (Gavin et al., 2006; Walsh et al., 2010). Synchronous shifts in CHAR across multiple sites (Gavin et al., 2003; Rius et al., 2011) have been attributed to changes in regional fire activity related to periods of warmer temperature and decreased effective moisture. Nonetheless, multi-site comparison of fire histories also suggests that the climate-fire relationship is not linear, and the relative importance of regional- vs. watershed-scale controls of ecosystem dynamics changes as critical thresholds are reached (Ali et al., 2009; Iglesias et al., 2012). At present, for example, topography and fuel type effectively limit fires in the coastal temperate rainforest of western North America (Gavin et al., 2003) and the Alps (Carcaillet et al., 2009). During the early Holocene, however, when Northern Hemisphere summer insolation was higher than present, warmer fire-season temperatures and extended droughts synchronized fire activity in both regions. Thus, cross-scale studies reveal that the fire regime is the product of non-stationary interactions of large- and small-scale drivers of ecosystem dynamics (Turner and Romme, 1994).

Charcoal and pollen data show that vegetation influences fire regimes by affecting the type, load and spatial distribution of fuel across the landscape. In humid regions, dense woody vegetation provides abundant and continuous fuel. However, due to the high water content of plant tissues and soil, protracted droughts are necessary for fuel to become flammable and fires to spread. Thus, as exemplified by the fire history of Xishuangbanna in southern China (Gu et al., 2008), infrequent high-intensity fires in tropical rainforests are strongly controlled by fuel moisture.

Conversely, fine fuels in seasonally-xeric environments desiccate quickly and are frequently dry enough to support fires, even during non-drought years. Areas where weather and climate are conducive to natural ignitions may not be those with the most active fire regimes because their climates may not support continuous flammable vegetation (Huber et al., 2004; Parisien and Moritz, 2009; Iglesias et al., 2014). In these ecosystems, frequent low-intensity fires are limited by fuel accumulation (Krawchuk et al., 2009). Postglacial composite charcoal records from the Americas (Power et al., 2008; Moreno et al., 2010; Iglesias and Whitlock, 2014), for example, show a direct correlation between moisture and fire at millennial time scales suggesting that, during glacial times, regional-to-continental-scale biomass burning was primarily limited by fuel availability, and rose during the Holocene, as vegetation density increased.

Comparison of pollen and charcoal from multiple sites shows that climate-vegetation-fire dynamics are complex and change as ecosystems evolve. In Alaska, the colonization of *Picea mariana* during the cool humid middle Holocene was associated with a pronounced increase in fire frequency, suggesting that the bottom-up controls of vegetation may not only amplify the effects of climate on fire regimes, but also dampen them (Higuera et al., 2009). Similarly, the fire history of subalpine forest in the Colorado Rockies showed little variation in fire frequency over the last 6000 years, but changes in biomass burned per fire (and inferred fire severity) indicated a response to climate-induced changes in forest density (Higuera et al., 2014).

Complex ecosystem dynamics resulting from climate-vegetation-fire feedbacks may be difficult to infer from proxy data alone and modeling efforts have helped explicitly incorporate assumptions and estimate uncertainties. Several studies have compared, and in some cases integrated, charcoal records and ecological modeling at regional-to-continental scales, allowing for an improved understanding of the evolution of the fire regime (e.g., Brubaker et al., 2009; Colombaroli et al., 2010; Molinari et al., 2013). In general, successful efforts have relied on conceptual and statistical models of fire behavior and charcoal deposition, at a range of scales from landscape to global. Flannigan et al. (2001), for instance, derived present and future fire weather maps from general circulation model and regional climate model simulations for two levels of atmospheric CO_2 in Canada. For each of these scenarios, they generated 6000 years of fire weather, and compared the simulated paleofire weather data to charcoal data. Good agreement between the charcoal record and ignition patterns under the paleoclimate scenario lent support to their estimates of changes in area burned throughout Canada under the future climate scenarios.

More complex, dynamic representations of vegetation have also been effectively compared with fire-history reconstructions. Brubaker et al. (2009)matched fire-return intervals from a dynamic landscape ecological model (ALFRESCO) and reconstructed fire return intervals developed from three charcoal records from the southern Brooks Range, Alaska, USA, over a 100 km^2 simulation area. They assumed a correspondence between fire occurrence in 1 km^2 model cells and the fire episodes–as inferred from the charcoal peaks in the CHAR record—in the CHAR record's source area. Fire size and frequency were calibrated to the modern record and then validated against historic records under varying climate and vegetation scenarios. Model-data integration thus allowed for spatially explicit reconstructions of the fire regime and the quantitative characterization of the associated environmental conditions.

Regional Fire Activity: The Role of Humans in Altering the Fire Regime

Human use of fire likely dates back over 1,000,000 years (Berna et al., 2012) and anthropogenic impact on natural fire regimes is likely to have changed as economies shifted from hunting-gathering to pastoralism, farming, and industrialization (Pyne, 2001). Charcoal composites from around the world have been used to examine human-fire relationships at continental scales over millennia (Marlon et al., 2008). At this coarse resolution, climate emerges as the dominant driver of fire activity (Daniau et al., 2010; Mooney et al., 2011; Power et al., 2013). Additional insight has come at the local to regional scale where fire history reconstructions can be compared with archeological, ethnographic and historical records of anthropogenic burning. Cross-scale studies of biomass burning in Patagonia, for instance, show very limited human impact at both the local- and regional scales before European arrival (Iglesias and Whitlock, 2014). Conversely, striking evidence of human-induced fires and deforestation comes from charcoal and pollen records from South Island New Zealand, where initial Mâori arrival is strongly associated with widespread burning and loss of native forest (McWethy et al., 2009, 2010). Population growth during the Holocene was also associated with increased fire activity in watersheds in the Alps (Carcaillet et al., 2009), the Mediterranean basin (Vannière et al., 2011), and southeastern Australia (Black and Mooney, 2006). In eastern North America, paleoecological and ethnohistorical records of land-use reveal that human impacts were widespread but their intensity was heterogeneous (Muñoz et al., 2014).

Paleoenvironmental reconstructions thus indicate that anthropogenic impact depends on scale, prevailing climate conditions, and vulnerability of the dominant plant species to fire. Human influences are most strongly felt at the wet and dry ends of the precipitation gradient (Whitlock et al., 2010; Moritz et al., 2012; McWethy et al., 2013). At the wet end of the gradient, where biomass is high and natural ignitions are either infrequent or do not occur at the time of fuel desiccation, people have increased ignition frequency, converting closed forests to open vegetation types (e.g., Perry et al., 2012). At the dry end, fire suppression in fuel-limited landscapes has shifted vegetation to more closed, fire-prone plant communities (e.g., Staver et al., 2011).

Simulation modeling experiments play a valuable role in testing these hypotheses and assessing whether changes in climate and/or human activity were both necessary and sufficient to generate shifts in vegetation and fire under different environmental conditions. Regional vegetation modeling studies have used experimental designs that allow for the inference of humans as active agents environmental change. For example, Colombaroli et al. (2010) employed a 2 × 2 design of climate and ignition scenarios in LandClim to simulate Holocene fire activity in the Alps. By comparing the output with pollen and charcoal data, they assessed the long-term role of humans in altering the probability of fire and highlighted the risk that modern societies pose to mountain environments. Simulations have also been undertaken to infer anthropogenic impact on vegetation dynamics. Models suggest the rapid rate of deforestation reconstructed for New Zealand after the initial Mâori arrival cannot be explained by climate alone and could only have occurred if fires were targeted to the most flammable vegetation on the landscape (Perry et al., 2012). Under these conditions, 40% of forests would have been lost within decades, an estimate that matches inferences from the charcoal record (McWethy et al., 2014). Studies nested in the paleoecology-modeling interface thus serve as a context for understanding the resilience of ecosystems to natural and human-induced environmental change.

Global Fire Activity

The growing geographic coverage of high-resolution charcoal data allows for reconstructions of past fire activity at continental and global scales. With the objective of archiving and sharing data for research and education, the Global Paleofire Workgroup (www.gpwg.org) supports a public-access database of fire records (e.g., charcoal, black carbon, levoglucosan) from around the world (Global Paleofire Working Group, 2014), as well as computer code for data analysis and synthesis (e.g., Blarquez et al., 2014). Compositing of charcoal records has allowed the reconstruction of biomass burning in the Americas (Marlon et al., 2009; Moreno et al., 2010) and Australasia (Mooney et al., 2011), the northern and southern hemispheres (Daniau et al., 2012) and the globe (Marlon et al., 2008; Power et al., 2008). Furthermore, composite charcoal data have been used to test hypotheses regarding the causes of large-scale environmental change. Composite records from North America, for example, suggest that the proposed asteroid impact on the Laurentide ice sheet at ca. 12,900 cal yr BP did not result in continent-wide burning (Marlon et al., 2009), as proposed by Firestone et al. (2007), but rather fire activity increased during the late-glacial to early-Holocene transition in association with rising temperatures. Long-term trends in global biomass burning have also been attributed to changes in climate, inasmuch as trends in mean temperatures correlate well with charcoal data (Daniau et al., 2012). A composite record from North American shows that during the last 200 years, however, there has been a decline in global fire activity despite favorable climate conditions (Marlon et al., 2008). In the western United States, this decline has been attributed to land cover fragmentation, fire management, and anomalous fuel accumulations (Marlon et al., 2012).

Syntheses of paleoenvironmental data disclose temporal variations in the relative amount of biomass burned at continental scales but often offer little insight into the characteristics of fire and its drivers. To tackle this issue, Brücher et al. (2014) developed a new carbon cycle model by combining existing models (i.e., Earth system carbon cycle model and land-cover carbon cycle model) and compared the output with global paleofire patterns inferred from charcoal data. Simulations under transient climate yielded fire activity that changed in concert with observed changes in the charcoal record; minor discrepancies between the simulations and the empirical data were attributed to anthropogenic influences and the effect of CO_2 fertilization on fuel bulk density. Overall, the results of this modeling research agree with previous interpretations (Daniau et al., 2012) that show increasing fire activity in the mid to late Holocene due to changes in vegetation, although the modeling research could not indicate whether anthropogenic changes to vegetation were the ultimate driver of this increased fire activity.

Conclusions and Future Directions

The use of sedimentary charcoal as a fire proxy has increased markedly in recent years, as has interest in the role of biomass burning as a driver of and response to past environmental change. Several approaches, including decomposition of charcoal time series, multi-proxy studies, multi-site comparisons within regions, and composite records at regional-to-global scales have been used to characterize fire regimes and test hypotheses regarding the relative influences of climate, vegetation, and humans on fire regimes. These insights from the past help provide important context for efforts to manage fire, conserve biodiversity, and restore ecosystems (Willis et al., 2005). Full realization of the sedimentary charcoal potential, however, requires continued research into charcoal taphonomy, and the relationship between fire activity and charcoal accumulation rates in different depositional settings (i.e., large vs. small lakes; mountainous vs. flat regions; forested ecosystems vs. grasslands). In addition, laboratory and analytical techniques employed in charcoal analysis and interpretation should continually be reviewed to validate methodological assumptions, assure replicability, and quantify uncertainties. There is also a need to interpret charcoal data in ways that are more closely comparable to simulations of fire. For example, relative abundances of charcoal in paleofire records are less useful than estimates of area burned, fire intensity or carbon emissions for data-model comparisons.

Recent advances in ecological modeling present opportunities to reduce some of these uncertainties. For example, simulations of charcoal accumulation rates have helped fill gaps in our understanding of charcoal taphonomy, which is an important step in establishing a two-way flow of information between disciplines. Modeling approaches can also be used to interpolate—spatially and temporally—between charcoal records. Furthermore, fire simulation results may be considered as independent evidence to support ecological dynamics inferred from charcoal records.

To date, paleoecological records have been used in model validation, but it is possible to integrate paleoecological data in all phases of model development and testing. Such integration opens the door to a new generation of hypotheses in fire research and a more realistic understanding of fire and its drivers at different spatial and temporal scales. Realistic representations of climate-vegetation-fire linkages can be interrogated with resource planning goals, inasmuch as they provide a sound basis for identifying trends, mechanisms and thresholds under different socio-environmental conditions (Dearing et al., 2012).

Conflict of Interest Statement

The authors declare that the research was conducted in the absence of any commercial or financial relationships that could be construed as a potential conflict of interest.

ACKNOWLEDGMENTS

The research discussed in the paper has been supported by grants from the National Science Foundation (ATM-0714061, OISE 09667472; EPS-1101342), and the US Geological Survey North-Central Climate Science Center. This manuscript benefitted from the comments of L. Stahle and two anonymous reviewers.

REFERENCES

1. Aleman, J. C., Blarquez, O., Bentaleb, I., Bonté, P., Brossier, B., Carcaillet, C., et al. (2013). Tracking land-cover changes with sedimentary charcoal in the Afrotropics. *Holocene* 23, 1853–1862. doi: 10.1177/0959683613508159
2. Ali, A. A., Blarquez, O., Girardin, M. P., Hély, C., Tinquaut, F., El Guellab, A., et al. (2012). Control of the multi-millennial wildfire size in boreal North America by spring climatic conditions. *Proc. Natl. Acad. Sci. U.S.A.*

109, 20966–20970. doi: 10.1073/pnas.1203467109

3. Ali, A. A., Carcaillet, C., and Bergeron, Y. (2009). Long-term fire frequency variability in the eastern Canadian boreal forest: the influences of climate vs. local factors. *Glob. Change Biol.* 15, 1230–1241. doi: 10.1111/j.1365-2486.2009.01842.x

4. Allué, E., Euba, I., Caceres, I., Esteban, M., and Perez, M. J. (2005). "Experimentacion sobre recogida de lena en el parque faunistico de los pirineos "Lacuniacha" (Huesca). Una aproximacioìn a la tafonomia del registro anthracologico," in *Avances en Arqueometriìa. Actas del VI Congreso Ibeìrico de Arqueometriìa*, eds J. Molera, J. Farjas, T. Pradell, and P. Roura (Girona: Universitat de Girona), 295–303.

5. Anderson, N. J., Bugmann, H., Dearing, J. A., and Gaillard, M.-J. (2006). Linking palaeoenvironmental data and models to understand the past and to predict the future. *Trends Ecol. Evol.* 21, 696–704. doi: 10.1016/j.tree.2006.09.005

6. Asouti, E., and Austin, P. (2005). Reconstructing woodland vegetation and its exploitation by past societies, based on the analysis and interpretation of archaeological wood charcoal macro-remains. *Environ. Archaeol.* 10, 1–18. doi: 10.1179/env.2005.10.1.1

7. Asselin, H., and Payette, S. (2005). Detecting local-scale fire episodes on pollen slides. *Rev. Palaeobot. Palynol.* 137, 31–40. doi: 10.1016/j.revpalbo.2005.08.002

8. Bartlein, P. J., Harrison, S. P., Brewer, S., Connor, S., Davis, B. A. S., Gajewski, K., et al. (2011). Pollen-based continental climate reconstructions at 6 and 21 ka: a global synthesis. *Clim. Dyn.* 37, 775–802. doi: 10.1007/s00382-010-0904-1

9. Bartlein, P. J., and Hostetler, S. W. (2004). *Modeling Paleoclimates*. Amsterdam: Elsevier.

10. Bégin, Y., and Marguerie, D. (2002). Characterization of tree macroremains production in a recently burned conifer forest in northern Quebec, Canada. *Plant Ecol.* 159, 143–152. doi: 10.1023/A:1015537021039

11. Behling, H. (1997). Late quaternary vegetation, climate and fire history from the tropical mountain region of Morro de Itapeva, SE Brazil. *Palaeogeogr. Palaeoclimatol. Palaeoecol.* 129, 407–422. doi: 10.1016/S0031-0182(97)88177-1

12. Bergeron, Y., Cyr, D., Drever, C. R., Flannigan, M., Gauthier, S., Kneeshaw, D., et al. (2006). Past, current, and future fire frequencies in Quebec's commercial forests: implications for the cumulative effects of harvesting and fire on age-class structure and natural disturbance-based management. *Can. J. For. Res.* 36, 2737–2744. doi: 10.1139/x06-177

13. Berna, F., Goldberg, P., Horwitz, L. K., Brink, J., Holt, S., Bamford, M., et al. (2012). Microstratigraphic evidence of *in situ* fire in the Acheulean strata of Wonderwerk Cave, Northern Cape province, South Africa. *Proc. Natl. Acad. Sci. U.S.A.* 109, E1215–E1220. doi: 10.1073/pnas.1117620109

14. Black, M. P., and Mooney, S. D. (2006). Holocene fire history from the Greater Blue Mountains World Heritage area, New South Wales, Australia: the climate, humans and fire nexus. *Reg. Environ. Change* 6, 41–51. doi: 10.1007/s10113-005-0003-8

15. Blarquez, O., Girardin, M. P., Leys, B., Ali, A. A., Aleman, J. C., Bergeron, Y., et al. (2013). Paleofire reconstruction based on an ensemble-member strategy applied to sedimentary charcoal. *Geophys. Res. Lett.* 40, 2667–2672. doi: 10.1002/grl.50504

16. Blarquez, O., Vannière, B., Marlon, J. R., Daniau, A.-L., Power, M. J., Brewer, S., et al. (2014). paleofire: an R package to analyse sedimentary charcoal records from the Global Charcoal Database to reconstruct past biomass burning. *Comput. Geosci.* 72, 255–261. doi: 10.1016/j.cageo.2014.07.020

17. Bowman, D. M. J. S., Balch, J., Artaxo, P., Bond, W. J., Cochrane, M. A., D'Antonio, C. M., et al. (2011). The human dimension of fire regimes on Earth. *J. Biogeogr.* 38, 2223–2236. doi: 10.1111/j.1365-2699.2011.02595.x

18. Bowman, D. M. J. S., Balch, J. K., Artaxo, P., Bond, W. J., Carlson, J. M., Cochrane, M. A., et al. (2009). Fire in the Earth system. *Science* 324, 481–484. doi: 10.1126/science.1163886

19. Boyle, J. F., Gaillard, M.-J., Kaplan, J. O., and Dearing, J. A. (2011). Modeling prehistoric land use and carbon budgets: a critical review. *Holocene* 21, 715–722. doi: 10.1177/0959683610386984

20. Braadbaart, F., and Poole, I. (2008). Morphological, chemical and physical changes during charcoalification of wood and its relevance to archaeological contexts. *J. Archaeol. Sci.* 35, 2434–2445. doi: 10.1016/j.jas.2008.03.016

21. Bradshaw, R. W. (2008). Detecting human impact in the pollen record using data-model comparison. *Veg. Hist. Archaeobot.* 17, 597–603. doi: 10.1007/s00334-007-0116-8

22. Brubaker, L. B., Higuera, P. E., Rupp, T. S., Olson, M. A., Anderson, P. M., and Hu, F. S. (2009). Linking sediment-charcoal records and ecological modeling to understand causes of fire-regime change in boreal forests. *Ecology* 90, 1788–1801. doi: 10.1890/08-0797.1

23. Pubmed Abstract | Pubmed Full Text | CrossRef Full Text | Google Scholar

24. Brücher, T., Brovkin, V., Kloster, S., Marlon, J. R., and Power, M. J. (2014). Comparing modelled fire dynamics with charcoal records for the Holocene. *Clim. Past* 10, 811–824. doi: 10.5194/cp-10-811-2014

25. Brunelle, A., Whitlock, C., Bartlein, P. J., and Kipfmuller, K. (2005). Postglacial fire, climate, and vegetation history along an environmental gradient in the Northern Rocky Mountains. *Quat. Sci. Rev.* 24, 2281–2300. doi: 10.1016/j.quascirev.2004.11.010

26. Carcaillet, C. (1998). A spatially precise study of fires, climate and human impact within the Maurienne valley, North French Alps. *J. Ecol.* 86, 384–396. doi: 10.1046/j.1365-2745.1998.00267.x

27. Carcaillet, C. (2007). "Charred particles analyses," in *Encyclopedia of Quaternary Science*, ed S. Elias (Oxford: Elsevier), 1582–1593.

28. Carcaillet, C., Ali, A. A., Blarquez, O., Genries, A.,

Mourier, B., and Bremond, L. (2009). Spatial variability of fire history in subalpine forests: from natural to cultural regimes. *Ecoscience* 16, 1–12. doi: 10.2980/16-1-3189
29. Carcaillet, C., Almquist, H., Asnong, H., Bradshaw, R. H. W., Carrión, J. S., Gaillard, M.-J., et al. (2002). Holocene biomass burning and global dynamics of the carbon cycle. *Chemosphere* 49, 845–863. doi: 10.1016/S0045-6535(02)00385-5
30. Carcaillet, C., and Muller, S. D. (2005). Holocene tree-limit and distribution of Abies alba in the inner French Alps: anthropogenic or climatic changes? *Boreas* 34, 468–476. doi: 10.1080/03009480500231377
31. Clark, J. S. (1988). Particle motion and the theory of charcoal analysis: source area, transport, deposition, and sampling. *Quat. Res.* 30, 67–80. doi: 10.1016/0033-5894(88)90088-9
32. Clark, J. S., and Royall, P. D. (1996). Local and regional sediment charcoal evidence for fire regimes in presettlement north-eastern North America. *J. Ecol.* 84, 365–382. doi: 10.2307/2261199
33. COHAMP members. (1998). Climatic changes of the last 18,000 years: observations and model simulations. *Science* 241, 1043–1052.
34. Colombaroli, D., Henne, P. D., Kaltenrieder, P., Gobet, E., and Tinner, W. (2010). Species responses to fire, climate and human impact at tree line in the Alps as evidenced by palaeo-environmental records and a dynamic simulation model. *J. Ecol.* 98, 1346–1357. doi: 10.1111/j.1365-2745.2010.01723.x
35. Conedera, M., Tinner, W., Neff, C., Meurer, M., Dickens, A. F., and Krebs, P. (2009). Reconstructing past fire regimes: methods, applications, and relevance to fire management and conservation. *Quat. Sci. Rev.* 28, 555–576. doi: 10.1016/j.quascirev.2008.11.005
36. Cordeiro, R. C., Turcq, B., Suguio, K., Oliveira da Silva, A., Sifeddine, A., and Volkmer-Ribeiro, C. (2008). Holocene fires in East Amazonia (Carajás), new evidences, chronology and relation with paleoclimate. *Glob. Planet. Change* 61, 49–62. doi: 10.1016/j.gloplacha.2007.08.005
37. Cwynar, L. C. (1987). Fire and the forest history of the North Cascade Range. *Ecology* 68, 791–802. doi: 10.2307/1938350
38. Daniau, A.-L., Bartlein, P. J., Harrison, S. P., Prentice, I. C., Brewer, S., Friedlingstein, P., et al. (2012). Predictability of biomass burning in response to climate changes. *Glob. Biogeochem. Cycles*26:GB4007. doi: 10.1029/2011GB004249
39. Daniau, A. L., Harrison, S. P., and Bartlein, P. J. (2010). Fire regimes during the last glacial. *Quat. Sci. Rev.* 29, 2918–2930. doi: 10.1016/j.quascirev.2009.11.008
40. Daniau, A.-L., Sánchez-Goñi, M. F., Beaufort, L., Laggoun-Défarge, F., Loutre, M.-F., and Duprat, J. (2007). Dansgaard–Oeschger climatic variability revealed by fire emissions in southwestern Iberia. *Quat. Sci. Rev.* 26, 1369–1383. doi: 10.1016/j.quascir ev.2007.02.005
41. Davis, M. B. (1994). Ecology and paleoecology begin to merge. *Trends Ecol. Evol.* 9, 357–358. doi: 10.1016/0169-5347(94)90049-3
42. Dearing, J. A., Bullock, S., Costanza, R., Dawson, T. P., Edwards, M. E., Poppy, G. M., et al. (2012). Navigating the perfect storm: research strategies for socialecological systems in a rapidly evolving world. *Environ. Manag.* 49, 767–775. doi: 10.1007/s00267-012-9833-6
43. Dufraisse, A. (2006). "Charcoal anatomy potential, wood diameter and radial growth," in *Charcoal Analysis: New Analytical Tools and Methods for Archaeology*, Papers from the Table- Ronde held in Basel 2004, BAR International Series S1483, ed A. Dufraisse (Oxford: Archaeopress), 47–60.
44. Dunnette, P. V., Higuera, P. E., McLauchlan, K. K., Derr, K. M., Briles, C. E., and Keefe, M. H. (2014). Biogeochemical impacts of wildfires over four millennia in a Rocky Mountain subalpine watershed. *New Phytol.* 203, 900–912. doi: 10.1111/nph.12828
45. Dupuy, J.-L., and Morvan, D. (2005). Numerical study of a crown fire spreading toward a fuel break using a multiphase physical model. *Int. J. Wildland Fire* 14, 141–151. doi: 10.1071/WF04028
46. Enache, M. D., and Cumming, B. F. (2007). Charcoal morphotypes in lake sediments from British Columbia (Canada): an assessment of their utility for the reconstruction of past fire and precipitation. *J. Paleolimnol.* 38, 347–363. doi: 10.1007/s10933-006-9084-8
47. Firestone, R. B., West, A., Kennett, J. P., Becker, L., Bunch, T. E., Revay, Z. S., et al. (2007). Evidence for an extraterrestrial impact 12,900 years ago that contributed to the megafaunal extinctions and the Younger Dryas cooling. *Proc. Natl. Acad. Sci. U.S.A.* 104, 16016–16021. doi: 10.1073/pnas.0706977104
48. Flannigan, M., Campbell, I., Wotton, M., Carcaillet, C., Richard, P. J. H., and Bergeron, Y. (2001). Future fire in Canada's boreal forest: paleoecology results and general circulation model - regional climate model simulations. *Can. J. For. Res.* 31, 854–864. doi: 10.1139/x01-010
49. Flannigan, M., Stocks, B., Turetsky, M., and Wotton, M. (2009). Impacts of climate change on fire activity and fire management in the circumboreal forest. *Glob. Change Biol.* 15, 549–560. doi: 10.1111/j.1365-2486.2008.01660.x
50. Froyd, C. A., and Willis, K. J. (2008). Emerging issues in biodiversity and conservation management: the need for a palaeoecological perspective. *Quat. Sci. Rev.* 27, 1723–1732. doi: 10.1016/j.quascirev.2008.06.006
51. Ganopolski, A., Kubatzki, C., Claussen, M., Brovkin, V., and Petoukhov, V. (1998). The influence of vegetation-atmosphere-ocean interaction on climate during the mid-Holocene. *Science* 280, 1916–1919. doi: 10.1126/science.280.5371.1916
52. Gardner, J. J., and Whitlock, C. (2001). Charcoal accumulation following a recent fire in the cascade range, northwestern USA, and its relevance for

fire-history studies. *Holocene* 11, 541–549. doi: 10.1191/095968301680223495

53. Gavin, D. G., Brubaker, L. B., and Lertzman, K. P. (2003). Holocene fire history of a coastal temperate rain forest based on soil charcoal radiocarbon dates. *Ecology* 84, 186–201. doi: 10.1890/0012-9658(2003)084[0186:HFHOAC]2.0.CO;2

54. Gavin, D. G., Hu, F. S., Lertzman, K., and Corbett, P. (2006). Weak climatic control of stand-scale fire history during the late Holocene. *Ecology* 87, 1722–1732. doi: 10.1890/0012-9658(2006)87[1722:WCCOSF]2.0.CO;2

55. Pubmed Abstract | Pubmed Full Text | CrossRef Full Text | Google Scholar

56. Gimmi, U., and Bugmann, H. (2013). Integrating historical ecology and ecological modeling. *Landsc. Ecol.* 28, 785–787. doi: 10.1007/s10980-013-9884-y

57. Global Paleofire Working Group. (2014). *paleofire: an R Package to Analyze Sedimentary Charcoal Records from the Global Charcoal Database to Reconstruct Past Biomass Burning*. Available online at: http://CRAN.R-project.org/package=paleofire

58. Gu, Y., Pearsall, D. M., Xie, S., and Yu, J. (2008). Vegetation and fire history of a Chinese site in southern tropical Xishuangbanna derived from phytolith and charcoal records from Holocene sediments. *J. Biogeogr.* 35, 325–341. doi: 10.1111/j.1365-2699.2007.01763.x

59. Haberle, S. G., and Ledru, M.-P. (2001). Correlations among charcoal records of fires from the past 16,000 years in Indonesia, Papua New Guinea, and Central and South America. *Quat. Res.* 55, 97–104. doi: 10.1006/qres.2000.2188

60. Heiri, C., Bugmann, H., Tinner, W., Heiri, O., and Lischke, H. (2006). A model-based reconstruction of Holocene treeline dynamics in the Central Swiss Alps. *J. Ecol.* 94, 206–216. doi: 10.1111/j.1365-2745.2005.01072.x

61. Henne, P., Elkin, C., Colombaroli, D., Samartin, S., Bugmann, H., Heiri, O., et al. (2013). Impacts of changing climate and land use on vegetation dynamics in a Mediterranean ecosystem: insights from paleoecology and dynamic modeling. *Landsc. Ecol.* 28, 819–833. doi: 10.1007/s10980-012-9782-8

62. Heyerdahl, E. K., Brubaker, L. B., and Agee, J. K. (2001). Spatial controls of historical fire regimes: a multiscale example from the interior west, USA. *Ecology* 82, 660–678. doi: 10.1890/0012-9658(2001)082[0660:SCOHFR]2.0.CO;2

63. Heyerdahl, E. K., Morgan, P., and Riser, J. P. (2008). Multi-season climate synchronized historical fires in dry forests (1650-1900), northern Rockies, U.S.A. *Ecology* 89, 705–716. doi: 10.1890/06-2047.1

64. Higuera, P. (2009). *CharAnalysis 0.9: Diagnostic and Analytical Tools for Sedimentcharcoal Analysis*. Bozeman, MT: User's Guide Montana State University.

65. Higuera, P. E., Briles, C. E., and Whitlock, C. (2014). Fire-regime complacency and sensitivity to centennial-through millennial-scale climate change in Rocky Mountain subalpine forests, Colorado, U.S.A. *J. Ecol.* 102, 1429–1441. doi: 10.1111/1365-2745.12296

66. Higuera, P. E., Brubaker, L. B., Anderson, P. M., Hu, F. S., and Brown, T. A. (2009). Vegetation mediated the impacts of postglacial climate change on fire regimes in the south-central Brooks Range, Alaska. *Ecol. Monogr.* 79, 201–219. doi: 10.1890/07-2019.1

67. Higuera, P. E., Gavin, D. G., Bartlein, P. J., and Hallett, D. J. (2011). Peak detection in sediment–charcoal records: impacts of alternative data analysis methods on fire-history interpretations. *Int. J. Wildland Fire* 19, 996–1014. doi: 10.1071/WF09134

68. Higuera, P. E., Gavin, D. G., Henne, P. D., and Kelly, R. F. (2010). Recent advances in the analysis and interpretation of sediment-charcoal records. *Pages News*. 18, 57–59.

69. Higuera, P. E., Peters, M. E., Brubaker, L. B., and Gavin, D. G. (2007). Understanding the origin and analysis of sediment-charcoal records with a simulation model. *Quat. Sci. Rev.* 26, 1790–1809. doi: 10.1016/j.quascirev.2007.03.010

70. Higuera, P. E., Sprugel, D. G., and Brubaker, L. B. (2005). Reconstructing fire regimes with charcoal from small-hollow sediments: a calibration with tree-ring records of fire. *Holocene* 15, 238–251. doi: 10.1191/0959683605hl789rp

71. Huber, U. M., Markgraf, V., and Schäbitz, F. (2004). Geographical and temporal trends in Late Quaternary fire histories of Fuego-Patagonia, South America. *Quat. Sci. Rev.* 23, 1079–1097. doi: 10.1016/j.quascirev.2003.11.002

72. Iglesias, V., and Whitlock, C. (2014). Fire responses to postglacial climate change and human impact. *Proc. Natl. Acad. Sci. U.S.A.* 101, E5545–E5554. doi: 10.1073/pnas.1410443111

73. Iglesias, V., Whitlock, C., Bianchi, M. M., Villarosa, G., and Outes, V. (2012). Climate and local controls of long-term vegetation dynamics in northern Patagonia (Lat 41° S). *Quat. Res.* 78, 502–512. doi: 10.1016/j.yqres.2012.07.004

74. Iglesias, V., Whitlock, C., Markgraf, V., and Bianchi, M. M. (2014). Postglacial history of the Patagonian forest/steppe ecotone (41–43° S). *Quat. Sci. Rev.* 94, 120–135. doi: 10.1016/j.quascirev.2014.04.014

75. Kaplan, J., Krumhardt, K., and Zimmermann, N. (2009). The prehistoric and preindustrial deforestation in Europe. *Quat. Sci. Rev.* 28, 3016–3034. doi: 10.1016/j.quascirev.2009.09.028

76. Kaplan, J. O., Krumhardt, K. M., Ellis, E. C., Ruddiman, W. F., Lemmen, C., and Klein Goldewijk, K. (2011). Holocene carbon emis- sions as a result of anthropogenic land cover change. *Holocene* 21, 775–791. doi: 10.1177/0959683610386983

77. Keane, R. E., Loehman, R. A., and Holsinger, L. M. (2011). *The FireBGCv2 Landscape Fire Succession Model: a Research Simulation Platform for Exploring Fire and Vegetation Dynamics*. Fort Collins, CO: US Department of Agriculture, Forest Service, Rocky Mountain Research Station.

78. Keller, F., Lischke, H., Mathis, T., Moehl, A., Wick, L., Ammann, B., et al. (2002). Effects of climate, fire, and

humans on forest dynamics: forest simulations compared to the palaeological record. *Ecol. Model.* 152, 109–127. doi: 10.1016/S0304-3800(02)00011-X

79. Kelly, R., Chipman, M. L., Higuera, P. E., Stefanova, I., Brubaker, L. B., and Hu, F. S. (2013). Recent burning of boreal forests exceeds fire regime limits of the past 10,000 years. *Proc. Natl. Acad. Sci. U.S.A.* 110, 13055–13060. doi: 10.1073/pnas.1305069110

80. Kitzberger, T., Araìoz, E., Gowda, J., Mermoz, M., and Morales, J. (2012). Decreases in fire spread probability with forest age promotes alternative community states, reduced resilience to climate variability and large fire regime shifts. *Ecosystems* 15, 97–112. doi: 10.1007/s10021-011-9494-y

81. Kleinen, T., Tarasov, P., Brovkin, V., Andreev, A., and Stebich, M. (2011). Comparison of modeled and reconstructed changes in forest cover through the past 8000 years: Eurasian perspective. *Holocene* 5, 723–734. doi: 10.1177/0959683610386980

82. Krawchuk, M. A., Moritz, M. A., Parisien, M.-A., Van Dorn, J., and Hayhoe, K. (2009). Global pyrogeography: the current and future distribution of wildfire. *PLoS ONE* 4:e5102. doi: 10.1371/journal.pone.0005102

83. Long, C. J., Whitlock, C., Bartlein, P. J., and Millspaugh, S. H. (1998). A 9000-year fire history from the Oregon Coast Range, based on a high-resolution charcoal study. *Can. J. For. Res.* 28, 774–787. doi: 10.1139/x98-051

84. Lorenzen, E. D., Nogués-Bravo, D., Orlando, L., Weinstock, J., Binladen, J., Marske, K., et al. (2011). Species- specific responses of late quaternary megafauna to climate and humans. *Nature* 479, 359–365. doi: 10.1038/nature10574

85. Lynch, J. A., Clark, J. S., and Stocks, B. J. (2004). Charcoal production, dispersal, and deposition from the fort providence experimental fire: interpreting fire regimes from charcoal records in boreal forests. *Can. J. For. Res.* 34, 1642–1656. doi: 10.1139/x04-071

86. MacDonald, G. M., Larsen, C. P. S., Szeicz, J. M., and Moser, K. A. (1991). The reconstruction of boreal forest fire history from lake sediments: a comparison of charcoal, pollen, sedimentological, and geochemical indices. *Quat. Sci. Rev.* 10, 53–71. doi: 10.1016/0277-3791(91)90030-X

87. Marguerie, D., and Thibaudeau, C. (2004). "Apport de l'anthracologie et de la dendrologie à l'analyse architecturale et paléoenvironnementale, " *Le site de la Hersonnais à Pléchatel, (Ille-et-Vilaine): un Ensemble de Bâtiments Collectifs du Néolithique Final*, Vol. 5, ed J.-Y. Tinevez (Paris: Travaux Société Préhistorique Française), 139–148.

88. Markgraf, V., Iglesias, V., and Whitlock, C. (2013). Late and postglacial vegetation and fire history from Cordón Serrucho Norte, northern Patagonia. *Palaeogeogr. Palaeoclimatol. Palaeoecol.* 371, 109–118. doi: 10.1016/j.palaeo.2012.12.023

89. Marlon, J. R., Bartlein, P. J., Carcaillet, C., Gavin, D. G., Harrison, S. P., Higuera, P. E., et al. (2008). Climate and human influences on global biomass burning over the past two millennia. *Nat. Geosci.* 1, 697–702. doi: 10.1038/ngeo313

90. Marlon, J. R., Bartlein, P. J., Daniau, A.-L., Harrison, S. P., Maezumi, S. Y., Power, M. J., et al. (2013). Global biomass burning: a synthesis and review of Holocene paleofire records and their controls. *Quat. Sci. Rev.* 65, 5–25. doi: 10.1016/j.quascirev.2012.11.029

91. Marlon, J. R., Bartlein, P. J., Gavin, D. G., Long, C. J., Anderson, R. S., Briles, C. E., et al. (2012). Long-term perspective on wildfires in the western USA. *Proc. Natl. Acad. Sci. U.S.A.* 109, 535–543. doi: 10.1073/pnas.1112839109

92. Marlon, J. R., Bartlein, P. J., Walsh, M. K., Harrison, S. P., Brown, K. J., Edwards, M. E., et al. (2009). Wildfire responses to abrupt climate change in North America. *Proc. Natl. Acad. Sci. U.S.A.* 106, 2519–2524. doi: 10.1073/pnas.0808212106

93. McWethy, D. B., Higuera, P. E., Whitlock, C., Veblen, T. T., Bowman, D. M. J. S., Cary, G. J., et al. (2013). A conceptual framework for predicting temperate ecosystem sensitivity to human impacts on fire regimes. *Glob. Ecol. Biogeogr.* 22, 900–912. doi: 10.1111/geb.12038

94. McWethy, D. B., Whitlock, C., Wilmshurst, J. M., McGlone, M. S., Fromont, M., Li, X., et al. (2010). Rapid landscape transformation in South Island, New Zealand, following initial Polynesian settlement. *Proc. Natl. Acad. Sci. U.S.A.* 107, 21343–21348. doi: 10.1073/pnas.1011801107

95. McWethy, D. B., Whitlock, C., Wilmshurst, J. M., McGlone, M. S., and Li, X. (2009). Rapid deforestation of South Island, New Zealand by early Polynesian fires. *Holocene* 19, 883–897. doi: 10.1177/0959683609336563

96. McWethy, D. B., Wilmshurst, J. M., Whitlock, C., Wood, J., and McGlone, M. S. (2014). High-resolution chronology of forest transitions following early human-set fires South Island New Zealand. *PLoS ONE* 9:e111328. doi: 10.1371/journal.pone.0111328

97. Mensing, S. A., Michaelsen, J., and Byrne, R. (1999). A 560-year record of Santa Ana fires reconstructed from charcoal deposited in the Santa Barbara Basin, California. *Quat. Res.* 51, 295–305. doi: 10.1006/qres.1999.2035

98. Meyer, G. A., Wells, S. G., and Jull, A. J. T. (1995). Climatic and intrinsic controls on Holocene geomorphic processes. *Geol. Soc. Am. Bull.* 107, 1211–1230.

99. Molinari, C., Lehsten, V., Bradshaw, R. H., Power, M. J., Harmand, P., Arneth, A., et al. (2013). Exploring potential drivers of European biomass burning over the Holocene: a data–model analysis. *Glob. Ecol. Biogeogr.* 22, 1248–1260. doi: 10.1111/geb.12090

100. Mooney, S. D., Harrison, S. P., Bartlein, P. J., Daniau, A.-L., Stevenson, J., Brownlie, K. C., et al. (2011). Late Quaternary fire regimes of Australasia. *Quat. Sci. Rev.* 30, 28–46. doi: 10.1016/j.quascirev.2010.10.010

101. Moreno, P. I. (1997). Vegetation and climate near Lago Llanquihue in the Chilean Lake District between 20 200 and 9500 14C yr BP. *J. Quat. Sci.* 12, 485–500.

102. Moreno, P. I., Kitzberger, T., Iglesias, V., and Holz, A. (2010). Paleofires in southern South America since the last glacial maximum. *Pages News* 2, 75–77.

103. Moritz, M.A., Parisien, M.-A., Batllori, E., Krawchuk, M. A., Van Dorn, J., Ganz, D. J., et al. (2012). Climate change and disruptions to global fire activity. *Ecosphere* 3, art49. doi: 10.1890/ES11-00345.1

104. Muñoz, S. E., Mladenoff, D. J., Schroeder, S., and Williams, J. W. (2014). Defining the spatial patterns of land use associated with the indigenous societies of eastern North America. *J. Biogeogr*. 41, 2195–2210. doi: 10.1111/jbi.12386

105. Mworia-Maitima, J. (1997). Prehistoric fires and land-cover change in western Kenya: evidences from pollen, charcoal, grass cuticles and grass phytoliths. *Holocene* 7, 409–417. doi: 10.1177/095968369700700404

106. Oris, F., Ali, A. A., Asselin, H., Paradis, L., Bergeron, Y., and Finsinger, W. (2014). Charcoal dispersion and deposition in boreal lakes from 3 years of monitoring: differences between local and regional fires. *Geophys. Res. Lett*. 41, 6743–6752. doi: 10.1002/2014GL060984

107. Paduano, G. M., Bush, M. B., Baker, P. A., Fritz, S. C., and Seltzer, G. O. (2003). A vegetation and fire history of Lake Titicaca since the Last Glacial Maximum. *Palaeogeogr. Palaeoclimatol. Palaeoecol*. 194, 259–279. doi: 10.1016/S0031-0182(03)00281-5

108. Parisien, M.-A., and Moritz, M. A. (2009). Environmental controls on the distribution of wildfire at multiple spatial scales. *Ecol. Monogr*. 79, 127–154. doi: 10.1890/07-1289.1

109. Patterson, W. A. III, and Backman, A. E. (1988). "Fire and disease history of forests," in *Vegetation History*, eds B. Huntley and T. Web III (Amsterdam: Springer), 603–632.

110. Patterson, W. A. III, Edwards, K. J., and Maguire, D. J. (1987). Microscopic charcoal as a fossil indicator of fire. *Quat. Sci. Rev*. 6, 3–23. doi: 10.1016/0277-3791(87)90012-6

111. Peng, C., Guiot, J., Wu, H., Jiang, H., and Luo, Y. (2011). Integrating models with data in ecology and palaeoecology: advances towards a model–data fusion approach. *Ecol. Lett*. 14, 522–536. doi: 10.1111/j.1461-0248.2011.01603.x

112. Perry, G. L. W., and Millington, J. D. A. (2008). Spatial modelling of succession-disturbance dynamics in forest ecosystems: concepts and examples. *Perspect. Plant Ecol. Evol. Syst*. 9, 191–210. doi: 10.1016/j.ppees.2007.07.001

113. Perry, G. L. W., Wilmshurst, J. M., McGlone, M. S., McWethy, D. B., and Whitlock, C. (2012). Explaining fire-driven landscape transformation during the Initial Burning Period of New Zealand's prehistory. *Glob. Change Biol*. 18, 1609–1621. doi: 10.1111/j.1365-2486.2011.02631.x

114. Peters, M. E., and Higuera, P. E. (2007). Quantifying the source area of macroscopic charcoal with a particle dispersal model. *Quat. Res*. 67, 304–310. doi: 10.1016/j.yqres.2006.10.004

115. Pfeiffer, M., Spessa, A., and Kaplan, J. O. (2013). A model for biomass burning in preindustrial time: LPJ-LMfire (v1.0). *Geosci. Model Dev*. 6, 643–685 doi: 10.5194/gmd-6-643-2013

116. Pitkänen, A., Lehtonen, H., and Huttunen, P. (1999). Comparison of sedimentary microscopic charcoal particle records in a small lake with dendrochronological data: evidence for the local origin of microscopic charcoal produced by forest fires of low intensity in eastern Finland. *Holocene* 9, 559–567. doi: 10.1191/095968399670319510

117. Ponader, K., Pienitz, R., Vincent, W., and Gajewski, K. (2002). Limnological conditions in a subarctic lake (northern Québec, Canada) during the late Holocene: analyses based on fossil diatoms. *J. Paleolimnol*. 27, 353–366. doi: 10.1023/A:1016033028144

118. Power, M. J., Marlon, J., Ortiz, N., Bartlein, P. J., Harrison, S. P., Mayle, F. E., et al. (2008). Changes in fire regimes since the last glacial maximum: an assessment based on a global synthesis and analysis of charcoal data. *Clim. Dyn*. 30, 887–907. doi: 10.1007/s00382-007-0334-x

119. Power, M. J., Mayle, F. E., Bartlein, P. J., Marlon, J. R., Anderson, R. S., Behling, H., et al. (2013). Climatic control of the biomass-burning decline in the Americas after AD 1500. *Holocene* 23, 3–13. doi: 10.1177/0959683612450196

120. Pyne, S. J. (2001). *Fire: a Brief History*. Seattle: Jeremy Mills Publishing.

121. Rius, D., Vannière, B., Galop, D., and Richard, H. (2011). Holocene fire regime changes from multiple-site sedimentary charcoal analyses in the Lourdes basin (Pyrenees, France). *Quat. Sci. Rev*. 30, 1696–1709. doi: 10.1016/j.quascirev.2011.03.014

122. Rothermel, R. C. (1972). *A Mathematical Model for Predicting Fire Spread in Wildland Fuels*. Ogden, UT: USDA Forest Service, Intermountain Forest and Range Experiment Station.

123. Rummery, T. A. (1983). The use of magnetic measurements in interpreting the fire histories of lake drainage basins. *Hydrobiologia* 103, 53–58. doi: 10.1007/BF00028427

124. Rummery, T. A., Bloemendal, J., Dearing, J., Oldfield, F., and Thompson, R. (1979). Persistance of fire-induced magnetic oxides in soils and lake-sediments. *Annal. Géophys*. 35, 103–107.

125. Scheller, R. M., Domingo, J. B., Sturtevant, B. R., Williams, J. S., Rudy, A., Gustafson, E. J., et al. (2007). Design, development, and application of LANDIS-II, a spatial landscape simulation model with flexible temporal and spatial resolution. *Ecol. Model*. 201, 409–419. doi: 10.1016/j.ecolmodel.2006.10.009

126. Scholze, M., Knorr, W., and Heimann, M. (2003). Modelling terrestrial vegetation dynamics and carbon cycling for an abrupt climate change event. *Holocene* 13, 327–333. doi: 10.1191/0959683603hl625rp

127. Scott, A. C., and Glasspool, I. J. (2006). The diversification of Paleozoic fire systems and fluctuations in atmospheric oxygen concentration. *Proc. Natl. Acad. Sci. U.S.A*. 103,

10861–10865. doi: 10.1073/pnas.0604090103

128. Scott, J. H., and Burgan, R., E. (2005). *Standard Fire Behavior Fuel Models: a Comprehensive Set for Use with Rothermel's Surface Fire Spread Model*. Fort Collins, CO: USDA Forest Service, Rocky Mountain Research Station.

129. Sitch, S., Smith, B., Prentice, I. C., Arneth, A., Bondeau, A., Cramer, W., et al. (2003). Evaluation of ecosystem dynamics, plant geography and terrestrial carbon cycling in the LPJ dynamic global vegetation model. *Glob. Change Biol.* 9, 161–185. doi: 10.1046/j.1365-2486.2003.00569.x

130. Staver, A. C., Archibald, S., and Levin, S. A. (2011). The global extent and determinants of savanna and forest as alternative biome states. *Science* 334, 230–232. doi: 10.1126/science.1210465

131. Sun, X., Li, X., and Chen, H. (2000). Evidence for natural fire and climate history since 37 ka BP in the northern part of the South China Sea. *Sci. China D Earth Sci.* 43, 487–493. doi: 10.1007/BF02875310

132. Swain, A. M. (1973). A history of fire and vegetation in northeastern Minnesota as recorded in lake sediments. *Quat. Res.* 3, 383–396. doi: 10.1016/0033-5894(73)90004-5

133. Swetnam, T. W. (1993). Fire history and climate change in giant sequoia groves. *Science (Washington)* 262, 885–889. doi: 10.1126/science.262.5135.885

134. Thery-Parisot, I., Chabal, L., and Chrzavzez, L. (2010). Anthracology and taphonomy, from wood gathering to charcoal analysis. A review of the taphonomic processes modifying charcoal assemblages. *Palaeogeogr. Palaeoclimatol. Palaeoecol.* 291, 142-153. doi: 10.1016/j.palaeo.2009.09.016

135. Thonicke, K., Spessa, A., Prentice, I. C., Harrison, S. P., Dong, L., and Carmona- Moreno, C. (2010). The influence of vegetation, fire spread and fire behaviour on biomass burning and trace gas emissions: results from a process-based model. *Biogeosciences* 7, 1991–2011. doi: 10.5194/bg-7-1991-2010

136. Tolonen, M. (1978). Palaeoecology of annually laminated sediments in Lake Ahvenainen, Southern Finland. I. Pollen and charcoal analyses and their relation to human impact. *Annal. Bot. Fennici* 15, 177–208.

137. Turner, M. G. (2001). *Landscape Ecology in Theory and Practice: Pattern and Process*. New York, NY: Springer.

138. Turner, M. G., and Romme, W. H. (1994). Landscape dynamics in crown fire ecosystems. *Landsc. Ecol.* 9, 59–77. doi: 10.1007/BF00135079

139. Umbanhowar, C. E., and Mcgrath, M. J. (1998). Experimental production and analysis of microscopic charcoal from wood, leaves and grasses. *Holocene* 8, 341–346. doi: 10.1191/095968398666496051

140. Vannière, B., Bossuet, G., Walter-Simonnet, A.-V., Gauthier, E., Barral, P., Petit, C., et al. (2003). Land use change, soil erosion and alluvial dynamic in the lower Doubs Valley over the 1st millenium AD (Neublans, Jura, France). *J. Archaeol. Sci.* 30, 1283–1299. doi: 10.1016/S0305-4403(03)00017-7

141. Vannière, B., Power, M. J., Roberts, N., Tinner, W., Carrión, J., Magny, M., et al. (2011). Circum-Mediterranean fire activity and climate changes during the mid-Holocene environmental transition (8500-2500 cal. BP). *Holocene* 21, 53–73. doi: 10.1177/0959683610384164

142. Walsh, M. K., Whitlock, C., and Bartlein, P. J. (2010). 1200 years of fire and vegetation history in the Willamette Valley, Oregon and Washington, reconstructed using high-resolution macroscopic charcoal and pollen analysis. *Palaeogeogr. Palaeoclimatol. Palaeoecol.* 297, 273–289. doi: 10.1016/j.palaeo.2010.08.007

143. Whitlock, C., and Bartlein, P. J. (2003). "Holocene fire activity as a record of past environmental change," in *The Quaternary Period in the United States*, eds A. R. Gillespie, S. C. Porter, and B. F. Atwater (Amsterdam: Elsevier), 479–490.

144. Whitlock, C., Bianchi, M. M., Bartlein, P. J., Markgraf, V., Marlon, J., Walsh, M., et al. (2006). Postglacial vegetation, climate, and fire history along the east side of the Andes (lat 41- 42.5°S), Argentina. *Quat. Res.* 66, 187–201. doi: 10.1016/j.yqres.2006.04.004

145. Whitlock, C., Higuera, P. E., McWethy, D. B., and Briles, C. E. (2010). Paleoecological perspectives on fire ecology: revisiting the fire-regime concept. *Open Ecol. J.* 3, 6–23. doi: 10.2174/1874213001003020006

146. Whitlock, C., and Larsen, C. (2001). "Charcoal as a fire proxy," in *Tracking Environmental Change Using Lake Sediments*, eds J. P. Smol, H. J. B. Birks, and W. M. Last (Dordrecht: Springer), 75–97.

147. Whitlock, C., and Millspaugh, S. H. (1996). Testing the assumptions of fire-history studies: an examination of modern charcoal accumulation in Yellowstone National Park, USA. *Holocene* 6, 7–15. doi: 10.1177/095968369600600102

148. Whitlock, C., Moreno, P. I., and Bartlein, P. (2007). Climatic controls of Holocene fire patterns in southern South America. *Quat. Res.* 68, 28–36. doi: 10.1016/j.yqres.2007.01.012

149. Wick, L., and Mohl, A. (2006). The mid-Holocene extinction of silver fir (*Abies alba*) in the Southern Alps: a consequence of forest fires? *Veg. Hist. Archaeobot.* 15, 435–444. doi: 10.1007/s00334-006-0051-0

150. Willis, K. J., Gillson, L., Brncic, T. M., and Figueroa-Rangel, B. L. (2005). Providing baselines for biodiversity measurement. *Trends Ecol. Evol.* 20, 107–108. doi: 10.1016/j.tree.2004.12.003

151. Zumbrunnen, T., Bugmann, H., Conedera, M., and Bürgi, M. (2009). Linking forest fire regimes and climate—a historical analysis in a dry inner alpine valley. *Ecosystems* 12, 73–86. doi: 10.1007/s10021-008-9207-3

CHAPTER 5 FUNCTIONAL MORPHOLOGY AND PALEOECOLOGY

Mark P. Witton, Darren Naish

INTRODUCTION

Azhdarchids were a highly successful Cretaceous pterosaur clade, distributed virtually worldwide [1] with a fossil record extending from perhaps the Aptian-Albian to the end of the Maastrichtian [2], [3]. Azhdarchid fossils are most abundant in the Upper Cretaceous and are best known for gigantic forms like *Quetzalcoatlus northropi* from the Maastrichtian of the USA (wingspan c. 10 m). Not all azhdarchids are enormous however, and the clade also includes smaller forms such as *Montanazhdarcho minor* from the Campanian of Montana (wingspan c. 2.5 m), *Zhejiangopterus linhaiensis* from the early Campanian of China (wingspan c. 3.5 m), and *Bakonydraco galaczi* from the Santonian of Hungary (wingspan c. 3.5 m) [4]–[6]. Some taxa apparently exceeded *Q. northropi* in size, with *Hatzegopteryx thambema* from the Maastrichtian of Romania having an estimated wingspan greater than 12 m [7]. Earlier remains of azhdarchids currently await re-evaluation: a possible Upper Jurassic azhdarchid from Tanzania has recently been suggested to be a ctenochamatoid [8], while the the Jiufotang Formation taxon *Eoazhdarcho liaoxiensis*, once thought to be a diminutive azhdarchid with a wingspan of 1.6 m [3], has since been suggested to represent a basal azhdarchoid [9] or pteranodontid [10].

Azhdarchids share a suite of characters with a number of smaller Cretaceous pterosaurs (the tapejarids from South America and China, and the tupuxuarids from South and possibly North America), and are united with them in the pterodactyloid clade Azhdarchoidea [11]–[13]. Azhdarchid fossils are relatively abundant compared to those of other azhdarchoids: while many are fragmentary, complete skeletons are known, but are yet to be adequately described. Despite these problems, we recognize that, while some minor morphological differences in skull and limb proportions can be recognised between azhdarchid genera (e.g. *Quetzalcoatlus*, *Zhejiangopterus*, *Hatzegopteryx*), broad morphological similarities can be identified across the group: all azhdarchids exhibit large skulls with long, edentulous rostra, elongate, cylindrical cervical vertebrae, proportionally short wings with an abbreviated fourth phalanx, and elongate hindlimbs [12]. These anatomical features, combined with the large size of some taxa, make azhdarchids one of the most striking and distinctive pterosaur groups (Figure 1).

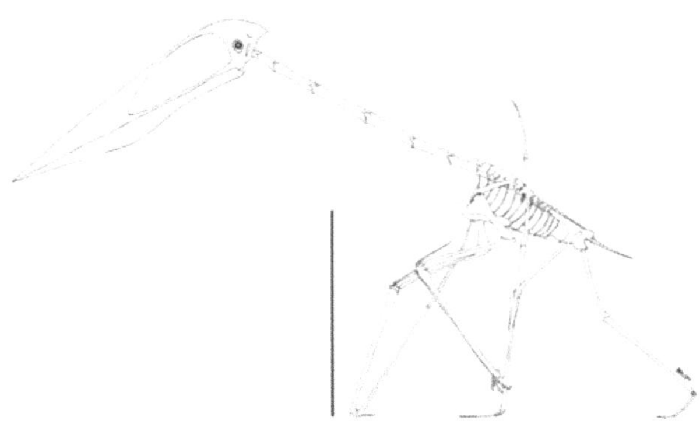

Figure 1. Reconstructed skeleton of *Zhejiangopterus linhaiensis* **based on [40]**and **[47].**

Scale bar represents 500 mm.

By far the greatest controversy in azhdarchid studies concerns their paleoecology, the resolution of which has remained enigmatic due to their unusual anatomy, the sedimentary settings in which their fossils are found, and a general lack of research into pterosaur biomechanics other than

Citation: Witton MP, Naish D (2008) Functional Morphology and Paleoecology. PLoS ONE 3(5): e2271. doi:10.1371/journal.pone.0002271.

those concerned with flight (e.g. [14]–[16]). Many authors have made speculations about azhdarchid paleoecology [2], [17]–[32], often showing a preference for a skim-feeding lifestyle. The only extant obligate skim-feeders are the skimmers, *Rynchops*(Laridae, Charadriiformes), a highly specialised group of gull-like birds that fly low over the water surface, trawling their unusually elongate, laterally compressed lower jaws through the water and grabbing animal prey that they contact. In contrast to that of many other birds, the feeding behaviour and cranial morphology of skimmers has been well described (e.g. [33]–[36]). Two tern species, the Royal tern *Thalasseus maximus* and Caspian tern *Hydroprogne caspia*, are capable of facultative skim-feeding but lack the many unusual features of *Rynchops* [37]. Despite the preference many pterosaur workers have for the concept of skim-feeding in azhdarchids, little work on the biological feasibility of this, or other, feeding strategies have been performed. While most azhdarchids are known from isolated fragments, the complete skeletons that are known provide us with enough information to assess paleoecological hypotheses for the group.

Proposed Azhdarchid Lifestyles

Pterosaurs as a whole are most often thought to have been shorebird-like piscivores, with some imagined as insectivores or durophagores [24]. Azhdarchid anatomy has been difficult to interpret (as evidenced by [38], [39]), and it was not until relatively complete specimens of the Javelina Formation form *Quetzalcoatlus* were discovered that paleoecological interpretations could be made [2]. The first proposed hypothesis of azhdarchid habits - prompted by the stiff, hyper-elongate neck and large size of *Quetzalcoatlus* combined with the sedimentology of the Javelina Formation - was that *Quetzalcoatlus* was a carrion feeder, using a vulture-like flight style and long neck to probe into dinosaur carcasses [2]. The traditional pterosaur diet of piscivory was ruled out due to the structure of the cervical vertebrae and large size of *Quetzalcoatlus*, and the small size of the streams that would have existed in the environment represented by the Javelina Formation [2].

Langston [17] also suggested that the Javelina paleoenvironment would prevent piscivory, with an absence of fish fossils cited as evidence against a piscivorous lifestyle. Langston's alternative scenario, based predominantly on the association of invertebrate trace fossils with *Quetzalcoatlus*, was that giant azhdarchids fed on burrowing arthropods by probing for them in the substrate [17]. The probing scenario was also favoured by Wellnhofer [24] and Lehman and Langston [26]. Paul [19], [20], Chatterjee and Templin [16] and Witton [32] regarded azhdarchids as terrestrial foragers, suggesting that they patrolled water courses, grabbing fish and other animals. Bennett [31] concluded, based on its relatively robust hindlimb elements, that *Quetzalcoatlus* might have been heron- or stork-like in its ecology, and a heron-like lifestyle was also intimated by Padian [21]. O″si et al. [6] imagined *Bakonydraco* to have had a diverse diet of both fishes and fruits, with the latter collected whilst the animal walked through sparsely forested environments.

An entirely different paleoecological role was imagined for azhdarchids by Nessov [18]. In describing *Azhdarcho lancicollis* from the Coniacian of Uzbekistan, Nessov [18] suggested that azhdarchids might have skim-fed in the manner of *Rynchops*, writing 'If it is assumed that the Azhdarchinae could have flown like the Ornithocheirinae and Pteranodontinae – that is, like the Recent skimmers…' (p. 42). Supposedly, the long neck would enable azhdarchids to reach food at depth whilst flying or swimming, thereby circumventing the need to dive [16], [18]. Mid-air predation of 'poorly-flying vertebrates' was also suggested [18].

Of these proposed lifestyles, in-flight piscivory appears to have gained the most acceptance [16], [18], [22], [25], [27]–[30], [40], with skim-feeding being a frequently suggested foraging method [18], [25], [27], [30]. According to these suggestions, the long azhdarchid neck would enable dip- or skim-feeding without damaging the wing-tips [27], [29], whilst the streamlined skull would reduce drag when skim-feeding [27]. Prieto [30] argued that a laterally compressed bill, poor terrestrial abilities, and wing shapes resembling those of highly aerial birds like frigatebirds, swallows and some kites indicated that azhdarchids were specialised for feeding on the wing. Similarities in the gape of *Quetzalcoatlus* and *Rynchops* have been cited as further evidence for skim-feeding behaviour [25].

It is noteworthy that many paleobiological interpretations of azhdarchid anatomy are contradictory. The long, stiff azhdarchid neck is interpreted by some authors as having *impaired*in-flight feeding [2], [6], [17], [32] but is taken to *suggest* in-flight feeding by others (e.g. [18], [25], [27], [30]). Similarly, the azhdarchid wing planform is evidence of a slow, soaring flight to some [2] but indicative of fast, dynamic flight to others [16], [30]. Such contradictions highlight the lack of research into azhdarchid functional morphology, a situation further hampered by inadequate descriptions of azhdarchid fossils. It is also notable that, rather than being based on details of azhdarchids themselves, some of these hypotheses rely strongly on comparisons with other pterosaurs: when proposing the azhdarchid skimming hypothesis, Nessov [18] directly compared *Azhdarcho* with 'Ornithocheirinae' and 'Pteranodontidae' (see quotation, above), and Kellner and Langston regarded skim-feeding as plausible for *Quetzalcoatlus* on the basis that it had been 'previously advocated for *Rhamphorhynchus* … and later assumed for many other pterosaurs, including the larger toothless pterosaurs' ([25] p. 231). Given that the lifestyles of other pterosaurs are no more extensively researched than those of azhdarchids, this type of argumentation is weak and ignores the many obvious anatomical distinctions between these pterosaur taxa. Oversimplified views that all pterosaurs were ecologically alike have undoubtedly added to the controversy surrounding azhdarchid feeding methods.

METHODS

We have assessed the likelihood of proposed azhdarchid lifestyles through both a functional analysis of the azhdarchid skeleton, and by employing comparisons with the anatomy of modern animals occupying similar niches to those inferred for azhdarchids. We acknowledge that azhdarchids may have possessed unique adaptations to particular lifestyles that are not seen in any modern taxa, but pterosaur anatomy is not so disparate from that of extant animals that some functional convergence should not be expected through adaptation to a similar lifestyle (e.g. large, weight-spreading feet for wading, robust jaw joints for skim feeders). An assessment of azhdarchid trace fossils was used to evaluate their terrestrial competence. Additionally, we tested the notion that azhdarchids were, as has been suggested for other pterosaurs, predominately shore-dwelling animals with a quantitative assessment of their geological context. Using sedimentary and other fossil remains as evidence of paleoenvironment, azhdarchid fossils were scored as occurring in settings that were fully terrestrial, coastal, marine with terrestrial input or fully marine with no terrestrial input. This dataset of 33 azhdarchid-bearing localities was then assessed along with consideration of the completeness and abundance of their remains to develop an understanding of preferred azhdarchid habitat.

RESULTS

Inferences from the azhdarchid fossil record

The notion that pterosaurs were predominantly analogues of seabirds seems to have arisen from their frequent occurrences in marine deposits [24], and this inference has been applied to azhdarchids by several authors [18], [25], [41], [42]. However, increasing numbers of pterosaurs are being recovered from inland deposits (e.g. [43]–[45]), implying that the supposed connection pterosaurs had to marine environments may reflect artefacts of preservation rather than actual habitat preference. Notably, most azhdarchids are found in continental fluvial deposits [44], [46], a condition perhaps best demonstrated by the occurrence of *Quetzalcoatlus* 400 km from the nearest contemporary shoreline [2]. At least 16 azhdarchid occurrences (52% of surveyed material) are from inland sediments (e.g., fluvial or alluvial sediments, overbank deposits: Table 1, Figure 2). Furthermore, all but five marine or coastal occurrences are associated with terrestrially-derived fossils such as non-avian dinosaurs, plants and amphibians (83% of surveyed literature: Table 1, Figure 2), and only terrestrial deposits preserve remains of associated azhdarchid individuals [2], [6], [40], [43], [47]. Moreover, the most complete, best preserved azhdarchid fossils are found in terrestrial settings, whereas fossils found in marine settings are generally isolated bones or bone fragments. Possible azhdarchid footprints are also only known from inland lacustrine settings [46].

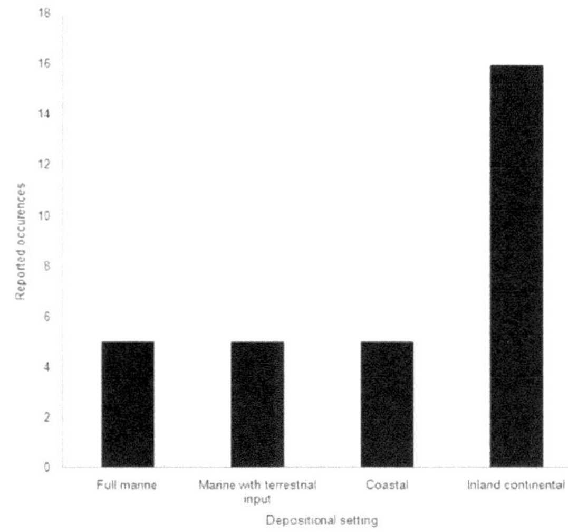

Figure 2. The terrestrial skew of azhdarchid fossils based on data in

Table 1. Geological context of azhdarchid fossils.

Taxon	Locality	Lithology	Associated fossils	Depositional setting	References
Aralazhdarcho	Shakh-Shahk locality, Kazakhstan	Red beds	Amphibians, dinosaurs, crocodiles, mammals	Continental	[52,42]
Arambourgiania	Balqa Group, Jordan	Phosphates	Fishes, turtles, marine reptiles, crocodiles, dinosaurs	Marine	[58]
Azhdarcho	Taykarshinskaya unit, Uzbekistan	Lenticular sands	Turtles, amphibians,	Coastal	[18]
cf. *Azhdarcho*	Laño Locality, Basque Country	Sands and clays	Fish, amphibians, reptiles and mammals	Continental	[121]
Bakonydraco	Csehbánya Formation, Hungary	Silts and sandstones	Turtles, amphibians, lizards, dinosaurs, crocodiles, fishes	Continental; fluvial	[6]
Hatzegopteryx	Densuş-Ciula Formation, Romania	Siltstone	Dinosaur eggs, amphibians, turtles, crocodiles, mammals	Continental; fluvial	[7]
Montanazhdarcho	Two Medicine Formation, USA	Oxidised, cross bedded sands	Dinosaurs, chelonians, freshwater molluscs, lizards	Continental; lacustrine	[122,123]
Phosphatodraco	Oulad Abdou Basin, Morocco	Phosphates, marls	Mosasaurs, sauropterygians, turtles, fish, sharks	Marine	[124]
Quetzalcoatlus	Javelina Formation, USA	Siltstone	Dinosaurs, plants	Continental; fluvial	[2]
Quetzalcoatlus(?)	Oldman Formation, Canada	Grey silt/white sandstone	Dinosaurs, plants, turtles, crocodiles	Continental; fluvial	[125]
cf. *Quetzalcoatlus*	Marnes d'Auzas Formation, France	Coarse sandstone	Turtles, crocodiles, dinosaurs	Freshwater-brackish	[44]
cf. *Quetzalcoatlus*	Hell Creek Formation, USA	Cross bedded sands	Angiosperm leaves	Continental; fluvial	[60]
Zhejiangopterus	Tangshang Formation, China	Volcanic tuffs	Dromaeosaur	Continental; lacustrine	[4]
Azhdarchidae indet.	Azizbek District, Armenia	Lenticular sands	Ammonites, leaf imprint	Marine*	[41]
Azhdarchidae indet.	Pudovkino Formation, Russia	Phosphates	Echinoderms, oysters, belemnites, mosasaurs	Marine	[1]
Azhdarchidae indet.	Volga Region, Russia	Phosphates	Plesiosaurs, mosasaurs	Marine	[126]
Azhdarchidae indet.	Miria Formation, Australia	Chalk	Marine invertebrates, saurichian humerus(?), reptilia indet.	Marine	[127]
Azhdarchidae indet.	Saint Foy site, France	'Non-marine sediments'	Dinosaurs, crocodiles	Continental	[128]
Azhdarchidae indet.	Montplasir site, France	Red and purple marls	Charophytes, dinosaurs, crocodiles, chelonians	Continental*	[129]
Azhdarchidae indet.	Sierra Perenchiza Formation, Spain	Mottled siltstone	Fishes, crocodiles, dinosaurs, frogs	Continental; lacustrine	[117]
Azhdarchidae indet.	Dinosaur Park Formation, Canada	Fine/medium sands	Dinosaurs, amphibians, mammals, chelonians	Continental; fluvial	[130]
Azhdarchidae indet.	Mifune Group, Japan	Coarse sands, muddy lenses	Dinosaurs, crocodiles, turtles, fishes, mammals	Continental	[118]
Azhdarchidae indet.	Hakobuchi Group, Japan	Sandstones, mudstones, lignites	Marine invertebrates, mosasaurs, cehlonians	Coastal	[28]
Azhdarchidae indet.	Ksar es Souk, Morocco	Coarse sands	Dinosaurs, fish, crocodiles, chelonians	Continental	[131]
Azhdarchidae indet.	Paki, Sénégal	Quartzic sandstones	Molluscs, echinoderms, angiosperms	Coastal	[132]
Azhdarchidae indet.	Glen Rose Formation, USA	Micrite	Logs, plant cuticles, ostracodes, fish	Marine	[133]
Azhdarchidae indet.	Kita-ama Formation, Japan	Coarse sandstone	Ammonites, inoceramid bivalves, chelonians, dinosaurs	Marine	[134]
Azhdarchidae indet.	Seidan Formation, Japan	Siltstone	Gastropods	Marine	[135]
Azhdarchidae indet.	Peedee Formation, USA	Alternating limestones and sands	Mardine invertebrates, marine reptiles, dinosaurs	Marine	[136]
Azhdarchidae indet.	Portezuelo Formation, Argentina	Siltstones, sandstones	Dinosaurs, crocodiles	Continental	[137]
?Azhdarchidae	Kem Kem Region, Morocco	Red beds	Fish, turtles, lizards, crocodiles, dinosaurs, pterosaurs	Coastal	[53]
?Azhdarchidae	Illd Formation, Japan	Shales	Bivalves	Marine	[138]

Although interpreting the ecology of extinct organisms from their depositional settings is problematic due to taphonomic influences [48], [49], a significant body of evidence indicates that azhdarchids were denizens of continental settings. Their relative abundance in terrestrial settings contradicts the expectation that marine settings are more conducive to fossilisation than continental environments, and we suggest that this reflects a genuine signal of higher azhdarchid populations inland [50]. This is supported by the relative concentration, completeness and articulation of continentally-preserved azhdarchids compared to their marine counterparts. While scavenging and decay are important taphonomic factors in both marine and continental settings [51], it is significant that marine azhdarchid remains are typically isolated limb bones or cervical vertebrae, as these are among the most readily transported skeletal components in modern vertebrates (Voorhies Groups 1 and 2 of [49] and references therein). Additionally, the inclusion of allochthonous terrestrial material in many azhdarchid-bearing marine horizons implies that the azhdarchid material preserved in these units could also be derived from terrestrial settings. These lines of evidence converge to suggest a strong continental bias for azhdarchid fossils which is most parsimoniously interpreted as an indication of preferred habitat.

An alternative hypothesis suggests that azhdarchids were migratory [16], [25], [41] and that their occurrences inland may result from deaths that occurred *en route* to other locations [25]. While the volant abilities of azhdarchids would certainly permit migratory behaviour, this hypothesis is speculative and, moreover, it is highly unlikely that the vast majority of azhdarchid fossils became associated with continental deposits through chance deaths of migrating animals. Moreover, this hypothesis requires an explanation as to why there are relatively few azhdarchids from coastal or marine deposits. It is far more parsimonious to interpret the overwhelming number of azhdarchid occurrences in terrestrial settings as representing the favoured habitat of these animals. Indeed, strong evidence that azhdarchids inhabited continental environments also comes from their anatomy: numerous details of the azhdarchid skeleton suggest greater terrestrial competence than that of many other pterosaurs.

Functional anatomy

Skull.

Azhdarchid skull material is rare, with complete or near-complete skulls only known for *Quetzalcoatlus* sp. and *Zhejiangopterus linhaiensis* (Figure 3B–C; [25], [40]). Unfortunately, these specimens have suffered crushing that obscures many details bar those of the skull profiles. Fragmentary but three-dimensional skull material is known, however, for *Azhdarcho lancicollis* [18], *Hatzegopteryx thambema* (Figure 3A; [7]), *Bakonydraco galaczi* (Figure 3D; [6]) and *Aralazhdarcho bostobensis* [42], [52]. An incomplete three-dimensional rostrum, probably from an azhdarchid, is also known from the Cretaceous of Morocco [53].

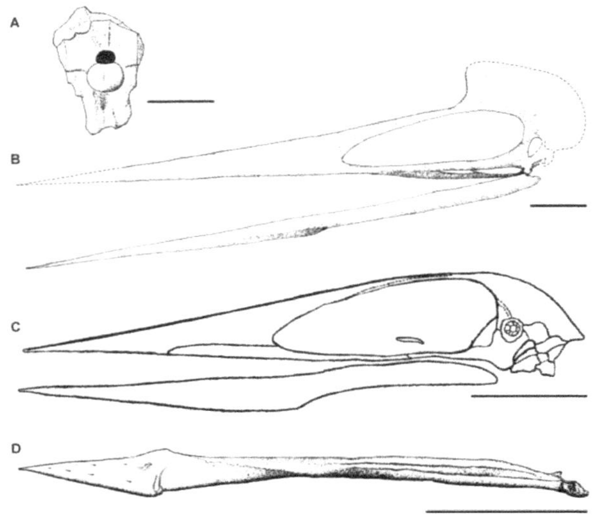

Figure 3. Azhdarchid skull material.

A, occipital region of *Hatzegopteryx* (modified from [7]), B, reconstruction of *Quetzalcoatlus* sp. based on photographs in [25]; C, *Zhejiangoperus* (modified from [40]); D, mandible of *Bakonydraco* (modified from [6]); Scale bars represent 100 mm.

The azhdarchid skull is long, lightly built and approximately triangular in lateral profile (Figure 3; [25], [40]). Azhdarchids appear to have variable jaw width-length ratios, ranging from 0.12 in *Quetzalcoatlus* sp. to 0.22 in *Bakonydraco galaczi*. The giant azhdarchid *Hatzegopteryx*appears to have a particularly wide skull (500 mm across the quadrates), but the relationship between this and skull length cannot be ascertained due to its fragmentary nature. Azhdarchid mandibles are slender in lateral view with a mandibular symphysis that extends for 45–66% the length of the jawline. The jaws are straight and edentulous [6], [18], [25], [40], although *Bakonydraco* possesses a low transverse ridge across the dorsal surface of the mandibular symphysis [6]. The occlusal surfaces of the jaws are flat save for low ridges along the jaw margins [6], [18], [25]. *Quetzalcoatlus* may have a ventrally convex palatal region beneath the posterior half of the nasoantorbital fenestra: four *Quetzalcoatlus* specimens with this region preserved have broken palatal elements that extend below the ventral margin of the maxilla and jugal bars (Figure 3B; [25]), suggesting that their palates were similar to those of the tupuxuarid azhdarchoid *Tupuxuara* [54]. The azhdarchid mandibular joint is typical of pterosaurs in being a simple hinge permitting slight lateral movement of the mandible during jaw extension. A process on the glenoid fossa of *Quetzalcoatlus* may have prevented lateral movement of the quadrates when the jaws were at maximum gape [25]. The cranium is relatively small with a ventrally oriented occiput set within a sculpted occipital region [7], [12]. The occipital condyle is only known in three-dimensions from *Hatzegopteryx*: it reveals a large, well developed hemispherical condyle with no obvious 'neck' [7].

Several lines of evidence suggest that azhdarchid jaw musculature was relatively weak. Attachment sites for jaw musculature on the mandible appear to be small: the mandibular rami show no dorsoventral expansion around the coronoid region, and the retroarticular process is no longer than 3% of the jaw length [6], [25], [40], indicating the small sizes of M. pterygoideus posterior, M. adductor mandibulae posterior and M. intramandibularis (assuming a jaw myology similar to that of extant archosaurs – see [55]). Similarly, the supratemporal fenestrae are reduced and positioned at a low angle to the palate in *Zhejiangopterus* [40], providing the already small M. pseudotemporalis with relatively little mechanical advantage during jaw adduction. The large subtemporal fenestra of *Quetzalcoatlus* and *Hatzegopteryx* indicates the most important jaw adductor was M. pterygoideus anterior, occupying approximately 8% of the jaw length in *Quetzalcoatlus* and 15% of the estimated jaw width in *Hatzegopteryx*. This simplified jaw myology is similar to that of birds in which M. pterygoideus anterior is also the dominant adductor [56] and would have provided azhdarchids with a relatively weak bite. Furthermore, the elongate azhdarchid rostrum would have resulted in low bite pressures delivered at the jaw tips and the straight jaw margins would provide no concentration of bite force. We therefore conclude that azhdarchid jaws were ill suited for demanding feeding techniques or for subduing large, struggling prey, and were better adapted for handling relatively small or immobile food items.

Cervical vertebrae.

Cervical vertebrae are perhaps the best known elements of azhdarchid anatomy, with detailed descriptions provided by Howse [57], Frey and Martill [58], Martill et al. [29], Pereda Suberbiola et al. [47], Godfrey and Currie [59] and Henderson and Peterson [60]. They are among the most common of azhdarchid remains, although complete cervical series are only known for *Quetzalcoatlus* sp., *Zhejiangopterus* and *Phosphatodraco* (Figure 4; [40], [47], [57]). The vertebrae are typically elongate and strongly procoelous with low, ridge-like neural spines and no transverse processes (but see below), although the neural spines are somewhat more developed on the anterior vertebrae. Vestigial cervical ribs occur in small, possibly immature individuals and are entirely fused to the ventral surfaces of the prezygapophyses, with the sutures between ribs and vertebrae absent in larger individuals [59]. Azhdarchid cervicals typically possess large zygapophyses and exapophyses that extend at low angles relative to the vertebral long axis [29], [57], [47]. The postzygapophyses have posteroventrally oriented articular surfaces, with opposing faces seen on the prezygapophyses. The centra extend posteriorly beyond the postzygapophyses but terminate well before the anterior extremity of the prezygapophyses [57]. The condyle is prominent and shows anterodorsal articular surfaces. The cotyle is dorsoventrally asymmetrical, being deeper and broader in the dorsal half. A posteroventrally projecting hypapophysis [59], [60] extends from the cotyle to insert between the exapophyses of the following vertebra. Transverse sections taken at mid-vertebral-length show sub-circular or slightly dorsoventrally flattened cross sections, but both cotyles and condyles are dorsoventrally compressed [29], [57], [60].

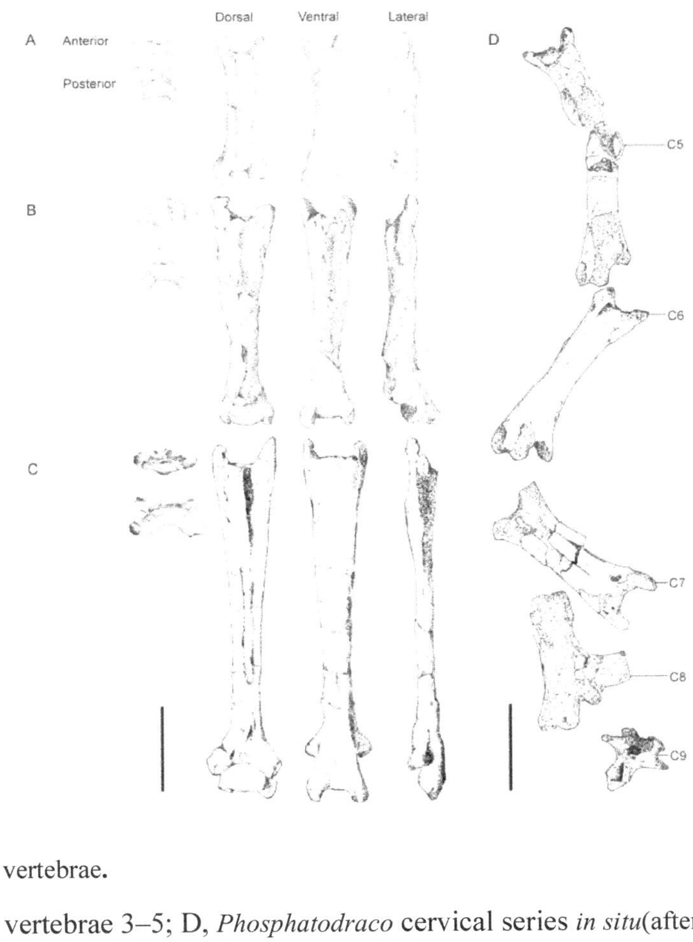

Figure 4. Azhdarchid cervical vertebrae.

A–C, *Quetzalcoatlus* cervical vertebrae 3–5; D, *Phosphatodraco* cervical series *in situ*(after [47]). Scale bars represent 100 mm.

The *Phosphatodraco* cervical series suggests that cervicals eight and nine are exceptions to the slender construction present in the rest of the cervical series, showing relatively large neural spines and, on cervical nine, prominent transverse processes (Figure 4D; [47]). Larger neural spines have also been reported on cervical eight of *Quetzalcoatlus* and *Zhejiangopterus* [8]. The postzygapophyses of cervical eight also project at higher angles relative to the long axis of the vertebrae, but the condition of the prezygapophyses is not known [47].

The crushed or fragmentary nature of most azhdarchid vertebrae has prevented accurate interpretation of their mechanics (e.g. [2], [24], [29], [58]), and, although well preserved azhdarchid cervicals have recently been described [47], [59] qualifying the degrees of articulation between vertebrae remains difficult. It appears that ventroflexion of the neck was limited by the presence of prominent exapophyses and hypapophyses [29], with flexibility decreasing posteriorly (Unwin, pers. comm. 2008). Rotation and lateral flex was hampered by the dorsoventrally compressed centra, laterally flared prezygapophyses and interlocking exapophyses and hypapophysis. Articulation was particularly restricted in the mid-cervicals where the zygapophses are longer and broader, and the absence of posteroventral articulatory faces between these vertebrae seems to confirm their limited ventroflexion.

By contrast, the condylar articular surfaces suggest that a moderate amount of dorsal extension was permitted between vertebrae. The bulbous posterodorsal convexity of the condyle would have facilitated moderate degrees of dorsal rotation during which the prezygapophyses would slide ventrolaterally beneath the postzygapophyses of the preceding vertebra (a motion confirmed by the articular surfaces of these processes). The dorsoventrally compressed but laterally expanded condyle and cotyle appear to have been hinge-like in that they permit flexion in one plane (sagittal) to the exclusion of others. The longer prezygapophyses of the mid-cervicals may have limited extension in this region, but the anterior-most vertebrae have shorter zygapophyses that may have permitted greater degrees of dorsal rotation. These observations are corroborated by comparisons of zygapophyseal shape with animals bearing flexible necks: forms with increased cervical flexibility have short, high-angle zygapophyses with mediodorsally inclined articulator facets, allowing the zygapophyses to slide as the neck flexes. Manipulation of the posterior *Azhdarcho* and *Quetzalcoatlus* cervical series suggests that very little extension was permitted at the neck base (Unwin, pers. comm. 2008).

The relative inflexibility of the mid-cervicals may have served to strengthen inter-vertebral joints, thereby forgoing the need for large epaxial ligaments. The reduced extent of such soft tissue is verified by the lack of sculpting for deep muscle or ligament attachment on all azhdarchid cervicals with the exception of cervicals eight and nine, which contrast with their predecessors in bearing prominently developed neural spines and, on cervical nine, transverse processes [47]. The prominent processes of cervical nine indicate it is probably a 'dorsalised' cervical akin to that seen in *Pteranodon* [31], leaving cervical eight to form the posterior-most 'neck element' of the cervical series. The relatively sculpted morphology of these vertebrae indicates that they anchored larger epaxial muscles than the preceding cervicals that were probably integral in controlling elevation of the neck. Additional anchorage for neck-elevating tissues would be located on the notarium, scapulacoracoid and sternum (as partially reconstructed for *Anhanguera*; [61]). This is in accordance with observations on modern animals demonstrating that soft tissues at the base of the neck control neck elevation [62].

The atlas-axis articulation may also have had limited flexion, despite proposals that this was a main source of flexibility in the azhdarchid neck [24], [29]. This part of the cervical column is not well known, but the atlas and axis are reported to be fused together and truncated in *Zhejiangopterus* [40] and similar to that of *Pteranodon* in *Quetzalcoatlus* and *Azhdarcho* [31]. Detailed descriptions of this region of the neck have yet to be published, however. A fragmentary *Aralazhdarcho* atlas-axis demonstrates a morphology similar to that of other azhdarchid cervical vertebrae with posterolaterally directed exapophyses and a dorsoventrally compressed condyle [42]. Presumably therefore, this morphology imposed the same limit on rotational and vertical articulation as seen between other cervicals. This observation is supported by the small size of the atlas-axis in *Zhejiangopterus* [40], which appears inadequately sized to anchor powerful skull musculature. The generally inflexible nature of the neck therefore appears to extend from the base of the skull throughout much of the neck.

Combined, these observations suggest a relatively limited range of motion in the azhdarchid cervical series. Azhdarchids probably held their necks in a manner similar to lizards and crocodiles with relatively little curvature of their vertebral columns [63]. Mid-series cervicals were probably held with only minor dorsoflexion, forming a gentle, dorsal arc from the torso. The greater length between the pre- and postzygapophyses compared to the length between cotyle and exapophyses dictates that a slight downturn in the cervical series may have occurred at cervical three. However, the vestigial nature of the neural spines indicates that the nuchal ligament did not play a role in maintaining a cervical 'S' shape like that of birds and some mammals [63], or in maintaining the strong dorsal arcs seen in some other pterosaurs [31]. Despite this, the *Hatzegopteryx* supraoccipital bears a well-developed medial ridge similar to that of ornithocheirids [64] and pteranodontids [31] for strong attachment of the nuchal ligament. This medial ridge is not seen in smaller forms such as *Tapejara* [65] or *Rhamphorhynchus* [24], suggesting that the development of a large ligament was an adaptation for the supporting of a heavier neck and head.

Wing structure and flight capability.

Research into pterosaur flight has principally focused on ornithocheiroids like *Pteranodon* [15], [66]–[69] and *Anhanguera* [70], with some studies investigating multiple pterosaur taxa [16], [71], [72]. Azhdarchid flight has yet to be researched in great detail and some controversy surrounds their flight capability. Paul [73] and Frey et al. [74] concluded that azhdarchids would be able to perform prolonged flapping flight using large flight muscles, with Paul [73] suggesting this for even the largest forms. Other workers have argued that the flight muscles of large pterosaurs were not sufficient to maintain flapping flight [16]–[18], [75] and that they were dynamic soarers akin to modern albatrosses [16], [18]. Similar conclusions were drawn by Lawson [2], but vulture-like static soaring was suggested rather than dynamic soaring. Drawing conclusions about the flight of pterosaurs is problematic due to our limited understanding of their paleobiology. Modelling flight is particularly difficult for the larger forms due to the lack of equivalently sized extant analogues [72]: the absence of such creates problems in estimating the masses of giant pterosaurs, a critical value in modelling even basic flight attributes such as wing loading and flight speed. Despite many attempts at estimating the masses of giant azhdarchids, little agreement has been achieved. Langston [17] suggested a *Quetzalcoatlus* with an 11–12 m wingspan may have weighed 86 kg and, with a 15.5 m wingspan, 136 kg. Citing Bramwell and Whitfield's [15] aeronautical work with a lightweight *Pteranodon*, Wellnhofer [24] suggested that an 86 kg estimate for a 10 m span *Quetzalcoatlus*could be too high and that its weight may have been comparable to modern ultra-light aircraft. Shipman [76] suggested that an azhdarchid of similar size would have a mass of 126 kg. The results of a multivariate analysis by Atanassov and Strauss [77] gave mass estimates of 90–120 kg depending on body density, but similar techniques used by Templin [78] and Chatterjee and Templin [16] produced estimates of 62–77 kg, a range of figures also cited by Witton [32]. A similar mass of 70 kg was produced by Brower and Veinus [71] using regression analysis of geometrically modelled pterosaurs. Considerably higher estimates are given by Marden [79]and Paul [73] at 200–250 kg, but calculations by Chatterjee and Templin [16] suggest a pterosaur of this magnitude would never become airborne. These calculations are contradicted by Marden [79], and other workers have criticised lower mass estimates for being impossibly low [66], [73], [80].

It is noteworthy that all mass estimates of azhdarchids have been based on methods for which pterosaur soft tissue density has to be estimated. Many pterosaurs exhibited extensive skeletal pneumaticity (e.g. [31], [81]) and azhdarchid vertebrae and humeri were clearly pneumatised [29], [40]. We therefore assume that azhdarchids exhibited pneumaticity in both their soft tissue anatomy as well as in their skeleton. However, given that pneumaticity has been shown to vary considerably among extant birds [82] and has a significant impact on mass estimates [83], we know too little about pterosaur anatomy to accurately predict their masses using density-dependent calculating techniques. A regression analysis of dry skeletal mass relative to total body mass [84], a technique that avoids the complications of estimating body density, generates a mass of 250 kg for a 10 m span azhdarchid, a figure matching the higher mass estimates of Paul [73], [80] and Marden [79]. Moreover, calculating the body volume of a giant azhdarchid suggests that soft tissue densities have to be less than 0.25 g/cm^3 in order to allow masses of under 125 kg (Witton, unpublished data). We note that the masses of not only giant azhdarchids but all pterosaurs have been grossly underestimated and suggest that the flight calculations based on these hyper-lightweight estimates be treated with caution.

Some conclusions on azhdarchid flight can be drawn from their wing morphology alone. Wing planform is highly diagnostic of flight style in extant animals [85], but controversy over the shape of the pterosaur brachiopatagium has resulted in multiple interpretations of azhdarchid flight style. Langston [17], Wellnhofer [24] and Chatterjee and Templin [16] reconstructed azhdarchids with narrow brachiopatagia extending to the top of the hindlimbs, whereas Frey et al. [74] suggested that the membrane extended to the ankle, forming a much broader wing. No fossilised azhdarchid wing membranes are known, but evidence from anurognathids, campylognathoidids, rhamphorhynchids, ctenochasmatoids and non-azhdarchid azhdarchoids [86]–[91] indicates that ankle-attached wing configurations are more accurate.

Many authors have noted the relatively short wing configuration of azhdarchids, with their abbreviated distal wing phalanges contrasting with their long forearms, the latter a result of hyper-elongate wing metacarpals [12], [16], [17], [74]. These short wings contrast with long hindlimbs, a condition resulting from proportionally elongate femora [12], [17], [74]. Consequently, the forelimb-hindlimb ratio (length of humerus+radius+wing metacarpal/length of femur+tibia) of azhdarchids is low: Chatterjee and Templin [16] suggested a ratio for *Quetzalcoaltus* of 1.13, one of the lowest limb ratios for any pterodactyloid. This contrasts with the limb ratio of 1.45 present in *Zhejiangopterus*: a more typical pterodactyloid limb ratio (unfortunately, a precise ratio for a complete wing to hindlimb length cannot be given as no complete azhdarchid wing fingers have yet been described). Consequently, with their abbreviated wing fingers and long hindlimbs incorporated into the wing membrane, azhdarchids possessed relatively short, broad wings. Reconstructing the azhdarchid wing with broad brachiopatagia (wing shape derived from the 'dark wing' *Rhamphorhynchus* [91]) generates an aspect ratio of 8.1 (Figure 5), a value comparable to the aspects of modern storks, raptors and bats that engage in static soaring [85], [92]–[94]. With the relatively low wing loading that such broad wings produce, it is likely that large azhdarchids were also static soarers, using the warmed, rising air of thermals to gain altitude before soaring cross-country. Smaller azhdarchids may have been more capable of complementing gliding with sustained flapping flight than larger forms due to their lower masses and less demanding energetic requirements [16], [72]. These observations are supported by principal component analysis of azhdarchid wing form, with 10 m and 3 m span taxa plotting in the same ecomorphospace as condors, ibises and other thermal soarers [see 85 for further details]. The shape of azhdarchid wings implies that, like modern static soarers, they would have had relatively small turning radii when soaring, but comparatively poor glide performance compared to longer, narrower-winged forms [85], [92]. However, their broad wing area would act in concert with the deep camber produced by the elongate pteroid [40] to generate greater lift than that present in the narrower wing planform of dynamic soarers [70], [92]. This may have been crucial for azhdarchids, allowing them to take off in cluttered inland habitats where wind and topography are too variable to always allow an assisted takeoff. Additionally, it is of obvious benefit to have short, broad wings when taking off in vegetated inland settings [85]. That azhdarchids appear to have wings well adapted for flight in terrestrial environments correlates well with the regular occurrence of azhdarchid fossils in terrestrial strata.

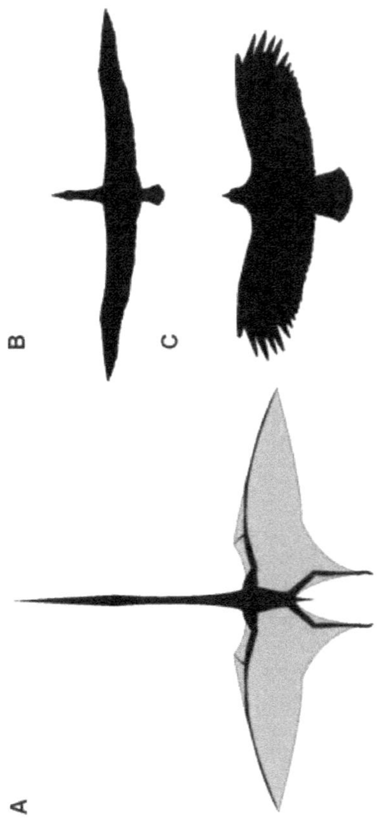

Figure 5. Azhdarchid wing shape.

A, reconstructed planform of *Quetzalcoatlus* (wing shape derived from the 'dark wing' *Rhamphorhynchus*: see [91]); B, planform of the dynamically soaring wandering albatross (*Diomedea exulans*); C, planform of the statically soaring Andean condor (*Vultur gryphus*). Images not to scale.

Terrestrial capability.

While the terrestrial abilities of pterosaurs were once regarded as non-existent to poor, reassessments of pterosaur trace fossils indicate that pterodactyloids were competent walkers and runners [23], [95], [96]. Azhdarchid tracks are rare, but large, Upper Cretaceous footprints from Korea (*Haenamichnus uhangriensis*) are thought to have been produced by an azhdarchid (Figure 6; [46]). *Haenamichnus* is distinguished from other pterosaur tracks by its slim shape, rounded heel and apparent absence of large pedal claws, although the manus prints are broadly similar to those of other pterosaur ichnotaxa. A comprehensive comparison between the *Haenamichnus* footprints and azhdarchid feet cannot be made because of the incomplete nature of azhdarchid foot remains [46]: but limited comparisons can be made between the tracks and the partial pes material known from *Zhejiangopterus* [40] and *Quetzalcoatlus* [31]. However, while the identification of *Haenamichnus* as an azhdarchid trace is not proven, it is supported by the Santonian-Campanian age of the tracks, their large size (no other pterosaurs were large enough to produce 35 cm long pes prints), and sub-equal digit lengths (a character shared with *Zhejiangopterus*). Tapejarid feet show some of these characters but bear large claws on all digits [91], while tupuxuarids bear extremely elongate fifth digits. Based on pes print morphology and a Campanian age, tracks referred to *Pteraichnus* sp. from Mexico [97] may also be *Haenamichnus*, albeit produced by a considerably smaller azhdarchid.

Figure 6. The probable azhdarchid trace fossil *Haenamichnus uhangriensis*.

A, the 7.3 m trackway CNUPH.P9; B, *H. uhangriensis* holotype (CNUPH.P2), manus (top) and pes (bottom) prints. Modified from [46]. Scale bars represent 1 m (A) and 100 mm (B).

Using *Zhejiangopterus* as a template, the *Haenamichnus* trackmaker can be estimated to have stood almost 3 m tall at the shoulder and to have had a wingspan of over 10 m. In concert with the large size of this trackmaker, one *Haenamichnus* trackway has a length of 7 m and is the longest pterosaur trackway yet known [46]. It records a pterosaur moving with an efficient, parasagittal gait [46] rather than in a sprawled posture as suggested by earlier studies (e.g. [98]). In fact, the *Haenamichnus* trackway demonstrates that the presumed azhdarchid trackmaker had a particularly narrow gait with pes prints regularly overlying manus prints, an observation suggesting particularly efficient terrestrial locomotion compared to that inferred from other pterosaur trackways. The *Haenamichnus* pes prints show that the feet possessed soft tissue pads on the digits, metatarsal heads and heel in the manner of some tapejarids [91], [99]) with webbing between the digits [46]. This webbing may also have been present between the digits of the manus [46].

Good descriptions of azhdarchid pes material are lacking, but Bennett [31] reported that the metatarsals of *Quetzalcoatlus* are relatively robust compared to those of *Pteranodon*. This contrasts with the relative pes length of azhdarchids, which was comparatively small for their body size. The pes of

Quetzalcoatlus is approximately 25% of tibial length, and that of *Zhejiangopterus* 30% tibial length [16], [40]. These figures contrast with the 47% pes-tibial length of *Pteranodon* and *Huaxiapterus* [100], 58% in *Germanodactylus*, 69% in *Pterodactylus*and 84% in *Pterodaustro*. In modern birds, long feet with large surface areas are often associated with swimming or wading behaviour [101], suggesting that the proportionally small feet of azhdarchids were poorly suited for these lifestyles. The hands of azhdarchids are similarly truncated with reduced digits and claws that would function poorly in support on soft substrates. We therefore conclude that azhdarchids were more competent at walking on firm substrates than in marshes, swamps or intertidal environments.

Further evidence of terrestrial competence in azhdarchids stems from their atypically long limbs. With the exception of ornithocheirids with their disproportionately long forearms and small bodies, *Zhejiangopterus* demonstrates longer limbs relative to body length than any other pterosaur. Increased forelimb length is mainly achieved through elongation of the wing metacarpal, creating limb bones proportions similar to those of cursorial ungulates. However, the plantigrade hindlimb does not resemble that of cursorial tetrapods, with a tibia only 20% longer than the femur [40]. Hence, although probably not cursorial, azhdarchids may have been relatively fast, energy efficient terrestrial locomotors merely thanks to the increased stride length allowed by their longer limbs. Further interpretation of long azhdarchid limbs is complicated by the poorly understood selection pressures for long limbs among modern animals [102]–[104], but their potential enhancement of terrestrial proficiency is noteworthy.

DISCUSSION

Evaluation of suggested azhdarchid lifestyles

Scavenging.

Lawson's [2] suggestion that *Quetzalcoatlus* was an obligate scavenger is partially based on the association of sauropod remains with the pterosaur material, but this was refuted by Martill [27]as circumstantial evidence. Some aspects of azhdarchid anatomy appear to support the scavenging hypothesis, particularly their possible adaptations for long-distance static soaring in the manner of vultures [2]. The correlation between high body mass and carcass dominance is well documented in extant scavengers (e.g. [105]–[107]) and the large size of many azhdarchids would almost certainly prove beneficial in this regard, and also permit them to swallow small animal carcasses whole. However, many authors have contrasted the flexible necks of scavenging birds with the stiff necks of azhdarchids and have suggested that the latter would limit any carcass-probing ability [6], [24], [27], [29], [58]. The lack of hooked jaw tips has also been cited as evidence against the scavenging hypothesis [24], [30], but scavenging storks and corvids manage to open carcasses quickly and bite off pieces of flesh without the aid of curved jaw tips [108]. Presumably, the lack of a hooked bill in these forms reflects the fact that scavenging is only part of a broader diet facilitated by a generalized bill morphology [101], [109]. Therefore, it seems almost certain that azhdarchids would have been capable of feeding upon at least some elements of large carcasses, although their long skulls and necks would inhibit their ability to obtain flesh from the deepest recesses of a corpse. However, although carrion was a likely component of azhdarchid diets, they possess no anatomical features to suggest they were obligate scavengers.

Probing.

The suggestion that azhdarchids may have probed into sediments in search of infaunal invertebrates is based on the association of *Quetzalcoatlus* remains with invertebrate trace fossils [17], [24], [26], but this evidence is also circumstantial [27], [29]. The restricted ventral flexion of the azhdarchid neck is problematic for this hypothesis [27], [29], as is the cross-sectional shape of the rostrum [27]. Extant probing birds, such as sandpipers, share long jaws and ventrally located occiputs with azhdarchids, but differ in that the margins of their rostra are parallel rather than tapering, and in that their jaws are sub-cylindrical in cross-section (Figure 7B; [110], [111]). Many probing birds possess batteries of closely packed pits on the premaxillary and dentary tips that house pressure-sensitive Herbst corpuscles (Figure 7A; [110]), but analogous features are absent in azhdarchids. Azhdarchids also lack the pleurokenetic bills of probers or the large retroarticular processes of 'gaping' birds [101]prompting the question as to how they could obtain infaunal prey once they had located it. The relatively small feet of azhdarchids suggest they were not adapted for supporting their weight on soft substrates where probing would be best facilitated, a point particularly pertinent for the largest forms given their potential masses of around 250 kg. The size of these forms also dictates that they would need to process enormous amounts of probed invertebrates to sustain themselves. This, in concert with the absence of cranial specialisations for probing, the relatively inflexible neck, and proportionally small feet lead us to conclude that the probing hypothesis can be rejected.

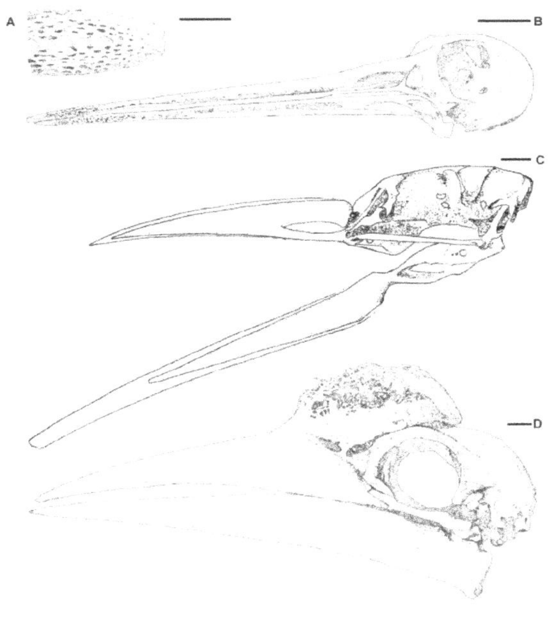

Figure 7. Suggested modern analogues of azhdarchids.

A, anterior premaxilla of the western sandpiper (*Calidris mauri*) showing densely packed Herbst corpuscles, dorsal view [after 110]; B, skull of the probing common snipe (*Gallinago gallinago*); C, skull of the black skimmer (*Rynchops nigra*) ; D, skull of the northern ground hornbill *(Bucorvus abyssinicus)*. Scale bars represent 1 mm (A) and 10 mm (B–D).

Mid-air predation.

The possibility that azhdarchids were aerial predators of smaller flying animals [18] has not gained acceptance among pterosaur workers. Unlike modern raptorial birds, pterosaurs hawking airborne prey would have to rely on their jaws for prey capture rather than their limbs: employment of any limb in mid-air prey capture would compromise the wing membrane and stall the wing. Correspondingly, azhdarchids do not bear raptorial claws on any appendage that could be used to subdue prey in this manner. Extant volant tetrapods that employ oral apprehension of aerial prey have short, wide skulls and often possess deep mandibular symphyses [112], [113], a condition that contrasts markedly with the elongate, narrow azhdarchid skull They also tend to be relatively small, fast and agile fliers with below-average wing loading and moderately high aspect ratios [85], but while the wing loadings of modern aerial hawkers are comparable to those restored for azhdarchids, their aspect ratios are far higher. Consequently, azhdarchids would be slow, cumbersome fliers in comparison. Members of the pterosaur clade Anurognathidae conform to these criteria far better and are hence usually regarded as having been aerial insectivores convergent with swifts, nightjars and some microbats [23], [24], [114]. Given that azhdarchids were large, relatively narrow-skulled animals with stiff necks and wings better adapted for gentle gliding than high-velocity pursuit, the hypothesis that they were capable of routine aerial predation of other volant animals can be rejected.

Swimming and diving.

Although many modern birds regularly swim or dive in pursuit of food, there is no anatomical evidence that azhdarchids did the same [contra. 16,18]. Extant birds demonstrate multiple approaches to feeding on and in water, including surface feeding, plunge diving, and surface diving [101]. Tetrapods that habitually swim possess limbs modified to greater or lesser degrees for propulsion through water, and those that regularly dive bear streamlined bodies to minimise drag [115]. With their relatively small, narrow feet, expansive wings and ventrally oriented skulls atop long, stiffened necks, azhdarchids lack both the propulsion mechanisms and streamlining for efficient movement through water and their limbs show no modifications (e.g. enlarged olecranon or cnemial processes) for swimming. Extant swimming and diving birds also hold their heads close to or above their centre of buoyancy when alighted on the water surface, but the anatomy of azhdarchid cervical vertebrae disallows the possibility of holding the neck at a high angle and may have created issues of stability if the animal were to alight on the water surface. Although other pterosaurs may have been competent swimmers [31], [116], there is no anatomical evidence that azhdarchids were suited for an aquatic existence. Rather, the elongate, slender limbs and proportionally large neck and skull of azhdarchids probably cast them as some of the least aquatically-adapted of all pterosaurs.

Skim-feeding.

Many authors have suggested that azhdarchids were airborne piscivores [16], [18], [22], [24], [25], [27], [29], [40], and skim-feeding is often suggested as the feeding method [18], [25], [27], [30]. Both Lawson [2] and Langston [17] suggested that the lack of fish fossils and large water bodies in the *Quetzalcoatlus*-bearing Javelina Formation might refute a piscivorous diet, but the occurrence of azhdarchids in a variety of depositional settings with associated fish fossils (e.g. [58], [117], [118]) demonstrates the circumstantial nature of this argument. It has been proposed that azhdarchids used the combined lengths of their heads and necks (perhaps up to 5 m in the largest forms) to feed whilst keeping their wingtips from contacting the water, with narrow jaws minimising drag [27]. Further evidence cited for skimming includes the similarity alleged between the 50° gape of *Quetzalcoatlus* and that of *Rynchops* [25]. However, the hypothesis that azhdarchids may have been skim-feeders fails to acknowledge the remarkable and highly distinctive specialisations necessary for skim-feeding and ignores the fact that, among extant vertebrates, habitual skimming is unique to *Rynchops* [34]–[36], [115], although Royal terns *Thalasseus maximus* and Caspian terns *Hydroprogne caspia* are known to perform facultative skim-feeding behaviour [37]. The head and neck of *Rynchops* has 30 skimming adaptations [36], the most obvious being the extreme streamlining and keratinous extension of the mandibular symphysis (Figure 7C), the reinforcement and secondary bracing of the jaw joint, and the robust nature of the cervical vertebrae [35], [36]. These adaptations reflect a lifestyle that is considerably more energetically demanding and specialised than previously appreciated [75].

No azhdarchid exhibits any of these adaptations nor any functional alternatives, and in many details azhdarchids appear maladapted for skim-feeding. While slender, the dorsal surfaces of the mandibular symphyses in *Bakonydraco* and *Quetzalcoatlus* are flattened, producing triangular (rather than blade-like) cross-sections [6], [25]. Humphries et al. [75] highlighted the importance of the knife-like mandibular symphysis of skimmers in drag reduction and suggested that even small (<2 m wingspan) pterosaurs might have struggled to skim with their dorsoventrally flattened mandibles. Modelling the skimming energetics of a 10 m span azhdarchid suggests that even an unfeasibly lightweight (50 kg) individual would lack the metabolic energy to skim-feed, and this assertion applies even more to a realistically estimated 200 kg individual [75]. Also noteworthy is that, in contrast to *Rynchops*, no portion of the azhdarchid mandibular symphysis extended beyond the premaxilla, and the rhamphothecae preserved in other azhdarchoid taxa [91] suggest there was no keratinous extension either.

Crucially, other than in their gape, there is little similarity between the jaw articulation in *Quetzalcoatlus* and *Rynchops*. Although the quadrate-articular joint of *Quetzalcoatlus* shows some additional lateral bracing [25], it is fundamentally under-developed compared to the robust, doubly-reinforced jaw joint of skimmers, a morphological adaptation critical to withstanding the impacts experienced in skim-feeding [35]. Similarly, although relatively short compared to other pterosaurs, the mandibular rami of azhdarchids remain long and slender in contrast to the deep, robust rami of *Rynchops*

[6], suggesting that azhdarchid jaw muscles were relatively weak compared to the enlarged, drag- and impact-resistant jaw musculature of skim-feeding birds. The necessity for these adaptations in skim-feeding cannot be understated: along with the impacts experienced on striking prey items, modern skimmers regularly strike submerged obstacles that can be severe enough to cause crashes [34], [36], [119]. Although azhdarchids could avoid these risks by trawling deeper water, modern skimmers regularly trawl shallow water bodies to maximise their chances of catching prey. We assume that azhdarchids would have had to skim in similar settings and hence would have faced similar risks: the lack of shock-absorbing structures in the azhdarchid jaw alone is compelling evidence against the skim-feeding hypothesis.

The azhdarchid neck also lacks the structures needed to cope with the drag forces and jarring impacts incurred during skim-feeding. Not only is the azhdarchid neck totally incapable of the considerable flexion needed for skim-feeding, it also lacks sculpting for strong ligament and muscle attachment that are unusually prominent in *Rynchops* [36]. The low neural spines of azhdarchid cervicals and stiffened neck structure result in very low mechanical advantage for any associated musculature. This contrasts strongly with the short neck of *Rynchops* that produces high mechanical advantage through relatively tall neural spines and pronounced dorsal curvature of the cervical series [36].

Furthermore, it is highly unlikely that ventral rotation of the azhdarchid skull at the occipital condyle could have absorbed skimming impacts in compensation for the stiffened neck. The apparent inflexibility of the axis-cervical three joint requires that all rotation took place at the occiput-axis and atlas-axis junctions, both of which are small and wholly insufficient to absorb jarring impacts. The position of the occiput also precludes extending the jaw tips to the combined length of the jaw and neck, as the skull cannot be rotated enough to align the jaws and neck on the same plane [contra. 27]. Equally, the suggested strategy of keeping the wings held at neck›s length from the water surface may compromise the wing-in-ground effect benefits that are known to be important in the flight of modern skimmers [120]. Azhdarchid wings also contrast with the high aspect wings of *Rynchops* that facilitate fast, efficient 'flap-gliding' [36], [75], [120] rather than static soaring [contra. 30]. Hence, despite regular mention by pterosaur workers, azhdarchids lack characters that support the skim-feeding hypothesis and it is an entirely unlikely foraging method for this group of pterosaurs. Additionally, we note that the concept of skim-feeding in other pterosaurs is no more secure than it is for azhdarchids (see [75] for further details).

Dip-feeding.

Many elements of the azhdarchid skeleton that preclude skim-feeding also apply to their inability to dip-feed (our use of the term dip-feeding here applies to the style of foraging practised by frigatebirds, gulls and terns where prey items at or near the water surface are picked up by the bird while it is on the wing). This feeding method has been proposed for azhdarchids by various authors [16], [22], [24], [29], [40], as it has for virtually all other pterosaurs. While the long necks and jaws of azhdarchids superficially resemble those of modern dip-feeders, the details of their neck and skull anatomy preclude such a foraging method. The azhdarchid neck is not flexible enough to allow the animal to reach beneath and behind the body in the manner practised by extant dip-feeders, an essential adaptation for dip-feeders given the momentum of their bodies in flight compared to their relatively stationary prey. Similarly, the ventrally-orientated occiput does not permit the azhdarchid skull to extend in line with the neck during the 'strike' phase of prey apprehension. Azhdarchids also lack the ventrally-curved jaw tips of many dip-feeders. Given that azhdarchids also lack the ability to apprehend prey in flight with their limbs (see discussion of mid-air predation, above), it seems highly unlikely that azhdarchids were capable of seizing prey from the water surface in flight.

Wading.

The wading ecology proposed for azhdarchids by several workers [16], [19], [21], [31], [32]agrees with many aspects of azhdarchid anatomy and with the sedimentological context of some specimens, but has been poorly explored by its proponents. Chatterjee and Templin [16]used the occurrence of azhdarchids in lacustrine deposits as evidence of a wading lifestyle, but such argumentation is circumstantial in light of the preservation bias afforded by aquatic settings compared to terrestrial environments. Furthermore, the azhdarchid skeleton suggests that, while competent walkers, they were poor waders. The elongate limbs, neck and jaws of azhdarchids appear well suited for wading, but their manus and pes anatomy and the *Haenamichnus* footprints suggest that their extremities had relatively small surface areas, a condition quite different to the splayed feet of wading birds [101]. Some storks with relatively small feet are known to wade [109], indicating that azhdarchids may have been capable of some wading activity, but the high masses of large azhdarchids may have limited their ability to wade on soft substrates. Moreover, other pterodactyloids with larger pedal surface areas (most notably ctenochasmatoids) were almost certainly better adapted waders than azhdarchids. In view of this evidence, we suggest that azhdarchids were not habitual, although perhaps faculatative, waders.

Terrestrial stalkers.

Our interpretation of the evidence has led us to conclude that azhdarchids severed the ties with aquatic foraging conventionally assumed for pterosaurs (e.g. [24]), and that they were instead terrestrial opportunists, finding much of their food via terrestrial, ground-level foraging. The number of bird lineages that have independently evolved to exploit such a niche suggests that it is highly plausible that some pterosaur groups could do the same. The skewed distribution of azhdarchid fossils in continental settings corroborates this hypothesis at least in part: with the possible exception of obligate scavenging, all other proposed azhdarchid ecologies would predict the occurrence of azhdarchids in other depositional contexts, but their continental preference suggests a lifestyle in which most of their time – and subsequent foraging – occurred inland. Moreover, this hypothesis most adequately explains the anatomical details that have proved problematic in other hypotheses, such as the structure and carriage of the neck and skull and relatively small feet. At the expense of swimming or wading ability, shorter feet decrease the out-lever arm of the foot during the flexion phase of the step cycle and, therefore, increase walking efficiency. Azhdarchids, therefore, possessed feet well adapted for walking: an observation that agrees with their relatively long limbs. The latter would not only increase stride length but also allow easier passage through dense vegetation and provide a high vantage point to spot prey, both adaptations reflected in modern avian terrestrial stalkers [104]. The robust foot skeleton and padded soles present further adaptations to a primarily-grounded lifestyle, providing both traction and cushioning when walking on hard substrates. An efficient standing and walking ability in azhdarchids is further verified by the unusually narrow-gauge *Haenamichnus* trackway: bringing the limbs closer to the midline allowed grounded azhdarchids to support their weight largely through compressive forces acting on sub-vertical limbs. Other trackways show that non-azhdarchid pterosaurs had partially abducted limbs when moving terrestrially [e.g. 96] and consequently would have had to exert some effort to counter bending forces acting on their limb bones and joints. That azhdarchids appear to have overcome this problem suggests that they had one of the most energy-efficient postures and gaits known in any pterosaur, and we speculate that strong specialisation to weight-bearing may have facilitated the evolution of exceptional size in the clade. Combined with their wings well adapted for flight around inland settings (see above), azhdarchids appear well suited for locomotion in cluttered terrestrial environments.

The unusual pterosaur manual morphology and lack of large claws suggests that azhdarchids did not employ their limbs in apprehension of prey, even when grounded. It is assumed, therefore, that the jaws were the primary agents of prey apprehension, and would need to be lowered to ground level to procure food. Due to the long azhdarchid hindlimb, relatively little flexion is required in the forelimbs to bring the skull and neck towards the ground (Figure 8). Moreover, the relatively simple neck and skull mobility required in terrestrial foraging means that, unlike most other purported feeding styles, the stiff azhdarchid neck and ventrally oriented occiput do not present mechanical problems for this feeding method, as only slight flexion of the anterior neck vertebrae will fully lower the jaws to the substrate. The perpendicular orientation of the skull to the neck decreases the cervical flexion required to lower the jaws and, augmented by the great length of the skull, the distance between the jaw tips and the ground. The elongate neck also serves to decrease the flexion necessary to lower the jaws, as relatively minor inter-vertebral rotations are exaggerated along the length of the series, although this in itself does not necessarily preclude the influence of other factors (e.g. counterbalance of long limbs, greater visual acuity gained by increased skull height, sexual selection) on the evolution of the azhdarchid neck. The limited mobility of the neck means azhdarchids could not, however, projectile-feed in the manner of many long-necked birds.

The azhdarchid skull has a similar construction to modern birds that habitually stalk terrestrial environments (such as marabou storks and ground hornbills; Figure 7D) in bearing a long but relatively deep rostrum that extends anteriorly without invasion of the nasoantorbital fenestra (as is typical of other azhdarchoids [e.g. 11], [12]). The long jaw and relatively small jaw muscles of azhdarchids would presumably limit them to small food items that would not require strong bite forces or high mechanical strength to subdue or process. As with extant avian terrestrial stalkers, the generalised bills of azhdarchids would enable them to have a broad carnivorous diet comprised of relatively small vertebrates and large invertebrates, possibly supplemented with fruit [6] and carrion (see above).

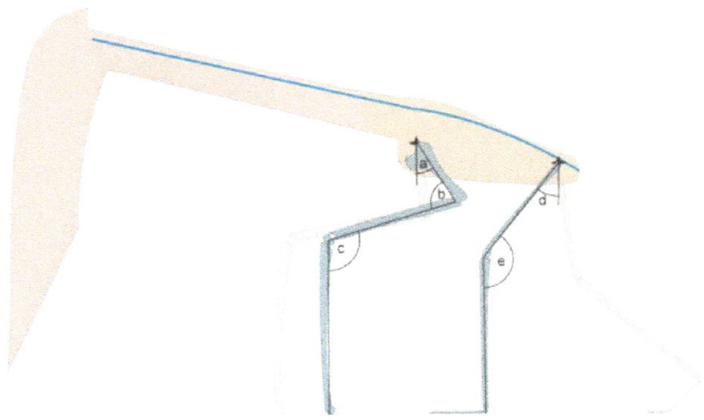

Figure 8. Reconstructed feeding posture of an azhdarchid with sagittally aligned limbs, as evidenced by [46].

The blue line indicates the dorsal and cervical column; note how the long jaws require little flexion of the forelimb to be lowered to the ground and how only moderate flexion of the anterior cervical series would lower the jaws fully. Letters denote approximate angles used in this reconstruction; a, 30°; b, 80°; c, 120°; d; 35°; e, 145°.

Hence, although azhdarchid anatomy is unique in a number of aspects, they appear to have been stork- or ground hornbill-like terrestrial stalkers (Figure 9), with the best modern analogues being the most generalized storks, such as the *Ciconia* species. Note that azhdarchids lack the specializations seen in some stork taxa, such as the *Mycteria* wood storks (which specialise in tactile feeding and wading), or *Anastomus*, the open-billed stork (which possesses scopate tomial edges and upper and lower jaws that bow away from each other. These are apparently specializations that assist in the holding of hard-shelled prey [109]).

Figure 9. Life restoration of a group of giant azhdarchids, *Quetzalcoatlus northropi*, foraging on a Cretaceous fern prairie.

A juvenile titanosaur has been procured by one pterosaur, while the others stalk through the scrub in search of small vertebrates and other foodstuffs.

Concluding remarks

Studies of pterosaur ecology have suffered from the dogmatic attitude that pterosaurs were predominately aerial piscivores living in coastal settings, in spite of steady accretion of evidence that they occupied a variety of ecological roles in a suite of environments. The unusual anatomy of azhdarchids strongly indicates that they had a unique ecology and inhabited unusual environments compared to many other pterosaurs: these details have been overlooked by most authors who have interpreted azhdarchids as marine piscivores occupying niches conventionally considered typical of pterosaurs as a whole. This unusual lifestyle may explain the resilience of azhdarchids to decline in contrast to other Cretaceous pterosaur lineages, few or none of which persisted to the late Maastrichtian as did azhdarchids. It is hoped that this re-revaluation of azhdarchid ecology will inspire much-needed descriptions of azhdarchid material, empirical testing of the hypotheses presented here, and further research into the lifestyles of pterosaurs beyond their flight capability.

ACKNOWLEDGMENTS

For access to specimens we thank S. Chapman (The Natural History Museum, London) and E. Frey (Staatliches Museum für Naturkunde, Karlsruhe). For valuable discussion and advice on the manuscript, we thank D. M. Martill (University of Portsmouth) D. M. Unwin (University of Leicester) and M. Wedel (University of California, Merced).

Author Contributions

Analyzed the data: MW DN. Wrote the paper: MW DN.

REFERENCES

1. Averianov AO, Arkhangelsky MS, Pervushov EM, Ivanov AV (2005) A new record of an azhdarchid (Pterosauria: Azhdarchidae) from the Upper Cretaceous of the Volga Region. Paleontological Journal 39: 433–439.
2. Lawson DA (1975) Pterosaur from the Latest Cretaceous of West Texas: Discovery of the largest flying creature. Science 185: 947–948.
3. Lü J, Ji Q (2005) New azhdarchid Pterosaur from the Early Cretaceous of Western Liaoning. Acta Geologica Sinica 79: 301–307.
4. Unwin DM, Lü J (1997) On Zhejiangopterus and the relationships of pterodactyloid pterosaurs. Historical Biology 12: 199–210.
5. McGowen MR, Padian K, de Sosa MA, Harmon RW (2002) Description of Montanazhdarcho minor, an azhdarchid pterosaur from the Two Medicine Formation (Campanian) of Montana. Paleobios 22: 1–9.
6. O″si A, Weishampel DB, Jianu CM (2005) First evidence of azhdarchid pterosaurs from the Late Cretaceous of Hungary. Acta Palaeontologica Polonica 50: 777–787.
7. Buffetaut E, Grigorescu D, Csiki Z (2003) Giant azhdarchid pterosaurs from the terminal Cretaceous of Transylvania (western Romania). In: Buffetaut E, Mazin JM, editors. Evolution and Palaeobiology of Pterosaurs, Geological Society Special Publication. 217. : 91–104.
8. Andres B, Ji Q (2008) A new pterosaur from the Liaoning Province of China, the phylogeny of the Pterodactyloidea, and the convergence in their cervical vertebrae. Palaeontology 51: 453–469.
9. Lü J, Ji Q (2006) Preliminary results of a phylogenetic analysis of the pterosaurs from western Liaoning and surrounding areas. Journal of the Paleontological Society of Korea 22: 239–261.
10. Wang X, Zhou Z (2006) Pterosaur assemblages of the Jehol Biota and their implication for the Early Cretaceous pterosaur radiation. Geological Journal, 41: 405–418.
11. Kellner AWA (2003) Pterosaur phylogeny and comments on the evolutionary history of the group. In: Buffetaut E, Mazin JM, editors. Evolution and Palaeobiology of Pterosaurs, Geological Society Special Publication,. 217. : 105–137.
12. Unwin DM (2003) On the phylogeny and evolutionary history of pterosaurs. In: Buffetaut E, Mazin JM, editors. Evolution and Palaeobiology of Pterosaurs, Geological Society Special Publication,. 217. : 139–190.
13. Martill DM, Naish D (2006) Cranial crest development in the azhdarchoid pterosaur Tupuxuara, with a review of the genus and tapejarid monophyly. Palaeontology 49: 925–941.
14. Hankin EH, Watson DMS (1914) On the flight of pterodactyls. Aeronautical Journal 18: 324–335.
15. Bramwell CD, Whitfield GR (1974) Biomechanics of Pteranodon. Philosophical Transactions of the Royal Society of London 267: 503–581.
16. Chatterjee S, Templin RJ (2004) Posture, locomotion and palaeoecology of pterosaurs. Geological Society of America Special Publication 376: 1–64.
17. Langston W Jr. (1981) Pterosaurs. Scientific American 244: 92–102.
18. Nessov LA (1984) Pterosaurs and birds of the Late Cretaceous of Central Asia. Paläontologische Zeitschrift 1: 47–57.
19. Paul GS (1987) Pterodactyl habits – real and radio controlled. Nature 328: 481.
20. Paul GS (1987) The science and art of restoring the life appearance of dinosaurs and their relatives - a rigorous how-to guide. In: Czerkas SJ, Olson EC, editors. Dinosaurs Past and Present Vol. II. Natural History Museum of Los Angeles County/University of Washington Press, Seattle and London. pp. 4–49.
21. Padian K (1988) The flight of pterosaurs. Natural History 97: 58–65.
22. Unwin DM (1988) The Azhdarchidae: giant pterosaurs from the upper Cretaceous. Abstracts of the Palaeontological Association Annual Conference: Aston, 16-17.
23. Unwin DM (2005) The Pterosaurs from Deep Time. New York: Pi Press.
24. Wellnhofer P (1991) The Illustrated encyclopaedia of Pterosaurs. London: Salamander Books Ltd.
25. Kellner AWA, Langston W Jr. (1996) Cranial remains of Quetzalcoatlus (Pterosauria, Azhdarchidae) from

Late Cretaceous sediments of Big Bend National Park. Journal of Vertebrate Paleontology 16: 222–231.

26. Lehman TM, Langston W Jr. (1996) Habitat and behaviour of Quetzalcoatlus: paleoenvironmental reconstruction of the Javelina Formation (Upper Cretaceous), Big Bend National Park, Texas. Journal of Vertebrate Paleontology 16: 48A.
27. Martill DM (1997) From hypothesis to fact in a flight of fancy: the responsibility of the popular scientific media. Geology Today 13: 71–73.
28. Unwin DM, Bakuhrina NN, Lockley MG, Manabe M, Lü J (1997) Pterosaurs from Asia. Paleontological Society of Korea Special Publication 2: 43–65.
29. Martill DM, Frey E, Sadaqah RM, Khoury HN (1998) Discovery of the holotype of the giant pterosaur Titanopteryx philadephiae Arambourg, 1959 and the status of Arambourgiania and Quetzalcoatlus. Neues Jahrbuch für Geologie and Paläontologie, Abhandlungen 207: 57–76.
30. Prieto IR (1998) Morfología funcional y hábitos alimentarios de Quetzalcoatlus (Pterosauria). Coloquios de Paleontologia 49: 129–144.
31. Bennett SC (2001) The osteology and functional morphology of the Late Cretaceous pterosaur Pteranodon. (parts 1 and 2). Paleontographica Abt. A 260: 1–153.
32. Witton MP (2007) Titans of the skies: azhdarchid pterosaurs. Geology Today 23: 33–38.
33. Arthur SC (1921) The feeding habits of the black skimmer. The Auk 38: 566–574.
34. Tomkins IR (1951) Method of feeding of the black skimmer, Rynchops nigra. The Auk 68: 236–239.
35. Bock WJ (1960) Secondary articulation of the avian mandible. Auk 77: 19–55.
36. Zusi RL (1962) Structural adaptations of the head and neck in the Black Skimmer, Rynchops nigra Linnaeus. Publications of the Nuttal Ornithological Club, 3,101.
37. Tomkins IR (1963) Skimmer-like behaviour in the royal and Caspian terns. The Auk 80: 549.
38. Arambourg C (1954) Sur la présence d'un ptérosaurien gigantesue dans les phosphates de Jordanie. Canadian Royal Academy of Science, Paris 283: 133–134.
39. Arambourg C (1959) Titanopteryx phildelphiae nov. gen., nov. sp., pterosaurien géant. Notes et Mémoires du Moyen-Orient 7: 229–234.
40. Cai Z, Wei F (1994) Zhejiangopterus linhaiensis (Pterosauria) from the Upper Cretaceous of Linhai, Zhejiang, China. Vertebrata PalAsiatica 32: 181–194.
41. Averianov AO, Atabekyan AA (2005) The first discovery of a flying reptile (Pterosauria) in Armenia. Paleontological Journal 38: 669–671.
42. Averianov AO (2007) New records of azhdarchids (Pterosauria, Azhdarchidae) from the Late Cretaceous of Russia, Kazakhstan, and Central Asia. Paleontological Journal 2: 73–79.
43. Bakhurina NA, Unwin DM (1995) A survey of pterosaurs from the Jurassic and Cretaceous of the former Soviet Union and Mongolia. Historical Biology 10: 197–245.
44. Buffetaut E, Laurent Y, Le Lœuff J, Bilotte M (1997) A terminal Cretaceous giant pterosaur from the French Pyrenees. Geological Magazine 134: 553–556.
45. Wang X, Kellner AWA, Zhou Z, Campos DA (2005) Pterosaur diversity and faunal turnover in Cretaceous terrestrial ecosystems in China. Nature 437: 875–879.
46. Hwang KG, Huh M, Lockley MG, Unwin DM, Wright JL (2002) New pterosaur tracks (Pteraichnidae) from the Late Cretaceous Uhangri Formation, S. W. Korea. Geological Magazine 139: 421–435.
47. Pereda Suberbiola X, Bardet N, Jouve S, Iarochéne M, Bouya B, et al. (2003) A new azhdarchid pterosaur from the Late Cretaceous phosphates of Morocco. In: Buffetaut E, Mazin JM, editors. Evolution and Palaeobiology of Pterosaurs. Geological Society Special Publication,. 217. : 79–90.
48. Western D (1980) Linking the ecology of past and present mammal communities. In: Behrensmeyer AK, Hill AP, editors. Fossils in the making – Vertebrate taphonomy and paleoecology,. Chicago: University of Chicago Press. pp. 41–54.
49. Lyman RL (1994) Vertebrate Taphonomy. Cambridge: Cambridge University Press.
50. Anderson LC (2001) Transport and spatial fidelity. In: Briggs DE, Crowther PR, editors. Palaeobiology II,. Oxford: Blackwell Publishing. pp. 289–292.
51. Behrensmeyer AK (2001) Terrestrial vertebrates. In: Briggs DE, Crowther PR, editors. Palaeobiology II,. Oxford: Blackwell Publishing. pp. 318–321.
52. Averianov AO (2004) New data on Cretaceous flying reptiles (Pterosauria) from Russia, Kazakhstan, and Kyrgyzstan. Paleontological Journal 38: 426–436.
53. Wellnhofer P, Buffetaut E (1999) Pterosaur remains from the Cretaceous of Morocco. Paläontologische Zeitschrift 73: 133–142.
54. Kellner AWA (2004) New information on the Tapejaridae (Pterosauria, Pterodactyloidea) and discussion of the relationships of this clade. Ameghiniana 41: 521–534.
55. Schumacher G (1973) The head muscles and hyolaryngeal skeleton of turtles and crocodilians. In: Gans C, Parsons TS, editors. Biology of the Reptilia Volume 4: Morphology D. London: Academic Press. pp. 101–200.
56. Witmer LM (1987) The nature of the antorbital fossa of archosaurs: shifting the null hypothesis. In: Currie PJ, Koster EH, editors. Fourth Symposium on Mesozoic Terrestrial Ecosystems, Short Papers. Drumheller: Boxtree Books. pp. 230–235.
57. Howse SCB (1986) On the cervical vertebrae of the Pterodactyloidea (Reptilia: Archosauria). Zoological Journal of the Linnean Society 88: 307–328.
58. Frey E, Martill DM (1996) A reappraisal of Arambourgiania (Pterosauria, Pterodactyloidea): one of the world›s largest flying animals. Neues Jahrbuch für Geologie und Paläontologie, Abhandlugen 199: 221–247.
59. Godfrey SJ, Currie PJ (2005) Pterosaurs. In: Currie PJ, Koppelhus EB, editors. Dinosaur Provincial Park: A

Spectacular Ancient Ecosystem Revealed. Bloomington and Indianapolis: Indiana University Press. pp. 292–311.

60. Henderson MD, Peterson JE (2006) An azhdarchid pterosaur cervical vertebra from the Hell Creek Formation (Maastrichtian) of southeastern Montana. Journal of Vertebrate Paleontology 26: 192–195.

61. Bennett SC (2003) Morphological evolution of the pectoral girdle in pterosaurs: myology and function. In: Buffetaut E, Mazin JM, editors. Evolution and Palaeobiology of Pterosaurs, Geological Society Special Publication. 217. : 191–215.

62. Graf W, de Waele C, Vidal PP (1995) Functional anatomy of the head-neck movement system of quadrupedal and bipedal animals. Journal of Anatomy 186: 55–74.

63. Vidal PP, Graf W, Berthoz A (1986) The orientation of the cervical vertebral column in unrestrained awake animals. Experimental Brain Research 61: 549–559.

64. Veldmeijer AJ (2003) Description of Coloborhynchus spielbergi sp. nov. (Pterodactyloidea) from the Albian (Lower Cretaceous) of Brazil. Scripta Geologica 125: 35–139.

65. Kellner AWA (1996) Description of the braincase of two Early Cretaceous pterosaurs (Pterodactyloidea) from Brazil. American Museum Novitates 3175: 1–34.

66. Heptonstall WB (1971) An analysis of the flight of the Cretaceous pterodactyl Pteranodon ingens Marsh. Scottish Journal of Geology 7: 61–78.

67. Stein RS (1975) Dynamic analysis of Pteranodon ingens: a reptilian adaptation to flight. Journal of Paleontology 49: 534–548.

68. Mateer NJ (1975) A study of Pteranodon. Bulletin of the Geological Institute of the University of Uppsala 6: 23–53.

69. Brower JC (1983) The aerodynamics of Pteranodon and Nyctosaurus, two large pterosaurs from the upper Cretaceous of Kansas. Journal of Vertebrate Paleontology 3: 84–124.

70. Wilkinson MT, Unwin DM, Ellington CP (2006) High lift function of the pteroid bone and forewing of pterosaurs. Proceedings of the Royal Society B 273: 119–126.

71. Brower JC, Venius J (1981) Allometry in pterosaurs. The University of Kansas Paleontological Contributions 105: 1–32.

72. Hazlehurst GA, Rayner JMV (1992) Flight characteristics of Triassic and Jurassic Pterosauria: an appraisal based on wing shape. Paleobiology 18: 447–463.

73. Paul GS (2002) Dinosaurs of the Air: The Evolution and Loss of Flight in Dinosaurs and Birds. Baltimore: John Hopkins University Press.

74. Frey E, Buchy MC, Martill DM (2003) Middle- and bottom-decker Cretaceous pterosaurs: unique designs in active flying vertebrates. In: Buffetaut E, Mazin JM, editors. Evolution and Palaeobiology of Pterosaurs, Geological Society Special Publication,. 217. : 267–274.

75. Humphries S, Bonser RHC, Witton MP, Martill DM (2007) Did pterosaurs feed by skimming? Physical modelling and anatomical evaluation of an unusual feeding method. PLoS Biology 5: e204 doi:10.1371/journal.pbio.0050204.

76. Shipman P (1998) Taking wing: Archaeopteryx and the Evolution of Bird Flight. London: Weidenfeld and Nicolson.

77. Atanassov MN, Strauss RE (2002) How much did Archeopteryx and Quetzalcoatlus weigh? Mass estimation by multivariate analysis of bone dimensions. Journal of Vertebrate Paleontology 22: 33A.

78. Templin RJ (2000) The spectrum of animal flight: from insects to pterosaurs. Progress in Aerospace Sciences 36: 393–436.

79. Marden JH (1994) From damselflies to pterosaurs: how burst and sustainable flight performance scale with size. American Journal of Physiology 266: 1077–1084.

80. Paul GS (1991) The many myths, some old, some new, of dinosaurology. Modern Geology 16: 69–99.

81. Bonde N, Christiansen P (2003) The detailed anatomy of Rhamphorhynchus: axial pneumaticity and its implications. In: Buffetaut E, Mazin JM, editors. Evolution and Palaeobiology of Pterosaurs, Geological Society Special Publication. 217. : 217–232.

82. Seamans TW, Hamershock DW, Bernhardt GE (1995) Determination of body density for twelve bird species. Ibis 137: 424–428.

83. Wedel MJ (2005) Postcranial skeletal pneumaticity in sauropods and its implications for mass estimates. In: Wilson JA, Curry-Rogers K, editors. The Sauropods: Evolution and Paleobiology. Berkeley: University of California Press. pp. 201–228.

84. Prange HD, Anderson JF, Rahn H (1979) Scaling of skeletal mass to body mass in birds and mammals. The American Naturalist 113: 103–122.

85. Rayner JM (1988) Form and function in avian flight. Current Ornithology 5: 1–66.

86. Wild R (1993) A juvenile specimen of Eudimorphodon ranzii Zambelli (Reptilia, Pterosauria) from the Upper Triassic (Norian) of Bergamo. Rivista del Museo Civico di Scienze Naturali 16: 95–120.

87. Unwin DM, Bakhurina NN (1994) Sordes pilosus and the nature of the pterosaur flight apparatus. Nature 371: 62–64.

88. Frey E, Martill DM (1998) Soft tissue preservation in a specimen of Pterodactylus kochi (Wagner) from the Upper Jurassic of Germany. Neues Jahrbuch für Geologie und Paläontologie, Abhandlugen 20: 421–441.

89. Wang X, Zhou Z, Zhang F, Xu X (2002) A nearly completely articulated rhamphorhynchoid pterosaur with exceptionally well-preserved wing membranes and "hairs" from Inner Mongolia, northeast China. Chinese Science Bulletin 47: 226–230.

90. Lü J (2002) Soft tissue in an Early Cretaceous pterosaur from Liaoning province, China. Memoir of the Fukui Prefectural Dinosaur Museum 1: 19–28.

91. Frey E, Tischlinger H, Buchy MC, Martill DM (2003) New specimens of Pterosauria (Reptilia) with soft parts with implications for pterosaurian anatomy and locomotion. In: Buffetaut E, Mazin JM, editors. Evolution and Palaeobiology of Pterosaurs, Geological

Society Special Publication,. 217. : 233–266.

92. Pennycuick CJ (1971) Gliding flight of the white-backed vulture Gyps africanus. Journal of Experimental Biology 55: 13–38.

93. Pennycuick CJ (1972) Soaring behaviour and performance of some east African birds, observed from a motor glider. Ibis 114: 178–218.

94. Norberg UML, Brooke AP, Trewhella WJ (2000) Soaring and non-soaring bats of the family Pteropodidae (flying foxes, Pteropus spp.): wing morphology and flight performance. The Journal of Experimental Biology 203: 651–664.

95. Unwin DM (1997) Pterosaur tracks and the terrestrial ability of pterosaurs. Lethaia 29: 373–386.

96. Mazin JM, Billon-Bruyat J, Hantzepergue P, Larauire G (2003) Ichnological evidence for quadrupedal locomotion in pterodactyloid pterosaurs: trackways from the late Jurassic of Crayssac. In: Buffetaut E, Mazin JM, editors. Evolution and Palaeobiology of Pterosaurs, Geological Society Special Publication,. 217. : 283–296.

97. Rodrigurez-de la Rosa RA (2003) Pterosaur tracks from the latest Campanian Cerro del Pueblo Formation of southeastern Coahuila, Mexico. In: Buffetaut E, Mazin JM, editors. Evolution and Palaeobiology of Pterosaurs, Geological Society Special Publication,. 217. : 275–282.

98. Wellnhofer P (1988) Terrestrial locomotion in pterosaurs. Historical Biology 1: 3–16.

99. Frey E, Tischlinger H (2000) Weichteilanatomie der Flugsaurierfüße und Bau der Scheitelkämme: Neue Pterosaurierfunde aus der Solnhofener Schichten (Bayern) und der Crato-Formation (Brasilien). Archaeopteryx 18: 1–16.

100. Lü J, Jin X, Unwin DM, Zhao L, Azuma Y, et al. (2006) A new species of Huaxiapterus (Pterosauria: Pterodactyloidea) from the Lower Cretaceous of Western Liaoning, China with comments on the systematics of tapejarid pterosaurs. Acta Geologica Sinica 80: 315–326.

101. Storer RW (1971) Adaptive radiation of birds. In: Farner DS, King JR, editors. Avian Biology 1,. London: Academic Press. pp. 150–188.

102. Harris MA, Steudel K (1997) Ecological correlates of hind-limb lengths in the Carnivora. Journal of Zoology 241: 381–408.

103. Christiansen P (1999) Scaling of limb long bones to body mass in terrestrial mammals. Journal of Morphology 239: 167–190.

104. Zeffer A, Johansson C, Marmebro A (2003) Functional correlation between habitat use and leg morphology in birds (Aves). Biological Journal of the Linnean Society 79: 461–484.

105. Wallace MP, Temple SA (1987) Competitive interactions within and between species in a guild of avian scavengers. The Auk 104: 290–295.

106. Travaini A, Donázar JA, Rodríguez A, Ceballos O, Funes M, et al. (1998) Use of European hare (Lepus europaeus) carcasses by an avian scavenging assemblage in Patagonia. Journal of Zoology 246: 175–181.

107. Shelley EL, Tanaka MYU, Ratnathicam AR, Blumstein DT (2004) Can Lanchester›s laws help explain interspecific dominance in birds? The Condor 106: 395–400.

108. Witmer LM, Rose KD (1991) Biomechanics of the jaw apparatus of the gigantic Eocene bird Diatryma: implications for diet and mode of life. Paleobiology 17: 95–120.

109. Kahl MP (1987) An overview of the storks of the world. Colonial Waterbirds 10: 131–134.

110. Piersma T, van Aelst R, Kurk K, Berkhoudt H, Maas LRM (1998) A new pressure sensory mechanism for prey detection in birds: the use of principles of seabed dynamics? Proceedings of the Royal Society B 265: 1377–1383.

111. Nebel S, Jackson DL, Elner RW (2005) Functional association of bill morphology and foraging behaviour in calidrid sandpipers. Animal Biology 55: 235–243.

112. Lederer RJ (1972) The role of avian rictal bristles. The Wilson Bulletin 84: 192–197.

113. Hertel F (1995) Ecomorphological indicators of feeding behaviour in recent and fossil raptors. The Auk 112: 890–903.

114. Bennett SC (2003) New insights in Anurognathus and anurognathids. Rivista del Museo Civico di Scienze Naturali "E. Caffi" Bergamo 22: 9–11.

115. Ashmole NP (1971) Sea bird ecology and the marine environment. In: Farner DS, King JR, editors. Avian Biology, Volume 1,. New York: Academic Press Inc. pp. 223–286.

116. Lockley MG, Wright JL (2003) Pterosaur swim tracks and other ichnological evidence of behaviour and ecology. In: Buffetaut E, Mazin JM, editors. Evolution and Palaeobiology of Pterosaurs, Geological Society Special Publication,. 217. : 297–313.

117. Company J, Ruiz-Omeñaca JI, Pereda Suberbiola X (1999) A long-necked pterosaur (Pterodactyloidea, Azhdarchidae) from the Upper Cretaceous of Valencia, Spain. Geologie en Mijinbouw 78: 319–333.

118. Ikegami N, Kellner AWA, Tomida Y (2000) The presence of an azhdarchid pterosaur in the Cretaceous of Japan. Paleontological Research 4: 165–170.

119. Potter JK (1932) Fishing ability of the black skimmer (Rynchops nigra nigra). The Auk 49: 477.

120. Withers PC, Timko PL (1977) The significance of ground effect to the aerodynamic cost of flight and energetics of the black skimmer (Rynchops nigra). Journal of Experimental Biology 70: 13–26.

121. Buffetaut E (1999) Pterosauria from the Upper Cretaceous of Laño (Iberian Peninsula): a preliminary comparative study. Estudios del Museo de Ciencias Naturales de Alava 14: 289–294.

122. Padian K, Ricqlés AJ, Horner JR (1995) Bone histology determines a new fossil taxon of pterosaur (Reptilia: Archosauria). Comptes Rendus de l›Academie des Sciences, Paris, Serie IIa 320: 77–84.

123. Padian K (1984) A large pterodactyloid pterosaur from the Two Medicine Formation (Campanian) of Montana. Journal of Vertebrate Paleontology 4: 516–524.

124. Pereda Suberbiola X, Company J, Ruiz-Omeñaca J (2007) Azhdarchid pterosaurs from the Late Cretaceous (Campanian-Maastrichtian) of the Iberian Penninsula: an update. Abstracts of the Flugsaurier Wellnhofer pterosaur meeting, Munich,27.

125. Currie PJ, Russell DA (1982) A giant pterosaur (Reptilia: Archosauria) from the Judith River (Oldman) Formation of Alberta. Canadian Journal of Earth Sciences 19: 894–897.

126. Unwin DM, Bakuhrina NN (2000) Pterosaurs from Russia, Middle Asia and Mongolia. In: Benton MJ, Shishkin MA, Unwin DM, Kurochkin EN, editors. The Age of Dinosaurs in Russia and Mongolia,. Cambridge: Cambridge University Press. pp. 420–433.

127. Bennett SC, Long JA (1991) A large pterodactyloid pterosaur from the Late Cretaceous (Late Maastrichtian) of Western Australia. Records of the Western Australian Museum 15: 435–443.

128. Buffetaut E (2001) An azhdarchid pterosaur from the Upper Cretaceous of Cruzy (Herault, France). Comptes Rendus de l›Academie des Sciences, Paris, Sciences de la Terre et des Planétes 333: 357–361.

129. Buffetaut E, Clarke JB, Le Lœuff J (1996) A terminal Cretaceous pterosaur from the Corbiéres (southern France) and the problem of pterosaur extinction. Bulletin de la Societe Geologique de France 167: 753–759.

130. Currie PJ, Jacobsen AR (1995) An azhdarchid pterosaur eaten by a velociraptorine theropod. Canadian Journal of Earth Sciences 32: 922–925.

131. Kellner AWA, Mader BJ (1996) First report of Pterosauria (Pterodactyloidea, Azhdarchidae) from Cretaceous rocks of Morocco. Journal of Vertebrate Paleontology 16: 45A.

132. Monteillet J, Lappartient JR, Taquet P (1982) Un ptérosaurien géant dans le Crétacé supérieur de Paki (Sénégal). Comptes Rendus de l›Academie des Sciences, Paris, Serie IIa, 295: 409–414.

133. Murry PA, Winkler DA, Jacobs LL (1991) An azhdarchid pterosaur humerus from the Lower Cretaceous Glen Rose Formation of Texas. Journal of Paleontology 65: 167–170.

134. Obata I, Shibata K, Matsujawa M, Unwin DM (2007) New record of a pterosaur from the Late Cretaceous Izumi Group, Awaji Island, Hyogo Prefecture, Japan. Annual Report of the Fukada Geological Institute 8: 149–162.

135. Saegusa H, Furutani H (2004) A cervical vertebra of an azhdarchid pterosaur from the Seidan Formation, Izumi Group. Abstracts of the 111st Annual Meeting of the Geological Society of Japan: 252.

136. Parris DC, Grandstaff BS, Clements D (2004) A pterosaur femur from the Upper Cretaceous of North Carolina. Southeastern Geology 43: 52–55.

137. Codorinú L, Gasparini Z (2007) Pterosauria. In: Gasparini Z, Coria RA, Salgado L, editors. Patagonian Mesozoic Reptiles. Bloomington: Indiana University Press. pp. 143–166.

138. Obata I, Hasegawa Y, Otsuka H (1972) Report on the Cretaceous reptile fossils from Hokkaido. Memoirs of the National Science Musuem, Tokyo 5: 213–222.

CHAPTER 6

THE LAST MARINE PELOMEDUSOIDS (TESTUDINES: PLEURODIRA): A NEW SPECIES OF BAIRDEMYS AND THE PALEOECOLOGY OF STEREOGENYINA

Gabriel S. Ferreira, Ascanio D. Rincón, Andrés Solórzano, Max C. Langer

INTRODUCTION

The Stereogenyina is a peculiar extinct podocnemidid lineage, promptly recognized among pleurodiran turtles by their unique secondary palate, with a midline cleft formed by the maxilla and the palatine (Gaffney et al., 2011). With a relatively rich fossil record, the oldest known stereogenyins are Cordichelys antiqua (Andrews, 1903) and Stereogenys cromeri (Andrews, 1901) from the late Eocene of the Birket el-Qurun Formation, in Egypt, and its most recent record is Shweboemys pilgrimi (Swinton, 1939) from the Plio-Pleistocene Irrawaddy Formation, in Burma. The broad geographical range of the group includes North and South America, North Africa, and Southeast Asia (Gaffney et al., 2011).

Six of the twelve known stereogenyin species have been formally assigned to Bairdemys (Weems & Knight, 2013), a genus known from the Oligocene of the USA to the Miocene of Puerto Rico and Venezuela. They are characterized by the curved edges of the secondary palate midline cleft, and the very small, slitlike opening of the antrum postoticum (Gaffney et al., 2008). The first known taxon of this group was described by Wood & Díaz de Gamero (1971) as Podocnemis venezuelensis, from the late Miocene Urumaco Formation, Venezuela. It was inferred as possibly corresponding to a marine turtle, on the basis of the depositional environment of the type locality (Wood & Díaz de Gamero, 1971), as supported by further evidence (Winkler & Sánchez-Villagra, 2006; Smith et al., 2010).

Here we describe a new species of Bairdemys, from the marine carbonatic deposits of the middle Miocene Capadare Formation, Venezuela (Fig. 1). We further explore the structure of the secondary palate of Stereogenyina and evidences of their marine adaptations, allowing a better understanding of the evolutionary history of these unique pleurodiran turtles.

Citation: Ferreira GS, Rincón AD, Solórzano A, Langer MC. (2015), The last marine pelomedusoids (Testudines: Pleurodira): a new species of Bairdemys and the paleoecology of Stereogenyina. PeerJ 3:e1063 https://doi.org/10.7717/peerj.1063.

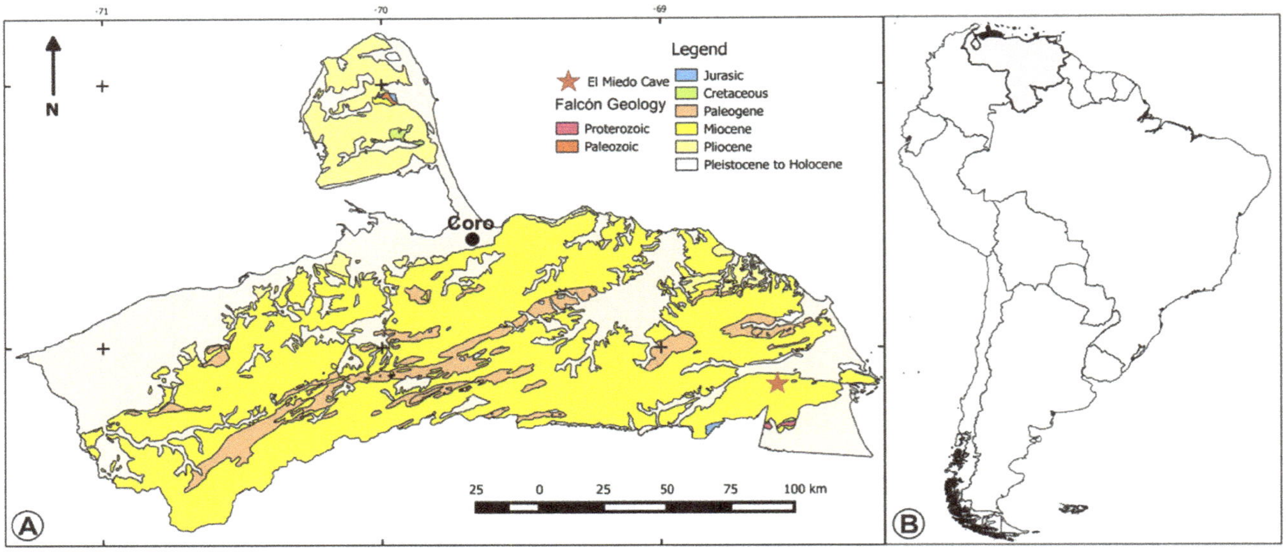

Figure 1: Maps of the type location of the holotype of Bairdemys thalassica.

(A) Surface geology map of the Falcón state, Venezuela, showing the location of the El Miedo Cave, where the holotype of Bairdemys thalassica was found, and (B) map of South America showing Venezuela (light gray) and Falcón state (black).

MATERIALS & METHODS

The fossil material was found partially encased into a limestone bed, and was prepared using acid solution. The material was described comparatively to other pelomedusoids, following the anatomical nomenclature used by Gaffney (1979), Gaffney, Tong & Meylan (2006) and Gaffney et al. (2011).

Nomenclatural acts

The electronic version of this article in Portable Document Format (PDF) will represent a published work according to the International Commission on Zoological Nomenclature (ICZN), so that the new names contained in the electronic version are effectively published under that Code from the electronic edition alone. This published work and the nomenclatural acts it contains have been registered in ZooBank, the online registration system for the ICZN. The ZooBank LSIDs (Life Science Identifiers) can be resolved and the associated information viewed through any standard web browser by appending the LSID to the prefix "http://zoobank.org/". The LSID for this publication is: [urn:lsid:zoobank.org:pub:F431F398-5442-4917-83E7-04CD0598225A]. The online version of this work is archived and available from the following digital repositories: PeerJ, PubMed Central and CLOCKSS.

Clade names follow, whenever possible, Phylogenetic Nomenclature principles and rules, as defined by the PhyloCode v. 4.0 (Cantino & de Queiroz, 2010). As such, suprageneric italicized names have been phylogenetically defined and those non-italicized refer to names defined in the context of the International Code of Zoological Nomenclature (ICZN, 1999).

The newly described skull was coded into a taxon-character matrix using Mesquite v. 3.0 (Maddison & Maddison, 2014). The matrix includes 57 characters (nine of which are new and 48 previously proposed by various authors) (SI1; Bona & de la Fuente, 2005; Cadena et al., 2012; Dumont Júnior, 2013; de la Fuente, 2003; Gaffney, 1977; Gaffney, Tong & Meylan, 2006; Gaffney et al., 2011; Meylan, Gaffney & Campos, 2009; Thomson & Georges, 2009, and 16 taxa. Podocnemis unifilis (Troschel, 1848), Peltocephalus dumerilianus (Schweigger, 1812), Erymnochelys madagascariensis (Grandidier, 1867) and Mogharemys blackenhorni(Dacqué, 1912) formed the outgroup and twelve Stereogenyina were included in the ingroup. The resulting matrix was analysed in TNT v. 1.0 (Goloboff, Farris & Nixon, 2008) using parsimony and implicit enumeration algorithms and bootstrap values (GC, 1,000 replicates, Fig. 5; Goloboff et al., 2003) were calculated using TNT implemented functions. A second analysis was run in TNT with a positive constraint (using the command force) for the monophyly of all species previously assigned to Bairdemys, and the resulting suboptimal topology was compared with the most parsimonious tree (MPT) using the Templeton test (Templeton, 1983) in PAUP* v. 4.0 (Swofford, 2002) (Table 1).

Table 1: Summary of the results of the Templeton test. Output of the Templeton test of the most parsimonious tree and one suboptimal tree with positive constraint for the monophyly of Bairdemys.

Templeton (Wilcoxon signed-ranks) test:	
N	11
Test statistic (T)	0
P	0.0021*

Two geometric morphometric analyses were performed to test if the morphology of the triturating surface of the upper and lower jaws of Stereogenyina corresponds to that of a durophagous turtle (Claude et al., 2004). The morphometric analysis of the upper jaw employed 14 extant taxa and three Stereogenyina with well preserved upper jaws. The lower jaw analysis employed 15 extant taxa plus three Stereogenyina with well preserved mandibles. The extant taxa used in both analyses were chosen based on the knowledge of their feeding habits (Moll , 1976; Meylan, 1988; Burke, Morreale & Standora, 1994; Teran, Vogt & Gomez, 1995; Pérez-Emán & Paolillo, 1997; Claude et al., 2004; Polovina et al., 2004; Salmon, Jones & Horch, 2004; Garcia & Lourenco, 2007). The list of employed specimens is seen in Supplemental Information 3. The skulls were photographed in ventral view and the lower jaws in dorsal view, and landmarks were digitized using the software tpsDig2 (Rohlf, 2005). Twelve landmarks for the upper and ten for the lower jaw were chosen (Fig. 6; description of the landmarks on Supplemental Information 3), aiming at representing the size and general shape of those structures. The shape data were extracted from the landmark dataset by a Generalized Procrustes Analysis (GPA, Rohlf, 1999) in the software MorphoJ v. 1.06d (Klingenberg, 2011), taking into account object symmetry (Klingenberg & McIntyre, 1998; Klingenberg, Barluenga & Meyer, 2002). Principal Component Analysis (PCA) (Table 2; Figs. 7A and 7C) were conducted for each dataset and components that explained morphological variation were retained after analysis. We tested the normality of shape components for upper and lower jaws with the Shapiro–Wilk's test and a MANOVA was performed to detect differences between durophagous, non-durophagous, and fossil species, using an a posteriori pairwise comparison test. All tests were performed using geomorph package (Adams & Otarola-Castillo, 2013; Adams, Collyer & Sherratt, 2015) in R environment software (R Development Core Team, 2013).

Table 2: Summary of the results of the geometric morphometric analyses. Results from MANOVA for the PCs of the palate and lower jaws and distance probabilities from the pairwise comparison tests for the palate and lower jaws using three predefined groups (durophagous, non-durophagous and fossil turtles).

MANOVA	**gl**	**SS/Wilks**	**F**	**P**	
Palate	2	0.43087	10.251	0.01*	
LowerJaws	2	0.67515	11.248	0.01*	

	Durophagous	**Non-durophagous**	**Fossils**
Palate			
Durophagous	1.00	0.01	0.32
Non-durophagous	0.01	1.00	0.01
Fossils	0.32	0.01	1.00
Lower Jaws			
Durophagous	1.00	0.01	0.30
Non-durophagous	0.01	1.00	0.01
Fossils	0.30	0.01	1.00

Systematic Paleontology

Pleurodira Cope, 1864

Pelomedusoides Cope, 1868

Podocnemididae Cope, 1868

Comment

Both Podocnemididae (Cope, 1868) and Podocnemidae (Baur, 1893) entered the literature in the 19th century and have been used since to designate the same biological group. The rules of the ICZN (1999) state that "A family-group name is formed by adding to the stem of the name [...] a suffix as specified in Article 29.2" (Article 29.1), that the suffix -IDAE is used for family names (Article 29.2), and that "If a generic name is or ends in a Greek or Latin word, or ends in a Greek or Latin suffix, the stem for the purposes of the Code is found by deleting the case ending of the appropriate genitive singular" (ICZN, 1999). For example, in the case of the genus Archaeopteryx, the genitive is Archaeopterygis and, thus, the stem for the purposes of the ICZN is Archaeopteryg-, giving the family name Archaeopterygidae; similarly, in the case of the genus Homo, the genitive is Hominis and the stem for the ICZN is Homin-, giving the family name Hominidae. In the case of the genus Podocnemis, formed by the greek words podos, foot, and cnemis, ocrea (Wagler, 1830), the Roman soldiers leg armour, the genitive case is Podocnemidos (D Amorim, F Lotti, E-E Kischlat, G Oliveira, pers. comm., 2015) and the stem for the purposes of the ICZN is Podocnemid-, giving the family name Podocnemididae. Additionally, Podocnemididae (Cope, 1868) has priority (Article 23.1) over Podocnemidae (Baur, 1893), and its usage (Article 29.5) is more common. As revealed by a Google Scholar search, "Podocnemididae" returns 1,020 and "Podocnemidae" 212 results. Accordingly, we chose to use the name Podocnemididae and advise future authors to do the same.

Stereogenyina (Gaffney et al., 2011)

Phylogenetic definition: Stereogenyina refers to the node-based clade that includes the most recent common ancestor of Stereogenys cromeri(Andrews, 1901), Bairdemys hartsteini (Gaffney & Wood, 2002), and Cordichelys antiqua (Andrews, 1903), and all its descendents.

Diagnosis (following Gaffney et al., 2011): Podocnemidids with a unique secondary palate among turtles, which is formed by maxillae and palatines and separated on the midline by a narrow cleft; no median maxillary ridge seen in other Podocnemididae; palate with a variable development of the rostral convexity and caudal concavity; palatine making up half or more of secondary palate; fossa orbitalis with extensive caudal pocket behind orbital rim enclosed by the septum orbitotemporale.

Bairdemys (Gaffney & Wood, 2002)

Derivatio nominis: For Donald Baird (see Gaffney & Wood, 2002).

Type species: Bairdemys hartsteini (Gaffney & Wood, 2002).

Diagnosis: Stereogenyins with an extremely small and slitlike antrum postoticum entrance; a ventral vertical flange on the squamosal; a distinct caudal projection of the squamosal; a deep convexity on the triturating surface; medial edges of palatal cleft curved; eustachian tube separated by bone from fenestra postotica; both foramina nervi-hypoglossi combined and recessed in a short canal opening on occipital surface; jugal-pterygoid contact; pectoral-abdominal sulcus not crossing mesoplastron; pectoral-humeral sulcus in cranial half of entoplastron; and intergular scales barely extending onto entoplastron.

Bairdemys thalassica sp. nov.

Derivatio nominis: From the Greek word thalassa (= sea).

Holotype: IVIC-P-2908 (IVIC-P: Colección de Paleontología, Instituto Venezolano de Investigaciones Científicas, Caracas, Venezuela), almost complete skull, lacking the pre-orbital region, the right portion of the skull roof, and the right lateral wall of the fossa temporalis.

Locality: El Miedo Cave (10°51′26″ N; 68°36′41″ W), located 15.5 km southwest of Yaracal town, Cerro Misión (Fig. 1), Falcón State, Venezuela.

Geology: The Capadare Formation is a thick sequence of massive carbonate layers deposited in shallows marine waters (Lorente, 1978), which crops out in northwestern Venezuela. It represents isolated carbonate platforms, without coastal or continental influence, in entirely open sea conditions, with well oxygenated and clear waters, normal salinity, and moderate energy in a tropical climate (Lorente, 1978; Díaz de Gamero, 1985). Development of karstic systems is common in the Capadare Formation, as exemplified by the well-known El Miedo and Zumbador caves. Based on its foraminifer assemblage, the Capadare Formation was assigned to the middle Miocene, Globorotalia fohsi fohsi-Globorotalia sjakensisBiozone (Renz, 1948; Lorente, 1978; Díaz de Gamero, 1985). Using strontium isotope analyses (ratio of $^{87}Sr/^{86}Sr$) from a marine mollusc shell (Ostrea sp.), (Solórzano & Rincón, in press) placed the Capadare Formation in the middle Miocene (~12.24 Ma). The record of vertebrates from the Capadare Formation includes Pelagornis cf. P. chilensis (Aves: Odontopterygiformes) and shark teeth (Rincón & Stucchi, 2003; Solórzano & Rincón, in press), but Pleistocene fossils were also recovered in the bottom of the El Miedo cave (Rincón, 2003; Rincón & White, 2007; Montellano-Ballesteros, Rincón & Solórzano, 2014).

Diagnosis: A Bairdemys with a short midline contact of the pterygoids; a longer basisphenoid-quadrate contact; a more rounded rostral tip of the basisphenoid; dermal scale iv expanding onto the postorbital, quadratojugal, and parietal; and a slitlike opening of the canal containing both foramina nervi hypoglossi significantly smaller than the foramen jugulare posterius.

Description

Skull—The holotype and only specimen assigned to Bairdemys thalassica, IVIC-P-2908 (Figs. 2 and 3), is very well preserved, with almost no taphonomic distortion. It is nearly complete, lacking only the rostral and the right dorsal regions. Only the vomeri, the premaxillae, and the prefrontals are completely absent. The frontal, jugal, quadratojugal, parietal, squamosal, maxilla, and palatine have one of the elements of the pair entirely or partially missing, but the other is complete, allowing a detailed description.

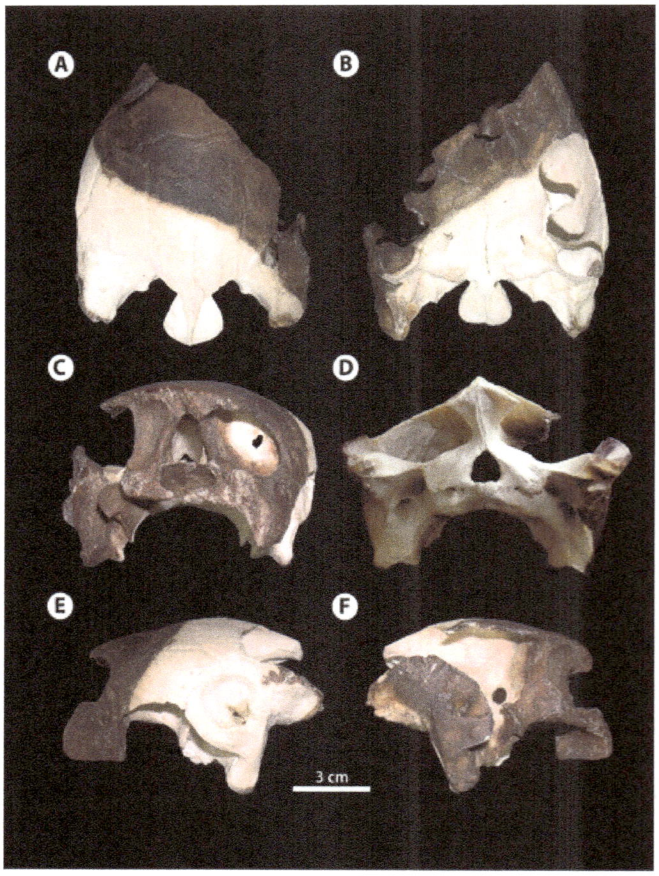

Figure 2: Bairdemys thalassica **sp.** nov. Holotype skull IVIC-P-2908 in (A) dorsal, (B) ventral, (C) rostral, (D) caudal, (E) left, and (F) right lateral views.

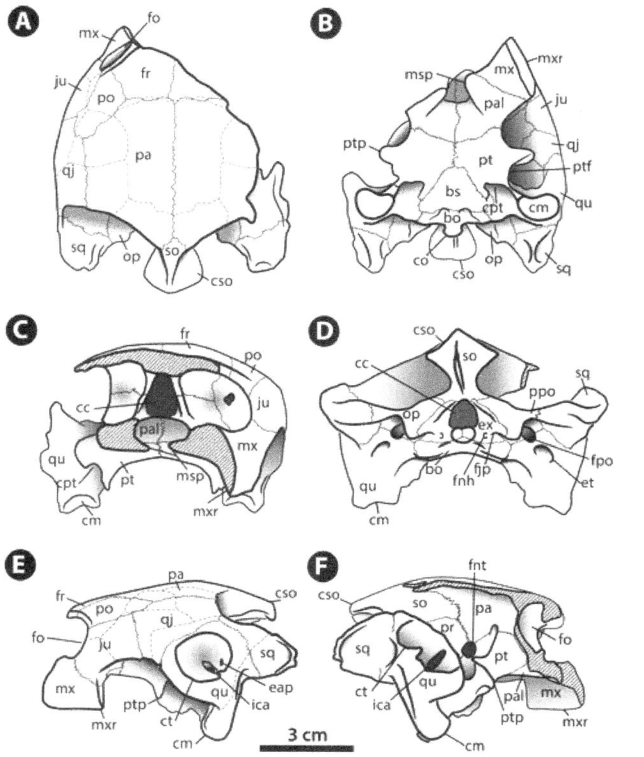

Figure 3: Bairdemys thalassica **sp.** nov.

Drawing of the holotype skull IVIC-P-2908 in (A) dorsal, (B) ventral, (C) rostral, (D) caudal, (E) left, and (F) right lateral views. **Abbreviations: bo,** basioccipital; **bs,**basisphenoid; **cc,** cavum cranii; **cm,** condylus mandibularis; **co,** condylus occipitalis; **cpt,**cavum pterygoidei; **cso,** crista supraoccipitalis; **ct,** cavum tympani; **eap,** entrance of the antrum postoticum; **et,** eustachian tube; **ex,** exoccipital; **fjp,** foramen jugulare posterius; **fnh,** foramina nervi hypoglossi; **fnt,** foramen nervi trigemini; **fo,** fossa orbitalis; **fpo,**fenestra postotica; **fr,** frontal; **ica,** incisura columellae auris; **ju,** jugal; **msp,** midline cleft of the secondary palate; **mx,** maxilla; **mxr,** maxillary ridge; **op,** opisthotic; **pa,** parietal; **pal,**palatine; **po,** postorbital; **ppo,** processus paroccipitalis; **pt,** pterygoid; **ptf,** pterygoid flange; **ptp,** processus trochlearis pterygoidei; **qj,** quadratojugal; **qu,** quadrate; **so,**supraoccipital; **sq,** squamosal.

Among Bairdemys species, the general shape of the skull of B. thalassica is more similar to that of B. venezuelensis, which is rounded, almost oval in dorsal outline (Fig. 2A), and relatively high in lateral view (Figs. 2C and 2D). The temporal emargination is not well developed, and rostrocaudally shallower than in B. hartsteini and B. sanchezi (Gaffney et al., 2008). The cheek emargination is as developed as in other Bairdemys, not reaching the dorsal edge of the cavum tympanii (Fig. 2E). As in other Bairdemys species, the orbits face laterally and the skull bone surface is smooth, lacking ornamentations except for the dermal scale sulcii.

Dermal scales of the skull—The scale sulcii of the skull are well defined in IVIC-P-2908. In comparison to those of Meiolaniformes, the only turtle lineage with a described and systematized scale pattern (Simpson, 1938; Gaffney, 1996; Sterli & de la Fuente, 2011; Sterli & de la Fuente, 2013; Rabi et al., 2014 also adopted this system to describe the scales of the eucryptodiran Annemys levensis), B. thalassica, as well as all pelomedusoids with preserved sulcii, shows a great reduction in the number of scale units. For example, Mongolochelys efremovi (Khosatzky, 1997), bears 14 types of scales on the skull (Sterli & de la Fuente, 2013), whereas pelomedusoids have only between 6 and 8 types. This hampers tracing scale homology between those groups, and a more in depth analysis, also including cryptodires and stem-Testudines, is needed to fully establish homology relations of the turtle skull dermal scales. For now, we decided to create a new system to identify the Pelomedusoides scales, using lower-case latin numbers. Potencial homologues with the proposed Meiolaniformes system (herein called SF13) are mentioned in the description.

The scales i, iii, v, vi and iv (Fig. 4) are paired in Bairdemys thalassica and in all studied Pelomedusoides. Scales iii can be divided in two units in some pelomedusoids, e.g., Pelomedusa subrufa (Bonnaterre, 1789), Erymnochelys madagascariensis, and Podocnemis vogli (Müller, 1935), or absent, e.g., Podocnemis sextuberculata (Cornalia, 1849), and Peiropemys mezzalirai (Gaffney et al., 2011). Scales ii, vii and viii are unpaired in B. thalassica, as well as in most pelomedusoids, e.g., Peltocephalus dumerilianus. In some taxa, e.g., Pelomedusa subrufa, Podocnemis unifilis, and Neochelys fajumensis (Andrews, 1903), scale ii is paired. Scale viii is absent in Pelomedusa subrufa and Cordichelys antiqua, the. Scale i forms the ventral-most portion of the rostral edge of the skull, covering the premaxilla, maxilla, and a portion of the jugal. It contacts scales ii, dorsally (caudal to the orbit), and iii, caudally. This scale can be the homologue of scales J plus I of SF13. A large and single ii scale covers the prefrontal, frontal, and portions of the postorbital and jugal, and contacts scales iii lateroventrally, iv laterocaudally, and vii caudomedially. We suggest that scale ii of Pelomedusoides can be a total or partial fusion of scales F, Z, and Y of SF13. Scale iii covers parts of the jugal, postorbital and quadratojugal in the lateral region of the skull. In contrast to the condition in Peltocephalus dumerilianus, E. madagascariensis, and Podocnemis vogli, scale iii in B. thalassica does not reach the orbital margin. This condition is shared with Bairdemys hartsteini and may represent a synapomorphy of Bairdemys. Scale iii contacts scales iv, medially, v, laterocaudally and vi, caudally, and we suggest that it corresponds to scale E in SF13. Over parts of the postorbital, quadratojugal and parietal of B. thalassica lies a semi-circular paired scale iv, that contacts scales vii medially, iii laterally and vi caudally. This scale is not found in any of the other analyzed pelomedusoids and may be the homologue to scale H of SF13. Scale v is relatively large and covers the quadrate, squamosal and parts of the quadratojugal, contacting scale vi medially. In this region, there are two scales in meiolaniforms, K and C (Sterli & de la Fuente, 2013), and we suggest that both are fused into scale v of Pelomedusoides. Due to the presence of scale iv, scale vi is reduced in B. thalassica compared to the condition of other pelomedusoids, e.g., Peltocephalus dumerilianus. It contacts scales vii, medially, and viii, caudomedially. Scale vii (identified otherwise as "interparietal scale") has parallel lateral margins and covers half of the parietals and a small portion of the frontals, contacting scale viii caudally. Scale viii is also unpaired and covers a small caudal portion of the parietals and supraoccipital. We suggest that scales vi, vii, and viii are respectively homologues of scales D, X, and A of SF13.

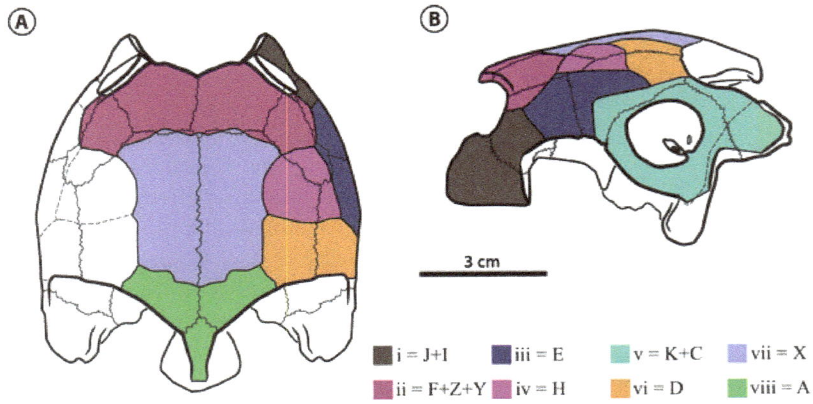

Figure 4: Bairdemys thalassica **sp.** nov.

Dermal scales of the skull in (A) dorsal and (B) lateral view. On the identification of each scale (in lower case latin numbers, from i to viii) there is after the "=" mark the preliminary homology assessment in relation to the Sterli & de la Fuente (2013) system (see Dermal scales of the skull section).

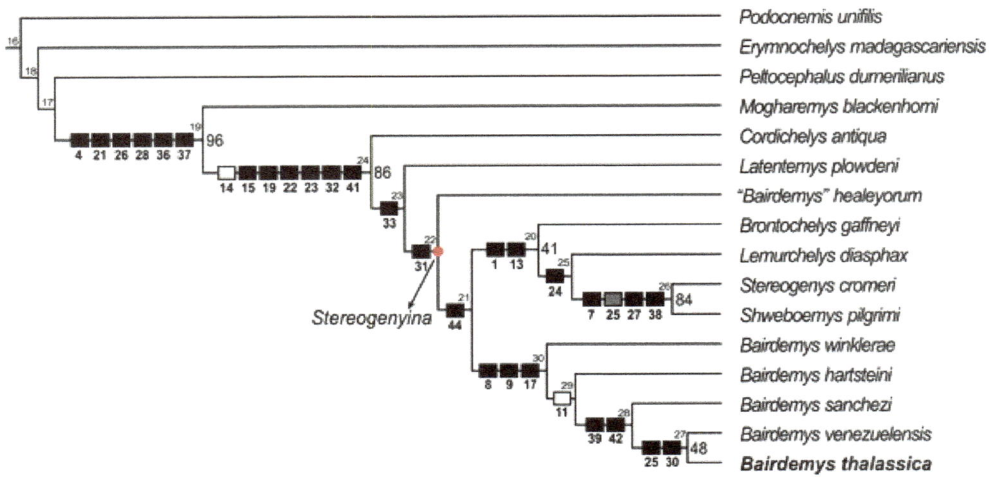

Figure 5: Phylogenetic relations of Stereogenyina **based on the single most parsimonious tree obtained in the present study.**

Node numbers are indicated above each node and GC bootstrap values above 40 in front of them. Squares indicate the synapomorphies of the respective clade (black = 0 to 1, gray = 0 to 2, white = 1 to 0).

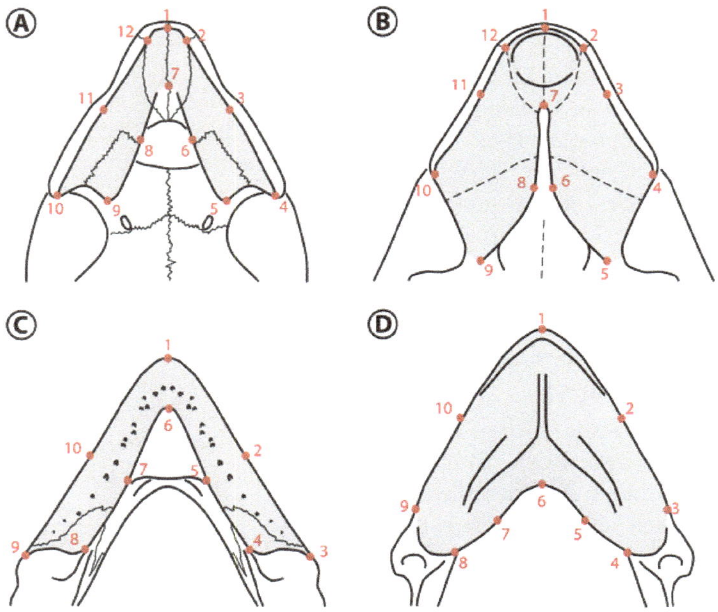

Figure 6: Landmarks used in the geometric morphometric analyses.

Landmarks shown in the palates of (A) Peltocephalus dumerilianus and (B) Bairdemys winklerae, and in the lower jaws of (C) Peltocephalus dumerilianus and (D) Bairdemys venezuelensis. For the description of the landmarks see Supplemental Information 3.

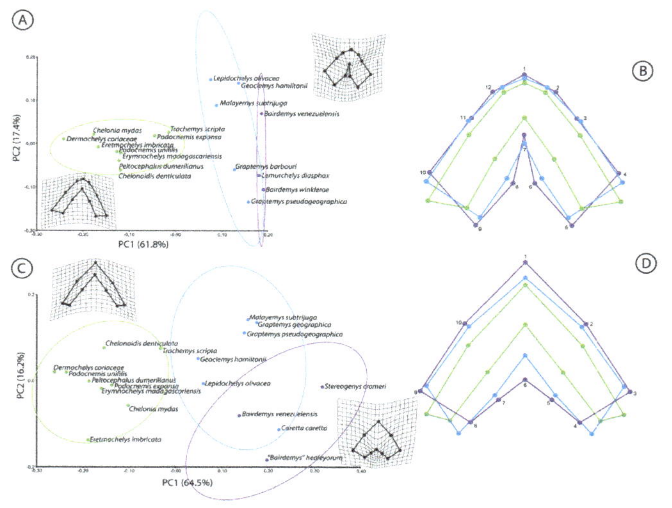

Figure 7: Results of the geometric morphometric analyses.

Principal component analysis (A and C) derived from the first two principal components (PC1 and PC2) and comparison of the mean values of the coordinates (B and D) for the upper (A and B) and the lower jaws (C and D). D'Arcy Thompson grids correspond to extreme shape variation. Durophagous taxa are represented in blue, non-durophagous in green, and fossil taxa with unknown diet in purple.

Frontal—Each frontal forms the caudomedial portion of the orbital ridge, and is sutured with the postorbital laterocaudally, parietal caudally, and with its counterpart medially. The curved skull outline in lateral view in the prefrontal-frontal contact region, characteristic of Bairdemys (Gaffney & Wood, 2002) cannot be seen in IVIC-P-2908, because it is usually located rostral to the preserved portion of the frontal. As in all Bairdemys species the frontal is very thick and also has a thick parasagittal process that contacts the processus inferior parietalis and the ascending process of the palatine (Fig. 3C); this separates the sulcus olfactorius from the fossa orbitalis.

Parietal—As in all Podocnemididae, each parietal is composed of a horizontal plate and a descending process, the processus inferior parietalis (Gaffney & Wood, 2002). In dorsal view, the horizontal plate contacts the frontal rostrally, postorbital laterorostrally, quadratojugal laterally, supraoccipital caudomedially, and the other parietal medially. The processus inferior parietalis is a very thick structure, as in Stereogenys cromeri, Shweboemys pilgrimi, and other Bairdemys species (Gaffney & Wood, 2002). It forms the roof and lateral wall of the cavum cranii and, together with the postorbital, the caudal surface of the septum orbitotemporale. The processus contacts the medial and lateral ascending processes of the pterygoid caudomedially, forming the roof of the sulcus palatinopterygoideus (Figs. 2F and 3F). Caudal to this sulcus, the parietal forms the dorsal margin of the foramen nervi trigemini.

Jugal—As in all pelomedusoid turtles, the jugal forms the laterocaudal portion of the orbital ridge, between the maxilla and the postorbital (Gaffney & Wood, 2002). The semi-circular dorsal margin of the cheek emargination is formed by the jugal rostrally and the quadratojugal caudally, and the contact between this two bones is an almost vertical line. The jugal-maxilla contact is nearly horizontal as in B. sanchezi and B. venezuelensis. As in most Stereogenyina, the jugal of B. thalassica does not enter the palatal triturating surface. In the septum orbitotemporale, it contacts the postorbital dorsomedially, the pterygoid medially, and the palatine ventrally, differing only from Stereogenys cromeri and Shweboemys pilgrim, among Stereogenyina in which the palatine prevents the jugal-pterygoid contact.

Quadratojugal—The quadratojugal contacts the jugal rostrally, the postorbital rostromedially, the parietal medially, the quadrate laterally, and the squamosal, in a very thin process (Fig. 3E). This arrangement is shared with all pelomedusoids, except for Podocnemis spp. and Cerrejonemys wayuunaiki (Cadena, Bloch & Jaramillo, 2010), in which there is no quadratojugal-postorbital contact, due to the great reduction of the postorbital (Gaffney & Wood, 2002; Cadena, Bloch & Jaramillo, 2010). As in all pelomedusoids the quadratojugal-quadrate suture is curved, following the margin of the cavum tympani (Fig. 3E).

Squamosal—The squamosal contacts the quadratojugal rostrodorsally, on the lateral edge of the temporal emargination, the opisthotic medially, and the quadrate rostrally, in a long suture. A dorsal process forms the laterocaudal portion of the wall of the fossa temporalis superior. As in all Testudines the squamosal of B. thalassica encloses a cavity caudodorsal to the cavum tympani, the antrum postoticum (Gaffney, 1979). In Bairdemysspp., Lemurchelys diasphax (Gaffney et al., 2011), and Stereogenys cromerithe opening of the antrum postoticum in the cavum tympani is extremely reduced and slit-shaped. The squamosal in Bairdemys spp. projects more caudally than in any other Stereogenyina. In B. thalassica its outline in dorsal view is more rounded caudally, aproaching the condition of B. hartsteini and differing from the squamosal of the other species of Bairdemys, e.g., B. winklerae (Gaffney et al., 2008), which have a sharp ending. The dorsomedial and ventrolateral surfaces of the caudal projection of the squamosal were probably attachment sites for the Musculus (M.) adductor mandibulae and the M. depressor mandibulae, respectively. Both of them are more developed in B. thalassica than in any other known Stereogenyina.

Postorbital—The postorbital is composed of a dorsal horizontal plate, very thick in Bairdemys spp., and a ventral vertical process that forms most of the septum orbitotemporale (Figs. 2C and 3C). The horizontal plate forms the caudal portion of the orbital ridge and contacts the frontal medially, parietal caudomedially, quadratojugal laterocaudally, and jugal laterally. The ventral process contacts the palatine rostrally and the pterygoid and jugal caudally, as seen in all other Bairdemys species, as well as in Lemurchelys diasphaxand Latentemys plowdeni (Gaffney et al., 2011).

Quadrate—The quadrate forms the entire cavum tympani which encloses the incisura columellae auris and the opening of the antrum postoticum (Figs. 2E and 3E). As in other Bairdemys species (Gaffney & Wood, 2002; Gaffney et al.,

2008) the fossa precolumellaris is absent in B. thalassica. The incisura columellae auris is completely surrounded by bone and encloses the stapes and the eustachian tube. In B. thalassica this bone forms a very thick caudal bar, as seen in all Stereogenyina that have this region preserved (Gaffney & Wood, 2002). The fenestra postotica is separated from the eustachian tube (Fig. 3D), as in all Bairdemys species and Latentemys plowdeni (Gaffney et al., 2011). The rostral and dorsal surfaces of the quadrate of B. venezuelensis and B. hartsteini, which form the caudal wall of the fossa temporalis inferior and the floor of the fossa temporalis superior, respectively, bears a small crest previously known only in those two species, in the same region where the processus trochlearis oticum of cryptodires is found (Gaffney & Wood, 2002). In B. thalassica, this structure is also present and, perhaps due to its excellent preservation, an additional crest is seen parallel to the former one, forming a smooth groove between them. The function of this structure is unknown.

In the wall and floor of the fossa temporalis, the quadrate contacts the pterygoid rostroventrally, the prootic rostrodorsally, the opisthotic caudomedially, and the squamosal caudally and laterocaudally. Together with the prootic, it forms the margins of the foramen stapedio-temporale, which opens rostrodorsally as in other podocnemidids. On its ventral surface, the quadrate contacts the basioccipital caudomedially, the basisphenoid medially, and the quadrate ramus of the pterygoid rostrally, close to the condylus mandibularis (Figs. 2B and 3B). The condylus has a concave rostral margin, as in other Bairdemys species, contrasting with the condition of Stereogenys cromeri and Lemurchelys diasphax, in which this margin is straight (Gaffney et al., 2011). The condylus mandibularis of B. thalassica is in the same line as the basisphenoid-basioccipital contact, a position closer to that seen in Cordichelys antiqua, Stereogenys cromeri, and B. venezuelensis than to that of other Bairdemys species. The quadrate forms the caudal portion of the roof of the cavum pterygoidei (Fig. 3B), which is also formed by the basisphenoid and pterygoid. On its lateral surface, around the cavum tympani, the quadrate contacts the quadratojugal rostrally and the squamosal caudally. On the caudal surface, it contacts the squamosal laterodorsally, the opisthotic dorsomedially, the exoccipital medially, and the basioccipital ventromedially. Finally, it forms the lateral and vetral margins of the fenestra postotica.

Maxilla—The maxila contacts the palatine caudomedially and the jugal dorsocaudally (Figs. 3B and 3E). The dorsal process that contacts the frontal in B. venezuelensis and B. hartsteini (Gaffney & Wood, 2002) is replaced by a process of the palatine in B. thalassica (Fig. 3C). The maxilla contributes less to the orbital floor in comparisson to that of other Bairdemys species, forming only its rostrolateral portion, where it contacts the jugal caudally and the palatine medially, as in Brontochelys gaffneyi (Wood, 1970) and Podocnemis expansa (Schweigger, 1812). On the palatal triturating surface, the suture with the palatine extends laterally, ending caudal to the end of the labial ridge (Fig. 3E). This condition is different from that of B. hartsteini, in which this suture ends more medially. In comparison to other Stereogenyina, the labial ridge in Bairdemys sp. is high and thin, as in Cordichelys antiqua.

Palatine—As in other Stereogenyina (Gaffney & Wood, 2002) the palatine of B. thalassica is a very complex bone. It bears two distinct horizontal plates forming the caudal portions of the primary and secondary palate, and two very thick vertical plates, one separating the fossa orbitalis from the sulcus palatinopterygoideus and another forming the rostromedial wall of the sulcus palatinopterygoideus and the laterorostral wall of the cavum cranii (Figs. 2Fand 3F). The horizontal plates are sutured to the maxilla rostrally and to the pterygoid caudally. These form the caudal portion of the midline cleft and the triturating surface, which extends towards the palatine-pterygoid contact, but does not reach it as in Stereogenys cromeri, Shweboemys pilgrimi, Brontochelys gaffneyi, and Lemurchelys diasphax. Close to this contact there is a tiny aperture identified as the foramen palatinum posterius. This is usually much reduced and sometimes absent in Stereogenyina. The palatine forms most of the fossa orbitalis floor, where it contacts the maxilla laterorostrally, the jugal laterocaudally, and the postorbital caudally. The horizontal plate also forms the rostral most part of the floor of the sulcus palatinopterygoideus.

The typical Stereogenyina secondary palate, (unique among Pleurodira) is well developed in B. thalassica, but less extensive than in Lemurchelys, Stereogenys, and Shweboemys (Gaffney et al., 2011). Although it is not completely preserved, the edges of its midline cleft are curved, as in other Bairdemys species, Cordichelys antiqua, and Latentemys plowdeni; in contrast to the straight/parallel edges found in Stereogenys, Shweboemys, Lemurchelys, and Brontochelys.

The lateral vertical plate is a dorsal process of the palatine, medial to the jugal and rostral to the ascendent process of the pterygoid. It contacts the ventral parasagittal process of the frontal in all species of Bairdemys, separating the fossa orbitalis from the sulcus palatinopterygoideus. As in all Stereogenyina, this process is well developed and becomes thicker rostrally. The medial vertical plate contacts the pterygoid caudally and the parietal dorsally, forming the wall that separates the cavum cranii from the sulcus palatinopterygoideus.

Pterygoid—As the palatine and quadrate, the pleurodiran pterygoid is a very complex bone. It is composed of a ventral horizontal plate that in Bairdemysspecies contacts the palatine rostrally, the quadrate laterocaudally, and the basisphenoid caudally. As is typical of Pleurodira (Gaffney, 1979), the pterygoid has a laterally expanding processus trochlearis pterygoidei, forming a right angle to the horizontal plate (Figs. 3B and 3F). Caudal to the base of the process, the pterygoid flange is well developed, covering the cavum pterygoideum as in all podocnemidids. This flange is a very thin sheet of bone and is frequently broken in fossils. In IVIC-P-2908 only parts of the left flange are preserved, but it is possible to infer that its shape and extension are similar to those of other species of Bairdemys.

Two dorsal processes are seen rostral and medial to the processus trochlearis pterygoidei. The rostral process contacts the jugal rostrally, the postorbital rostrodorsally and the parietal dorsally. The medial process is parallel to the processus trochlearis pterygoidei, forming the caudoventral portion of the sulcus palatinopterygoideus. It contacts the parietal rostrodorsally and the prootic caudodorsally, which together form the margins of the foramen nervi-trigemini opening laterally to the cavum cranii (Fig. 3F).

As typical of Podocnemoidae, the cavum pterygoidei of B. thalassica lies on the caudal portion of the ventral surface of the pterygoid. As in other Bairdemys species, Cordichelys antiqua, and Stereogenys cromeri, this structure is formed mainly by the pterygoid and quadrate with a small laterocaudal contribution of the basisphenoid. The extension of the medial contact between the pterygoids in B. thalassica is similar to that of B. venezuelensis but shorter than in B. sanchezi. The sutures of this region are not preserved in B. hartsteini and B. winklerae.

Supraoccipital—The supraoccipital forms the dorsal margin of the foramen magnum and its contacts are very similar in all Pelomedusoides: rostrodorsally to the parietals, lateroventrally to the prootic, caudally and lateroventrally to the opisthotic, and caudoventrally to the exoccipital, the latter on a narrow suture. In B. thalassica it has well developed horizontal plates that extend laterally to the crista supraoccipitalis, a common feature among Stereogenyina (Gaffney et al., 2011). In Bairdemys spp. those plates are even more pronounced and form a bulbous structure on the caudal edge of the crista supraoccipitalis (Figs. 2A and 3A). The crista supraoccipitalis of B. thalassica is not as expanded as in B. venezuelensis and B. winklerae, and does not reach the line formed by the caudal edges of the squamosals.

Exoccipital—The exoccipitals form the lateral and ventral margins of the foramen magnum and meet the supraoccipital dorsally. They also form the laterodorsal portion of the condylus occipitalis, where they contact the basioccipital ventrally (Fig. 3D). Each exoccipital contacts the opisthotic laterodorsally, the basioccipital along its entire ventral margin and its counterpart on the bottom of the foramen magnum. The foramen jugulare posterius (Fig. 3D) is formed mainly by the exoccipitals with small contributions of the basioccipital ventrally and opisthotics laterally. Dorsal to this foramen, each exoccipital and opisthotic form a covering. This develops into a trough extending laterally from this foramen (Gaffney et al., 2008), as seen in B. venezuelensis and B. sanchezi. As in all Stereogenyini (sensuGaffney et al., 2011) both foramina nervi-hypoglossi are combined and recessed into a single short canal, which opens laterocaudally, and is located between the condylus occipitalis and the foramen jugulare posterius. In B. thalassica the single opening of the canal containing both foramina nervi-hypoglossi is slit-shaped and substantially smaller than the foramen jugulare posterius , an autapomorphy of the new species.

Basioccipital—The basioccipital of B. thalassica contacts the basisphenoid rostroventrally, the quadrate lateroventrally, the opisthotic laterodorsally, and the exoccipital caudodorsally. It forms the ventral portion of the condylus occipitalis, parts of the foramina jugulare posterius and anterius, as well as part of the cavum labyrinthicum floor. The contact between the basioccipital and basisphenoid is convex rostrally, as in other species of Bairdemys(Gaffney & Wood, 2002) and Stereogenyina, except in Latentemys plowdeni(Gaffney et al., 2011).

Prootic—The prootic contacts the parietal rostrodorsally, the supraoccipital caudodorsally, the opisthotic caudally, the quadrate laterally, and the pterygoid rostroventrally. It forms the medial edge of the foramen stapedio-temporale, which in B. thalassica and other Stereogenyina (Gaffney & Wood, 2002) opens more rostrally than in Podocnemis spp. At its rostral margin, between the pterygoid and parietal contact, it forms the caudal edge of the foramen nervi trigemini (Fig. 3F).

Opisthotic—The opisthotic contacts the prootic rostrodorsally, the supraoccipital laterodorsally, the quadrate laterorostrally, and the exoccipital caudomedially. At its laterocaudal margin, the processus paraoccipitalis meets the squamosal dorsally and, ventral to this process, the opisthotic contacts the basioccipital. It encloses the hemispherical cavity of the recessus labyrinthicus opisthoticus (Gaffney, 1979) and forms the caudal and dorsal walls of the cavum labyrinthicum and the rostral edge of the foramen jugulare anterius. As in other species of Bairdemys (Gaffney & Wood, 2002; Gaffney et al., 2008), the opisthotic of B. thalassica also forms the roof of the cavum acustico-jugulare, the dorsomedial margin of the fenestra postotica, and the laterodorsal margin of the foramen jugulare posterius. Lateral to the later foramen, a trough extends laterally towards the fenestra postotica. In B. thalassica, as well as in B. venezuelensis and B. sanchezi, this trough is deep, in contrast to the shallower trough of other Stereogenyina (Gaffney et al., 2008).

Basisphenoid—The basisphenoid of B. thalassica is subtriangular and has a rounded rostral edge (Figs. 2B and 3B), as seen in B. venezuelensis, and in contrast to the angular edge of B. hartsteini and B. sanchezi (Gaffney et al., 2008). The lateral margins of the bone are straight, differing from the curved lateral margins as seen in B. sanchezi. It contacts the pterygoid rostrally, the quadrate laterally, and the basioccipital caudally. The basisphenoid forms a small part of the medial surface of the cavum pterygoidei, where the foramen anterius canalis carotici interni opens laterally.

RESULTS

Phylogenetic analysis

A single MPT, 83 steps long, with CI = 0.723 and RI = 0.826, was found (Fig. 5) and it supports the assignment of B. thalassica to the genus Bairdemys(Fig. 5). A Bairdemys clade was recovered (node 30, Fig. 5) exclusive of "Bairdemys" healeyorum (Weems & Knight, 2013). The latter taxon was positioned as sister to the Stereogenyita + Bairdemys clade (node 21, Fig. 5), as supported by a smooth rostral tip of the lower jaw, lacking a rostral hook (car. 44, 0 → 1). In a recent phylogenetic analysis (Menegazzo, Bertini & Manzini, 2015), the only one to include "B." healeyorum, that taxon was placed as sister to B. venezuelensis + B. hartsteini. In order to test this possibility, we enforced the position of "B." healeyorum inside Bairdemys, resulting in 30 MPTs 84 steps long, the strict consensus of which includes a politomy with all the Bairdemys species. The comparison of the two topologies using the Templeton test (Templeton, 1983) yields a significant pvalue of 0.0021 (Table 1), supporting the position of Bairdemys healeyorumoutside the Bairdemys clade. However, we chose not to propose a new genus for this taxon, waiting for further analyses to confirm our results. Because of its very incomplete condition, we did not include Bairdemys miocenica (Collins & Lynn, 1936), the new combination of "Taphrosphys" miocenica proposed by Weems & Knight (2013), in the analysis. Yet, the authors used comparisons to the plastron morphology of Bairdemys venezuelensis and "B." healeyorum to include that taxon in the Bairdemysclade, and our analysis supports a more widespread distribution of those characteristics across the Stereogenyina lineage. Accordingly, we consider "B." miocenica a Stereogenyina incertae sedis.

Bairdemys thalassica is closer to Bairdemys venezuelensis (node 27, Fig. 5) than to the other species of Bairdemys, sharing a moderately developed (caudally) palatine that reduces the contact between the pterygoids (car. 25, state 0 → 1) and a condylus mandibularis caudal or at the same line of the basioccipital-basisphenoid contact (car. 30, state 0 → 1). B. hartsteini is the sister taxon to this clade, followed by B. sanchezi and B. winklerae. This arrangement implies an early Miocene (Fig. 8A), rather than Oligocene (Weems & Knight, 2013) origin of Bairdemys.

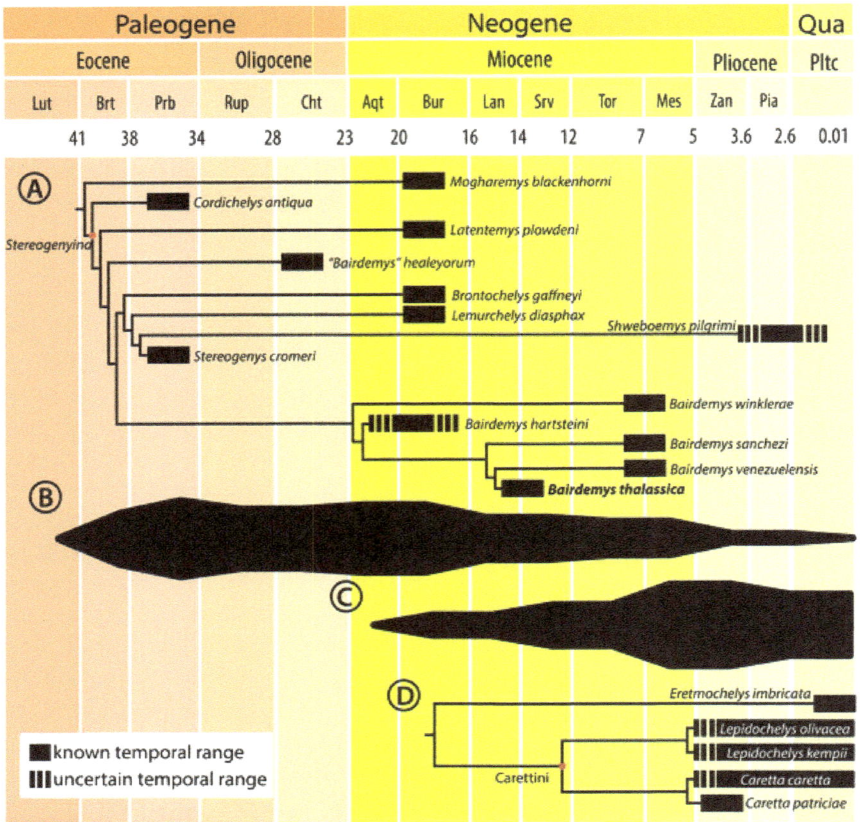

Figure 8: Cladogram calibrated on a time scale and species diversity plot.

Cladogram calibrated on a time scale (A and D) and species diversity plot based on the count of lineages on each age (B and C), from the Eocene to the present, for Stereogenyina (A and B) and Carettini turtles (C and D, cladogram based on Parham & Pyenson, 2010). Layout modified from Romano et al. (2014)).

The Stereogenyita clade (node 20, Fig. 5) was recovered with the same arrangement as in Gaffney et al. (2011), with Brontochelys gaffneyi and Lemurchelys diasphax as successive sister-taxa to the Stereogenys cromeriplus Shweboemys pilgrimi clade. Stereogenyita is supported by a very wide pre-orbital region of the skull (car. 1, state 0 → 1) and a low labial ridge (car. 13, state 0 → 1). In contrast to Gaffney et al. (2011), Cordichelys antiquaand Latentemys plowdeni were not recovered as closer to Bairdemys than to Stereogenyita, but as successive sister taxa to the "B." healeyorum + (Stereogenyita + Bairdemys) clade (node 22, Fig. 5). The latter clade is supported by a condylus mandibularis slightly wider than longer, with a "kidney bean" shape and a convex caudal margin (car. 31, state 0 → 1), and implies that the secondary palate with parallel edged midline cleft, synapomorphic for Stereogenyita clade, is derived from a secondary palate with curved-edged midline cleft, as in Cordichelys antiqua and Latentemys plowdeni.

Morphometric analysis

The estimated shape measurement error was 2.4% for the upper and 5.2% for the lower jaws total variance. The coordinate data from the upper and lower jaws are normally distributed (Shapiro–Wilk test resulted in $w = 0.9778$ and 0.9772, respectively). The shape variation was decomposed in ten and eight Principal Components (PCs) for the upper and lower jaws, respectively. For the upper jaw, the first two PCs accounted for 79.2% of the variation, and for the lower jaw the first two PCs explained 80.7% of the total variation in shape.

The main variation (PC1 = 61.8%) in the shape of the upper jaw in the analyzed species is related to the expansion to the midline or the lateral reduction (Fig. 7A) of the caudomedial edges of the triturating surfaces (landmarks 5, 6, 8, and 9). The variation accounted for in the PC2 (17.4%) is related to the rostrocaudal displacement of the caudomedial edge (landmark 7). The shape of the lower jaws changed mainly (PC1 = 64.5%) in the rostrocaudal displacement of the caudomedial edge (landmark 6) and the caudomedial-rostrolateral displacement of the medial edge of the triturating surface (landmarks 5 and 7). The PC2 (16.2%) accounted for the variation in the caudal edge of the medial portion of the triturating surface (landmarks 4 and 8) in the rostromedial-caudolateral axis (Fig. 7C). Considering the a priori defined groups (non-durophagous, durophagous and unknown), the changes in the upper and lower jaws shape of the durophagous and the unknown (Stereogenyina) groups when compared to the non-durophagous is characterized by a medial and caudal expansion of the triturating surfaces.

DISCUSSION

Feeding habits of Stereogenyina

A greater development of the triturating surface of the jaws is characteristic of the Stereogenyina among podocnemidids. Similar conditions have been investigated for other turtle groups and related to their feeding habits. Analyses of the skull and lower jaw morphology of bothremydids from the Dakhla Formation, Late Cretaceous of Egypt, suggest that the well-developed maxillary region of those turtles were related to a durophagous diet, composed mainly of hard shell/carapace preys, such as ammonites, oysters, and arthropods (Lapparent de Broin & Werner, 1998; Gaffney, Tong & Meylan, 2006; Rabi, Vremir & Tong, 2013). More recently, Claude et al. (2004) analyzed the skull morphology of 85 species of Testudinoidea using geometric morphometrics, and found that a durophagous diet significantly changes the morphology of the rostral part of the skull, independently of their phylogenetic relations. Species with this feeding habit are "characterized by a wide expansion of the triturating surface and secondary palate" (Claude et al., 2004).

Aiming at exploring this correlation for Stereogenyina turtles, we conducted two geometric morphometric analyses, for the upper and lower jaws, including, respectively, 14 and 15 living taxa with known feeding habits (Moll , 1976; Meylan, 1988; Burke, Morreale & Standora, 1994; Teran, Vogt & Gomez, 1995; Pérez-Emán & Paolillo, 1997; Claude et al., 2004; Polovina et al., 2004; Salmon, Jones & Horch, 2004; Garcia & Lourenco, 2007), plus three extinct Stereogenyina. Our results agreed with those found by Claude et al. (2004), suggesting that the triturating surfaces of the upper and lower jaws of durophagous turtles are significantly more expanded caudally and medially than those of non-durophagous species (Fig. 7; Table 2). This difference in morphology implies an increase in the area of both the upper and lower jaws used for processing food. At the same time, the morphology of the three included Stereogenyina is very similar to that of the durophagous turtles (Figs. 7B and 7C) and significantly different from those of the non-durophagous species (Table 2), supporting the hypothesis that stereogenyins were adapted to a durophagous feeding strategy.

Palaeohabitat of Stereogenyina

In the original description of "Podocnemis" venezuelensis, Wood & Díaz de Gamero (1971) raised the possibility that it could represent a marine turtle. Yet, they at the time asserted that: "to be able to do so (i.e., confirm that Podocnemis venezuelensis was a marine turtle) would be particularly interesting because, while all living pelomedusids (Pelomedusoides) are inhabitants of fresh waters, in the past some were marine (...). If P. venezuelensis were, in fact, marine, it would be the last recorded pelomedusid so adapted".

More recently, in the systematics and morphology review of Podocnemididaeturtles by Gaffney et al. (2011), the authors also suggested that "many or all (the taxa) of the subtribe Stereogenyina were marine or near-shore marine". Indeed, the shell morphology of the Urumaco Formation eggs assigned to B. venezuelensis by Winkler & Sánchez-Villagra (2006), which differ from that of other podocnemidids, as well as the depositional environment where they were found, indicate that they were deposited in a beach facing a saline body of water, possibly open-sea (Winkler & Sánchez-Villagra, 2006). This led the authors to suggest that at least some species of Bairdemys were colonial nesters, laying eggs in beaches and having a marine lifestyle.

The limb morphology of stereogenyin turtles also suggests a similar lifestyle. In the description of "B." healeyorum, Weems & Knight (2013) analysed the morphology of the proximal limb bones, specially the humerus. They found that its elongated head, with the long axis almost parallel to the shaft axis, restricted the movements of the fore limb to the vertical plan. Turtles with this kind of humeral head, e.g., extant marine cheloniid sea turtles, usually have strong aquatic adaptations, with limbs functioning as paddles. These are very capable swimmers, living in large water bodies; and this can be also inferred for stereogenyins.

The paleoenvironments inferred for the deposits where stereogenyins are found also suggest marine habitats. With the exception of Brontochelys gaffneyi, the type-locality of which is unknown (Wood, 1970; Gaffney et al., 2011), and Shweboemys pilgrimi from the Irrawaddy Beds, in Burma, which is considered of fluvial origin (Thein, Nu & Pyone, 2012), all other Stereogenyina come from marine coastal or open sea depositional environments. The Qasr el Sagha Formation is interpreted to include "nearshore marine and alluvial deposits" (Holroyd et al., 1996), and the horizons where Cordichelys antiqua and Stereogenys cromeri were found also bear shark and sea cow fossils (Andrews, 1906). The Moghra Formation, where Lemurchelys diasphax and possibly Latentemys plowdeniwere found (Gaffney et al., 2011), is interpreted as an estuarine deposit (Hassan et al., 2012) and the Urumaco Formation, where B. venezuelensis, B. sanchezi, and B. winklerae were collected, is currently considered a succession of marine paleoenvironments, maybe related to a lagoon or bay (Smith et al., 2010). The deposits of the Cibao Formation, where B. hartsteiniwas found, are interpreted as a sequence of different marine environments, with different sea levels, including open sea and coastal paleoenvironments (MacPhee & Wyss, 1990). The Chandler Bridge Formation, from North Carolina, USA, where "B." healeyorum was found, is also interpreted as a succession of marine deposits (Sanders, Weems & Lemon, 1982), with little fluvial input (Weems & Knight, 2013). It is dominated by a marine fauna, with sharks, cetaceans, and cheloniid sea turtles, suggesting an open sea environment (Cicimurri & Knight, 2009). Additionally, the Cerro La Cruz deposits, Castillo Formation, which yielded Bairdemys aff. MBLUZ-P-5045 (Sánchez-Villagra et al., 2004), also correspond to mainly near-shore marine environments (Wheeler, 1960; Rincón et al., 2014). These are mostly coastal marine records, whereas the Capadare Formation, which yielded the new Bairdemys thalassica, was deposited in entirely open sea conditions, with no influence from coastal or continental environments (Lorente, 1978; Díaz de Gamero, 1985). Thus, until now, B. thalassica provides the strongest evidence for the marine habits of Stereogenyina.

Salt glands in pleurodiran turtles?

Turtles that invaded saltwater environments, e.g., cheloniid sea turtles (sensu Joyce, Parham & Gauthier, 2004), possess orbital salt glands, which function as an osmoregulatory organ, helping their kidneys to excrete the excess of salt (Schmidt-Nielsen & Fange, 1958). Similar glands are also found in all other reptile lineages adapted to marine environments, e.g., birds, crocodiles, and snakes, having independent evolutionary origins (Schmidt-Nielsen & Fange, 1958; Babonis & Brischoux, 2012). These multiple origins are supported by the uneven phylogenetic distribution of the character, as well as by its various anatomical positions. In living turtles, however, the salt gland is always placed behind the orbits (Baccari, Matteo & Minucci, 1992; Babonis & Brischoux, 2012).

The presence of salt glands has been inferred for some extinct reptile. Mesosaurs have a unique foramen nariale obturatum, located behind the external nares, that has been interpreted as an osteological correlate of a salt gland duct (Huene, 1941; Piñeiro et al., 2012). In metriorhynchids, a crocodylomorph lineage highly adapted to marine lifestyle, the presence of salt glands was inferred based on lobulated structures rostral to the orbits (Fernández & Gasparini, 2000; Fernández & Gasparini, 2008; Gandola et al., 2006; Fernández & Herrera, 2009). In Santanachelys gaffneyi(Hirayama, 1998), an early cheloniid sea turtle, and the Jurassic plesiochelyids, an extinct lineage of marine turtles, the enlarged foramen interorbitale is used to infer the presence of salt glands (Hirayama, 1998). In all this cases, besides the characteristic morphology of the structures, their anatomical position is also used to suggest the relation to salt glands.

Given the compelling evidence that stereogenyin turtles were adapted to life on sea, they might bear osteological correlates of salt glands. A very good candidate is an excavation, identified by Gaffney, Tong & Meylan (2006)(character 27) as synapomorphic for Bothremydidae, and by Gaffney et al.(2011; character 27) as the "fossa orbitalis posterior pocket in septum orbitotemporale", seen in all Stereogenyina in which that area is exposed (Fig. 9A). In fact, the authors speculate that it could contain "eyeball attachments or orbital glands". Indeed, its location is comparable to that of salt glands in extant turtles (Baccari, Matteo & Minucci, 1992; Babonis & Brischoux, 2012), and the prevalence of such an organ in extant marine reptiles suggests the "fossa orbitalis posterior pocket in septum orbitotemporale" as a possible osteological correlate of salt glands in Stereogenyina turtles.

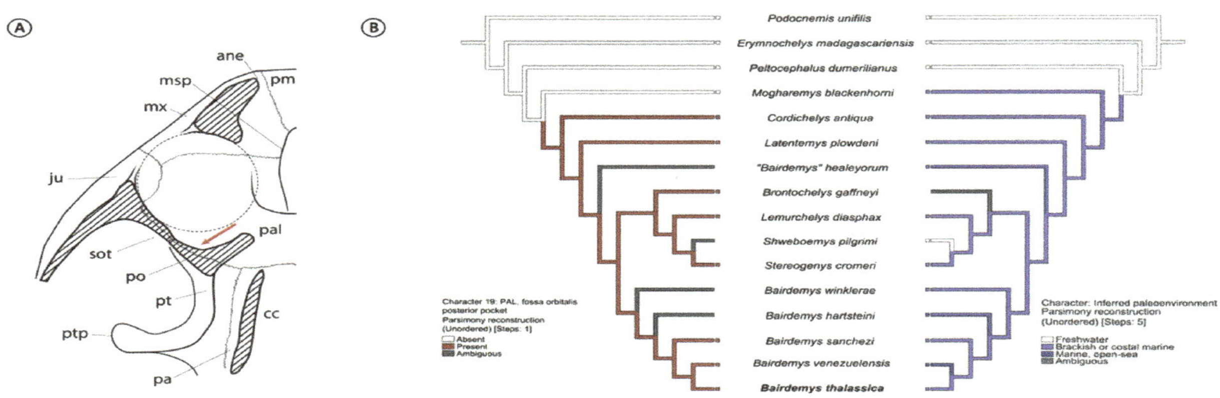

Figure 9: Inference for salt glands in Stereogenyina **turtles.**

(A) cross section of a generalized Stereogenyina skull, indicating (red arrow) the position of the "posterior pocket on septum orbitotemporale"; (B) distribution of character 19 "fossa orbitalis caudal pocket" (left tree) and inferred paleoenvironment (right tree) in the phylogenetic hypothesis presented here. The coding of the character 19 is based on personal observations and on Gaffney et al. (2011) coding. **Abbreviations: ane,** apertura narium externa; **cc,** cavum cranii; **ju,** jugal; **msp,** midline cleft of the secondary palate; **mx,** maxilla; **pa,** parietal; **pal,** palatine; **pm,** premaxilla; **po,** postorbital; **pt,** pterygoid; **ptp,** processus trochlearis pterygoidei; **sot,** septum orbitotemporale.

CONCLUSIONS

A turtle skull from the middle Miocene Capadare Formation, Venezuela, is described as the holotype of the new species, Bairdemys thalassica. Two autapomorphies were recognized for this new taxon: 1—dermal scale iv expanding onto the postorbital, quadratojugal, and parietal; 2—slitlike opening of the canal containing both foramina nervi hypoglossi significantly smaller than the foramen jugulare posterius. Recovered as the sister-taxon to Bairdemys venezuelensis, B. thalassica adds new information on the morphology and phylogeny of Stereogenyina. The addition of the new taxon and of new characters to a phylogenetic data matrix changed the position of some stereogenyins, e.g., Cordichelys antiqua, Latentemys plowdeni, and "Bairdemys" healeyorum, the latter being recovered outside Bairdemys. The new set of relations suggest that some traits were more widely distributed among Stereogenyina, including a secondary palate with curved medial edges.

Morphometric analyses revealed that Stereogenyina turtles probably had a durophagous diet, and a review of the morphology and geological provenance of Stereogenyina indicates that those pelomedusoids had a marine lifestyle. These two features are also seen in the extant Carettini sea turtles, which began their diversification around the middle Miocene (Dodd & Morgan, 1992; Parham & Fastovsky, 1997; Parham & Pyenson, 2010) and may have since competed with the stereogenyins. This is consistent with the decline of the latter group from the late Miocene onwards (Figs. 8B and 8C), which may have been partially driven by the rise of Carettini, although a much better fossil record of both groups is necessary to test that hypothesis.

COMPETING INTERESTS

The authors declare there are no competing interests.

AUTHOR CONTRIBUTIONS

Gabriel S. Ferreira conceived and designed the experiments, performed the experiments, analyzed the data, contributed reagents/materials/analysis tools, wrote the paper, prepared figures and/or tables, reviewed drafts of the paper.

Ascanio D. Rincón and Max C. Langer conceived and designed the experiments, analyzed the data, contributed reagents/materials/analysis tools, reviewed drafts of the paper.

Andrés Solórzano conceived and designed the experiments, analyzed the data, contributed reagents/materials/analysis tools, prepared figures and/or tables, reviewed drafts of the paper.

Data Deposition

The following information was supplied regarding the deposition of related data:

MorphoBank: http://www.morphobank.org/index.php/Projects/ProjectOverview/project_id/2194.

New Species Registration

The following information was supplied regarding the registration of a newly described species:

New taxon LSID: urn:lsid:zoobank.org:act:D2192A69-5019-410A-A875-48097A50F8CD.

Funding

This research was supported by Fundação de Amparo à Pesquisa do Estado de São Paulo (FAPESP) scholarships to GSF (2012/11604-1) and research funding to MCL (2014/03825-3), and by Venezuelan education university, science, and technology Ministry (MEUCT), research fundings IVIC-1096 and PEII2012-456 to ADR and AS. The funders had no role in study design, data collection and analysis, decision to publish, or preparation of the manuscript.

ACKNOWLEDGEMENTS

The authors gratefully acknowledge Juliana Sterli, Marta Fernández and Gustavo R. Oliveira for the discussions of some aspects of this work, Fábio C. de Barros for the help with the morphometric analyses, and the editor Mark Young and the two reviewers, Gustavo R. Oliveira and Márton Rabi, for their useful comments on the manuscript.

REFERENCES

1. Adams DC, Collyer ML, Sherratt E. 2015. R package. (version 2.1.4). Available at http://cran.r-project.org/web/packages/geomorph/index.html software
2. Adams DC, Otarola-Castillo E. 2013. geomorph: an R package for the collection and analysis of geometric morphometric shape data. Methods in Ecology and Evolution 4:393-399
3. Andrews CW. 1901. Preliminary note on some recently discovered extinct vertebrates from Egypt (Part II) Geological Magazine IV(8):436-444
4. Andrews CW. 1903. On some pleurodiran chelonians from the Eocene of the Fayum, Egypt. Annals and Magazine of Natural History Series 7(11):115-122
5. Andrews CW. 1906. A descriptive catalog of the Tertiary Vertebrata of the Fayum, Egypt. London: Trustees of the British Museum (Natural History). 324
6. Babonis LS, Brischoux F. 2012. Perspectives on the convergent evolution of tetrapod salt glands. Integrative and Comparative Biology 52:1-12
7. Baccari GC, Matteo LD, Minucci S. 1992. The orbital glands of the chelonians Pseudemys scripta and Testudo graeca: comparative histological, histochemical and ultrastructural investigations. Journal of Anatomy. 180:1-13
8. Baur G. 1893. Notes on the classification of Cryptodira. American Naturalist 27:672-675
9. Bona P, de la Fuente MS. 2005. Phylogenetic and paleobiogeographic implications of Yaminuechelys maior (Staesche, 1929) new comb., a large long-necked chelid turtle from the early Paleocene of Patagonia, Argentina. Journal of Vertebrate Paleontology 25(3):569-582
10. Bonnaterre PJ. 1789. Tableau encyclopédique et méthodique des trois règnes de la nature: dédié et présenté a M. Necker, ministre d'État, & directeur général des Finances . In: Erpétologie. Paris: Panckoucke. 71
11. Burke VJ, Morreale SJ, Standora EA. 1994. Diet of the Kemp's ridley sea turtle, Lepidochelys kempii, in New York waters. Fishery Bulletin 96:26-32
12. Cadena EA, Bloch JI, Jaramillo CA. 2010. New Podocnemidid turtle (Testudines: Pleurodira) from the middle-upper Paleocene of South America. Journal of Vertebrate Paleontology 30(2):367-382
13. Cadena EA, Ksepka DT, Jaramillo CA, Bloch JI. 2012. New pelomedusoid turtles from the late Palaeocene Cerrejón Formation of Colombia and their implications for phylogeny and body size evolution. Journal of Systematic Palaeontology 10(2):313-331
14. Cantino PD, de Queiroz K. 2010. International code of phylogenetic nomenclature (Version 4c).
15. Cicimurri DJ, Knight JL. 2009. Late Oligocene sharks and rays from the Chandler Bridge Formation, Dorchester County, South Carolina, USA. Acta Palaeontologica Polonica 54(4):627-647

16. Claude J, Pritchard P, Tong H, Paradis E, Auffray J-C. 2004. Ecological correlates and evolutionary divergence in the skull of turtles: a geometric morphometric assessment. Systematic Biology 53(6):933-948

17. Collins RL, Lynn WG. 1936. Fossil turtles from Maryland. Proceedings of the American Philosophical Society 76:151-174

18. Cope ED. 1864. On the limits and relations of the Raniformes. Proceedings of the Academy of Natural Sciences of Philadelphia 1864:181-183

19. Cope ED. 1868. On the origin of genera. Proceedings of the Academy of Natural Sciences of Philadelphia 1868:242-300

20. Cornalia E. 1849. Vertebratorum Synopsis in Musaeo Mediolanense extantium quae per novum Orbem Cajetanas Osclati collegit Annis 1846-47-48. Speciebus novis vel minus cognitus adjectis, nec non Descriptionibus atque Iconibus illustratis, curante Aemilio Cornalia . In: Mediolani. 16

21. Dacqué E. 1912. Die fossilen Schildkröten Aegyptens. Geologische und Palaeontologische Abhandlungen 14(10):275-337

22. de la Fuente MS. 2003. Two new pleurodiran turtles from the Portezuelo Formation (Upper Cretaceous) of Northern Patagonia, Argentina. Journal of Paleontology 77(3):559-575

23. Díaz de Gamero ML. 1985. Estratigrafia de Falcón Nororiental. VI Congreso Geológico de Venezuela. Memoria I:454-502

24. Dodd CK, Morgan GS. 1992. Fossil sea turtles from the early Pliocene Bone Valley Formation, Central Florida. Journal of Herpetology 26:1-8

25. Dumont Júnior MV. 2013. Um novo podocnemídeo fóssil de grande porte da Formação Solimões (Mioceno-Plioceno), Acre, Brasil e as relações filogenéticas entre os Podocnemidae . Unpublished M.Sc. dissertation, Universidade de Brasília, Instituto de Geociências, Brasília, DF, Brazil thesis

26. Fernández MS, Gasparini Z. 2000. Salt glands in a Tithonian metriorhynchid crocodyliform and their physiological significance. Lethaia 33:269-276

27. Fernández MS, Gasparini Z. 2008. Salt glands in the Jurassic metriorhynchid Geosaurus: implications for the evolution of osmoregulation in Mesozoic marine crocodyliforms. Naturwissenschaften 95:79-84

28. Fernández MS, Herrera Y. 2009. Paranasal sinus system of Geosaurus araucanensis and the homology of the antorbital fenestra of metriorhynchids (Thalattosuchia: Crocodylomorpha) Journal of Vertebrate Paleontology 29(3):702-714

29. Gaffney ES. 1977. The side-necked turtle family Chelidae: a theory of relationships using shared derived characters. American Museum Novitates 2620:1-28

30. Gaffney ES. 1979. Comparative cranial morphology of recent and fossil turtles. Bulletin of the American Museum of Natural History 164(2):65-376

31. Gaffney ES. 1996. The postcranial morphology of Meiolania platyceps and a review of the Meiolaniidae. Bulletin of the American Museum of Natural History 229:1-165

32. Gaffney ES, Meylan PA, Wood RC, Simons E, Campos DA. 2011. Evolution of the side-necked turtles: the Family Podocnemididae. Bulletin of the American Museum of Natural History 350:1-237

33. Gaffney ES, Scheyer TM, Johnson KG, Bocquentin J, Aguilera OA. 2008. Two new species of the side necked turtle genus, Bairdemys (Pleurodira, Podocnemididae), from the Miocene of Venezuela. Paläontologische Zeitschrift 82(2):209-229

34. Gaffney ES, Tong H, Meylan PA. 2006. Evolution of the side-necked turtles: the families Bothremydidae, Euraxemydidae, and Araripemydidae. Bulletin of the American Museum of Natural History 300:1-700

35. Gaffney ES, Wood RC. 2002. Bairdemys, a new side-necked turtle (Pelomedusoides: Podocnemididae) from the Miocene of Caribbean. American Museum Novitates 3359:1-28

36. Gandola R, Buffetaut E, Monaghan N, Dyke G. 2006. Salt glands in the fossil crocodile Metriorhynchus. Journal of Vertebrate Paleontology 26(4):1009-1010

37. Garcia G, Lourenco W. 2007. Erymnochelys madagascariensis—diet. Herpetological Review 38(3):330-331

38. Goloboff PA, Farris JS, Källerjö M, Oxelman B, Ramírez M, Szumik C. 2003. Improvements to resampling measures of group support. Cladistics 19:324-332

39. Goloboff PA, Farris JS, Nixon KC. 2008. TNT, a free-program for phylogenetic analysis. Cladistics 24:774-786

40. Grandidier A. 1867. Liste des reptiles nouveaux découverts, en 1866, sur la côte sud-ouest de Madagascar. Revue et Magasin de Zoologie Pure et Appliquée 19(2):232-234

41. Hassan SM, Steel RJ, El Barbooky A, Hamdan M, Olariu C, Helper MA. 2012. Stacked, Lower Miocene tide-dominated estuary deposits in a trangressive succession, Western Desert, Egypt. Sedimentary Geology 282:241-255

42. Hirayama R. 1998. Oldest known sea turtle. Nature 392:705-708

43. Holroyd PA, Simons EL, Bown TM, Polly PD, Kraus MJ. 1996. New records of terrestrial mammals from the Upper Eocene Qasr el Sagha Formation, Fayum Depression, Egypt. Palaeovertebrata 25(2–4):175-192

44. Huene F. 1941. Osteologie und systematische Stellung von Mesosaurus. Palaeontographica, Abteilung A 92:45-58

45. ICZN. 1999. International code of zoological nomenclature (4th edition). London: The International Trust for Zoological Nomenclature. 306

46. Joyce WG, Parham JF, Gauthier JA. 2004. Developing a protocol for the conversion of rank-based taxon names to phylogenetically defined clade names, as exemplified by turtles. Journal of Paleontology 78(5):989-1013

47. Khosatzky LI. 1997. Big turtle of the Late Cretaceous of Mongolia. Russian Journal of Herpetology 4(2):148-154

48. Klingenberg CP. 2011. MorphoJ: an integrated software package for geometric morphometrics. Molecular Ecology Resources 11(2):353-357

49. Klingenberg CP, Barluenga M, Meyer A. 2002. Shape analysis of symmetric structures: quantifying variation among individuals and asymmetry. Evolution 56:1909-1920

50. Klingenberg CP, McIntyre GS. 1998. Geometric morphometrics of developmental instability: analyzing patterns of fluctuanting asymmetry with procrustes methods. Evolution 52:1363-1375

51. Lapparent de Broin F, Werner C. 1998. New Late Cretaceous turtles from the Western Desert, Egypt. Annales de Paléontologie 84(2):131-214

52. Lorente MA. 1978. La Caliza de Capadare y sus relaciones con la cuenca de Agua Salada, estado Falcón. Caracas: Universidad Central de Caracas, Tese de Engenharia. 161

53. MacPhee RDE, Wyss AR. 1990. Oligo-Miocene vertebrates from Puerto Rico, with a catalog of localities. American Museum Novitates 2965:1-45

54. Maddison WP, Maddison DR. 2014. Mesquite: a modular system for evolutionary analysis. (Version 3.01). Available at http://mesquiteproject.org software

55. Menegazzo MC, Bertini RJ, Manzini FF. 2015. A new turtle from the Upper Cretaceous Bauru Group of Brazil, updated phylogeny and implications for age of the Santo Anastácio Formation. Journal of South American Earth Sciences 58:18-32

56. Meylan A. 1988. Spongivory in Hawksbill turtles: a diet of glass. Science 239(4838):393-395

57. Meylan PA, Gaffney ES, Campos DA. 2009. Caninemys, a new side-necked turtle (Pelomedusoides: Podocnemididae) from the Miocene of Brazil. American Museum Novitates 3639:1-26

58. Moll D. 1976. Food and feeding strategies of the Ouachita Map Turtle (Graptemys pseudogeographica ouachitensis) American Midland Naturalist 96(2):478-482

59. Montellano-Ballesteros M, Rincón AD, Solórzano A. 2014. Record of tayassuids in ?late Pliocene to Quaternary deposits in Venezuela. Revista Brasileira de Paleontología 17(2):169-182

60. Müller L. 1935. Über eine neue Podocnemis-Art (Podocnemis vogli) aus Venezuela nebst ergänzenden Bemerkungen über die systematischen Merkmale der ihr nächstverwandten Arten. Zoologischer Anzeiger 110(5–6):97-109

61. Parham JF, Fastovsky DE. 1997. The phylogeny of cheloniid sea turtles revisited. Chelonian Conservation and Biology 2:548-554

62. Parham JF, Pyenson ND. 2010. New sea turtle from the Miocene of Peru and the iterative evolution of feeding ecomorphologies since the Cretaceous. Journal of Paleontology 84(2):231-247

63. Pérez-Emán JL, Paolillo OA. 1997. Diet of the pelomedusid turtle Peltocephalus dumerilianus in the Venezuelan Amazon. Journal of Herpetology 31(2):173-179

64. Piñeiro G, Ramos A, Goso C, Scarabino F, Laurin M. 2012. Unusual environmental conditions preserve a Permian mesosaur-bearing Konservat-Lagerstätte from Uruguay. Acta Palaeontologica Polonica 57(2):299-318

65. Polovina JJ, Balazs GH, Howell EA, Parker DM, Seki MP, Dutton PH. 2004. Forage and migration habitat of loggerhead (Caretta caretta) and olive ridley (Lepidochelys olivacea) sea turtles in the central North Pacific Ocean. Fisheries Oceanography 13(1):36-51

66. R Development Core Team. 2013. R: a language and environment for statistical computing. Viena: R Foundation for Statistical Computing.

67. Rabi M, Sukhanov VB, Egorova VN, Danilov I, Joyce WG. 2014. Osteology, relationships, and ecology of Annemys (Testudines, Eucryptodira) from the Late Jurassic of Shar Teg, Mongolia, and phylogenetic definitions of Xinjiangchelyidae, Sinemydidae, and Macrobaenidae. Journal of Vertebrate Paleontology 34(2):324-352

68. Rabi M, Vremir M, Tong H. 2013. Preliminary overview of Late Cretaceous turtle diversity in Eastern Central Europe (Austria, Hungary, and Romania) In: Brinkman D, Holroyd P, Gardner J, eds. Morphology and evolution of turtles, Vertebrate Paleobiology and Paleoanthropology Series. Netherlands: Dordrecht. 307-336

69. Renz HH. 1948. Stratigraphy and fauna of the Agua Salada Group, State of Falcón, Venezuela. Geological Society of America Memoir 32:1-219

70. Rincón AD. 2003. Los mamíferos fósiles del Pleistoceno de la Cueva del Zumbador (Fa. 116), Estado Falcón, Venezuela. Boletin de la Sociedad Venezolana de Espeleologia 37:18-26

71. Rincón AD, Solórzano A, Benammi M, Vignaud P, McDonald HG. 2014. Chronology and geology of an Early Miocene mammalian assemblage in North of South America, from Cerro La Cruz (Castillo Formation), Lara State, Venezuela: implications in the 'changing course of Orinoco River' hypothesis. Andean Geology 41(3):507-528

72. Rincón AD, Stucchi M. 2003. Primer registro de la familia Pelagornithidae (Aves: Pelecaniformes) para Venezuela. Boletin de la Sociedad Venezolana de Espeleologia 37:27-30

73. Rincón AD, White RS. 2007. Los Xenarthra Cingulata del Pleistoceno Tardio (Lujanense) de Cerro Misión, Estado Falcón, Venezuela. Boletin de la Sociedad Venezolana de Espeleologia 41:2-12

74. Rohlf FJ. 1999. Shape statistics: Procrustes superimpositions and tangent spaces. Journal of Classification 16:197-223

75. Rohlf FJ. 2005. tpsDig. (versão 2.04). Stony Brook: Department of Ecology and Evolution, State University of New York. software

76. Romano PSR, Gallo V, Ramos RRC, Antonioli L. 2014. Atolchelys lepida, a new side-necked turtle from the Early Cretaceous of Brazil and the age of crown Pleurodira. Biology Letters 10(7) 20140290

77. Salmon M, Jones TT, Horch KW. 2004. Ontogeny of diving and feeding behavior in juvenile seaturtles: Leatherback seaturtles (Dermochelys coriacea L) and Green seaturtles (Chelonia mydas L) in the Florida Current. Journal of Herpetology 38(1):36-43

78. Sánchez-Villagra MR, Asher RJ, Rincón AD, Carlini AA, Meylan P, Purdy RW. 2004. New faunal reports for the Cerro La Cruz locality (lower Miocene), north-western Venezuela. Special Papers in Palaeontology 71:105-112

79. Sanders AE, Weems RE, Lemon EM. 1982. The Chandler Bridge Formation; a new Oligocene stratigraphic unit in the lower Coastal Plain of South Carolina, Contributions to stratigraphy. US Geological Survey Bulletin. 1529-H:105-124

80. Schmidt-Nielsen K, Fange R. 1958. Salt glands in marine reptiles. Nature 182(4638):783-785

81. Schweigger AF. 1812. Prodromus Monographiae Cheloniorum. Königsberger Archiv für Naturwissenschaft und Mathematik 1(271–368):406-458

82. Simpson GG. 1938. Crossochelys, Eocene horned turtle from Patagonia. Bulletin of the American Museum of Natural History 74:221-254

83. Smith CJ, Collins LS, Jaramillo C, Quiroz LI. 2010. Marine paleoenvironments of Miocene-Pliocene formations of north-central Falcón State, Venezuela. Journal of Foraminiferal Research 40(3):266-282

84. Solórzano A, Rincón AD. The earliest record of a bony-toothed bird of South America from the Early Miocene of Venezuela, with a reexamination of Venezuelan pelagornithids. Journal of Vertebrate Paleontology In Press

85. Sterli J, de la Fuente MS. 2011. Re-description and evolutionary remarks on the Patagonian horned turtle Niolamia argentina Ameghino, 1899 (Testudinata; Meiolaniidae) Journal of Vertebrate Paleontology 31(6):1210-1229

86. Sterli J, de la Fuente MS. 2013. New evidence from the Palaeocene of Patagonia (Argentina) on the evolution and palaeo-biogeography of Meiolaniformes (Testudinata, new taxon name) Journal of Systematic Palaeontology 11(7):835-852

87. Swinton WE. 1939. A new fresh-water tortoise from Burma. Records of the Geological Survey of India 74(4):548-551

88. Swofford DL. 2002. PAUP*. Phylogenetic analysis using parsimony (*and other methods). (Version 4.0b). Suderland: Sinauer Associates. software

89. Templeton AR. 1983. Phylogenetic inference from restriction endonuclease cleavage site maps with particular reference to the evolution of humans and the apes. Evolution 37(2):221-244

90. Teran AF, Vogt RC, Gomez MFS. 1995. Food habits of an assemblage of five species of turtles in the Rio Guapore, Rondonia, Brazil. Journal of Herpetology 29(4):536-547

91. Thein ZMM, Nu TT, Pyone WW. 2012. Lithofacies and depositional environment of the Irrawaddy Formation in Myogintha Area, Pauk Township, Magway region. Universities Research Journal 5(6):69-82

92. Thomson S, Georges A. 2009. Myuchelys gen. nov. - a new genus for Elseya latisternum and related forms of Australian freshwater turtle (Testudines: Pleurodira: Chelidae) Zootaxa 2053:32-42

93. Troschel FH. 1848. Amphibien. In: Schomburgk, ed. Reisen in Britisch-Guiana in den Jahren 1840-44. Im Auftrage Majestät des Königs von Preussen ausgeführt. Versuch einer Zusammenstellung der Fauna und Flora von Britisch-Guiana. Leipzig: Weber. 645-661

94. Wagler J. 1830. Natürliches System der Amphibien mit vorangehender Classification der Säugethiere und Vögel. München: J. G. Cotta'sche Buchhandlung. 354

95. Weems RE, Knight JL. 2013. A new species of Bairdemys (Pelomedusoides: Podocnemididae) from the Oligocene (Early Chattian) Chandler Bridge Formation of South Carolina, USA, and its paleobiogeographic implications for the genus. In: Brinkman D, Holroyd P, Gardner J, eds. Morphology and evolution of turtles, Vertebrate paleobiology and paleoanthropology series. Netherlands: Dordrecht. 289-303

96. Wheeler CB. 1960. Estratigrafía del Oligoceno y Mioceno inferior de Falcón occidental y nororiental. Memorias, III Congreso Geológico Venezolano 1:407-465

97. Winkler JD, Sánchez-Villagra MR. 2006. A nesting site and egg morphology of a Miocene turtle from Urumaco, Venezuela: evidence of marine adaptations in Pelomedusoides. Palaeontology 49(3):641-646

98. Wood RC, Díaz de Gamero ML. 1971. Podocnemis venezuelensis, a new fossil pelomedusid (Testudines, Pleurodira) from the Pliocene of Venezuela and a review of the history of Podocnemis in South America. Breviora 376:1-23

99. Wood RC. 1970. A review of the fossil Pelomedusidae (Testudines, Pleurodira) of Asia. Breviora 357:1-24

CHAPTER 7

PALEOECOLOGY OF THE LATE CRETACEOUS MULTI-TAXA VERTEBRATE ASSEMBLAGE

Laura Domingo, Fernando Barroso-Barcenilla, Oscar Cambra-Moo

INTRODUCTION

The Mid-Cretaceous thermal maximum, which peaked in Turonian times and constituted the warmest climate warming of the last 144 Ma [1–2] was followed by a long-term cooling trend beginning at the early Campanian and detected in both the terrestrial and marine realms [3–8]. This trend was characterized by alternating cooling and warming episodes across the Campanian and Maastrichtian [9–11]. In spite of this cooling pattern, overall warmer conditions than today persisted until the end of the Cretaceous, as suggested by high atmospheric CO_2concentrations and a lack of permanent ice at the poles [11–14]. During this time, enhanced ocean heat transport along with the maintainance of a low albedo at high latitudes, due to the presence of forests, contributed to a reduced latitudinal thermal gradient [15].

Although stable isotope studies on Cretaceous terrestrial vertebrate taxa can help to better understand continental climatic and environmental conditions, they are not as common as those performed on Cenozoic vertebrate taxa due to the scarcity of adequate sample sizes of fossil material, the destructive nature of the stable isotope technique, as well as pervasive diagenetic effects observed in fossils of this age. Some stable isotope studies on Cretaceous vertebrate assemblages have focused on the analysis of $\delta^{18}O$ on the bioapatite with the aim of characterizing paleoclimate [16–18], faunal behavior and/or paleohydrology [18–20], and metabolism with special interest on dinosaurian body temperature [21–23]. $\delta^{13}C$ values have been used to make inferences on paleoecology and to study the fractionation between diet (vegetation) and dinosaur enamel [24–26]. Domingo *et al.* [27] carried out a preliminary stable isotope study on the "Lo Hueco" vertebrate paleofauna (central eastern Spain) as a first approach to evaluate the potential, validity and degree of preservation of the bioapatite from this locality and as a first approach to characterize paleoclimatic ($\delta^{18}O_{H2O}$, temperature) and paleoecological/paleoenvironmental (diet, type of vegetation) variables existing during the late Campanian-early Maastrichtian in the southeastern area of the Iberian region. Albeit the Late Cretaceous Iberian geological record shows a good representation of vertebrate localities and outcrops, there are very scarce stable isotope studies on this area and time period [27–28]. Therefore, the isotopic investigation of the "Lo Hueco" vertebrate assemblage provide valuable data, which may shed light on climatic and environmental conditions that occurred in this subtropical setting located in the Tethys realm.

The main objectives of our study are three-fold: 1) to determine seasonal thermal variability in the Iberian Late Cretaceous, 2) to characterize the isotopic offset between $\delta^{13}C$ values of dinosaur tooth enamel and their diet ($\Delta^{13}C_{enamel-diet}$), and 3) to unveil dietary and habitat affinities for "Lo Hueco" crocodyliforms and turtles. The richness and diversity of the Late Cretaceous "Lo Hueco" Fossil-Lagerstätte [29–30] have allowed us to approach the aforementioned questions by performing multi-taxa stable isotope analyses ($\delta^{18}O_{PO4}$, $\delta^{18}O_{CO3}$ and $\delta^{13}C$) on dinosaurs, crocodyliforms, turtles, gars and bulk organic matter. Most isotopic studies dealing with Mesozoic vertebrates focused on one or two groups [16–17, 19, 25–26, 31–32], so the advantage of considering co-existing taxa relies on the fact that they can shed light on differences in habitat and resource utilization, informing about climatic, environmental and ecological variables such as hydrology of the area, seasonality, type of vegetation, resource/habitat overlap and/or partitioning.

While different paleontological, geochemical and modelling studies offered a good characterization of long-term climatic variability for the Middle and Late Cretaceous (e.g., [1–3, 6, 16, 33], among others), the evaluation of seasonal thermal variability has proven challenging on account of the lack of proxies capable of recording intra-annual information. Controversy arises when discerning whether Middle and

Citation: Domingo L, Barroso-Barcenilla F, Cambra-Moo O (2015) Paleoecology of the Late Cretaceous Multi-Taxa Vertebrate Assemblage. PLoS ONE 10(3): e0119968. doi:10.1371/journal.pone.0119968.

Late Cretaceous climate was more seasonally thermally equable (low mean annual range of temperatures) [34–39], although, there seems to be an agreement indicating a higher seasonal thermal variability (i.e., lower equability) during the colder Late Cretaceous in comparison to the warmer Middle Cretaceous [38].

$\delta^{18}O$ values in the phosphate and carbonate fractions of vertebrate bioapatite record the $\delta^{18}O$ value of their body water ($\delta^{18}O_{bw}$). In the case of terrestrial endothermic species, $\delta^{18}O_{bw}$ is a reflection of oxygen uptake (inspired O_2 and water vapor, drinking water, dietary water, oxygen in food dry matter) and loss (excreted water and solids, expired CO_2, and water vapor). Dinosaur metabolism is a controversial issue that remains under debate [21, 40–45]. Fricke and Rogers [22] performed $\delta^{18}O_{PO4}$ analyses on Late Cretaceous theropods and crocodilians from fossil sites located at different latitudes and found that the reconstructed theropod slope of the $\delta^{18}O_{PO4}$-latitude relationship was steeper than the one observed for crocodilians (ectotherms) and similar to that estimated for present-day endotherms such as mammals and birds. Amiot *et al.* [23] used a similar approach to that of Fricke and Rogers [22] on a wider suite of Cretaceous dinosaurs and obtained body temperatures of 36–38°C for dinosaurs inhabiting high and low latitudes. Other authors such as O'Connor and Dodson [46] and Gillooly *et al.* [47] argued that dinosaur body temperature depended on body mass (inertial homeothermy), ranging from 25°C at 12 kg to 41°C at 13,000 kg [47]. However, Fricke and Rogers [22] observed similar metabolisms for Late Cretaceous theropods showing different body mass (e.g., *Albertosaurus*: 1.3–1.7 tons; *Majungasaurus*: 1.5 tons; *Saurornitholestes*: 10 kg). Amiot *et al.* [23] observed a widespread endothermy in different groups of Late Cretaceous dinosaurs (theropods, sauropods, ornithopods, and ceratopsians). Eagle *et al.* [48] performed clumped isotope analyses on sauropods and estimated body temperatures between 4 and 7°C lower (36 to 38°C) than those proposed by Gillooly *et al.* [47], arguing that sauropods were able to regulate their body temperatures, preventing overheating. Ectothermic semi-aquatic taxa, such as crocodiles and turtles, form their bioapatite within a narrow thermal window, and their bioapatite $\delta^{18}O$ is a reflection of the $\delta^{18}O_{H2O}$ value of warm months when conditions are favorable for apatite synthesis [20, 49–50]. In the case of ectothermic aquatic taxa (gars), bioapatite forms in isotopic equilibrium with ambient water and since they do not thermoregulate, their bioapatite $\delta^{18}O$ values record both $\delta^{18}O_{H2O}$ and ambient temperature independently of body temperature [51]. Therefore, it is possible to estimate temperature values from gar $\delta^{18}O$ values if the $\delta^{18}O_{H2O}$ value is independently estimated [20, 51–52].

Studies dealing with dinosaur $\delta^{13}C$ values have mainly focused on herbivore taxa (e.g., [24–26, 53]). These authors obtained relatively high $\delta^{13}C$ values for hadrosaurian and ceratopsian dinosaurs for a typical C_3 environment (C_4 plant expansion took place in the late Miocene–Pliocene). Plants following the C_3 or Calvin-Benson photosynthetic pathway (trees, shrubs, forbs, and cool-season grasses) strongly discriminate against ^{13}C during fixation of CO_2, yielding tissues with $\delta^{13}C$ values averaging ~ -27.5‰ (VPDB) (ranging from -36‰ to -22‰). The most negative $\delta^{13}C$ values of this range (-36‰ to -30‰) reflect closed canopy conditions due to recycling of ^{13}C-depleted CO_2 and low irradiance. The highest values (-25‰ to -22‰) correspond to C_3 plants from high insolated, arid, or water stressed environments [54–56]. When considering fossil taxa, it is necessary to account for shifts in the $\delta^{13}C$ value of atmospheric CO_2 ($\delta^{13}C_{atmCO2}$), the source of plant carbon, including anthropogenic modification due to fossil fuel burning, which has decreased the $\delta^{13}C$ value of atmospheric $CO_2$from a value of −6.0‰ in the Late Cretaceous [57] to a modern value of −8.0‰ [58–59]. Accounting for these shifts in baseline and assuming the modern fractionation of ~ -19.5‰ between $\delta^{13}C_{atmCO2}$ and C_3 vegetation $\delta^{13}C$ values [60], Late Cretaceous mean $\delta^{13}C$ value for C_3 plants would be ~ -25.5‰ (ranging from -34‰ to -20‰), with the most negative $\delta^{13}C$ values (-34‰ to -28‰) reflecting closed canopy conditions and the highest values (-23‰ to -20‰) reflecting high insolated, arid, or water stressed environments. Studies of $\delta^{13}C$ offsets between terrestrial vertebrate enamel and diet ($\Delta^{13}C_{enamel\text{-}diet}$) have been mainly carried out on modern mammalian taxa, which allows for extrapolation on paleontological studies dealing with mammals, so that a consistent $\Delta^{13}C_{enamel\text{-}diet}$ of +12 to +14‰ has been observed between herbivore mammalian tooth enamel and vegetation $\delta^{13}C$ values [61–63], while the difference between mammalian carnivore and herbivore tooth enamel $\delta^{13}C$ values ($\Delta^{13}C_{carnivore\text{-}herbivore}$) has been proposed to be ~ -1.3‰ [64]. Johnson *et al.* [65] and Angst *et al.* [66] also investigated this tissue-diet fractionation on modern birds by analyzing ostrich eggshells and diet, obtaining a $\Delta^{13}C_{eggshell\text{-}diet}$ value of +16‰ and +13.4‰, respectively. The lack of modern counterparts in the case of dinosaurs makes it difficult to assess this offset and we can just rely on the fossil record geochemical imprint to determine tissue-diet fractionation values. In this vein, analyzing herbivore dinosaur bioapatite and bulk organic matter $\delta^{13}C$ values, Fricke and Pearson [25] and Fricke *et al.* [26] argued that $\Delta\delta^{13}C_{enamel\text{-}diet}$ for ornithischian dinosaurs was ~ +18‰, whereas Tütken [32] estimated a $\Delta\delta^{13}C_{enamel\text{-}diet}$ value of ~ +16‰ for sauropods.

GEOGRAPHIC AND GEOLOGICAL SETTING

The Late Cretaceous "Lo Hueco" Fossil-Lagerstätte is located in the province of Cuenca (central eastern Spain: 2° 02'50"W, 39° 57'15"N) (Fig. 1). It was fortuitously discovered in 2007, while constructing the Madrid-Levante high-speed railway. More than 10,000 macrofossils of different taxonomic groups of flora and fauna were unearthed. The stratigraphic position and associated fauna support a late Campanian–early Maastrichtian age [29, 67–68]. The "Lo Hueco" site outcrops in Garumn facies, which is the informal term that designates marl, clay, and gypsum, mainly of red color, deposited in shallow marine, coastal, and/or continental environments of southwestern Europe during the latest Cretaceous and the earliest Paleogene [69]. It corresponds to the upper part of the Villalba de la Sierra Formation (Fig. 2A), a lithological unit interpreted as a coastal marsh with distributary channels and sporadic establishment of sabkhas [29]. Six stratigraphic levels were defined at the "Lo Hueco" outcrop named V, G1, R1, G2, R2, and M from bottom to top [29] (Fig. 2B). This succession appears slightly modified laterally by a lower sulphate interval (S1) that cuts the V level in the eastern area of the outcrop, by a sandy channel structure (C) that interrupts V, G1, and R1 levels in the southern area, and by an upper sulphate interval (S2) that cuts part of G2 in the northeastern area (Fig. 2B). Four bonebeds have yielded the majority of fossils: the C structure, the G1 and G2 levels and the lower part of the R2 level (Fig. 2B) [29, 68]. The G1 level is interpreted as a proximal muddy floodplain (close to the distributary channels) and vertebrate fossils usually appear complete and associated. More than 14 partially articulated sauropod skeletons were recovered from this level [28, 68]. Exceptionally well-preserved plant remains have been also described from this lithosome [30]. The G2 level corresponds to the distal part of a poorly drained muddy floodplain (distal from the distributary channels) and points to a relatively calm aquatic environment (i.e., marsh/swamp) exposed to partial or total desiccation, although there is a low degree of articulation of preserved skeletons [29, 68]. Recent fluid inclusion and geochemical analyses on the sulphates from S1 and S2 have pointed to a near coast playa-lake environment with mainly brackish and freshwater influence in the area, while not totally ruling out some extent of marine influence [70]. During the late Campanian-early Maastrichtian, the "Lo Hueco" locality was placed at a subtropical paleolatitude of ~ 31°N.

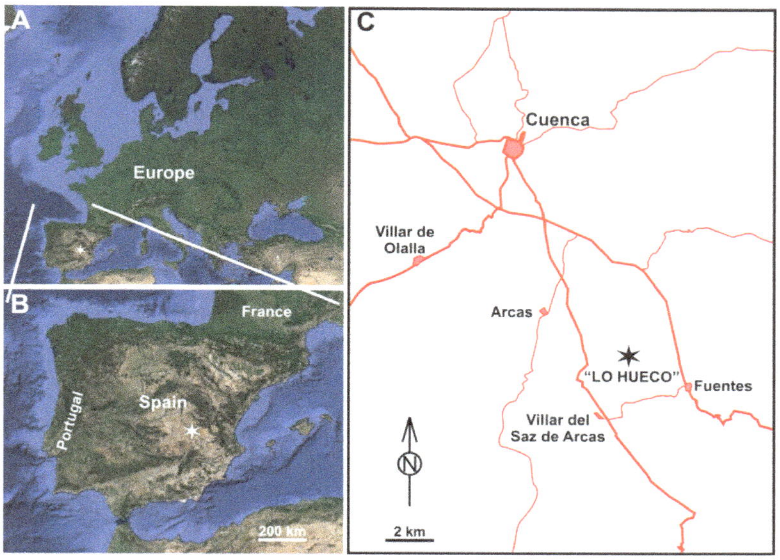

Figure 1. Geographic setting of the "Lo Hueco" fossil site.

Satellite images of A) Europe, and B) Iberian Peninsula; the star shows the situation of the "Lo Hueco" fossil site. C) Close-up view of "Lo Hueco" near the town of Fuentes showing other towns, and the main and secondary roads. Europe and Iberian Peninsula satellite images from USGS EROS (Earth Resources Observatory and Science (EROS) Center) http://eros.usgs.gov/#.

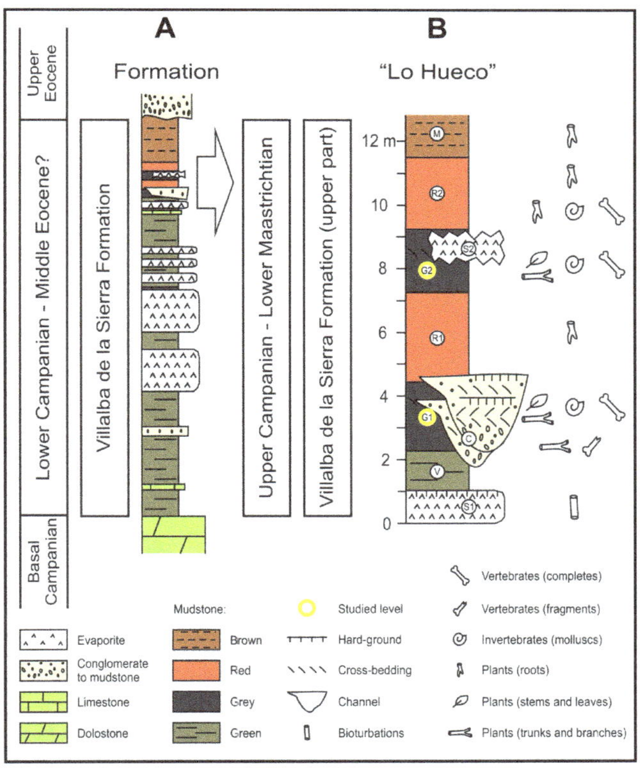

Figure 2. Stratigraphic context of the "Lo Hueco" fossil site.

A) The Villalba de la Sierra Formation. B) Detailed stratigraphic section of "Lo Hueco" showing the studied levels G1 and G2.

MATERIALS AND METHODS

The rich and diverse vertebrate assemblage of the "Lo Hueco" fossil site allowed us to collect multi-taxa samples including dinosaur (theropods: n = 11, sauropods: n = 4) and crocodyliform (n = 36) tooth enamel, turtle shell (n = 9) and gar ganoine (n = 19) from levels G1 and G2. Paleontological prospection and excavation undertaken at the "Lo Hueco" fossil site were authorized by the Dirección General de Patrimonio y Museos of the Junta de Comunidades de Castilla-La Mancha (Spain) (permit number: 04–0392-P11). "Lo Hueco" vertebrate remains are housed at Museo de las Ciencias de Castilla-La Mancha (Cuenca, Spain). Complete repository information including specimen number, stratigraphic situation and geographic location. Dinosaurs, crocodyliforms and turtles were analyzed for their $\delta^{18}O_{PO4}$, $\delta^{18}O_{CO3}$ and $\delta^{13}C$ values, whereas gars were analyzed for their $\delta^{18}O_{PO4}$ values. Dinosaur dentine was also analyzed in order to check for diagenetic effects. Although three major groups of dinosaurs (theropods, sauropods and ornithopods) are represented in "Lo Hueco", in this study we had access to theropods belonging to Dromaeosaurinae and Velociraptorinae Eumaniraptora and to sauropods, which are presumably representing two titanosaurian species, one of them close to the genus *Ampelosaurus* [71–73]. Eusuchian crocodile remains are very abundant at this locality, although the record is monotonous and seems to be represented by two genera close to *Allodaposuchus* [74]. Most of the turtle material probably belongs to two forms of Pleurodira, specifically to members of Bothremydidae [29, 75–76]. Finally, actinopterygians from "Lo Hueco" are mainly represented by lepisosteids, commonly known as gars [29]. We also performed $\delta^{13}C$ analyses on bulk organic matter from the "Lo Hueco" G1 (n = 10) and G2 (n = 8) levels with the aim of characterizing the type of vegetation that was present in the area.

The carbon and oxygen isotope results are reported in the δ-notation. $\delta^{H}X_{sample} = [(R_{sample}—R_{standard})/R_{standard}] \times 1000$, where X is the element, H is the mass of the rare, heavy isotope, and R = $^{13}C/^{12}C$ or $^{18}O/^{16}O$. Vienna Pee Dee Belemnite (VPDB) is the standard for $\delta^{13}C$ values, whereas $\delta^{18}O$ values for both bioapatite carbonate and phosphate are reported relative to Vienna Standard Mean Ocean Water (VSMOW).

Sample chemical treatments followed those described in Bassett *et al.* [77] for phosphate in bioapatite, Koch *et al.* [78] for carbonate in bioapatite and Domingo *et al.* [79] for bulk organic matter. All samples were analyzed at the Stable Isotope Laboratory of the University of California Santa Cruz (USA). Bioapatite $\delta^{18}O_{PO4}$ values were measured using a Thermo Finnigan Delta plus XP isotope ratio mass spectrometer coupled via continuous flow to a high temperature conversion elemental analyzer (TCEA). Bioapatite $\delta^{13}C$ and $\delta^{18}O_{CO3}$ analyses were conducted on a Thermo MAT253 dual-inlet isotope-ratio mass spectrometer coupled to a Kiel IV carbonate device. Bulk organic matter $\delta^{13}C$ analyses were performed using a Carlo Erba 1108 elemental analyzer interfaced to a ThermoFinningan Delta Plus XP isotope ratio mass spectrometer. The standards used in the case of phosphates were Fisher Standard ($\delta^{18}O$ = 8.4‰), UCSC Low standard ($\delta^{18}O$ = 11.3‰), UCSC High standard ($\delta^{18}O$ = 19.0‰) and Kodak standard ($\delta^{18}O$ = 18.1‰) (all of them are silver phosphate). Standard deviations for repeated measurements of Fisher (n = 52), UCSC Low (n = 13), UCSC High (n = 16), and Kodak (n = 7) standards were 0.51‰, 0.55‰, 0.21‰, and 0.36‰ for $\delta^{18}O_{PO4}$. As suggested by Suarez *et al.* [20], we used NIST 120c as a quality control standard and not as a calibration standard since its value is highly dependent on both pretreatment and analysis methods. NIST 120c value (n = 8) was 21.5±0.4. Replicate $\delta^{18}O_{PO4}$ analyses were carried out in ~80% of the samples. The average absolute difference for $\delta^{18}O_{PO4}$ was 0.02‰ and the standard deviation of this average difference was 0.20‰. The standards used in the case of carbonates were Carrara Marble (CM, $\delta^{13}C$ = 2.05‰ and $\delta^{18}O$ = -1.91‰, both VPDB), NBS-18 ($\delta^{13}C$ = -5.03‰ and $\delta^{18}O$ = -23.01‰, VPDB) and NBS-19 ($\delta^{13}C$ = 1.95‰ and $\delta^{18}O$ = -2.20‰, VPDB). The standard deviations for repeated measurements of CM (n = 16), NBS-18 (n = 19), and NBS-19 (n = 10) were 0.03‰, 0.33‰, and 0.04‰ for $\delta^{13}C$, respectively, and 0.08‰, 0.07‰, and 0.09‰ for $\delta^{18}O$, respectively. Duplicate analyses were performed for ~50% of the samples. The average absolute difference for $\delta^{13}C$ and $\delta^{18}O_{CO3}$ was 0.03‰ and 0.14‰, respectively, and the standard deviations of these average differences were 0.02‰ and 0.05‰ for $\delta^{13}C$ and $\delta^{18}O_{CO3}$, respectively. The standards used in the case of bulk organic matter analyses were Pugel ($\delta^{13}C$ = -12.60‰) and Acetanilide ($\delta^{13}C$ = -29.53‰). The standard deviations for repeated measurements of Pugel (n = 11) and Acetanilide (n = 3) were 0.15‰ and 0.02‰, respectively. Statistical tests were performed using the program SPSS PASW Statistics 18.0 software.

We used the following equations to calculate $\delta^{18}O_{H2O}$ values from "Lo Hueco" vertebrates:

Dinosaurs: $\delta^{18}O_{H2O} = 1.11\delta^{18}O_{PO4}–26.44$ [16] (1)

Crocodyliforms: $\delta^{18}O_{H2O} = 0.82\delta^{18}O_{PO4}–19.13$ [80] (2)

Turtles: $\delta^{18}O_{H2O} = 1.06\delta^{18}O_{PO4}–21.6$ [81] (3)

Equation (1) was calculated by Amiot *et al.* [16] constructing a database with modern mammalian $\delta^{18}O_{PO4}$ values and the $\delta^{18}O$ value of meteoric water from IAEA stations. Since most of dinosaur samples from "Lo Hueco" belong to theropods, we considered them as endotherms following most authors [22–23, 45], and therefore we applied equation (1) as Amiot *et al.* [16] suggested in their study. Equation (2) was experimentally determined by Amiot *et al.* [80] analyzing modern crocodilian $\delta^{18}O_{PO4}$ values and the $\delta^{18}O$ value of the water in which they live. Equation (3) was determined by Coulson *et al.* [81] for freshwater and marine extant turtle datasets combining experimental and literature results. This equation was adjusted to correct for the difference in NIST 120c value obtained by Coulson *et al.* [81] (22.6‰) and in our study (21.5±0.4).

Once dinosaur, crocodyliform and turtle $\delta^{18}O_{H2O}$ values were estimated, we used gar ganoine $\delta^{18}O_{PO4}$ values as an independent proxy to calculate temperature values and to eventually determine temperature offsets related to seasonal patterns (see Discussion). Following Domingo *et al.* [82], we selected lepisosteid middle flank scales since they record the greatest number of layers of ganoine per unit of time and grow all year round. In the case of dinosaurs, we have only used theropod $\delta^{18}O_{PO4}$ values on account of 1) the low number of sauropod samples, and 2) the uncertainty about whether sauropods obtained water mainly from drinking (i.e., obligate drinkers) or from vegetation (i.e., non-obligate drinkers).

Different phosphate-water oxygen isotope fractionation equations have been proposed as paleothermometers based mainly on studies carried out on extinct and extant fish and invertebrate phosphates. The utilization of a given equation may yield significant differences in calculated temperature values when compared to other equations, which can be as large as 8–9°C [83–84]. Here, we applied three different equations in order to check whether they can be used to calculate temperature offsets related to seasonal patterns independently of the absolute temperature values yielded by each of them (see Discussion). We considered the following equations:

$$T(°C) = 119.3\ (\pm 12.9) - 4.38\ (\pm 0.54)\ (\delta^{18}O_{PO4} - \delta^{18}O_{H2O})\ [51] \qquad (4)$$

This equation was rescaled for a NBS 120b value of 21.4‰ (see [84])

$$T(°C) = 118.7\ (\pm 4.9) - 4.22\ (\pm 0.20)\ (\delta^{18}O_{PO4} - \delta^{18}O_{H2O})\ [85] \qquad (5)$$

$$T(°C) = 117.4\ (\pm 9.5) - 4.50\ (\pm 0.43)\ (\delta^{18}O_{PO4} - \delta^{18}O_{H2O})\ [84] \qquad (6)$$

RESULTS

Table 1 shows dinosaur, crocodyliform, turtle and gar mean $\delta^{18}O_{PO4}$, $\delta^{13}C$, $\delta^{18}O_{CO3}$, and $\delta^{18}O_{H2O}$ values. Individual values per sample and statistical tests are given in S1 and S2Datasets, respectively.

Table 1. Mean and standard deviation (SD) $\delta^{18}O_{PO4}$ **(‰ VSMOW),** $\delta^{18}O_{H2O}$ **(‰ VSMOW),** $\delta^{13}C$ (‰ VPDB), and $\delta^{18}O_{CO3}$ **(‰ VSMOW) values of "Lo Hueco" theropods, sauropods, crocodyliforms, turtles and gars for the total dataset (TOTAL) and independently for G1 and G2 levels.**

	$\delta^{18}O_{PO4}$ (‰ VSMOW)		$\delta^{18}O_{H2O}$ (‰ VSMOW)		$\delta^{13}C$ (‰ VPDB)		$\delta^{18}O_{CO3}$ (‰ VSMOW)	
	Mean	SD	Mean	SD	Mean	SD	Mean	SD
TOTAL								
Theropods	20,8	0,9	-3,5	1,0	-10,7	0,8	30,0	1,1
Sauropods	20,9	0,4	-	-	-10,5	0,8	29,0	0,1
Crocodyliforms	19,4	0,9	-3,2	0,7	-10,4	1,9	28,2	1,0
Turtles	18,3	0,7	-2,2	0,7	-11,8	0,6	27,1	0,3
Gars	19,8	1,0						
G1								
Theropods	20,2	1,2	-4,1	1,4	-10,7	1,2	31,2	0,2
Sauropods	20,6	-	-	-	-10,6	1,0	29,0	0,02
Crocodyliforms	19,0	0,7	-3,6	0,6	-10,3	2,1	28,1	1,1
Turtles	17,9	0,6	-2,7	0,6	-12,5	0,1	27,0	0,3
Gars	19,1	1,5						
G2								
Theropods	20,7	0,5	-3,5	0,5	-10,9	0,6	29,7	1,1
Sauropods	21,1	-	-	-	-10,3	0,9	29,0	0,1
Crocodyliforms	20,0	0,7	-2,7	0,6	-10,6	1,8	28,2	1,1
Turtles	18,7	0,5	-1,8	0,5	-11,3	0,2	27,2	0,2
Gars	19,7	0,9	-	-	-	-	-	-

$\delta^{18}O_{PO4}$ results

Fig. 3 shows mean $\delta^{18}O_{PO4}$ values for theropods, sauropods, crocodyliforms, turtles and gars. Mean $\delta^{18}O_{PO4}$ value for theropods is 20.8±0.9‰, with a non-significant increase of ~0.5‰ between G1 and G2 ($t = -1.146$, $p = 0.276$). Mean $\delta^{18}O_{PO4}$ value for sauropods is 20.9 ± 0.4‰ and there is also an increase of ~0.5‰ in $\delta^{18}O_{PO4}$ between G1 and G2, although in this case no statistical tests were done since only one sauropod per level could be sampled. Mean $\delta^{18}O_{PO4}$ value for crocodyliforms is 19.4 ± 0.9‰ with a significant increase of ~1.0‰ between G1 and G2 ($t = -3.491$, $p = 0.002$). Turtles show a mean $\delta^{18}O_{PO4}$ value of 18.3‰ ± 0.7‰, with a non-significant increase of ~0.8‰ between G1 and G2 ($t = -2.208$, $p = 0.063$). Crocodyliforms and turtles show statistically consistent lower $\delta^{18}O_{PO4}$ values than dinosaurs for the total dataset (Table 1). Mean $\delta^{18}O_{PO4}$ value for gars is 19.8 ± 1.0‰, with a non-significant decrease of ~0.2‰ between G1 and G2 ($t = 0.475$, $p = 0.641$). ANOVA tests show significant differences for $\delta^{18}O_{PO4}$ values when comparing all taxa for the total dataset and for G1 and G2 independently.

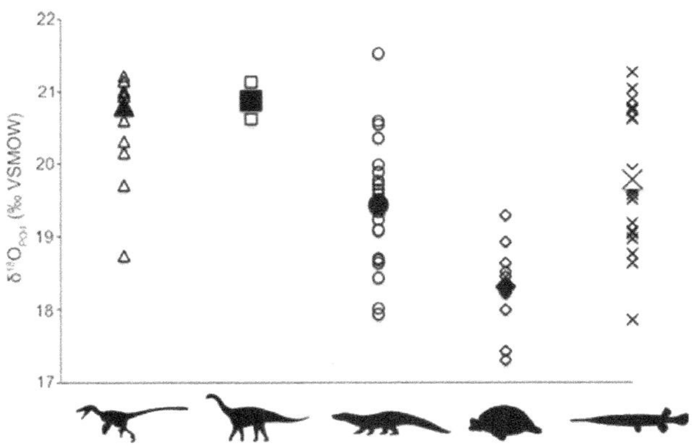

Figure 3. $\delta^{18}O_{PO4}$ **(‰ VSMOW) values of "Lo Hueco" vertebrates.**

Theropods (triangles), sauropods (squares), crocodyliforms (circles), turtles (diamonds), and gars (crosses). Large black symbols represent mean $\delta^{18}O_{PO4}$ values.

$\delta^{18}O_{H2O}$ results

Mean calculated $\delta^{18}O_{H2O}$ value for theropods is -3.5 ± 1.0‰, with a non-significant increase of ~0.6‰ between

G1 and G2 ($t = -0.670$, $p = 0.566$). Mean $\delta^{18}O_{H2O}$ value for crocodyliforms is -3.2 ± 0.7‰, with a significant increase of ~0.9‰ between G1 and G2 ($t = -2.541$, $p = 0.019$). Turtle mean $\delta^{18}O_{H2O}$ value is -2.2 ± 0.7‰, with a non-significant increase of ~0.9‰ between G1 and G2 ($t = -2.215$, $p = 0.062$). ANOVA test perfomed for the total dataset and for levels G1 and G2 showed significant differences in $\delta^{18}O_{H2O}$ value among theropods, crocodyliforms and turtles.

$\delta^{13}C$ and $\delta^{18}O_{CO3}$ results

Fig. 4 shows a biplot $\delta^{13}C$- $\delta^{18}O_{CO3}$ graph for the "Lo Hueco" vertebrate assemblage. Theropod mean $\delta^{13}C$ and $\delta^{18}O_{CO3}$ values are -10.7 ± 0.8‰ and 30.0 ± 1.1‰, respectively, with a decrease in both values between G1 and G2, although only significant in the case of $\delta^{18}O_{CO3}$ ($t = 2.603$, $p = 0.022$). Sauropod mean $\delta^{13}C$ and $\delta^{18}O_{CO3}$ values are -10.5 ± 0.8‰ and 29.0 ± 0.1‰, respectively, with a non-significant increase in $\delta^{13}C$ between G1 and G2 ($t = -0.334$, $p = 0.770$) and with the same $\delta^{18}O_{CO3}$ value in both levels. Crocodyliform mean $\delta^{13}C$ and $\delta^{18}O_{CO3}$ values are -10.4 ± 1.9‰ and 28.2 ± 1.0‰, respectively, with a non-significant decrease in $\delta^{13}C$ between G1 and G2 ($t = 0.214$, $p = 0.834$) and a non-significant increase in $\delta^{18}O_{CO3}$ between G1 and G2 ($t = -0.204$, $p = 0.842$). Finally, turtle mean $\delta^{13}C$ and $\delta^{18}O_{CO3}$values are -11.8 ± 0.6‰ and 27.1 ± 0.3‰, respectively, with increases in both values between G1 and G2, although only significant in the case of $\delta^{13}C$ ($t = -14.300$, $p < 0.001$). $\delta^{13}C$ values do not show statistically significant differences among taxa for the total dataset, G1 and G2, whereas $\delta^{18}O_{CO3}$ values do show significant differences in all cases.

Bulk organic matter mean $\delta^{13}C$ value is -25.1±1.4‰, with a non-significant increase of ~0.2‰ between G1 and G2 ($t = -0.302$, $p = 0.767$).

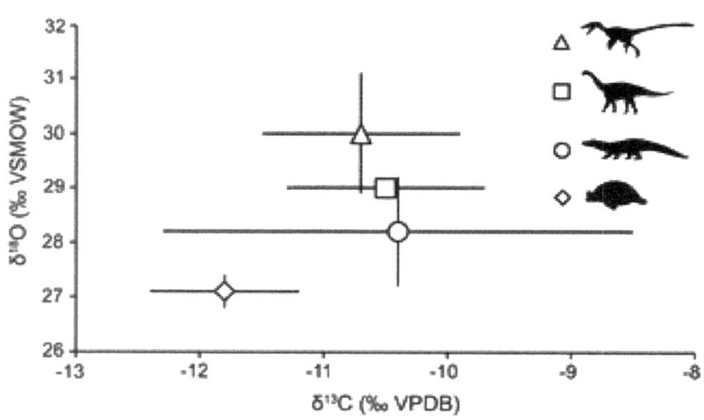

Figure 4. Mean ± 1 SD $\delta^{13}C$ (‰ VPDB) and $\delta^{18}O_{CO3}$ **(‰ VSMOW) values of the "Lo Hueco" vertebrates.**

Theropods (triangles), sauropods (squares), crocodyliforms (circles), and turtles (diamonds).

Diagenesis

Several lines of evidence point to a low degree of diagenetic alteration for the "Lo Hueco" fossil vertebrates:

1) Tooth enamel was the selected tissue for dinosaurs, crocodyliforms, whereas ganoine was analyzed in the case of gars. Relative to dentine, bone and isopedine, tooth enamel and ganoine have larger apatite crystals, a lower content of organic matter, and a low porosity, making them the least prone tissues to undergo diagenetic alteration [86].
2) Several studies of living organisms point to a difference between bioapatite $\delta^{18}O_{CO3}$ and $\delta^{18}O_{PO4}$ values ($\Delta^{18}O_{CO3-PO4}$) of ~8.6–9.1‰ [87–88]. Obtaining a $\Delta\delta^{18}O_{CO3-PO4}$ value near this range in fossil bioapatite has been viewed as an indication that both phases retain pristine isotopic values [87, 89–91]. The mean $\Delta^{18}O_{CO3-PO4}$ value for "Lo Hueco" dinosaurs is 9.1 ± 1.7‰ and thus, it is in the range of the expected equilibrium difference supporting a low degree of diagenetic alteration (Fig. 5). Although this relationship was established for endotherms, "Lo Hueco" turtle $\Delta^{18}O_{CO3-PO4}$ value is also in that range (8.8 ± 0.5‰) (Fig. 5), probably reflecting sustained and constant body temperature during mineral growth (preferentially during summer months: [49–50]). The fact that we only sampled compact bone from the outermost part of turtle shells reduces the possibility of diagenetic alteration. $\Delta^{18}O_{CO3-PO4}$ value could not be calculated for crocodyliforms since $\delta^{18}O_{CO3}$ and $\delta^{18}O_{PO4}$ values did not come from the same samples.

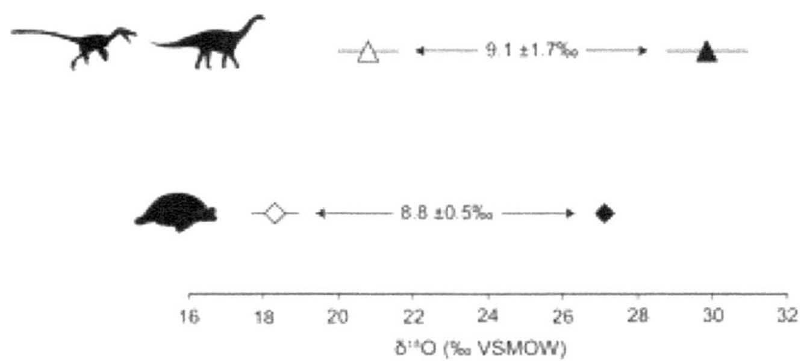

Figure 5. $\Delta^{18}O_{CO3-PO4}$ **values for dinosaurs (theropods and sauropods) and turtles.**

Both values are within the range of the expected equilibrium difference for modern bioapatites (~8.6–9.1‰) [87–88]. Open symbols represent $\delta^{18}O_{PO4}$ values and solid symbols correspond to $\delta^{18}O_{CO3}$.

3) Dinosaur enamel and dentine $\delta^{18}O_{PO4}$ analyses were performed in order to evaluate the response of these tissues against postburial alteration. Several authors have observed higher isotopic values in dinosaur dentine compared to enamel (e.g., [24–26]). Fricke and Pearson [25] argued that this difference was due to a variable exposure of dentine to diagenetic fluids, as well as more isotopic exchange and/or secondary mineral formation in dentine compared to enamel. In the case of the "Lo Hueco" dinosaurs, we also observed higher isotopic values in dentine with theropods showing $\Delta^{18}O_{PO4dentine-enamel} = 1.04$, and sauropods showing $\Delta^{18}O_{PO4dentine-enamel} = 1.34$. Significant differences have been detected between both tissues when considering the total dataset ($t = -2.983$, $p = 0.007$).
4) An additional line of evidence to assess the preservation of original $\delta^{18}O_{PO4}$ values was outlined by Fricke and Rogers [22] when trying to determine the type of metabolism of dinosaurs. According to these authors, differences in body temperatures between

ectotherms and endotherms give rise to differences in $\delta^{18}O_{PO4}$ values as a function of latitude in such a way that at high latitudes, ectotherms have higher $\delta^{18}O_{PO4}$ values than endotherms (due to lower body temperature for ectotherms). Conversely, at low latitudes, endotherms show higher $\delta^{18}O_{PO4}$ values than ectotherms. Amiot *et al.* [16] stated that diagenetic processes would bring about the homogenization of isotopic values and therefore, the aforementioned differences would not be observable. The "Lo Hueco" fossil site was situated at a low paleolatitude (~31°N) during the Late Cretaceous [92]. Theropod dinosaurs, which have been proposed to be endotherms [22–23, 93], show consistently higher $\delta^{18}O_{PO4}$ values in comparison to crocodyliforms and turtles (ectotherms) at the "Lo Hueco" fossil site when considering the total dataset as well as G1 and G2 levels separately (Table 1), in agreement with the results observed by Fricke and Rogers [22] and Amiot *et al.* [16]. In addition to the difference observed in $\delta^{18}O_{PO4}$ values between endotherms and ectotherms, lower $\delta^{18}O_{PO4}$ values for aquatic gars and semi-aquatic crocodyliforms and turtles, and higher $\delta^{18}O_{PO4}$ values for terrestrial theropods constitute another line of evidence pointing to the preservation of the original isotopic signal, as previously stated by other authors [19–20].

DISCUSSION

Seasonal thermal variability in the Iberian Late Cretaceous

Controversy arises when discerning whether Middle and Late Cretaceous climate witnessed a lower mean annual range of temperatures (i.e., seasonal thermal equability) than today [34–39], although it seems that during the colder Late Cretaceous there was a higher seasonal thermal variability (ie., lower equability) in comparison to the warmer Middle Cretaceous [38]. Multi-taxa studies as the one carried out here allow us to investigate this issue on account of differences in the timing of bioapatite growth, thus tracking different moments throughout the year. While in general, $\delta^{18}O_{H2O}$ values calculated for theropods, crocodyliforms and turtles from "Lo Hueco" correspond to typical precipitation waters in subtropical locations [94], theropods recorded consistently lower $\delta^{18}O_{H2O}$ and temperature values than crocodyliforms and turtles in G1 and G2 levels as well as in the total dataset (Table 1). Lower $\delta^{18}O_{H2O}$ values of terrestrial endothermic taxa such as theropods would correspond to ingested water probably consumed during the whole year and hence, recording all seasons. In contrast, $\delta^{18}O_{H2O}$ data of semiaquatic ectothermic taxa such as crocodyliforms and turtles would represent local meteoric waters over a shorter time scale when the conditions are favorable for apatite synthesis (i.e., growth during the warm season and higher $\delta^{18}O_{H2O}$ values) [21, 49, 80]. Temperature values calculated from theropod $\delta^{18}O_{H2O}$ values would track mean annual temperature (MAT), whereas temperature values estimated from crocodyliform and turtle $\delta^{18}O_{H2O}$ values would record temperature of the warmest months (TWMs), when apatite synthesis is more likely to occur. Albeit the seasonal thermal amplitude is usually calculated as the difference between temperature of the warmest months (TWMs) *minus* temperature of the coldest months (TCMs), in our approach it is not feasible to work out this latter value with the available proxies. We opted to infer the semi-seasonal thermal variability characterized as the difference between TWMs and MAT (ΔTWMs-MAT). Since the "Lo Hueco" site was located in a coastal subtropical setting, we created a database with modern meteorological information from coastal stations situated within the subtropics in both hemispheres (25°–35°) compiling MAT and TWMs (i.e., July, August, September) data and calculating ΔTWMs-MAT. We also considered seasonal data provided by Tethyan Cretaceous rudists [36, 38]. These authors argued that during the colder Early Cretaceous (late Barremian-middle Albian), seasonal thermal variability was more intense than during warmer Cretaceous episodes (late Albian-early Campanian). Fig. 6 shows ΔTWMs-MAT *vs* subtropical latitude range (25°–35°) for data calculated from vertebrate taxa from "Lo Hueco", modern subtropical stations, and Santonian-Campanian eastern Tethyan rudists (Greece and Turkey: [36, 38]). $\Delta TWMs_{crocodyliforms}$-$MAT_{theropods}$ values from "Lo Hueco" are 2.2±0.1°C, 3.5±0.1°C, and 2.7±0.1°C for G1, G2, and the total dataset, respectively, whereas $\Delta TWMs_{turtles}$-$MAT_{theropods}$ yielded higher values of 6.1±0.2°C, 7.4±0.20°C, and 6.7±0.2°C for G1, G2, and the total dataset, respectively. Similar $\Delta TWMs_{crocodyliforms}$-$MAT_{theropods}$ and $\Delta TWMs_{turtles}$-$MAT_{theropods}$ values were obtained when using equations (4), (5) and (6). ΔTWMs-MAT values from modern data vary between minimum values of ~2.8±0.4°C (Easter Island, Chile, 27.17°S) and maximum values of ~8.1±0.3°C (Fuzhou, China, 26.08°N). Finally, the mean sea-surface ΔTWMs-MAT value from Late Cretaceous Tethyan rudists is ~5.0±2.6°C (Greece and Turkey, ~30°N). No significant statistical differences have been found between the "Lo Hueco" ΔTWMs-MAT values and those shown by modern meteorological stations and Cretaceous rudists indicating that the climatic conditions in the subtropical western area of the Tethys during the late Campanian-early Maastrichtian were not more significantly equable than those observed today. Previous studies have suggested that during mid-Cretaceous greenhouse conditions the latitudinal thermal gradient was weaker than the one observed today, whereas during the cooler late-Cretaceous, the latitudinal thermal gradient has been proposed to be either weaker or similar to that observed today [6, 95–96]. Cool greenhouse periods, as Huber *et al.* [6] refer to mid-Campanian-Maastrichtian, showed warm temperatures in high latitudes and cool temperatures in the subtropics and tropics (e.g., [95]). Cooler conditions have been traditionally associated to a lower thermal equability and *vice versa* [37, 97] and in this vein, Steuber *et al.* [38], studying seasonal thermal variability on Tethyan rudists, observed a lower thermal seasonality during the warmer Cretaceous episodes, whereas cooler Cretaceous episodes show a higher thermal seasonality. "Lo Hueco" temperature data do not show a significant different seasonal thermal variability when compared to modern data pointing to a similar seasonal thermal amplitude between central eastern Iberia during the late Campanian-early Maastrichtian and today. The determination of past seasonality remains a difficult issue to determine due to the limitation of the seasonal information yielded by paleoproxies and the shortage of these indicators in terrestrial settings. Isotopic studies of multi-taxa vertebrate assemblages as the one from the "Lo Hueco" locality help to characterize past seasonality. Future clumped isotope analyses on soil carbonate and invertebrate carbonate from the "Lo Hueco" locality may allow us to double-check TWMs values and compare them with those provided by crocodyliforms and turtles.

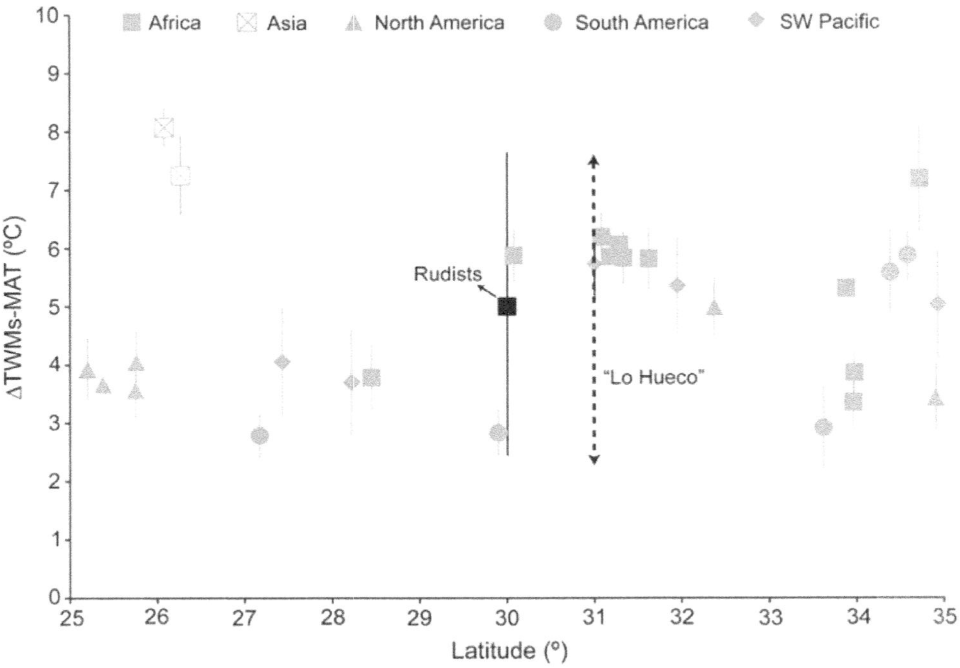

Figure 6. ΔTWMs-MAT (°C) *vs* latitude (°) for "Lo Hueco" (black dotted arrow), Late Cretaceous eastern Tethyan rudists from Steuber [36] and Steuber *et al.* [38] (black square), and modern meteorological stations located at latitudes 25° to 35° in both hemispheres from Africa (grey square), Asia (crossed square), North America (grey triangle), South America (grey circle), and Southwest Pacific (grey diamond).

Absolute latitude is shown including North and South hemispheres.

"Lo Hueco" vertebrate isotopic paleoecology

Carbon isotope data have not been widely investigated in Cretaceous multi-taxa assemblages and studies have mainly focused on herbivore terrestrial taxa (e.g., [24–26]). We have analyzed $\delta^{13}C$ values of sauropods, theropods, crocodyliforms and turtles as well as $\delta^{13}C$ values of the bulk organic matter from the "Lo Hueco" sediments. Sauropod mean $\delta^{13}C$ value is -10.5±0.8‰ and bulk organic matter mean $\delta^{13}C$ value is -25.1±1.4‰ (Fig. 7). While we are aware that the low number of "Lo Hueco" sauropod tooth enamel samples may prevent us from attaining conclusive results, we still believe that it is worth discussing them in the context of other isotopic studies focused on herbivore dinosaurs. The "Lo Hueco" bulk organic matter mean $\delta^{13}C$ value is almost identical to the estimated Late Cretaceous C_3 vegetation mean $\delta^{13}C$ value (-25.5‰), assuming a $\delta^{13}C_{atmCO2}$ value of ~-6‰ for the late Campanian-early Maastrichtian [57] and the modern fractionation value of ~-19.5‰ between $\delta^{13}C_{atmCO2}$ and C_3 vegetation $\delta^{13}C$ values [60]. This suggests a pure C_3 environment for the "Lo Hueco" locality for that time period. Assuming that the "Lo Hueco" bulk organic matter mean $\delta^{13}C$ value is representative of the $\delta^{13}C$ value of the vegetation present in the area, then the fractionation between the "Lo Hueco" sauropod enamel and diet ($\Delta^{13}C_{enamel-diet}$) would be ~15‰. This offset lies between the one estimated for mammals and ostriches (12 to 14‰: [61–63, 66]) and the one estimated for sauropod dinosaurs and ostriches (~16‰: [32, 65], respectively), and it is narrower when compared to the one estimated for ornithischian dinosaurs (i.e., 18‰: [25–26]). Stanton-Thomas and Carlson [24], Fricke and Pearson [25] and Fricke *et al.* [26] obtained higher $\delta^{13}C$ values for hadrosaur and ceratopsian tooth enamel (-4‰ to -5.9‰) from different Late Cretaceous North American localities located in the vicinity of the Western Interior Seaway when compared to "Lo Hueco" sauropod $\delta^{13}C$ values. Since bioapatite $\delta^{13}C$ values of terrestrial vertebrates are ultimately controlled by $\delta^{13}C$ values of diet giving rise to consistent $\Delta^{13}C_{enamel-diet}$ values [63, 86], the difference between the fractionation values obtained in our study and those from Stanton-Thomas and Carlson [24], Fricke and Pearson [25] and Fricke *et al.* [26] might be likely associated to metabolic differences among taxonomic groups (sauropods *vs.* ceratopsians and hadrosaurs) (Fig. 8) [22–23, 46–48]. Bulk organic matter mean $\delta^{13}C$ values showed by Fricke and Pearson [25] are ~-24.0‰, just 1‰ higher than our bulk organic matter mean $\delta^{13}C$ values, supporting differences in dinosaur metabolism in a scenario with similar isotopic values of vegetation (Fig. 7). Fricke and Pearson [25] pointed out that differences in the methane production in herbivore dinosaur stomachs during digestion along with different utilization of plant organic compounds (e.g., carbohydrates, proteins, lipids) and/or plant parts (e.g., leaves, seeds, wood) may also explain the differences observed in $\Delta^{13}C_{enamel-diet}$. The larger ceratopsian and hadrosaur $\Delta^{13}C_{enamel-diet}$ may be due to their ability to incorporate low quality, fibrous vegetation [25], whereas the smaller sauropod $\Delta^{13}C_{enamel-diet}$ value may be indicative of consumption of more digestible food items (e.g., soft leaves), fact also supported by the dental morphology of "Lo Hueco" titanosaur sauropods (i.e., chisel-like teeth) more adapted to leave behind the least digestible tissues [98–99]. The 3‰ difference observed in $\Delta^{13}C_{enamel\text{-}diet}$ values for ornithischians from Fricke and Pearson [25] and Fricke *et al.* [26] and sauropods from our study is statiscally significant ($t = -4.481$, $p = 0.003$), and therefore, even considering the variability yielded by dinosaur and bulk organic matter $\delta^{13}C$ values, these two sets of data do not significantly overlap. Stanton-Thomas and Carlson [24] stated that higher $\delta^{13}C$ values observed in hadrosaurs might be related to ingestion of vegetation subject to high salinity levels and/or consumption of gymnosperms (which show an enrichment of ~1.1 to 2‰ compared to mean C_3 $\delta^{13}C$ values). "Lo Hueco" sauropods may have relied heavily on angiosperms as revealed by the palynological content of the "Lo Hueco" sediments showing pollen assemblages dominated by angiosperms (93%: [100]). Specifically, the palynological assemblage from "Lo Hueco" is dominated by freshwater palynomorphs, spores and pollen grains related to swamp or local wetland vegetation [100]. The preferential incorporation of angiosperms by "Lo Hueco" sauropods may also lie behind the difference observed between our $\Delta^{13}C_{enamel\text{-}diet}$ value and the one estimated by Tütken [32]. Interestingly, in the compilation of sauropod $\delta^{13}C$ values carried out by this author, *Ampelosaurus* showed the lowest $\delta^{13}C$ values, and according to Knoll *et al.* [73], one of the "Lo Hueco" sauropod taxa remains close to this genus. These uncertainties regarding herbivore dinosaur isotopic paleoecology open up new lines of investigation dealing with the question of the type of dietary behavior and physiology of different genera giving rise to different $\Delta^{13}C_{enamel\text{-}diet}$ offsets.

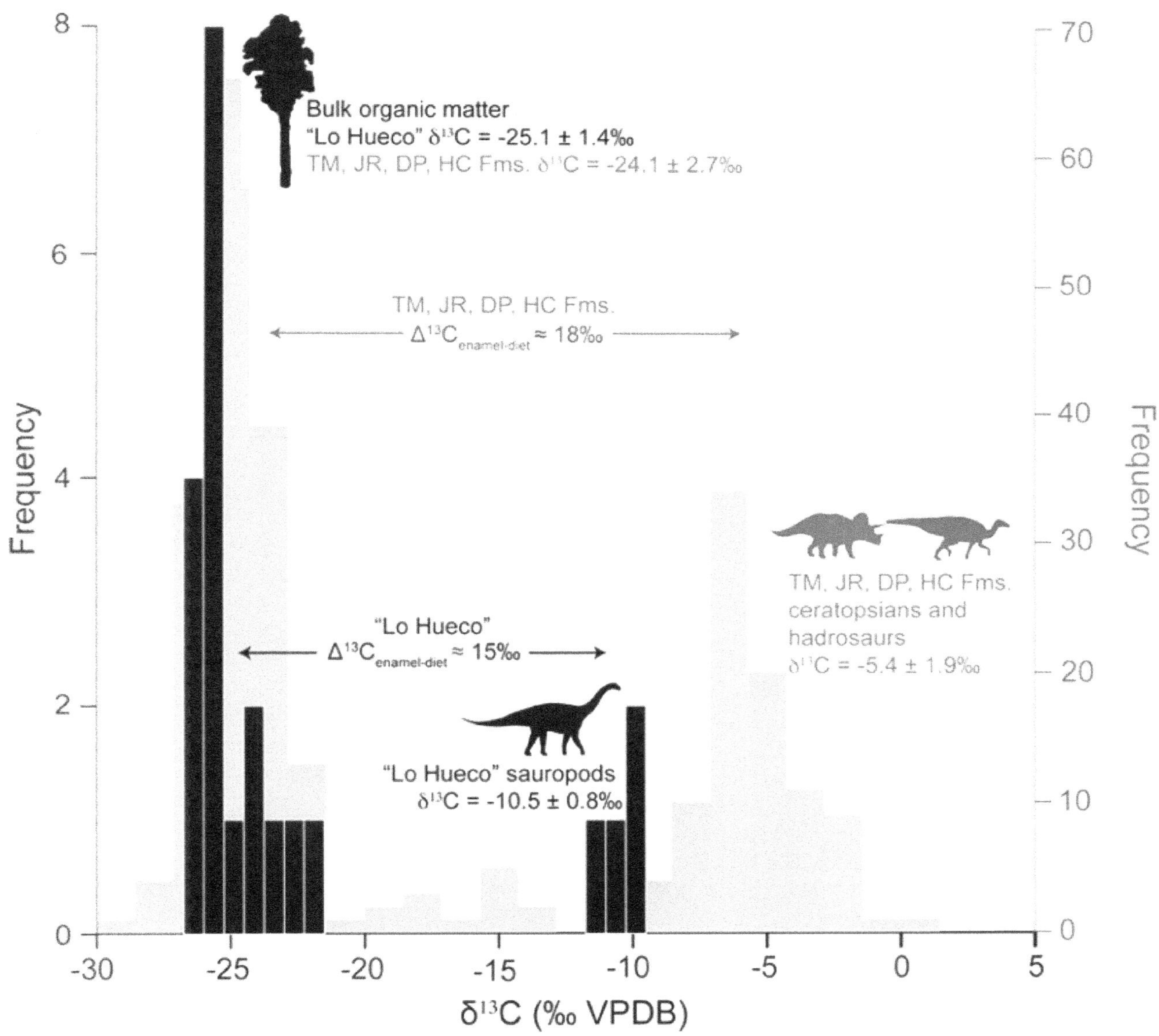

Figure 7. Fractionation between herbivore enamel and herbivore diet ($\Delta^{13}C_{enamel\text{-}diet}$).

"Lo Hueco" bulk organic matter and sauropod mean ± 1 SD $\delta^{13}C$ (‰ VPDB) (black). Two Medicine (TM), Judith River (JR), Dinosaur Park (DP), and Hell Creek (HC) Formations bulk organic matter and ceratopsian and hadrosaur mean ± 1 SD $\delta^{13}C$ (‰ VPDB) (grey) [25–26]. "Lo Hueco" $\Delta^{13}C_{enamel\text{-}diet}$ is 15‰, when analysing sauropods, whereas Fricke and Pearson [25] and Fricke *et al.* [26] obtained a value of 18‰ when analysing ceratopsians and hadrosaurs. Note different frequency scale for "Lo Hueco" (black) and Fricke and Pearson [25] and Fricke *et al.* [26]´s data (grey).

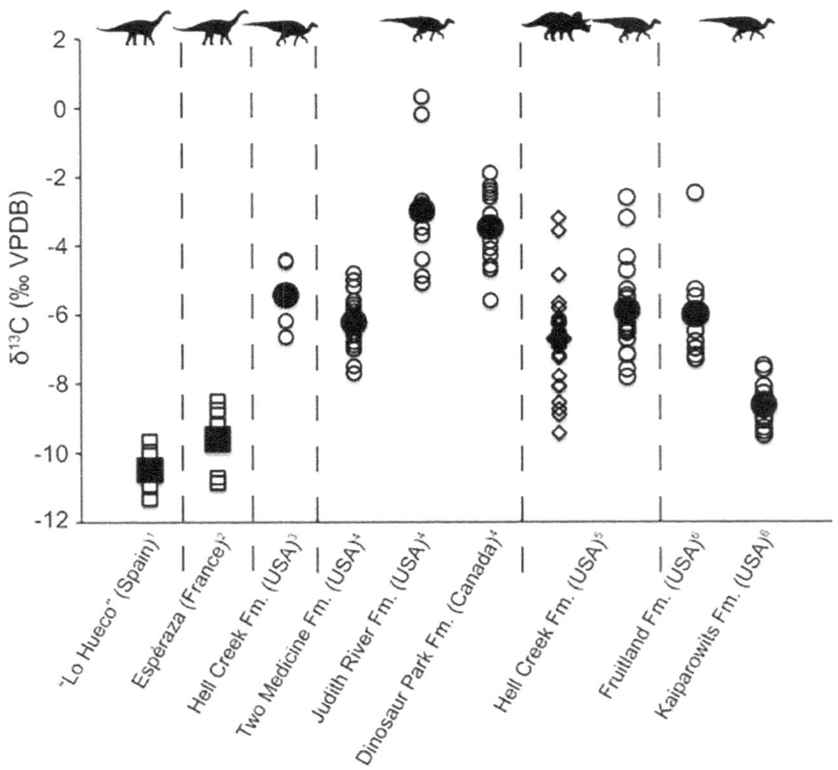

Figure 8. $\delta^{13}C$ (‰ VPDB) values for Campanian-Maastrichtian herbivore dinosaurs: titanosaur sauropods (squares), hadrosaurs (circles), and ceratopsians (diamonds).

Large black symbols represent mean $\delta^{13}C$ values. [1]This study; [2]Tütken [32]; [3]Stanton-Thomas and Carlson [24]; [4]Fricke *et al.* [26]; [5]Fricke and Pearson [25]; [6]Fricke *et al.* [53].

"Lo Hueco" theropod mean $\delta^{13}C$ value is -10.7±0.8‰, statistically indistinguishable from the sauropod mean $\delta^{13}C$ value (-10.5±0.8‰) (Fig. 4). To the best of our knowledge, there are no isotopic studies dealing with herbivore and carnivore dinosaur feeding paleoecology from the same locality. Isotopic studies on ancient vertebrate predator-prey systems have mainly focused on Quaternary mammals [64, 101–103] and to a lesser extent on Miocene mammals [104]. Clementz *et al.* [64] observed a $\delta^{13}C$ offset between predators (wolves) and their prey (moose and elks) ($\Delta^{13}C_{carnivore\text{-}herbivore}$) of ~ -1.3‰. These authors argued that lower tooth enamel $\delta^{13}C$ values for carnivores might be due to differences in their digestive physiology in comparison to herbivores. The lack of significant differences between "Lo Hueco" carnivore and herbivore dinosaur tooth enamel $\delta^{13}C$ values ($\Delta^{13}C_{carnivore\text{-}herbivore}$= - 0.2‰,) may be due to differences in metabolic and/or physiological mechanisms between dinosaurs and mammals, however a more plausible explanation is that titanosaur sauropods were not likely prey for theropods belonging to Dromaeosaurinae and Velociraptorinae subfamilies on account of body mass differences. Interestingly, the lack of significant differences in $\delta^{13}C$ values between both theropod subfamilies ($t = 0.516$, $p = 0.618$), along with the similarity in their body masses and likely hunting strategies point to some degree of resource overlap. Also, the lack of significant differences in variance of $\delta^{13}C$ values between these subfamilies of theropods (Levene test; $F = 1.582$, $p = 0.240$) suggests that none was more versatile than the other in resource utilization.

In the case of aquatic ectotherms (crocodyliforms and turtles), $\delta^{13}C$ values are a reflection of food oxidation (respiration) and ambient water (e.g., disolved inorganic carbon or DIC can constitute an additional source of carbon) [105–106]. "Lo Hueco" crocodyliforms belong to forms close to the genus *Allodaposuchus* [74]. Since it is not possible to determine physiological tolerance to salinity from the morphological standpoint in the basal Eusuchia from "Lo Hueco", (J.L. Sanz and F. Ortega, personal communication), the information supplied by stable isotopes may shed light about salt tolerance, dietary behavior and habitat occupancy. We have compared the isotopic values of "Lo Hueco" crocodyliforms to those provided by Wheatley *et al.* [107] for Louisiana and Florida modern crocodylians. We are aware that extrapolation between extinct and extant organisms is not straightforward due to unknown physiological mechanisms in ancient taxa as well as different environmental conditions. However, we still believe that a comparison can be made since 1) the marine-freshwater isotopic threshold values are well established not only in modern studies, but also in the past, 2) the marine $\delta^{18}O$ value has remained fairly constant throughout the geological time and 3) the Wheatley *et al.* [107]´s study considered modern crocodylians from Louisiana and Florida with a similar latitude to the one suggested for "Lo Hueco" (25–30°N for Louisiana and Florida *vs*31°N for "Lo Hueco"). "Lo Hueco" crocodyliforms show a mean $\delta^{13}C$ value, which is statistically indistinguishable from the value observed by Wheatley *et al.* [107] for Louisiana and Florida modern coastal *Alligator mississippiensis* ($p = 0.999$) and Florida modern coastal *Crocodylus acutus* ($p = 0.868$), while it is significantly different from Florida modern inland *A. mississippiensis* ($p < 0.001$), whereas "Lo Hueco" crocodyliforms show a significantly lower mean $\delta^{18}O_{CO3}$ value, when compared to coastal and inland *A. mississippiensis* ($p = 0.019$ and $p = 0.001$, respectively) and coastal *C. acutus* ($p < 0.001$) (Fig. 9A). These results pose interesting issues concerning the salinity discrimination and osmoregulation capacity of the "Lo Hueco" eusuchian crocodyliforms. Jackson *et al.* [108] in a study of salinity tolerance and osmoregulation mechanisms by modern crocodilians argued that reptiles inhabiting marine and estuarine waters keep a constant plasma osmolality by behavioral modifications (avoiding drinking seawater) and/or morphological adaptation (salt-secreting glands and reduced integumental permeability). Crocodyliforms from "Lo Hueco" are under study and with the current information we cannot determine their physiological tolerance to salinity from a morphological standpoint (e.g., presence or absence of salt-secreting glands) (JL Sanz and F Ortega, personal communication, 2014). While waiting for these morphological studies, isotopic results suggest that the "Lo Hueco" crocodyliforms may have incorporated food items from brackish waters as shown by their $\delta^{13}C$ values, whereas they avoided ingesting this water and consumed preferentially freshwater, as suggested by their $\delta^{18}O_{CO3}$ values. Clementz and Koch [109] and Wheatley *et al.* [107] argued that animals incorporating marine food items and drinking seawater show a lower variability in their $\delta^{18}O_{CO3}$ values. Even though "Lo Hueco" crocodyliforms likely drank freshwater, its $\delta^{18}O_{CO3}$ variability (1.0‰) is not as high as that shown by either inland (2.2‰) and coastal (1.8‰) *A. mississippiensis*, and it is more similar to the one shown by saltwater tolerant *C. acutus* (0.8‰) (Fig. 9B). This might indicate that $\delta^{18}O$ values of the water ingested by the "Lo Hueco" crocodyliforms may have remained homogeneous during the time window in which bioapatite mineralized (i.e., warmest months of the year).

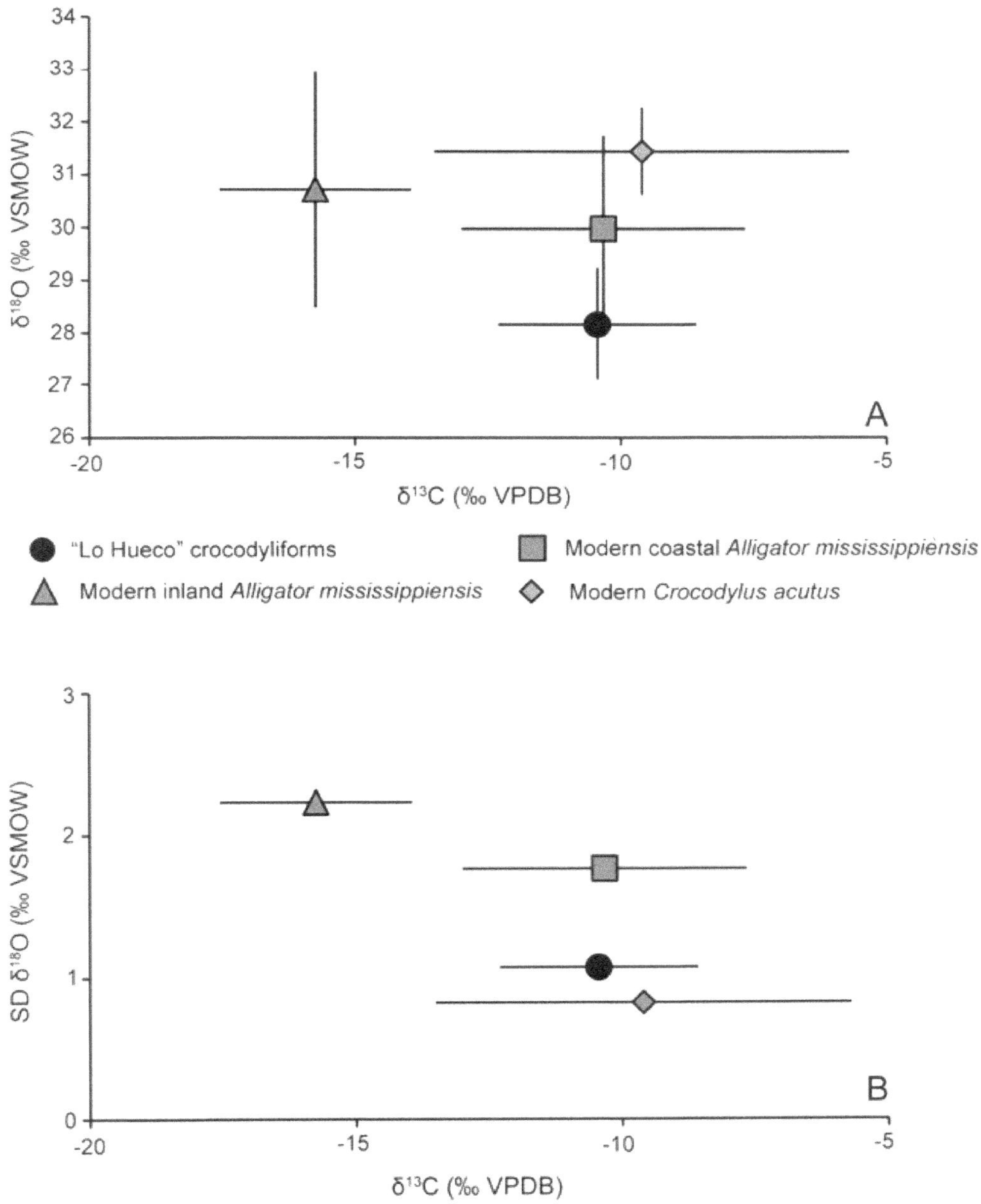

Figure 9. Comparison between isotopic values of "Lo Hueco" crocodyliforms and modern crocodilians.

A) $\delta^{18}O_{CO3}$ (‰ VSMOW) *vs* $\delta^{13}C$ (‰ VPDB) mean ± 1 standard deviation (SD) values of the "Lo Hueco" crocodyliforms (black circle), and modern inland *Alligator mississippiensis* (grey triangle), coastal *Alligator mississippiensis* (grey square) and *Crocodylus acutus* (grey diamond) [107]. B) SD $\delta^{18}O_{CO3}$ (‰ VSMOW) *vs* $\delta^{13}C$ (‰ VPDB) values of the "Lo Hueco" crocodyliforms and modern crocodilians (symbols as in A).

"Lo Hueco" turtle mean $\delta^{13}C$ and $\delta^{18}O_{CO3}$ values are the lowest when compared to dinosaurs and crocodyliforms (Fig. 4) (although significant differences are only observed between turtles and dinosaur $\delta^{18}O_{CO3}$ values). "Lo Hueco" turtles belong to two forms of Pleurodira, specifically to members of Bothremydidae [29, 75–76]. They are omnivore freshwater turtles, so low $\delta^{13}C$ values may be indicative of a diet based on a mixture of aquatic and terrestrial C_3 vegetation and/or invertebrates, whereas low $\delta^{18}O$ values may indicate ingestion of water with an inland source. Unlike crocodyliforms, there are not isotopic studies dealing with bone carbonate of freshwater turtles from the subtropics and they focused on marine turtles [105–106]. As observed in the case of "Lo Hueco" crocodyliforms, "Lo Hueco" turtles show a low variability in their mean $\delta^{18}O_{CO3}$ value, in spite of inhabiting freshwater environments, when compared to modern marine loggerhead turtles reported by Wheatley *et al.* [107] (0.3‰ *vs.* 0.6‰, for "Lo Hueco" and marine loggerhead turtles, respectively) pointing to a low variability of $\delta^{18}O_{CO3}$ values and presumably of the $\delta^{18}O$ values of the ingested water for the time of bioapatite mineralization (i.e., warmest months of the year).

CONCLUSIONS

Isotopic analyses on the multi-taxa terrestrial vertebrate assemblage of the "Lo Hueco" locality (central eastern Spain) provides valuable information about climatic and environmental conditions existing in southeastern Iberia during the late Campanian-early Maastrichtian (Late Cretaceous). Seasonal thermal varibility has been inferred as the difference between temperature of the warmest months (TWMs), supplied by crocodyliforms and turtles (whose bioapatite mineralizes during the warm season) and mean annual temperature (MAT), provided by theropods (that record ingested water throughout the year). "Lo Hueco" ΔTWMs-MAT results do not point to a significant different seasonal thermal variability to that observed today. From the paleoecological standpoint, $\delta^{13}C$ values of the "Lo Hueco" taxa point to consumption of pure C_3 vegetation, fact that agrees well with bulk organic matter $\delta^{13}C$ values from the "Lo Hueco" sediments. The estimated fractionation between sauropod enamel and diet ($\Delta^{13}C_{enamel-diet}$) is ~15‰, lower than other fractionation values calculated for sauropods (~16‰) and ornithischians (~18‰), and likely indicating differences in metabolic and/or physiological processes or different utilization of plant organic compounds and/or plant parts. Since "Lo Hueco" titanosaur sauropods may have not been likely prey for theropods belonging to Dromaeosaurinae and Velociraptorinae subfamilies on account of differences in their body mass, no conclusive information concerning dinosaurian predator-prey $\delta^{13}C$ offset could be attained. Although "Lo Hueco" crocodyliform material remains under study and no paleoecological conclusions have been drawn from a morphological standpoint, isotopic results indicate that they may have incorporated food items from brackish waters as shown by their $\delta^{13}C$ values, whereas they avoided ingesting saline water and consumed preferentially freshwater, as suggested by their $\delta^{18}O_{CO3}$ values, when compared with isotopic values of modern crocodilians inhabiting subtropical regions. "Lo Hueco" turtles show the lowest $\delta^{13}C$ and $\delta^{18}O_{CO3}$ values of the vertebrate assemblage likely indicating a diet based on a mixture of aquatic and terrestrial C_3 vegetation and/or invertebrates and ingestion of water with an inland source, a fact that agrees well with their taxonomic designation.

ACKNOWLEDGMENTS

We thank JL Sanz (Universidad Autónoma de Madrid), F Ortega (Universidad Nacional de Educación a Distancia), A Pérez-García (Universidad Complutense de Madrid) and A Torices (University of Alberta) for the access to the analyzed material and suggestions that helped to improve the manuscript. We are indebted to PL Koch (University of California Santa Cruz) for valuable comments and access to his laboratory, and to D Andreasen (University of California Santa Cruz) for his help with stable isotope analyses. O Sanisidro is acknowledged for illustrations. We thank editor P Dodson and reviewers C Suarez and R Amiot for their valuable comments.

Author Contributions

Conceived and designed the experiments: LD FBB OCM. Performed the experiments: LD. Analyzed the data: LD FBB OCM. Contributed reagents/materials/analysis tools: LD FBB OCM. Wrote the paper: LD FBB OCM.

REFERENCES

1. Poulsen CJ, Gendaszek AS, Jacob RL. Did the rifting of the Atlantic Ocean cause the Cretaceous thermal maximum? Geology. 2003; 31: 115–118. doi: 10. 1130/0091-7613(2003)031<0115:dtrota>2. 0. co;2
2. Forster A, Schouten S, Baas M, Sinninghe Damsté JS. Mid-Cretaceous (Albian–Santonian) sea surface temperature record of the tropical Atlantic Ocean. Geology. 2007; 35: 919–922. doi: 10. 1130/g23874a. 1
3. Parrish JT, Spicer RA. Late Cretaceous terrestrial vegetation:Anearpolartemperaturecurve. Geology. 1988; 16: 22–25. doi: 10. 1130/0091-7613(1988)016<0022:lctvan>2. 3. co;2
4. Jenkins HC, Gale AS, Corfield RM. Carbon- and oxygen-isotope stratigraphy of the English chalk in Italian Scaglia and its paleoclimatic significance. Geol Mag. 1994; 131: 1–34. doi: 10. 1017/s0016756800010451
5. Huber BT, Hodell DA, Hamilton CP. Middle–Late Cretaceous climate of the high southern latitudes: Stable isotope evidence for minimal equator to pole thermal gradients. Geol Soc Am Bull. 1995; 107: 1164–1191. doi: 10. 1130/0016-7606(1995)107<1164:mlccot>2. 3. co;2
6. Huber BT, Norris RD, MacLeod KG. Deep-sea paleotemperature record of extreme warmth during the Cretaceous. Geology. 2002; 30: 123–126. doi: 10. 1130/0091-7613(2002)030<0123:dsproe>2. 0. co;2

7. Friedrich O, Herrle JO, Wilson PA, Cooper MJ, Erbacher J, Hemleben C. Early Maastrichtian carbon cycle perturbation and cooling event: Implications from the South Atlantic Ocean. Paleoceanography. 2009; 24: PA2211. doi: 10. 1029/2008pa001654
8. Jenkyns HC, Forster A, Schouten S, Sinninghe Damsté JS (2004) High temperatures in the Late Cretaceous Arctic Ocean. Nature 432: 888–892. pmid:15602558 doi: 10. 1038/nature03143
9. Nordt L, Atchley S, Dworkin S. Terrestrial evidence for two greenhouse events in the latest Cretaceous. GSA Today. 2003; 13: 4–9. pmid:14655765 doi: 10. 1130/1052-5173(2003)013<4:teftge>2. 0. co;2
10. MacLeod KG, Huber BT, Isaza-Londoño C. North Atlantic warming during global cooling at the end of the Cretaceous. Geology. 2005; 33: 437–440. doi: 10. 1130/g21466. 1
11. Royer DL. CO_2-forced climate thresholds during the Phanerozoic. Geochim Cosmochim Acta. 2006; 70: 5665–5675. doi: 10. 1016/j. gca. 2005. 11. 031
12. Frakes LA, Francis JE, Syktus JI. Climate Modes of the Phanerozoic. Cambridge: Cambridge University Press; 1992.
13. Markwick PJ, Rowley DB. The geological evidence for Triassic to Pleistocene glaciations: Implications for eustasy. In: Pindell JL, Drake CL, editors. Paleogeographic Evolution and Non-Glacial Eustasy, Northern South America: Society for Sedimentary Geology (SEPM) Special Publication, Tulsa, Oklahoma. 1998; 58: 17–43.
14. DeConto RM, Brady EC, Berggren J, Hay WW. Late Cretaceous climate, vegetation, and ocean interactions. In: Huber BT, MacLeod KG, Wing SL, editors. Warm Climates in Earth History. Cambridge: Cambridge University Press; 2000. pp. 275–296.
15. DeConto RM, Hay WW, Thompson SL, Bergengren JC. Late Cretaceous climate and vegetation interactions: Cold continental paradox. In: Barrera E, Johnson C, editors. The evolution of Cretaceous ocean/climate systems. Geological Society of America Special Paper. 1999; 332: 391–432.
16. Amiot R, Lécuyer C, Buffetaut E, Fluteau F, Legendre S, Martineau F. Latitudinal temperature gradient during the Cretaceous upper Campanian–middle Maastrichtian: $\delta^{18}O$ record of continental vertebrates. Earth Planet Sci Lett. 2004; 226: 255–272. doi: 10. 1016/j. epsl. 2004. 07. 015
17. Amiot R, Wang X, Zhou Z, Wang X, Buffetaut E, et al. Oxygen isotopes of East Asian dinosaurs reveal exceptionally cold Early Cretaceous climates. Proc Natl Acad Sci USA. 2011; 108: 5179–5183. doi: 10. 1073/ pnas. 1011369108. pmid:21393569
18. Straight WH, Barrick RE, Eberth DA. Reflections of surface water, seasonality and climate in stable oxygen isotopes from tyrannosaurid tooth enamel. Palaeogeogr Palaeoclimatol Palaeoecol. 2004; 206: 239–256. doi: 10. 1016/j. palaeo. 2004. 01. 006
19. Amiot A, Buffetaut E, Lécuyer C, Wang X, Boudad L, et al. Oxygen isotope evidence for semi-aquatic habits among spinosaurid theropods. Geology. 2010; 38: 139–142. doi: 10. 1130/g30402. 1
20. Suarez CA, Gonzalez LA, Ludvigson GA, Cifelli RL, Tremain E. Water utilization of the Cretaceous Mussentuchit Member local vertebrate fauna, Cedar Mountain Formation, Utah, USA: Using oxygen isotopic composition of phosphate. Palaeogeogr Palaeoclimatol Palaeoecol. 2012; 313–314: 78–92. doi: 10. 1016/j. palaeo. 2011. 10. 011
21. Barrick RE, Showers WJ, Fisher AG. Comparison of thermoregulation of four ornithischian dinosaurs and a varanid lizard from the Cretaceous Two Medicine Formation: Evidence from oxygen isotopes. Palaios. 1996; 11: 295–305.
22. Fricke HC, Rogers RR. Multiple taxon-multiple locality approach to providing oxygen isotope evidence for warm-blooded theropod dinosaurs. Geology. 2000; 28: 799–802. doi: 10. 1130/0091-7613(2000)28<799:mtlatp>2. 0. co;2
23. Amiot R, Lécuyer C, Buffetaut E, Escarguel G, Fluteau F, Martineau F. Oxygen isotopes from biogenic apatites suggest widespread endothermy in Cretaceous dinosaurs. Earth Planet Sci Lett. 2006; 246: 41–54. doi: 10. 1016/j. epsl. 2006. 04. 018
24. Stanton-Thomas KJ, Carlson SJ. Microscale $\delta^{18}O$ and $\delta^{13}C$ isotopic analysis of an ontogenetic series of the hadrosaurid dinosaur *Edmontosaurus*: Implications for physiology and ecology. Palaeogeogr Palaeoclimatol Palaeoecol. 2004; 206: 257–287. doi: 10. 1016/j. palaeo. 2004. 01. 007
25. Fricke HC, Pearson DA. Stable isotope evidence for changes in dietary niche partitioning among hadrosaurian and ceratopsian dinosaurs of the Hell Creek Formation, North Dakota. Paleobiology. 2008; 34: 534–552. doi: 10. 1666/08020. 1
26. Fricke HC, Rogers RR, Backlund R, Dwyer CN, Echt S. Preservation of primary stable isotope signals in dinosaur remains, and environmental gradients of the Late Cretaceous of Montana and Alberta. Palaeogeogr Palaeoclimatol Palaeoecol. 2008; 266: 13–27. doi: 10. 1016/j. palaeo. 2008. 03. 030
27. Domingo L, Barroso-Barcenilla F, Cambra-Moo O. Paleoenvironmental reconstruction of the "Lo Hueco" fossil site (Upper Cretaceous, Cuenca, Spain): Preliminary stable isotope analyses on crocodilians and dinosaurs. Palaios. 2013; 28: 195–202. doi: 10. 3989/ egeol. 41148. 244
28. Lécuyer C, Bogey C, Garcia JP, Grandjean P, Barrat JA, et al. Stable isotope composition and rare earth element content of vertebrate remains from the Late Cretaceous of Northern Spain (Laño): Did the environmental record survive? Palaeogeogr Palaeoclimatol Palaeoecol. 2003; 193: 457–471. doi: 10. 1016/s0031-0182(03)00261-x
29. Barroso-Barcenilla F, Cambra-Moo O, Escaso F, Ortega F, Pascual A, et al. New and exceptional discovery in the Upper Cretaceous of the Iberian Peninsula: The palaeontological site of "Lo Hueco", Cuenca, Spain. Cretaceous Res. 2009; 30: 1268–1278. doi: 10. 1016/j. cretres. 2009. 07. 010

30. Cambra-Moo O, Barroso-Barcenilla F, Coruña F, Postigo-Mijarra JM. Exceptionally well-preserved vegetal remains from the Upper Cretaceous “Lo Hueco” site, Cuenca, Spain. Lethaia. 2013; 46: 127–140. doi: 10. 1111/j. 1502-3931. 2012. 00331. x
31. Tütken T, Pfretzschner HU, Vennemann T, Sun G, Wand YD. Paleobiology and skeletochronology of Jurassic dinosaurs: implications from the histology and oxygen isotope compositions of bones. Palaeogeogr Palaeoclimatol Palaeoecol. 2004; 206: 217–238. doi: 10. 1016/j. palaeo. 2004. 01. 005
32. Tütken T. The diet of sauropod dinosaurs: Implications from carbon isotope analysis of teeth, bones, and plants. In: Klein N, Remes K, Sander M, editors. Biology of the Sauropod Dinosaurs: Understanding the Life of Giants. Bloomington, IN: Indiana University Press; 2011. pp. 57–79.
33. Markwick PJ. Fossil crocodilians as indicators of Late Cretaceous and Cenozoic climates: implications for using palaeontological data in reconstructing palaeoclimate. Palaeogeogr Palaeoclimatol Palaeoecol. 1998; 137: 205–271. doi: 10. 1016/s0031-0182(97)00108-9
34. Sloan LC, Barron E. “Equable” climates during Earth history? Geology. 1990; 18: 489–492. doi: 10. 1130/0091-7613(1990)018<0489:ecdeh>2. 3. co;2
35. Bush ABG, Philander SGH. The late Cretaceous: Simulation with a coupled atmosphere-ocean general circulation model. Paleoceanography. 1997; 12: 495–516. doi: 10. 1029/97pa00721
36. Steuber T. Isotopic and chemical intra-shell variations in low-Mg calcite of rudist bivalves (Mollusca: Hippuritacea)—disequilibrium fractionations and Late Cretaceous seasonality. Int J Earth Sci. 1999; 88: 551–570. doi: 10. 1007/s005310050284
37. Golovneva LB. The Maastrichtian (Late Cretaceous) climate in the Northern Hemisphere. In: Hart MB, editor. Climates: Past and Present. Geological Society, London, Special Publications. 2000; 181: 43–54.
38. Steuber T, Rauch M, Masse JP, Graaf J, Malkoc M. Low-latitude seasonality of Cretaceous temperatures in warm and cold episodes. Nature. 2005; 437: 1341–1344. pmid:16251961 doi: 10. 1038/nature04096
39. Hunter SJ, Valdes PJ, Haywood AM, Markwick PJ. Modelling Maastrichtian climate: investigating the role of geography, atmospheric CO_2 and vegetation. Clim Past. 2008; 4: 981–1019. doi: 10. 5194/cpd-4-981-2008
40. Weaver JC. The improbable endotherm: the energetics of the sauropod *Brachiosaurus*. Paleobiology. 1983; 9: 173–182.
41. Dunham AE, Overall KL, Porter WP, Forster CA. Implications of ecological energetics and biophysical and developmental constraints for life-history variation in dinosaurs. In: Farlow JO, editor. Paleobiology of the dinosaurs. Geological Society of America, Special Paper. 1989; 238: 1–19.
42. Seebacher F. Dinosaur body temperatures: the occurrence of endothermy and ectothermy. Paleobiology. 2003; 29: 105–122. doi: 10. 1666/0094-8373(2003)029<0105:dbttoo>2. 0. co;2
43. McNab BK. Resources and energetics determined dinosaur maximal size. Proc Natl Acad Sci USA. 2009; 106: 12184–12188. doi: 10. 1073/pnas. 0904000106. pmid:19581600
44. Sander PM, Christian A, Clauss M, Fechner R, Gee CT, et al. Biology of sauropod dinosaurs: the evolution of gigantism. Biol Rev. 2011; 86: 117–155. doi: 10. 1111/j. 1469-185X. 2010. 00137. x. pmid:21251189
45. Clarke A. Dinosaur Energetics: Setting the bounds on feasible physiologies and ecologies. Am Nat. 2013; 182: 283–297. doi: 10. 1086/671259. pmid:23933721
46. O’Connor MP, Dodson P. Biophysical constraints on the thermal ecology of dinosaurs. Paleobiology. 1999; 25: 341–368.
47. Gillooly JF, Allen AP, Charnov EL. Dinosaur fossils predict body temperatures: PLOS BIOL. 2006; 4: e248. pmid:16817695 doi: 10. 1371/journal. pbio. 0040248
48. Eagle RA, Tütken T, Martin TS, Tripati AK, Fricke HC, et al. Dinosaur body temperatures determined from isotopic (^{13}C-^{18}O) ordering in fossil biominerals. Science. 2011; 333: 443–445. doi: 10. 1126/science. 1206196. pmid:21700837
49. Barrick RE, Fisher AG, Showers WJ. Oxygen isotopes from turtle bone: applications for terrestrial paleoclimates? Palaios. 1999; 14: 186–191.
50. Pough FH, Janis CM, Heiser JB. Vertebrate Life, 6th ed. Upper Saddle River, NJ: Prentice Hall; 2002.
51. Kolodny Y, Luz B, Navon O. Oxygen isotope variations in phosphate of biogenic apatites, I. Fish bone apatite rechecking the rules of the game. Earth Planet Sci Lett. 1983; 64: 398–404 doi: 10. 1016/0012-821x(83)90100-0
52. Grimes ST, Mattey DP, Hooker JJ, Collinson ME. Paleogene paleoclimate reconstruction using oxygen isotopes from land and freshwater organisms: the use of multiple paleoproxies. Geochim Cosmochim Acta. 2003; 67: 4033–4047. doi: 10. 1016/s0016-7037(03)00173-x
53. Fricke HC, Rogers RR, Gates TA. Hadrosaurid migration: Inferences based on stable isotope comparisons among Late Cretaceous dinosaur localities. Paleobiology. 2009; 35: 270–288. doi: 10. 1666/08025. 1
54. Farquhar GD, Ehleringer JR, Hubick KT. Carbon isotopic discrimination and photosynthesis. Annu Rev Plant Biol. 1989; 40: 503–537. doi: 10. 1146/annurev. pp. 40. 060189. 002443
55. Ehleringer JR, Sage RF, Flanagan LB, Pearcy RW. Climate change and the evolution of C_4 photosynthesis. Trends Ecol Evol. 1991; 6: 95–99. doi: 10. 1016/0169-5347(91)90183-X. pmid:21232434
56. Hayes JM. Fractionation of carbon and hydrogen isotopes in biosynthetic processes. Rev Mineral Geochem. 2001; 43: 225–277. doi: 10. 2138/gsrmg. 43. 1. 225
57. Ekart DD, Cerling TE, Montañez IP, Tabor NJ. A 400 million year carbon isotope record of pedogenic carbonate: implications for paleoatmospheric carbon dioxide. Am J Sci. 1999; 299: 805–827. doi: 10. 2475/ajs. 299. 10. 805

58. Friedli H, Lotscher H, Oeschger H, Siegenthaler U, Stauver B. Ice core record of the $^{13}C/^{12}C$ ratio of atmospheric CO_2 in the past two centuries. Nature. 1986; 324: 237–238. doi: 10. 1038/324237a0

59. Marino BD, McElroy MB. Isotopic composition of atmospheric CO_2 inferred from carbon in C_4 plant cellulose. Nature. 1991; 349: 127–131. doi: 10. 1038/349127a0

60. Koch PL. Isotopic reconstruction of past continental environments. Annu Rev Earth Planet Sci. 1998; 26: 573–613. doi: 10. 1146/annurev. earth. 26. 1. 573

61. Cerling TE, Harris JM. Carbon isotope fractionation between diet and bioapatite in ungulate mammals and implications for ecological and paleoecological studies. Oecologia. 1999; 120: 347–363. doi: 10. 1007/ s004420050868

62. Passey BH, Robinson TF, Ayliffe LK, Cerling TE, Sponheimer M, et al. Carbon isotope fractionation between diet, breath CO_2, and bioapatite in different mammals. J Archaeol Sci. 2005; 32: 1459–1470. doi: 10. 1016/j. jas. 2005. 03. 015

63. Koch PL. Isotopic study of the biology of modern and fossil vertebrates. In: Michener R, Lajtha K, editors. Stable Isotopes in Ecology and Environmental Science, 2nd Edition. Boston: Blackwell Publishing; 2007. pp. 99–154.

64. Clementz MT, Fox-Dobbs K, Wheatley PV, Koch PL, Doak DF. Revisiting old bones: coupled carbon isotope analysis of bioapatite and collagen as an ecological and palaeoecological tool. Geol J. 2009; 44: 605–620. doi: 10. 1002/gj. 1173

65. Johnson BJ, Fogel M, Miller GH. Stable isotopes in modern ostrich eggshell: A calibration for paleoenvironmental applications in semi-arid regions of southern Africa. Geochim Cosmochim Acta. 1998; 62: 2451–2461. doi: 10. 1016/s0016-7037(98)00175-6

66. Angst D, Lécuyer C, Amiot R, Buffetaut E, Fourel F, et al. Isotopic and anatomical evidence of an herbivorous diet in the Early Tertiary giant bird *Gastornis*. Implications for the structure of Paleocene terrestrial ecosystems. Naturwissenschaften. 2014; 101: 313–322. doi: 10. 1007/s00114-014-1158-2. pmid:24563098

67. Ortega F, Sanz JL, Barroso-Barcenilla F, Cambra-Moo O, Escaso F, et al. El yacimiento de macrovertebrados fósiles del Cretácico Superior de "Lo Hueco" (Fuentes, Cuenca). Palaeontologica Nova, Publicaciones del Seminario de Paleontología de Zaragoza. 2008; 8: 119–131.

68. Cambra-Moo O, Barroso-Barcenilla F, Berreteaga A, Carenas A, Coruña F, et al. Preliminary taphonomic approach to "Lo Hueco" palaeontological site (Upper Cretaceous, Cuenca, Spain). Geobios. 2012; 45: 157–166. doi: 10. 1093/jncimonographs/lgs029. pmid:23271767

69. Gutiérrez G, Robles F. Consideraciones sobre la utilización del término "Garumniense" en la Cordillera Ibérica: Cuadernos de Geología Ibérica. 1979; 5: 127–214.

70. González-Acebrón L, Barroso-Barcenilla F, Cambra-Moo O, Carenas B, Segura M. Environmental significance of gypsum-bearing layers at the «Lo Hueco» palaeontological site (Upper Cretaceous, Cuenca, Spain): petrography, fluid inclusions and isotopic relations. Facies. 2014; 60: 755–771. doi: 10. 1007/s10347-014-0402-8

71. Díez Díaz V, Ortega F, Sanz J. Microwear patterns of the titanosaurian teeth from the Late Cretaceous of "Lo Hueco" (Cuenca, Spain). Program and Abstracts of the 73rd Meeting of the SVP. Journal of Vertebrate Paleontology. 2013; 116.

72. Knoll F, Witmer LM, Ridgely RC, Ortega F, Sanz JL. A new titanosaurian sauropod neurocranium from the Late Cretaceous of Spain. Program and Abstracts of the 73rd Meeting of the SVP. Journal of Vertebrate Paleontology. 2013; 155. doi: 10. 1371/journal. pone. 0054991

73. Knoll F, Ridgely RC, Ortega F, Sanz JL, Witmer LM. Neurocranial osteology and neuroanatomy of a Late Cretaceous titanosaurian sauropod from Spain (*Ampelosaurus*sp.). PLOS ONE. 2013; 8: e54991. doi: 10. 1371/journal. pone. 0054991. pmid:23355905

74. Narváez I, Ortega F, Brochu C, Escaso F. A new basal eusuchian crocodile from the Late Cretaceous of "Lo Hueco" (Cuenca, Spain). In: Torcida Fernández-Baldor F, Huerta P, editors. VI International Symposium about Dinosaur Paleontology and their Environment, abstract book; 2013. pp. 96–97.

75. Pérez-García A, Ortega F, Murelaga X. A new genus of Bothremydidae (Chelonii, Pleurodira) in the Cretaceous of southwestern Europe. Geobios. 2012; 45: 219–229. doi: 10. 1016/j. geobios. 2011. 03. 001

76. Pérez-García A, Ortega F, Murelaga X. Two synchronic and sympatric Bothremydidae taxa (Chelonii, Pleurodira) in the Late Cretaceous site of "Lo Hueco" (Cuenca, Spain). In: Morphology and Evolution of Turtles. Vertebrate Paleobiology and Paleoanthropology. Brinkman; 2013. pp. 251–259.

77. Bassett D, Macleod KG, Miller JF, Ethington RL. Oxygen isotopic composition of biogenic phosphate and the temperature of Early Ordovician seawater. Palaios. 2007; 22: 98–103.

78. Koch PL, Tuross N, Fogel ML. The effects of sample treatment and diagenesis on the isotopic integrity of carbonate in biogenic hydroxylapatite. J Archaeol Sci. 1997; 24: 417–429. doi: 10. 1006/jasc. 1996. 0126

79. Domingo L, López-Martínez N, Leng MJ, Grimes ST. The Paleocene–Eocene Thermal Maximum record in the organic matter of the Claret and Tendruy continental sections (South-central Pyrenees, Lleida, Spain). Earth Planet Sci Lett. 2009; 281: 226–237. doi: 10. 1016/j. epsl. 2009. 02. 025

80. Amiot R, Lécuyer C, Escarguel G, Billon-Bruyat JP, Buffetaut E, et al. Oxygen isotope fractionation between crocodilian phosphate and water: Palaeogeogr Palaeoclimatol Palaeoecol. 2007; 243: 412–420. doi: 10. 1016/j. palaeo. 2006. 08. 013

81. Coulson AB, Kohn MJ, Shirley MH, Joyce WG, Barrick RE. Phosphate–oxygen isotopes from marine turtle bones: Ecologic and paleoclimatic applications. Palaeogeogr Palaeoclimatol Palaeoecol. 2008; 264: 78–84. doi: 10. 1016/j. palaeo. 2008. 04. 008

82. Domingo L, Grimes ST, Soler-Gijón R, López-Martínez N. Analysis of the geochemical variability in lepisosteid scales from the Fontllonga-3 site (early Danian, Tremp Formation, South Central Pyrenees, Spain): implications for palaeoenvironmental studies. Palaeogeogr Palaeoclimatol Palaeoecol. 2009; 274: 204–214. doi: 10. 1016/j. palaeo. 2009. 02. 004

83. Coulson AB, Kohn MJ, Barrick RE. Isotopic evaluation of ocean circulation in the Late Cretaceous North American seaway. Nat Geosci. 2011; 4: 852–855. doi: 10. 1038/ngeo1312

84. Lécuyer C, Amiot R, Touzeau A, Trotter J. Calibration of the phosphate $\delta^{18}O$ thermometer with carbonate–water oxygen isotope fractionation equations. Chem Geol. 2013; 347: 217–226. doi: 10. 1016/j. chemgeo. 2013. 03. 008

85. Pucéat E, Joachimski MM, Bouilloux A, Monna F, Bonin A, et al. Revised phosphate–water fractionation equation reassessing paleotemperatures derived from biogenic apatite. Earth Planet Sci Lett. 2010; 298: 135–142. doi: 10. 1016/j. epsl. 2010. 07. 034

86. Kohn MJ, Cerling TE. Stable isotope compositions of biological apatite. Rev Mineral Geochem. 2002; 48: 455–488. doi: 10. 2138/rmg. 2002. 48. 12

87. Bryant JD, Koch PL, Froelich PN, Showers WJ, Genna BJ. Oxygen isotope partitioning between phosphate and carbonate in mammalian apatite. Geochim Cosmochim Acta. 1996; 60: 5145–5148. doi: 10. 1016/s0016-7037(96)00308-0

88. Iacumin P, Bocherens H, Mariotti A, Longinelli A. Oxygen isotope analyses of coexisting carbonate and phosphate in biogenic apatite: A way to monitor diagenetic alteration of bone phosphate? Earth Planet Sci Lett. 1996; 142: 1–6. doi: 10. 1016/0012-821x(96)00093-3

89. Fricke HC, Clyde WC, O'Neil JR, Gingerich PD. Evidence for rapid climate change in North America during the latest Paleocene thermal maximum: Oxygen isotope compositions of biogenic phosphate from the Bighorn Basin (Wyoming). Earth Planet Sci Lett. 1998; 160: 193–208. doi: 10. 1016/s0012-821x(98)00088-0

90. Tütken T, Vennemann TW, Janz H, Heizmann EPJ. Palaeoenvironment and palaeoclimate of the middle Miocene lake in the Steinheim basin, SW Germany: A reconstruction from C, O, and Sr isotopes of fossil remains. Palaeogeogr Palaeoclimatol Palaeoecol. 2006; 241: 457–491. doi: 10. 1016/j. palaeo. 2006. 04. 007

91. Domingo L, Cuevas-González J, Grimes ST, Hernández Fernández M, López-Martínez N. Multiproxy reconstruction of the paleoclimate and paleoenvironment of the middle Miocene Somosaguas site (Madrid, Spain) using herbivore tooth enamel. Palaeogeogr Palaeoclimatol Palaeoecol. 2009; 272: 53–68. doi: 10. 1016/j. palaeo. 2008. 11. 006

92. Hay WW, DeConto R, Wold CN, Wilson KM, Voigt S, et al. Alternative global Cretaceous paleogeography. In: Barrera E, Johnson C, editors. The Evolution of Cretaceous Ocean/Climate Systems. Geological Society of America Special Paper. 1999; 332: 1–47.

93. Pontzer H, Allen V, Hutchinson JR. Biomechanics of running indicates endothermy in bipedal dinosaurs. PLOS ONE. 2009; 4: e7783. doi: 10. 1371/journal. pone. 0007783. pmid:19911059

94. Rozanski K, Araguás-Araguás L, Gonfiantini R. Isotopic patterns in modern global precipitation. In: Swart PK, Lohmann KC, McKenzie J, Savin S, editors. Climate Change in Continental Isotopic Records. Geophysical Monograph Series. 1993; 78: 1–36.

95. Pucéat E, Lécuyer C, Sheppard SMF, Dromart G, Reboulet S, et al. Thermal evolution of Cretaceous Tethyan marine waters inferred from oxygen isotope composition of fish tooth enamels. Paleoceanography. 2003; 18: 1029. doi: 10. 1029/2002pa000823

96. Pucéat E, Lécuyer C, Donnadieu Y, Naveau P, Cappetta H, et al. Fish tooth $\delta^{18}O$ revising Late Cretaceous meridional upper ocean water temperature gradients. Geology. 2007; 35: 107–110. doi: 10. 1130/g23103a. 1

97. Greenwood DR, Wing SL. Eocene continental climates and latitudinal temperature gradients. Geology. 1995; 23: 1044–1048. doi: 10. 1130/0091-7613(1995)023<1044:eccalt>2. 3. co;2

98. Hummel J, Clauss M. Sauropod feeding and digestive physiology. In: Klein N, Remes K, Sander M, editors. Biology of the Sauropod Dinosaurs: Understanding the Life of Giants. Bloomington, IN: Indiana University Press; 2011. pp. 11–33.

99. Calvo J. Jaw mechanics in sauropod dinosaurs. Gaia. 1994; 10: 183–193.

100. Peyrot D, Barroso-Barcenilla F, Cambra-Moo O. Paleoecology of the late Campanian/early Maastrichtian Fossil-Lagerstätte of "Lo Hueco" (Cuenca, Spain): Palynological insights. Palaeogeogr Palaeoclimatol Palaeoecol. 2013; 387: 27–39. doi: 10. 1016/j. palaeo. 2013. 07. 005

101. Fox-Dobbs K, Leonard JA, Koch PL. Pleistocene megafauna from Eastern Beringia: paleoecological and paleoenvironmental interpretations of stable carbon and nitrogen isotope and radiocarbon records. Palaeogeogr Palaeoclimatol Palaeoecol. 2008; 261, 30–46. doi: 10. 1016/j. palaeo. 2007. 12. 011

102. Yeakel JD, Guimarães PR Jr, Novak M, Fox-Dobbs K, Koch PL. Probabilistic patterns of interaction: the effects of link-strength variability on food web structure. J R Soc Interface. 2012; 9: 3219–3228. doi: 10. 1098/rsif. 2012. 0481. pmid:22832361

103. Yeakel JD, Guimarães PR Jr, Bocherens H, Koch PL. The impact of climate change on the structure of Pleistocene food webs across the mammoth steppe. Proc R Soc B. 2013; 280: 20130239. doi: 10. 1098/rspb. 2013. 0239. pmid:23658198

104. Domingo MS, Domingo L, Badgley C, Sanisidro O, Morales J. Resource partitioning among top predators in

a Miocene food web. Proc R Soc B. 2013; 20122138. doi: 10. 1098/rspb. 2012. 2138

105. Vennemann TW, Hegner E, Cliff G, Venz GW. Isotopic composition of recent shark teeth as a proxy for environmental conditions. Geochim Cosmochim Acta. 2001; 65: 1583–1599. doi: 10. 1016/s0016-7037(00)00629-3

106. Biasatti DM. Stable carbon isotopic profiles of sea turtle humeri: implications for ecology and physiology. Palaeogeogr Palaeoclimatol Palaeoecol. 2004; 206: 203–216. doi: 10. 1016/j. palaeo. 2004. 01. 004

107. Wheatley PV, Peckham H, Newsome SD, Koch PL. Estimating marine resource use by the American crocodile *Crocodylus acutus* in southern Florida, USA. Mar Ecol Prog Ser. 2012; 447: 211–229. doi: 10. 3354/meps09503

108. Jackson K, Butler DG, Brooks DR. Habitat and phylogeny influence salinity discrimination in crocodilians: implications for osmoregulatory physiology and historical biogeography. Biol J Linn Soc Lond. 1996; 58: 371–383. doi: 10. 1111/j. 1095-8312. 1996. tb01441. x

109. Clementz MT, Koch PL. Differentiating aquatic mammal habitat and foraging ecology with stable isotopes in tooth enamel. Oecologia. 2001; 129: 461–472. doi: 10.1007/s004420100745

CHAPTER 8

THE PALEOECOLOGY, HABITATS, AND STRATIGRAPHIC RANGE OF THE ENIGMATIC CRETACEOUS BRACHIOPOD PEREGRINELLA

Steffen Kiel, Johannes Glodny, Daniel Birgel, Luc G. Bulot, Kathleen A. Campbell, Christian Gaillard, Roberto Graziano, Andrzej Kaim, Iuliana Lazaˇ r, Michael R. Sandy, Jörn Peckmann

INTRODUCTION

The discovery of highly specialized animal communities around hydrothermal vents and methane seeps in the late 1970s and 1980s [1], [2] and the question of their origin spurred the search for fossil examples of these ecosystems [3]. The major players of the modern, mollusk-dominated fauna, such as vesicomyid bivalves and bathymodiolin mussels, can be traced into the Eocene, and minor bivalve groups and various gastropods range into the Cretaceous [4]–[8]. In contrast, many Paleozoic and Mesozoic seep communities were dominated by organisms that are virtually absent from modern vents and seeps: rhynchonellid brachiopods [9]. Perhaps the most notable of these is the Early Cretaceous genus Peregrinella (Figure 1), the largest Mesozoic rhynchonellid brachiopod. Peregrinella long puzzled paleontologists because of its large size, its mass occurrence in isolated lenses and its widespread, yet disjunct, distribution [10]–[16].

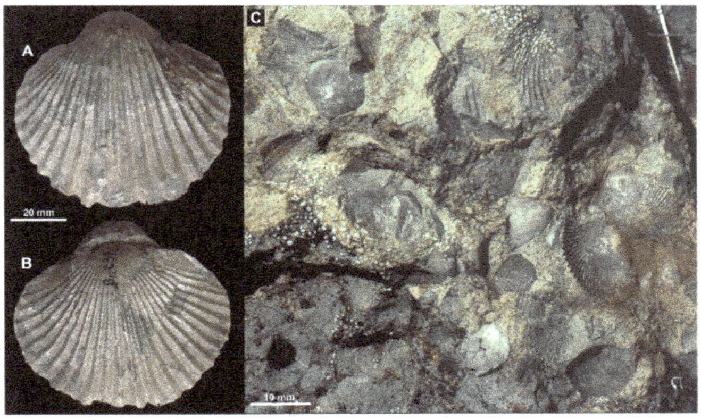

Figure 1. Examples of Peregrinella.

A, B. Peregrinella multicarinata (Lamarck, 1819) from Rottier, southeastern France (EM 36190). C. Field image showing the mass occurrence of Peregrinella whitneyi (Gabb, 1866) at the Wilbur Springs seep site in California, USA.

Early in the 1980s various Peregrinella occurrences in the Vocontian basin in southern France were discussed as analogs to modern hydrothermal vent communities [17], [18]. But it was only after the discovery of methane-seep communities in the Gulf of Mexico that Peregrinella-bearing deposits were identified as ancient methane seeps [19], based on stable carbon isotope analyses of a Californian example. These authors listed most occurrences of Peregrinella and noted that all are found as isolated carbonate lenses within deep-water sediments

Citation: Kiel S, Glodny J, Birgel D, Bulot LG, Campbell KA, Gaillard C, et al. (2014) The Paleoecology, Habitats, and Stratigraphic Range of the Enigmatic Cretaceous Brachiopod Peregrinella. PLoS ONE 9(10): e109260. doi:10.1371/journal.pone.0109260.

similar to the Californian example. Therefore they suggested that Peregrinella lived exclusively at ancient methane seeps like the major taxa at vents and seeps today [19]. However, brachiopods are rare at modern vents and seeps and are only found in the marginal areas of these habitats. Thus, there are no modern analogs to the large and superabundant Peregrinella at fossil methane seeps (Figure 1C) and its paleoecologic and evolutionary traits remain poorly understood. Campbell and Bottjer [19] listed 18 occurrences world-wide, another one was subsequently reported from Alaska [20], and a summary of the biostratigraphic evidence for Peregrinella suggested a Berriasian to Hauterivian range for the genus [21].

Here we investigate strontium-isotope stratigraphy, petrography, stable carbon and oxygen isotopes, and the associated fauna of samples from 11 Peregrinella occurrences worldwide (Figure 2) that span its (suggested) stratigraphic range, together with a review of the available literature, to provide new insights into the nature of this enigmatic brachiopod. Peregrinellabelongs to the superfamily Dimerelloidea, a clade that includes several seep-inhabiting or potentially seep-restricted brachiopods [22]–[27]: therefore, our results may have bearings on the role of brachiopods at vents and seeps throughout most of the Phanerozoic [9].

Figure 2. Distribution of the Peregrinella **occurrences discussed herein, plotted on a paleogeographic map of the late Berriasian (140 m.y.a., [98]).**

Gray stars indicate occurrences for which new data are available: 1, Bonanaza Creek, Alaska; 2, Rice Valley and Wilbur Springs, California; 3, Incoronata, Italy; 4, Châtillon-en-Dois, Curnier and Rottier, southern France; 5, Musenalp, Switzerland; 6, Raciborsko, Poland; 7, Zizin Valley, Romania; 8, Planerskoje, Crimean peninsula. White stars indicate occurrences known from the literature: 9, Guanajuato seamount, Mexico; 10, Bohrung Werle, northern Germany; 11, Kuban, Russia; 12, Xainza County, Tibet.

MATERIAL

Most localities sampled for this work have been described in detail elsewhere; here we just provide essential information and references in which further information can be found. The material is deposited in the Geowissenschaftliches Museum, Georg-August-Universität Göttingen, Germany (GZG), Institute of Paleobiology at the Polish Academy of Sciences, Warsaw, Poland (ZPAL), the collection of the Université Claude Bernard, Lyon-1, France (FSL) and the collection of the Ecole des Mines (EM) that is also housed at the Université Lyon-1, the University of Bucharest, Laboratory of Palaeontology, Romania (LPB), and the Smithsonian Natural History Museum, Washington, DC, USA (USNM).

Bonanza Creek (Alaska)

Peregrinella chisania shells are embedded in a carbonate matrix, a rather unusual lithology for the Chisana Formation that is composed mainly of volcanic rocks and mudstones and is considered as Valanginian-Hauterivian in age [20], [28]. Two small specimens (USNM 487761 and 603602) were available for thin sectioning and isotope analyses (C, O, Sr).

Châtillon-en-Dois (France)

This locality in the Vocontian Basin is frequently mentioned in the literature on Peregrinella and was considered as Hauterivian in age [13], [19], [29], but the original locality has disappeared under urban development. A well-preserved shell was available for isotope analyses (C, O, Sr) from the collection of the Université Claude Bernard, Lyon-1 (FSL 425076).

Curnier (France)

A limestone block of about 1 m^3 with Peregrinella multicarinata was found as float in a dry creek near Curnier in southeastern France (coordinates: 44°23′23.4″N, 05°15′17.4″E). The associated macrofauna consists of the lucinid bivalve Tehamatea vocontiana, the gastropod Humptulipsia macsotayi [30], and various ammonites. Samples were collected by SK and LGB in 2010 and the material was used for thin sectioning, isotope analyses (C, O, Sr), and biomarker analysis (GZG. INV.82725-29). This occurrence was considered as Hauterivian in age [19].

Incoronata (Italy)

Abundant material was collected in 2011 by SK and RG at the base of the Maiolica Formation, which onlaps the flank of the Apulia carbonate platform near the village of Mattinata on the Gargano promontory in southeastern Italy (coordinates: 41°43'25.8"N, 16°02'29.2"E), and was used for thin sectioning and isotope analyses (C, O, Sr; GZG.INV.82745, 46, 54). The locality and its geologic setting have been described elsewhere and the age of the Peregrinella-bearing sediments was considered as latest Valanginian-Early Hauterivian (NC4a Subzone) [31]–[33].

Musenalp (Switzerland)

A small block (30×20×20 cm) with abundant Peregrinella subsilvana was found about 10 m NE of the Schwändi hut (coordinates 46°56'26.4"N, 8°27'13.5"E) by SK in 2010, and was used for thin sectioning, isotope analyses (C, O, Sr), and biomarker analysis (GZG.INV.82730-38). The scattered Peregrinella-bearing blocks on the small and steep pasture around the Schwändi hut are from a nappe of the Préalpes Médianes and are derived from a post-glacial landslide [14]. This site was previously considered as the oldest occurrence of Peregrinella, being possibly of late Berriasian age [34].

Planerskoje (Crimean peninsula)

An isolated carbonate lens with Peregrinella multicarinata was found within well-bedded mudstone of the Planerskoje section in the southeastern Crimean peninsula (coordinates: 44°58'08.1"N, 35°12'01.5"E), considered as Hauterivian in age [35], and was sampled by SK in 2007. The petrography, stable isotopes, biomarkers, and fauna of this locality have been described in detail elsewhere [35]–[37]; here we investigate oxygen and strontium isotopes of Peregrinella shells (GZG. INV.82739-44).

Raciborsko (Poland)

Isolated, pyrite-rich carbonate blocks with Peregrinella multicarinata occur within dark, marly shales of the Grodziszcze beds in the Silesian unit in the western Carpathians, considered as being of 'middle Neocomian' age [13]. Despite various efforts, this locality has not been re-located in recent years but several small specimens of Biernat's collection (ZPAL Bp.III) at the Institute of Paleobiology at the Polish Academy of Sciences, Warsaw, were available for thin sectioning and isotope analyses (C, O, Sr).

Rice Valley (California)

Isolated limestone lenses with Peregrinella whitneyi occur within the Rice Valley outlier of the Great Valley Group in Lake County, California, USA [38]; approximate coordinates: 39°21′02″N, 122°52′01″W). A hand sample collected by KAC was used for thin sectioning, and isotope analyses (C, O, Sr) of Peregrinella shells (GZG.INV.82747-52).

Rottier (France)

Peregrinella-bearing limestone crops out near the village of Rottier in southeastern France [29]and were sampled by SK and CG in 2006 and by SK and LGB in 2010, but only a few Peregrinella specimens were found. Additional Peregrinella-rich material for thin sectioning and isotope analyses (C, O, Sr) was selected from the collection of the Université Claude Bernard, Lyon-1 (FSL 425077-79).

Wilbur Springs (California)

This is a large, isolated carbonate lens with Peregrinella whitneyi enclosed in serpentinite of the Great Valley Group. The petrography, stable isotopes, biomarkers and fauna of this locality have been described in detail elsewhere [39]–[41]. Its age was considered as Valanginian in the older literature but more recently as Hauterivian [19]. A hand sample collected by KAC was used for thin sectioning and for isotope analyses (C, O, Sr) of Peregrinella shells (GZG.INV.82753, 82755, 82756); four specimens from the USNM collection were used for stable C and O isotope analyses of Peregrinella shells (USNM 603596-603599).

Zizin Valley (Romania)

Shells of Peregrinella multicarinata are found here both within isolated carbonate blocks as well as scattered through the surrounding flysch deposits of the Sinaia Formation, which is reportedly of late Hauterivian to early Barremian age [12], [42]. The Peregrinella-bearing carbonates were identified as seep deposits using petrographic, stable isotope, and biomarker investigations [42]. Shells from both types of occurrence were used for isotope analyses (C, O, Sr).

Two shells from active methane seeps in the northern Gulf of Mexico were available for comparative Sr isotope work, one of the bathymodiolin mussel "Bathymodiolus" childressicollected at brine pool NR-1 (site GC233) in 640 m depth [43], and one of a terebratulid brachiopod, Ecnomiosa gerda? Cooper, from Garden Banks block 647 in about 950–1000 m depth [44]. Specimens of a putative Peregrinella from a drill core in northeastern Germany [45]were also investigated, but their affinity to Peregrinella is questionable.

METHODS

Thin sections of ca. 60 micrometer thickness were produced from the available material and the surfaces of the counterparts were polished to facilitate the selection of sampling sites for isotope analyses. Samples for stable carbon and oxygen isotope analyses were extracted either from the counterparts of the thin sections or from the brachiopod shells using a hand-held microdrill. Carbonate powders were reacted with 100% phosphoric acid at 75°C using a Finnigan Kiel IV Carbonate Device attached to a Finnigan DELTA V PLUS mass spectrometer. All values are reported in per mil relative to the PDB standard by assigning a $\delta^{13}C$ value of +1.95‰ and a $\delta^{18}O$ value of −2.20‰ to NBS19. Reproducibility was checked by replicate analysis of laboratory standards and is better than ±0.05‰. Paleotemperatures were calculated from the $\delta^{18}O$ values of Peregrinella shells using the formula provided by [46] and assuming an ice-free ocean with a $\delta^{18}O_{seawater}$ value of −1.2. [47] and ignoring artifacts that were possibly introduced due to isotope exchange during late diagenesis and rock alteration.

For strontium isotopes we analyzed two different Peregrinella shells (where available) from each locality listed in the 'Material' section to check for potential inhomogeneity, distributed alteration or recrystallization of the original calcite. The microstructure of shell fragments was examined optically and with scanning electron microscopy (LEO 1530 SEM at 3.8 KV; specimens coated with 14 nm of platinum: Figure 3) to check for possible diagenetic alteration or intergrowth with siliciclastic material. Fragments devoid of any indications of recrystallization or contamination were selected and cleaned in distilled water and in pure ethanol. Wherever feasible we also cleaned the fragments by rinsing them in 1 N HCl for about 5 seconds to remove any secondary calcareous material and contaminants [48]–[50]. After another rinse in ultrapure water the shell fragments were dissolved in 2.5 N HCl. Strontium was isolated and recovered from the solutes by ion exchange using Dowex AG-50 cation exchange resin.

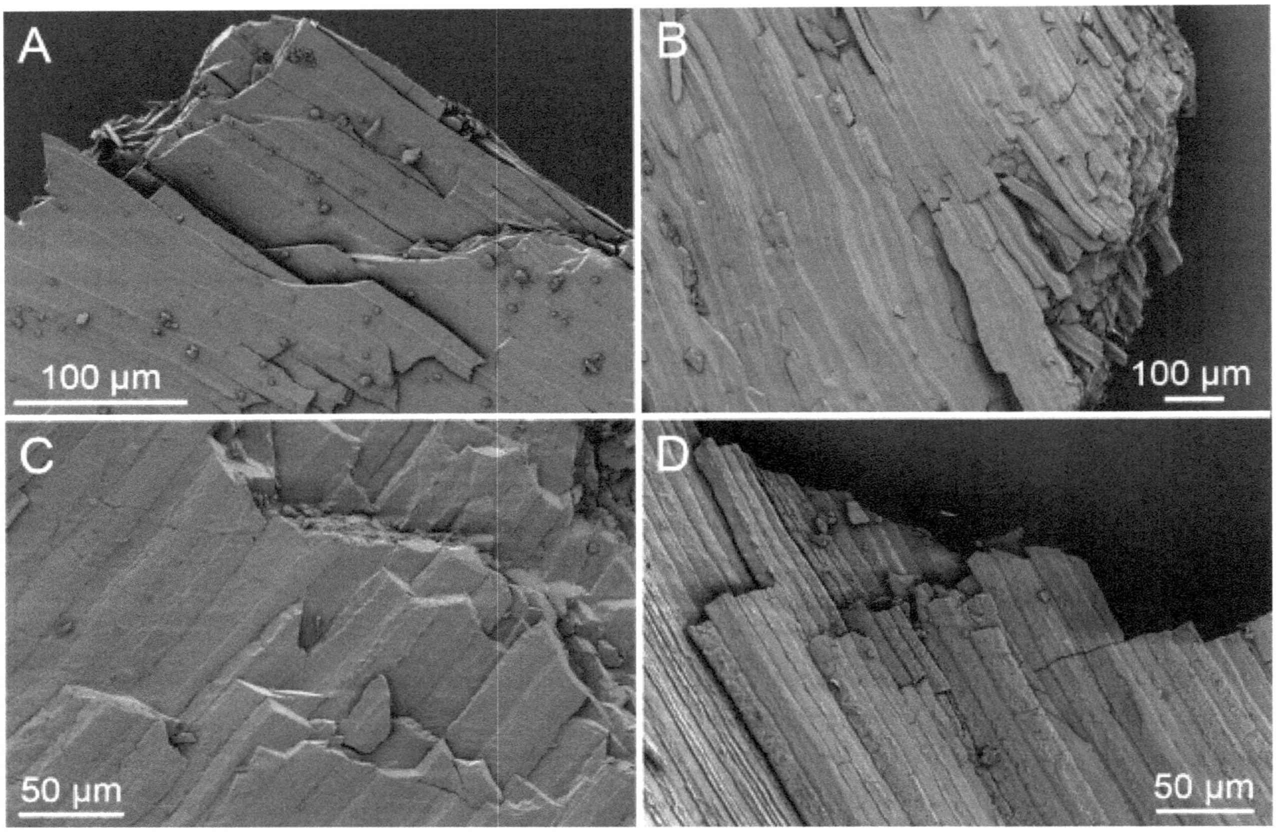

Figure 3. SEM images of well preserved Peregrinella **shells with unaltered shell microstructure.**

The examples are from: A: Incoronata (GZG.INV.82745); B: Planerskoje (GZG.INV.82739); C: Musenalp (GZG. INV.82737); D: Curnier (FSL 425078).

Mass spectrometry analyses of Sr isotopic compositions were carried out at GFZ Potsdam, using a Thermo-Finnigan Triton TIMS (thermal ionization mass spectrometry) instrument. Strontium was analyzed in dynamic multicollection mode. For the duration of the study, the mean value obtained for $^{87}Sr/^{86}Sr$ of the NBS standard SRM 987 (now NIST 987) was 0.710261±0.000006 (n=10, 2σ SD). The $^{87}Sr/^{86}Sr$ value for SRM 987 used for construction of the LOWESS dataset for conversion of Sr isotope data into ages is 0.710248 [51], which requires application of a normalization factor of 0.999982 for our Sr isotope ratio data. Calculation of ages and age errors was performed using the 'Look-Up Table Version 5: 04/13' [52]. The error intervals calculated for our Sr isotope stratigraphy age data (see Table 1) are based on typical 2σ absolute uncertainties for single-analysis, first-measurement $^{87}Sr/^{86}Sr$ values of ±0.000020. This uncertainty is estimated from the overall analytical range of values obtained for the NBS standard SRM 987. The only exception is sample PS1804. Here we used the measured $2\sigma_m$ uncertainty since it was, because of small sample size, slightly higher than the standard-derived uncertainty. Analytical blanks for Sr are regarded negligible compared to the amounts of sample Sr in each analysis; hence, no blank correction was applied. To check for possible radiogenic in-growth of ^{87}Sr after shell deposition, we analyzed a few samples for Rb concentrations using a Thermo-Neptune multicollection ICP-MS instrument (GFZ Potsdam). Rubidium concentrations in the carbonate material are very low, in the range of less than 30 ppb. With typical Sr concentrations of at least several hundred ppm in both recent and fossil shell material (compilation of data in [53], [54]), Rb/Sr ratios are low enough to rule out any significant post-depositional radiogenic in-growth of ^{87}Sr since shell growth.

Table 1. Sr-isotope data and possible stratigraphic ages derived from the LOWESS 5.0 curve [52].

Sample	$^{87}Sr/^{86}Sr \pm 2\sigma_m$ meas.	$^{87}Sr/^{86}Sr$ norm.	age 1 ($\pm 2\sigma$) (in Ma)	age 2 ($\pm 2\sigma$) (in Ma)	age 3 ($\pm 2\sigma$) (in Ma)
Bonanza Creek, Alaska (USNM 603602)	0.707416+/−0.000028	0.707403	**133.20** (+1.85/−0.95)	**125.40** (+1.10/−1.25)	**108.90** (+1.90/−>6.65
Bonanza Creek, Alaska (USNM 487761)	0.707437+/−0.000017	0.707424	**132.60** (+0.70/−0.60)	**126.10** (+0.80/−0.85)	(**104.05** or **96.65**)
Chatillon, France (FSL 425076)	0.707414+/−0.000005	0.707401	**133.25** (+1.10/−0.65)	**125.30** (+0.90/−0.90)	**109.10** (+1.50/−5.70)
Chatillon, France (FSL 425076)	0.707447+/−0.000005	0.707434	**132.40** (+0.60/−0.80)	**126.40** (+0.80/−0.75)	**approx. 100**
Curnier, France (GZG.INV.82729)	0.707440+/−0.000016	0.707427	**132.55** (+0.65/−0.65)	**126.20** (+0.80/−0.80)	(**103.30** or **97.00**)
Incoronata, Italy (GZG.INV.82745)	0.707365+/−0.000006	0.707352	**136.70** (+1.20/−2.40)	**123.30** (+1.00/−1.40)	**111.25** (+0.80/−0.95)
Incoronata, Italy (GZG.INV.82746)	0.707363+/−0.000006	0.707350	**136.80** (+1.20/−2.20)	**123.20** (+1.05/−1.40)	**111.30** (+0.80/−0.95)
Musenalp, Switzerland (GZG.INV.82737)	0.707271+/−0.000007	0.707258	**141.15** (+1.45/−1.20)	**117.50** (+1.75/−1.15)	**113.50** (+0.65/−0.55)
Musenalp, Switzerland (GZG.INV.82738)	0.707270+/−0.000004	0.707257	**141.20** (+1.45/−1.20)	**117.45** (+1.75/−1.15)	**113.50** (+0.65/−0.55)
Planerskoje, Crimea (GZG.INV.82739)	0.707298+/−0.000005	0.707285	**139.75** (+1.25/−1.05)	**119.10** (+1.60/−1.85)	**113.00** (+0.60/−0.65)
Planerskoje, Crimea (GZG.INV.82740)	0.707293+/−0.000007	0.707280	**140.00** (+1.30/−1.05)	**118.75** (+1.65/−1.65)	**113.10** (+0.60/−0.65)
Raciborsko, Poland (ZPAL Bp.III)	0.707399+/−0.000007	0.707386	**133.70** (+2.25/−0.70)	**124.70** (+1.00/−0.90)	**110.05** (+1.05/−4.70)
Rice Valley, California (GZG.INV.82747)	0.707387+/−0.000007	0.707374	**134.65** (+2.15/−1.30)	**124.25** (+0.95/−1.00)	**110.55** (+0.90/−1.55)
Rottier, France (FSL 425077)	0.707454+/−0.000008	0.707441	**132.20** (+0.60/−0.85)	**126.65** (+0.85/−0.75)	
Wilbur Springs, California (GZG.INV. 82755)	0.707389+/−0.000013	0.707376	**134.40** (+2.30/−1.10)	**124.30** (+0.95/−0.95)	**110.45** (+0.95/−1.70)
Wilbur Springs, California (GZG.INV. 82756)	0.707417+/−0.000009	0.707404	**133.20** (+0.95/−0.65)	**125.45** (+0.80/−0.95)	**108.80** (+1.70/−6.00)
Zizin Valley, Romania (LPB III br 387)	0.707392+/−0.000006	0.707379	**134.15** (+2.35/−0.95)	**124.45** (+0.95/−0.95)	**110.35** (+0.95/−2.25)
Zizin Valley, Romania (LPB III br 385)	0.707399+/−0.000005	0.707386	**133.70** (+2.25/−0.70)	**124.70** (+1.00/−0.90)	**110.05** (+1.05/−4.70)
"Bathymodiolus" childressi, Gulf of Mexico (GZG.INV.82764)	0.709189+/−0.000004	0.709176	**0.00** (+0.65/−0.00)		
Ecnomiosa gerda?, Gulf of Mexico (GZG.INV.82765)	0.709170+/−0.000008	0.709157	**0.55** (+0.45/−0.55)		
Ecnomiosa gerda?, Gulf of Mexico	0.709181+/−0.000004	0.709168	**0.24** (+0.58/−0.24)		

For biomarker analysis, limestone samples from Curnier and Musenalp were prepared and decalcified after a method described previously [40]. After saponification with 6% potassium hydroxide in methanol, the samples were extracted with a microwave extraction system (CEM Discovery) at 80°C and a maximum of 250 W with a mixture of dichloromethane:methanol (3:1). The resulting total lipid extracts were pre-separated into an n-hexane soluble and dichloromethane soluble fraction. For further separation of the n-hexane fraction by column chromatography, the samples were separated into four fractions of increasing polarity [55]. Only the hydrocarbon fractions were found to contain indigenous compounds. The polar fractions did not yield genuine lipid biomarker patterns, and were found to be severely affected by thermal maturation and biodegradation. Hydrocarbon fractions were analyzed by coupled gas chromatography- mass spectrometry (GC-MS) with an Agilent 7890 A GC system coupled to an Agilent 5975 C inert MSD mass spectrometer at the Department of Geodynamics and Sedimentology, University of Vienna. The GC-MS was equipped with a 30 m HP-5 MS UI fused silica capillary column (0.25 mm i.d., 0.25 µm film thickness). The carrier gas was helium. The GC temperature program was as follows: 60°C (1 min); from 60°C to 150°C at 10°C/min; from 150°C to 320°C at 4°C/min, 25 min isothermal. Identification of compounds was based on retention times and published mass spectral data. Compound-specific carbon isotope analysis was performed with a Thermo Fisher Trace GC Ultra connected via a Thermo Fisher GC Isolink interface to a Thermo Fisher Delta V Advantage spectrometer at the Department of Terrestrial Ecosystem Research, University of Vienna. Conditions for gas chromatography were the same as described above. Stable carbon isotopic compositions are given as δ values in per mil relative to the PDB standard. Each measurement was calibrated using several pulses of carbon dioxide with known isotopic composition at the beginning and the end of the run. Instrument precision was checked with a mixture of n-alkanes (C_{14} to C_{38}) of known isotopic composition. Analytical standard deviation was below 0.9‰.

Statistic analyses of faunal data were conducted using the software package PAST v. 2.13 [56].

Sr-Isotope Stratigraphy of Peregrinella Occurrences

Using strontium isotope ratios to correlate the Sr isotopic signature of marine fossils with the global Sr seawater evolution curve [52], we dated the available Peregrinella occurrences. As outlined above, the sample preparation procedure was such that alteration or contamination of the analyzed samples is highly unlikely, so that the $^{87}Sr/^{86}Sr$ ratios obtained herein can reliably be converted into Sr isotope stratigraphic ages. Reliability of the calculated ages is also supported by the observation that pairs of two independent analyses of different shells from one locality in all cases revealed consistent results (Table 1).

Because all known occurrences of Peregrinella are Early Cretaceous (145.0 to 100.5 Ma) in age, we focused on the Early Cretaceous Sr seawater evolution curve. For the Early Cretaceous the Sr seawater evolution curve displays two maxima and one marked minimum (Figure 4), therefore, a specific Sr isotopic composition may correspond to different stratigraphic ages, and age determinations may be ambiguous. To resolve this ambiguity, we used available biostratigraphic and geologic information from the individual Peregrinellaoccurrences to constrain the definite Sr isotope stratigraphic ages. Results, including Sr isotopic compositions of specific samples and their corresponding ages, are presented in Table 1. For comparison, we also present Sr isotopic data on Recent shells of a terebratulid brachiopod and of the mussel

"Bathymodiolus" childressi from methane seeps in the Gulf of Mexico, which are, as expected, indistinguishable from today's Sr seawater composition ($^{87}Sr/^{86}Sr$ =0.709175, [52]). This is important because the Sr-isotope signature of the seeping fluids at these sites often differs substantially from that of seawater [57], [58] and it shows that mollusks and brachiopods incorporate the seawater Sr isotope signature in their shells, not that of the seeping fluids. Therefore we assume that the Sr-isotope signature in the shells of Peregrinella reflects that of the seawater in which they lived.

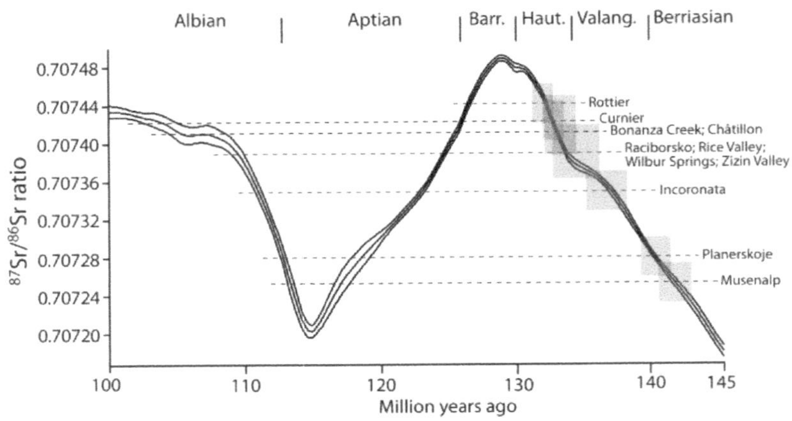

Figure 4. Strontium isotope curve for the Early Cretaceous and its upper and lower confidence interval [52].

Dashed lines indicate our measurements on Peregrinella shells; shaded areas indicate the error margins (vertical: Sr isotope ratio; horizontal: geologic age) of those ages that fall within the biostratigraphically possible age range.

The Sr-isotope ages of two of the fossil sites that are biostratigraphically well-dated are in agreement with biostratigraphy. In particular, the occurrence at the Swiss Musenalp was dated as late Berriasian using tintinnids (calpionellids) [34], consistent with its Sr-isotope age [141.20 (+1.45/−1.20) Ma]; the occurrence at Rottier, France, was dated as early Hauterivian (Jeannoti-zone) using ammonites [29], consistent with its Sr-isotope age [132.20 (+0.60/−0.85) Ma] (Table 1).

Of interest is the Sr-isotope age of the Romanian occurrence in the Sinaia Formation. Peregrinella shells from this formation were transported, deriving from turbidites or isolated, allochthonous blocks of limestone enclosed in deep-water strata. These Peregrinella-bearing rocks were considered the geologically youngest occurrence, ranging into the early Barremian. This inference, however, was based on foraminifera from the surrounding mudstone of the Sinaia Formation, not from the Peregrinella limestone itself. Both investigated shells from this site, one from a carbonate, the other from a siltstone, yielded Sr-isotope values of 0.707379 to 0.707386, indicating a latest Valanginian to earliest Hauterivian age [133.70 (+2.25/−0.70) to 134.15 (+2.35/−0.95) Ma]. Therefore, it seems likely that the Romanian Peregrinella shells were eroded from their original latest Valanginian to earliest Hauterivian sediments and reworked into the late Hauterivian-early Barremian Sinaia Formation. The two younger ages that could also be derived from the measured Sr-isotope ratios (123 Ma, early Aptian, and 109 Ma, mid-Albian; see Table 1) can be ruled out because they are younger than the biostratigraphic age of the Sinaia Formation.

The Californian Peregrinella whitneyi at Wilbur Springs was considered as Valanginian in the earlier literature [13], [21], [59], [60], but more recently a Hauterivian age was suggested due to the position of the locality just above the last occurrence of the index fossil Buchia pacifica [19]. Our Sr isotope data indicate a latest Valanginian to earliest Hauterivian age [133.20 (+0.95/−0.65) to 134.40 (+2.30/−1.10) Ma].

A Hauterivian age was suggested for the occurrence in Rice Valley, California, based on the presence of Peregrinella whitneyi and a mollusk assemblage that is now known from many Early Cretaceous seep deposits in California [38]. The Sr-isotope age of this locality [134.65 (+2.15/−1.30) Ma (Table 1)] indicates a latest Valanginian age, similar to the Wilbur Springs site mentioned above. The sediments directly overlying the Peregrinella-bearing unit at Rice Valley are dated as Albian-Cenomanian based on palynomorphs [38], thus ruling out the equally consistent Albian Sr-isotope age (Table 1). However, we cannot entirely discard the equally consistent early Aptian Sr-isotope age, but the Hauterivian age is more likely because two of the mollusk species present in the Rice Valley seep carbonate (Paskentana paskentaensis and Retiskenea? tuberculata) are known from other Valanginian-Hauterivian seep deposits, but an Aptian seep deposit from this area (Cold Fork of Cottonwood Creek) contains a different set of taxa [41], [61].

The Polish Peregrinella site at Raciborsko is situated roughly at the boundary between the Grodziszcze and Wierzowice (Verovice) shales in the Silesian Nappe and was considered as 'Neocomian' [13]. The age of this boundary is diachronous, being oldest (Hauterivian) in the west and becoming younger (as young as Aptian) in the east [62]. Raciborsko is located in the western part of the Silesian Nappe, and our Sr isotope data indicate a basal Hauterivian age for this locality [133.70 (+2.25/−0.70) Ma], consistent with its geographic location.

The Alaskan species Peregrinella chisania was previously considered as Valanginian-Hauterivian [20]; the Sr isotope data indicate an early Hauterivian age [133.20 (+1.85/−0.95) to 132.60 (+0.70/−0.60) Ma]. Because the Chisana Formation, from which the species name P. chisania is derived, hosts Valanginian to Barremian bivalves [28], the equally possible younger (early Aptian and Albian, see Table 1) Sr-isotope ages can be ruled out.

In summary, our Sr-isotope stratigraphic investigation confirms the so far oldest and youngest biostratigraphically well-dated occurrences as late Berriasian and early Hauterivian, respectively. It also shows that some occurrences of Peregrinella are slightly older than previously considered, but still within the range late Berriasian–late Hauterivian. Furthermore, it is evident that Peregrinella shells and even Peregrinella–bearing carbonate blocks can be reworked from their original sediment into younger sediments. Finally, we can refine ages of previously poorly dated occurrences (those dated as Valanginian-Hauterivian or 'Neocomian'), and show that they are confined within the range late Berriasian–late Hauterivian. An occurrence of Peregrinella possibly as young as Albian was reported from sediments of the Guanajuato seamount, central Mexico, an age that was based on co-occurring, reworked fossils of 'Neocomian' to Albian age [63]. However, considering that all specimens investigated here are of 'Neocomian' age and that the Romanian example documented here indicates that Peregrinella can be reworked into younger sediments, it seems likely that also the Mexican occurrence falls within the stratigraphic interval of the other localities, rather than being of Albian age. Thus, Peregrinella originated in the late Berriasian and most likely became extinct at the end of the early Hauterivian, giving it a geologic range of 9.0 (+1.45/−0.85) million years according to the Gradstein et al. 2012 time scale [64].

The Habitats of Peregrinella

To date, three Peregrinella-bearing limestones have unequivocally been identified as ancient methane-seep deposits using a combination of petrographic, stable isotope, and biomarker analyses [19], [35], [37], [40], [42]. Appropriate material to carry out at least some of these analyses was available from a further seven localities: Bonanza Creek, Curnier, Incoronata, Musenalp, Raciborsko, Rice Valley, and Rottier. Characteristic petrographic features of seep carbonates are authigenic, early diagenetic carbonate phases, such as microcrystalline calcite (micrite), yellow calcite, and banded and botryoidal aggregates of fibrous cement [39], and the carbon isotope signature of these early diagenetic carbonate phases typically have very negative $\delta^{13}C$ signatures, significantly lower than marine bicarbonate and occasionally even lower than −50‰ [65]. A detailed description and discussion of the petrographic and stable isotope data of each locality would exceed the scope of the present paper. Therefore we provide a summary of the characteristics of the investigated sites (Table 2) and discuss their implications for reconstruction of the paleoenvironmental conditions under which Peregrinellalived. Representative thin section images are shown in Figure 5, and the stable isotope data are summarized in Table 3A and plotted in Figure 6.

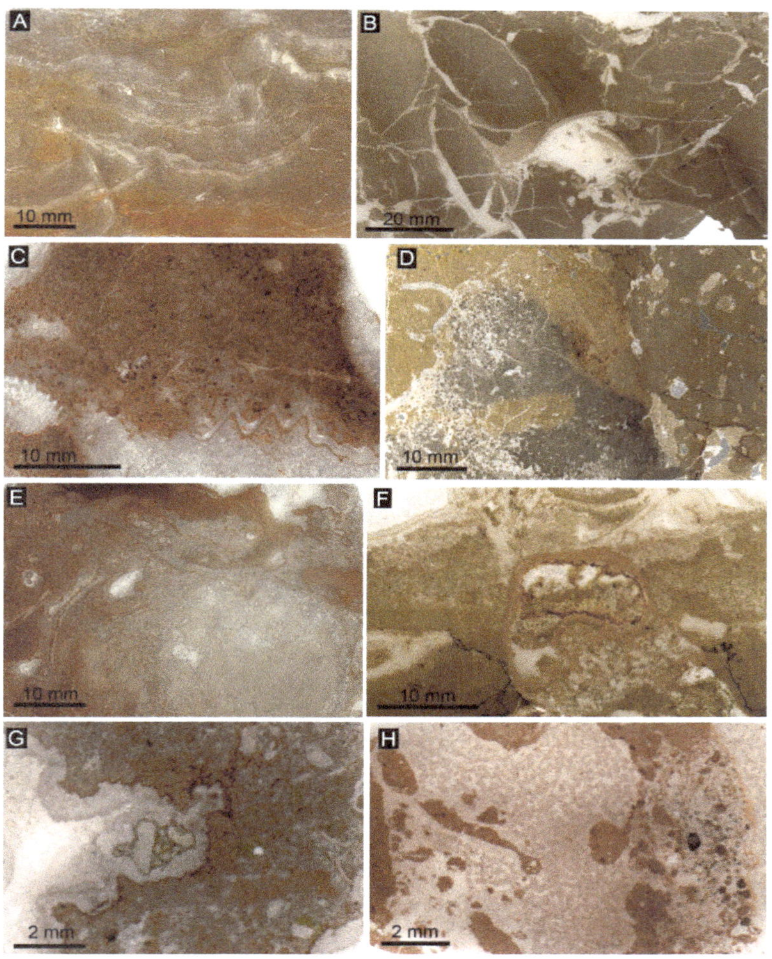

Figure 5. Petrography of Peregrinella **limestones.**

Images are ordered by an increasing proportion of seep cements (light colored), compared to microcrystalline carbonate (micrite; dark colored). A: Incoronata (GZG. INV.82754); B: Musenalp (GZG.INV.82734); C: Rottier (FSL 425077); D: Curnier (GZG.INV.82728); E: Rice Valley (GZG. INV.82748); F: Wilbur Springs (GZG.INV.82753); G: Bonanza Creek (USNM 603602), H: Raciborsko (ZPAL Bp.III).

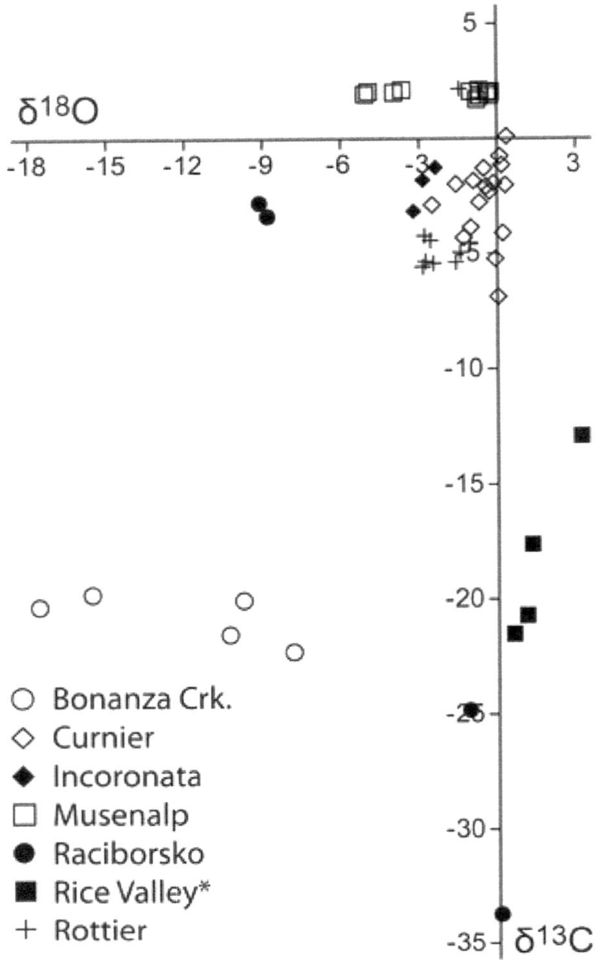

Figure 6. Carbon and oxygen isotope values of Peregrinella-bearing limestones.

All values are reported relative to the PDB standard. * Data from [99].

Table 2. Summary of ecologic and environmental characteristics of Peregrinella-bearing localities, compiled from our own data and the literature.

Locality	Species	Max. size (mm)	Type of tectonic margin	Range of δ^{13}C-values [‰]*1	Seep cement (%)	Mean δ^{18}O-paleotemperature (°C)	References
Bonanza Creek, Alaska	*P. chisania*	30	active	−22.3 to −19.8	~20	nd	[20] and herein
Yongzhu bridge	*P. dongoensis, P. baingoinensis, P. cheboensis*	49*2	active	nd	nd	nd	[21]
Rice Valley, USA	*P. whitneyi*	40	active	−21.5 to −12.9	~20	12.2	[99] and herein
Wilbur Springs, USA	*P. whitneyi*	60	active	−24.3 to −19.3	10	13.3	[39] and herein
E of Lhasa, China	*P. gongboxueensis*	62.8	active	nd	nd	nd	[21]
Koniakov Castle, Czech Republic	*P. multicarinata*	26	passive	nd	nd	nd	[11]
Musenalp, Switzerland	*P. subsilvana*	56	passive	1.7 to 2.1	0	17.3	[14] and herein
Raciborsko, Poland	*P. multicarinata*	59	passive	−33.7 to −2.8	<10	nd	[13] and herein
Curnier, France	*P. multicarinata*	70	passive	−6.9 to 0.1	0	14	herein
Zizin Valley, Romania	*P. multicarinata*	85	passive	−29.7 to −20.4	5	12.2	[42]
Incoronata, Italy	*P. garganica*	85	passive	−3.1 to −1.2	0	nd	[32] and herein
Chatillon, France	*P. multicarinata*	85	passive	nd	nd	15.1	herein
Rottier, France	*P. multicarinata*	88	passive	−5.6 to 2.2	0	13.8	[29] and herein
Planerskoje, Crimea	*P. multicarinata*	90	passive	−13.6 to 2.9	<5	11.7	[35]
Koniakov, Czech Republic	*P. silesica*	90	passive	nd	nd	nd	[11]
Kuban, Russia	*P. pinguis*	102	passive	nd	nd	nd	[102]

nd = no data,
*1: only the values of carbonate mineral phases are given here, not those of *Peregrinella* shells;
*2: size of largest species.

Table 3. Summary of the stable carbon and oxygen isotope data.

A - Limestone samples			B - *Peregrinella* shells			
Sample	δ¹³C(PDB) [‰]	δ¹⁸O(PDB) [‰]	Sample	δ¹³C (PDB) [‰]	δ¹⁸O (PDB) [‰]	Paleotemperature [°C]
Bonanza Creek, micrite	−19.8	−15.6	Châtillon, FSL 425076	0.0	−0.5	14
Bonanza Creek, micrite	−20.3	−17.6	Châtillon, FSL 425076	−0.1	−0.9	16
Bonanza Creek, rim cement	−20.0	−9.8	Châtillon, FSL 425076	0.1	−0.6	14
Bonanza Creek, rim cement	−22.3	−7.9	Châtillon, FSL 425076	0.2	−1.0	16
Bonanza Creek, rim cement	−21.5	−10.3	Curnier, shell 1, GZG.INV.82725	0.5	−0.3	13
Curnier, micrite, GZG.INV.82728	−4.3	−1.3	Curnier, shell 1, GZG.INV.82725	0.7	−0.2	13
Curnier, micrite, GZG.INV.82728	−3.9	−1.0	Curnier, shell 2, GZG.INV.82726	−0.3	−1.6	18
Curnier, micrite, GZG.INV.82728	−2.8	−0.7	Curnier, shell 2, GZG.INV.82726	−0.3	−1.1	17
Curnier, micrite, GZG.INV.82728	−2.0	−1.6	Curnier, shell 3, GZG.INV.82727	0.3	0.1	12
Curnier, micrite, GZG.INV.82728	−1.3	−0.5	Curnier, shell 3, GZG.INV.82727	0.9	0.2	11
Curnier, micrite, GZG.INV.82728	−6.9	0.0	Musenalp, shell 1, GZG.INV.82735	0.8	−1.3	17
Curnier, micrite, GZG.INV.82728	−5.2	−0.1	Musenalp, shell 2, GZG.INV.82736	1.1	−1.7	19
Curnier, micrite, GZG.INV.82728	−2.9	−2.5	Musenalp, shell 3, GZG.INV.82730	0.4	−1.1	16
Curnier, micrite, GZG.INV.82728	−1.8	−0.9	Musenalp, shell 4, GZG.INV.82731	1.1	−0.8	15
Curnier, micrite, GZG.INV.82728	−2.0	0.3	Musenalp, shell 5, GZG.INV.82732	1.0	−1.3	17
Curnier, micrite, GZG.INV.82728	−1.9	−0.1	Musenalp, shell 5, GZG.INV.82732	0.9	−1.7	19
Curnier, micrite, GZG.INV.82728	−4.1	0.2	Planerskoje, shell 1, GZG.INV.82739	−1.4	0.1	11
Curnier, micrite, GZG.INV.82728	−2.3	−0.3	Planerskoje, shell 2, GZG.INV.82740	−0.1	0.1	12
Curnier, micrite, GZG.INV.82728	−2.1	−0.4	Planerskoje, shell 3, GZG.INV.82741	−0.2	0.0	12
Curnier, rim micrite, GZG.INV.82728	0.1	0.4	Planerskoje, shell 4, GZG.INV.82742	−0.8	0.1	12
Curnier, rim micrite, GZG.INV.82728	−1.2	0.2	Planerskoje, shell 5, GZG.INV.82743	−0.8	0.0	12
Curnier, rim micrite, GZG.INV.82728	−0.8	0.1	Planerskoje, shell 6, GZG.INV.82744	−2.5	0.0	12
Incoronata, Italy, GZG.INV.82754	−3.2	−3.3	Rice Valley, GZG.INV.82748	−1.7	−0.5	14
Incoronata, Italy, GZG.INV.82754	−1.8	−2.9	Rice Valley, GZG.INV.82749	−3.0	−0.3	13
Incoronata, Italy, GZG.INV.82754	−1.3	−2.4	Rice Valley, GZG.INV.82750	−0.6	−0.1	12.
Musenalp, micrite, GZG.INV.82733	2.1	−4.9	Rice Valley, GZG.INV.82751	0.8	0.2	11
Musenalp, micrite, GZG.INV.82733	1.9	−5.1	Rice Valley, GZG.INV.82752	−2.8	0.4	10
Musenalp, micrite, GZG.INV.82733	1.7	−0.8	Rottier, FSL 425077	0.2	−0.2	13
Musenalp, micrite, GZG.INV.82733	2.1	−0.2	Rottier, FSL 425077	0.0	−0.3	13
Musenalp, micrite, GZG.INV.82733	2.0	−0.4	Rottier, FSL 425078	0.3	−1.3	18
Musenalp, micrite, GZG.INV.82733	2.0	−3.9	Rottier, FSL 425078	−2.6	−1.2	17
Musenalp, micrite, GZG.INV.82733	2.1	−3.6	Rottier, FSL 425079	−0.1	0.1	12
Musenalp, micrite, GZG.INV.82733	1.9	−0.7	Rottier, FSL 425079	0.0	0.2	11
Musenalp, micrite, GZG.INV.82733	2.0	−0.3	Wilbur Springs, GZG.INV.82753	−1.1	−0.4	13
Musenalp, sparitic rim in shell, GZG.INV.82734	1.9	−0.6	Wilbur Springs, USNM 603596	−1.4	−0.4	14
Musenalp, sparitic rim in shell, GZG.INV.82734	1.9	−0.3	Wilbur Springs, USNM 603597	−0.6	−0.2	13
Musenalp, sparitic rim in shell, GZG.INV.82734	2.0	−1.0	Wilbur Springs, USNM 603598	−2.5	−1.0	16
Musenalp, sparitic rim in shell, GZG.INV.82734	2.1	−0.7	Wilbur Springs, USNM 603599	−0.4	0.2	11
Raciborsko, micrite, ZPAL Bp.III	−33.7	0.1	Zizin Valley, LPB III br 364	0.1	0.1	12
Raciborsko, peloidal micrite, ZPAL Bp.III	−2.8	−9.1	Zizin Valley, LPB III br 372	0.0	−0.3	13
Raciborsko, micrite, ZPAL Bp.III	−24.8	−1.1	Zizin Valley, LPB III br 381	0.3	0.1	12
Raciborsko, peloidal micrite, ZPAL Bp.III	−3.4	−8.8	Zizin Valley, LPB III br 381	0.3	0.1	11
Rice Valley*	−21.5	0.6	Zizin Valley, LPB III br 389	0.4	−0.4	13
Rice Valley*	−12.9	3.2				
Rice Valley*	−20.7	1.1				
Rice Valley*	−17.6	1.3				
Rottier, micrite, FSL 425077	−5.3	−2.8				
Rottier, micrite, FSL 425077	−5.4	−2.5				
Rottier, micrite, FSL 425077	−5.6	−2.9				
Rottier, micrite, FSL 425077	−5.4	−1.6				
Rottier, micrite, FSL 425077	−4.9	−1.4				
Rottier, micrite, FSL 425077	−4.5	−1.0				
Rottier, rim micrite, FSL 425077	−4.2	−2.8				
Rottier, rim micrite, FSL 425077	−4.4	−2.6				

The investigated occurrences of Peregrinella in this study fall within three categories:

1) Occurrences where Peregrinella has clearly been transported away from its original living place: Incoronata and Musenalp. Both show bedding and alignment of Peregrinella shells parallel to bedding (although less obvious in the case of the Musenalp). Both lack typical seep cements (Figures 5A, B), and both have $\delta^{13}C$-signatures of marine carbonate (Figure 6). No conclusions can be drawn about the original habitat of Peregrinella at these sites. In the case of the Incoronata locality, which overlies the mid-

Valanginian drowning surface between the Maiolica pelagic limestones and the underlying coarse-grained slope debrites, Peregrinella has been considered as a disaster taxon that bloomed due to the environmental disruptions linked to the late Weissert Oceanic Anoxic Event [33], [66].

2) In situ occurrences within carbonate settings with only minor ^{13}C depletion: Curnier and Rottier. The limestones of both sites show evidence for bioturbation, they lack typical seep cements, but authigenic micrites are present (Figures 5C, D). Their carbon isotope signatures (as low as −6.9 and −5.6‰, respectively, Figures 6, Table 2) are below that of marine carbonate, but clearly not low enough to rule out carbon sources other than methane. However, considering that they were deposited in an carbonate setting, a strong input of marine carbonate is obvious, which would have diluted a potentially much lower original signal [37], [65]. We were not able to reproduce the extremely high $\delta^{13}C$ values of up to +20‰, as reported earlier from Rottier [18]. The accompanying mollusk fauna at Curnier and Rottier consists of taxa that are otherwise found exclusively at methane-seep deposits [30], [67]. Furthermore, the presence of ^{13}C-depleted PMI and phytane in the Curnier limestone (see below) indicates that anaerobic oxidation of methane occurred in the depositional environment [65], agreeing with methane seepage. The Peregrinella-bearing limestones at Curnier and Rottier are therefore interpreted as ancient methane-seep deposits, probably reflecting diffuse seepage sites.

3) In situ occurrences in carbonate bodies with ^{13}C-depleted early diagenetic cements: Bonanza Creek, Raciborsko, and Rice Valley. Authigenic micrites, and typical seep cement such as banded and botryoidal cement are common in these limestones (Figures 5F-H, Table 2). The carbon isotope signatures of these cements are as low as −33.7 to −21.5‰ (Figure 6, Table 2) and are thus well within the range of other Peregrinella-bearing seep limestones [19], [35], [42]. The associated fauna, where present, consists largely of seep-endemic mollusks [41]. Therefore, Bonanza Creek, Raciborsko, and Rice Valley are here also identified as ancient seep deposits.

Lipid biomarker data have been reported for Peregrinella-bearing seep limestones from the Wilbur Springs [40], Planerskoje [37], and Zizin Valley [42] localities. For this study two new limestone samples suited for biomarker analyses were available. The hydrocarbon fraction of the Curnier limestone is typified by an unresolved complex mixture (Figure 7), reflecting a crude oil pattern and indicating pronounced biodegradation. The timing of possible oil ingress and biodegradation cannot be constrained with certainty. Thus, the possibility of syndepositional oil seepage cannot be ruled out. Although the observed compound pattern has obviously been affected by biodegradation and thermal maturity, two compounds and their carbon isotopic compositions point to syndepositional methane seepage, having fueled the anaerobic oxidation of methane involving methanotrophic archaea. Irregular, ^{13}C-depleted pentamethylicosane (PMI) is a stable and particularly long-lasting biomarker of methanotrophic archaea [65], [68]. Its $\delta^{13}C$ value of −48‰ is much higher than values of PMI typically observed at methane seeps; yet, it is also much lower than the values of the regular isoprenoid pristane in the same sample, the precursors of which are produced by phototrophs [69]. Most likely, the relatively high $\delta^{13}C$ value of PMI is due to masking of its pristine isotopic signature caused by the unresolved complex mixture made up of compounds with higher ^{13}C content, on average, than PMI. Hence, it is difficult to determine whether the observed intermediate value of PMI represents a pure methanotrophic or mixed source (methanotrophic and methanogenic archaea) signal. Moreover, the isotopic composition of the regular isoprenoid phytane (−44‰) is much lower than that of pristane (−35‰), revealing that phytane did not exclusively derive from phototrophic organisms. The greater ^{13}C-depletion of phytane is best explained by input from methanotrophic archaea (archaeal diethers), in addition to input from phototrophic organisms (chlorophyll), resulting in a carbon isotopic composition that falls between the compositions commonly observed for lipids of methanotrophs and phototrophs. To sum up, in combination with the lowest $\delta^{13}C_{carbonate}$ values of −6.9‰ found for the Curnier limestone, the observed lipid patterns are best explained by anaerobic oxidation of methane at a seafloor seep. The seepage fluids also may have included crude oil or, alternatively, oil may have intruded the limestone at a later stage. Unlike the Curnier limestone, no biomarker evidence for seepage was found for the Musenalp limestone, agreeing with its marine $\delta^{13}C_{carbonate}$ values (Figure 6). No irregular isoprenoids were detected in the Musenalp sample, and the regular isoprenoids pristane and phytane as well as n-alkanes yielded $\delta^{13}C$ values (−31 to −30‰) that are typical for lipids resulting from marine primary production in the photic zone [70].

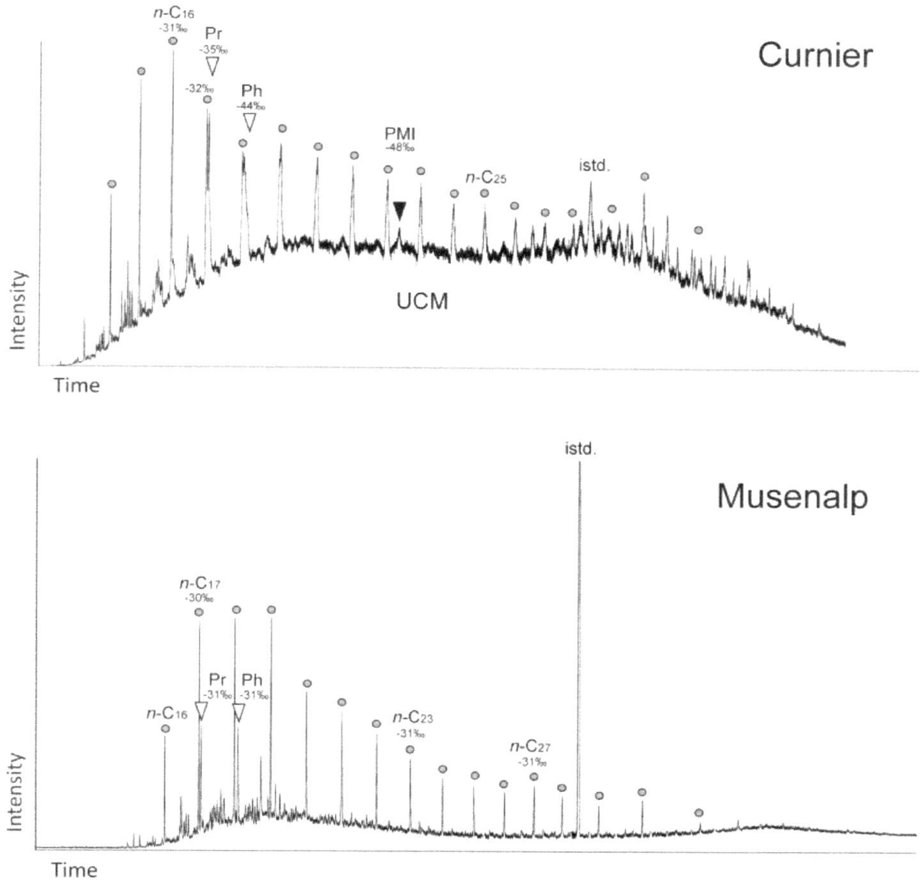

Figure 7. Hydrocarbon fractions (total ion currents) of the Curnier and Musenalp limestones with Peregrinella.

Circles: n-alkanes; white triangles: regular isoprenoids; black triangles: irregular isoprenoids; Pr: pristane; Ph: phytane; PMI: pentamethylicosane; istd.: internal standard; UCM: unresolved complex mixture; compound-specific $\delta^{13}C$ values are indicated in per mil relative to the Peedee belemnite standard.

Impact of Peregrinella on Other Seep Inhabitants

The sheer abundance of Peregrinella at the localities it colonized (Figure 1) has led to suggestions that it had a negative impact on at least some of the co-occurring taxa, in particular infaunal and semi-infaunal bivalves [19], [35], but these suggestions were based only on observations for individual Peregrinella occurrences. We therefore attempted to address this question in a more comparative way. The mollusks in methane-seep communities of the late Mesozoic show an overall similarity in faunal composition, although with a notable turnover during the mid-Cretaceous, and are often dominated by the modiomorphid bivalve Caspiconcha, several seep-restricted lucinid bivalves, and the gastropod Hokkaidoconcha [8], [30], [71]–[75]. Therefore, our comparisons of specimen size and diversity of the seep mollusk fauna in the presence or absence of Peregrinella focused on two time intervals: first the late Jurassic through late Cretaceous (henceforth referred to as the 'late Mesozoic'), to assess possible impacts of Peregrinella on the late Mesozoic seep fauna in general, and second only the time interval spanning the stratigraphic range of Peregrinella (henceforth referred to as the 'Peregrinella interval'). The faunal data for these analyses and their sources are summarized in Table 4; included are only sites where we had some confidence that the fauna had been adequately sampled, i.e., reports of Peregrinella from collections with unknown sampling strategies (such as Bonanza Creek or Châtillon-en-Dois reported here) are not included.

Table 4. Peregrinella and chemosymbiotic bivalves at Cretaceous seep deposits.

Locality (age)	*Peregrinella*	Solemyidae, size (mm)	Thyasiridae, size (mm)	Lucinidae, size (mm)	References
Beauvoisin (Oxfordian)	a			*Beauvoisina carinata*, 200	[103]
Gateway Pass (Tithonian)	a			lucinid, 50	[104]
NW Berryessa (Tithonian)	a			*Tehamatea ovalis*	[41]
Paskenta (Tithonian)	a	*Acharax stantoni*, 58		*Tehamatea ovalis*, 60	[41,105]
				Tehamatea colusaensis, 95	
Stony Creek (Tithonian)	a	*Acharax stantoni*		*Tehamatea colusaensis*	[41]
				Tehamatea ovalis	
*** Sassenfjorden (Berriasian)**	a	solemyid, 77	thyasirid, 50	lucinid, 94	[76]
*** Planerskoje (Berriasian)**	p			*Tehamatea vocontiana*, 70	[30,35]
*** Musenalp (Berriasian)**	p				[14] and herein
*** Bear Creek (Valanginian)**	a	*Acharax stantoni*, 60		*Tehamatea ovalis*, 25	[30,101]
*** Little Indian Valley (Valanginian)**	a	*Solemya* sp.			[41]
*** Rocky Creek (Valanginian)**	a	*Acharax stantoni*, 56		*Tehamatea colusaensis*, 145	[30]
*** Raciborsko (Hauterivian)**	p				[13] and herein
*** Koniakov (Hauterivian)**	p			*Lucina* sp., 22	[11,106]
*** Koniakover Schloss (Hauterivian)**	p			*Lucina valentula*, 13	[11,106]
				Lucina obliqua, 11	
				lucinid sp. ind., 70	
*** Curnier (Hauterivian)**	p			*Tehamatea vocontiana*, 50	herein
*** Rottier (Hauterivian)**	p			*Tehamatea vocontiana*, 200	[30]
*** Wilbur Springs (Hauterivian)**	p	*Acharax stantoni*, 70		*Tehamatea colusaensis*, 77	[30,105]
Baska (Barremian)	a			lucinid indet., 30	[106]
Eagle Creek (Barremian)	a	*Acharax stantoni*, 47		*Tehamatea ovalis*	[101]
Kuhnpasset Beds (Barremian)	a	*Solemya* sp., 50		*Amanocina kuhnpassetensis*, 129	[107]
Awanui I (Albian)	a			*Cubathea awanuiensis*, 27	[30,78]
				Amanocina raukumara, 33	
Awanui II (Albian)	a			*Cubathea awanuiensis*, 50	[30,78]
CFCC (Albian)	a	*Acharax stantoni*, 35		*Lucina* sp., 22	[30]
Ispaster (Albian)	a			*Tehamatea agirrezabalai*, 135	[30,108]
Ponbetsu (Albian)	a	*Acharax mikasaensis*, 55	*Thyasira tanabei*, 11.7	"*Nipponothracia*" *ponbetsensis*, 120	[77]
Awanui GS 688 (Cenomanian)	a		thyasirid, 6	*Amanocina raukumara*, 80	[30,78]
				Cubathea awanuiensis, 48	
				Ezolucina ? sp. "ridge", 37	
Obira-cho (Cenomanian)	a	solemyid indet.	*Thyasira tanabei*, 10.4	lucinid indet., 43	[77]
				Miltha sp.	
				Amanocina yezoensis, 100	
Omagari lens (Campanian)	a		*Thyasira tanabei*, 50	lucinid *ponbetsensis*, 60	[77]
				Tehamatea sp.?, 50	
Romero Creek (Campanian)	a	*Solemya* sp.	*Thyasira cretacea*		[41]
Sada Limestone (Campanian)	a	solemyid, 70	*Thyasira hataii*, 80	*Myrtea*? sp., 35	[30,109]
				lucinid, 71	
Waipiro I (Campanian)	a			*Ezolucina* sp., 120	[30,78]
Waipiro III (Campanian)	a	solemyid, 20	thyasirid, 10	lucinid, 30	[30,78]
Yasukawa (Campanian)	a	*Acharax cretacea*, 60	*Thyasira tanabei*, 10.4	*Miltha* sp., 18	[110]
				Myrtea sp., 10	

We performed a set of tests with each of the three infaunal chemosymbiotic bivalve families Solemyidae, Thyasiridae and Lucinidae. First, for the late Mesozoic we tested whether they were more commonly (1) present, and (2) larger, at seep sites outside the Peregrinella interval than at seep sites during the Peregrinella interval. Second, for the Peregrinella interval we tested whether they were (1) more frequently absent from, and (2) smaller at sites with Peregrinella than at sites without Peregrinella. For the size tests, only the largest record was used when more than one species of a family was present at a locality.

During the late Mesozoic, all three bivalve families were neither more common nor larger at seep sites outside the Peregrinella interval than during the Peregrinella interval (Figure 8). However, in the case of the thyasirids these comparisons are problematic: they were reported from shallow-water seeps from the Berriasian on Svalbard [76], are entirely absent from seeps during the remainder of the lower half of the Early Cretaceous, and re-appear in the Albian of Japan and New Zealand [77], [78]. An alternative explanation to a potential impact of Peregrinella is that thyasirids did not colonize deep-water seeps until Albian time.

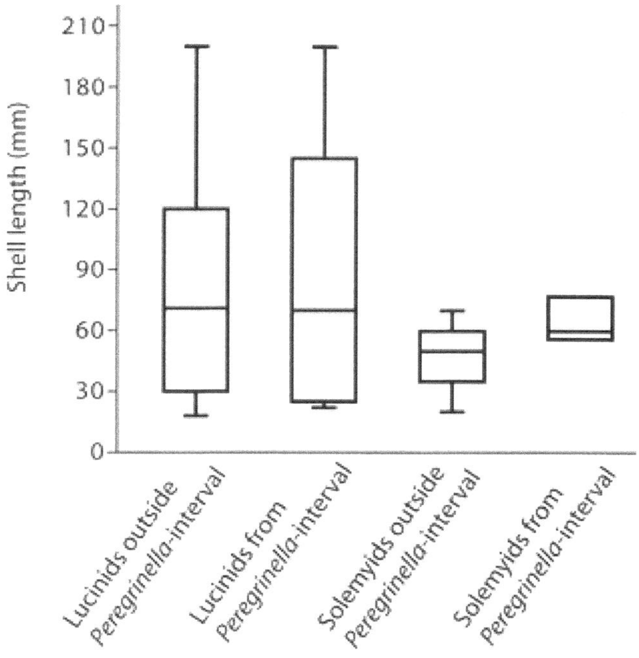

Figure 8. Shell sizes of chemosymbiotic bivalves at late Mesozoic methane seeps.

When only the Peregrinella interval was considered, lucinids were neither more frequently absent from, nor smaller at sites with Peregrinella than at sites without Peregrinella. In a rigorous statistical sense this is also true for the solemyids (Fisher›s exact test, p=0.23), although it is remarkable that solemyids occur at all non-Peregrinella sites but only at a single site with Peregrinella (Table 4). In summary, contrary to previous suggestions, we found little empirical evidence that Peregrinella had a negative impact on the presence or size of infaunal bivalves at seeps. However, the observation from the seep deposit at Wilbur Springs that Peregrinella is more abundant than the co-occurring modiomorphid bivalves [19] (Figure 1) also applies to the infaunal chemosymbiotic bivalves.

At localities where Peregrinella occurs, it is always more abundant than any other taxon. Therefore, we were interested in whether the presence of Peregrinella affected species diversity at the sites it colonized. For the Peregrinella interval we tested whether seep sites with fewer than 10 associated species (i.e., excluding Peregrinella) are more common among the Peregrinella-bearing deposits than seep sites with more than 10 associated species. Sites without associated species were excluded. This analysis showed no significant difference (Fisher›s exact test, p=0.56, Table 5). At its various occurrences, Peregrinella coexisted with virtually all coeval seep mollusks, as well as occasional tube worms and sponges. Neither the origin nor the extinction of Peregrinella coincides with the origin or extinction of any coeval seep-inhabiting mollusk (Figure 9).

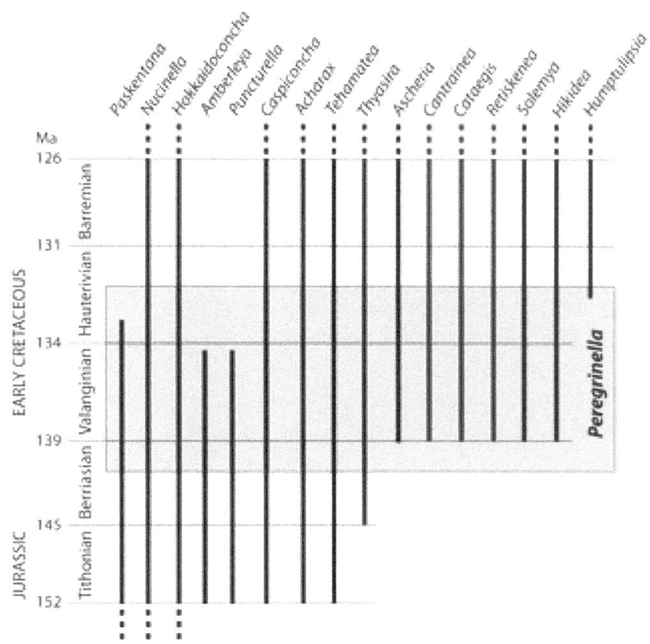

Figure 9. Geologic ranges of seep-inhabiting mollusk genera and Peregrinelladuring the Late Jurassic – Early Cretaceous.

Data from [8], [25], [30], [41], [61], [67], [74], [100], [101]; absolute ages from [64].

Table 5. Numbers of species at seep deposits from the Peregrinella interval; a=absent, p=present.

Locality	# associated species	*Peregrinella*	Reference
Musenalp, Switzerland	1	p	[14]
Curnier, France	2	p	herein
Rottier, France	2	p	[29] and herein
East Berryessa, USA	3	a	[41]
Foley Canyon, USA	3	p	[41]
Gravelly Flat, USA	3	p	[41]
Rice Valley, USA	4	p	[41]
West Berryessa, USA	4	a	[41]
Little Indian Valley, USA	5	a	[41]
Koniakov, Czech Republic	5	p	[11,106]
Planerskoje, Crimean peninsula	7	p	[35]
Wilbur Springs, USA	10	p	[41,71]
Rocky Creek, USA	13	a	[41,71,101]
Koniakov Castle, Czech Republic	17	p	[11,106]
Bear Creek, USA	18	a	[41,71,101]

Environmental Impacts on Peregrinella

Peregrinella reached significantly greater maximum sizes at passive continental margins than at active margins (Mann-Whitney U test, p=0.04; Table 2). Assuming that environmental disturbances such as turbidites are more common at active margins than in deep shelf settings at passive margins, it may be inferred that Peregrinella was sensitive toward such environmental disturbances.

The intensity of seep fluid flow plays a major role in structuring seep communities today because some taxa prefer sites with focused, advective fluid flow while others prefer sites of slow, diffusive seepage [79]. The apparent restriction of Peregrinella to seeps suggest that it benefited, in as-yet unknown ways, from the seeping fluids and the reduced compounds in them. Most likely, Peregrinella benefited from chemotrophic bacteria using methane and/or sulfide, but whether Peregrinella was filter-feeding on free-living bacterioplankton, for example like modern vent and seep-inhabiting barnacles [80], or had a symbiotic relationship with such bacteria, remains unknown and cannot be resolved with the data presented here. We were interested whether Peregrinella had a preference for a certain paleo-seep fluid intensity and compared the maximum size of Peregrinella to two proxies for seepage intensity: (i) the carbon isotope signature of the enclosing limestone, and (ii) the abundance of the typical seep cements such as banded or botryoidal cement, quantified based on thin section observations; the localities with uncertain original habitat type, Incoronata and Musenalp, were not used in this comparison. There is no correlation between the maximum size of Peregrinella and the carbon isotope signature of the enclosing limestone, but we found a significant negative correlation between the maximum size of Peregrinella and the abundance of seep cements (i.e. fibrous, banded and botryoidal cement rather than micrite): Peregrinella is on average larger at sites with fewer seep cements ($p<0.003$, Table 2), suggesting that it was better adapted to diffusive rather than advective seepage [37].

Paleo-temperatures calculated from $\delta^{18}O$ values of Peregrinella shells indicate that these specimens lived at temperatures ranging from 10 to 19°C (Tables 2, 3B). At some localities different shells showed a remarkably uniform range of values (i.e., Planerskoje: 11 to 12°C) while at other localities different shells showed a wide scatter of values (i.e., Rottier: 11 to 18°C; Curnier: 11 to 18°C). The data are summarized in Figure 10. These absolute values should be treated cautiously because some rhynchonellid brachiopods exert vital effects when incorporating oxygen isotopes into their shells [46]. But because we investigated only specimens of a single genus, we infer that all investigated specimens lived within a temperature range spanning about 9°C from the coldest to the warmest. Does this reflect the preferred temperature range of the genus Peregrinella in general? Unfortunately, the available material from the northernmost occurrence, the Alaskan Bonanza Creek site, was too sparse for oxygen isotope analyses. Therefore Peregrinella might have lived also at lower temperatures than inferred here and thus might have had a broader temperature range than the 9°C estimated from the available specimens.

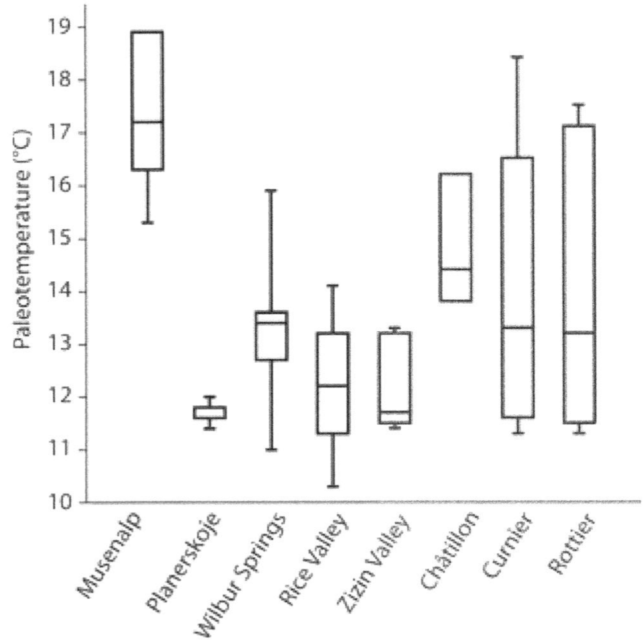

Figure 10. Paleotemperatures derived from $\delta^{18}O$ values of Peregrinella shells.

Localities ordered from oldest (left) to youngest (right).

Brachiopod Geologic Ranges

The 9 m.y. long geologic range of Peregrinella appears short when compared to genera of seep-inhabiting mollusks, including chemosymbiotic bivalves, which may reach geologic ranges of 50 m.y. or more [4], [30]. But how does this compare to the geologic ranges of other brachiopod genera? Virtually all brachiopod genera that dominated vent and seep deposits in the geologic past belong to a single rhynchonellid superfamily, the Dimerelloidea [22], [24]–[27], [81] and their geologic ranges seem to never exceed 10 m.y. in duration (see fig. 9.6 in [23]). To investigate how the geologic ranges of Peregrinella and other dimerelloids compare to those of rhynchonellid brachiopod genera in general, we summarized the geologic ranges of genera of the Rhynchonellida listed in Sepkoski's compendium [82], using the following criteria: only taxa with their first appearances (FAs) and last appearances (LAs) given at least to series level were included. Only taxa having their FAs in Devonian through Cretaceous were included. Subgenera and genera with "?" were excluded. Taxa with questionable FAs or LAs, as indicated by a "?" were excluded. Stratigraphic subdivisions such as lower, middle, and upper were ignored. Taxa with FAs or LAs in the Leonardian (Permian) were excluded because this stage could not be correlated with the Gradstein et al. 2012 time scale [64]. The geologic ranges of the remaining 483 genera were calculated as follows: numerical age of base of FA minus numerical age of end of LA. This resulted in a median geologic range of 12.7 m.y., and 25 and 75 percentiles of 6.6 and 17.3 m.y., respectively. Thus the 9 m.y. range of Peregrinellaand the geologic ranges of dimerelloid genera in general, are somewhat shorter than the median, but certainly not abnormal for rhynchonellid brachiopods. In this respect, dimerelloids differ from mollusk genera at seeps which are, on average, longer-lived than marine mollusks in general [4]. These observations fit with two macroevolutionary patterns among brachiopods and mollusks. First, on low taxonomic levels (e.g., genus, species) brachiopods show higher origination and extinction rates than mollusks [83], [84], as seen here also among their seep-inhabiting members. Second, on the family/superfamily level brachiopods appear to persist longer in deep-water environments than gastropods do [85], consistent with the long persistence of the Dimerelloidea in the vent/seep environment.

DISCUSSION

The oldest occurrences of Peregrinella are the Musenalp site in Switzerland and the Planerskoje site in the eastern Crimean peninsula; both are of late Berriasian age. The genus thus apparently originated in the Tethys Ocean [21], in contrast to a previous hypothesis of an origin in California and subsequent dispersal to Europe [13], [45]. In the Valanginian Peregrinella had spread eastward as far as Tibet [21] and by Hauterivian time it had spread to California and Alaska. Whether this spread to the American West coast was eastward or westward from the Tethys remains unclear. The present understanding of Early Cretaceous ocean currents [86] makes a westward spread possible, although an eastward spread along the active continental margins around the North Pacific Ocean also seems plausible [87]. The youngest occurrences are those from southern France and Alaska, and are of late early Hauterivian age.

The timing of the appearance and disappearance of Peregrinella is puzzling, because neither does it coincide with the appearance and disappearance of any other seep-inhabiting taxon, nor with any of the major oceanic disturbances such as oceanic anoxic events (OAEs). Its extinction at the end of the early Hauterivian roughly coincides with the 'Faraoni event', a presumed, short-lived anoxic event in the late Hauterivian [88], but this event was apparently restricted to the Tethyan Realm and does not explain the disappearance of Peregrinella from California and Alaska.

The implied concurrent demise of Peregrinella and the large, seep-inhabiting bivalve Caspiconcha at Cretaceous seeps in figure 15 of [74] is not supported by our Sr-isotope stratigraphy results and by the finding of a mid-Cenomanian Caspiconcha-dominated seep deposit in New Zealand [78]. There are about 36 m.y. between the extinction of Peregrinella at the end of the early Hauterivian and the last Caspiconcha-dominated seep deposit.

The apparent preference of Peregrinella for diffusive seepage rather than for focused, advective fluid flow is surprising considering that taxa dominating modern vents and seeps prefer strong fluid flow and resulting high hydrogen sulfide concentrations [79], [89], [90]. A potential implication of this pattern is that the link of Peregrinella to seeps is related to methane rather than to hydrogen sulfide. Because hydrogen sulfide is quickly oxidized in the presence of oxygen, it only reaches the seafloor when fluid flow is strongly advective. When fluid flow is slow and diffusive, hydrogen sulfide is oxidized before it reaches the sediment-water interface, as shown by modeling approaches and in situ measurements [91], [92]. Methane, on the other hand, is only oxidized biologically in marine environments and therefore has a greater chance to reach the sediment-water interface under diffusive flow conditions than hydrogen sulfide. In case of the Planerskoje and Zizin Valley seep deposits, the presence of molecular fossils of aerobic methane-oxidizing bacteria [37], [42] reveals that at least some of the seeping methane was not oxidized in the zone of anaerobic oxidation of methane but entered the oxic zone. Based on the low preservation potential of lipids of endosymbiotic bacteria, such a source can be ruled out for the observed molecular fossils, but their presence reveals that oxygen-dependent consumption of methane indeed occurred at the ancient seep sites [93]. Thus, assuming that Peregrinella relied – in as-yet unknown ways – on chemotrophic bacteria, then its apparent preference for diffusive seepage may suggest a more prominent role of methane-oxidizing bacteria, rather than sulfide-oxidizing bacteria, in the diet of Peregrinella.

We infer the preference of Peregrinella for diffuse seepage from the larger average sizes of Peregrinella at sites with fewer seep cements. This may have implications for other seep-inhabiting dimerelloids from different geologic ages. In contrast to Peregrinella, certain dimerelloid genera including the Devonian Dzieduszyckia [26] and the Jurassic Sulcirostra [27], occur at sites where the proportion of seep cements of the limestone exceeds that seen in any Peregrinella site by a great margin, and may even constitute the majority of the limestone. These brachiopods potentially lived at seep sites with strong, advective fluid flow. The early Jurassic dimerelloid Anarhynchia cf. gabbi is common at a hydrothermal vent deposit [94]where sulfide most likely was the dominant reduced chemical compound. Therefore, it seems possible that the different dimerelloids developed various strategies for living at vent and seep environments. However, most of these genera have been documented from one or two seep deposits only, making a direct comparison with the trend seen in Peregrinella impossible.

CONCLUSIONS

All in situ occurrences of Peregrinella investigated here were confirmed as ancient seep deposits. This supports the view that Peregrinella lived exclusively at seeps [19] and calls for further investigations to determine whether other dimerelloid genera that have been reported from seep or hydrothermal-vent deposits (i.e., Anarhynchia, Cooperrhynchia, Dzieduszyckia, Halorella, Ibergirhynchia, Sulcirostra) were also restricted to this type of habitat, or had broader ecologic plasticity [26].

While the question whether Peregrinella was chemosymbiotic or not remains unresolved, the comparative approach used here provides new insights into the paleoecology of this enigmatic brachiopod, notably the apparent preference for living at diffusive rather than advective seepage sites. Expanding this line of work to other seep-inhabiting dimerelloids could show whether this preference is shared by dimerelloids in general or whether different dimerelloid genera had adapted to vents and seeps in different ways.

Compared to other seep-inhabiting groups, especially the well-studied mollusks, Peregrinellahas two remarkable traits: first, its quite short geologic range of only 9 m.y., and second, its extremely gregarious mode of occurrence. However, from a rhynchonellid brachiopod›s perspective, these traits may not be unusual: Peregrinella›s stratigraphic range is not very different from the average rhynchonellid›s range, and given a suitable habitat, rhynchonellid brachiopods tend to be gregarious [95]–

[97]. Thus while seep mollusks clearly differ from other marine mollusks in their geologic longevity, this seems not to be the case for Peregrinella and perhaps also for other seep-inhabiting dimerelloids back to the Devonian.

ACKNOWLEDGMENTS

We thank Chuck Fisher (Penn State University) for providing shells from Recent seeps for comparison, Tina Kollaske (BGR Berlin) for the loan of a putative Peregrinella specimen from northeastern Germany, Abel Prieur and Emmanuel Robert (Lyon) for arranging the loan of specimens from southern France, Amanda Millhouse (Washington, DC) for arranging the loan of specimens from the USNM collection, Andreas Pack (Göttingen) for stable C and O isotope analyses, Axel Hackmann (Göttingen) for thin section preparation, Dorothea Hause-Reitner (Göttingen) for assistance with the SEM, Birgit Wild and Andreas Richter (Vienna) for aid measuring compound-specific stable carbon isotopes, Beatrix Bethke (Vienna) for biomarker sample extractions, John M. McArthur (London) for providing the LOWESS 5.0 look-up table, Przemysław Gedl (ING PAN, Cracow) for discussion of Carpathian stratigraphy, and two anonymous reviewers for their insightful comments that helped to improve this manuscript.

Author Contributions

Conceived and designed the experiments: SK. Performed the experiments: SK JG DB JP. Analyzed the data: SK JG DB JP. Wrote the paper: SK JG DB JP. Conducted field work and/or contributed specimens: SK LGB KAC CG RG AK IL MRS.

REFERENCES

1. Corliss JB, Dymond J, Gordon LI, Edmond JM, Von Herzen RP, et al. (1979) Submarine thermal springs on the Galápagos Rift. Science 203: 1073–1083. doi: 10.1126/science.203.4385.1073
2. Paull CK, Hecker B, Commeau R, Freeman-Lynde RP, Neumann C, et al. (1984) Biological communities at the Florida Escarpment resemble hydrothermal vent taxa. Science 226: 965–967. doi: 10.1126/science.226.4677.965
3. Campbell KA (2006) Hydrocarbon seep and hydrothermal vent paleoenvironments and paleontology: Past developments and future research directions. Palaeogeogr Palaeoclimatol Palaeoecol 232: 362–407. doi: 10.1016/j.palaeo.2005.06.018
4. Kiel S, Little CTS (2006) Cold seep mollusks are older than the general marine mollusk fauna. Science 313: 1429–1431. doi: 10.1126/science.1126286
5. Amano K, Kiel S (2007) Fossil vesicomyid bivalves from the North Pacific region. Veliger 49: 270–293.
6. Kaim A, Jenkins RG, Warén A (2008) Provannid and provannid-like gastropods from Late Cretaceous cold seeps of Hokkaido (Japan) and the fossil record of the Provannidae (Gastropoda: Abyssochrysoidea). Zool J Linn Soc 154: 421–436. doi: 10.1111/j.1096-3642.2008.00431.x
7. Kiel S, Amano K (2013) The earliest bathymodiolin mussels: Evaluation of Eocene and Oligocene taxa from deep-sea methane seep deposits in western Washington State, USA. J Paleont 87: 589–602. doi: 10.1666/12-135
8. Kaim A, Jenkins RG, Hikida Y (2009) Gastropods from Late Cretaceous hydrocarbon seep deposits in Omagari and Yasukawa, Nakagawa area, Hokkaido, Japan. Acta Palaeontol Pol 54: 463–690. doi: 10.4202/app.2009.0042
9. Campbell KA, Bottjer DJ (1995) Brachiopods and chemosymbiotic bivalves in Phanerozoic hydrothermal vent and cold seep environments. Geology 23: 321–324. doi: 10.1130/0091-7613(1995)023<0321:bacbip>2.3.co;2
10. Gabb WM (1866–69) Cretaceous and Tertiary fossils. Calif Geol Surv Paleontol 2: 1–299.
11. Ascher E (1906) Die Gastropoden, Bivalven und Brachiopoden der Grodischter Schichten. Beitr Paläontol Geol Österreich-Ungarns Orients 19: 135–172.
12. Toula F (1911) Über Rhynchonella (Peregrinella Öhlert) multicarinata Lamk. sp. (1819) = Terebratula peregrina L. v. Buch (1833) von Zajzon bei Kronstadt. Abh kk Geol Reichsanst 20: 27–35.
13. Biernat G (1957) On Peregrinella multicarinata (Lamarck)(Brachiopoda). Acta Palaeontol Pol 2: 19–52.
14. Trümpy R (1956) Notizen zur mesozoischen Fauna der innerschweizerischen Klippen (I-II). Eclogae geol Helvetiae 49: 573–591.
15. Ager DV (1965) The adaptation of Mesozoic brachiopods to different environments. Palaeogeogr Palaeoclimatol Palaeoecol 1: 143–172. doi: 10.1016/0031-0182(65)90011-8
16. Remeš M (1903) Rhynchonella peregrina bei Freiberg in Mähren. Verh kk geol Reichsanst 11: 223–225.
17. Macsotay O (1980) Molluscques benthiques du Crétacé inférieur. Une méthode de corrélation entre la Téthys mésogéenne et le domaine paléocaraïbe (Vénézuela). PhD thesis, Université Claude-Bernard, Lyon. 158 p.
18. Lemoine M, Arnaud-Vanneau A, Arnaud H, Létolle R, Mevel C, et al. (1982) Indices possibles de paléo-hydrothermalisme marin dans le Jurassique et le Crétacé des Alpes occidentales (océan téthysien et sa marge continentale européenne): essai d'inventaire. Bull Soc Géol France 24: 641–647.
19. Campbell KA, Bottjer DJ (1995) Peregrinella: an Early Cretaceous cold-seep-restricted brachiopod. Paleobiology 24: 461–478.
20. Sandy MR, Blodgett RB (1996) Peregrinella (Brachiopoda; Rhynchonellida) from the Early Cretaceous Wrangellia Terrane, Alaska. In: Copper P, Jin J, editors. Brachiopods. Rotterdam: A.A. Balkema. pp. 239–242.
21. Sun D (1986) Discovery of Early Cretaceous Peregrinella (Brachiopoda) in Xizang (Tibet) and its significance. Palaeontol Cathayana 2: 211–227.
22. Sandy MR (1995) A review of some Palaeozoic and Mesozoic brachiopods as members of cold seep chemosynthetic communities: "unusual" palaeoecology and anomalous palaeobiogeographic pattern explained. Földtani Közlöny 125: 241–258.

23. Sandy MR (2010) Brachiopods from ancient hydrocarbon seeps and hydrothermal vents. In: Kiel S, editor. The Vent and Seep Biota. Heidelberg: Springer. pp. 279–314.

24. Gischler E, Sandy MR, Peckmann J (2003) Ibergirhynchia contraria (F.A. Roemer, 1850), an Early Carboniferous seep-related rhynchonellide brachiopod from the Harz Mountains, Germany-a possible successor to Dzieduszyckia? J Paleont 77: 293–303. doi: 10.1666/0022-3360(2003)077<0293:icfara>2.0.co;2

25. Peckmann J, Kiel S, Sandy MR, Taylor DG, Goedert JL (2011) Mass occurrences of the brachiopod Halorella in Late Triassic methane-seep deposits, Eastern Oregon. J Geol 119: 207–220. doi: 10.1086/658199

26. Peckmann J, Campbell KA, Walliser OH, Reitner J (2007) A Late Devonian hydrocarbon-seep deposit dominated by dimerelloid brachiopods, Morocco. Palaios 22: 114–122.

27. Peckmann J, Sandy MR, Taylor DG, Gier S, Bach W (2013) An Early Jurassic brachiopod-dominated seep deposit enclosed by serpentinite, eastern Oregon, USA. Palaeogeogr Palaeoclimatol Palaeoecol 390: 4–16. doi: 10.1016/j.palaeo.2013.01.003

28. Short EJ, Snyder DC, Trop J, Hart WK, Layer PW (2005) New findings on Early Cretaceous volcanism within the allochthonous Wrangellia Terrane, south-central Alaska: stratigrpahic, geochronogic, and geochemical data from the Chisana Formation, Nutzotin Mountains. Geol Soc Am Abstr Program 37: 81.

29. Thieuloy J-P (1972) Biostratigraphie des lentilles a peregrinelles (Brachiopodes) de l'Hauterivien de Rottier (Drôme, France). Geobios 5: 5–53. doi: 10.1016/s0016-6995(72)80004-4

30. Kiel S (2013) Lucinid bivalves from ancient methane seeps. J Moll Stud 79: 346–363. doi: 10.1093/mollus/eyt035

31. Bosellini A, Morsilli M (1997) A Lower Cretaceous downing unconformity on the eastern flank of the Apulia Platform (Gargano Promontory, southern Italy). Cret Res 18: 51–61. doi: 10.1006/cres.1996.0049

32. Posenato R, Morsilli M (1999) New species of Peregrinella (Brachiopoda) from the Lower Cretaceous of the Gargano Promontory (southern Italy). Cret Res 20: 641–654. doi: 10.1006/cres.1999.0171

33. Graziano R, Taddei Ruggiero E (2008) Cretaceous brachiopod-rich facies of the carbonate platform-to-basin transitions in southern Italy: stratigraphic and paleoenvironmental significance. Boll Soc Geol Ital 127: 407–422.

34. Boller K (1963) Stratigraphische und mikropaläontologische Untersuchungen in Neocom der Kippendecke (östlich der Rhone). Eclogae geol Helvetiae 56: 15–102.

35. Kiel S, Peckmann J (2008) Paleoecology and evolutionary significance of an Early Cretaceous Peregrinella-dominated hydrocarbon-seep deposit on the Crimean Peninsula. Palaios 23: 751–759.

36. Kiel S (2008) Parasitic polychaetes in the Early Cretaceous hydrocarbon seep-restricted brachiopod Peregrinella multicarinata. J Paleont 82: 1214–1216. doi: 10.1666/08-003.1

37. Peckmann J, Birgel D, Kiel S (2009) Molecular fossils reveal fluid composition and flow intensity at a Cretaceous seep. Geology 37: 847–850. doi: 10.1130/g25658a.1

38. Berkland JO (1973) Rice Valley outlier - new sequence of Cretaceous-Paleocene strata in northern Coast Ranges, California. Geol Soc Am Bull 84: 2389–2406. doi: 10.1130/0016-7606(1973)84<2389:rvosoc>2.0.co;2

39. Campbell KA, Farmer JD, Des Marais D (2002) Ancient hydrocarbon seeps from the Mesozoic convergent margin of California: carbonate geochemistry, fluids and palaeoenvironments. Geofluids 2: 63–94. doi: 10.1046/j.1468-8123.2002.00022.x

40. Birgel D, Thiel V, Hinrichs K-U, Elvert M, Campbell KA, et al. (2006) Lipid biomarker patterns of methane-seep microbialites from the Mesozoic convergent margin of California. Org Geochem 37: 1289–1302. doi: 10.1016/j.orggeochem.2006.02.004

41. Kiel S, Campbell KA, Elder WP, Little CTS (2008) Jurassic and Cretaceous gastropods from hydrocarbon-seeps in forearc basin and accretionary prism settings, California. Acta Palaeontol Pol 53: 679–703. doi: 10.4202/app.2008.0412

42. Sandy MR, Lazar I, Peckmann J, Birgel D, Stoica M, et al. (2012) Methane-seep brachiopod fauna within turbidites of the Sinaia Formation, Eastern Carpathian Mountains, Romania. Palaeogeogr Palaeoclimatol Palaeoecol 323–325: 42–59. doi: 10.1016/j.palaeo.2012.01.026

43. MacDonald IR (1998) Stability and change in Gulf of Mexico chemosynthetic communities: Interim Report. Prepared by the Geochemical and Environmental Research Group, Texas A&M University. U.S. Dept. of the Interior, Minerals Mgmt. Service, Gulf of Mexico OCS Region, New Orleans. 116 p.

44. Brooks JM, Fisher CR, Roberts HH, Bernard B, MacDonald IR, et al. (2009) Investigations of chemosynthetic communities on the lower continental slope of the Gulf of Mexico: Interim Report 2. U.S. Dept. of the Interior, Minerals Mgmt. Service, Gulf of Mexico OCS Region, New Orleans. 360 p.

45. Chryploff G (1958) Ein Fund von Peregrinella cf. peregrina d'Orbigny in der Unteren Kreide Norddeutschlands (Gebiet Werle in Mecklenburg). Z angew Geol 4: 70.

46. Brand U, Logan A, Hiller N, Richardson J (2003) Geochemistry of modern brachiopods: applications and implications for oceanography and paleoceanography. Chem Geol 198: 305–334. doi: 10.1016/s0009-2541(03)00032-9

47. Lear CH, Elderfield H, Wilson PA (2000) Cenozoic deep-sea temperatures and global ice volumes from Mg/Ca in benthic foraminiferal calcite. Science 287: 269–272. doi: 10.1126/science.287.5451.269

48. McArthur JM (1994) Recent trends in strontium isotope stratigraphy. Terra Nova 6: 331–358. doi: 10.1111/j.1365-3121.1994.tb00507.x

49. Bailey TR, McArthur JM, Prince H, Thirlwall MF (2000) Dissolution methods for strontium isotope stratigraphy: whole rock analysis. Chem Geol 167: 313–319. doi: 10.1016/s0009-2541(99)00235-1
50. Reinhardt EG, Cavazza W, Patterson RT, Blenkinsop J (2000) Differential diagenesis of sedimentary components and the implication for strontium isotope analysis of carbonate rocks. Chem Geol 164: 331–343. doi: 10.1016/s0009-2541(99)00147-3
51. McArthur JM, Howarth RJ, Bailey TR (2001) Strontium isotope stratigraphy: LOWESS version 3: Best fit to the marine Sr-isotope curve for 0–509 Ma and accompanying look-up table for deriving numerical age. J Geol 109: 155–170. doi: 10.1086/319243
52. McArthur JM, Howarth RJ, Shields GA (2012) Strontium isotope stratigraphy. In: Gradstein FM, Ogg JG, Schmitz M, Ogg G, editors. The Geologic Time Scale 2012. Elsevier. pp. 127–144.
53. Carroll M, Romanek CS (2008) Shell layer variation in trace element concentration for the freshwater bivalves Elliptio complanata. Geo-Mar Let 28: 369–381. doi: 10.1007/s00367-008-0117-3
54. Nielsen SN, Glodny J (2009) Early Miocene subtropical water temperatures in the southeast Pacific. Palaeogeogr Palaeoclimatol Palaeoecol 280: 480–488. doi: 10.1016/j.palaeo.2009.06.035
55. Birgel D, Elvert M, Han X, Peckmann J (2008) 13C-depleted biphytanic diacids as tracers of past anaerobic oxidation of methane. Org Geochem 39: 152–156. doi: 10.1016/j.orggeochem.2007.08.013
56. Hammer Ø, Harper DAT, Ryan PD (2001) PAST: Palaeontological Statistics software package for education and data analysis. Palaeontol Electronica 4: 9 pp.
57. Paull CK, Chanton JP, Martens CS, Fullagar PD, Neumann AC, et al. (1991) Seawater circulation through the flank of the Florida platform: evidence and implications. Mar Geol 102: 265–279. doi: 10.1016/0025-3227(91)90011-r
58. Aharon P, Schwarcz HP, Roberts HH (1997) Radiometric dating of submarine hydrocarbon seeps in the Gulf of Mexico. Geol Soc Am Bull 109: 568–579. doi: 10.1130/0016-7606(1997)109<0568:rdoshs>2.3.co;2
59. Anderson FM (1938) Lower Cretaceous deposits in California and Oregon. Geol Soc Am Spec Pap 16: 1–339. doi: 10.1130/spe16-p1
60. Gabb WM (1869) Cretaceous and Tertiary fossils: Palaeontology, v. II. Geological Survey of California 1–299.
61. Campbell KA, Peterson D, Alfaro AC (2008) Two new species of Retiskenea? (Gastropoda: Neomphalidae) from Lower Cretaceous hydrocarbon seep-carbonates of northern California. J Paleont 82: 140–153. doi: 10.1666/05-025.1
62. Slaczka A, Kaminski MA (1998) A guidebook to excursions in the Polish Flysch Carpathians. Grzybowski Foundation Special Publication 6: 1–173.
63. Ortiz-Hernández LE, Acevedo-Sandoval OA, Flores-Castro K (2003) Early Cretaceous intraplate seamounts from Guanajuato, central Mexico: geochemical and mineralogical data. Rev Mex Sci Geol 20: 27–40.
64. Gradstein FM, Ogg JG, Schmitz M, Ogg G, Eds., The Geologic Time Scale 2012 (Elsevier, 2012), pp. 1176.
65. Peckmann J, Thiel V (2004) Carbon cycling at ancient methane-seeps. Chem Geol 205: 443–467. doi: 10.1016/j.chemgeo.2003.12.025
66. Graziano R (1999) The Early Cretaceous drowning unconformities of the Apulia carbonate platform (Gargano Promontory, southern Italy): local fingerprints of global palaeoceanographic events. Terra Nova 11: 245–250. doi: 10.1046/j.1365-3121.1999.00256.x
67. Kiel S, Campbell KA, Gaillard C (2010) New and little known mollusks from ancient chemosynthetic environments. Zootaxa 2390: 26–48.
68. Birgel D, Himmler T, Freiwald A, Peckmann J (2008) A new constraint on the antiquity of anaerobic oxidation of methane: Late Pennsylvanian seep limestones from southern Namibia. Geology 36: 543–546. doi: 10.1130/g24690a.1
69. Goossens H, de Leeuw JW, Schenck PA, Brassell SC (1984) Tocopherols as likely precursors of pristane in ancient sediments and crude oils. Nature 312: 440–442. doi: 10.1038/312440a0
70. Hayes JM, Strauss H, Kaufman AJ (1999) The abundance of 13C in marine organic matter and isotopic fractionation in the global biogeochemical cycle of carbon during the past 800 Ma. Chem Geol 161: 103–125. doi: 10.1016/s0009-2541(99)00083-2
71. Campbell KA, Carlson C, Bottjer DJ (1993) Fossil cold seep limestones and associated chemosymbiotic macroinvertebrate faunas, Jurassic-Cretaceous Great Valley Group, California. In: Graham SA, Lowe DR, editors. Advances in the Sedimentary Geology of the Great Valley Group, Sacramento Valley, California. Los Angeles: Pacific Section of the Society of economic Paleontologists and Mineralogists. pp. 37–50.
72. Kiel S, Wiese F, Titus AL (2012) Shallow-water methane-seep faunas in the Cenomanian Western Interior Seaway: No evidence for onshore-offshore adaptations to deep-sea vents. Geology 40: 839–842. doi: 10.1130/g33300.1
73. Kiel S (2010) On the potential generality of depth-related ecologic structure in cold-seep communities: Cenozoic and Mesozoic examples. Palaeogeogr Palaeoclimatol Palaeoecol 295: 245–257. doi: 10.1016/j.palaeo.2010.05.042
74. Jenkins RG, Kaim A, Little CTS, Iba Y, Tanabe K, et al. (2013) Worldwide distribution of modiomorphid bivalve genus Caspiconcha in late Mesozoic hydrocarbon seeps. Acta Palaeontol Pol 58: 357–382. doi: 10.4202/app.2011.0026
75. Kaim A, Kelly SRA (2009) Mass occurrence of hokkaidoconchid gastropods in the Upper Jurassic methane seep carbonate from Alexander Island, Antarctica. Antarct Sci 21: 279–284. doi: 10.1017/s0954102009001813
76. Hammer Ø, Nakrem HA, Little CTS, Hryniewicz K, Sandy MR, et al. (2011) Hydrocarbon seeps close to the

Jurassic-Cretaceous boundary, Svalbard. Palaeogeogr Palaeoclimatol Palaeoecol 306: 15–26. doi: 10.1016/j.palaeo.2011.03.019

77. Kiel S, Amano K, Jenkins RG (2008) Bivalves from Cretaceous cold-seep deposits on Hokkaido, Japan. Acta Palaeontol Pol 53: 525–537. doi: 10.4202/app.2008.0310

78. Kiel S, Birgel D, Campbell KA, Crampton JS, Schiøler P, et al. (2013) Cretaceous methane-seep deposits from New Zealand and their fauna. Palaeogeogr Palaeoclimatol Palaeoecol 390: 17–34. doi: 10.1016/j.palaeo.2012.10.033

79. Sahling H, Rickert D, Lee RW, Linke P, Suess E (2002) Macrofaunal community structure and sulfide flux at gas hydrate deposits from the Cascadia convergent margin, NE Pacific. MEPS 231: 121–138. doi: 10.3354/meps231121

80. Newman WA, Yamaguchi T, Southward AJ (2006) Arthropoda, Crustacea, Cirripedia. Denisia 18: 359–368.

81. Sandy MR, Campbell KA (1994) New rhynchonellid brachiopod genus from Tithonian (Upper Jurassic) cold seep deposits of California and its paleoenvironmental setting. J Paleont 68: 1243–1252.

82. Sepkoski JJ (2002) A compendium of fossil marine animal genera. Bull Am Paleont 363: 1–563.

83. Sepkoski JJ (1998) Rates of speciation in the fossil record. Philosophical Transactions of the Royal Society of London B 353: 315–326. doi: 10.1098/rstb.1998.0212

84. Stanley SM (2007) An analysis of the history of marine animal diversity. Paleobiology 33: 1–55. doi: 10.1666/06020.1

85. Thuy B, Kiel S, Dulai A, Gale AS, Kroh A, et al. (2014) First glimpse into Lower Jurassic deep-sea biodiversity: in-situ diversification and resilience against extinction. Proc Biol Sci 281: 20132624. doi: 10.1098/rspb.2013.2624

86. Pucéat E, Lécuyer C, Reisberg L (2005) Neodymium isotope evolution of NW Tethyan upper ocean waters throughout the Cretaceous. EPSL 236: 705–720. doi: 10.1016/j.epsl.2005.03.015

87. Ager DV (1986) Migrating fossils, moving plates and an expanding Earth. Modern Geol 10: 377–390.

88. Bodin S, Godet A, Föllmi KB, Vermeulen J, Arnaud H, et al. (2006) The late Hauterivian Faraoni oceanic anoxic event in the western Tethys: Evidence from phosphorus burial rates. Palaeogeogr Palaeoclimatol Palaeoecol 235: 245–264. doi: 10.1016/j.palaeo.2005.09.030

89. Cordes EE, Bergquist DC, Fisher CR (2009) Macro-ecology of Gulf of Mexico cold seeps. Ann Rev Mar Sci 1: 143–168. doi: 10.1146/annurev.marine.010908.163912

90. Van Dover CL (2000) The ecology of deep-sea hydrothermal vents. Princeton: Princeton University Press. 424 p.

91. Niemann H, Fischer D, Graffe D, Knittel K, Montiel A, et al. (2009) Biogeochemistry of a low-activity cold seep in the Larsen B area, western Weddell Sea, Antarctica. Biogeosciences 6: 2383–2395. doi: 10.5194/bg-6-2383-2009

92. Luff R, Wallmann K (2003) Fluid flow, methane fluxes, carbonate precipitation and biogeochemical turnover in gas hydrate-bearing sediments at Hydrate Ridge, Cascadia Margin: Numerical modeling and mass balances. Geochim Cosmochim Acta 67: 3403–3421. doi: 10.1016/s0016-7037(03)00127-3

93. Birgel D, Peckmann J (2008) Aerobic methanotrophy at ancient marine methane seeps: A synthesis. Org Geochem 39: 1659–1667. doi: 10.1016/j.orggeochem.2008.01.023

94. Little CTS, Herrington RJ, Haymon RM, Danelian T (1999) Early Jurassic hydrothermal vent community from the Franciscan Complex, San Rafael Mountains, California. Geology 27: 167–170. doi: 10.1130/0091-7613(1999)027<0167:ejhvcf>2.3.co;2

95. Thayer CW (1981) Ecology of living brachiopods. In: Broadhead TW, editor. Lophophorates. Notes for a Short Course. Knoxville: University of Tennessee. pp. 110–126.

96. Manceñido MO, Owen EF (2001) Post-Palaeozoic Rhynchonellida (Brachiopoda): classification and evolutionary background. In: Brunton CH, Cocks LRM, Long SL, editors. Brachiopods Past and Present. The Systematics Association, Special Volume series 63. London and New York: Taylor and Francis. pp. 189–200.

97. Lazar I, Panaiotu CE, Grigore D, Sandy MR, Peckmann J (2011) An unusual brachiopod assemblage in a Late Jurassic (Kimmeridgian) stromatactis mud-mound of the Eastern Carpathians (Haghimas Mountains), Romania. Facies 57: 627–647. doi: 10.1007/s10347-010-0261-x

98. Scotese CR (2004) A continental drift flipbook. J Geol 112: 729–741. doi: 10.1086/424867

99. Keenan KE (2010) Hydrocarbon seeps of the Mesozoic Great Valley Group forearc strata and Franciscan Complex, northern and central California, U.S.A. PhD thesis, University of California, Riverside. 273 p.

100. Amano K, Jenkins RG, Hikida Y (2007) A new gigantic Nucinella (Bivalvia: Solemyoida) from the Cretaceous cold-seep deposit in Hokkaido, northern Japan. Veliger 49: 84–90.

101. Kaim A, Jenkins RG, Tanabe K, Kiel S (2014) Mollusks from late Mesozoic seep deposits, chiefly in California. Zootaxa. 3861: 401–440 doi: 10.11646/zootaxa.3861.5.1

102. Renngarten V (1924) O Kavkaykikh Peregrinellae (Sur les Pérégrinelles du Caucase). Izvestiia Geol Komiteta, Leningrad 42: 119–128.

103. Gaillard C, Rio M, Rolin Y (1992) Fossil chemosynthetic communities related to vents or seeps in sedimentary basins: the pseudobioherms of southeastern France compared to other world examples. Palaios 7: 451–465.

104. Kelly SRA, Ditchfield PW, Doubleday PA, Marshall JD (1995) An Upper Jurassic methane-seep limestone from the Fossil Bluff Group forearc basin of Alexander Island, Antarctica. J Sediment Res A65: 274–282. doi: 10.1306/d426809a-2b26-11d7-8648000102c1865d

105. Stanton TW (1895) Contributions to the Cretaceous paleontology of the Pacific coast: the fauna of the Knoxville beds. US Geol Surv Bull 133: 1–132.

106. Kaim A, Skupien P, Jenkins RG (2013) A new Lower Cretaceous hydrocarbon seep locality from Czech Carpathians and its fauna. Palaeogeogr Palaeoclimatol Palaeoecol 390: 42–51. doi: 10.1016/j.palaeo.2013.03.010

107. Kelly SRA, Blanc E, Price SP, Withham AG (2000) Early Cretaceous giant bivalves from seep-related limestone mounds, Wollaston Forland, Northeast Greenland. In: Harper EM, Taylor JD, Crame JA, editors. The evolutionary biology of the Bivalvia. London: Geological Society of London, Special Publication. pp. 227–246.

108. Agirrezabala LM, Kiel S, Blumenberg M, Schäfer N, Reitner J (2013) Outcrop analogues of pockmarks and associated methane-seep carbonates: a case study from Lower Cretaceous (Albian) of the Basque-Cantabrian Basin, western Pyrenees. Palaeogeogr Palaeoclimatol Palaeoecol 390: 94–115. doi: 10.1016/j.palaeo.2012.11.020

109. Nobuhara T, Onda D, Kikuchi N, Kondo Y, Matsubara K, et al. (2008) Lithofacies and fossil assemblages of the Upper Cretaceous Sada Limestone, Shimanto City, Kochi Prefecture, Shikoku, Japan. Fossils Palaeont Soc Japan 84: 47–60.

110. Jenkins RG, Kaim A, Hikida Y, Tanabe K (2007) Methane-flux-dependent lateral faunal changes in a Late Cretaceous chemosymbiotic assemblage from the Nakagawa area of Hokkaido, Japan. Geobiology 5: 127–139. doi: 10.1111/j.1472-4669.2007.00106.x

CHAPTER 9

EVOLUTION OF ENDEMIC SPECIES, ECOLOGICAL INTERACTIONS AND GEOGRAPHICAL CHANGES IN AN INSULAR ENVIRONMENT

Antonella Cinzia Marra

Keywords: Sicily, mammals, Quaternary, endemism, island paleobiogeography, paleoecology, biochronology

INTRODUCTION

In the last few decades, knowledge of the Quaternary mammals of the island of Sicily has increased, allowing new considerations of time periods and systems of faunal dispersals from the surrounding continental areas [1,2,3,4,5,6]. The first mammal community, from the Early Pleistocene age, has been recorded in only one site. Its composition is very interesting and needs more investigation [1,2]. Later, Sicily experienced four different faunal complexes, characterized by the extinction of some species followed by the dispersal of newcomers from peninsular Italy [1,2,5]. Three faunal complexes include endemic species, evolved in insular conditions. Changes in faunal composition are related to dispersal events, followed by isolation phases, which induced processes of endemism in the pioneer species [2,5]. Dispersals depend on paleogeographical conditions and on dispersal abilities of the precursor species [2,5,6]. While the main dispersal direction came from Peninsular Italy, some of the earliest arrivals (dubitatively attributed to Late Miocene) from North Africa also need to be considered [4].

The spreading of mammals from the Italian mainland to Sicily has not only been shaped by the sea barrier of the Strait of Messina, but also by the physiography of Calabria, the southernmost region of peninsular Italy (Figure 1; [6]). During Early Pleistocene, Calabria and Sicily were fragmented in small islands, constituting an archipelago [6]. In the course of Middle Pleistocene, the uplift of inner and coastal plains determined a geographical setting more and more similar to present day [6]. Calabria has been the main filter to mammal spreading, especially in the first half of Pleistocene. During late Middle and Late Pleistocene, Calabria acted as a filter for its complex physiography and latitudinal extension. As a matter of fact, all mammals present in late Middle and Late Pleistocene of Calabria spread to Sicily, with the exception of a few species of prairie or mountain environments [6]. Moreover, faunal assemblages of Calabria include a lower number of species with respect to central and northern Italy [6]. Some species of Sicily are present on Malta [5]. Maltese faunal associations are impoverished with respect to Sicilian ones and include mammals which are thought to easily cross sea barriers.

Citation: Cinzia Marra, Evolution of Endemic Species, Ecological Interactions and Geographical Changes in an Insular Environment, doi:10.3390/geosciences3010114.

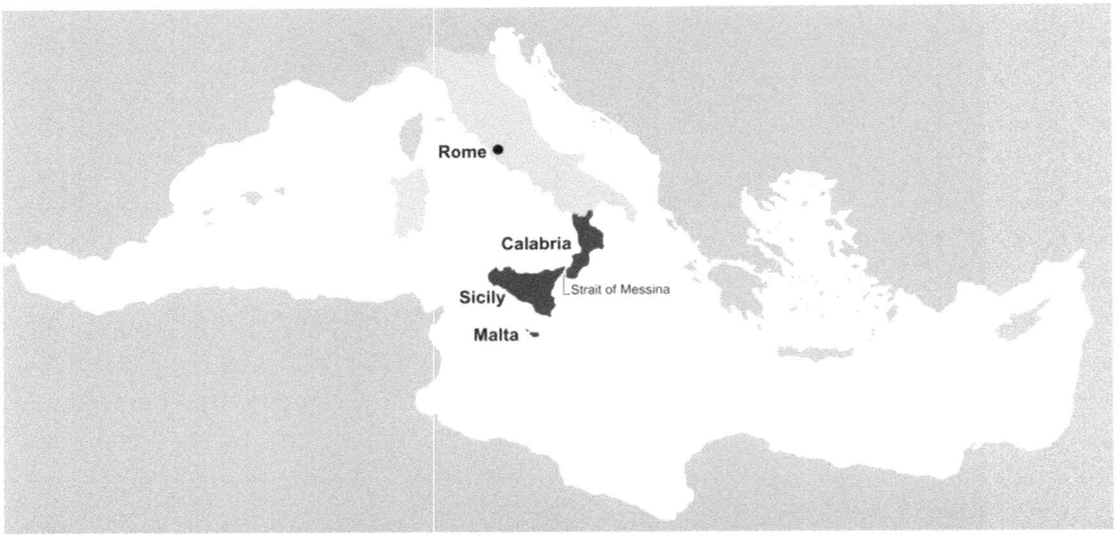

Figure 1. Map of Mediterranean area with locations of the regions quoted in the text.

Dispersal of mammals from the mainland to the islands depends on their body structure and behavior, in addition to geographical setting [7,8]. Good fliers and good swimmers clearly have better chances of reaching islands far from the mainland, while animals with more or less pronounced dispersal ability can disperse with some limitations. On the island, pioneer species can evolve adaptations, such as size changes and allometries, giving rise to endemic species or subspecies. They can lose their dispersal ability (e.g., flying attitude) or their defensive adaptations. The character of insular faunas needs to be investigated, not only as changing in species attributes, but also as features of the whole faunal association [7,8]. Usually, faunal associations on an island are affected by impoverishment in terms of biodiversity and by ecological disharmony [7,8].

According to Simpson [7], Dermitzakis and Sondaar [8] and Sondaar *et al.* [9], the composition of insular faunas is related to the way of dispersal: an ecologically balanced fauna, similar to the mainland one, results from a broad land connection, which acts as a corridor, allowing interchanges from one region to another. An impoverished but balanced fauna with absent or slight endemism is related to a land connection with obstacles, acting as a filter to dispersal, where spread is probable for some taxa only. An unbalanced fauna with mainland elements is related to a filter ("pendel route"), such as a narrow sea strait, which is easily crossed by good swimmers and flyers; an endemic fauna with endemic features is related to accidental ("sweepstake") transit, as across a wide sea barrier, and spread occurs accidentally and for some taxa only.

In the case of the Quaternary mammals of Sicily, it is possible to indicate the dispersal ability of mainland ancestors and the endemism degree of insular species, as well as impoverishment and ecological balance of each faunal complex [5]. As can be inferred by paleogeographical reconstructions, physiographic changes of Calabria and Sicily determined different possibilities of mammal spreading through time [2,6,10]. Reconstruction of times and ways of dispersal from the mainland, with observations on paleogeography and endemism, makes up the larger part of the literature. However, paleoecology has been considered only in terms of species autoecology and ecological balance of associations [1,4,5]. Very few studies about ecological relationships among species and stability of communities have been conducted.

The aim of this paper is to present the current state of knowledge and to raise some critical observations and possible re-interpretations. The first step is a critical review of mammal biochronology, being a reference to investigate paleobiogeography, endemism and paleoecology of Quaternary mammals of Sicily. Then, available data are used to give a picture of community structure and interspecific interactions on an island subject to relevant changes in geography and environment. The review of data sourcing from different branches of geology and paleontology, as well as from genetics, enables interesting considerations on time periods and systems of dispersals, on evolution of endemism, and on ecological interactions. It is worth noting that paleobiogeography of Sicily can be ascribed to different types of islands, as a consequence of its deep changes in geography, source availability, and possibilities of mammal dispersals.

REVIEW OF BIOCHRONOLOGY

The Quaternary vertebrate fauna of Sicily is presently arranged into five Faunal Complexes (FCs), according to Bonfiglio *et al.* [1,2] and subsequent integrations [4,5,11]. Data from the cited literature are listed below (obsolete scientific names are rectified in square brackets). Mammal associations and ages are reported; endemic species are indicated by "E". Although it is not applicable to island associations, the correspondence to the biochronological chart of Italy [12,13] is indicated, being significant to timespan and changes in mainland faunal associations.

- Monte Pellegrino Faunal Complex

Mammal association: *Asoriculus burgioi* (E), *Mustelercta* (=*Pannonictis*) *arzilla* (E), *Apodemus maximus* (E), *Leithia* sp. (E), *Maltamys* sp. (E), *Pellegrinia panormensis* (E), *Hypolagus peregrinus*(E). The Monte Pellegrino FC is based on a single faunal assemblage coming from a karst cavity with a problematic depositional setting.

Age: Pliocene–Early Pleistocene according to Thaler [14]; Early Pleistocene, according to Burgio and Fiore [15]; Late Villafranchian Mammal Age of the biochronological chart of Italy.

- *Elephas falconeri* Faunal Complex

Mammal association: *Crocidura esuae* (E), *Lutra trinacriae* (E), *Vulpes* sp. (dubitative), *Ursus* sp. ("small bear," dubitative), *Elephas falconeri* [recte *Palaeoloxodon falconeri*] (E), *Leithia cartei* (E), *Leithia melitensis* (E), *Maltamys gollcheri* (E), undetermined bats. The mammal fauna is associated with diversified herpetofauna and avifauna.

Age: early Middle Pleistocene, probably MIS 22–11; Middle and Late Galerian of the biochronological chart of Italy.

- *Elephas mnaidriensis* Faunal Complex

Mammal association: *Crocidura esuae* (E), *Erinaceus europaeus* (E), *Lutra trinacriae* (E), *Crocuta crocuta spelaea*, *Panthera leo*, *Canis lupus*, *Ursus arctos*, *Elephas mnaidriensis* [recte *Palaeoloxodon mnaidriensis*] (E), *Sus scrofa*, *Hippopotamus pentlandi* (E), *Cervus elaphus siciliae* (E), *Dama carburangelensis* (E), *Bos primigenius siciliae* (E), *Bison priscus siciliae* (E), *Leithia melitensis* (E), *Maltamys wiedincitensis* (E), *Microtus* (*Terricola*) sp. In the opinion of Petruso *et al.* [11], *Bison priscus siciliae* is erroneously attributed to this faunal complex and must be considered present only in the "Grotta San Teodoro-Pianetti" FC.The mammal fauna is associated with diversified herpetofauna and avifauna.

Age: late Middle Pleistocene, probably MIS 10–4; Aurelian Mammal Age of the biochronological chart of Italy.

- Grotta San Teodoro—Pianetti Faunal Complex

Mammal association: *Erinaceus europaeus*, *Crocidura* cf. *sicula* (E), *Crocuta crocuta spelaea*, *Canis lupus*, *Vulpes vulpes*, *Ursus arctos*, *Elephas mnaidriensis* [recte *Palaeoloxodon mnaidriensis*] (E), *Sus scrofa*, *Cervus elaphus siciliae* (E), *Bos primigenius siciliae* (E), *Bison priscus siciliae* (E), *Equus hydruntinus*, *Apodemus* cf. *A. sylvaticus*, *Microtus* (*Terricola*) *savii* (E).

Mammal fauna is associated with herpetofauna and avifauna.

Age: Late Pleistocene; Grotta San Teodoro: 32,000 ± 4000 yr BP (^{230}Th/^{234}U dating on a concretion, Bonfiglio *et al.* [16]; late Aurelian (Last Glacial cycle, more precisely) according to the biochronology of peninsular Italy.

- Castello Faunal Complex

Mammal association: *Erinaceus europaeus*, *Crocidura* cf. *C. sicula* (E), *Canis lupus*, *Vulpes vulpes*, *Equus hydruntinus*, *Sus scrofa*, *Cervus elaphus*, *Bos primigenius*, *Equus ferus*, *Microtus* (*Terricola*) *savii*, *Apodemus* cf. *sylvaticus*, *Lepus europaeus*.

Age: late MIS 3—early MIS 2; late Aurelian (late Glacial) Mammal Age of the biochronological chart of Italy.

The arrangement of the vertebrate fauna of Sicily proposed in several papers by the research teams coordinated by Bonfiglio and Masini [1,2,4] started from new finds and stratigraphical data and followed, among the others, the significant papers by Vaufrey [17], Kotsakis [18] and Burgio [19].

Vaufrey [17] distinguished a first upper level attributed to Upper Paleolithic and a second lower level named "*terre rouge à elephants*" (=red soil with elephant remains). At Luparello cave, Vaufrey excavated elephant remains having two different sizes, interpreted as two subsequent steps of dwarfism in the Sicilian elephants, moving from the forerunner *Elephas antiquus*. This theory has been accepted for almost 50 years.

In 1979, Tassos Kotsakis [18] provided the first modern chronological study of the fossil vertebrates of Sicily. Kotsakis [18] grouped the known local faunal associations into four stages, named by the most representative site, and correlated them to the peninsular Italy biochronology, at that time still in progress [20] and here reported only for "historical" purposes. Kotsakis [18] defined five faunal "Stages", from the oldest to the youngest one (obsolete scientific names are rectified in square brackets):

-Monte Pellegrino Stage (fauna characterized by the presence of *Pellegrinia panormensis*associated with *Leithia* sp., a murid, a leporid, and a mustelid); type site: Monte Pellegrino (Palermo); Age: Ruscinian and/or Villafranchian and/or Galerian.

- Capo Tindari Stage: fauna characterized by the presence of a large sized *Hippopotamus* sp. associated with *Megaceros messinae* (probably attributable to *Dama carburangelensis*) and *Leithia*; type site: Capo Tindari (Messina); Age: Galerian or Flaminian or Pariolin.
- Maccagnone Stage (fauna characterized by the presence of *Elephas mnaidriensis* [recte *Palaeoloxodon mnaidriensis*] associated with *Cervus siciliae* [recte *Cervus elaphus siciliae*], *Leithia melitensis*, *Leithia cartei*), divided into two phases by the presence or absence of *Hippopotamus pentlandi*:

(a) *Hippopotamus pentlandi* is present; type site: Maccagnone Cave (Palermo); Age: Pariolin, Rianin, Maspinian.

(b) *Hippopotamus pentlandi* is absent; type site: Carburangeli Cave (Palermo); Age: end of Maspinian, Pontinian.

- Spinagallo Stage (fauna characterized by the presence of *Elephas falconeri* [recte *Palaeoloxodon falconeri*] associated with *Leithia melitensis* and *Leithia cartei*); type site: Spinagallo cave (Syracuse); Age: Middle Pontinian.
- Castello Stage (fauna characterized by the presence of *Equus hydruntinus* associated with *Cervus elaphus* and *Homo sapiens;* elephants are absent); type site: Castello cave (Palermo); Age: Upper Pontinain.

New stratigraphic evidences from Alcamo [21] and Comiso [22], also supported by absolute dating [23], assigned to the Spinagallo Stage an earlier age than the Maccagnone one. The "classical" interpretation was based on the derivation of *Elephas falconeri* [recte *Palaeoloxodon falconeri*] from *Elephas mnaidriensis* [recte *Palaeoloxodon mnaidriensis*], through an evolutionary process characterized by a progressive reduction in size. On the contrary, the new data were consistent with two different arrivals of elephants and two different dwarfing processes on the forerunners.

In light of these new data, Burgio [19] redefined the stages by Kotsakis [18] (obsolete scientific names are rectified in square brackets; locations of type sites are in Figure 2).

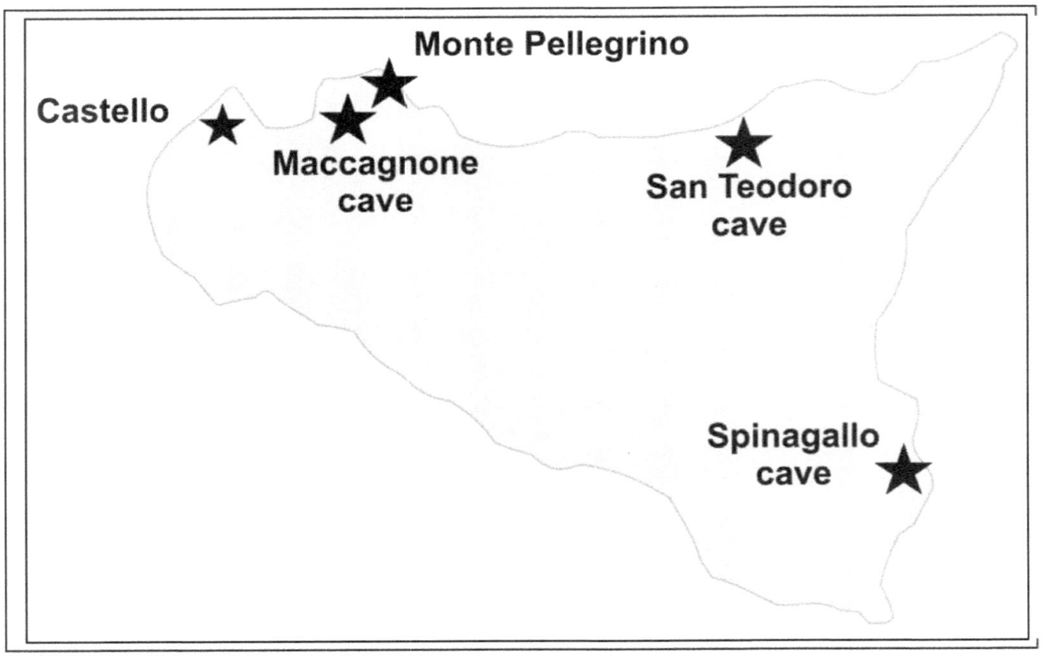

Figure 2. Locations of type-sites representative of faunal complexes of Sicily.

- Monte Pellegrino Stage (fauna characterized by the presence of *Pellegrinia panormensis*). Faunal associations: *Apodemus maximus*, *Maltamys* cf. *gollcheri*, *Pellegrinia panormensis*, *Hypolagus* sp. [recte *Hypolagus peregrines*], *Soricidae* gen. et sp. indet., *Mustelercta arzilla* [recte *Mustelercta* (= *Pannonictis*) *arzilla*]
- Spinagallo Stage (fauna characterized by the presence of *Elephas falconeri* [recte *Palaeoloxodon falconeri*]). Faunal association: *Leithia melitensis*, *Leithia cartei*, *Crocidura esuae*, *Lutra trinacriae*, *Elephas falconeri* [recte *Palaeoloxodon falconeri*]. Age: 500 ka by absolute dating [23]; Mindel-Riss.
- Maccagnone Stage (fauna characterized by the presence of *Elephas mnaidriensis* [recte *Palaeoloxodon mnaidriensis*]). Faunal association: *Leithia melitensis*, *Maltamys wiedinticitensis*, *Crocidura esuae*, *Crocuta crocuta spelaea*, *Panthera leo spelaea*, *Canis lupus*, *Ursus arctos*, *Elephas antiquus* [recte *Palaeoloxodon antiques*], *Elephas mnaidriensis* [recte *Palaeoloxodon mnaidriensis*], *Elephas* sp., *Sus scrofa*, *Hippopotamus pentlandi*, *Hippopotamus* sp., *Cervus elaphus siciliae*, *Dama dama* [recte *Dama carburangelensis*], Megaceros (*s.l.*) *carburangelensis* [recte *Dama carburangelensis*], *Bos primigenius siciliae*, *Bison priscus siciliae*. Age: 200 ka by absolute dating [23]; attributable to MIS 5.
- Castello Stage (fauna characterized by the presence of *Equus hydruntinus*). Faunal association: *Hystrix cristata*, *Pytimis savii* [recte *Microtus* (*Terricola*) savii], *Lepus europaeus*, *Erinaceus europaeus*, *Crocidura* sp., *Lynx lynx*, *Felis silvestris*, *Canis lupus*, *Vulpes vulpes*, *Equus caballus*[recte *Equus ferus*], *Equus hydruntinus*, *Sus scrofa*, *Cervus elaphus elaphus*, *Bos primigenius primigenius*, *Homo sapiens sapiens*. Age: some millennia before Holocene (10 ka), Upper Paleolithic.

PALEOBIOGEOGRAPHY, ENDEMISM AND ECOLOGICAL INTERACTIONS

The historical review of biochronology is helpful in understanding the current biochronological framework of Sicily. Papers by Bonfiglio, Masini and colleagues [1,2,3,4] greatly improved the current knowledge and represent a good reference, as the studies were based on broad datasets. However, some criticism does exist. The authors correctly abandoned the name "stage" and used the most appropriate term "Faunal Complex" (FC), which is synonymous with "faunal unit." Faunal complexes are represented by non-overlapping faunal associations well defined in time, space, and ecological attributes (according to the definition of faunal units by Tedford [24]). The authors used new denominations for faunal complexes that, in spite of new data, were substantially coincident with the old "stages" stated by Burgio [19] in 1997: *Elephas falconeri* FC for the Spinagallo Stage; *Elephas mnaidriensis* FC for the Maccagnone Stage. However, these denominations are inadequate for several reasons:

- faunal complexes are usually stated on the basis of Local Faunal Units (LFAs) and are named by the most representative site [12,13];
- denominations by species names are unusual; moreover, in this case, scientific names are obsolete, being *Elephas falconeri* and *Elephas mnaidriensis* currently named *Palaeoloxodon falconeri* and *Palaeoloxodon mnaidriensis*, respectively [25];
- the Grotta San Teodoro-Pianetti FC proposed in 2001 [26] and updated in 2008 [16] includes *Palaeloxodon mnaidriensis*; as a consequence, this latter species is not exclusive of the previous FC;
- the names indicated by Burgio [19] are maintained in the other FCs: Monte Pellegrino FC and Castello FC;

- Spinagallo and Maccagnone caves are particularly significant in the fossil record of mammals of Sicily.

The old names of the "stages" by Burgio, widely coincident to the updated faunal complexes, can be considered still valid: the *Elephas falconeri* FC can be renamed Spinagallo FC, and the *Elephas mnaidriensis* FC can be renamed Maccagnone FC. These denominations do respect criteria of priority and uniformity, and avoid the use of obsolete scientific names and of a non-exclusive species to indicate a faunal complex, as in the case of *Elephas mnaidriensis* FC. Moreover, the denomination Grotta San Teodoro-Pianetti faunal complex, named by the two LFAs known at present, is unusual and could be simplified as San Teodoro FC, the San Teodoro cave being the most representative site. Another criticism is in the use of both small and large mammals to state faunal complexes and will be a matter of discussion in this paper.

Figure 3 represents a new proposal for a biochronological chart based on the large mammals of Sicily, correlated with the standard mammal biochronology of Italy, the marine isotopic signal, and geochronology. Monte Pellegrino FC includes mainly small mammals and is not included in Figure 3. The proposed denominations of faunal complexes will be used in the rest of the paper, with replaced ones in brackets (only in headings).

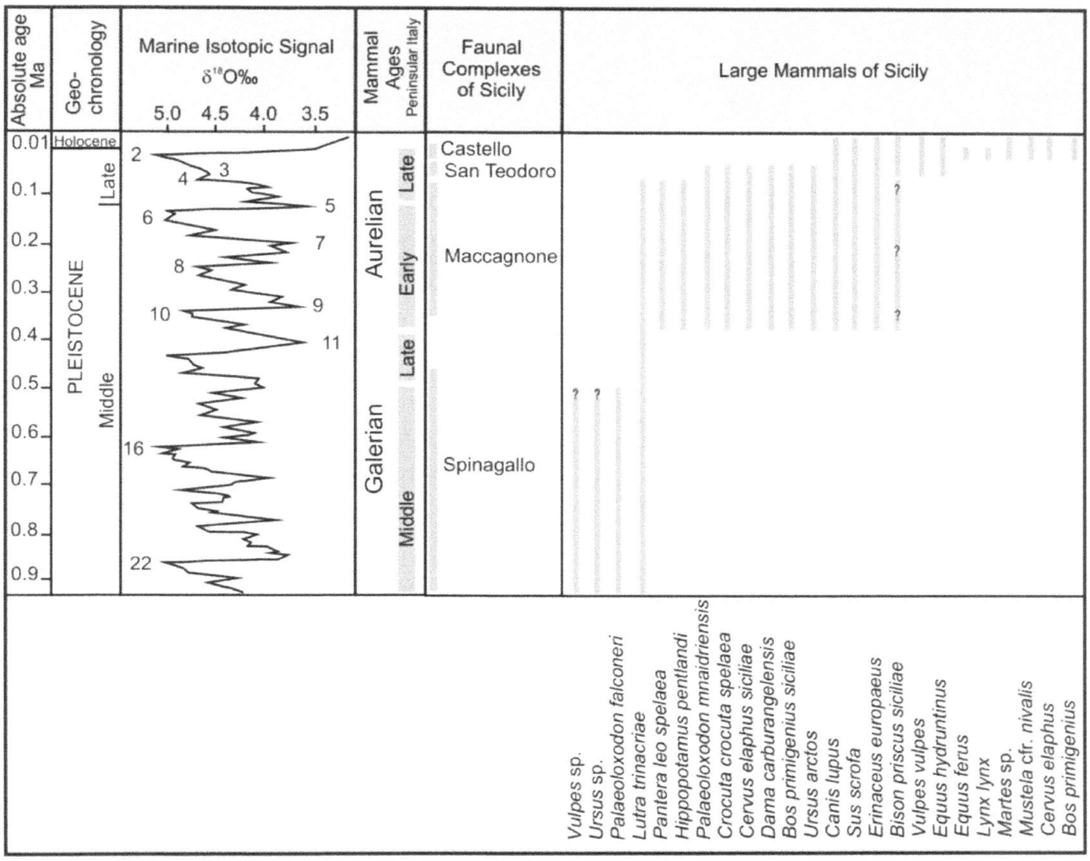

Figure 3. Biochronological chart of the faunal complexes of Sicily, correlated with the standard mammal biochronology of Italy, the marine isotopic signal, and geochronology.

Monte Pellegrino Faunal Complex

The Monte Pellegrino faunal complex is likely from the early Pleistocene age and its fauna is endemic, unbalanced, and poorly diversified, mainly composed by small mammals (Figure 4).

During Early Pleistocene, Sicily was made up of two islands, and the volcano Etna had not yet emerged; Calabria also consisted of two small islands separated by narrow sea straits, while the Strait of Messina was wider than today (Figure 4) [6]. Malta was emergent [27].

Different degrees of endemism, mainly consisting of enlargements of size, have been observed in species composing the FC. *Asoriculus burgioi*, *Apodemus maximus* and *Maltamys* sp. have marked endemic features and are dubitatively considered relics of an undocumented previous fauna, dubitatively spread during Messinian [28,29]. Thaler [14] reported also *Leithia* sp. from Monte Pellegrino. Daams and De Bruijn [30] considered an ancient form of *Leithia* as the precursor of *Leithia* and *Maltamys* species present in Middle and Late Pleistocene of Sicily and Malta. The authors identified the ancestor of *Leithia* (and *Maltamys*) in a form of *Eliomys* sp. probably dispersed during Messinian (late Miocene). *Pellegrinia panormensis* has very marked endemic features and has been compared to African Ctenodactilids [14]. *Hypolagus peregrinus* and *Mustelercta* (=*Pannonictis*) *arzilla* show moderate endemism, and are considered affine to European forms, dubitatively related to an Early Pleistocene dispersal [15,31]. There are no fossil records of mammals from the Early Pleistocene of Malta and Calabria [6,27].

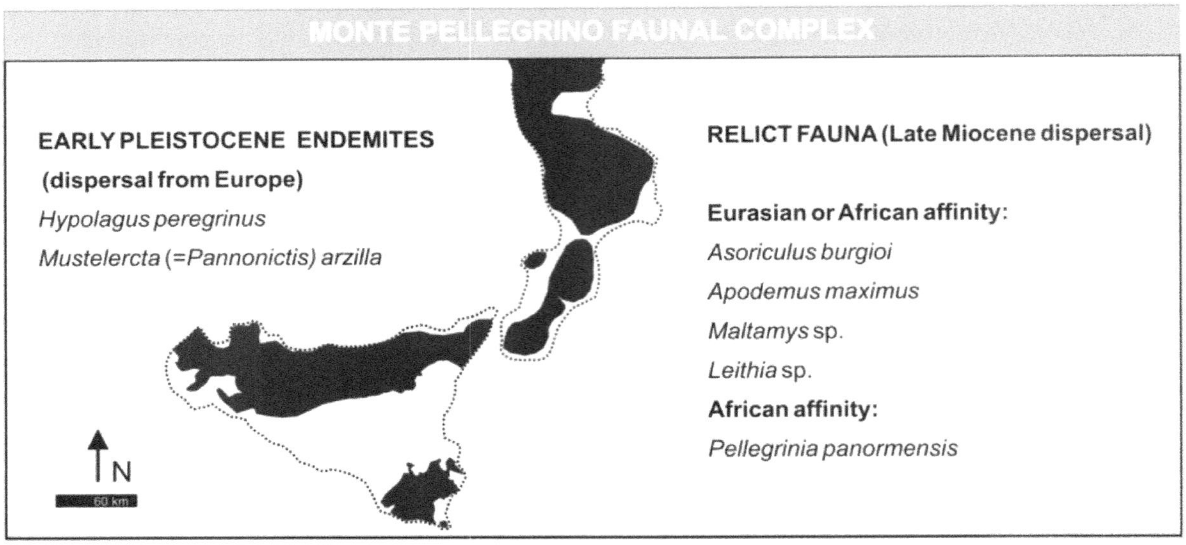

Figure 4. Paleobiogeography of the Monte Pellegrino faunal complex.

Spinagallo Faunal Complex (Formerly Elephas falconeri FC)

The Spinagallo faunal complex is early Middle Pleistocene in age and is strongly impoverished in number of species (Figure 3), as well as unbalanced in ecological composition [2,5].

During early Middle Pleistocene (Figure 5), Sicily was still consisting of two small islands [32]; the Strait of Messina persisted as a sea strait for the entire Quaternary, with some variations in wideness [10,33]. During Middle Pleistocene, a general uplift influenced all of Calabria, which progressively attained its present shape ([6] and references therein). During early Middle Pleistocene, Calabria was a land not easy to cross, characterized by high mountains and few extended plains [6]. Malta was emerged [27].

Palaeoloxodon antiquus, probable ancestor of *Palaeoloxodon falconeri* [25,34], is recorded in northern Calabria, in the "Mercure A" site [35], attributed to the Middle Galerian. *Praemegaceros* cf. *P. verticornis* and *Dama* cf. *D. clactoniana*, recorded at "Mercure A" site, did not disperse to Sicily [6]. There are no other records of mammals from Calabria. On Malta, *Palaeoloxodon falconeri* is associated with *Maltamys gollcheri*, *Maltamys wiedincitensis* (dubitative), *Crocidura esuae* and *Lutra euxena*. The Maltese otter is the only species not shared in common with Sicily. Moreover, in Sicily, *Maltamys wiedincitensis*, considered a descendant of *M. gollcheri*, is present in the younger Maccagnone FC. Zammit Maempel and De Bruijn [36] consider *Maltamys gollcheri* as a junior synonym of *Leithia cartei*, while Masini *et al.* [4] consider *Maltamys* sp.–*wiedincitensis*–*gollcheri* as chronospecies and *Leithia cartei* as a distinct species. In both cases, endemic dormice (*Leithia* and *Maltamys*) seem to be derived by taxa recorded at Monte Pellegrino, it follows, therefore, that their ancestors reached Sicily and Malta before Quaternary. The probable ancestor might have been a form of *Eliomys* because the genus was present in both continents during Miocene, though this is not enough to indicate African or European origin [37]. Since dormice are susceptible to passive transport on natural rafts and since their phylogeny is uncertain, it is still not clear if their forerunners reached Sicily from Africa or from Italy.

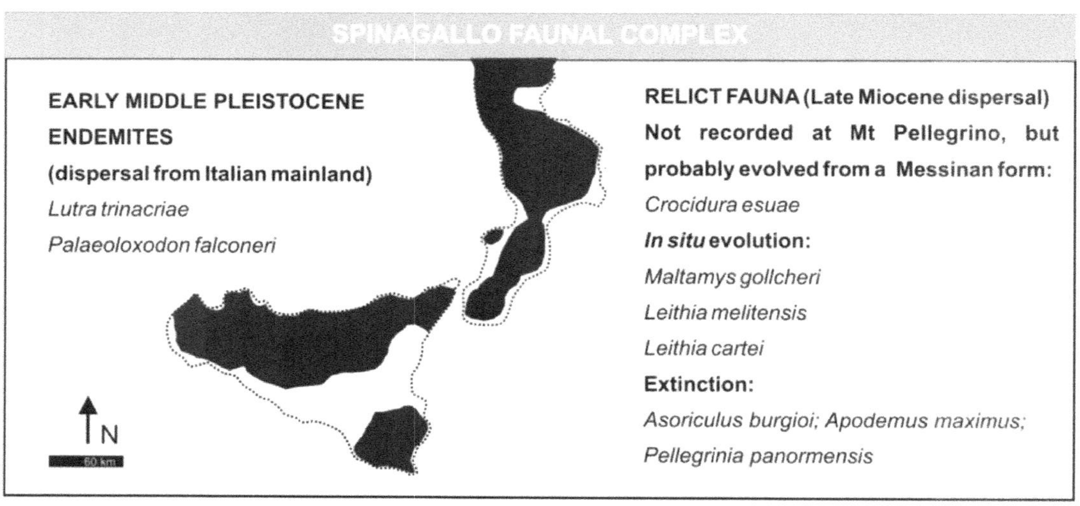

Figure 5. Paleobiogeography of the Spinagallo faunal complex.

The shrew *Crocidura esuae* is considered by Kotsakis [38], Esu *et al.* [39], and Sarà [40] to be the ancestor of *Crocidura sicula*, a species present in the fossil record since Late Pleistocene (San Teodoro FC) and still living in Sicily, while it is considered to be the descendant of a newcomer that arrived in the Late Pleistocene, according to Masini *et al.* [4]. The debate is still open, but genetic and morphological data indicate marked differences among *Crocidura sicula* and species from the Italian peninsula and, alternatively, strong affinities with living afro-tropical forms [41,42]. Studies on DNA provided by Dubey *et al.* [43] consider *Crocidura sicula* to be the descendant of the Saharan shrew *Crocidura tarfayaensis*, with a split between Saharan and Siculo-Maltese clades dated to Messinian. Taxa possibly related to *Crocidura esuae* have not been recorded in the Monte Pellegrino FC, although a Late Miocene spreading from North Africa cannot be excluded. Therefore, this small mammal fauna has been long persistent on Sicily and its spreading, probably by floating on natural rafts, might have occurred before Quaternary in a quite different paleogeographical setting.

Large mammals reached Sicily by swimming, most likely from peninsular Italy. The forerunners of the dwarf elephant, *Palaeoloxodon falconeri,* and the endemic otter, *Lutra trinacriae,* were good swimmers. Living elephants have been observed swimming the seas, mainly in the Indian Ocean, in order to reach islands [44,45]. They can swim for long distances, searching a new habitat, and are able to see islands on the horizon and to smell edible food resources [44]. Otters have good dispersal ability and they can spread to islands as far as 30 km [46]. Therefore, it can be deduced that the ancestors of *Palaeoloxodon falconeri* and *Lutra trinacriae* reached Sicily and small islands of Calabria by swimming. *Palaeoloxodon falconeri* underwent a strong reduction in size, being an adult male high about 120 cm at the scapula, and heavy at about 170 kg [47]. Moreover, the dwarf elephant shows important changes in skull morphology and body proportions [47]. *Lutra trinacriae*was moderately reduced in size, probably as a consequence of the permanence of its populations on the island [48]. Chiroptera are recorded but still undetermined [2]. Bats, birds, reptiles, and amphibians associated with mammals enable some paleoecological considerations.

Small vertebrates other than mammals that probably arrived by transport on natural rafts are: Amphibians (*Bufo* cf. B. *viridis*, *Hyla* sp., and *Discoglossus* cf. *D. pictus*); Reptiles (Chelonia: *Emys orbicularis*, *Testudo hermanni*, and a giant form of *Geochelone* sp.); Squamata: (*Coluber* cf. *viridiflavus*, *Natrix* sp., *Lacerta viridis* and the giant endemic *Lacerta siculomelitensis*) [2].

Avifauna was diversified, with 66 species [2,49]. Among birds, 24 migrant species have spent winters in Sicily [49,50]. Fresh waters have been available, as indicated by the presence of Gruiformes and Anseriformes [50]. The presence of sedentary birds such as Piciformes, Colombiformes and Passeriformes indicates permanent availability of fruits, seeds and invertebrates [50]. Some endemites were present: the giant crane, *Grus melitensis*, the dwarf swan *Cygnus equitum*, the Strigiformes *Tyto mourerchauvireae* and *Aegolius martes*, both giant species, and *Athene trinacriae*, which had lengthened forelimbs [49]. *Tyto mourerchauvireae* seem to have evolved from *T. balearica*, an owl common in the Mediterranean area, and was at the top of the food chain [51]. This species preyed upon the three species of Gliridae, the shrew *Crocidura*, and several birds [51]. The presence in Sicily of large-sized Gliridae could have favored the evolution of *T. mourerchauvireae*, which became extinct in the Maccagnone FC, probably for the arrival of *Bubo bubo*[51].

Guglielmo and Marra [50] proposed a reconstruction of the food chain. During the night, Strigiformes preyed on small vertebrates; in daylight, Colombiformes, Passeriformes, and small vertebrates were preyed upon by Falconiformes. Birds' eggs and nestlings might have been eaten by reptiles, dormice and, occasionally, by otters. Elephants did not have had predators and intra-guild competitors. Their strong reduction in size could be related to these peculiar paleocological conditions, and perhaps it occurred through anticipation of the sexual maturity and shortening of the gestational period [52]. In the opinion of Palombo [47], the strong reduction in size of *Palaeoloxodon falconeri* could also be related to the availability of niches typical of medium- and large-sized herbivores. The reduced size allowed the persistence of *P. falconeri* populations with a number of individuals adequate to avoid extinction. Moreover, a smaller body mass reduced the deep modification of vegetation typically produced by elephants.

Maccagnone Faunal Complex (Formerly Elephas mnaidriensis FC)

The Maccagnone faunal complex is late Middle Pleistocene in age and almost completely renewed with respect to the Spinagallo FC (Figure 3, Figure 6). Among small mammals, *Leithia melitensis* and *Crocidura esuae* persist from the previous FC; *Maltamys gollcheri* evolves in *Maltamys wiedincitensis* [2,4,5]. *Microtus* (*Terricola*) sp. has "subterraneomorph" characteristics, similar to the European *Microtus* (*Terricola*) *subterraneus* and could have spread to Sicily during MIS 6 [53]. The dwarf elephant *Palaeoloxodon falconeri* became extinct, while *Lutra trinacriae* persists. Large mammal fauna is renewed and represented by an association impoverished but balanced and diversified, with a moderate endemism.

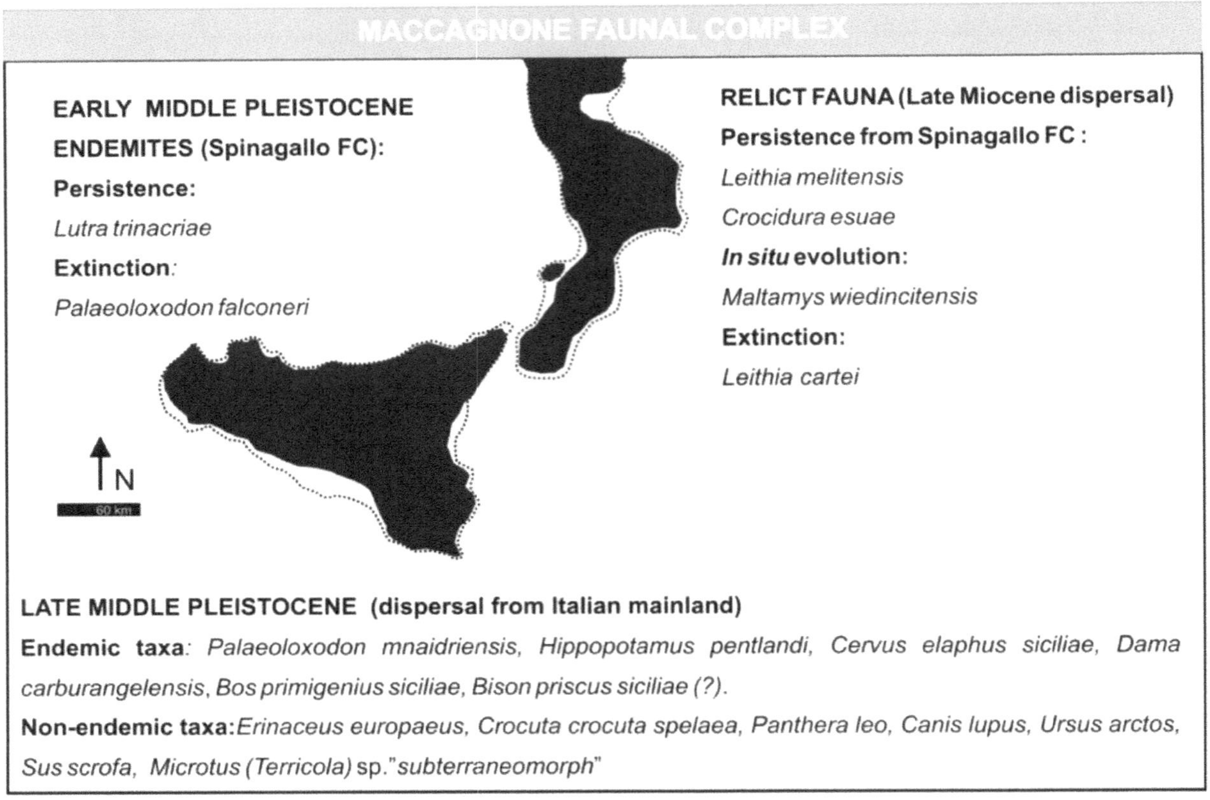

Figure 6. Paleobiogeography of the Maccagnone faunal complex.

In the late Middle Pleistocene, the geography of Sicily, Malta and Calabria was attaining its present-day shape, with an extension of inner and coastal plains (Figure 6). The arrival of several new taxa of European origin can be related to favorable conditions of dispersal through the Strait of Messina [2,5,47]. Dispersals might be related to several events that occurred during the stadial oscillations of the sea level in late Middle Pleistocene (MIS 10, MIS 8, and MIS 6 [47]), or to a single event that occurred 0.3 Ma (MIS 8 [4]). A combination of both the tectonic uplift of Calabria and Sicily and low-stand phases of the sea level likely reduced the relative distance between the two coastlines of the Messina Strait (Figure 6). The strait has been crossable by large mammals with different dispersal abilities, ranging from good swimmers (elephant, cervids) to animals with more or less limited swimming ability, such as bovids and carnivores.

The long-term persistence of the small mammal fauna has been attributed to the absence of new competitors, due to a geographical filter active in the Strait of Messina such as a partially emerged sea floor or a swampy lagoon system, crossable only by large mammals [2,4,5]. However, this hypothesis is not easily sustainable, because of the fact that small mammals can follow more dispersal ways than large ones can; they can disperse simply by walking, and also on natural rafts. Furthermore, there is no reason why a partially emerged bridge cannot be crossed by small mammals. And there are no geological data to support the existence of this peculiar land connection. Paleoecology, discussed in the following, can be the key.

The poor quality of the fossil record in the Middle Pleistocene of Calabria allows limited investigations on the dispersal of the precursors of the Maccagnone FC [6]. However, the latest Middle Pleistocene fossil assemblages from Calabria include all the large mammals considered precursors of Middle Pleistocene endemic species of Sicily [6]. Their presence in Calabria, when they spread to Sicily thereby originating the Maccagnone FC, can only be hypothesized. More than one dispersal might have occurred. As a matter of fact, "normal" sized elephants found in Sicily at Contrada Fusco (MIS 5) can be interpreted as newcomers, or at least as strong intraspecific variations [47]. It is not difficult to imagine elephants that swam several times across the Messina Strait. Among mammals of the Maccagnone FC, only elephant and cervids had swimming attitude, while hippopotamus, wild boar and bear had a good attitude to humid/freshwater environments. Other mammals did not swim, and therefore paleogeography might have been favorable to their dispersal.

On the basis of the available geological and paleontological data, Marra [6] hypothesized a partially emerged "stepping stone" way in the Strait of Messina, available during regressive phases. Mammals could have reach the island crossing the "sill," an area of the Strait of Messina that is currently deep (80–100 m offshore Sicily) and irregular in its western portion, though shallower (60–100 m offshore Calabria) and smoothed in its eastern part [54]. In correspondence to the "sill," the Strait of Messina is at present 3 km wide. The sea strait had a maximum wideness in Upper Pliocene–Lower Pleistocene; later, it underwent a sensible narrowing, with morphology very similar to present day [10,33]. The invertebrate fossils and extant forms of marine environment indicate the persistence of elevated hydrodynamics in the Strait of Messina, which exclude the emergence of a land bridge for all Quaternary until now [10,33]. The tectono-sedimentary evolution of the "sill" indicates a sensible uplift from the late Middle Pleistocene

[10]. The emergence of small areas of the "sill" during sea-lowering phases occurred since late Middle Pleistocene would be consistent with geopaleontological data and could have allowed "stepping stone" dispersals [6]. Also, the possibility of a "pendel route" (*sensu* Dermitzakis and Sondaar [8]) has to be taken into account, considering that the Strait of Messina was a narrow sea barrier that enabled the transit of mammals with reduced dispersal abilities. Low-stand phases might be responsible of the faunal renewal on Malta. The faunal association recorded in the late Middle Pleistocene of Malta is impoverished with respect to Sicily and includes *Leithia cartei*, *Maltamys wiedincitensis*, *Lutra euxena*, *Palaeoloxodon mnaidriensis*, *Hippopotamus melitensis*, and *Crocidura esuae* [5,27]. Small mammals persist, while the dwarf elephant is replaced by the new species spread from Sicily. The Maltese hippopotamus is considered the descendant of *Hippopotamus pentlandi* of Sicily [5,27]. Connections between Sicily and Malta were occasional and submitted to a strong filter, and only two species with aquatic attitude (elephant, hippopotamus) reached the island. On the other hand, the small island did not offer to possible newcomers source availability comparable to that of Sicily.

The increasing biodiversity, the moderate endemism of the herbivores, and the presence of non-endemic carnivores (with top predators among them) observed in the Maccagnone FC are indicative of a large island with diversified environments in addition to a paleogeography enabling the dispersion of species with different dispersal abilities [6].

On the island, species evolved endemic features. Large herbivores were moderately modified with respect to the mainland ancestors, showing a slight reduction in size, while carnivores did not evolve endemic features and included top predators [1,2,5]. Medium-sized ruminants (*Cervus elaphus siciliae* and *Dama carburangelensis*) reduced their size (intended as linear dimensions of the skeleton) by about 20%–25% with respect to their ancestors (*Cervus elaphus* and *Dama dama* cf *tiberina*); the large ruminant *Bos primigenius siciliae* reduced its size by about 20% with respect to *Bos primigenius* [1,2,55,56]. Endemic features observed in non-ruminant herbivores are of different degrees. *Hippopotamus pentlandi* had not marked endemic features and a reduction of about 10%–15% in linear dimensions with respect to the ancestor *H. amphibius* [57]. *Palaeoloxodon mnaidriensis*had marked allometries and a strong reduction in size, about 50%–60% with respect to the ancestor *Palaeoloxodon antiquus* [25]; some morphological modifications are related to dwarfing processes, but some others might be related to ecological adaptations (*i.e.*, fusion of radius and ulna; [25]). Omnivorous (*Sus scrofa* and *Ursus arctos*) and carnivorous (*Canis lupus*, *Crocuta crocuta spelaea*, *Panthera leo spelaea*) large mammals did not modify their size [2,4,5,58].

Small mammal fauna persisted from the Spinagallo FC, with only one new arrival. The reason for this long persistence might be found in paleoecology. Although sustaining the existence of a filtering land bridge, which prevented the spreading of small mammals, Masini *et al.* [4] did admit that longer survivorship of small mammals might be due to the equilibrium they established with predators, or to their prolificacy and food requirements. Guglielmo and Marra [50] argued that small mammals could have established deep-rooted populations that defeated possible newcomer competition. Moreover, they might have not been affected by the extinction of their primary predator *Tyto mourerchauvireae*, replaced by *Bubo bubo*, having similar body size and predatory attitude [49]. The stability of small mammal populations can also be attributable to the absence of possible secondary predators (*i.e.*, small carnivores) that might have perturbed their ecological equilibrium [50]. It is worth noting that endemic Gliridae disappeared when *Vulpes vulpes*, their possible secondary predator, entered the island and were replaced by species smaller in size (San Teodoro FC). The life of secondary predators did not depend on the survival of their secondary prey, so the red fox might have caused the extinction of endemic Gliridae [50]. Moreover, small mammals can risk overpopulation (especially in conditions of equilibrium with predators), and the evolution of large body size, probably accompanied by lower metabolic rate and greater life duration, might have been a good life strategy [50]. Furthermore, ecological niches available for possible newcomers were stably occupied.

Among small vertebrates, the amphibian *Discoglussus* cf. *pictus* and the Reptiles *Emys orbicularis*, *Testudo hermanni* and *Lacerta siculomelitensis* persisted, while the giant turtle *Geochelone* sp. became extinct [2]. The other amphibians (*Bufo* cf. *viridis*, *Hyla* sp.) and Reptiles (*Coluber* cf. *viridiflavus*, *Natrix* sp., *Lacerta viridis*) present in the previous fauna have not been recorded. However, molecular data by their living relatives seem to indicate their long persistence in Sicily [50].

The avifauna underwent several changes [49]. The number of migrant birds increased, with taxa preferring fresh waters. Podicipediformes and Pelicaniformes, adapted to marine waters, dispersed to Sicily. Anseriformes were represented by new species and Ciconiformes increased, with some species of swamp environment. Among Caradriformes, *Numenius phaeopus*, a species adapted to humid environments, was present. Passeriformes sensitively decreased. Guglielmo and Marra [50] argued for a good availability of humid and fresh water environments. It is likely that a good ichthyofauna was present in the freshwaters, sufficient to sustain the rich avifauna. Endemic Strigiformes disappeared, replaced by *Bubo bubo*. Among Falconiformes, it is worthy of noting the presence of *Aquila heliaca*, a bird able to prey on animals with a weight up to 10 kg (mainly mammals and birds). The eagle was probably a menace for the cubs of many mammal's species, while the other Falconiformes preyed upon small vertebrates, such as amphibians, reptiles, birds and dormice [50].

Top predators (*Canis lupus*, *Crocuta crocuta spelaea*, and *Panthera leo spelaea*) fed on populations of herbivores, mainly cervids (*Cervus elaphus siciliae* and *Dama carurangelensis*). Bovids, small mammals, and birds have probably been occasional prey, as observed in the behavior of the extant wolves, hyenas, and lions. As scavengers, large predators also fed on dead animals. In conditions of isolation and reduced areas, the slight reduction in size of herbivores might have been a strategy to maintain the number of individuals adequate to avoid extinction. Furthermore, herbivore populations had been numerous enough to feed three species of social predators, which did not evolve endemic features because the size of endemics fell into the variability range of usual prey.

In the opinion of Palombo [47], *Hippopotamus pentlandi* and *Palaeoxodon mnaidriensis*, thanks to their reduction in size, occupied ecological niches vacant on the island, that are respectively those of large horses (grazers) and rhinoceroses (mixed-feeders).

Omnivorous mammals (*Sus scrofa* and *Ursus arctos*) exploited various food resources, without suffering the effects of isolation.

Large mammal association is indicative of a composite environment, with prairies and open woods, typical of Mediterranean "*macchia*" (Table 1).

Table 1. Habitat, Feeding Behavior and Body Mass of endemic large mammals of Sicily. Categories according to Palombo [5]. **Habitat**: O = grassland, steppe or savanna; W = forests and closed woodland; Wc = bushland, Mediterranean "macchia" and open woodland. **Feeding behavior**: Br = browser; Gr = grazer; MF = mixed-feeder; Om = omnivore; P = predator; Sc = scavenger; sP = small predator; SP = superpredator. **Body mass**: <1 kg (class I); 1–10 kg (class II); 10–60 kg (class III); 60–120 kg (class IV); 120–350 kg (class V); 350–850 kg (class VI); >850 kg (class VII).

Large mammal	Habitat	Feeding behavior	Body mass	Faunal complexes			
				Spinagallo	Maccagnone	S. Teodoro	Castello
Palaeoloxodon falconeri	O-Wc	Br	V	X			
Lutra trinacriae	F	sP	II	X	X		
Panthera leo spelaea	O	SP	V		X		
Hippopotamus pentlandi	O-F	Gr	VII		X		
Dama carburangelensis	O-Wc	MF	IV		X		
Crocuta crocuta spelaea	O-Wc	Sc-SP	III/IV		X	X	
Cervus elaphus siciliae	Wc	MF	V		X	X	
Bison priscus siciliae	O	Gr	VII		X	X	
Bos primimigenius siciliae	O-Wc	Gr	VII		X	X	
Palaeoloxodon mnaidriensis	O-Wc	Br	VII		X	X	
Ursus arctos	Wc	Om	IV		X	X	
Canis lupus	O-W-Wc	P	III		X	X	X
Vulpes vulpes	O-Wc	sP	II		X	X	X
Sus scrofa	Wc	Om	IV		X	X	X
Equus hydruntinus	O	Gr	V			X	X
Equus ferus	O	Gr	VI/VII				X
Lynx lynx	W	P	III				X
Martes sp.	O-Wc	sP	II				X
Mustela cf. *nivalis*	Wc	sP	I				X
Cervus elaphus	Wc	MF	V				X
Bos primigenius	O-Wc	Gr	VII				X

San Teodoro Faunal Complex (Formerly Grotta S. Teodoro—Pianetti Faunal Complex)

In the San Teodoro Faunal Complex (Figure 3), some large mammals of the previous fauna became extinct (*Panthera leo spelaea*, *Hippopotamus pentlandi*, *Dama carburangelensis*) and the newcomers (*Vulpes vulpes*, and *Equus hydruntinus*) did not evolve endemic features [2]. *Bison priscus siciliae* might be considered as a newcomer with moderate endemic adaptations, according to Petruso *et al.* [11]. As already discussed, the arrival of *Vulpes vulpes* might have been decisive for the extinction of the endemic small mammals' fauna and the entrance of a renewed stock on the island.

Among small mammals, the endemites *Leithia melitensis* and *Maltamys wiedincitensis* became extinct, while the new presence of *Apodemus sylvaticus* (non endemic), *Crocidura sicula*, and *Microtus*(*Terricola*) *savii* is recorded. As previously mentioned, *Crocidura sicula* has been considered a descendant of a new immigrant species [4] and of *Crocidura esuae* ([38,39,40]. *Microtus* (*Terricola*) *savii* is considered a newcomer, likely dispersing during MIS 4–3, and not a descendent of the *Microtus* (*Terricola*) "subterraneumorph" of the Maccagnone FC [53], and probably evolved in the endemic *Microtus* (*Terricola*) *savii nebrodensis*, currently present in Sicily [11].

Paleogeography of Sicily and Calabria was similar to nowadays, with some local variations in the coastal lines (Figure 7). The arrival in Sicily of a new stock of small mammals and some large mammals has been related by Bonfiglio *et al.* [2] to the opening of a land bridge as a consequence of a eustatic low stand, and by Marra [6] to a "stepping stone" dispersal way. Considerations made for the Maccagnone FC are valid also for the San Teodoro FC: if the hypothesis of a "stepping stone" way can be accepted, small mammals could have been easily spread by natural rafts and/or by swimming crossing a partially emerged "sill".

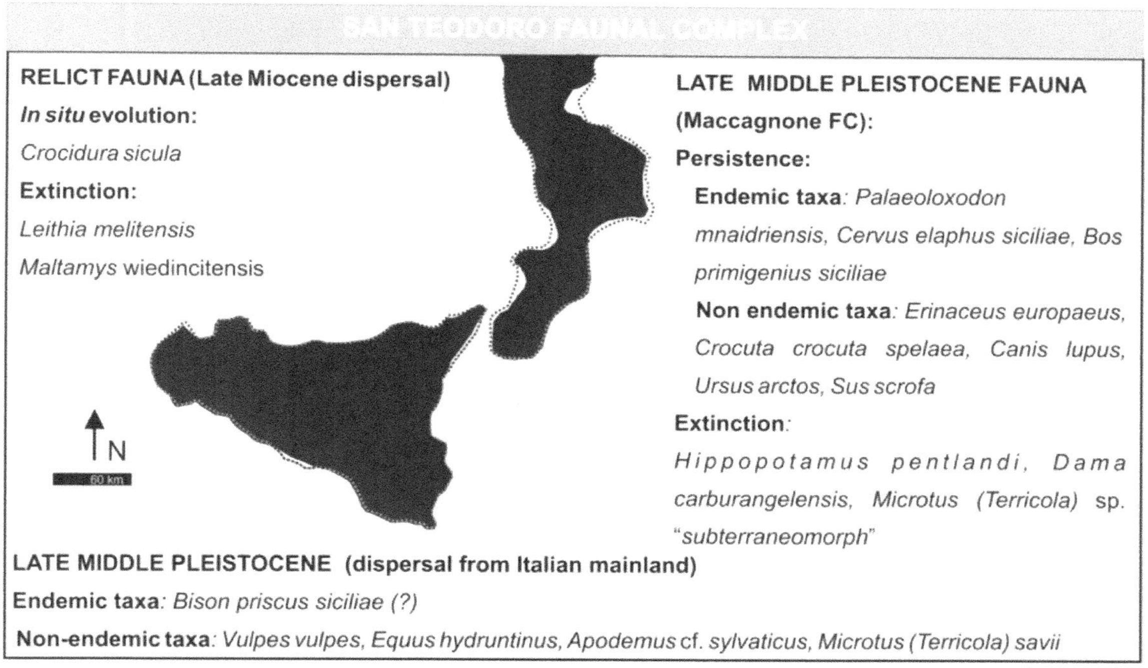

Figure 7. Paleobiogeography of the San Teodoro faunal complex.

The fossil record of Calabria is richer than before and comes from different depositional settings, demonstrating important geographical and environmental changes, such as a wider extension of inner and coastal plains [6]. During Late Pleistocene, Calabria was easier to cross than before, and the Strait of Messina was not a strong barrier. As a matter of fact, precursors of the species of the San Teodoro FC have been recorded in Calabria, in assemblages attributable to the early Aurelian and to the late Aurelian [6]. Only few mammals that adapted to prairie or mountain environments were present in Calabria and not in Sicily [6]: *Equus ferus*, *Stephanorhinus kirchbergensis*, *Capreolus capreolus*, *Capra ibex*, *Rupicapra rupicapra*, and *Meles meles*. *Homo neanderthalensis* was sporadically present in Calabria, but its presence is not recorded in Sicily.

At Malta, dormice persisted and three new species, strongly endemic (*Cervus* sp., *Microtus*(*Pitimys*) *melitensis*, *Crocidura* sp.), have been recorded [27]. Unfortunately, these species are poorly studied. It is evident that the filter between Sicily and Malta was strong.

Paleoecology of the large mammal association of Sicily is indicative of Mediterranean "*macchia*" (Table 1). The arrival of *Equus hydruntinus*, and possibly of *Bison priscus siciliae*, is corroborated by vegetation reconstructions based upon pollen from coprolites of *Crocuta crocuta spelaea* recorded in the San Teodoro cave, indicating pre-Late Glacial conditions [59]. Vegetation was dominated by steppic taxa, but also arboreal taxa, and small percentages of mesophilus taxa were present [59]. During the Last Glacial cycle, more precisely, 32,000 ± 4,000 yr BP in Sicily, a wooded steppe biome, with elements representing a variety of local conditions, was present [59]. A paleobiome of southern Italy shows the predominance of wooded steppes during Upper Pleistocene [60]. The character of the steppe was Mediterranean, indicating relatively cool summers.

Avifauna was less diversified then earlier, with prevailing migrant species and some Passeriformes [2,61]. The disappearance of birds typical of humid environments is coherent with the wooded steppe biome.

Castello Faunal Complex

The Castello faunal complex is represented by a mammal fauna low diversified, without endemites and with *Homo sapiens*. These features indicate easier communication between Italy and Sicily: all mammals recorded in Calabria, except for *Martes* sp., are present in the Castello FC (Figure 8) [6].

According to Barrier [10] and Di Geronimo [33] the Strait of Messina has always had an elevated hydrodynamics, while according to Bonfiglio *et al.* [2] and Lambeck *et al.* [62] a land bridge emerged 20,000 yr BP (Figure 8).

The Castello FC has an ecological assessment still unbalanced but closer to modern ones. Presence of mammals of wooded (*Cervus elaphus*, *Sus scrofa*) and open (*Equus ferus*, *Equus hydruntinus*, *Bos primigenius*, *Lepus europaeus*) environments (Table 1) can be related to the dry conditions of Late Glacial. Amphibians and reptiles are absent from the fossil record. Avifauna was diversified, with species typical of wooded and open environments and with a large number of migrant species.

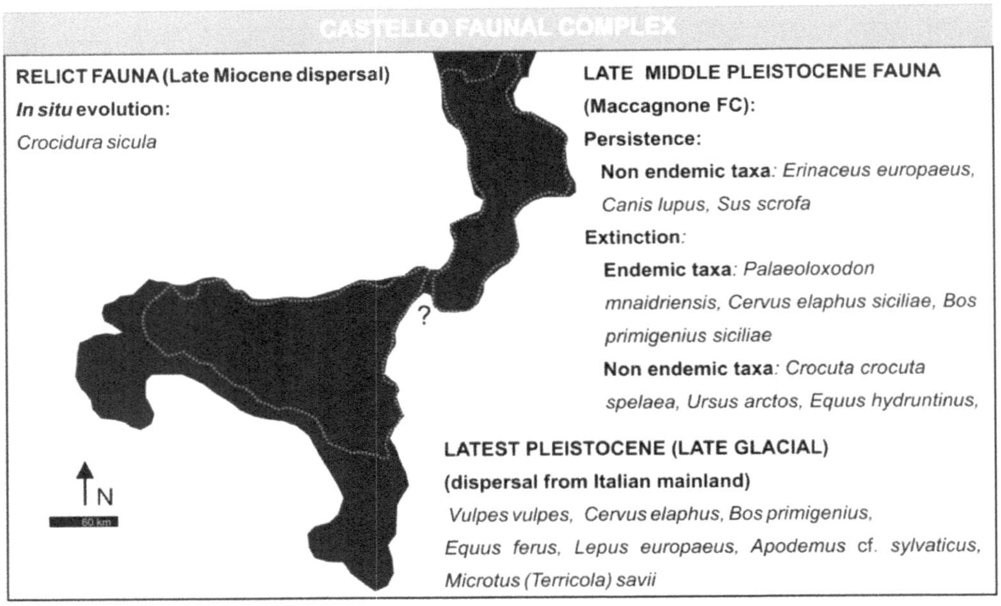

Figure 8. Paleobiogeography of the Castello faunal complex.

DISCUSSION

The present review started from a reconsideration of the current biochronology of mammals of Sicily. The first part has been devoted to some nomenclature problems, but after the discussion of other available data, it becomes evident that something more substantial is needed. In particular, a formalization of faunal complexes as faunal units having biochronological significance is still lacking. In order to state a formal biochronology, large mammals have to be considered apart from small mammals. The current faunal complexes are based on numerous sites and some of them were formerly excavated without screening for small mammals; it therefore follows that large mammals have to be used to formalize faunal units, as do peninsular ones [12]. The biochronological chart of Quaternary large mammals of Sicily presented here (Figure 3) is correlated to geochronology, the marine isotopic signal, and the mammal ages of Italy, and thus represents a first step to formalization of faunal units.

Small mammals of the Pleistocene of Italy are currently arranged in zones correlated to small mammal ages of Europe [63]. A framework for Sicily is still lacking.

Review of data on paleogeography and the possibility of dispersals allowed the elaboration of paleobiogeographical maps of Sicily and Calabria (Figure 4, Figure 5, Figure 6, Figure 7, Figure 8). The Strait of Messina (Figure 1) played an important role during the Pleistocene period, being a more or less wide sea barrier between southern Italy (Calabria) and Sicily, controlled by tectonics and eustatism [2,3,5,6]. In Early Pleistocene, two sea-straits divided Calabria in three parts, filtering mammal dispersals until their emersion during Early and Middle Pleistocene [6].

For a mammal, the possibility to cross a composite environment is related to its dispersal ability that depends on its anatomy and physiology. Moreover, its settlement on the island depends on available ecological niches and environmental resources, as well as competition and predation. Old and new data suggest an integrated approach, in which the insular species attributes (*i.e.*, dispersal ability, size change, endemism, and adaptive radiation) and the community attributes (*i.e.*, ecological balance and biodiversity) are considered in a biochronological, palaeoecological and paleogeographical framework. Furthermore, comparisons with peninsular Italy (in particular Calabria, the southernmost region of peninsular Italy) and the island of Malta, which was part of the insular system, are necessary.

The possibility that Malta could have represented a dispersal way from Africa might have been considered. However, this hypothesis is not sustainable and not only for the distance from the African mainland. On the basis of the presence of taxa with African affinity in the Monte Pellegrino FC and their degree of endemism, the possible dispersal from northern Africa occurred probably in Late Miocene (Messinian), as already discussed. Malta emerged during Late Miocene and has probably been connected to the small islands of Comino and Sicily until Early Pliocene; fossils of that age have not been recorded [27]. No records occur in Calabria, where African fauna might have dispersed. The Quaternary fossil record of Malta is coherent with filtered dispersals from Sicily, which in its turn received faunas from Calabria.

The most reliable reconstruction is that relics of one or more ancient faunas persisted on Sicily from late Miocene to Pleistocene. In Sicily, a late Miocene faunal assemblage including non-endemic large mammals has been found at Messina (Gravitelli site), but has been lost during the 1908 earthquake. Papers published before the terrible event [64,65] allowed the review and determinations of species related to Eurasian forms. The assemblage, recovered in lignite deposits, has been attributed to Messinian for a long time, but Rook *et al.* [66] attributed it to late Turolian, proposing the existence of a Late Miocene Calabrian–Sicilian bio-province without endemics. These data cannot be used to clarify the character of the fauna of Monte Pellegrino FC and its age. Moreover, the site of Monte Pellegrino is problematic by a stratigraphic point of view, being represented by fissure-filling deposit. However, the fauna of Monte Pellegrino FC could open a window on Late Miocene scenarios: *Asoriculus burgioi*, *Apodemus maximus*, *Maltamys* sp., *Leithia* sp. and *Pellegrinia panormensis* could be descendant of forms spread during Messinian, likely coming from north Africa. Also, *Crocidura esuae* might have been part of this faunal stock. Although the species is not recorded at Monte Pellegrino, its living relative, *Crocidura sicula*, has a consistent genetic distance from European forms and has affinity to African populations dated to Messinian.

During Quaternary, faunas spread from the Italian peninsula through Calabria and the Strait of Messina (Figure 4, Figure 5, Figure 6, Figure 7, Figure 8); Sicily has been an island becoming more and more "continental," changing its paleogeography from an archipelago made of two small islands, to a large single island. Moreover, Calabria underwent similar paleogeographic conditions, attaining its territorial continuity and extending its plains thanks to the emersion of some areas. Moreover, Calabria has a very elongated shape, and the main part of its territory is mountainous, with reduced inner and coastal plains. For these reasons, and for its latitudinal extension, Calabria represented a strong filter to the spreading of fauna from north to south. Mammals who challenged these lands underwent endemic processes on the island, mainly consisting in enlargements of size for small mammals, and dwarfing of large mammals, with allometries.

The role of the Strait of Messina in controlling dispersal to Sicily has been greatly overestimated, just because insular associations have been roughly compared to Italian peninsula ones. However, it is widely assumed that, in southern Italy, mammal associations were less diverse than in central or northern Italy, for climatic and geographic reasons. Paleobiography does demonstrate that the sea-strait has been not difficult to cross for mammals, in particular during low-stand phases of the sea level, but they needed to spread to Calabria at first.

Debates on the full emersion of the Strait of Messina can be passed over by considering a "stepping stone" way. It is coherent with geological data, as well as with paleontological ones, being crossable by the majority of recorded mammals and other vertebrates (amphibians included). It is worth noting that elephants could have swam the Strait of Messina several times. The occurrence of "normal" sized elephants in some Middle and Late Pleistocene sites has been recently emphasized by Herridge [67]. Preliminary data indicate the occurrence of large-sized elephants, temporary named "*Palaeoloxodon* sp. -Sicily 3" at Puntali cave, Zà Minica cave (both refereable to Maccagnone FC) and San teodoro cave.

Dispersal by swimming and natural rafts probably occurred many times during Quaternary and sometimes might have been helped by the emersion of small pieces of land, thanks to the lowering of the sea level during cold stages. From the oldest to the youngest, faunal complexes show increasing biodiversity and decreasing endemism, they progressively lost the character of strong isolation and became more uniform to Calabria, indicating easier connections with mainland.

In the course of these changes in geography, Sicily offered different environments to pioneer faunas. The vertebrate association of the Maccagnone FC indicates a relatively humid environment, with availability of fresh waters. It appears that wooded areas were alternated to open landscapes and Mediterranean "*macchia.*" In the later association (San Teodoro FC), the climate was more dry, as a consequence of glaciation. Vertebrate associations and vegetation are, according to a steppe biome, related to dry conditions. In terms of interspecific ecological relationships, faunal associations became more complex through time.

The long persistence of small mammals in Sicily, also in the presence of relevant changes of paleogeography, events of extinctions and new arrivals among large mammals and other vertebrates, is worth noting. As already discussed, the long persistence of small mammals on Sicily could have had ecological reasons: species were deeply rooted on the islands and their niche probably did not change through time; possible competitors found giant relatives and presumably lost the struggle for life; main predators of dormice and shrew have been birds of prey of similar sizes. The extinction of endemic small mammals can be due to the environmental changes of Late Pleistocene, shifted to dry conditions, and to the entrance on the island of *Vulpes vulpes*. The canid was likely the secondary predator of small endemites and did not establish an equilibrium with its occasional prey, thereby causing their extinction.

CONCLUSIONS

Quaternary mammals of Sicily represent a good case study for island biogeography. The wide dataset reviewed in this paper enables considerations on biochronology, bioevents, evolution of endemites, paleoecology and times and ways of dispersals (Figure 9). The current biochronological framework is well documented, but informal. The names of faunal complexes here proposed avoid ambiguous denominations and open discussions on formalization of biochronology of Sicily.

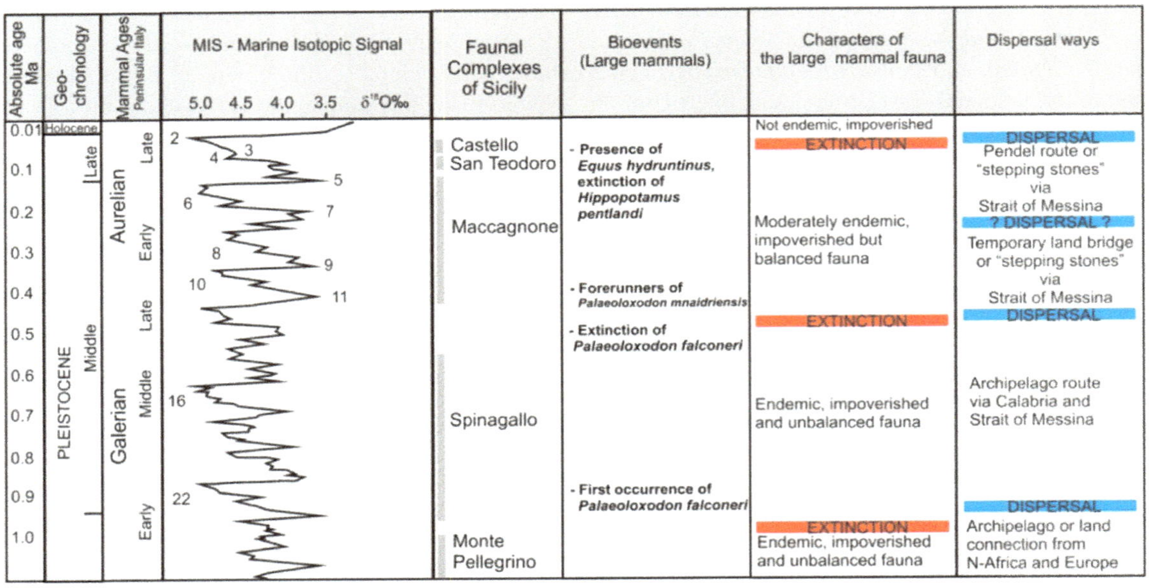

Figure 9. Quaternary mammals of Sicily: bioevents, endemic characters, times and ways of dispersals.

Bioevents (*i.e.*, extinctions, turnovers) are strictly related to paleogeography of Sicily and surrounding areas, which changed from an archipelago to a large island (Figure 4, Figure 5, Figure 6, Figure 7, Figure 8). The role of the Strait of Messina in controlling dispersals from mainland has thus far been overestimated, while Calabria can be considered a first filter. Large mammals have been influenced by relevant bioevents, while small mammals persisted unchanged until Late Pleistocene. As a consequence, faunal complexes can be distinguished on the basis of complete or partial renewal of large mammal faunas. Long persistence of small mammals can be hardly related to selective palaeogeographic filters, while it is more reliably due to a well-established equilibrium with predators, disturbed by the spreading of the red fox during Late Pleistocene. Differences in times and ways of dispersals and in reproduction rates, as well as in occurrence in the fossil record existing between large and small mammals require distinct biochronological frameworks. Paleoecology of vertebrate associations is indicative of changes in the availability of environmental resources. Complexity of interspecific relationships and of ecological niches increased through time.

Finally, the data reviewed here enable a paleobiogeographical characterization of Sicily. By a biogeographical point of view, islands can be characterized as oceanic islands (never connected to mainland; a strongly impoverished, unbalanced and endemic fauna is expected) or as continental islands (broadly connected to continents in the past; their faunas are impoverished but similar to continental ones, with low endemism and high diversity) [68]. However, intermediate types of islands can be defined. Alcover *et al.* [68] defined oceanic-like islands as an island connected to mainland in the distant past; the fauna is similar but more diversified with respect to oceanic islands. Marra [5] indicated a second type of oceanic-like island, which received diversified faunas through occasional routes from the near mainland. Louchart [69] proposed an intermediate type of island, between oceanic and continental ones. In this "intermediate" type, terrestrial carnivorous mammals are missing, while non-flying mammals, mainly rodents, are present and Strigiformes represent the top of the food chain.

Sicily experienced different types of insular conditions. The Monte Pellegino FC is problematic, but it can be tentatively related to an oceanic-like island *sensu* Alcover *et al.* [68], which hosted taxa coming from different lands, maybe from a past land connection in the late Miocene. The Spinagallo FC is probably related to the intermediate type of island suggested by Louchart [69]. The MaccagnoneFC and the San Teodoro FC can be indicative of an island that received the faunas spread through varying, sporadic means (natural rafts, active dispersal by swimming or flying), or maybe even a "stepping stone" way. It might correspond to the oceanic-like island of the second type, suggested by Marra [5].

Sicily became more and more "continental" during Late Pleistocene, when exchanges with the mainland were more frequent and the faunal composition indicated a large island, with abundant resources and a well-developed trophic net.

ACKNOWLEDGMENTS

The author is deeply indebted to: the anonymous reviewers, whose comments improved this paper; Donald R. Prothero, for his critical reading of the paper; the staff of Geosciences. A.C. Marra dedicates this paper to her daughter Gaia. Founds are provided by University of Messina (PRA 2008-2009).

REFERENCES

1. Bonfiglio, L.; Marra, A.C.; Masini, F. The contribution of Quaternary vertebrates to the paleoenvironmental and paleoclimatic reconstructions in Sicily. *Spec. Pubbl. Geol. Soc. London***2000**, *181*, 169–182.
2. Bonfiglio, L.; Mangano, G.; Marra, A.C.; Masini, F.; Pavia, M.; Petruso, D. Pleistocene Calabrian and Sicilian bioprovinces. *Geobios. Spec. Mem.* **2002**, *24*, 29–39.

3. Bonfiglio, L.; di Maggio, C.; Marra, A.C.; Masini, F.; Petruso, D. Biochronology of Pleistocene vertebrate faunas of Sicily and correlation of verterate bearing deposits with marine deposits. *Ital. J. Quat. Sci.* **2003**, *16*, 107–114.
4. Masini, F.; Petruso, D.; Bonfiglio, L.; Mangano, G. Origination and extinction patterns of mammals in three central western Mediterranean islands from the Late Miocene to Quaternary. *Quat. Int.* **2008**, *182*, 63–79.
5. Marra, A.C. Pleistocene Mammals of Mediterranean Islands. *Quat. Int.* **2005**, *129*, 5–14.
6. Marra, A.C. Pleistocene mammal faunas of Calabria. *Boll. Soc. Paleontol. Ital.* **2009**, *48*, 113–122.
7. Simpson, G.G. *The Geography of Evolution*; Chilton Books: New York, NY, USA, 1965.
8. Dermitzakis, M.D.; Sondaar, P.Y. The importance of fossil mammals in reconstructing palaeogeography wight special reference to the Pleistocene Aegean Archipelago. *Ann. Geolog. Pays Ellen.* **1978**, *29*, 808–840.
9. Sondaar, P.Y.; Dermitzakis, M.D.; de Vos, J. The paleogeography and faunal evolution of the land mammals of Crete. In *Pleistocene and Holocene Fauna of Crete and Its First Settlers*; Reese, D.S., Ed.; Prehistory Press: Madison, WI, USA, 1996; pp. 61–67.
10. Barrier, P. Stratigraphie des dépôts pliocènes et quaternaires du Dètroit de Messine [in French]. In *Le Détroit de Messine (Italie) Évolution Tectono-Sédimentaire Récente (Pliocène et Quaternaire et Environmnt Actuel)*; Barrier, P., di Geronimo, I., Montenat, C., Eds.; Institut Géologique Albert de Lapparent (IGAL): Paris, France, 1987; pp. 59–81.
11. Petruso, D.; Sarà, M.; Surdi, G.; Masini, F. Le faune a mammiferi della Sicilia tra il tardo-glaciale e l'Olocene [in French]. *Biogeographia* **2011**, *30*, 27–39.
12. Gliozzi, E.; Abbazzi, L.; Ambrosetti, P.; Argenti, P.; Azzaroli, A.; Caloi, L.; Capasso Barbato, L.; di Stefano, G.; Esu, D.; Ficcarelli, G.; *et al.* Biochronology of selected Mammals, Molluscs, Ostracods from the Middle Pliocene to the Late Pleistocene in Italy. The state of the art. *Riv. Ital. Palentol. Strat.* **1997**, *103*, 369–388.
13. Palombo, M.R. Biochronology of Plio-Pleistocene mammalian faunas of the Italian peninsula: Knowledge, problems and perspectives. *Ital. J. Quat. Sci.* **2004**, *17*, 565–582.
14. Thaler, L. Les rongeurs (Rodentia et Lagomorpha) du Monte Pellegrino et la question des anciens isthmus de la Sicilie [in French]. *Comptes Rendues de l'Académie de Sciences* **1972**, *274*, 188–190.
15. Burgio, E.; Fiore, M. *Pannonictis arzilla* (DE GREGORIO, 1886) a "Villafranchian" element in the fauna of Monte Pellegrino (Palermo, Sicily). *Ital. J. Quat. Sci.* **1997**, *10*, 65–74.
16. Bonfiglio, L.; Esu, D.; Mangano, G.; Masini, F.; Petruso, D.; Soligo, M.; Tuccimei, P. Late Pleistocene vertebrate-bearing deposits at San Teodoro Cave (Northe-Eastern Sicily): Preliminary data on faunal diversification and chronology. *Quat. Int.* **2008**, *190*, 26–37.
17. Vaufrey, R. Les Elephants nains des iles mediterrannéennes et la question des Isthmes Pléistocène [in French]. *Archives de l'Institut de Paléontologie Humaine* **1929**, *6*, 1–220.
18. Kotsakis, T. Sulle mammalofaune quaternarie siciliane. *Boll. Serv. Geol. Ital.* **1979**, *99*, 263–276.
19. Burgio, E. Le attuali conoscenze sui Mammiferi terrestri quaternary della Sicilia. [in Italian]; In *Prima Sicilia:Alle Origini Della Società Siciliana*; Tusa, S., Ed.; Edilprint: Palermo, Italy, 1997; pp. 55–72.
20. Ambrosetti, P.; Azzaroli, A.; Bonadonna, F.P.; Follieri, M. A scheme of Pleistocene chronology for the Thyrrenian side of centra Italy. *Boll. Soc. Geol. Ital.* **1972**, *91*, 169–184.
21. Burgio, E.; Cani, M. Sul ritrovamento di elefanti fossili ad Alcamo (Trapani, Sicilia). *Il Naturalista Sicil.* **1988**, *12*, 87–97.
22. Bonfiglio, L.; Insacco, G. Palaeoenvironmental, paleontologic and stratigraphic significante of vertebrate remains in Pleistocene limnic and alluvial deposits from South Eastern Sicily. *Palaeogeogr. Palaeoclimatol. Paleoecol.* **1992**, *95*, 195–208.
23. Bada, J.L.; Belluomini, G.; Bonfiglio, L.; Branca, M.; Burgio, E.; Delitala, L. Isoleucine Epimerization ages of Quaternary Mammals of Sicily. *Ital. J. Quat. Sci.* **1991**, *4*, 5–11.
24. Tedford, R.H. Principles and Practices of Mammalian Geochronology in North America. In Proceedings of the North American Paleontological Convention, Chicago, IL, USA, 5–7 September 1969; pp. 666–703, Part F.
25. Ferretti, M.P. he dwarf elephant *Palaeoloxodon mnaidriensis* from Puntali cave, Carini (Sicily; late Middle Pleistocene): Anatomy, systematics and phylogenetic relationships. *Quat. Int.***2008**, *182*, 90–108.
26. Bonfiglio, L.; Mangano, G.; Marra, A.C.; Masini, F. A new Late Pleistocene vertebrate faunal complex from Sicily (S. Teodoro cave. North-eastern Sicily, Italy). *Boll. Soc. Paleontol. Ital.***2001**, *40*, 149–1587.
27. Van der Geer, A.; Lyras, G.; de Vos, J.; Dermitzakis, M. *Evolution of Island Mammals Adaptation and Extinction of Placental Mammals on Islands*; Wiley-Blackwell: Hoboken, NJ, USA, 2010.
28. Masini, F.; Sarà, M. *Asoricolus burgioi* sp. nov. (Soricidae, Mammalia) from the Monte Pellegrino faunal complex (Sicily). *Acta Zool. Crac.* **1998**, *41*, 111–124.
29. Petruso, D. Il Contributo dei Micro Mammiferi Alla Stratigrafia e Paleogeografia del Quaternario Continentale Siciliano. [in Italian]. Ph.D. Thesis, Universities of Naples and Palermo, Palermo, Italy, 2003.
30. Daams, R.; de Bruijn, H. A classification of the Gliridae (Rodentia) on the basis of dental morphology. *Hystrix* **1995**, *6*, 3–50.
31. Fladerer, F.A.; Fiore, M. The Early Pleistocene insular hare *Hypolagus pellegrinus* sp. nov. from Northern Sicily. *Palaeontogr. Ital.* **2002**, *89*, 37–63.
32. Agnesi, V.; Macaluso, T.; Masini, F. L'ambiente e il clima della Sicilia nell'ultimo milione di anni. [in Italian]; In

Prima Sicilia:Alle Origini Della Società Siciliana; Tusa, S., Ed.; Edilprint: Palermo, Italy, 1997; pp. 31–53.

33. Di Geronimo, I. Bionomie des peuplements benthiques des sustrats meubles et rocheux plio-quaternaires du Dètroit de Messine [in French]. In *Le Détroit de Messine (Italie) Évolution Tectono-Sédimentaire Récente (Pliocène et Quaternaire et Environmnt Actuel)*; [in Italian]; Barrier, P., di Geronimo, I., Montenat, C., Eds.; Institut Géologique Albert de Lapparent (IGAL): Paris, France, 1987; pp. 153–169.
34. Palombo, M.R. Climate Change *versus* biotic interaction: A case study of large mammal faunal complexes on the Italian Peninsula from the Pliocene to the Late Pleistocene. New methodological approaches. *Cour. Forsch. Senckenberg* **2007**, *259*, 13–46.
35. Cavinato, G.P.; Petronio, C.; Sardella, R. The Mercure River Basin (Southern Italy): Quaternary stratigraphy and large mammal biochronology. In The World of Elephants, Proceedinggs of the 1st International Congress, Rome, 16–20 October 2001; Cavaretta, G., Gioia, P., Mussi, M., Palombo, M.R., Eds.; Consiglio Nazionale delle Ricerche: Rome, Italy, 2001; pp. 187–190.
36. Zammit Maempel, G.; de Bruijn, H. The Plio-Pleistocene Gliridae grom the Mediterranean Islands reconsidered. *Proc. K. Ned. Akad. Ser. B Phys.* **1982**, *85*, 113–128.
37. Filippucci, M.G.; Kotkakis, T. Biochemical systematic and evolution of Myoxidae. *Hystrix***1994**, *6*, 77–97.
38. Kotsakis, T. *Crocidura esui* n. sp. (Soricidae, Insectivora) du Plèistocène de Spinagallo (Sicile orientale, Italie) [In French]. *Geol. Rom.* **1986**, *23*, 51–64.
39. Esu, D.; Kotsakis, T.; Burgio, E. Vertebrati e Molluschi continentali fossili del Pleistocene di Poggio Schinaldo (Palermo, Sicilia). [in Italian]. *Boll. Soc. Geol. Ital.* **1986**, *105*, 233–241.
40. Sarà, M. The Sicilian (*Crocidura sicula*) and the Canary (*C.canariensis*) shrew (Mammalia, Soricidae): Peripheral isolate formation and geographic variation. *Boll. Zool.* **1995**, *62*, 173–182.
41. Maddalena, T.; Ruedi, M. Chromosomal evolution in the genus Crocidura. In *Advances in the Biology of Shrews*; Merritt, F.J., Kirkland, G.L., Rose, R.K., Eds.; Carnegie Museum of Natural History Special Publication: Pittsburgh, PA, USA, 1994; Volume 18, pp. 334–344.
42. Vogel, P.; Cosson, J.P.; Lòpez Jurado, L.F. Taxonomic status and origin of the shrews (Soricidae) from the Canary islands inferred from mtDNA comparison with the European *Crocidura* species. *Mol. Phyl. Evol.* **2003**, *27*, 271–282.
43. Dubey, S.; Koyasua, K.; Parapanova, R.; Ribic, M.; Hutterere, R.; Vogela, P. Molecular phylogenetics reveals Messinian, Pliocene, and Pleistocene colonizations of islands by North African shrews. *Mol. Phylogenet. Evol.* **2007**, *47*, 877–882.
44. Johnson, D.L. Problems in the Land Vertebrate Zoogeography of certain islands and the swimming powers of elephants. *J. Biogeog.* **1980**, *7*, 383–398.
45. Quammen, D. *The Song of the Dodo. Island Biogeography in an Age of Extinctions*; Pimlico: London, UK, 1996; pp. 1–702.
46. Alcover, J.A.; McMinn, M. Predators of vertebrates on islands. *BioScience* **1994**, *44*, 12–18.
47. Burgio, E.; Fiore, M. *Nesolutra trinacriae* n. sp. lontra quaternaria della Sicilia. *Boll. Soc. Paleontol. Ital.* **1988**, *23*, 259–275.
48. Palombo, M.R. How can endemic proboscideans help us to understand the "island rule"? A case study of Mediterranean islands. *Quat. Int.* **2007**, *169–170*, 105–124.
49. Pavia, M. The evolution dynamics of the Strigiformes in the mediterranean islands with the description of *Aegolius martae* n. sp. (Aves, Strigidae). *Quat. Int.* **2008**, *182*, 80–89.
50. Guglielmo, M.; Marra, A.C. Le due Sicilie del Pleistocene Medio: osservazioni paleogeografiche. [in Italian]. *Biogeographia* **2011**, *30*, 11–25.
51. Pavia, M. A new large Barn Owl (Aves, strigiformes, Tytonidae) from the Middle Pleistocene of Sicily, Italy, and its taphonomical significance. *Geobios* **2004**, *37*, 631–641.
52. Raia, P.; Barbera, C.; Conte, M. The fast life of a dwarfed elephant. *Evol. Morph.* **2003**, *17*, 293–312.
53. Petruso, D.; Locatelli, E.; Dalla Valle, C.; Masini, F.; Sala, B. Phylogeny and biogeography of fossil and extant Microtus (Terricola) (Mammalia, Rodentia) of Sicily and the southern Italian peninsula based on current. *Quat. Int.* **2011**, *243*, 192–203.
54. Colantoni, P. Marine geology of the Strait of Messina. In *Le Détroit de Messine (Italie) Évolution Tectono-Sédimentaire Récente (Pliocène et Quaternaire et Environmnt Actuel)*; Barrier, P., Di Geronimo, I., Montenat, C., Eds.; Institut Géologique Albert de Lapparent (IGAL): Paris, France, 1987; pp. 191–210.
55. Abbazzi, L.; Bonfiglio, L.; Marra, A.C.; Masini, F. A revision of medium and small sized deer from the Middle and Late Pleistocene of Calabria and Sicily. *Boll. Soc. Paleontol. Ital.* **2001**, *40*, 115–126.
56. Brugal, J.P. Cas de "Nanisme" Insulaire dans L'Auroche [in Franch]. In Proceedings of 112th Congrès National des Sociétés Savantes, Lyon, France, 1987; pp. 53–66.
57. Marra, A.C. Pleistocene Hippopotamuses of Mediterranean Islands: Looking for Ancestors. In Proceedings of the International Symposium Insular Vertebrate Evolution: The Palaeontological Approach, Mollorca, Spain, 16–19 September 2003.
58. Marra, A.C. *Ursus arctos* from selected Pleistocene sites of eastern Sicily. *Boll. Soc. Paleontol. Ital.* **2003**, *42*, 145–150.
59. Yll, R.; Carriòn, J.S.; Marra, A.C.; Bonfiglio, L. Vegetation reconstruction on the basis of pollen in late Pleistocene hyena coprolites from San Teodoro Cave (Sicily, Italy). *Palaeogeogr. Palaeoclim. Palaeoecol.* **2006**, *237*, 32–39.

60. Allen, J.R.M.; Watts, W.A.; Huntley, B. Weichselian palynostratigraphy, palaeovegetation and palaeoenvironment: The record from Lago Grande di Monticchio, southern Italy. *Quat. Int.*.**2000**, *74*, 91–110.

61. Pavia, M. The Middle Pleistocene Fossil Avifauna from the "*Elephas mnaidriensis* Faunal Complex" of Sicily (Italy). Preliminary Results. In The World of Elephants—International Congress, Proceedings of the First International Congress, Rome, Italy, 16–20 October 2001; pp. 497–501.

62. Lambeck, K.; Antonioli, F.; Purcell, A.; Silenzi, S. Sea-level change along the Italian coasts for the past 10,000 years. *Quat. Sci. Rev.* **2004**, *23*, 1567–1598.

63. Masini, F.; Sala, B. Large- and small-mammal distribution patterns and chronostratigraphic boundaries from the Late Pliocene to the Middle Pleistocene of the Italian peninsula. *Quat. Int.* **2007**, *160*, 43–56.

64. Seguenza, G. I vertebrati fossili della provincia di Messina: Parte II, Mammiferi e Geologia del piano Pontico. [in Italian]. *Boll. Soc. Geol. Ital.* **1902**, *21*, 115–175.

65. Seguenza, G. Nuovi resti di mammiferi pontici di Gravitelli presso Messina. [in Italian]. *Boll. Soc. Geol. Ital.* **1907**, *26*, 7–122.

66. Rook, L.; Mazza, P.; Rustioni, M.; Torre, D. Lands and Endemic Mammals in the Late Miocene of Italy: Paleogeographic Outlines of Tyrrhenian and Adriatic Areas between 11–9 and 7–4 Ma. In Proceedings of Environments and Ecosystem Dynamics of the Eurasian Neogene (EEDEN), State of the Art Workshop, Lyon, France, 16–18 November 2000; pp. 56–58.

67. Herridge, V.L. Dwarf Elephants on Mediterranean Islands: A Natural Experiment in Parallel Evolution. Ph.D. Thesis, University College London, London, UK, 2010.

68. Alcover, J.A.; Sans, A.; Palmer, M. The extent of extinction of mammals on islands. *J. Biogeogr.***1998**, *25*, 913–918.

69. Louchart, A. Integrating the Fossil Record in the Study of Insular Body Size Evolution: Examples in Owls (Aves: Strigiformes). In Proceedings of the Internation Symposyum Insular Vertebrate Evolution: the Palaeontological Approach, Mollorca, Spain, 16–19 September 2003; pp. 155–174.

CHAPTER 10

A REEVALUATION OF THE MORPHOLOGY, PALEOECOLOGY, AND PHYLOGENETIC RELATIONSHIPS OF THE ENIGMATIC WALRUS PELAGIARCTOS

Robert W. Boessenecker and Morgan Churchill

INTRODUCTION

The extant walrus *Odobenus rosmarus* represents one of the most morphologically aberrant members of the Pinnipedia, and is the sole extant member of the family Odobenidae. Despite the low taxonomic diversity of walruses today, the fossil record indicates that odobenids over the past 16 Ma showed a wider variety of morphological adaptations and body sizes than at present, exploited a greater number of resources based on inferred feeding ecology, and inhabited a wider variety of marine environments including subtropical shallow marine settings [1]–[7].

Despite the extensive fossil record of walruses during the Neogene (see table 3.2 of Deméré et al., [8]), the phylogenetic relationships of walruses remain poorly understood. Non-cladistic studies have placed walruses as the sister taxon of the Otariidae (fur seals and sea lions) within a monophyletic Otarioidea (= Otariidae of Barnes [9]. In contrast, phylogenetic analyses of morphological data support a sister taxon relationship between walruses and phocoid pinnipeds (Desmatophocidae+Phocidae), forming the clade Phocomorpha (i.e. Odobenidae+Phocoidea; Berta and Wyss [10]). However, virtually all molecular analyses of extant pinnipeds have supported a monophyletic Otarioidea [11]–[16].

The "Imagotariinae" are a group of stem walruses considered by Barnes [9], Barnes and Raschke [1] and Kohno et al. [6] to be a subfamily of "walrus-like" pinnipeds; Deméré [2], however, found the group to be paraphyletic, a conclusion supported by additional phylogenetic analyses [2], [4], [5], [17], and to represent a stem grade of walruses ancestral to Dusignathinae and Odobeninae. "Imagotariinae" includes *Proneotherium*, *Prototaria*, *Neotherium*, *Kamtschatarctos*, *Imagotaria*, *Pelagiarctos*, *Pseudotaria*, and possibly *Pontolis*([5]; other analyses include *Pontolis* as a dusignathine walrus; [2]). The earliest diverging walruses – *Prototaria*, *Proneotherium*, and *Neotherium* – are generally similar in size and cranial, mandibular, and dental morphology to early diverging pinnipedimorphs (="Enaliarctinae" of Mitchell and Tedford [18]), while geologically younger and later diverging taxa such as *Pelagiarctos* and *Imagotaria* are much larger; *Imagotaria, Pseudotaria,* and all later diverging odobenids (Dusignathinae and Odobeninae) share numerous derived cranial features, such as a transversely arched palate, a reduced pseudosylvian sulcus, and less than three roots on the M[1] [2], [5].

Pelagiarctos thomasi was described on the basis of an isolated and fragmentary pair of fused mandibles and several isolated teeth from the middle Miocene Sharktooth Hill Bonebed in Kern County, California [19]. Due to the apparently large body size of *Pelagiarctos*, the morphology of its postcanine teeth, robust mandible, and rarity in the Sharktooth Hill Bonebed, Barnes [19]concluded that *Pelagiarctos* was a marine apex predator. *Pelagiarctos* was hypothesized to be adapted toward macrophagy and fed upon warm blooded prey such as marine birds and small marine mammals, "in addition to (or instead of) the expected diet of fishes" ([19]:9). Since the discovery of the *Pelagiarctos thomasi* holotype, only a few isolated teeth referable to the same taxon have been discovered at Sharktooth Hill ([19]; see Table 1) and we only know of a

Citation: Boessenecker RW, Churchill M (2013) A Reevaluation of the Morphology, Paleoecology, and Phylogenetic Relationships of the Enigmatic Walrus Pelagiarctos. PLoS ONE 8(1): e54311. doi:10.1371/journal.pone.0054311.

few isolated teeth that have been collected since the taxon was named by Barnes [19]; these newly collected specimens remain uncurated. Description of a new fossil (SDNHM 131041) from the "Topanga" Formation (Fig. 1) referable to *Pelagiarctos* serves to expand our knowledge of the anatomy of this poorly known walrus (Fig. 2). Although fragmentary, this new specimen offers the opportunity to include *Pelagiarctos* within a phylogenetic analysis for the first time, and to reevaluate its hypothesized feeding ecology.

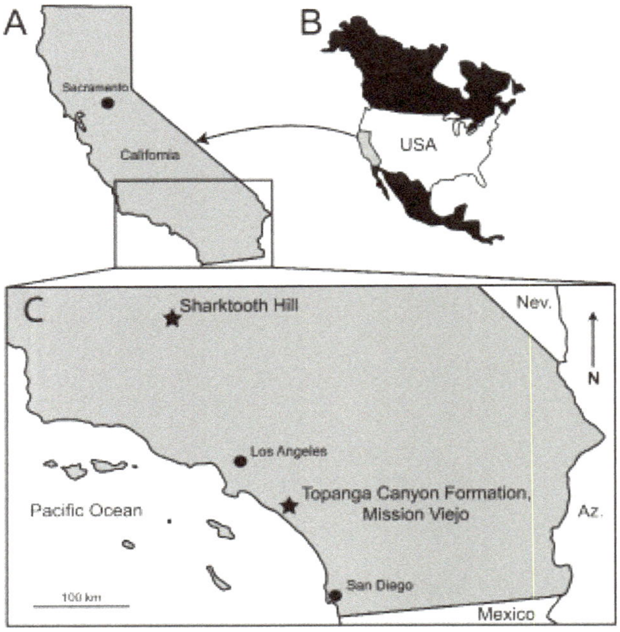

Figure 1. Location of fossil occurrences of *Pelagiarctos*.

Map of A) California and B) North America, and C) Map of southern California showing the location of Sharktooth Hill (Round Mountain Silt) and the "Topanga" Formation. Abbreviations: Az., Arizona; Nev., Nevada.

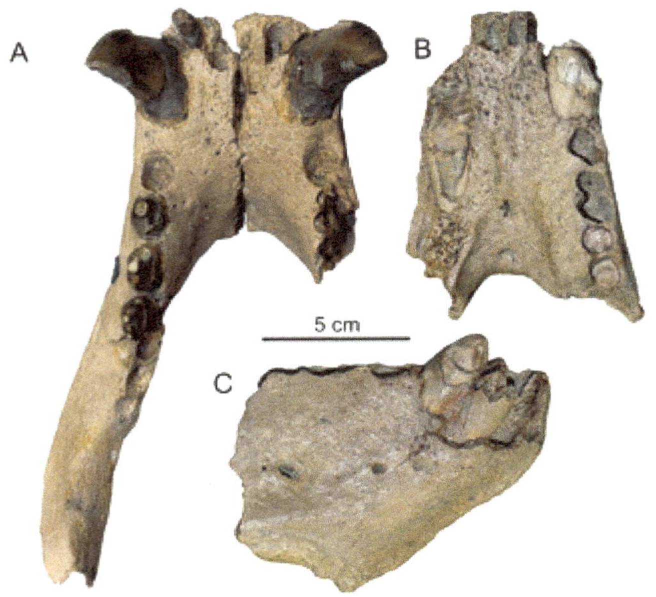

Figure 2. Mandibles of *Pelagiarctos*.

Comparison of A) *Pelagiarctos* sp. (SDNHM 131041) in dorsal aspect, and the holotype specimen of *Pelagiarctos thomasi* (LACM 121501) in B) dorsal and C) lateral aspect.

Table 1. Table of known fossils of *Pelagiarctos*. *Pelagiarctos* **sp. is from the "Topanga" Formation (This study), while all other material listed is from the Sharktooth Hill Bonebed, Round Mountain Silt (Barnes, 1988).**

Specimen	Taxonomic Referral	Element
LACM 121501	*Pelagiarctos thomasi* holotype	symphyseal portion of mandibles, damaged I_2, C_1, P_3
LACM 123415	*Pelagiarctos thomasi*	right P^3 or P^4
LACM 38812	*Pelagiarctos thomasi*	right P_1
LACM 72856	*Pelagiarctos thomasi*	right P_3 or P_4
LACM 122310	*Pelagiarctos thomasi*	left P_3 or P_4
TATE 2694	*Pelagiarctos thomasi*	left P_1 or right P^1
UCMP 93058	*Pelagiarctos thomasi*	left P_2
SDNHM 131041	*Pelagiarctos* sp.	Partial right and left dentary with left I_3, right and left C_1, and left P_2–P_4

doi:10.1371/journal.pone.0054311.t001

MATERIALS AND METHODS

The new specimen of *Pelagiarctos* (SDNHM 131041) was compared with the *Pelagiarctos thomasi* holotype and referred specimens described by Barnes (Table 1; [19]), in addition to all odobenids (fossil and modern) for which mandibles are known. With the exception of *Ontocetus emmonsi* and *Kamtschatarctos sinelnikovae* (represented by casts), we examined the original fossil and modern specimens of these taxa (published and unpublished). Anatomical terminology used in this study follows Mead and Fordyce [20] for mammals.

Phylogenetic Methods

To understand the evolutionary relationships of *Pelagiarctos* with other walruses, a phylogenetic analysis was performed using maximum parsimony in TNT [21] and Bayesian inference in MrBayes 3.1.2 [22]. Sixteen walrus taxa were included within our analysis, including representatives of the later diverging Odobeninae and Dusignathinae. Seven outgroup taxa were included for comparison: the "enaliarctine" taxa *Enaliarctos emlongi* [23] and *Pteronarctos goedertae* [9]; two members of the extinct Desmatophocidae (*Desmatophoca oregonensis* [24] and *Allodesmus gracilis* [25]); one otariid taxon (*Callorhinus ursinus*), and two phocid taxa, representing Monachinae (*Monachus tropicalis*) and Phocinae (*Erignathus barbatus*). The otariid and phocid taxa selected were chosen as they represent the earliest diverging extant lineages within their respective families and subfamilies. As part of this study, we produced a matrix of 90 cranial, dental, mandibular, and postcranial characters to examine the phylogenetic position of *Pelagiarctos*. These characters are fully illustrated on Morphobank. These characters were then included within a larger dataset employing the cranial and postcranial characters of Kohno [5], resulting in a matrix of 90 morphological characters. Only minor changes were made to the original matrix, these mostly pertain to using narrower taxonomic units for coding, whenever possible using species as our terminal units, not genera, subfamilies, or families. The only exception for this rule was *Ontocetus*, in which neither author has been able to view mandibular and cranial material referable to the same species. In this case, *O. emmonsi* forms the basis of our coding for mandibular and lower dental characters, while coding of other features is largely based on that of Kohno [5]. The phylogenetic matrices for the mandible and dental characters with associated media is archived at http://morphobank.org. Characters used in this analysis were based on personal observation as well as prior phylogenetic studies [2], [9], [10], [26]. We include and code for the first time a referred mandible of *Proneotherium repenningi* (USNM 314628), which is referable based on size, similarity with *Neotherium mirum* (e.g. LACM 123000, UCMP 81665, 116018, 191874), and lack of a carnassial-like lower molar that characterizes Astoria Formation enaliarctine pinnipeds.

All parsimony analyses were run using the new technology options with 10,000 random addition replicates and an implied weighting scheme of K=2–5. Using implied weighting allowed us to downweight characters prone to homoplasy, reducing their impact on the phylogeny [27]. Bootstrap values were calculated using symmetric resampling with 1000 replicates. Bayesian Inference analyses were run using 10 million generations, 4 chains, and a sample frequency of 1000. A conservative burn-in of 5000 was discarded. Two separate runs were created with each analysis, and a consensus of both runs was used to produce the most probable tree. Bayes factors were used to determine the morphological model of evolution employed in the analysis, with the GTR model [28] selected. Bootstrap support (BS) and Bayesian posterior probabilities (PP) are reported for all nodes.

RESULTS

Systematic Paleontology

Mammalia Linnaeus, 1758.

Carnivora Bowdich, 1821.

Odobenidae Allen, 1880.

Pelagiarctos Barnes, 1988.

Emended diagnosis.

Pelagiarctos is distinguished from the early diverging odobenids *Proneotherium, Neotherium*, and *Kamtschatarctos* by the following derived characters: rugose, vascularized bone at mandibular terminus; mandible that is dorsoventrally much deeper anteriorly, C_1 with lateral sulci, inflated postcanine crowns, less prominent metaconid cusps, and medially swollen and crenulated lingual cingulum. *Pelagiarctos* differs from the later diverging odobenids *Imagotaria*and *Pontolis* in retaining the following primitive characters: double-rooted P_{2-4}; metaconid cusps; sinuous ventral margin of mandible. *Pelagiarctos* is distinguished from all dusignathine and odobenine walruses in the retention of the following primitive characters: well developed paraconid, metaconid, and hypoconid cusps on postcanine crowns; double-rooted lower postcanines; lower canine with posterior crista; retention of M_2.

Pelagiarctos sp

Horizon and age.

SDNHM 131041 was collected in 1997 by H. Lozana from SDNHM locality 4984 in Orange County, California (Fig. 1). SDNHM locality 4984 was a temporary exposure of a rock unit currently mapped as the "Topanga" Formation, exposed during construction within the city of Mission Viejo (Fig. 1). Locally, the "Topanga" Formation overlies the lower Miocene Vaqueros Formation, and is overlain by the middle-upper Miocene Monterey Formation. Detailed locality information is available on request. Howard and Barnes [29] noted that the Topanga Canyon Group is probably not an appropriate stratigraphic name to be applied to this unit in Orange County; the type section in Los Angeles County has been renamed and elevated to group status [30]. Because no stratigraphic revision of this unit has been published in the time being, we retain the current applicable name, although it will likely change in the future due to concerns addressed by Howard and Barnes [29], and we follow Whistler and Lander [31] in referring to this unit in Orange County as the "Topanga" Formation to reflect its difference from the type section of the Topanga Canyon Group in Los Angeles County. Fossil vertebrates from SDNHM locality 4984 occur in a marine bonebed within a gray to yellow, friable, medium to coarse-grained sandstone, exposed about 2 meters above a prominent oyster-rich shell bed. SDNHM locality 4984 constitutes the uppermost 3 meters of the "Topanga" Formation, below its contact with the overlying Monterey Formation. A nearby locality in the "Topanga" Formation has yielded a diverse marine vertebrate assemblage, including sharks, marine birds, pinnipeds, and odontocetes [29], [32]. Howard and Barnes [29] reported a meager marine mammal assemblage, including an archaic walrus (*Neotherium* sp.), a desmatophocid seal (*Allodesmus* sp.), an allodelphinid odontocete (cf. *Zarhinocetus errabundus*), and a kentriodontid dolphin (*Kentriodon* sp.). Although this assemblage still awaits description, all of these taxa occur within the middle Miocene (16.1–14.5 Ma) Sharktooth Hill Bonebed, suggesting a similar age of the two units [29]. At a nearby locality in the "Topanga" Formation, Aranda-Manteca et al. [33] reported a partial skull of the dugongid sirenian *Metaxytherium arctodites*; more complete material of this taxon was described from the Los Indios Member of the Rosarito Beach Formation at La Mision in Baja California. The Rosarito Beach Formation also shares a number of marine mammal taxa (aff. *Neotherium*, *Allodesmus* sp., aff. *Tiphyocetus temblorensis*, aff. *Parietobalaena* sp., aff. *Kentriodon* sp., *Liolithax* aff. *L. kernensis*, and aff. *Desmostylus* sp.) with the Sharktooth Hill assemblage [34], [35]. Several terrestrial mammals were reported from the "Topanga" Formation by Raschke [32] and Howard and Barnes [29], including an indeterminate paleomerycid, a camel (*Aepycamelus* sp.), and a horse (*Merychippus* sp.), suggesting correlation with the Barstovian North American Land Mammal Age (NALMA; [29], [32]). Further work on the terrestrial mammals of the "Topanga" Formation in this region was published by Whistler and Lander [31], who reported a much more diverse assemblage; they noted that the presence of *Copemys* and *Merychippus* indicated a slightly older age, being correlative with the early late Hemingfordian NALMA. The late Hemingfordian is approximately 17.5–15.9 Ma in age [36]. Relizian and/or Luisian benthic foraminifera and mollusks assignable to the Temblor Provincial Marine Molluscan Stage have also been reported from these localities in the "Topanga" Formation [31]. Further to the west, in the San Joaquin Hills, an andesitic lava flow at the base of the Paulerino Member of the "Topanga" Formation has been K/Ar dated at 15.8+– 1.3 Ma [37]; recalculated with revised decay constants by Whistler and Lander [31]. Taken in full, "Topanga" Formation in Orange County is probably late early to early middle Miocene in age (~17.5–15 Ma), and is broadly correlative with the Round Mountain Silt Member of the Temblor Formation and the Los Indios Member of the Rosarito Beach Formation. Detailed locality information is available on request.

Referred material.

SDNHM 131041, a pair of partial associated left and right mandibles, preserving left I_3, left and right C_1, left P_{2-4} and partial root of right P_2. SDNHM 131041 was collected from SDNHM locality 4984 in the middle Miocene "Topanga" Formation. Because the left and right mandibles were found in close association, and tightly articulate at the symphysis when placed together (Fig. 2a), they represent one individual.

Description of SDNHM 131041

Mandible.

The referred specimen includes portions of the left and right mandibles (Fig. 2, 3, 4). The more incomplete right mandible is only preserved anterior to the P_3, and there is some slight breakage of the dorsal portion medial to the right canine. The less damaged left mandible is missing most of the ascending ramus, and just the anteriormost portion of the coronoid process is preserved (Fig. 4). Unless stated otherwise, the description is based on the left mandible. Measurements of SDNHM 131041 are presented in Table 2.

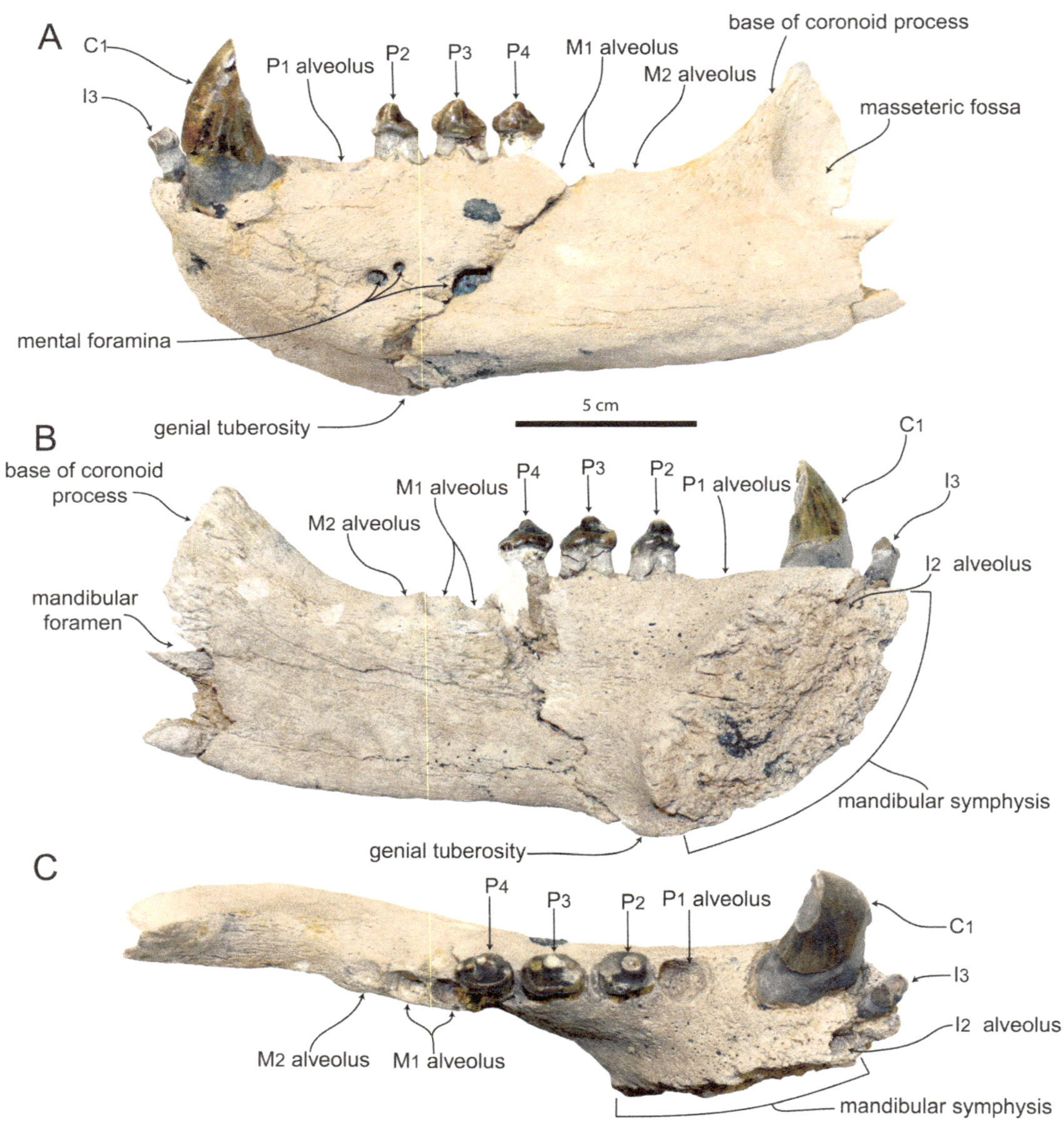

Figure 3. Left mandible of SDNHM 131041, *Pelagiarctos* **sp. SDNHM in A) lateral, B) medial, and C) dorsal aspect.**

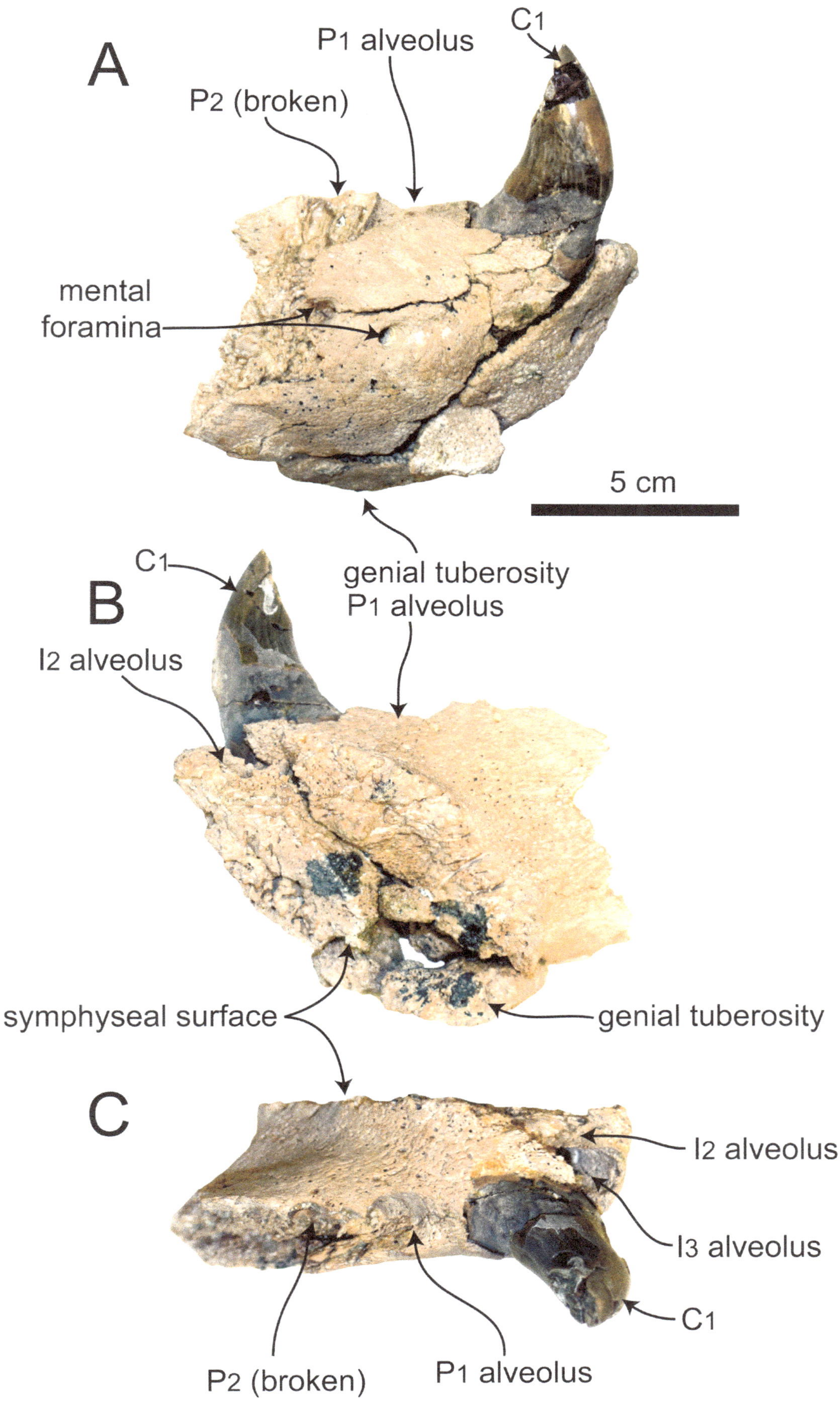

Figure 4. Right mandible of SDNHM 131041, *Pelagiarctos* sp. **Mandible of SDNHM 131041 in A) lateral, B) medial, and C) dorsal aspect.**

Table 2. Measurements of *Pelagiarctos* **sp., SDNHM 131041.**

Variable	Measurement in cm
Total length (as preserved)	20.57
Anterior tip to base of ascending ramus	15.33
Greatest length of mandibular symphysis	8.78
Anteroposterior length of symphysis	7.38
Depth of mandibular symphysis	4.83
Depth of ramus at C1	5.41
Depth of ramus at P1	6.46
Depth of ramus at P2	6.95
Depth of ramus at P3	6.56
Depth of ramus at P4	6.45
Depth of ramus at M1	5.48
Depth of ramus at M2	5.46
Length of postcanine tooth row	8.98
Length of C1–P1 diastema	0.86
Anterior tip to M1	10.86
M1 to base of ascending ramus	1.76
M2 to base of ascending ramus	1.18
Width across canines (apical)	8.83
Width across canines (basal)	6.43
Transverse width of ramus at P1	3.65
Transverse width of ramus at P2	4.40
Transverse width of ramus at P3	2.71
Transverse width of ramus at P4	2.21
Transverse width of ramus at M1	2.26
Transverse width of ramus at M2	2.24
I3 height/width	0.64/0.84
Canine height (right/left)	4.15/419
Canine anteroposterior length (right/left)	2.6/2.96
P2 crown height/width/length	0.85/1.06/1/31
P3 crown height/width/length	0.97/1.03/1.52
P4 crown height/width/length	0.86/1.02/1/44

doi:10.1371/journal.pone.0054311.t002

The mandibles are robust, dorsoventrally deep, and transversely wide. When in articulation, the anterior portions of the mandibles are transversely widest at the position of the canines (Fig. 2A). In lateral aspect, the dorsal and ventral margins of the horizontal ramus are nearly parallel. The dorsal margin of the coronoid process meets the horizontal ramus at an angle of 130°. Although most of the posterior portion of the mandible is missing, what is preserved of the anteriormost portion of the masseteric fossa indicates it was deep and sharply defined anteriorly. The medial surface of the coronoid process and horizontal ramus are generally flat. Along the long axis of the horizontal ramus, the vertical orientation changes so that near the position of the M_1, the medial surface of the ramus faces slightly ventromedially, while further anterior at the position of the P_2, the medial surface faces slightly dorsomedially. On the medial surface of the horizontal ramus there is a broad shelf immediately adjacent to the mandibular symphysis.

The symphyseal surface is rugose and pitted, but unlike the holotype specimen, the mandibles are not firmly ankylosed at the mandibular symphysis (Fig. 2A, 3B). In medial aspect, the symphyseal surface is oval-shaped, with convex anterior and posterior margins. The bone surface at the anterior end of the mandibles is rough and vascularized, with many small foramina, and most evident near the mandibular symphysis. Around the base of the canines, the lateral surface of the mandible bulges out around the canine root. Anteromedial to this, a slight medial keel exists adjacent to the mandibular symphysis. Adjacent to the symphysis and below the I_3, a large foramen occurs on each mandible, as in the holotype specimen.

The lateral surface of the horizontal ramus is slightly convex. Two large mental foramina occur on the lateral surface of the left mandible below the P_2, and a third large mental foramen occurs below the P_3. On the right mandible, a small single mental foramen occurs below the P_1. A well-developed genial tuberosity is situated at the posteroventral termination of the mandibular symphysis, although it does not sit on a ventral ridge as in the holotype specimen. Along the postcanine tooth row, adjacent alveoli and tooth roots are separated by small triangular extensions of bone.

Lower Dentition

On all teeth preserved, the basal margin of the crown is elevated nearly 10 mm above the alveolar margin, ranging from 8–10 mm on the P_{2-3} and nearly 20 mm on the anterior portion of the canine (Fig. 5). The I_2 is missing, but its alveolar morphology indicates it was small with a cylindrical, anterodorsally oriented root. The I_2 alveolus is medially confluent with the symphyseal surface, and there is no medial wall of bone preserved between the alveolus and the symphysis, indicating the lower second incisors were closely appressed; there are no I_1alveoli. The I_2 alveolus is positioned posteromedially to the I_3, resulting in a posteriorly V-shaped incisor alveolar row as in many early diverging odobenids and otariids. The left I_3 is preserved, and is a peg-like tooth with a small, bulbous crown and smooth enamel. The crown is roughly circular in occlusal aspect, and set on a column-like root. The crown is obliterated by two anteromedial and posterolateral wear facets, formed by occlusion with the I^2 and I^3, respectively.

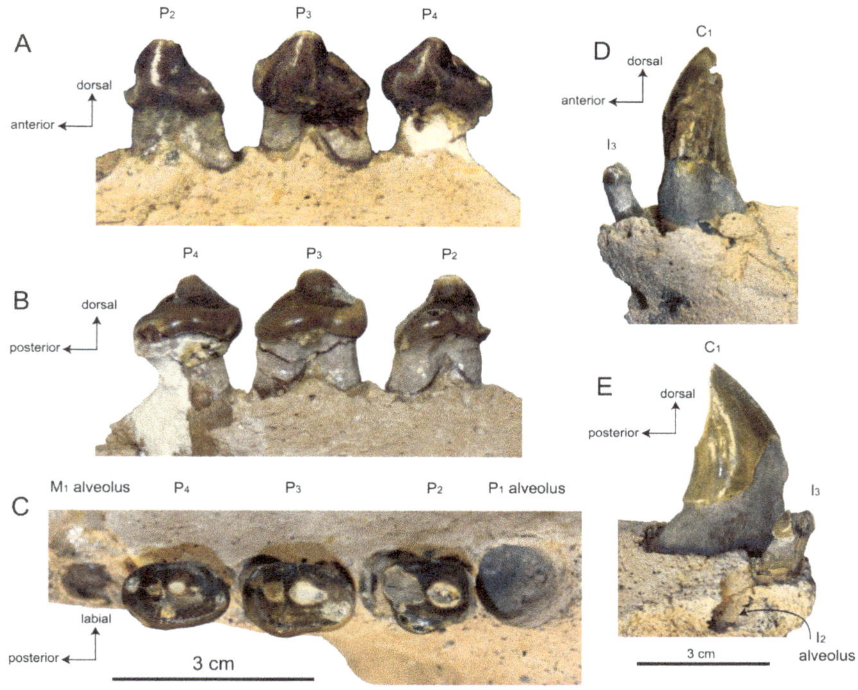

Figure 5. Lower dentition of SDNHM 131041, *Pelagiarctos* sp. **Lower premolars in A) labial, B) lingual, and C) occlusal aspect.** The M_1 and P_1 alveoli, and C_1 and I_3 are depicted in D) labial and E) mesial aspect.

Both left and right C_1 are present and well preserved. The canines are robust, vertically oriented, and slightly posteriorly curved (Fig. 5D, E). The crowns are conical and laterally compressed with an oval-shaped cross section, and thick, smooth enamel. There are two large wear facets on the crowns: a small anteromedial wear facet on the base of the crown from occlusion with the I^3, and a large, posterior facing, oval-shaped apical wear facet from occlusion with the C_1. The canines have shallow, longitudinal sulci on the lateral surface of the root, like the holotype of *Pelagiarctos thomasi* (Fig. 2, 5D, E). The sulci terminate below the crown, and there are no sulci on the medial surfaces of the canine roots.

Although both first lower premolars are missing, their alveoli are well preserved and indicate that the P_1 had a single, cylindrical root with a circular cross-section (Fig. 5C). The P_1 alveolus is notably wider than any other postcanine alveolus. The left P_2 (Fig. 5A–C, 6) appears to be either double-rooted or have a bilobate root, with a longitudinal sulcus on the lateral side, and a stronger sulcus on the medial surface. The broken right P_2 preserves root fragments in their alveoli, and no bony septum separates the roots, indicating the right P_2 was bilobate. In occlusal aspect, the P_2 crown is oval shaped. The P_2 has a large principal cusp (protoconid), while the anterior cusp (paraconid) and posterior cusp (hypoconid) morphology is unknown because of two large wear facets on the anterior and posterior edges of the crown (Fig. 5A, B). A faint labial cingulum is developed along the posterolabial margin of the crown. The posterior wear facet is so extensive that it is not clear if a metaconid was present. A third apical wear facet occurs at the tip of the protoconid. A strong lingual cingulum is present, with at least three cuspules (=crenations of Deméré and Berta [38]), although these are slightly worn. The cingulum is thickest posteromedially, where it forms a slight shelf and concavity, probably homologous to the talonid basin. The base of the crown is dorsally arched lingually and labially.

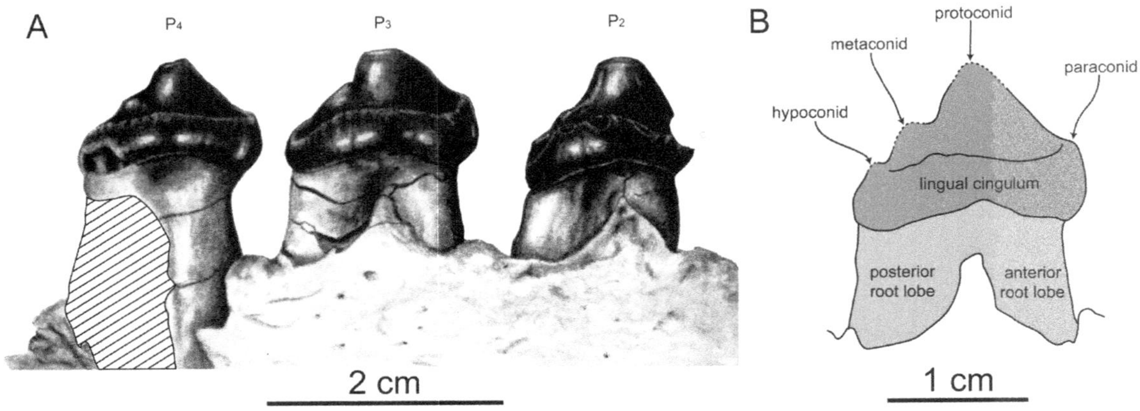

Figure 6. Dentition of SDNHM 131041, *Pelagiarctos* **sp. A) Illustration of lower premolars in lingual aspect.** B) line drawing of P_2 in lingual aspect, showing cusp homologies.

The P_3 (Fig. 5A–C, 6) root is double-rooted above the alveolar margin, and the root lobes appear to be widely divergent. The base of the P_3 crown is also dorsally arched lingually and labially, and the crown itself is bulbous. The crown of the P_3 is slightly anteroposteriorly longer than the P_2, and oval shaped in occlusal aspect. The protoconid cusp is very prominent, and there is a minute and worn paraconid cusp is present near the base of the crown and anterior to the protoconid. The tip of the protoconid has an apical wear facet; a blunt longitudinal crista connects the protoconid and paraconid. A small and worn metaconid cusp is located on the blunt posterior crista of the protoconid; the hypoconid cusp is positioned posterior and slightly labial to the metaconid, and is also worn. The labial surface is convex, with smooth enamel. A faintly developed labial cingulum is present posterolabially, which merges posteriorly with the hypoconid; anteriorly, the labial cingulum merges into a smooth surface. On the lingual surface there is a well developed lingual cingulum with fine cuspules along the crest. Between the protoconid and the cingulum is a small posterolingual shelf-like talonid basin. Due in part to the well-developed cingulum, the crown bears a posteromedial extension. A poorly developed labial cingulum occurs on the posterolabial portion of the crown, and terminates posterior to the midpoint of the crown.

Although the P_4 (Fig. 5A–C, 6) root is damaged, the anterior root lobe is present and cylindrical, and a fragment of the posterior root lobe is preserved, clearly indicating a double-rooted condition for this tooth. The basal margin of the crown is dorsally arched lingually and labially, as in the P_2 and P_3 crowns. The labial surface of the crown is convex with smooth enamel, and there is a poorly defined and discontinuous labial cingulum that is positioned posterolabially. The labial cingulum extends anteriorly from the hypoconid, and extends roughly 3/4 of the way to the paraconid cusp. The bulbous crown is oval-shaped in occlusal view, and the medial surface is strongly convex. The P_4 crown is transversely narrower than the P_3crown. The crown has a strong protoconid, and a small paraconid is developed on the anterior portion of the crown; a blunt longitudinal crista connects the paraconid and protoconid. A small accessory cusp occurs on the crest just anterior to the protoconid wear facet. A small metaconid cusp occurs on the posterior crista, and the hypoconid cusp occurs slightly labially to this crista. The tips of the paraconid, protoconid, metaconid, and hypoconid cusps have small circular apical wear facets (Fig. 5C). The lingual cingulum of the P_4 is rough and bears many small cuspules (Fig. 5B, C, 6A); the cuspules are more strongly developed posterolingually on the cingulum. The cingulum defines a slight talonid and trigonid basin, and is transversely wider anterolingually.

Both M_1 and M_2 are missing (Fig. 2, 3). Three alveoli occur posterior to the P_4; because most other pinnipedimorphs bearing a second lower molar show a tendency for the M_2 (rather than the M_1) to possess a single-rooted morphology (i.e. *Allodesmus*, *Enaliarctos*, *Imagotaria*, *Pontolis* and *Pteronarctos*) the first two molar alveoli are identified as the M_1 alveoli, and the third alveolus is identified as a single alveolus for the M_2. The M_1 alveoli are deep and separated by a well-developed bony septum, indicating widely divergent root lobes. The M_2alveolus is shallow, and along with the relatively long root exposure above the alveolar margin in SDNHM 131041, suggests that the root of this tooth may have been primarily embedded in the gums.

Comparisons

Among modern and fossil pinnipeds, SDNHM 131041 shares the most similarities with early diverging odobenids (*sensu* [2]) in retaining a large and caniniform canine, an M_2, metaconid cusps on P_{2-4}, double-rooted postcanine teeth, and two lower incisors (Fig. 7). In particular, the robustness of the mandible, canines with strongly developed longitudinal sulci, bulbous postcanine tooth crowns, and presence of crenulated lingual cingulum in SDNHM 131041 are most similar to *Pelagiarctos thomasi* and *Imagotaria downsi* (Fig. 2, 7). The combination of morphological characteristics preserved in SDNHM 131041 indicates it is most similar to the fused holotype mandibles and referred postcanine teeth described for *Pelagiarctos thomasi*[19].

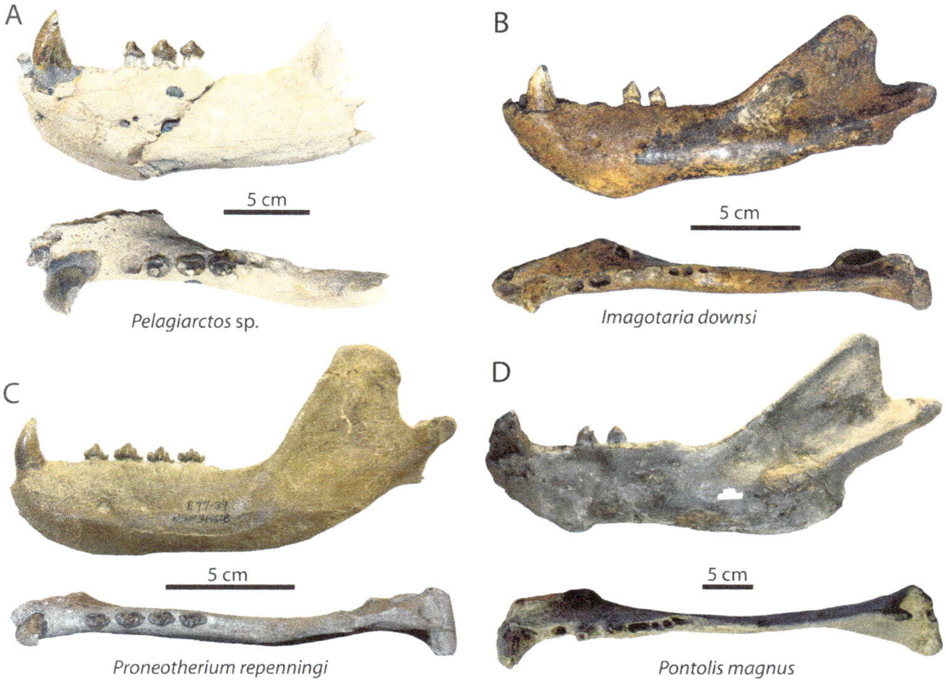

Figure 7. Comparison of *Pelagiarctos* **sp. mandible with other Miocene "imagotariine" odobenids.**

A) Mandible of *Pelagiarctos* sp. (SDNHM 131041) in lateral (top) and dorsal (bottom) aspect; "Topanga" Formation, Orange County, California. B) Mandible of *Imagotaria downsi* (USNM 23858) in lateral (top) and dorsal (bottom) aspect; reflected image of right mandible; Santa Margarita Sandstone, Santa Cruz County, California. C) Mandible of *Proneotherium repenningi* (USNM 314628) in lateral (top) and dorsal (bottom) aspect; Astoria Formation, Lincoln County, Oregon. D) Mandible of *Pontolis magnus* (USNM 335563), in lateral (top) and dorsal (bottom) aspect; reflected image of right mandible; Empire Formation, Coos County, Oregon. Scale bars equal 5 cm.

A few minor differences exist between *Pelagiarctos thomasi* from the Sharktooth Hill Bonebed and the new specimen (SDNHM 131041) from the "Topanga" Formation (Fig. 2). All postcanine teeth from the Sharktooth Hill Bonebed show a more strongly developed labial cingulum, and SDNHM 131041 lacks a fused mandibular symphysis, which characterizes the holotype of *Pelagiarctos thomasi*, as well as *Dusignathus seftoni*, *Valenictus*, and *Odobenus* (Fig. 2B). Because of these minor differences, SDNHM 131041 cannot be referred to *Pelagiarctos thomasi*. Although symphyseal fusion is not typically known to be variable within individual pinnipedimorph species and in this case may be a useful diagnostic criterion, we refrain from naming a new taxon due to the incompleteness of this specimen and lack of other differing characteristics, and identify SDNHM 131041 as *Pelagiarctos* sp. Ray (*in* [2]) reported that adult female *Odobenus rosmarus* occasionally retain an unfused mandibular symphysis, indicating that there is some degree of variation within the modern relative of *Pelagiarctos*; thus it is unclear whether or not this feature could be just as variable within *Pelagiarctos*. The combined possession of a fully erupted dentition, large wear facets, and large size suggest that SDNHM 131041 was an adult. The apparent maturity of this specimen thus suggests that symphyseal fusion is not an ontogenetic feature which had not yet developed (i.e. prior to the death of SDNHM 131041). Due to the small sample size, it is unclear whether the symphyseal fusion of the holotype specimen of *Pelagiarctos thomasi* is characteristic of that species, is variable, or represents a pathologic condition; further study of symphyseal fusion variability within *Odobenus* and other pinnipeds is required for further evaluation. *Pelagiarctos* sp. shares with all fossil and modern odobenids a mandible that is dorsoventrally deepest anteriorly, and differs from "enaliarctines" which are characterized by having a mandible deepest near the M_1. Rugose and vascularized bone at the mandibular terminus distinguishes *Pelagiarctos* from *Imagotaria* and all other earlier diverging odobenids, *Dusignathus santacruzensis*, *Aivukus*, and *Protodobenus*.

A well-developed and knob-like genial tuberosity further distinguishes *Pelagiarctos* from "enaliarctines", early diverging odobenids (*Proneotherium*, *Neotherium*), *Gomphotaria*, *Protodobenus*, and *Odobenus*. *Pelagiarctos* differs from dusignathines and odobenines in lacking an elongate mandibular symphysis and divergent mandibular rami. *Pelagiarctos* lacks an edentulous mandibular terminus, which is present in the Odobenini; furthermore, a dorsal longitudinal symphyseal furrow is also absent, unlike *Valenictus* and *Odobenus*. The horizontal symphyseal portion of the mandibular ramus in *Pelagiarctos* differs from the condition in *Ontocetus* and *Valenictus*, where it is upturned. *Pelagiarctos* exhibits a straight ventral margin of the mandible, unlike the sinuous condition in *Dusignathus*, *Pontolis*, and some specimens of *Imagotaria*.

The dentition of *Pelagiarctos* is generally plesiomorphic and close to the morphology of other early diverging odobenids (Fig. 7). For example, it differs from dusignathines, odobenines, *Pontolis*, and *Imagotaria* in the retention of a metaconid cusp and thick postcanine enamel. The presence of paraconid and hypoconid cusps, an M_2, and double-rooted P_{2-4} further distinguishes *Pelagiarctos* from dusignathines and odobenines. *Pelagiarctos* differs from the dusignathines in retaining a vertically oriented lower canine with posterior crista and lacks longitudinal fluting. The odobenines differ from *Pelagiarctos* in exhibiting a premolariform lower canine, retaining an M_1, lacking tooth enamel, and lacking a posterolingual shelf; *Pelagiarctos*further differs from *Odobenus* and *Valenictus* in retaining lower incisors and a P_4. *Pelagiarctos*is distinguished from earlier diverging odobenids and "enaliarctines" in possessing several derived features. *Pelagiarctos* exhibits a short postcanine toothrow, unlike *Enaliarctos*; *Pelagiarctos* is further distinguished from *Enaliarctos*, *Pteronarctos*, and *Proneotherium* in having a reduced metaconid and an expanded talonid or posterolingual shelf. *Pelagiarctos*differs from "enaliarctines", *Proneotherium*, *Neotherium*, and *Kamtschatarctos* in exhibiting inflated, bulbous postcanine crowns. *Pelagiarctos* sp. differs from *Proneotherium* in lacking a bicuspidate paraconid on the P_{3-4}, and lacking a deep notch between the paraconid and protoconid cusps (Figure 7); the latter feature also characterizes *Enaliarctos emlongi* and *Enaliarctos barnesi* [23], but not *Pteronarctos* [39].

Phylogenetic Analysis

Phylogenetic analysis of the complete character set using implied weighting (K=3) recovers one most parsimonious tree (tree length 48.31, CI=0.564, RI=0.730), which is shown in Figure 8. Application of different weights (K=2–5) produced no changes in topology. When no weighting scheme was applied, the resolution of the tree is reduced, with members of Dusignathinae recovered in an unresolved polytomy with Odobeninae, and *Pelagiarctos* sp., *Imagotaria, Neotherium*, and *Kamtschatarctos* occurred in an unresolved polytomy with a clade consisting of *Pontolis*, Dusignathinae, and Odobeninae. Bayesian analysis of the character set produced a similar topology, with the main differences including the recovery of a monophyletic Phocoidea (Desmatophocidae+Phocidae) with moderate support, which in turn was sister to Otariidae; *Pelagiarctos* as the sister taxon to a clade comprised of *Imagotaria*, *Pontolis*, and all later diverging odobenids; and members of *Dusignathus* and *Gomphotaria* in an unresolved polytomy with a monophyletic Odobeninae.

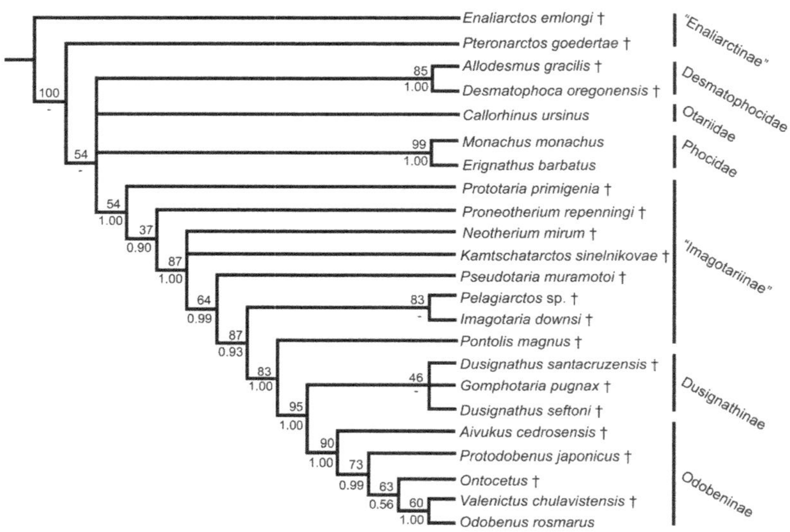

Figure 8. Strict consensus tree of odobenid relationships.

Bootstrap support and Bayesian posterior probabilities are labeled adjacent to nodes (above and below, respectively). Odobenidae and subfamilies and tribes within Odobenidae are labeled to the right. Extinct taxa are labeled with '†'.

The "enaliarctines" *Enaliarctos* and *Pteronarctos* are recovered as the earliest diverging taxa within this study, with *Pteronarctos* sister to crown Pinnipedia with strong support (BS=100). However, resolution within crown Pinnipedia remains poor, although the monophyly of Desmatophocidae (BS=85, PP=1.00) and Phocidae (BS=99, PP=1.00) is recovered with strong support. Otariidae, Desmatophocidae, and Phocidae are recovered as unresolved within crown Pinnipedia.

Within Odobenidae (BS=54, PP=1.00), *Prototaria* is recovered as the earliest diverging walrus, the sister taxon to a poorly supported (BS=37, PP=0.90) clade comprised of the remainder of Odobenidae. Odobenidae is characterized by five unequivocal synapomorphies: anterior narial opening large, thick-margined and dorsoventrally elliptical (character 2); a dorsoventrally thick and laterally broad pterygoid strut (character 13); possession of a large epitympanic recess (character 23); Bony tentorium appressed to petrosal (character 29); and a triple-rooted M^1(character 77); One equivocal synapomorphy was also suggested, presence of well developed cuspules on the P^{1-2} cingulum (character 71).

Proneotherium was recovered as the next diverging lineage of walrus, the sister taxa to a poorly resolved clade comprised of all later diverging walruses. This clade is supported by one unequivocal synapomorphy, a short and slender zygomatic process of the squamosal (character 19) and one equivocal synapomorphy, absence of the fossa muscularis on the orbital wall (character 14). *Neotherium* and *Kamtschatarctos* are found in an unresolved polytomy with a clade comprised of all later diverging walruses. This clade is supported by the following six synapomorphies: ascending process of maxilla along nasal short with no contact between ascending process and frontal (character 3); palatine long and posterolaterally expanded (character 11); a narrow and parallel-sided interorbital bar (character 17); a flattened and plate-like paroccipital process (character 34); presence of a talonid basin in the form of a slight concavity or shelf in the lower postcanine dentition (character 69); and a reduced metaconid (character 70). Five equivocal synapomorphies included absence of a supraorbital process (character 16); humerus with the medial lip of the distal trochlea greater in diameter than the distal capitulum (character 84); the distal end of the radius expanded with a large radial process (character 85); insertion of the pollicle extensor and metacarpal I forms a pit (character 86); and a scapholunar forms a well-formed pit for the magnum (character 87).

Pseudotaria is found as the earliest diverging lineage of a clade comprised of *Pelagiarctos, Imagotaria, Pontolis*, Dusignathinae, and Odobeninae, with moderate to strong support (BS=64, PP=0.99). Characters that supported this topology include the following synapomorphies: a transversely arched palate (character 9); a reduced or absent pseudosylvian sulcus (character 28); and a double-rooted M^1. The next diverging lineage identified as a strongly supported (BS=87, PP=0.93) clade comprised of *Pelagiarctos* sp. and *Imagotaria downsi* and all later diverging walruses. This clade is characterized by six unequivocal synapomorphies, including presence of a shallow glenoid fossa of the squamosal (character 26); a broad and pentagonal basioccipital (character 30); a large mastoid process of the squamosal (character 33); lateral lower incisors greater in size than medial incisors (character 54); bulbous postcanine crowns (character 63); and single and bilobed P^3 roots (character 74).

Within this clade, *Pelagiarctos* sp. and *Imagotaria downsi* are recovered as sister taxa with strong support (BS=83). Unequivocal synapomorphies for this clade included lower canine roots bilobate in cross-section (character 59) and presence of rough or crenulated postcanine lingual cingulum (character 67).

The next diverging lineage is *Pontolis*, which is recovered with strong support (BS=83, PP=1.00) as the sister taxon to a Dusignathinae+Odobeninae clade. Unequivocal synapomorphies for this clade included termination of mandibular symphysis at the level of P_2 (character 42); a sinuous ventral border of the mandible (character 45); an enlarged digastric insertion (character 46); mandibular condyle elevated above the tooth row (character 47); absence of posterior crista on C_1 (character 57); presence of thin or patchy enamel (character 62); and absence of lower postcanine paraconid cusps (character 66); and single-rooted P^4 roots (character 76).

Dusignathinae and Odobeninae are recovered as a clade, with strong support (BS=95, PP=1.00). Unequivocal synapomorphies for this clade included a very broad and short interorbital bar (character 17); presence of two upper incisors (character 51); absence of hypoconid cusps on lower premolars (character 68); cingulum on P^{1-2} weak and bulbous (character 71); single-rooted P_2 (character 72) and P_{3-4} (character 73); single and cylindrical P^3 roots (character 74); reduced or absent P^4 protocone shelf (character 75); single-rooted M^1 (character 77) and M_1(character 79); and absence of M_2 (character 81). Equivocal synapomorphies included a long mandibular symphysis (character 36) and a sharply divergent mandibular arch (character 44).

Dusignathinae is also recovered as monophyletic, but with poor support (BS=46). Unequivocal synapomorphies for Dusignathinae included tusk-like upper canines without globular osteodentine (character 55); procumbency of lower canines (character 60); and presence of a single-rooted M_1 (character 37). One equivocal synapomorphy is also found, a single-rooted M^2 (character 80).

Odobeninae is recovered as monophyletic with strong support (BS=90, PP=1.00). Unequivocal synapomorphies for Odobeninae included: a very short ascending process of the maxilla (character 3); a dorsally projected postorbital process of the jugal (character 18); a reduced squamosal fossa on the external auditory meatus (character 25); premolariform I^3(character 52); absence of enamel in adults (character 62); and absence of M^2 (character 80). Within Odobeninae, *Aivukus* is the sister taxon to a moderately supported (BS=73, PP=0.99) clade comprised of *Protodobenus* and Odobenini. Unequivocal synapomorphies that supported this topology included a palate that is arched transversely and longitudinally (character 9); tusk-like upper canines with globular osteodentine (character 55); C_1 reduced in size and premolariform (character 56); and absence of M_1 (character 78). Equivocal synapomorphies included a telescoped palatine that underlies the alisphenoid (character 11); a large and broad hamular process of the pterygoid (character 12); a sharply divergent mandibular arch (character 44); a mandibular condyle well elevated above the tooth row (character 47); root lobes of postcanines equal or narrower in width than the crowns (character 64); and absence of a talonid basin (character 69).

Odobenini is recovered as monophyletic but poorly supported (BS=63, PP=0.56). Unequivocal synapomorphies for Odobenini included: a symphyseal region composed of rugose, vascular bone (character 38) and a mandibular terminus which possesses an edentulous pad (character 41). Equivocal synapomorphies included presence of a posteriorly positioned orbital vacuity (character 20); a funnel-shaped optic foramen and orbitosphenoid (character 21); a flattened anterodorsally projected lambdoidal crest (character 32); very large mastoid processes (character 33); presence of an upturned horizontal mandibular ramus (character 39) and presence of the deltoid tubercle of the humerus on the lateral edge of the crest (character 83). Within Odobenini, *Valenictus* and *Odobenus* are recovered as sister taxa, with moderate to strong support (BS=60, PP=1.00). Unequivocal synapomorphies for this clade included: fusion of the mandibular symphysis (character 35); genial tuberosity developed as a small tubercle or process (character 37); presence of a longitudinal furrow on the anterior portion of the mandible (character 40); loss of lower incisors (character 53); absence of P^4 (character 76); and M^1(character 77); and deltoid tubercle of the humerus located off crest (character 83).

DISCUSSION

The Phylogenetic Relationships of *Pelagiarctos*

The phylogenetic analysis performed in this study is the first to include *Pelagiarctos,* and confirms that this taxon is a stem "imagotariine" walrus, in agreement with Barnes [19]. "Imagotariine" walruses are also found to be a paraphyletic assemblage of early-diverging stem taxa, a finding consistent with prior phylogenetic analyses of Odobenidae [2], [4], [5], [17]. Monophyly of Odobeninae is confirmed, with the phylogenetic relationships within this clade consistent with prior studies [2], [5].

A sister relationship between *Imagotaria* and *Pelagiarctos* is identified with excellent support on the basis of two synapomorphies. These synapomorphies included possession of lower canines bilobate in cross-section and presence of rough or crenulated postcanine lingual cingulum. Neither feature is found in other odobenids, although ontogenetic age and tooth wear may have impacted the coding of the latter character.

Our analysis identified *Pontolis* as a stem walrus sister to a clade composed of Dusignathinae and Odobeninae. Identification of *Pontolis* as a stem "imagotariine" is consistent with phylogenetic analyses of Kohno [5], [17], although the latter study found *Pontolis* to be the sister taxon to *Imagotaria*. Characters in Kohno [5] which were used support a sister taxa relationship included absence of a ventral tuberosity on the zygomatic root, and a sinuous ventral margin of the mandible. Our study interprets the absence of a ventral tuberosity as instead a character that was lost in later stem "imagotariines" and regained in the clade comprising Dusignathinae and Odobeninae. Our study also confirms the possession of a sinuous ventral margin of the mandible in *Pontolis*, but this feature is also possessed by *Dusignathus* and *Protodobenus*. This character was originally found to unite *Dusignathus* and *Pontolis* within the Dusignathinae [2]. The character is also found to be variable within *Imagotaria*, with both character states represented in the taxon.

The recovery of *Pontolis* as a stem walrus is contrary to the phylogenetic analysis presented by Deméré [2], which placed *Pontolis* as the earliest diverging lineage of a dusignathine clade. The latter study recognizes the *Pontolis* as a dusignathine walrus on the basis of one character, development of tusk-like lower canines, and three potential equivocal cranial synapomorphies. We agree that *Pontolis* possesses enlarged lower canines, but can find no other characters of the mandible or dentition that support inclusion of this genus within or as sister taxon to the Dusignathinae. Given the poor support found for Dusignathinae as well as the possibly non-monophyly of *Dusignathus*, further research on odobenid phylogeny is needed.

Kamtschatarctos has previously only been included within a phylogenetic analysis by Kohno [17]. *Kamtschatarctos* was recovered as a stem "imagotariine" walrus (Fig. 8), although the placement of this species within basal walruses still remains uncertain; it has been generally thought to have been an "imagotariine" walrus [4]–[6], [9], [17], [40]. Bayesian likelihood and maximum parsimony analysis supported placement of this taxon as close to *Neotherium*, and within a grade of stem walruses intermediate between *Proneotherium* and *Pelagiarctos*-*Imagotaria*. *Kamtschatarctos* is a fragmentary taxon [41] with little re-examination since its description, and we were only able to examine casts and photographs of the holotype jaw. Re-examination of the holotype and incorporation of additional cranial and postcranial characters may improve the resolution of this taxon within Odobenidae.

Odobenid Dental and Mandibular Evolution

Walruses have long been recognized as being more diverse in the geologic past both in terms of numerical richness and morphology [2], [7], [19]. While the earliest walruses closely resembled "enaliarctine" pinnipedimorphs in overall morphology, by the latest Miocene and Pliocene Dusignathine and Odobenine walruses had diversified into a number of derived morphologies. Among these were strange double-tusked suction feeding walruses (*Dusignathus* and *Gomphotaria*, [42]), tuskless and possibly piscivorous odobenines (*Aivukus*and *Protodobenus*; [40]), and tusked odobenine walruses which were dedicated suction-feeding molluskivores (*Ontocetus*, *Valenictus*, and *Odobenus*; [1]–[3], [40], [42]). The ecological diversity of walruses in the geologic past is reflected in the diverse dental and cranial modifications of these various fossil walruses. Previous authors have summarized changes in cranial morphology and the dentition [2], [42], and we provide a supplementary discussion of the evolution of the mandible and lower dentition in odobenids.

Pelagiarctos is distinctive in being the earliest known pinniped to evolve symphyseal fusion, a condition which is limited to the odobenids (among pinnipedimorphs). Examination of this character in a phylogenetic context suggested that symphyseal fusion evolved three times (Fig. 9): once in *Pelagiarctos thomasi* [19], once in *Dusignathus seftoni* [3], and once in the common ancestor of *Valenictus* and *Odobenus* [2], [3].

Dental characters shown on left cladogram, and mandibular characters shown on right cladogram, with diagrams of mandibles adjacent to taxon names; white indicates unknown morphology. Character acquisition and loss mapped directly from results of cladistic analysis, with the exception of "short tooth row", which was mapped *a posteriori*.

In lateral view, some "enaliarctine" mandibles (*Enaliarctos emlongi*, *Pteronarctos goedertae*) are dorsoventrally deepest posteriorly near the molars, reflecting the primitive carnivoran condition where the mandible is thickest near the large carnassials, and becomes shallower nearer the anterior dentition ([23], fig. 7a therein). *Pelagiarctos*, and all other odobenids (and most otariids) instead possess a mandible that is dorsoventrally deepest anteriorly near the canines and symphysis (Fig. 7, 9). *Enaliarctos, Pteronarctos*, and the possible early pinnipedimorph *Puijila* retain the primitive 'fissiped' condition where the middle of the mandible is most robust adjacent to the carnassials [23], [43]. The carnassial morphology was lost early in pinniped evolution during the transition from chewing and shearing to pierce feeding [42]; additionally, pinnipeds possess larger lower canines and robust anterior portions of the mandibles relative to terrestrial carnivorans. Early diverging walruses such as *Proneotherium*and *Neotherium* have completely lost carnassial-like postcanines [4] and have an anteriorly deep mandible (Fig. 9).

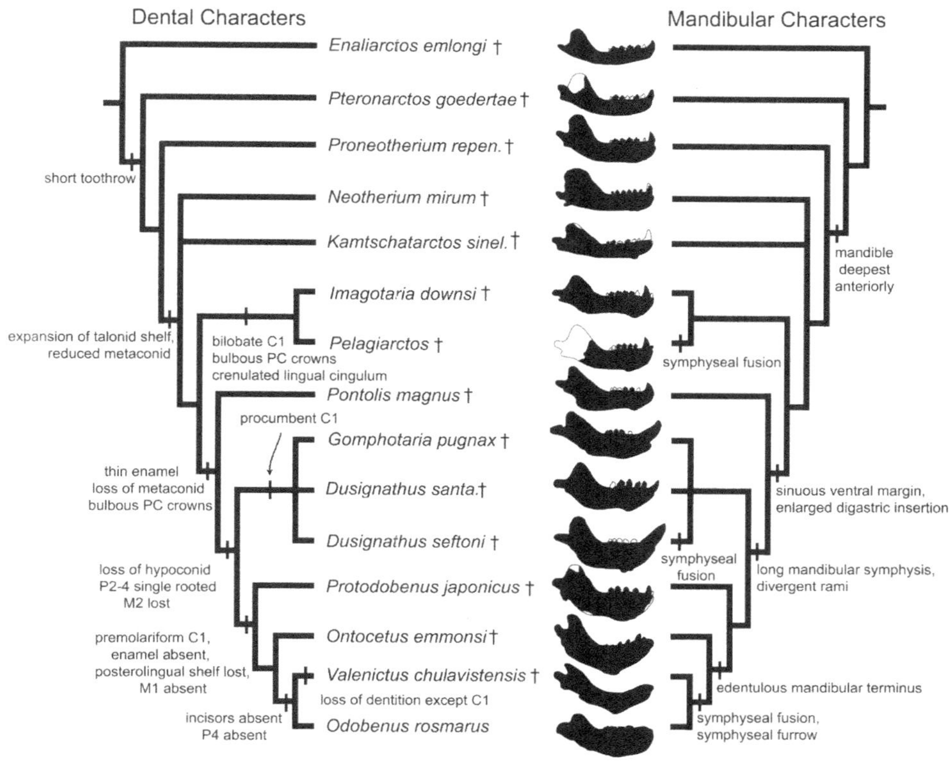

Figure 9. Hypothesized sequence of mandibular and dental character transformations during odobenid evolution.

Other mandibular changes occurred within the Dusignathinae and the Odobeninae. Dusignathines are characterized by a sinuous lower margin of the mandible, owing to the posteriorly enlarged digastric insertion; it is unclear whether this is related to stronger gular muscles associated with suction feeding. A posteriorly enlarged digastric insertion is also present in some phocoids (e.g. *Allodesmus*, *Acrophoca*, *Erignathus*, *Mirounga*, *Piscophoca*) and an otariid (*Otaria*). Odobenine mandibles show further modifications from the ancestral pinniped condition, including a mandibular symphysis that is greatly elongated and widely flaring mandibular rami, the latter relating to the extreme shortening of the rostrum [2]. The anterior part of the horizontal ramus is dorsally upturned within *Ontocetus* and *Valenictus* (Fig. 9), and in *Valenictus* and *Odobenus*, the mandibular terminus exhibits a dorsal furrow and forms a small, edentulous mandibular "pad". Extant *Odobenus rosmarus* and the extinct *Odobenus mandanoensis* share a hypertrophied and pachyostotic symphyseal region that is transversely thickest ventrally and narrower dorsally, whereas the primitive condition possessed by all other pinnipeds is a mandible that is widest dorsally [2]; rostral and cranial pachyostosis [44] possibly functions as ballast during benthic foraging.

Changes in the dentition of odobenids can generally be placed into five categories: 1) postcanine crown simplification, 2) tooth root fusion, 3) tooth reduction, 4) changes in tooth placement, and 5) tusk development. The postcanine dentition of extant walruses is remarkably simple in comparison to earlier diverging taxa like *Neotherium* and *Pelagiarctos*. Simplification of postcanine crowns was achieved by stepwise loss of the paraconid, metaconid, and hypoconid cusps (Fig. 9); these are all present (in addition to the protoconid or principal cusp) in *Proneotherium*, *Neotherium*, and *Pelagiarctos*. The metaconid was lost in *Imagotaria*, and the paraconid and hypoconid cusps disappeared within the dusignathines and odobenines, resulting in a simple, peg-like crown (Fig. 9).

Fusion of tooth roots was a major dental transition within the Odobenidae [2], [7], and the early diverging walrus *Imagotaria downsi* was noted to have a stage of intermediate root fusion where the anterior postcanines were single-rooted or fused (but bilobate), and the posterior postcanines retained double-roots [7]. *Pontolis* exhibits a similar (but slightly more derived) pattern; dusignathines and all odobenines possess single-rooted postcanine teeth [2], [7], [40]. This same pattern of root fusion, beginning with fusion of the anterior premolar roots and progressing posteriorly, also occurs during the evolution of otariids [45], [46]. Adam and Berta [42] suggested that the anterior migration of the M_1 in early pinnipedimorphs reflects an adaptation towards increasing functional gape. Because double-rooted teeth are anteroposteriorly more elongate than single-rooted teeth, Boessenecker [46] further argued that fusion of postcanine tooth roots was also an adaptation for increasing functional gape by crowding teeth anteriorly. Such a transition has occurred at least eight times, including twice within Otariidae (*Callorhinus*, as well as the clade composed of "Otariinae"+southern fur seals), twice within Odobenidae, once within Desmatophocidae, and three times within Phocidae (*Mirounga*, *Cystophora*, and *Halichoerus*; using the phylogeny of Fulton and Strobeck [15]; modified from Boessenecker [46]). The existence of unpublished latest Miocene specimens of *Imagotaria* with single-rooted teeth [2] and an unpublished early late Miocene *Gomphotaria* with some double-rooted teeth [47] suggests that root fusion may have occurred three times within odobenids (once in dusignathines, once in odobenines, and once in *Imagotaria*, totaling nine times within pinnipeds).

Reduction of teeth began with geochronologically late "imagotariines" like *Imagotaria*, with the incipient loss of the M_2; although the M_2 is present in *Pontolis* [2], it is present in some specimens of *Imagotaria downsi* and absent in others [7], [48]. The lower incisors were reduced and lost in dusignathines, with one incisor present in *Dusignathus seftoni*, and none in *Gomphotaria*; incisor loss independently occurred in the common ancestor of *Odobenus* and *Valenictus*. Incisor loss in odobenines may be related to allowing an oral pathway for suction feeding [42], whereas in *Gomphotaria*, it is probably related to lower canine enlargement. Further postcanine reduction occurred in odobenines with the loss of the M_1 in *Ontocetus* and *Protodobenus*, and further loss of the P_4 in *Odobenus*. Also within odobenines, the C_1 is reduced and premolariform. Extreme tooth reduction characterizes the most derived walrus, *Valenictus chulavistensis*, which lost all teeth with the exception of the C^1 [3]. Further tooth reduction within odobenines and dusignathines includes thinning and complete loss of tooth enamel [2], a feature which has also evolved within several cetacean lineages [49]–[51] and may reflect reduced reliance on dentition in the processing of prey.

Numerous changes occurred regarding the placement of anterior teeth, probably as a result of upper canine enlargement. The lateral upper and lower incisors (I^3 and I_3) are enlarged in many pinnipeds, and the medial incisors are often smaller and within odobenids are the first to be lost [2]. Decreased size of the medial incisor (I_2) in early diverging odobenids is also associated with it being positioned posteromedial to the lateral incisor, so that the canines and incisors do not form a transverse arc, but instead are medially indented; this condition is present in *Pelagiarctos*, *Neotherium*, *Imagotaria*, and *Ontocetus*, and also many otariids and phocids [26].

Tusk development occurs within the dusignathine and odobenine walruses, with the development of modest upper and lower tusks (Dusignathinae) and the development of very elongate upper tusks (Odobenini). Within Dusignathinae, the tusks of *Dusignathus seftoni* and *Gomphotaria* are not only greatly enlarged, but also procumbent ([1], [3]; Fig. 9). The function of upper and lower tusks within these taxa remains unclear. Elongation of tusks in the Odobenini was achieved by the addition of a globular dentine column within the tusks, a dental tissue that is unique to the Odobenini, and lacking in dusignathines [1]–[3]. Deméré [3]suggested that tusk evolution was driven by social and sexual display-related behavior; in modern *Odobenus* tusks are prominently displayed during aggressive confrontations [52] as well as used in intraspecific combat [53]. Muizon et al. [54] however, argued that the convergent evolution of tusks in the odontocete *Odobenocetops* suggests tusk evolution is related instead to feeding. Tusks in walruses may function as ‘sled runners’ on the seafloor, allowing the animal to orient its body with the sea floor by resting on its tusks [54]. This behavior has since been observed in the wild [55] and is also supported by wear patterns on the tusks. However, it is important to note that tusks evolved twice within odobenids, and within the dusignathines and basal odobenines, short procumbent tusks could scarcely serve such a function. Early evolution of tusks was thus likely driven by social interaction, and only later was exapted for use in feeding.

The Feeding Ecology of *Pelagiarctos* Revisited

With the description of the new material herein, *Pelagiarctos* (Fig. 2) has now been documented from both the “Topanga” Formation and the Round Mountain Silt (Fig. 1), two formations roughly correlative in age (Howard and Barnes, 1987). Although only 320 km south of exposures of the Round Mountain Silt at Sharktooth Hill, adjustment for strike-slip offset since deposition of the “Topanga” Formation suggests the two were originally separated by 620 km [29], indicating that fossils of *Pelagiarctos* have been found along a long section of the eastern North Pacific coastline. Within the Round Mountain Silt, three genera of pinnipeds are recognized, the desmatophocid *Allodesmus* [25], [56], [57], the early “imagotariine” *Neotherium*[17], [56], and *Pelagiarctos* [19]. Two other taxa may also be present within the Round Mountain Silt (Desmatophocine “A” and Desmatophocine “B”), although they are poorly known [58]. The “Topanga” Formation pinniped assemblage remains poorly known; however, material referable to *Allodesmus*, *Neotherium*, and an undescribed pinniped have been noted as occurring within the fauna [29].

Allodesmus is the most abundant pinniped taxon represented within the Round Mountain Silt, with possibly one to three species represented [25], [58], [59]. *Allodesmus* was a large pinniped with enlarged orbits, simplified dentition, and strong sexual dimorphism [57]. These traits are found in modern elephant seals (*Mirounga*; [57]), and may suggest that *Allodesmus*was a deep-diving pelagic pinniped [60]. The second most common pinniped taxon is *Neotherium mirum*, a small walrus lacking many of the specialized suction-feeding adaptations of later walruses, and has been previously interpreted as a piscivorous generalist feeder [17]. Barnes [19] suggested that niche partitioning or inhabitation of this region in different seasons may have reduced competition and fostered such a diverse pinniped assemblage.

Although known only from mandibles and isolated teeth, Barnes [19] suggested that the unique morphology preserved in both the holotype and referred material of *Pelagiarctos* indicated occupation of a different niche than contemporary pinnipeds. Barnes [19] proposed a novel interpretation of the feeding ecology of *Pelagiarctos thomasi* as a macrophagous predator, based on four lines of evidence: 1) The robust morphology of the mandible; 2) the perceived similarity of the postcanine teeth with scavenging carnivorans (extant hyaenids and extinct borophagine canids); 3) The inferred large body size of *Pelagiarctos*; and 4) the apparent rarity of *Pelagiarctos* within the Sharktooth Hill Bonebed.

The mandible of *Pelagiarctos* is far larger and more robust than that of *Neotherium*, and the holotype mandible possesses a large and fused bony symphysis, unlike other contemporary pinnipeds. Symphyseal fusion is not present in the “Topanga” Formation specimen, although the mandible is similarly robust. Fusion and ankylosis of the mandibular symphysis, among pinnipeds, is only typically found in *Pelagiarctos thomasi*, *Dusignathus seftoni*, and the odobenines *Odobenus* and *Valenictus*; in odobenines it is hypothesized to reduce torsion during suction feeding [42]. Among terrestrial carnivores, mandibular symphyseal fusion occurs in a variety of taxa, including big cats (*Panthera*), aardwolves (*Proteles*), bears (Ursidae), kinkajou (*Potos*) and a variety of mustelid taxa [61]. Within terrestrial taxa, this feature is thought to reduce strain at the symphyseal joint and focus force along the horizontal midline of the mandibles [62]. Within the Feliformia symphyseal fusion is strongly correlated with taxa that regularly consume large struggling prey, however diet and symphyseal fusion appear to be only poorly correlated in caniforms [61]. Barnes [19] argued that symphyseal fusion might be related to a durophagous diet and an adaptation to processing the bones of large prey. However, most durophagous taxa such as borophagine canids, hyenas, and sea otters have loose, unfused mandibular symphyses, which may relate to the need to rotate the jaws when applying force to particularly tough foods such as shellfish or bone, to reduce tooth breakage [62]. Fusion of the mandibular symphysis would have probably increased tooth breakage if *Pelagiarctos* had a durophagous diet, something not observed in the new specimen or the previously described material. Given the poor correlation between diet and symphyseal fusion between caniforms, use of this character to determine the ecology of *Pelagiarctos* is probably premature, and cannot be used as evidence for hypercarnivory.

The dentition of both specimens of *Pelagiarctos* is similar to that of *Neotherium* and other earlier diverging "imagotariines", with a prominent protoconid and less prominent metaconid and paraconid cusps (Fig. 5–6). Simplification of dentition as seen in later diverging suction-feeding odobenids (Dusignathinae and Odobeninae [42]) is not present in *Pelagiarctos* sp. The number of incisors has not been reduced as seen in later diverging odobenids, a feature that may be related to accommodating the movement of food while the mouth is closed during suction feeding. Most of the characters of the dentition associated with a macropredatory lifestyle cannot be identified within the holotype or referred material of *Pelagiarctos*, including enlarged procumbent incisors and canines [42], [63], sharp postcanine cusps (*Hydrurga*; [42], [63], and presence of carinae on the anterior dentition [63]. Prominent anterior wear facets are present, as seen in taxa such as *Dusignathus santacruzensis* and *Pontolis magnus*, and imply that the teeth are still used in the capture and processing of prey. This is in contrast to the wear observed in later odobenines such as *Odobenus*, in which the teeth are polished as a result of inhalation of grit and movements of the tongue [64].

Overall, the morphology of *Pelagiarctos* suggests that *Pelagiarctos* was probably not a specialized suction feeder, given the distinct wear facets, complex dentition, and retention of incisors. A durophagous diet is also unlikely. Barnes [19] suggested that *Pelagiarctos* may have been "an active marine predator with strong crushing and biting abilities" ([19]:9), based on superficial similarities of the cheek teeth (i.e. shape, size, and position on the mandible) with hyenas and borophagine canids, as well as fusion of the mandibular symphysis. We do not find any convincing evidence of this in the dentition of *Pelagiarctos*, as there is little overall similarity between its teeth and borophagine canids, or hyaenids. Although the cheek teeth of *Pelagiarctos* are large (in terms of absolute size), they are small relative to the size of the mandible. *Pelagiarctos* postcanine teeth are superficially similar to anterior premolars of borophagines and hyaenids, and although Barnes [19] noted sharp anterior and posterior cristae on these teeth, we were unable to identify any such cristae on teeth referred to *Pelagiarctos thomasi* by Barnes [19]; instead, they appear to be rather blunt. *Pelagiarctos*generally lacks sharp cusps or cristae of any sort, unlike the aforementioned borophagous fissipeds. In general, the dentition of *Pelagiarctos* is similar to other early odobenids which have been interpreted as generalist piscivores.

The body size of *Pelagiarctos* may also provide useful information on its ecology. Based on length of the lower tooth row of SDNHM 131041 and linear regression equations based on measurements of modern taxa (Churchill unpublished data), we estimate that *Pelagiarctos*weighed ~350 kg, about the size of an adult male South American Sea Lion (*Otaria*). *Pelagiarctos* was the first truly large odobenid, and started a trend towards ever larger body sizes within later diverging odobenids. The contemporary pinnipeds *Neotherium* and *Allodesmus* would have had weighed ~280 and up to 1400 kg respectively, making *Pelagiarctos* intermediate in body size between the two taxa. Large body size has been indicated to be correlated with trophic level [65], [66], with larger carnivores being able to feed on larger size classes of prey. To test whether body size relates to trophic level for pinnipeds, we plotted log body weight data from Lindenfors [67] versus trophic level data from Pauly et al. [68]. Trophic level data for pinnipeds is based on categorization of pinniped diet based on stomach contents and behavioral and morphological data. All taxa with strong sexual dimorphism are treated separately, but with the same trophic level. No correlation is found between log body weight and trophic level ($\alpha=0.05$, $p=0.554$, adjusted $R^2=-0.01211$; GLM analysis performed in R 2.15; Fig. 10). When comparing trophic level and body size, the largest pinnipeds are found to occupy the highest trophic levels (e.g. *Mirounga*) and the lowest trophic levels (e.g. *Odobenus* and *Erignathus*; Fig. 10). In part this is because pinnipeds are generalist feeders; even taxa with fairly derived morphologies are likely to feed on a range of prey items. A significant component of the diet of the "macropredatory" *Hydrurga* is krill [69], while the specialized molluskivore *Odobenus* has been documented to kill and eat other seals [70]. The estimated body size of *Pelagiarctos* places this taxon within range of body sizes reported for generalist feeders, suggesting that this taxon did not occupy a low trophic level (as reported for the much larger *Odobenus* and other molluskivores) nor within the range of body size reported for the high trophic level *Mirounga*.

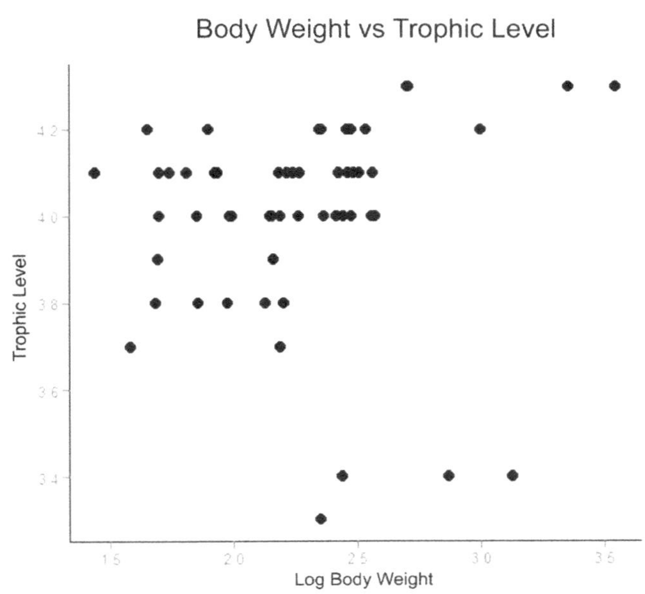

Figure 10. Relationship between body weight and calculated trophic level among modern pinnipeds.

Taxa with strong sexual dimorphism are represented by separate points for male and female taxa. Note that the greatest body weight is associated with the highest and the lowest trophic levels, and the lack of a clear trophic level trend for lower body weights.

The rarity of *Pelagiarctos* within the fossil record has also been cited as a possible indicator of its ecology [19]. Although the Sharktooth Hill Bonebed, the type locality for *Pelagiarctos*, preserves one of the largest and most diverse marine vertebrate assemblages of the Neogene and has been prospected for decades, remains of *Pelagiarctos* from this locality remain limited to six isolated teeth and a partial mandible. Given the abundance of fossil material referable to *Neotherium* and *Allodesmus*, this rarity appears to be a real phenomenon. Barnes [19] argued that the rarity as well as size of *Pelagiarctos* may correlate with a fairly high trophic level, and that a significant portion of its diet may have consisted of other pinnipeds, limiting the population to a level far below that of contemporary pinnipeds. He identified the holotype as a male individual, based on the presence of large broken canines and a fairly rugose mandible; because some isolated cheek teeth fit into alveoli of the holotype mandible, Barnes [19] further identified these isolated teeth as representing males. He used this identification to argue that *Pelagiarctos* possessed a partially allopatric distribution of genders, with females not ranging into the Temblor Sea. Given the limited amount of material of *Pelagiarctos* available for study and scarcity of studies quantifying sexual dimorphism in modern and fossil pinnipeds, evaluation of sexual dimorphism is probably premature and we question gender identification of isolated teeth. Other explanations for the rarity of *Pelagiarctos* within the Round Mountain Silt also exist. Many seals are prone to vagrancy [71]–[74], and the rarity of an individual taxon within a specific formation may reflect only occasional occurrence within the given region. In addition, the Sharktooth Hill bonebed was formed over a protracted period of time, possibly as long as 700 ka [75]. The rarity of *Pelagiarctos* within this fossil locality may indicate the existence of brief periods of unusual climate, which in turn could have led to shifts in the species composition of the local fauna and preservation of taxa not normally present along this section of coastline.

Conclusions

New fossil material of the enigmatic early walrus *Pelagiarctos* includes a pair of well preserved mandibles from the lower-middle Miocene "Topanga" Formation of Orange County, California. This fossil is more complete than the fragmentary holotype of *Pelagiarctos thomasi*, and confirms the referral of isolated teeth from the Sharktooth Hill Bonebed to this taxon. Because of minor differences including the lack of a fused mandibular symphysis, this new specimen is not referable to *P. thomasi* and is instead identified as *Pelagiarctos* sp. Owing to the more complete preservation of this new specimen, inclusion of *Pelagiarctos* within a phylogenetic analysis was possible for the first time, clearly establishing it as an early diverging odobenid, and as a sister taxon of *Imagotaria*. Although previously hypothesized to be a macrophagous predator, reevaluation of the evidence failed to support such an interpretation. Previously identified lines of evidence may have other explanations (rarity of *Pelagiarctos* within the Sharktooth Hill Bonebed), do not apply to pinnipeds (correlation of body size to trophic level), or were not verified in this study (borophagine/hyaenid like postcanine dentition with sharp cristae). The robust and fused mandibular symphysis may genuinely support the macrophagy hypothesis, but other evidence is instead suggestive of a generalist piscivore diet for *Pelagiarctos*.

ACKNOWLEDGMENTS

We are indebted to T. A. Deméré for inviting us to study this specimen, and providing detailed locality information. We thank two anonymous reviewers and the editor L. Viriot for their careful comments and criticism. This study benefited from multiple discussions with L. G. Barnes, A. Berta, T. A. Deméré, H.W. Howell, N. Kohno, and Y. Tanaka. Y. Tanaka reviewed an earlier draft of this manuscript, and we appreciate his comments. We would like to thank the following, who facilitated access to museum collections: M. Flannery (CAS); S. McLaren and J. Wibble (CM); M. Motokawa (KUZ); L. G. Barnes, S. McLeod, and V. Rhue (LACM); C. Conroy (MVZ); N. Kohno and T. Yamada (NSMT), T. A. Deméré, J. El Adli, and K. Randall (SDNHM); J.P. Cavigelli (TATE); M. Goodwin and P. Holroyd (UCMP); D. Lunde, C. Potter, D. Bohaska, and N. Pyenson (USNM); and J. Bradley (UWBM). We appreciate the assistance of T. A. Deméré (SDNHM) and L. G. Barnes (LACM) who allowed us to study fossils loaned from other institutions. We benefited from and appreciate the warm hospitality of J. El Adli (SDNHM), who hosted us during this research.

AUTHOR CONTRIBUTIONS

Conceived and designed the experiments: RWB MC. Performed the experiments: MC RWB. Analyzed the data: MC RWB. Contributed reagents/materials/analysis tools: MC RWB. Wrote the paper: RWB MC.

REFERENCES

1. Barnes LG, Raschke RE (1991) *Gomphotaria pugnax*, a new genus and species of Late Miocene dusignathine otariid pinniped (Mammalia: Carnivora) from California. Contributions in Science, Natural History Museum of Los Angeles County 426: 1–16.
2. Deméré TA (1994) The family Odobenidae: a phylogenetic analysis of living and fossil forms. In: Berta A, Deméré TA, editors. Contributions in Marine Mammal Paleontology honoring Frank C Whitmore, Jr: Proceedings of the San Diego Society of Natural History. 99–123.
3. Deméré TA (1994) Two new species of fossil walruses (Pinnipedia: Odobenidae) from the Upper Pliocene San Diego Formation, California. In: Berta A, Deméré TA, editors. Contributions in Marine Mammal Paleontology Honoring Frank C Whitmore, Jr: Proceedings of the San Diego Natural History Society. 77–98.
4. Deméré TA, Berta A (2001) A reevaluation of *Proneotherium repenningi* from the Miocene Astoria Formation of Oregon and its position as a basal odobenid (Pinnipedia: Mammalia). Journal of Vertebrate Paleontology 21: 279–310. doi: 10. 1671/0272-4634(2001)021[0279:aroprf]2. 0. co;2

5. Kohno N (2006) A new Miocene odobenid (Mammalia: Carnivora) from Hokkaido, Japan, and its implications for odobenid phylogeny. Journal of Vertebrate Paleontology 26: 411–421. doi: 10. 1671/0272-4634(2006)26[411:anmomc]2. 0. co;2

6. Kohno N, Barnes LG, Hirota K (1995) Miocene fossil pinnipeds of the genera *Prototaria*and *Neotherium* (Carnivora; Otariidae; Imagotariinae) in the North Pacific Ocean: evolution, relationships, and distribution. The Island Arc 3: 285–308. doi: 10. 1111/j. 1440-1738. 1994. tb00117. x

7. Repenning CA, Tedford RH (1977) Otarioid seals of the Neogene. Geological Survey Professional Paper 992: 1–87.

8. Deméré TA, Berta A, Adams P (2003) Pinnipedimorph evolutionary biogeography. Bulletin of the American Museum of Natural History 13: 32–76. doi: 10. 1206/0003-0090(2003)279<0032:c>2. 0. co;2

9. Barnes LG (1989) A new enaliarctine pinniped from the Astoria Formation, Oregon, and a classification of the Otariidae (Mammalia: Carnivora). Contributions in Science, Natural History Museum of Los Angeles County 403: 1–26.

10. Berta A, Wyss AR (1994) Pinniped Phylogeny. In: Berta A, Deméré TA, editors. Contributions in Marine Mammal Paleontology Honoring Frank C Whitmore, Jr: Proceedings of the San Diego Society of Natural History. 33–56.

11. Flynn JJ, Nedbal MA (1998) Phylogeny of the Carnivora (Mammalia): congruence vs incompatibility among multiple data sets. Molecular Phylogenetics and Evolution 9: 414–426. doi: 10. 1006/mpev. 1998. 0504

12. Bininda-Emonds ORP, Gittleman JL, Purvis A (1999) Building large trees by combining phylogenetic information: a complete phylogeny of extant Carnivora. Biological Review 74: 143–175. doi: 10. 1111/j. 1469-185x. 1999. tb00184. x

13. Árnason U, Gullberg A, Janke A, Kullberg M, Lehman N, et al. (2006) Pinniped phylogeny and a new hypothesis for their origin and dispersal. Molecular Phylogenetics and Evolution 41: 345–354. doi: 10. 1016/j. ympev. 2006. 05. 022

14. Delisle I, Strobeck C (2005) A phylogeny of the Caniformia (order Carnivora) based on 12 complete protein-coding mitochondrial genes. Molecular Phylogenetics and Evolution 37: 192–201. doi: 10. 1016/j. ympev. 2005. 04. 025

15. Fulton TL, Strobeck C (2009) Multiple markers and multiple individuals refine true seal phylogeny and bring molecules and morphology back in line. Proceedings of the Royal Society of London B 277: 1065–1070. doi: 10. 1098/rspb. 2009. 1783

16. Sato JJ, Hosoda T, Wolsan M, Suzuki H (2004) Molecular phylogeny of arctoids (Mammalia: Carnivora) with emphasis on phylogenetic and taxonomic positions of the ferret-badgers and skunks. Zoological Science 21: 111–118. doi: 10. 2108/zsj. 21. 111

17. Kohno N (1994) A new Miocene pinniped in the genus *Prototaria* (Carnivora: Odobenidae) from the Moniwa Formation, Miyagi, Japan. Journal of Vertebrate Paleontology 13: 414–426. doi: 10. 1080/02724634. 1994. 10011568

18. Mitchell ED, Tedford RH (1973) The Enaliarctinae: a new group of extinct aquatic Carnivora and a consideration of the origin of the Otariidae. Bulletin of the American Museum of Natural History 151: 203–284.

19. Barnes LG (1988) A new fossil pinniped (Mammalia: Otariidae) from the middle Miocene Sharktooth Hill Bonebed, California. Contributions in Science, Natural History Museum of Los Angeles County 396: 1–11.

20. Mead JG, Fordyce RE (2009) The therian skull: a lexicon with emphasis on the odontocetes. Smithsonian Contributions to Biology 627: 1–248. doi: 10. 5479/si. 00810282. 627

21. Goloboff PA, Farris JS, Nixon KC (2008) TNT, a free program for phylogenetic analysis. Cladistics 24: 774–786. doi: 10. 1111/j. 1096-0031. 2008. 00217. x

22. Huelsenbeck JP, Ronquist F (2001) MrBayes: Bayesian inference of phylogenetic trees. Bioinformatics 17: 754–755. doi: 10. 1093/bioinformatics/17. 8. 754

23. Berta A (1991) New *Enaliarctos** (Pinnipedimorpha) from the Oligocene and Miocene of Oregon and the role of “Enaliarctids” in pinniped phylogeny. Smithsonian Contributions to Paleobiology 69: 1–33. doi: 10. 5479/si. 00810266. 69. 1

24. Condon T (1906) A new fossil pinniped (*Desmatophoca oregonensis*) from the Miocene of the Oregon coast. University of Oregon Bulletin 3: 1–14.

25. Barnes LG, Hirota K (1995) Miocene pinnipeds of the otariid subfamily Allodesminae in the North Pacific Ocean: Systematics and Relationships. The Island Arc 3: 329–360. doi: 10. 1111/j. 1440-1738. 1994. tb00119. x

26. Bininda-EmondsORP, RussellAP(1996)Amorphological perspective on the phylogenetic relationships of the extant phocid seals (Mammalia: Carnivora: Phocidae). Bonner, Zoologische Monographien 41: 1–256.

27. Goloboff PA (1993) Estimating character weights during tree search. Cladistics 9: 83–91. doi: 10. 1006/clad. 1993. 1003

28. Tavaré S (1986) Some probabilitistic and statistical problems in the analysis of DNA sequences. Lectures on Mathematics in the Life Sciences 17: 57–86.

29. Howard H, Barnes LG (1987) Middle Miocene marine birds from the foothills of the Santa Ana Mountains, Orange County, California. Contributions in Science, Natural History Museum of Los Angeles County 383: 1–9.

30. Yerkes RF, Campbell RH (1979) Stratigraphic nomenclature of the central Santa Monica Mountains, Los Angeles County, California. With a section on age of the Conejo Volcanics by D. L. Turner and R. H. Campbell. United States Geological Survey Bulletin 1457: E1–E31.

31. Whistler DP, Lander EB (2003) New late Uintan to early Hemingfordian land mammal assemblages from the Undifferentiated Sespe and Vaqueros Formations,

Orange County, and from the Sespe and equivalent marine formations in Los Angeles, Santa Barbara, and Ventura counties, southern California. Bulletin of the American Museum of Natural History 279: 231–268. doi: 10. 1206/0003-0090(2003)279<0231:c>2. 0. co;2

32. Raschke RE (1984) Early and middle Miocene vertebrates from the Santa Ana Mountains, California. Memoirs of the Natural History Foundation of Orange County 1: 61–67.

33. Aranda-Manteca FJ, Domning DP, Barnes LG (1994) A new middle Miocene sirenian of the genus *Metaxytherium* from Baja California and California: relationships and paleobiogeographic implications. Proceedings of the San Diego Society of Natural History 29: 191–204.

34. Barnes LG (1998) The sequence of fossil marine mammal assemblages in Mexico. In: Carranza-Castañeda O, Córdoba-Méndez DA, editors. Avances en Investigación Paleontología de Vertebrados Publicación Especial 1. Hidalgo, Mexico: Instituto de Investigaciones en Ciencias de la Tierra, Universidad Autónoma del Estado de Hidalgo. 26–79.

35. Deméré TA, Roeder MA, Chandler RM, Minch JA (1984) Paleontology of the middle Miocene Los Indios Member of the Rosarito Beach Formation, northwestern Baja California, Mexico. In: Minch JA, Ashby JR, editors. Miocene and Cretaceous Depositional Environments, Northwestern Baja California, Mexico. Los Angeles: Pacific Section of the American Association of Petroleum Geology. 47–56.

36. Tedford RH, Albright LB, Barnosky AD, Ferrusquia-Villafranca I, Swisher CC, et al. . (2004) Mammalian biochronology of the Arikareean through Hemphillian interval (late Oligocene through early Pliocene epochs). In: Woodburne MO, editor. Late Cretaceous and Cenozoic Mammals of North America: Biostratigraphy and Geochronology. New York: Columbia University Press. 169–231.

37. Turner DL (1970) Potassium-argon dating of the Pacific Coast Miocene foraminiferal stages. Geological Society of America Special Paper 124: 91–129. doi: 10. 1130/spe124-p91

38. Deméré TA, Berta A (2005) New skeletal material of *Thalassoleon* (Otariidae: Pinnipedia) from the Late Miocene-Early Pliocene (Hemphillian) of California. Bulletin of the Florida Museum of Natural History 45: 379–411.

39. Berta A (1994) New specimens of the pinnipediform *Pteronarctos* from the Miocene of Oregon. Smithsonian Contributions to Paleobiology 78: 1–30. doi: 10. 5479/si. 00810266. 78. 1

40. Horikawa H (1995) A primitive walrus of Early Pliocene age from Japan. The Island Arc 3: 309–328. doi: 10. 1111/j. 1440-1738. 1994. tb00118. x

41. Dubrovo IA (1981) A new subfamily of fossil seals (Pinnipedia, Kamtschatarctinae subfam. nov. (Doklady Earth Science Sections 256: 202–206.

42. Adam PJ, Berta A (2002) Evolution of prey capture strategies and diet in the Pinnipedimorpha (Mammalia, Carnivora). Oryctos 4: 83–107.

43. Rybczynski N, Dawson MR, Tedford RH (2009) A semi-aquatic Arctic mammalian carnivore from Miocene epoch and origin of Pinnipedia. Nature 458: 1021–1024. doi: 10. 1038/nature07985

44. De Muizon C (1993) *Odobenocetops peruvianus*: una remarcable convergencia de adaptacion alimentaria entre morsa y delfin. Institut Francais d'Etudes Andines 22: 671–683.

45. Berta A, Deméré TA (1986) *Callorhinus gilmorei* n. sp. , (Carnivora: Otariidae) from the San Diego Formation (Blancan) and its implications for otariid phylogeny. Transactions of the San Diego Society of Natural History 21: 111–126.

46. Boessenecker RW (2011) New records of the fur seal *Callorhinus* (Carnivora: Otariidae) from the Plio-Pleistocene Rio Dell Formation of Northern California and comments on otariid dental evolution. Journal of Vertebrate Paleontology 31: 454–467. doi: 10. 1080/02724634. 2011. 550362

47. Berkoff M, Barnes LG (1998) The evolution of the dusignathines: pseudo-walruses of the late Miocene. PaleoBios 18: 1–2.

48. Barnes LG (1971) *Imagotaria* (Mammalia: Otariidae) from the late Miocene Santa Margarita Formation near Santa Cruz, California. PaleoBios 11: 1–10.

49. Ishiyama M (1987) Enamel microstructure in odontocete whales. Scanning Microscopy 1: 1071–1079.

50. Davit-Béal T, Tucker AS, Sire J (2009) Loss of teeth and enamel in tetrapods: fossil record, genetic data, and morphological adaptations. Journal of Anatomy 214: 447–501. doi: 10. 1111/j. 1469-7580. 2009. 01060. x

51. Meredith RW, Gatesy J, Murphy WJ, Ryder OA, Springer MS (2009) Molecular decay of the tooth gene enamelin (ENAM) mirrors the loss of enamel in the fossil record of placental mammals. PLOS Genetics. pp. e1000634.

52. Miller EH (1975) Walrus ethology. I. The social role of tusks and applications of multidimensional scaling. Canadian Journal of Zoology 53: 590–613. doi: 10. 1139/z75-073

53. Fay FH (1982) Ecology and biology of the Pacific walrus, *Odobenus rosmarus divergens* Illiger. North American Fauna 74 1–279. doi: 10. 3996/nafa. 74. 0001

54. De Muizon C, Domning DP, Ketten DR (2002) *Odobenocetops peruvianus,* the Walrus-convergent delphinoid (Mammalia:Cetacea) from the Early Pliocene of Peru. Smithsonian Contributions to Paleobiology 93: 223–262.

55. Levermann N, Galatius A, Ehlme G, Rysgaard S, Born EW (2003) Feeding behavior of free-ranging walruses with notes on apparent dextrality of flipper use. BMC Ecology 3: 1–13.

56. Kellogg R (1922) Pinnipeds from the Miocene and Pleistocene deposits of California. University of California Publications in Geological Sciences 13: 23–132.

57. Mitchell ED (1966) The Miocene pinniped *Allodesmus*. University of California Publications in Geological Sciences 61: 1–46.

58. Barnes LG (1972) Miocene Desmatophocinae (Mammalia: Carnivora) from California. University of California Publications in Geological Sciences 89: 1–69.

59. Kohno N (1996) Miocene pinniped *Allodesmus* (Mammalia: Carnivora); with special reference to the "Mito seal" from Ibaraki Prefecture, Central Japan. Transactions and Proceedings of the Palaeontological Society of Japan, New Series 181: 388–404.

60. Berg LM, Pyenson ND (2012) Osteological correlates and phylogenetic analysis of deep diving in living and extinct pinnipeds: What good are big eyes? Marine Mammal Science DOI: 10. 1111/j. 1748-7692. 2011. 00545. x.

61. Scott JE, Hogue AS, Ravosa MJ (2012) The adaptive significance of mandibular symphyseal fusion in mammals. Journal of Evolutionary Biology 25: 661–673. doi: 10. 1111/j. 1420-9101. 2012. 02457. x

62. Scapino R (1981) Morphological investigations into functions of the jaw symphysis in carnivorans. Journal of Morphology 167: 339–375. doi: 10. 1002/jmor. 1051670308

63. Fitzgerald EM (2006) A bizarre new toothed mysticete (Cetacea) from Australia and the early evolution of baleen whales. Proceedings of the Royal Society B: Biological Sciences 273: 2955–2963. doi: 10. 1098/rspb. 2006. 3664

64. Gordon KR (1984) Models of tongue movement in the walrus (*Odobenus rosmarus*). Journal of Morphology 182: 179–196. doi: 10. 1002/jmor. 1051820206

65. Memmott J, Martinez ND, Cohen JE (2000) Predators, parasitoids, and pathogens: species richness, trophic generality, and body sizes in a natural food web. Journal of Animal Ecology 69: 1–15. doi: 10. 1046/j. 1365-2656. 2000. 00367. x

66. Cohen JE, Pimm SL, Yodzis P, Saldaña J (1993) Body sizes of animal predators and animal prey in food webs. Journal of Animal Ecology 62: 67–78. doi: 10. 2307/5483

67. Lindenfors P, Tullberg BS, Biuw M (2002) Phylogenetic analyses of sexual selection and sexual size dimorphism in pinnipeds. Behavioral Ecology and Sociobiology 52: 188–195. doi: 10. 1007/s00265-002-0507-x

68. Pauly D, Christensen V, Dalsgaard J, Froese R, Torres F (1998) Fishing down marine food webs. Science 279: 860–863. doi: 10. 1126/science. 279. 5352. 860

69. Lowry LF, Testa JW, Calvert W (1988) Notes on winter feeding of crabeater and leopard seals near the Antarctic Peninsula. Polar Biology 8: 475–478. doi: 10. 1007/bf00264724

70. Lowry LF, Fay FH (1984) Seal eating by walruses in the Bering and Chukchi Seas. Polar Biology 3: 11–18. doi: 10. 1007/bf00265562

71. Harris DE, Gupta S (2006) GIS-based analysis of ice-breeding seal strandings in the Gulf of Maine. Northeastern Naturalist 13: 403–420. doi: 10. 1656/1092-6194(2006)13[403:gaoiss]2. 0. co;2

72. Nores C, Pérez C (1988) The occurence of walrus (*Odobenus rosmarus*) in southern Europe. Journal of Zoology 216: 593–596. doi: 10. 1111/j. 1469-7998. 1988. tb02457. x

73. Machado FJ, Rosa de Oliveira L, Wynen LP, Bester MN, Guinet C, et al. (2008) Multiple origins of vagrant subantarctic fur seals: a long journey to the Brazilian coast detected by molecular markers. Polar Biology 31: 303–308. doi: 10. 1007/s00300-007-0358-z

74. Wilson JW, Burle MH, Bester MN (2006) Vagrant Antarctic pinnipeds at Gough Island Polar Biology. 29: 905–908. doi: 10. 1007/s00300-006-0187-5

75. Pyenson ND, Irmis RB, Lipps JH, Barnes LG, Mitchell ED, et al. (2009) The origin of a widespread marine bonebed deposited during the Middle Miocene Climatic Optimum. Geology 37: 519–522. doi: 10.1130/g25509a.1

CHAPTER 11

ABSENCE OF SUCTION FEEDING ICHTHYOSAURS AND ITS IMPLICATIONS FOR TRIASSIC MESOPELAGIC PALEOECOLOGY

Ryosuke Motani, Cheng Ji, Taketeru Tomita, Neil Kelley, Erin Maxwell, Da-yong Jiang and Paul Martin Sander

INTRODUCTION

Many large predators in the modern marine ecosystem are mammals. A similar role was probably played by marine reptiles in the Mesozoic, until their last members became extinct 65.5 million years ago during the end-Cretaceous mass extinction [1], [2], leaving the niches open for marine mammals. However, it is not known how similar their feeding ecology was to that of modern marine mammals. Massare [3] was the first to identify feeding guilds among Mesozoic marine reptiles using tooth morphology and stomach contents. However, limited progress has been made in this field since her pioneering study, although some studies are being conducted [4], [5].

Ichthyopterygia must be considered when comparing modern marine mammals with Mesozoic marine reptiles because they were the Mesozoic analog of cetaceans. Being derived from four-legged reptiles, they evolved a fish-shaped body profile about 200 million years before cetaceans. Also, their fossil record is more complete than that of other marine reptile groups [6]. They were probably the most abundant marine reptiles in the Triassic and Jurassic seas, although they became extinct in the mid-Cretaceous [7]. They are the only group of Mesozoic marine reptiles for which deep diving has been proposed [8]. If an abrupt change in their feeding ecology is inferred from the fossil record, it probably reflects the changes in their prey community and its environment. Therefore, it is important to reconstruct the evolution of ichthyosaurian feeding ecology through geologic time.

Among the major questions in the evolution of feeding ecology in marine reptiles is the evolution of suction feeding. Suction feeding is the most common strategy for prey capture among extant marine vertebrates [9], yet its evolution among marine reptiles has attracted limited attention. Recently, it was suggested that Triassic shastasaurid ichthyosaurs contained three species of suction feeders based on their superficial resemblance to beaked whales (Ziphiidae) [10]. This expanded an earlier suggestion [11] that one of the three species, *Shonisaurus sikanniensis*, was a suction feeder analogous to beaked whales. If true, the feeding mode interpretation would have many ecological implications. Most importantly, it raises a question of whether a suitable environment for beaked-whale-type predators, including their prey community in the meso-/bathypelagic zones, was available in the Triassic. This question has remained unaddressed until now. Prior to these two studies, suggestions of suction feeding among Mesozoic marine reptiles were made for various stem sauropterygians including the placodont *Henodus* [12], and for the protorosaur *Dinocephalosaurus* [13]. *Henodus* is known only from the Upper Triassic of Germany in a lagoonal setting, whereas *Dinocephalosaurus* is a coastal species that is known only from the Middle Triassic of China. The other species are also coastal dwellers (*sensu* [14]) and have little relevance to mesopelagic ecology; therefore, they will not be mentioned hereafter.

The proposal of suction feeding in these shastasaurid ichthyosaurs was based on their resemblance to beaked whales, namely 'short' snout, tooth reduction, and 'enlarged' hyobranchial rod [10], [11]. These features were loosely defined and never tested quantitatively. However, a functional

Citation: Motani R, Ji C, Tomita T, Kelley N, Maxwell E, Jiang D-y, et al. (2013) Absence of Suction Feeding Ichthyosaurs and Its Implications for Triassic Mesopelagic Paleoecology. PLoS ONE 8(12): e66075. doi:10.1371/journal.pone.0066075.

inference based on superficial resemblance, or "intuitive functional morphology" *sensu* [15], is often misleading. As with most functional questions, suction feeding in ichthyosaurs should be examined based on mechanical reasoning and quantitative tests. Re-evaluation of hyobranchial morphology is especially wanting because a recent study [16] showed that the hyobranchial size of *Guanlingsaurus* was likely overestimated by the previous study [10]. Also, new studies of hyobranchial structure in modern suction feeding vertebrates have been published since [10], e.g., [17], [18]. The hyobranchial apparatus is considered the most important hard-tissue correlate in the discussion of suction feeding in vertebrates [17]–[19] because it underlies the mechanism of subambient pressure generation for suction (see below).

The purpose of the present study is twofold: first to re-evaluate the hypothesis of suction feeding in ichthyosaurs through quantification of osteological features related to the mechanisms of suction; and second to discuss if a modern-style meso-/bathypelagic animal community existed in the Triassic and Early Jurassic, from the perspective of the large predators of the time. We will also review the feeding ecology of extant suction feeders among air-breathing marine vertebrates to make comparisons.

MATERIALS AND METHODS

Data Collection

We measured 18 ichthyosaurian specimens in which at least one hyobranchial rod is preserved. The small sample size is a consequence of the scarcity of such specimens. The samples span 15 genera and 18 species. They are *Chaohusaurus geishanensis* (AGM, Anhui Geological Museum, Hefei, China CH-628-22), *Eurhinosaurus longirostris* (SMNS, Staatliches Museum für Naturkunde, Stuttgart, Germany 57922), *Guanlingsaurus liangae* (Guanling National Geopark, Guanling, China dq-50), *Guizhouichthyosaurus tangae* (Wuhan Institute of Geology and Mineral Resources of China, Wuhan, China TR00001), *Hauffiopteryx typicus*(SMNS 81962), *Ichthyosaurus communis* (NHMUK, Natural History Museum, London, UK 36256), *Leptonectes moorei* (NHMUK R14370), *L. tenuirostris* (NHMUK 24300), *Mixosaurus cornalianus* (Paläontologisches Institut und Museum der Universität, Zurich T2414), *Qianichthyosaurus zhoui* (Institute of Vertebrate Paleontology and Paleoanthropology, Beijing, China 11838), *Shastasaurus alexandrae* (UCMP, University of California Museum of Paleontology, Berkeley, US 9017), *Shonisaurus sikanniensis* (Royal Tyrrell Museum of Paleontology, Drumheller, Canada 94.378.2), *Stenopterygius quadriscussus* (SMNS 50165), *St. triscissus* (SMNS 55074), *Suevoleviathan integer* (SMNS 4652), *Temnodontosaurus platyodon* (Oxford University Museum, Oxford, UK J29170), *Te. trigonodon* (SMNS 50000), and *Toretocnemus zitteli* (UCMP 8099). All but one specimen were in the existing collection of each institution and studied on-site in respective collections with permission. AGM CH-628-22 was excavated by a joint team of AGM, Peking University, University of California, Davis, and Università degli studi di Milano, with permits from the Ministry of Land and Resources of the People's Republic of China. The current study did not involve any purchase, loan, or donation of a specimen. Of the 18 specimens, mandibular width could reasonably be estimated in 15, and mandibular length in 12 specimens (Table 1)–mandibular rami were sometimes splayed beyond their natural angle and it was necessary to estimate the original width using the skull width when possible. Measurements taken are: hyobranchial rod length (HL), hyobranchial rod width at mid length (HW), mandibular ramus length (ML), mandibular width (MW), and mandibular width at the end of tooth row (TW). Fig. 1 summarizes the measurements taken, and Table 1 their values. TW for edentulous taxa was estimated by the mandibular width at the anterior margin of the orbit. This scheme may lead to a slight underestimation of the true values if applied to tooth-bearing taxa, and slight overestimation of suction feeding ability. However, such a bias does not affect the outcome of the study. Digital calipers were used for most measurements, which were recorded to the nearest 0.1 mm. Larger measurements were made by large calipers or tape measures and recorded to the nearest 1 mm. The mechanical importance of these measurements will be discussed later.

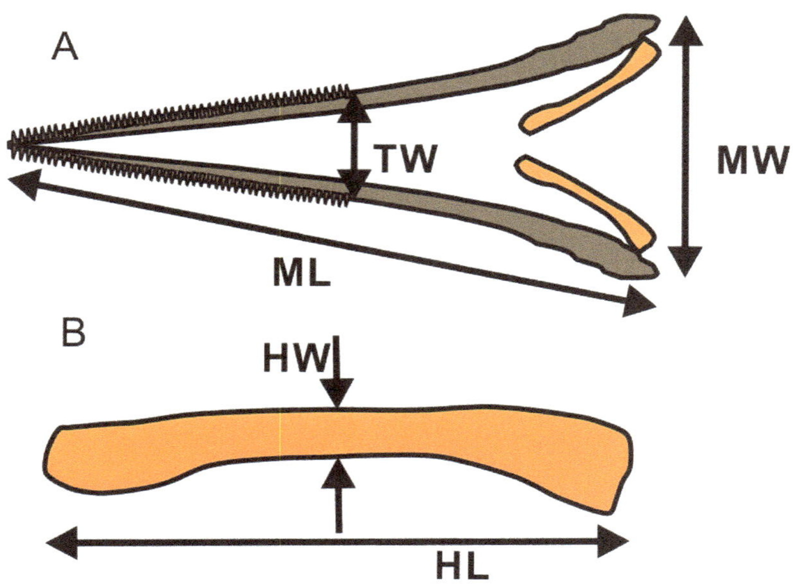

Figure 1. Measurements taken from ichthyopterygian specimens.

a, schematic drawing of mandible with a pair of ceratobranchial I. b, magnified view of a single ceratobranchial I. HL: hyobranchial rod length; HW, hyobranchial rod median width; ML, mandibular ramus length; MW, mandibular width; TW, mandibular width at the end of tooth row. Brown, mandible; orange, hyobranchial rod.

Table 1. Measurements (in mm) of ichthyopterygian mandible and hyobranchial rod.

	ML	MW	HL	HW	TW
Chaohusaurus geishanensis	103.4	—	11.7	1.3	—
Eurhinosaurus longirostris	650	200	143.0	13.1	45.0
Guanlingsaurus liangae	—	230	108.9	7.1	115.5
Guizhouichthyosaurus tangae	877	—	130.9	10.5	—
Hauffiopteryx typicus	—	134.0	98.0	9.0	—
Ichthyosaurus communis	376	83.1	88.1	6.0	44.2
Leptonectes tenuirostris	549	164.8	77.4	6.8	86.3
Leptonectes moorei	335	132.3	87.9	9.0	49.8
Mixosaurus cornalianus	184	—	30.0	2.5	—
Qianichthyosaurus zhoui	223	100.6	61.0	5.2	41.9
Shastasaurus alexandrae	—	205	227	16.7	—
Shonisaurus sikanniensis	—	1451	1220	78.8	1365
Stenopterygius quadriscissus	547	124.6	81.0	9.9	17.3
Stenopterygius triscissus	426	130.0	63.6	8.4	17.2
Suevoleviathan integer	537	192.9	103.2	9.9	104.1
Temnodontosaurus platyodon	1333	577	265	23.6	401
Temnodontosaurus trigonodon	—	420	365	21.9	190
Toretocnemus zitteli	—	82.3	57.4	5.2	35.5

See Fig. 1 for abbreviations and text for specimen number.
doi:10.1371/journal.pone.0066075.t001

Data for sharks were taken from [18], which used CT scans of sharks for measurements. Cetacean hyobranchial metrics were derived from [17], whose measurements were taken from dry specimens. Mandibular measurements of extant marine tetrapods were taken from [4].

Identity of Ossifications in Ichthyosaurian Hyobranchial Apparatus

The hyobranchial apparatus is usually a composite of hyoid and branchial elements, some of which are ossified/calcified while the rest are not. In mammals and sharks, the hyoid part is more dominant than the branchial part whereas in reptiles, the branchial part tends to be more dominant [18], [21], [30]. The apparatus comprises a median body (hyoid corpus) and at least one pair of horns (hyobranchial cornua). Each horn often has multiple segments, some of which may be ossified/calcified. There is usually a pair of ossified/calcified horn segments that is dominant within a given hyobranchial apparatus, and it tends to play a major role in the suspension of the entire apparatus. We will refer to these dominant horn segments as hyobranchial rods in this contribution.

The hyobranchial apparatus of ichthyosaurs is usually preserved as a pair of curved hyobranchial rods. These rods have variously been referred to as ceratohyals or hyoid rods [20], hypohyals [11] or simply hyoids, e.g., [6]. In the present contribution, the ossified rod is identified as the first ceratobranchial (CB1), given that when there is only one pair of rods ossified in the hyobranchial apparatus of extant reptiles, it is usually considered CB1 [21] in squamates [22], [23], crocodylians [24], and turtles [25]; the homology of the elements with those of fish, however, has been questioned [26].

Functionally, ichthyosaurian CB1 probably played two important roles. Given its position, it is likely that CB1 was involved in the suspension of hyobranchial apparatus, with its postero-lateral margin fixed, maybe via soft tissue, to the otic region of the cranium. The identity of the hyobranchial rod differs from clade to clade–it is the stylohyal in cetaceans, CB1 in reptiles, and the ceratohyal in sharks. We therefore compare the morphology of these hyobranchial rods across taxa because of functional similarity in suspension although they are not developmentally homologous with each other. CB1 of ichthyosaurs likely had other functions in addition to suspension: it most likely provided attachment surfaces for muscles that linked between the tongue anteriorly (M. hyoglossus, M. geniohyoideus, and M. mandibulohyoideus) and shoulder girdle posteriorly (M. sternohyoideus and M. omohyoideus), as in many extant reptiles [19], [27]–[29]. A similar role is played by thyrohyals in odontocete cetaceans [30], [31], although the stylohyal also provides attachment for an extrinsic tongue muscle (M. styloglossus). In sharks, it is the basihyal that provides attachment for the muscle to the shoulder (M. coracohyoideus) [32], whereas sharks generally do not have a muscular tongue [33].

Definition of Suction Feeding and 'Ram Feeding'

Despite the prevalence of suction feeding among aquatic vertebrates [9], the definition of the term "suction feeding" has been taxon-dependent and variable (Table 2), preventing broad comparisons across major vertebrate groups, such as sharks and cetaceans. When a jawed vertebrate (gnathostome) opens its mouth in water, a pressure gradient is necessarily produced, usually leading to subambient pressure in the mouth [34], [35]. Therefore, some level of 'suction' pressure is incurred during feeding in virtually all aquatic gnathostomes. However, the term suction feeder *sensu stricto* is applied only to those aquatic gnathostomes that draw prey closer by suction prior to capture. In contrast, those predators that capture their prey mainly by moving their body toward prey are called 'ram feeders' [36], even when the pressure gradient around the mouth helps the prey from being pushed away [34], [37].

Table 2. Taxon dependence of definition of suction and 'ram' feeding.

Suction is usually used in combination with various degrees of ram feeding [38]–[40]. The relative contribution of suction versus ram during feeding is often quantified by the Ram Suction Index (RSI), which compares the relative contribution of prey versus predator movements in regard to the total distance of movements by the two [41]. Despite its weaknesses, e.g., [39], [42], use of the RSI is common in the study of feeding kinematics in chondrichthyans, osteichthyans, and chelonians.

Suction feeding in the literature on whales has been defined less strictly: it variously contains suction feeding *sensu stricto* as defined above, as well as feeding where prey is captured by ram, followed by intraoral transpiration of prey for swallowing with help from subambient pressure induced by tongue movements [43]. This broader definition, however, obscures the status of suction feeding in cetaceans because it contains two mechanisms of prey capture. It also prevents comparisons with suction feeders among other major vertebrate groups.

Recent kinematic studies are shedding light on the status of cetacean suction feeding in comparison to that in other major vertebrate groups. A study of the feeding kinematics of the pygmy and dwarf sperm whales (*Kogia sima* and *K. breviceps*) and bottlenose dolphin (*Tursiops truncatus*) found that *Kogia* was a suction feeder and *Tursiops* a 'ram feeder' based on RSI [44]. The average suction distance was negative for *Tursiops* (i.e., the prey was pushed away, indicating that it barely used suction for prey capture). Also, an analysis of the feeding kinematics of three species of cetaceans, namely the beluga whale (*Delphinapterus leucas*), Pacific white-sided dolphin (*Lagenorhynchus obliquidens*), and long-finned pilot whale (*Globicephala melas*) found that all were capable of drawing prey toward them through suction by about 2.8 to 4.1 cm, whereas these cetaceans also used various degrees of 'ram feeding' in combination with suction [35]. In terms of RSI values, only *Delphinapterus* had feeding sequences that qualified as suction feeding, although many of the feeding sequences of the same species represented 'ram feeding'.

The use of the word 'ramming' for the movement of the predator toward its prey during capture is not universal across taxa (Table 2). Despite its use in a part of the cetacean literature [35], [44], it has been suggested that it is inappropriate to use this term for vertebrates without gills given its derivation [45], [46]–the term 'ram feeding' was first used in reference to ram ventilation, where water runs through the gills of certain fishes through swimming, rather than pumping. In this paper, we use the term 'ram feeding' to describe any feeding mode where movement of the body or jaws is mainly used to overtake prey, in order to permit a simple comparison across major vertebrate groups.

Statistical Analyses

We used R 2.15.2 for all calculations. Bivariate regressions are based on Model II regression [47], which is also called Standardized Major Axis (SMA) regression. The package smatr [48] of R was used for this purpose. All variables were log-transformed before statistical analyses to account for scaling effects. All ratios were log-transformed because ratios between two normally distributed variables with different means are highly skewed. Analysis of covariance (ANCOVA) is based on the aov function of R.

MECHANICALLY IMPORTANT HARD-TISSUE FEATURES

Hyoid Corpus Ossification/Calcification

The mechanism to produce subambient pressure in the oral cavity for suction feeding is almost uniform among jawed vertebrates, that is, posterior retraction and depression of the hyobranchial apparatus allowing expansion of the volume of the pharyngeal region, e.g., [18], [19], [49]. The muscles that connect the apparatus to the shoulder girdle, such as M. coracohyoideus (via M. coracoarcualis) or M. sternohyoideus, are recruited during the retraction, and the stress is applied to components of the apparatus, especially the central element (hyoid corpus) and the cornua that suspend the apparatus from the cranium (CB1 in reptiles, ceratohyal in sharks, and stylohyal in mammals). This common mechanism poses similar mechanical constraints on the morphology of the hyobranchial apparatus across taxa, allowing inference of suction capability based on hyobranchial morphology. One such mechanism-related feature is found in the hyoid corpus, or the central element of the hyobranchial apparatus. Suction-feeding jawed vertebrates have an ossified hyoid corpus that rigidly integrates with the right and left hyoid horns, in whales [50], turtles [19], [24], [42], [51], and sharks (TT pers. obs.; see also Fig. 2). It was specifically stated [24] that the absence of an ossified hyoid corpus in *Sternotherus* impaired the suction ability of this turtle. This is mechanically expected because the hyoid corpus is the central element of the hyobranchial apparatus that maintains its integrity and rigidity.

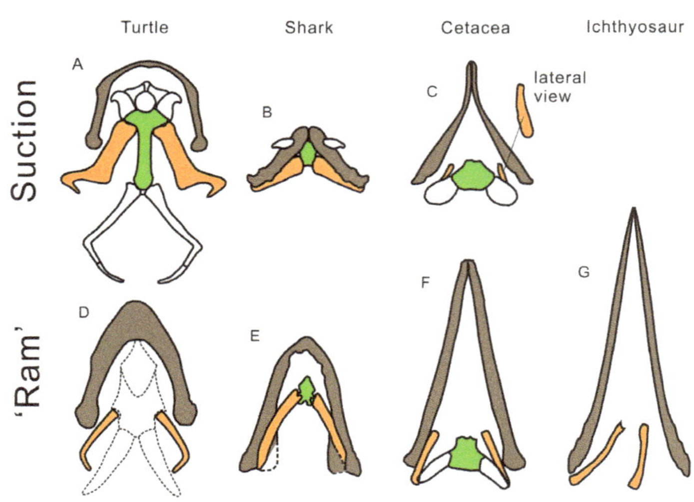

Figure 2. Mandible and hyobranchial apparatus of selected vertebrate groups.

Suction feeders are in the top row, and 'ram' feeders bottom row. Brown fill, mandible; green fill, hyoid corpus; orange fill , ossified/calcified hyobranchial rods that are discussed; white fill with real outline, other ossified hyobranchial elements; white fill with dotted outline: cartilaginous hyobranchial element. Taxa: a, Mata Mata Turtle *Chelus fimbriatus*; b, Japanese Angel Shark *Squatina japonica*; c, Pygmy Sperm Whale *Kogia breviceps*; d, Common Musk Turtle *Sternotherus odoratus*; e, Sharpnose Sevengill Shark *Heptranchias perlo*; f, Bottlenose Dolphin Tursiops truncatus; and g, Triassic Ichthyosaur *Qianichthyosaurus zhoui*. Derivations: a based [19]; b and e from CT data; c and f based on [52]; d based on [24]; and g based on IVPP 11838. Not to scale.

Hyobranchial Rod Robustness

The second feature that is rooted in the same mechanical constraints concerns the hyobranchial cornua, especially the suspensory element. The suspensory 'rods' of suction-feeding jawed vertebrates are known to be more robust compared to those of closely-related 'ram feeders' [17]–[19], [42], [52]. This observation is deeply rooted in suction mechanics because, as pointed out earlier, suspensory hyobranchial elements of suction-feeding jawed vertebrates need to withstand the stress incurred by abrupt rotation that ultimately results from posterior pulling by M. coracohyoideus or M. sternohyoideus muscles during suction pressure generation [18], [19]. It should be noted that this value is biased by change in the degree of ossification/calcification through ontogeny. Juvenile CB1 of ichthyosaurs, for example, is expected to have had cartilaginous ends that are not preserved in fossils in addition to the ossified center part that is preserved. As a result, the CB1 of juveniles may appear more robust in fossils than it was in life. We tried to avoid this bias by measuring mature specimens; however, it was not possible for taxa that are known only from juvenile specimens. However, the effect of this bias is probably small, judging from the uniformity of CB1 robustness across ichthyosaurs that is reported later in this paper.

Mandibular Pressure Concentration Index

Subambient pressure produced in the oral cavity is projected beyond the gape plane to suck prey toward the predator. The propagation of pressure beyond the gape needs to be controlled so that it is concentrated and directed toward the prey [37], [49]. It has been pointed out that shorter and smaller gapes perform better in achieving this goal. Many teleost fish, for example, have mouths that can be protruded to form a small semi-circular margin with minimal gape incision that enables concentrated projection of pressure [37]. The suction system may be compared to a syringe, with the mouth opening corresponding to its aperture and the hyobranchial apparatus to the cylinder. Given that the square of the ratio between aperture and cylinder diameters of a syringe describes the pressure concentration that occurs between the pressure generator (cylinder) and propagator (aperture), a similar ratio for the predator's feeding system may be used to approximately describe the pressure concentration that occurs during suction. One way to calculate such a ratio is to divide the mandibular width at the jaw articulation by that at the end of the toothrow, the former approximating the width of hyobranchial apparatus, and the latter the mouth width at the gape angle. This ratio is here called the Mandibular Pressure Concentration Index (MPCI). MPCI is expected to be higher in suction feeders than in 'ram feeders'. Measurements from extant marine mammals confirm this expectation (Fig. 3A).

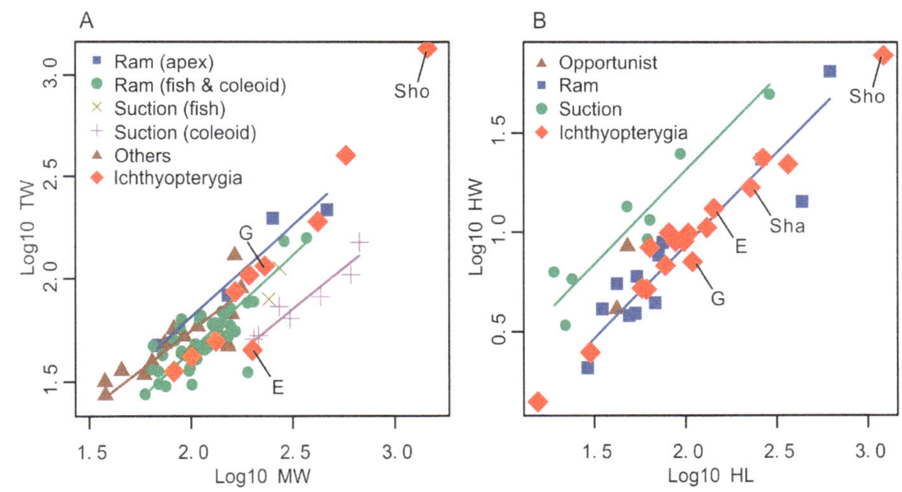

Figure 3. Bivariate SMA regression of features related to Mandibular Pressure Concentration Index and Hyobranchial Robustness.

a. Mandibular Pressure Concentration Index; b. Mandibular Robustness. Lower intercept values in a and b indicate: a, higher pressure concentration within the oral cavity; and b, less robust suspensory element of hyobranchial apparatus. Ichthyopterygians has limited pressure concentration (high intercept in a) and slender hyobranchial rod (low intercept in b). E, G, Sha, and Sho denote *Eurhinosaurus*, *Guanlingsaurus*, *Shastasaurus*, and *Shonisaurus*, respectively. Note that the latter three were considered suction feeders by [10].

Mandibular Bluntness

In odontocete cetaceans, it was found that shorter snouts were better suited for concentrating pressure projection based on measurements of suction pressure from the heads of three odontocete species with varying snout lengths [49]. For this mechanical reason, it is expected that mandibular bluntness [17], [49] is a feature of suction-feeding jawed vertebrates. It should be noted that mandibular bluntness, as measured by mandibular width/mandibular ramus length, has its limitation for being a simple ratio. Mandibular width exhibits a negative interspecific allometry against mandibular ramus length; therefore the larger the animal, the less blunt the mandible. However, given that the cetaceans and ichthyosaurs in question span similar size ranges, the range of mandibular bluntness in each group is still a useful indicator of average suction ability.

Features Not Considered

Tooth reduction is another feature that has been discussed in the literature on whales [17], [43]and sharks [53]. The word 'reduction', however, is used with different connotations in the two clades. In sharks, it is the reduction of size, as seen in the bottom-feeding nurse shark (*Ginglymostoma cirratum*), whitespotted bamboo shark (*Chiloscyllium plagiosum*), and the planktivorous whale shark (*Rhincodon typus*) [53], [54]. In whales, reduction usually refers to the reduction of tooth count [43], which may or may not be associated with decrease in tooth size. Therefore, it is difficult to define the term uniformly across taxa. To complicate the problem further, Werth [43](fig. 4) showed that tooth count reduction was clearly correlated with shortening of the mandible; this is expected because shorter mandibles have less space for teeth. Given that mandible bluntness has a hydromechanical reason to affect suction ability, this correlation has to be removed before the contribution of tooth reduction to suction ability can be tested. Also, extreme reduction in tooth count is only known among beaked whales (apart from *Tasmacetus*), mysticetes, and in *Monodon*, whose suction feeding has not been directly observed (i.e., a large contribution of 'ramming' may be possible, as in *Globicephala* and most feeding sequences of *Delphinapterus*). The two cetacean genera for which strict suction feeding has directly been established through kinematic studies (*Kogia* and some feeding sequences of *Delphinapterus*) have multiple pairs of teeth at least in the lower jaws. Furthermore, filter feeding mysticetes also exhibit completely reduced dentition; however only one mysticete, the gray whale *Eschrichtius robustus*, is a reported suction feeder [55], [56]. Thus tooth reduction alone is not an unequivocal indicator of suction feeding.

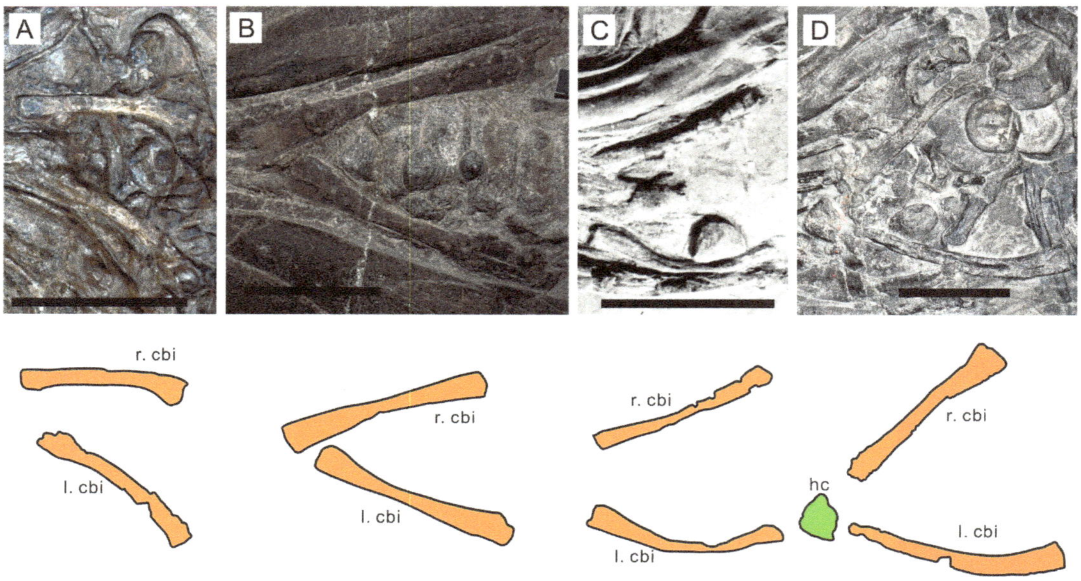

Figure 4. Hyobranchial apparatus of selected ichthyopterygians.

a., *Qianichthyosaurus zhoui* (IVPP11838); *Guanlingsaurus liangae* (GNG dq-50);c, *Ichthyosaurus communis* (OUM J10313); and d, *Hauffiopteryx typicus* (SMNS 81962). cbi, ceratobranchial I; hc, hyoid corpus; l, left; r, right. Scale bars are five centimeters.

RESULTS

Hyoid Corpus Ossification/Calcification

The fossilized hyobranchial apparatus of ichthyosaurs comprises a pair of CB1, whose posterior ends are located near that of the mandible in virtually all specimens examined, although the bones were preserved more anteriorly or posteriorly in some specimens than in others (Fig. 4). The only exception is *Shonisaurus sikanniensis*, which seems to have a small second pair of bones [11]. There was no evidence for ossified hyoid corpus in all but one specimen examined. The exception is *Hauffiopteryx* from the Toarcian (Lower Jurassic), in which a bone exists in front of the pair of CB1 (Fig. 4D). Given its symmetrical shape and position, it is tentatively identified as the hyoid corpus. As preserved, there is no evidence to suggest that the hyoid corpus and CB1 were strongly integrated in this taxon. The preservation suggests that a large part of the hyobranchial apparatus remained cartilaginous in ichthyosaurs, as in many reptiles [20].

Hyobranchial Rod Robustness

The shape of CB1 varies from strongly curved to almost straight, probably reflecting different degrees of compression during preservation rather than taxonomic differences. The effect of compaction is especially evident in one specimen of *Ichthyosaurus* (Fig. 4C), where the right CB1 is straight while the left one is curved. Virtually all CB1 were expanded toward both ends, with a slight constriction near the mid-shaft region. The anterior expansion is usually larger than the posterior one when the bone is well preserved, but there were some exceptions depending on the angle of fossil exposure. Many of them were somewhat sigmoidal in shape.

The CB1 of ichthyosaurs are more slender compared to suspensory hyobranchial elements of suction-feeding turtles (CB1), sharks (ceratohyal), and cetaceans (stylohyal) (Fig. 2; Fig. 5A). This can be quantified by comparing the mid-shaft diameter and length (Fig. 3B), although this ratio scales with size to some extent, revealing weakly negative interspecific allometry with a slope of 0.93 (Fig. 3B). It was also observed that the slenderness of the ichthyosaurian CB1 is similar to that of ceratohyals in 'ram-feeding' sharks (Fig. 3B). Notably, ichthyosaurian CB1 slenderness does not vary as widely as in sharks (Fig. 3B), suggesting that the mechanical function of the hyobranchial apparatus was largely uniform among ichthyosaurs. ANCOVA revealed a significant difference between regressions for suction-feeding sharks and ichthyosaurs ($p=9.60\times10^{-8}$, df=1, F=62.78) but not between those of 'ram-feeding' sharks and ichthyosaurs (p=0.777, df=1, F=0.082). 'Ram-feeding' cetaceans have more robust suspensory hyobranchial elements than do 'ram-feeding' sharks or ichthyosaurs (Fig. 5A). This anomaly, however, can be explained by suction feeding being plesiomorphic for cetaceans [17]. The close similarity in robustness of hyobranchial rods across clades is surprising given the expected difference in material properties of calcified cartilage of sharks and bones of tetrapods. Note, however, that the hyobranchial rod is slightly more slender in ichthyosaurs than in 'ram-feeding' sharks based on the median value, possibly reflecting the difference in material property. A careful literature search for the material property of calcified shark cartilage proved unsuccessful, and it is currently difficult to investigate this aspect.

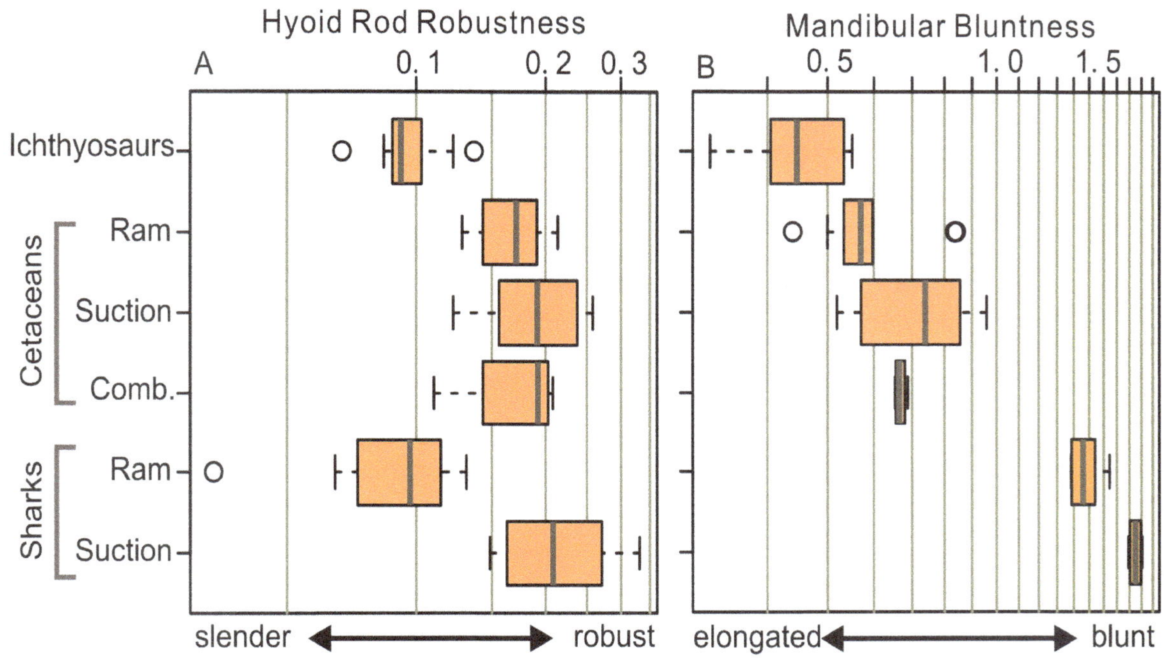

Figure 5. Boxplots of two ratios across taxa and feeding types.

a, hyobranchial robustness; b, mandibular bluntness. The thick line in the center denotes the median, the box surrounding it contains the middle 50% of the data points, and the whiskers extend to the most extreme data point which is no more than 1.5 times the interquartile range from the box. The data outside the whisker are considered outliers, plotted as small circles.

Mandibular Pressure Concentration Index

The MPCI of ichthyosaurs has a small variation, but it fits in the range of 'ram-feeding' marine mammals, with one exception of *Eurhinosaurus* (Fig. 3A). This observation is supported by ANCOVA: when plotting the mandibular width at the end of the tooth row against the maximum mandibular width as in Fig. 3A, there was a significant difference between regressions of suction-feeding marine mammals and ichthyosaurs excluding *Eurhinosaurus* ($p=4.57\times10^{-3}$, df=1, F=10.9), but not between those of 'ram-feeding' marine mammals and ichthyosaurs (p=0.122, df=1, F=2. 46).

Mandibular Bluntness

Ichthyosaurs are generally longer-snouted and therefore have low mandibular bluntness compared to odontocete cetaceans or sharks (Fig. 5B). Even the shortest-snouted ichthyosaurs in our data set, such as *Qianichthyosaurus* and *Guanlingsaurus*, have mandibular bluntness of about 0.42 or lower. These low bluntness values are found only in some beaked whales among suction-feeding odontocetes, whereas all others have much higher bluntness (Fig. 5B). Note that relatively low mandibular bluntness of beaked whales is compensated for by the precoronoid crest of the mandible that shortens the effective gape size.

DISCUSSION

Possibility of Suction Feeding

The hyobranchial apparatus of Triassic and Early Jurassic ichthyosaurs strongly suggests that these marine reptiles were incapable of suction feeding. All but one species (*Hauffiopteryx typicus*) lacked an ossified hyoid corpus, whereas robust integration between ossified hyoid corpus and cornua has been identified as essential for amniote suction feeders as pointed out earlier. Even in *H. typicus*, such a robust integration is questionable (Fig. 4D). Also, the CB1 of ichthyosaurs, including *H. typicus*, is not very robust compared to suspensory hyobranchial elements of suction-feeding turtles (CB1), whales (stylohyals), and sharks (ceratohyals) (Figs. 2, 3, and 5). Therefore, it seems reasonable to conclude that all ichthyosaurs examined were 'ram feeders' because they lacked hyobranchial adaptation toward suction feeding. Notably, two of the presumed suction feeders [11] have the most slender CB1 among ichthyosaurs examined; this makes them the least likely candidates for suction feeders among ichthyosaurs.

Lack of variation in the slenderness of the CB1 in ichthyosaurs is noteworthy. The ceratohyals of sharks show a broad range of slenderness depending on their prey capture ecology (Fig. 3B). The contrasting absence of such variation among ichthyosaurs seems to indicate that the use of the hyobranchial apparatus during feeding did not vary much among different types of ichthyosaurs, whether small or large, long- or short-snouted. We infer that the hyobranchial apparatus played a similar role during feeding across ichthyosaurian species.

The 'ram-feeding' hypothesis for ichthyosaurs is further strengthened by two independent pieces of morphological evidence that are rooted in the mechanics of suction feeding. First, as evident from MPCI values, syringe-like suction pressure concentration that occurs between the throat and mouth opening was not very high in ichthyosaurs, unlike in suction-feeding cetaceans. The MPCI values of ichthyosaurs are similar to those of 'ram-feeding' marine mammals, again suggesting that ichthyosaurs were 'ram feeders'. One exception is *Eurhinosaurus* that is discussed later. Second, ichthyosaurian mandibles are much more elongated than those of odontocete cetaceans on average (Figs. 2 and 5B). The most elongated mandibles among suction-feeding cetaceans are found in beaked whales, which are about as acute as the least elongated mandible of ichthyosaurs. Despite the unusually elongated mandibles for a suction feeder, beaked whales are able to suction feed thanks to the presence of superficial tissues around the corners of the gape, which in effect shorten the span of gape, while allowing a more rounded opening that is suitable for concentrating pressure projection [50]. This effective decrease in gape also has an osteological basis: beaked whales have the precoronoid crest of the mandible, a unique structure that enables the mandible to overlap the upper jaw far anterior to the jaw joint. There is no evidence for such soft- or hard-tissues in ichthyosaurs.

A previous study [10] mentioned the presence of dorsally convex coronoid region in *Guanlingsaurus*. This feature, however, is not well exposed in the specimens that they described. The newer specimen reported in [16] has a complete exposure of the mandible, where no special structure analogous to the precoronoid crest of beaked whales exists. Mandibular morphology of *Guanlingsaurus* is not very different from those of some deep-jawed ichthyosaurs of the Middle and Upper Triassic except its relative shortness. Also, even if a special structure existed in the region, it probably could not shorten the gape and narrow the mouth opening effectively because of its location (see below).

A part of the reason why suction feeding did not evolve in ichthyosaurs may be the basic skull design of the Ichthyosauria, which is a clade within the Ichthyopterygia [6]. In comparison with cetaceans, the jaw joint of this clade tends to be located much more posteriorly relative to the snout, at or behind the occiput that is often inclined to contribute to the posterior displacement of the joint as in *Guanlingsaurus*. Because of this posterior location, ichthyosaurs are destined to have more elongated mandibles than cetaceans even when the snout lengths are similar. Also, this design leads to posterior location of the coronoid region in ichthyosaurs, which is usually found at the level of the occipital condyle. Thus, the region is too far away from the snout to function as the gape-shortening apparatus–note that the precoronoid crest of beaked whales is located at the posterior part of the snout. This apparent constraint from the jaw-joint location is not applicable to basal Ichthyopterygia, so it may be possible to find a suction feeder among basal members of the clade in the future.

Eurhinosaurus merits a separate discussion because of its uniquely low MPCI value. There is an anatomical reason why this taxon exhibits an anomalous value in this index. The genus is known for extreme shortening of the mandible, leading to an extensive overbite that is analogous to that seen in swordfish [6]. A large part of its upper dentition is anterior to the mandible, which starts to widen immediately posterior to the tip but is still not very wide at the point where the dentition ends. Apart from the overall shortening, the mandibular morphology is not unusual for an ichthyosaur. For example, the jaw symphysis is short, unlike in suction-feeding cetaceans that have low MPCI values. Given that the CB1 of *Eurhinosaurus* is slender as in other ichthyosaurs and that its hyoid corpus is unossified, it lacked the 'equipment' to generate suction. Therefore, it was most likely a 'ram feeder' as with other ichthyosaurs.

Paleoecological Implications

Suction feeders among extant air-breathing marine vertebrates have a limited range of feeding ecology. This is partly because suction is only effective over a short distance [39], which is usually less than 6 cm in the cetaceans measured so far [35], [44]. These air-breathing suction feeders almost exclusively feed on stationary or slow-moving prey and capture them with limited pursuit. For example, suction-feeding pinnipeds and sharks feed from the sea floor [57]. With the exception of *Delphinapterus*, which use suction feeding during shallower benthic foraging [35], 'true' suction feeders among cetaceans feed in the mesopelagic and bathypelagic zones [58]–[60]) to catch squid and fish [61]–[63]). Most mesopelagic and bathypelagic squid and fish are considered to be slow moving [64], with the exception of some large squids [65]–[67]. Beaked whales and sperm whales tend to swim through concentrations of these prey items in the lower DSL (Deep Scattering Layer) and benthopelagic layers during feeding dives [60], [68], although sperm whales reportedly pursue large squids from time to time [68]. If some shastasaurid ichthyosaurs were suction feeders resembling beaked whales as once suggested [10], then they would have needed concentrations of slow-moving, soft-bodied prey, most likely coleoid cephalopods, as in the modern mesopelagic/ bathypelagic zones. None of the benthic feeders among suction-feeding, air-breathing marine predators is edentulous except the gray whale that uses side-suction [55], so benthic feeding is probably inappropriate for these ichthyosaurs.

Evidence for the presence of deep-water coleoid communities in the Triassic is scant, although its possibility cannot be completely excluded. The time period predates the common ancestor of extant decabrachian coleoids in the Late Jurassic or Early Cretaceous [69], [70]–this clade has extant members that are vertical migrants [69]. The oldest belemnites are known from the Carnian of Sichuan, China [71], [72], although they did not spread worldwide until the Early Jurassic [73], [74]. This clade has evidence for vertical migration reaching at least below the mixed surface layers in the Jurassic, while most belemnites are considered epipelagic organisms [75]–[77]. The habitat depth of the only Triassic belemnites is unknown; however they appear to be shallow water forms as with most belemnites, judging from published cross-section photographs [71] (pls.III and IV) and the premises of [75]. The Triassic saw two other lineages of coleoids (Fig. 6). Aulacocerids, which spread worldwide in the Late Triassic, especially along the Tethys Sea [78], may have inhabited deepwater [75]. It has even been suggested that they were not very active [78], as expected for deepwater organisms [64], [65]. The other coleoid lineage in the Triassic, namely phragmoteuthids, had a limited geographic distribution [78] and, judging from the phragmocone angle [79] and generalization by [75], probably lived in shallower waters. Thus, aulacocerids would have been the only possible prey coleoid group for deep-diving air-breathers. At his point, it is not known if the biomass of this group was sufficiently large to support deep-diving, air-breathing predators. Moreover, aulacocerids are considered to have had an extensive shell with a 'living chamber' as in nautiloids but unlike modern coleoids or belemnites [80]; thus, they may not have been suitable prey of suction feeders, and certainly were not analogous to modern deepwater coleoids in terms of their characteristics as prey.

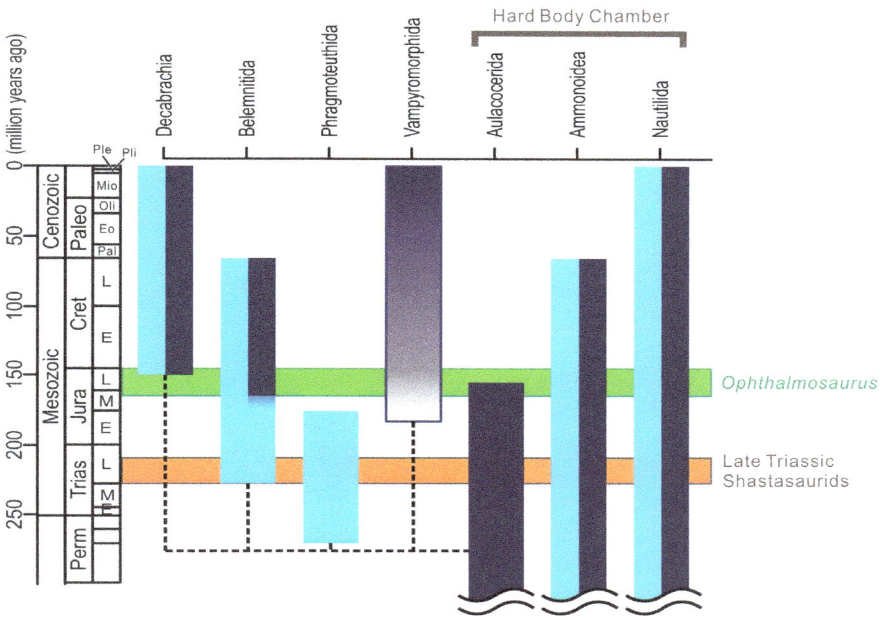

Figure 6. Stratigraphic ranges of major coleoid and key ichthyosaur groups being discussed.

Divergence time and tree topology is based on [70]. Ranges of fossil coleoid groups are based on [79], [89]. The shastasaurid ichthyosaurs of the Late Triassic, which were previously interpreted as suction feeders resembling beaked whales [10], did not co-exist with slow-moving and soft-bodied coleoid prey suitable for such suction feeders. The deep-diving ichthyosaur *Ophthalmosaurus* was coeval with some soft-bodied coleoid vertical migrants. Dark blue indicates deep habitat (reaching the mesopelagic zone) and light blue shallow (epipelagic). The color gradation for vampyromorphs indicates uncertainty in habitat depths of early forms. The upper range of vertically migrating belemnites is extended to the level indicated by [90].

Hypothetically, it may be possible that some slow moving, soft-bodied coleoids, comparable to today›s deepwater forms, inhabited shallow waters in the Triassic and could have served as potential prey for suction feeding marine reptiles. However, such a form has not been recognized to date, despite the fact that the fossil record of shallow water cephalopods is much richer than that of their deepwater counterparts. Absence of such slow moving forms in shallow water may be reasonable because the slowness of extant deepwater coleoids has been linked to the relaxed visually mediated predator/prey interactions in light-limited environments [65].

Another question is whether Triassic ichthyosaurs were capable of deep diving. Deep diving ability has been suggested for ichthyosaurs in the past but only for more derived ichthyosaurs in the Jurassic [8], [15], which had large scleral ring apertures that enabled dim-light adapted vision and a thunniform body plan that facilitated cruising ability. It is unlikely that shastasaurid ichthyosaurs were thunniform, judging from complete skeletons from China. Also, their eyes were not particularly large for ichthyosaurs, with apertures that were small for the eye size, indicating a lack of dim-light adaptation. It is thus likely that these shastasaurids were not deep divers.

Hypothetically, it may be possible that some slow-moving, soft-bodied coleoids, comparable to today›s deepwater forms, inhabited shallow waters in the Triassic, providing a viable prey source for suction feeders. However, such a form has not been recognized to date, even though shallow water fossils are far more abundant than their deepwater counterparts. Absence of such shallow water forms may be reasonable because the slowness of extant deepwater coleoids has been linked to the paucity of nutrients in deep sea layers [65].

A recent study suggested that ichthyosaurian eyes evolved in response to large predators rather than deep diving [81]. However, this suggestion is not well-supported optically [82] or by fossil evidence. First, the authors mainly discussed *Temnodontosaurus* and its possible predators, whereas deep diving was suggested for *Ophthalmosaurus* and possibly other thunnosaurs that are more derived than *Temnodontosaurus* [8]. Second, *Temnodontosaurus*, when it first appeared in the Hettangian (earliest Jurassic), was by far the largest of the coeval marine reptiles despite the claim of [81]. The very large *Rhomaleosaurus* that [81] mentioned, which was comparable in body length to *Temnodontosaurus* but was much smaller in gape size than the latter, is known from the Toarcian [83], some ten million years later. Predator-driven evolution has also been proposed for deep diving in ichthyosaurs [84]; however the discussion remains qualitative, especially concerning the involvement of predators. We suggest that the availability of deepwater prey (see below) can explain the data presented in [84] as well.

The appearance of deep-diving ichthyosaurs, such as *Ophthalmosaurus*, may predate that of the common ancestors of the Decabrachia [69] (Fig. 6). However, isotopic records suggest that belemnites that were coeval with *Ophthalmosaurus* included vertical migrants [77]. Also, one belemnite that co-occurs with *Ophthalmosaurus*, namely *Cylindrotheuthis*, is thought to have been capable of withstanding shallow mesopelagic water pressure [75]. The interpretation of [75] has been challenged in the past but the basic principle of his study remains justified [85]–[88]. Additionally, coleoids belonging to the lineage of vampyromorphs are known as early as the Toarcian of the Early Jurassic [88] (Fig. 6). Although the habitat depths of these early forms are debatable, the only extant member of the lineage, vampire squid *Vampyroteuthis infernalis*, is a mesopelagic inhabitant.

CONCLUSIONS

Triassic and Early Jurassic ichthyosaurs were most likely 'ram-feeders' based on functional inference from hyobranchial and mandibular morphology. Together with the inferred lack of deep-diving ability and dim-light vision in suspected suction feeders among Triassic ichthyosaurs, it is unlikely that these ichthyosaurs were meso-/bathypelagic feeders resembling beaked whales. Therefore, the evolutionary history of ichthyosaurs does not necessitate the formation of deepwater soft-bodied coleoid and fish communities in the Triassic. The coleoid fossil record also suggests the lack of deepwater coleoids with soft bodies in the Triassic. Such communities may have been available by the time *Ophthalmosaurus*, the postulated deep diving ichthyosaur, emerged. Hyobranchial morphology of marine reptiles has been largely understudied, despite its importance in inferring feeding ecology of these animals. It will be important to study it further, to understand the early evolution of modern marine ecosystems in the Mesozoic. Such a study is underway.

ACKNOWLEDGMENTS

We thank Christopher McGowan who provided photographs of some Liassic ichthyosaurs, and Mike Taylor for the discussion of Hettangian predators. Geerat Vermeij read an earlier version of this paper and made suggestions. Jin-ling Li and Chun Li (Institute of Vertebrate Paleontology and Paleoanthropology, Beijing, China), Sandra Chapman and Angela Milner (NHMUK), Heinz Furrer (Paläontologisches Institut und Museum der Universität, Zurich), Brandon Strilisky (Royal Tyrrell Museum of Paleontology, Drumheller, Canada), Rainer Schoch (SMNS), Pat Holroyd and Kevin Padian (UCMP) and Xiao-hong Chen and Long Cheng (Wuhan Institute of Geology and Mineral Resources of China, Wuhan, China) facilitated access to the specimens in their care.

AUTHOR CONTRIBUTIONS

Conceived and designed the experiments: RM CJ TT DJ. Performed the experiments: RM CJ TT NPK EM. Analyzed the data: RM. Contributed reagents/materials/analysis tools: RM DJ. Wrote the paper: RM. Manuscript revision: RM CJ TT NPK EM DJ PMS.

REFERENCES

1. Motani R (2009) The evolution of marine reptiles. Evolution: Education and Outreach 2: 224–235. doi: 10.1007/s12052-009-0139-y
2. Motani R (2010) Warm blooded sea dragons? Science 328: 1361–1362. doi: 10.1126/science.1191409
3. Massare JA (1987) Tooth morphology and prey preference of Mesozoic marine reptiles. J Vertebr Paleontol 7: 121–137. doi: 10.1080/02724634.1987.10011647
4. Kelley NP (2012) Trophic ecomorphology of Triassic marine reptiles. Unpubl Ph.D. dissertation, University of California, Davis.
5. Kelley N, Motani R, Jiang D, Rieppel O, Tintori A (2009) Rapid diversification of dental and jaw morphology among marine reptiles during the Triassic Recovery. J Vertebr Paleontol 29: 125A.
6. McGowan C, Motani R (2003) Ichthyopterygia. Handbuch der Paläoherpetologie Part 8. Verlag Dr. Friedrich Pfeil, München. 175 pp.
7. Motani R (2005) Ichthyosauria: evolution and physical constraints of fish-shaped reptiles. Ann Rev Earth Planet Sci 33: 395–420. doi: 10.1146/annurev.earth.33.092203.122707
8. Motani R, Rothschild BM, Wahl W Jr (1999) Large eyes in deep diving ichthyosaurs. Nature 402: 747. doi: 10.1038/45435
9. Ferry-Graham LA, Lauder GV (2001) Aquatic prey capture in Ray-finned fishes: a century of progress and new directions. J Morphol 248: 99–119. doi: 10.1002/jmor.1023

10. Sander PM, Chen X, Cheng L, Wang X (2011) Short-snouted toothless ichthyosaur from China suggests Late Triassic diversification of suction feeding ichthyosaurs. PLoS One 6: e19480. doi: 10.1371/journal.pone.0019480
11. Nicholls EL, Manabe M (2004) Giant ichthyosaurs of the Triassic–a new species of Shonisaurus from the Pardonet Formation (Norian: Late Triassic) of British Columbia. J Vertebr Paleontol 24: 838–849. doi: 10.1671/0272-4634(2004)024[0838:giottn]2.0.co;2
12. Rieppel O (2002) Feeding mechanics in Triassic stem-group sauropterygians: the anatomy of a successful invasion of the Mesozoic seas. Zool J Linn Soc 135: 33–63. doi: 10.1046/j.1096-3642.2002.00019.x
13. Li C, Rieppel O, LaBarbera MC (2004) A Triassic aquatic protorosaur with an extremely long neck. Science 305: 1931. doi: 10.1126/science.1100498
14. Kelley NP, Motani R, Jiang D, Rieppel O, Schmitz L (2012) Selective extinction of Triassic marine reptiles during long-term sea-level changes illuminated by seawater strontium isotopes. Palaeogeogr Palaeoclimatol Palaeoecol Available: http://dx.doi.org/10.1016/j.palaeo.2012.07.026.
15. Motani R (2008) Combining uniformitarian and historical data to interpret how Earth environment influenced the evolution of Ichthyopterygia. Paleontol Soc Pap 14: 147–164.
16. Ji C, Jiang D, Motani R, Hao W, Sun Z, et al. (2013) A new juvenile specimen of Guanlingsaurus (Ichthyosauria: Shastasauridae) from the Upper Triassic of Southwestern China. J Vertebr Paleontol 33: 1–9. doi: 10.1080/02724634.2013.723082
17. Johnston C, Berta A (2011) Comparative anatomy and evolutionary history of suction feeding in cetaceans. Mar Mamm Sci 27: 493–513. doi: 10.1111/j.1748-7692.2010.00420.x
18. Tomita T, Sato K, Suda K, Kawauchi J, Nakaya K (2011) Feeding of the Megamouth Shark (Pisces: Lamniformes: Megachasmidae) predicted by its hyoid arch: a biomechanical approach. J Morphol 272: 513–524. doi: 10.1002/jmor.10905
19. Lemell P, Lemell C, Sneldervaard P, Gumpenberger M, Wocheslãnder R, et al. (2010) Feeding patterns of Chelus fimbriatus (Pleurodira: Chalidae). J Exp Biol 205: 1495–1506.
20. Sollas WJ (1916) The skull of Ichthyosaurus, studied in serial sections. Philos Trans R Soc Lond B Biol Sci 208: 63–126. doi: 10.1017/s0016756800208510
21. Romer AS (1957) Osteology of the Reptiles. Chicago: University of Chicago Press. p. 772.
22. Rieppel O (1981) The hyobranchial skeleton in some little known lizards and snakes. J Herpetol 15: 433–440. doi: 10.2307/1563533
23. Reese AM (1947) The laryngeal region of Alligator mississipiensis. Anat Rec 92: 273–277. doi: 10.1002/ar.1090920308
24. Natchev N, Heiss E, Singer K, Kummer S, Salaberger D, et al. (2011) Structure and function of the feeding apparatus in the common musk turtle Sternotherus odoratus (Chelonia, Kinosternidae). Contrib Zool 808: 143–156.
25. Langerbartel DA (1968) The hyoid and its associated muscles in snakes. Illinois Biol Monogr 38: 1–156.
26. Kundrát M, Janáček J, Russell AP (2009) Developmental patterns of the crocodilian and avian columella auris: reappraisal of interpretations of the derivation of the dorsal hyoid arch in archosaurian tetrapods. Zool J Linn Soc 156: 384–410. doi: 10.1111/j.1096-3642.2009.00537.x
27. Smith KK (1984) The use of tongue and hyoid apparatus during feeding in lizards (Ctenosaura similis and Tupinambis nigropunctatus). J Zool 202: 115–143. doi: 10.1111/j.1469-7998.1984.tb04290.x
28. Smith KK (1986) Morphology and function of the tongue and hyoid apparatus in Varanus (Varanidae, Lacertilia). J Morphol 187: 261–287. doi: 10.1002/jmor.1051870302
29. Meyers JJ, Herrel A, Nishikawa KC (2002) Comparative study of the innervation patterns of the hyobranchial musculature in three iguanian lizards: Sceloporus undulatus, Pseudotrapelus sinaitus, and and Chamaeleo jacksonii. Anat Rec 267: 177–189. doi: 10.1002/ar.10096.abs
30. Reidenberg JS, Laitman JT (1994) Anatomy of the hyoid apparatus in Odontoceti (Toothed Whales): specializations of their skeleton and musculature compared with those of terrestrial mammals. Anat Rec 240: 598–624. doi: 10.1002/ar.1092400417
31. Werth AJ (2007) Adaptations of the cetacean hyolingual apparatus for aquatic feeding and thermoregulation. Anat Rec 290: 546–568. doi: 10.1002/ar.20538
32. Motta PJ, Wilga CAD (1995) Anatomy of the feeding apparatus of the Lemon Sharks, Negaprion bervirostris. J Morphol 226: 309–329. doi: 10.1002/jmor.1052260307
33. Dean MN, Wilga CD, Summers AP (2005) Eating without hands or tongue: specialization, elaboration and the evolution of prey processing mechanisms in cartilaginous fishes. Biol Lett 2005 1: 357–361. doi: 10.1098/rsbl.2005.0319
34. Lauder GV, Prendergast T (1992) Kinematics of aquatic prey capture in the snapping turtle Chelydra serpentina. J Exp Biol 164: 55–78.
35. Kane EA, Marshall CD (2009) Comparative feeding kinematics and performance of odontocetes: belugas, Pacific white-sided dolphins and long-finned pilot whales. J Exp Biol 212: 3939–3950. doi: 10.1242/jeb.034686
36. Liem KF (1980) Acquisition of energy by teleosts: adaptive mechanisms and evolutionary patterns. In: Ali MA, editor. Environmental Physiology of Fishes. New York: Plenum Press. pp. 299–334.
37. Lauder GV, Liem KF (1981) Prey capture by Luciocephalus pulcher: implications for models of jaw protrusion in teleost fishes. Evol Biol Fishes 6: 257–268. doi: 10.1007/bf00005755
38. Alexander RM (1967) Functional Design in Fishes. London: Hutchinson.

39. Wainwright PC, Ferry-Graham LA, Waltzek TB, Carroll AM, Hulsey CD, et al. (2001) Evaluating the use of ram and suction during prey capture by cichlid fishes. J Exp Biol 204: 3039–3051.

40. Higham TE, Day SW, Wainwright PC (2005) Sucking while swimming: evaluating the effects of ram speed on suction generation in bluegill sunfish Lepomis macrochirus using digital particle image velocimetry. J Exp Biol 208: 2653–2660. doi: 10.1242/jeb.01682

41. Norton SF, Brainerd EL (1993) Convergence in the feeding mechanics of ecomorphologically similar species in the Centrarchidae and Cichlidae. J Exp Biol 176: 11–29.

42. Lemell P, Beisser CJ, Gumpenberger M, Snelderwaard P, Gemel R, et al. (2002) The feeding apparatus of Chelus fimbriatus (Pleurodira; Chelidae)–adaptation perfected? Amphib-reptl 31: 97–107. doi: 10.1163/156853810790457803

43. Werth AJ (2006) Mandibular and dental variation and the evolution of suction feeding in Odontoceti. J Mammal 87: 579–588. doi: 10.1644/05-mamm-a-279r1.1

44. Bloodworth BE, Marshall CD (2005) Feeding kinematics of Kogia and Tursiops (Odontoceti: Cetacea): characterization of suction and ram feeding. J Exp Biol 208: 3721–3730. doi: 10.1242/jeb.01807

45. Van Damme J, Aerts P (1997) Kinematics and functional morphology of aquatic feeding in Australian snake-necked turtles (Pleurodira; Chelodina). J Morphol 233: 113–125. doi: 10.1002/(sici)1097-4687(199708)233:2<113::aid-jmor3>3.0.co;2-7

46. Summers AP, Darouian KF, Richmond AM, Brainerd EL (1998) Kinematics of aquatic and terrestrial prey capture in Terrapene carolina, with implications for the evolution of feeding in cryptodire turtles. J Exp Zool 281: 280–287. doi: 10.1002/(sici)1097-010x(19980701)281:4<280::aid-jez4>3.0.co;2-k

47. Sokal RR, Rohlf FJ (1995) Biometry, 3rd Ed. New York: W. H. Freeman and Company. 887 p.

48. Warton DI, Wright IJ, Falster DS, Westoby M (2006) Bivariate line-fitting methods for allometry. Biol Rev 81: 259–291. doi: 10.1017/s1464793106007007

49. Werth AJ (2006) Odontocete suction feeding: experimental analysis of water flow and head shape. J Morphol 267: 1415–1428. doi: 10.1002/jmor.10486

50. Heyning JE, Mead JG (1996) Suction feeding in beaked whales: morphological and observational evidence. Contrib Sci, Nat Hist Mus Los Angeles County 464: 1–12.

51. Bramble D (1973) Media dependent feeding in turtles. Am Zool 13: 1342.

52. Bloodworth BE, Marshall CD (2007) A functional comparison of the hyolingual complex in pygmy and dwarf sperm whales (Kogia breviceps and K. sima), and bottlenose dolphis (Tursiops truncatus). J Anat 211: 78–91. doi: 10.1111/j.1469-7580.2007.00755.x

53. Motta PJ, Hueter RE, Tricas TC, Summers AP, Huber DR, et al. (2008) Functional morphology of the feeding apparatus, feeding constraints, and suction performance in the Nurse Shark Ginglymostoma cirratum. J Morphol 269: 1041–1055. doi: 10.1002/jmor.10626

54. Wilga CD, Motta PJ, Sanford CP (2007) Evolution and ecology of feeding in elasmobranchs. Integr Comp Biol 47: 55–69. doi: 10.1093/icb/icm029

55. Ray GC, Schevill WE (1974) Feeding of a captive Gray Whale, Eschrichtius robustus. Mar Fisher Pap 36: 31–38.

56. Feyrer LJ, Duffs DA (2011) Predatory disturbance and prey species diversity: the case of gray whale (Eschrichtius robustus) foraging on a multi-species mysid (family Mysidae) community. Hydrobiologia 678: 37–47. doi: 10.1007/s10750-011-0816-z

57. Marshall CD, Kovacs KM, Lydersen C (2008) Feeding kinematics, suction and hydraulic jetting capabilities in bearded seals (Erignathus barbatus). J Exp Biol 211: 699–708. doi: 10.1242/jeb.009852

58. Hooker SK, Baird RW (1999) Deep-diving behaviour of northern bottlenose whale, Hyperoodon ampullatus (Cetacea: Ziphiidae). Proc R Soc Lond B Biol Sci 266: 671–676. doi: 10.1098/rspb.1999.0688

59. Tyack PL, Johnson M, Soto NA, Sturlese A, Madsen PT (2006) Extreme diving of beaked whales. J Exp Biol 209: 4238–4253. doi: 10.1242/jeb.02505

60. Arranz P, de Soto NA, Madsen PT, Brito A, Bordes F, et al. (2011) Following a foraging fish-finder; diel habitat use of Blainville's beaked whales revealed by echolocation. PLoS One 6: e28353. doi: 10.1371/journal.pone.0028353

61. Santos MB, Pierce GJ, Smeenk C, Addink MJ, Kinze CC, et al. (2001) Stomach contents of northern bottlenose whales Hyperoodon ampullatus stranded in the North Sea. J Mar Biol Assoc U. K. 81: 143–150. doi: 10.1017/s0025315401003484

62. Santos MB, Pierce GJ, Herman J, López A, Guerra A, et al. (2001) Feeding ecology of Cuvier's beaked whale (Ziphius cavirostris): a review with new information on the diet of this species. J Mar Biol Assoc U. K. 81: 687–694.

63. Santos MB, Martin V, Arbelo M, Fernández A, Pierce GJ (2007) Insights into the diet of beaked whales from the atypical mass stranding in the Canary Islands in September 2002. J Mar Biol Assoc U. K. 87: 243–251. doi: 10.1017/s0025315407054380

64. Childress JJ (1995) Are there physiological and biochemical adaptations of metabolism in deep-sea animals? Trends Ecol Evol 10: 30–36. doi: 10.1016/s0169-5347(00)88957-0

65. Seibel BA, Thuesen EV, Childress JJ (2000) Light-limitation of predator-prey interactions: consequences for metabolism and locomotion of deep-sea cephalopods. Biol Bull 198: 284–298. doi: 10.2307/1542531

66. Kubodera T, Koyama Y, Mori K (2007) Observations of wild hunting behaviour and bioluminescence of a large deep-sea, eight-armed squid, Taningia danae. Proc Biol Sci 274: 1029–1034. doi: 10.1098/rspb.2006.0236

67. Kubodera T, Mori K (2005) First-ever observations of a live giant squid in the wild. Proc Biol Sci 272: 2583–2586. doi: 10.1098/rspb.2005.3158

68. Aoki K, Amano M, Mori K, Kourogi A, Kubodera T, et al. (2012) Active hunting by deep-diving sperm whales: 3D dive profiles and maneuvers during bursts of speed. Mar Ecol Progr Ser 444: 289–301. doi: 10.3354/meps09371

69. Lindberg DR, Pyenson ND (2007) Things that go bump in the night: evolutionary interactions between cephalopods and cetaceans in the tertiary. Lethaia 40: 335–343. doi: 10.1111/j.1502-3931.2007.00032.x

70. Kröger B, Vinther J, Fuchs D (2011) Cephalopod origin and evolution: a congruent picture emerging from fossils, development and molecules. Bioessays 33: 602–613. doi: 10.1002/bies.201100001

71. Zhu K, Bian Z (1984) Sinobelemnitidae, a new Family of belemnitida from the Upper Triassic of Longmenshan, Sichuan. Gu Sheng Wu Xue Bao 23: 300–317 [in Chinese with English Summary].

72. Iba Y, Sano S, Mutterlose J, Kondo Y (2012) Belemnites originated in the Triassic—A new look at an old group. Geology 40: 911–914. doi: 10.1130/g33402.1

73. Doyle P, Donovan DT, Nixon M (1994) Phylogeny and systematic of the Coleoidea. Univ Kansas Paleontol Contrib 5: 1–15.

74. Doyle P, Kelly SRA, Pirrie D, Riccardi AC (1997) Jurassic belemnite distribution patterns: implications of new data from Antarctica and Argentina. Alcheringa 21: 219–228. doi: 10.1080/03115519708619175

75. Westermann GEG (1973) Strength of concave septa and depth limits of fossil cephalopods. Lethaia 6: 383–403. doi: 10.1111/j.1502-3931.1973.tb01205.x

76. Dutton A, Huber BT, Lohmann KC, Zinsmeister WJ (2007) High-resolution stable isotope profiles of a dimitobelid belemnite: implications for paleodepth habitat and Late Maastrichtian climate seasonality. Palaios 22: 642–650. doi: 10.2110/palo.2005.p05-064r

77. Price GD, Page KN (2008) A carbon and oxygen isotopic analysis of molluscan faunas from the Callovian-Oxfordian boundary at Redcliff Point, Weymouth, Dorset: implications for belemnite behaviour. Proc Geol Assoc 119: 153–160. doi: 10.1016/s0016-7878(08)80315-x

78. Jeletzky JA (1966) Mollusca 7. Comparative morphology, phylogeny, and classification of fossil Coleoidea. Univ Kansas Paleontol Contrib 42: 1–162.

79. Donovan DT (2006) Phragmoteuthida (Cephalopoda: Coleoidea) from the Lower Jurassic of Dorset, England. Palaeontology 49: 673–684. doi: 10.1111/j.1475-4983.2006.00552.x

80. Bandel K (1985) Composition and ontogeny of Dictyoconites (Aulacocerida, Cephalopoda). Paläontol Z 59: 223–244. doi: 10.1007/bf02988810

81. Nilsson D-E, Warrant EJ, Johnsen S, Hanlon R, Shashar N (2012) A unique advantage for giant eyes in Giant Squid. Curr Biol 22: 1–6. doi: 10.1016/j.cub.2012.02.031

82. Schmitz L, Motani R, Oufiero CE, Martin CH, McGee MD, et al. (2013) Allometry indicates giant eyes of giant squid are not exceptional. BMC Evolutionary Biology 13: 45. doi: 10.1186/1471-2148-13-45

83. Smith AS, Dyke GJ (2008) The skull of the giant predatory pliosaur Rhomaleosaurus cramptoni: implications for plesiosaur phylogenetics. Naturwissenschaften 95: 975–980. doi: 10.1007/s00114-008-0402-z

84. Rothschild BM, Xiaoting Z, Martin LD (2012) Adaptations for marine habitat and the effect of Triassic and Jurassic predator pressure on development of decompression syndrome in ichthyosaurs. Naturwissenschaften 99: 443–448. doi: 10.1007/s00114-012-0918-0

85. Westermann GEG, Ward P (1980) Septum morphology and bathymetry in cephalopods. Paleobiology 6: 48–50.

86. Hewitt RA, Westermann GEG (1988) Nautiloid septal strength: revisited and revised concepts. Alcheringa 12: 123–128. doi: 10.1080/03115518808619000

87. Hassan MA, Westermann GEG, Hewitt RA, Dokainish MA (2002) Finite-element analysis of simulated ammonoid septa (extinct Cephalopoda): septal and sutural complexities do not reduce strength. Paleobiology 28: 113–126. doi: 10.1666/0094-8373(2002)028<0113:feaosa>2.0.co;2

88. Ritterbush KA, Bottjer DJ (2012) Westermann morphospace displays ammonoid shell shape and hypothetical paleoecology. Paleobiology 28: 424–446. doi: 10.1666/10027.1

89. Fuchs D, Larson N (2011) Diversity, Morphology, and Phylogeny of coleoid cephalopods from the Upper Cretaceous Plattenkalks of Lebanon–Part II: Teudopseina. J Paleontol 85: 815–834. doi: 10.1666/10-159.1

90. Zakharov YD, Pletnev SP, Mel'nikov ME, Smyshlyaeva OP, Khudik VD, et al. (2007) The first find of Cretaceous belemnites from the Magellan Rise, Pacific Ocean. Russ J Pac Geol 1: 29–41. doi: 10.1134/s1819714007010058

CHAPTER 12

BROAD-SCALE PATTERNS OF LATE JURASSIC DINOSAUR PALEOECOLOGY

Christopher R. Noto, Ari Grossman

INTRODUCTION

Over the past twenty years new fossil discoveries, novel technologies, and a proliferation of analytical techniques have greatly increased our knowledge of dinosaur morphology, phylogeny, and behavior. These data, when combined with our growing knowledge of dinosaur biogeography, not only make it possible to address complex questions about changing dinosaur distributions, but also broad-scale ecological questions about the nature of dinosaur-dominated communities.

Advances in computer software enabled the creation of large, accessible databases recording data from an ever-increasing number of localities as new discoveries are made [1]–[3]. Coupled with geographic information system (GIS) technology, this allows us to examine patterns of dinosaur distribution at broad–regional to global–scales. There have been numerous studies of this type on dinosaur biogeography, focusing on vicariance, dispersal, and extinction patterns [4]–[10], which have helped identify possible areas of endemism, directions of dispersal, and even test the validity of different paleogeographic reconstructions. However, these data and methods have not yet been applied to ecological questions.

Historically, the ecology and behavior of dinosaurs has been reconstructed with reference to modern analogs, living animals that contain similar physical attributes or are closely related. Depending on the researcher, the modern analogs used to describe dinosaur ecology and behaviors have ranged from crocodilians and birds to mammals [11]–[21]. These models for dinosaur ecology and behavior are valuable starting points, providing a necessary conceptual framework from which to evaluate the unusual morphology of these extinct creatures. In many ways such thinking has been highly informative, as it is impossible to explain the biology of extinct taxa without first studying how living organisms operate. On the other hand it may lead to erroneous or unrealistic reconstructions based on constraints imposed by the organismal model used in the analogy, especially when those comparisons exist in the absence of quantitative data. Also, the focus on reconstructing the ecology of particular dinosaur taxa has hampered exploration of dinosaur-dominated ecosystems as a whole. While understanding the ecology of individual taxa is important, this information is insufficient outside of a broader ecosystem context.

An excellent example of a whole-ecosystem study is that of Foster [22], who carried out a detailed ecological analysis of the Late Jurassic Morrison Formation in the western United States. Such studies are few due to the immense effort involved in bringing together the multiple lines of data necessary to carry out such a comprehensive analysis. Therefore, detailed work of this nature is currently lacking for the large number of dinosaur-fossil bearing formations around the world. To gain a broad-scale view of dinosaur-dominated ecosystems a different approach is needed.

An ecological approach that reconstructs general habitat characteristics and ecological diversity among dinosaur faunas (as interpreted from fossil assemblages) in different regions of

Citation: Noto CR, Grossman A (2010) Broad-Scale Patterns of Late Jurassic Dinosaur Paleoecology. PLoS ONE 5(9): e12553. doi:10.1371/journal.pone.0012553.

the globe is an important tool for understanding some of the many forces shaping dinosaur distribution patterns. In addition, deciphering changes in the ecological structure of dinosaur communities over time is a powerful tool for elucidating the role of global changes in climate, continental arrangement, and land area in shaping dinosaur biogeography, ecology, and evolution. Various dynamics, such as Milankovich cycles (which impact solar energy distribution across the Earth's surface), or plate tectonics (which largely determine the location, size, and geology of terrestrial landmasses), affect climate patterns and determine the abiotic conditions to which ecosystems are exposed [23]. Changes in these processes over time are likely to be reflected in community structure and can be recorded in the fossil record [24]–[26].

This study utilizes dinosaur taxa from several Late Jurassic (161–145 Mya) fossil localities from around the world. The Late Jurassic is notable for its extremely warm and equable climate, which was dominated by a monsoonal circulation pattern [27]–[30]. These extreme conditions played a large role in the distribution and diversity of Late Jurassic biota [31]. Furthermore, throughout the Jurassic many dinosaur clades diversified [32], leading to the evolution of many extreme morphologies characteristic of these groups. While often speculated upon, the ecological role of these adaptations remains poorly understood, particularly how these adaptations were integrated to form stable, operational communities. How were dinosaur-dominated communities structured and what role did climatic conditions play? It is well known that biological communities evolving under similar environmental conditions often contain convergent adaptations [33]–[36]. We therefore predict that dinosaur fossil assemblages falling under similar climatic conditions will exhibit convergent community structure irrespective of their individual phylogenic or biogeographic history.

MATERIALS AND METHODS

Ecological Structure Analysis

The reconstruction of dinosaurs as living animals has greatly benefited from the increased emphasis on biomechanics and functional morphology as applied to fossil organisms. Use of biomechanical principals and a better understanding of the relationships between function and form substantially improve our ability to generate hypotheses about the behavior of extinct taxa [37]–[43]. These studies have led to many new insights about dinosaur paleoecology.

Traditionally, community analysis (as represented by fossil assemblages) uses indices of species richness [44] and/or taxonomic diversity indices (e.g., Simpson's Diversity Index). Species richness only analyzes the abundance of taxa at a site, and becomes extremely problematic when we lack tight control over the rate(s) of fossil accumulation at a site. In many cases, dinosaur fossil assemblages are a time-averaged collection representing a prolonged period of accumulation. Taxonomic diversity indices are more often used to determine the relative age of sites [45], [46] but may sometimes be used to make inferences about habitat as well [47]. However, on its own taxonomy is not sufficient to determine ecology. Different taxa may converge onto similar ecological niches, or alternatively closely related taxa may be ecologically diverse, especially when looking across broad spans of time. Thus, neither of these methods is sufficient for reconstructing the ecosystem represented at a fossil site.

Here we employ Ecological Structure Analysis (ESA), which uses functional morphology, to produce ecological profiles for different Late Jurassic localities. This method allows us to compare dinosaur assemblages, with varying taxonomic profiles and diversity, from a large number of localities. In applying ESA, taxa are classified using ecological criteria. Therefore, it is important to note that this is not a "taxon-free" method but merely a different way of classifying taxa. Reed [48]–[50] demonstrated that modern mammal communities located in different habitats differ significantly in the percentages of taxa found in trophic, locomotor, and body size categories (using both Kruskal-Wallis and the Mann-Whitney U tests), while those in similar habitats are more alike. She also demonstrated that the same ecological categories would correctly classify sites with different environments (using discriminant function analysis). Thus, it is possible to compare fossil mammal sites with modern ones and classify fossil localities according to habitat differences.

For this study, ESA was applied in the following way. A database of fossil localities and taxon lists was compiled from the Paleobiology Database [51] and Weishampel et al. [52]. Each locality was plotted on a paleogeographic map (150 mya reconstruction, Mollweide projection) using the mapping function available on the Paleobiology Database. Geographically close localities or those forming natural clusters were grouped together to form composite assemblages (Figure 1). This was done to provide the necessary sample size for analysis. The twelve assemblages include (Table 1): six from North America (M1–M6), two from Europe (E1–E2), one from Africa (A1), one from South America (S1) and two from Asia (C1–C2). Complete data are available in Table S1.

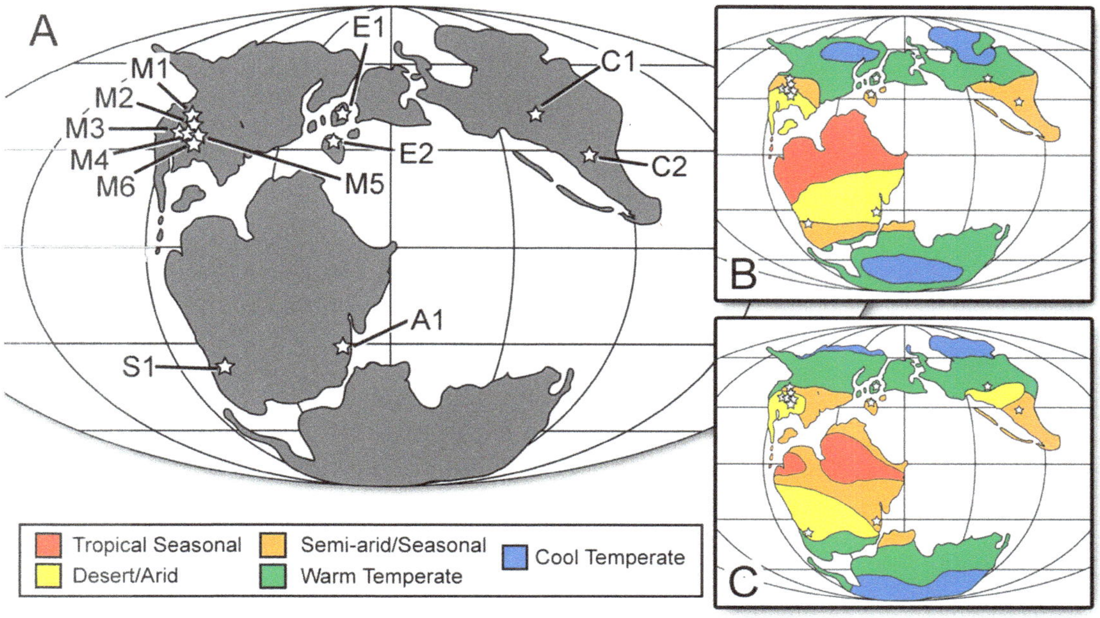

Figure 1. Late Jurassic paleogeographic map and reconstructed biomes.

A) Positions of fossil assemblages (stars): A1=Africa, C1–C2=China, E1–E2=Europe, M1–M6=North America, and S1=South America. Base map adapted from [27]. B) Biome reconstruction based on Rees et al. [27] model. C) Biome reconstruction based on Sellwood and Valdes [30] model.

Table 1. Location, biome assignments(s), and formations constituting each assemblage.

Assem.	Biome R[1]	Biome SV[2]	Region	Countries/States	Formation(s)
E1	2	1	Europe	England, France	Sables de Glos, Argiles d'Octeville, Marnes de Bléville, Kimmeridge Clay, Calcareous Grit, Corallian Oolite, Oxford Clay, Portland Stone
E2	2	2	Europe	Portugal, Spain	Villar del Arzobis, Alcobaça, Guimarota, Sobral, Amoreira-Porto Novo, Bombarral, Freixial, Louriñha
M1	2	3	North America	Wyoming, Montana, South Dakota	Morrison
M2	2	3	North America	Wyoming	Morrison
M3	3	3	North America	Colorado, Utah	Morrison
M4	3	3	North America	Colorado, Utah	Morrison
M5	3	3	North America	Colorado	Morrison
M6	3	3	North America	Colorado, New Mexico, Oklahoma	Morrison
C1	1	1	Asia	China (Xinjiang)	Shishugou, Kalazha
C2	2	2	Asia	China (Sichuan)	Shangshaximiao (Upper Shaximiao)
S1	2	3	South America	Chile, Argentina	Toqui, Cañadón Cálcereo
A1	2	3	Africa	Tanzania	Tendaguru

Biome Key: 1 = temperate, 2 = semi-arid/seasonally wet, 3 = desert/arid.
[1]Climate model of Rees et al. [27].
[2]Climate model of Sellwood and Valdes [30].
doi:10.1371/journal.pone.0012553.t001

For each assemblage genera were counted only once and higher taxa (i.e., generically indeterminate remains) were counted only if the group was not represented by generic remains, unless published reports provided a reasonable case that the remains represent distinct taxa. Isolated teeth were used only if they were assigned to a taxon for which a specific size range is known. Each taxon was classified using three separate ecological categories: Body Size, Locomotor Mode, and Trophic Mode, themselves divided into multiple classes (Table 2). Classification within each category was determined using morphological correlates as determined from the fossil itself, published reconstructions (if available), and closely related taxa (especially if the taxa are fragmentary). While other categories may also be useful, we prefer these particular three because osteological and dental evidence is potentially available in the fossil record for all taxa studied. For each assemblage the percentage of taxa in each class was calculated, thus generating an ecological profile. It is important to note that this analysis deals with paleoecological diversity and assumes a degree of correspondence between the diversity and relative abundance of ecomorphs in an assemblage. A detailed description of all ecological categories and classes is given below.

Table 2. Description of ecological categories and constituent classes used in this study.

Ecological Category[1]	Characteristics	
Trophic Mode		**Code**
Carnivore	Eats meat	C
Herbivore	High browsing: above 5 meters	HH
	Intermediate browsing: up to 5 meters	HI
	Low browsing up to 2 meters	HL
	Ground foraging: up to 1 meter	HG
Locomotor Mode		
Terrestrial Quadruped	Moves quadrupedally on the ground	TQ
Terrestrial Biped	Moves bipedally on the ground	TB
Facultative Biped	Capable of quadru- and bipedal motion	TF
Terrestrial/Arboreal Biped	Moves and feeds on the ground and in trees	TA
Body Size[2]		
Tiny	<1 kg	T
Very Small	1–10 kg	VS
Small	10–100 kg	S
Medium	100–1000 kg	M
Large	1000–10,000 kg	L
Very Large	>10,000 kg	VL

[1]Ecological categories after [49].
[2]Mass calculations from [63].
doi:10.1371/journal.pone.0012553.t002

Body Size.

Body size is perhaps the single most important ecological character. It not only influences lifestyle and behavior in an individual species, but is also responsible for driving ecosystem dynamics at higher levels of community organization. This occurs through the scaling of metabolic rate with body size, which influences population dynamics and species diversity, ultimately determining the flow of energy between trophic levels [53]–[57].

For individuals and species, body size can determine such ecologically meaningful characters as techniques of predator avoidance, the type of substrate an animal may utilize during locomotion [58], and the size of its home and day range [59]. In addition, body size generally affects overall mass and body proportions due to the mechanical constraints inherent in biological materials [60]. The forces that bodies of different size generate affect performance and therefore strategies in feeding, locomotion, and reproduction [61], [62]. In dinosaurs and other extinct organisms lacking modern analogs, many of these traits will remain difficult to determine directly from available fossil evidence. Use of body size estimates therefore can provide a first-order approximation of many important ecological characters when comparing taxa and assemblages.

Dinosaur body mass can be determined through estimate only. Each method has its own strengths and weaknesses, but discussion of these is outside the scope of this paper. More important is to provide a relative ranking of taxa even if absolute values are inaccurate. For continuity, we used body mass estimates based on Seebacher [63], which provides a wide list of taxa. For taxa not found on this list, we used the mass estimation equations given in the paper. We used only adult representatives where possible for mass estimation. Six size classes were used: Tiny (<1 kg), Very Small (1–10 kg), Small (10–100 kg), Medium (100–1000 kg), Large (1000–10 000 kg), and Very Large (>10 000 kg). Size classes are broad enough that expected body size variation among adults and estimate uncertainty is included for most taxa.

Trophic Mode.

Trophic mode refers to the diet and food processing strategies of an animal, the two primary categories being carnivore and herbivore. Further divisions depend on our knowledge of the morphology and behavior of the group in question. Morphology of the teeth is most useful in determining what type of food the animal processed during its life. While it is true that the type of prey consumed and the manner in which it is captured varies considerably among carnivores, these specializations are often much harder to decipher from available fossil material. Nevertheless, among non-mammalian carnivores, teeth tend to be elongate, sharp, and pointed; sometimes laterally compressed with a blade-like edge. In addition, many non-mammalian carnivores have long, laterally compressed skulls and possess sharp, recurved claws that aid in prey capture, dispatch, and processing. In contrast, non-mammalian herbivores tend to possess shorter, blunter teeth with grinding, slicing, or shearing surfaces. Their skulls tend to be broad and short, and the limbs lack the kinds of claws typically seen in carnivores.

Due to the unknown nature of plant preference in herbivorous dinosaurs, we assign herbivores to categories of browse height rather than plant type. We assign only an upper browsing limit based on characters such as neck posture and limb length. Ground level forms the lower browsing limit. These classes are (from lowest to highest feeding height): Ground Forager (<1 m), Low Browser (<2 m), Intermediate Browser (<5 m), and High Browser (>5 m). The neck posture of sauropods remains contentious given their extreme and unusual morphology [38], [42]. However, general differences in vertical feeding position between sauropod taxa are supported by morphological, biomechanical, and tooth wear analysis [38], [42], [64], [65].

Locomotor Mode.

We divide locomotor mode into two major categories, quadruped and biped. All locomotor modes used in this analysis are terrestrial in nature. Terrestrial quadrupeds are distinguished by having forelimbs that are closer in length to the hindlimbs and a robust radius locked with the ulna in a pronated position, increasing the stability of the front leg for weight-bearing during locomotion. In contrast, the forelimbs of bipeds are relatively shorter than the hind limbs. In addition, since the forelimbs are not necessary for weight-bearing the elbow joint is more mobile and the forelimb is not permanently locked in a pronated position. An additional division, facultative biped, was used for those taxa with intermediate axial and limb morphologies, such as shorter forelimbs capable of weight-bearing, suggesting that these taxa were capable of both bipedal and quadrupedal locomotion (e.g., some ornithopods) [66]–[68]. Some groups of dinosaurs evolved a quadrupedal stance secondarily, having descended from bipedal ancestors; however, these patterns in limb proportion and structure generally still hold.

The division Arboreal Biped was introduced because the locomotor behavior of some small maniraptorans and early birds (i.e, *Archaeopteryx*) is still debated [69]–[73]. For this analysis, we expect arboreal bipeds to have structures adapted for a combination of climbing, flight (gliding or active), and terrestrial locomotion, indicating they spent some amount of time both in trees and on the ground.

We acknowledge the great morphological variability in dinosaurs within the biped and quadruped categories, signaling important differences in locomotor strategy. More realistically, the locomotor strategies of taxa fall along a continuum, with discrete categories serving to delineate, rather than fully describe, how organisms move [74]. Again, further biomechanical work is necessary to determine how different limb and axial morphologies affected locomotor performance in dinosaurs. This work is currently underway and should lead to greater refinement of locomotor categories in the future [43], [75]–[78].

Comparison with Late Jurassic Climate

Biomes were used over individual locality-level environmental reconstructions to observe how well independent climate data reflect ecological differences between assemblages. Biomes are characterized by different groups of plants, which help form the basic structure of every environment and are related to the individual climatic tolerance of each species. This structure plays an important role in directing the subsequent evolution of constituent species inhabiting these areas. Therefore, biomes provide an estimate of the general environmental and climatic conditions that prevailed over the regions where different fossil localities formed. Furthermore, like ecological categories, biomes can be extended and applied to other time periods because they in part rely on morphological categories for assigning plant taxa [27], [79].

Each assemblage was assigned a biome using reconstructions from Rees et al. [27] or Sellwood and Valdes [30]. The former is based primarily on the distribution and morphology of fossil plant taxa and the distribution of climate sensitive sediments (e.g., evaporites and coals), similar to methods used in delineating modern biomes. The latter is based on a general circulation model (GCM), delineating each biome from model-predicted temperature and precipitation patterns (Table 1). While generally similar, each reconstruction differs in their interpretation of certain dinosaur habitats and should therefore be considered when comparing the assemblages.

Statistical Analyses

In order to compare ecological profiles, data were first arcsin transformed prior to analysis to approximate normality [80]. Similarity between assemblages was assessed with cluster analysis, utilizing the unweighted pair-group average (UPGMA) algorithm and Euclidean similarity measure. The stability of each node was assessed with a bootstrap test of 1000 replicates. In order to explore which ecological classes are most responsible for separating assemblages, a principal components analysis (PCA) was performed utilizing a covariance matrix. The number of significant components was determined using the eigenvalue bootstrapping method described in Peres-Neto et al. [81], in which we used the minimum recommended number of 1000 replicates. All analyses were run with PAST v 1.85 [82].

RESULTS

Cluster Analysis

Each assemblage exhibited a distinct ecological profile (Table 3). Cluster analysis separated the assemblages into several groups (Figure 2). S1 was found distinct from all other assemblages, while C1 was the next most distinct. The remaining ten assemblages are split into two groups, with the North American assemblages (M1–M6) split evenly between them. A1 and C2 are associated with one of these groups, while the other contains both E1 and E2. However, support for most nodes does not surpass 50%, except for those linking M1, M5, and M6 (73%) and E2, M2, M3, and M4 (52%). The relationship of C2 to either of the large nodes appears especially uncertain given its very low support (5%). Due to its lack of relationship with any other assemblage, S1 was excluded from further analysis.

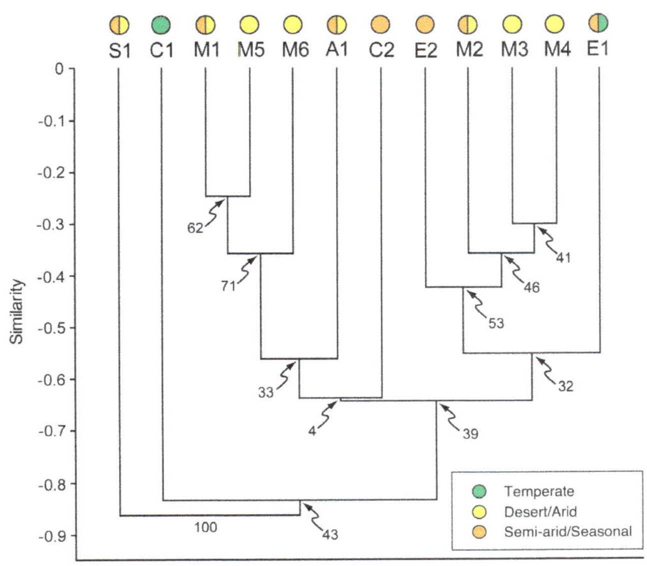

Figure 2. Results of cluster analysis (UPGMA with Euclidean similarity) for twelve Late Jurassic dinosaur fossil assemblages.

Numbers indicate support for node (bootstrap: N=1000). Labels refer to the fossil assemblage (see Table 1). Symbols denote the climate conditions assigned from the biome map. Two colors are shown if the biome assignment differs between the climate models.

Table 3. Percentages of taxa within ecological categories used in this study.

		Trophic Mode					Locomotor Mode				Body Size					
Assem.	**Total Taxa**	**C**	**HH**	**HI**	**HL**	**HG**	**TQ**	**TB**	**TF**	**TA**	**T**	**VS**	**S**	**M**	**L**	**VL**
E1	12	25.0	16.7	8.3	41.7	8.3	41.7	41.7	16.7	0.0	0.0	8.3	8.3	33.3	25.0	25.0
E2	24	37.5	12.5	25.0	20.8	4.2	50.0	41.7	4.2	4.2	4.2	12.5	8.3	16.7	29.2	29.2
M1	14	28.6	14.3	35.7	21.4	0.0	57.1	35.7	7.1	0.0	0.0	0.0	7.1	21.4	28.6	42.9
M2	22	36.4	9.1	27.3	18.2	9.1	45.5	50.0	4.5	0.0	0.0	9.1	22.7	18.2	13.6	36.4
M3	14	42.9	14.3	14.3	14.3	14.3	35.7	57.1	7.1	0.0	7.1	7.1	21.4	21.4	14.3	28.6
M4	20	40.0	10.0	20.0	25.0	5.0	45.0	50.0	5.0	0.0	5.0	5.0	15.0	25.0	20.0	30.0
M5	14	35.7	14.3	21.4	28.6	0.0	50.0	42.9	7.1	0.0	0.0	0.0	7.1	28.6	28.6	35.7
M6	10	30.0	20.0	30.0	20.0	0.0	60.0	30.0	10.0	0.0	0.0	0.0	0.0	20.0	30.0	50.0
C1	10	30.0	0.0	20.0	20.0	30.0	40.0	60.0	0.0	0.0	0.0	10.0	30.0	10.0	30.0	20.0
C2	11	27.3	9.1	27.3	27.3	9.1	63.6	36.4	0.0	0.0	0.0	9.1	0.0	27.3	27.3	36.4
S1	6	33.3	33.3	16.7	16.7	0.0	66.7	33.3	0.0	0.0	0.0	0.0	16.7	0.0	33.3	50.0
A1	11	45.5	9.1	27.3	18.2	0.0	45.5	54.5	0.0	0.0	0.0	0.0	18.2	36.4	18.2	27.3

Trophic Mode: C = carnivore, HH = high browser, HI = intermediate browser, HL = low browser, HG = ground foraging. Locomotor Mode: TQ = quadruped, TB = biped, TF = facultative biped, TA = arboreal biped. Body Size: VS = very small, S = small, M = medium, L = large, VL = very large. See text and Table 1 for explanation.

doi:10.1371/journal.pone.0012553.t003

Principal Components Analysis

Two significant principal components were recovered, which together account for 64.6% of the variance in the data (Figure 3). PC 1 accounts for 47.8% and PC 2 accounts for 16.8% of the variance, respectively. Positive values of PC 1 are associated with Ground Foragers, Small and Very Small sizes, and Bipeds, while negative values are associated with High Browsers, Very Large taxa, and Quadrupeds (Table 4). Positive values of PC 2 indicate a larger proportion of Very Small taxa while negative values suggest a greater proportion of Small taxa and Carnivores. The assemblages do not group tightly together but instead appear to fall along the continuum of PC 1, with members of the two major groups noted in the cluster analysis falling to either side of the origin. In order of increasing PC 1 score: M6, M1, M5, C2, A1, E1, E2, M4, M2, M3, and C1. Along PC 2 the greatest outliers are C2, E1, and A1. With the exception of assemblages M2, M3, and M4, each assemblage generally falls along PC 1 according to its biome. Assemblages with greater aridity have negative values, semi-arid assemblages remain near the origin, and assemblages representing moister environments are more positive.

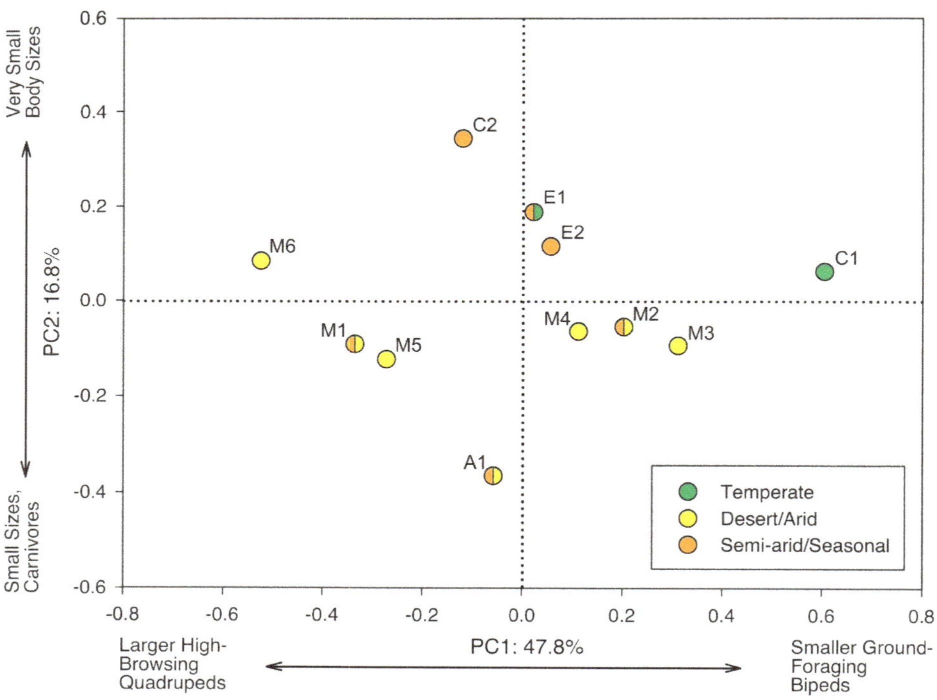

Figure 3. Results of Principal Components Analysis on twelve Late Jurassic dinosaur fossil assemblages.

Symbols denote the climate conditions assigned from the biome map. Two colors are shown if the biome assignment differs between the climate models. PC 1 corresponds to differences in the proportion of herbivore body sizes and feeding strategies, while PC 2 follows other environmental, taphonomic, and historical differences between the assemblages.

Table 4. Variable correlation values for the two major principal component axes.

Variable	Code	PC Axis 1	PC Axis 2
Carnivore	C	0.239	−0.7485
High Browser	HH	−0.7404	−0.01778
Intermediate Browser	HI	−0.4834	−0.183
Low Browser	HL	−0.2177	0.4712
Ground Forager	HG	0.8879	0.4218
Quadruped	TQ	−0.7707	0.3876
Biped	TB	0.864	−0.467
Facultative Biped	TF	−0.3563	0.07475
Arboreal Biped	TA	0.05912	0.2034
Tiny	T	0.3478	−0.07956
Very Small	VS	0.7331	0.5727
Small	S	0.8023	−0.545
Medium	M	−0.4491	−0.2363
Large	L	−0.3843	0.4739
Very Large	VL	−0.8327	0.06462

doi:10.1371/journal.pone.0012553.t004

DISCUSSION

Relationship with Climate Reconstructions

Based on the correlations between ecological categories and principal components, general characteristics are ascribed to the biota and climatic conditions of each assemblage based on its position in the component plot. PC 1 represents climatic differences between the assemblages: negative scores indicate aridity and positive scores more temperate conditions. PC 2 is discussed below.

Ultimately, assemblages did not group together by biome as expected. This discrepancy is related to 1) the different climate reconstructions used and 2) small-scale environmental variation between the localities comprising each assemblage. The biome assignments for each assemblage from Rees et al. [27] and Sellwood and Valdes [30] largely agree. Key differences occur in regions that are transitional zones between adjacent biomes. The actual boundaries between biomes are gradational and much less distinct and likely migrated in response to long-term shifts in climate patterns. This mainly affects the assessment of assemblages M1, M2, E1and A1, which are alternatively assigned semi-arid or desert/arid conditions (Table 1). In many of the assemblages, published environmental reconstructions differ from their assigned biome and may help interpret the results. These discrepancies point to major differences in the way climate models and ecological data reconstruct environmental conditions. Therefore, the model-assigned designations may not conform to the modern conception of these environments and instead relate to relative differences specific to the equable climate of the Late Jurassic [Foster, pers. comm.]. The conditions assessed for each assemblage are briefly discussed below.

Europe (E1 and E2).

Throughout the Late Jurassic western Europe became progressively more arid with a strongly seasonal, Mediterranean-type climate [28], [83]–[86]. E1 likely represents a drier environment, making it more similar to E2, and in accordance with the Rees et al. [27] biome model. This is supported by both the cluster analysis and PCA results, which place E1 and E2 close to the middle of the continuum. Both assemblages are therefore considered to have had a semi-arid climate.

North America (M1–M6).

The Morrison Fm. is reconstructed as a seasonally dry, savannah-like environment; much moister than both models, which indicate greater aridity. In addition these assemblages do not consistently group together. The spatial extent and topography of the Morrison depositional basin sheds some light on this pattern. Recent work suggests two spatial gradients exist in the Morrison Fm.: a south to north, arid to temperate climate gradient and an east to west precipitation/drainage gradient [87], [88]. These conditions created a greater proportion of lakes and wetlands towards the center of the depositional basin [89]. This work supports the observed division between Morrison Fm. assemblages, with M3 as a possible exception. M3 is expected to group with "drier" Morrison assemblages due to its southwestern position, although this relationship may not be resolvable with the current arrangement of assemblages and/or using dinosaurs only. Therefore, assemblages M2, M3 (tentatively), and M4 are considered to have been semi-arid or more seasonally wet, while M1, M5, and M6 were likely more arid or strongly seasonal.

Asia (C1 and C2).

Climate reconstructions for many formations in China indicate semi-arid and seasonal conditions [90]–[92], however Hallam [28], [84], [86] reconstructs eastern Eurasia as being moister than the west. Each biome model places C1 as a temperate assemblage and C2 as seasonally semi-arid. Both analyses find C1 is quite different from the other assemblages, which could be indicative of a temperate climate, although this is unlikely. C1 occurs close to the boundary making its biome assignment tenuous at best. Sedimentary indicators of seasonality further refute the temperate nature of this assemblage [91], [92]. Tectonic uplift throughout the Jurassic increased seasonality throughout the region, which was not accounted for in either biome model. Therefore C1 was likely semi-arid, but perhaps less seasonal or experienced more intense moist periods than other areas. C2, on the other hand, lies between the two major groupings making it semi-arid and more seasonal than C1.

Africa (A1).

The Tendaguru Fm. has been reconstructed with a semi-arid climate with coastal influences that maintained somewhat higher moisture levels than seen inland [93], supporting the biome model of Sellwood and Valdes [30]. The intermediate position of A1 in both analyses agrees with the assignment of a semi-arid climate to this assemblage.

South America (S1).

The region surrounding S1 may have shared a similar coastal semi-arid climate [94], [95, Rauhut, pers. comm.], but the number of dinosaur fossils from this area remains too sparse to allow a full comparison here. We will have to wait until further dinosaur fossils from the area are described.

Assemblage-Level Patterns

Results suggest that climatic and ecological factors played an important role in the distribution of Late Jurassic dinosaurs. In mammals, ecological preferences are often shared at the genus level or above [96], and the same was likely true of dinosaurs. The climatic and ecological similarity found here between North America, Europe, and southern Africa supports the biogeographic connections between these regions [5], [97], [98]. C2 shares relatively few taxa in common with the other semi-arid assemblages [5], [98], yet its general ecological similarity points to an ecosystem that evolved convergently under the same climatic conditions.

Generally, the faunas of the Morrison Fm. have been treated as a single unit, however it appears they were more ecologically diverse than previously thought, despite the relatively uniform distribution of dinosaur taxa [99], [100]. The ecological subdivisions present within the Morrison Fm. may also be related to its long history of intense study, which led to hundreds of described localities. This number is unmatched by the other assemblages. Future work may find further ecological subdivisions as more fossil localities are added in other regions.

C1, as the northernmost assemblage represented here, remains taxonomically and ecologically distinct. C1 may be like the central Morrison Fm. (M2–M4), maintaining greater moisture levels despite a seasonally semi-arid climate, or it may represent an entirely different semi-arid fauna that has yet to be encountered elsewhere.

PC2 appears related to assemblage-specific differences that contributed to the low bootstrap support in the cluster analysis. The wide distribution of scores among assemblages suggests PC2 cannot be directly attributed environmental variation. Instead, PC2 most likely represents some mixture of environmental, taphonomic, and sampling effects. For example, A1, E1, and C2 have high PC2 scores and also have a small number of taxa. Alternatively, this may be some relative measure of error in the spatial and/or temporal grouping of localities in an assemblage. Unfortunately the current database does not allow for more complex analysis, but the connection of PC2 to the ecology of these assemblages is certainly worthy of further inquiry.

Relationship with Dinosaur Paleoecology

The distribution of ecological categories varies accordingly with the proposed climatic differences between assemblages (Figure 4). Assemblages from more arid conditions (M1, M5, M6) tend to have a greater proportion of large-bodied taxa and lack anything smaller than 10 kg. High- and intermediate-browsing herbivorous sauropods and large carnivorous theropods such as *Allosaurus* dominate these assemblages. Smaller carnivores and ground-foraging herbivores are rare or absent. This pattern may reflect a lack of low-lying vegetation for a period of the season that restricted the numbers of smaller dinosaurs. Larger dinosaurs, on the other hand, were better able to cope with lower resource density and quality [101]–[104]. Assemblages representing semiarid or seasonally wet conditions usually contain representatives from each size class, though with no discernable decrease in the proportion of larger size classes. Ground Foragers are present as a larger portion of the herbivore population and High Browsers are less prevalent, indicating that more ground cover was likely available capable of supporting a greater diversity of herbivores. With increasingly moist conditions, assemblages exhibit greater proportions of Ground Foragers and Low- and Intermediate Browsers. In the extreme case of C1 High Browsers are conspicuously absent, reflecting perhaps a lack of suitable habitat or resources.

Figure 4. Schematic representation of variation in herbivore types and inferred habitat structure along PC 1.

Relatively arid conditions occur to the left (negative values), where very large, high-browsing herbivores dominate among sparser foliage. Assemblages with semi-arid/seasonal conditions are towards the center (low values), which includes a greater diversity of feeding modes, including high, intermediate, and low browsers among increased ground cover. To the right are more moist conditions (positive values), where smaller, ground-foraging herbivores are more prevalent within a more densely vegetated environment. Carnivorous theropods appear largely independent of this pattern. Green=high browser, orange=intermediate browser, blue=low browser, red=ground forager.

The environmental dependence of certain ecological categories demonstrates the importance of habitat structure in driving the relative abundance of dinosaur herbivore guilds. As observed in modern ecosystems, this is due to differences in habitat structure and resource availability. In the PCA results none of the herbivore classes are strongly positively correlated with each other. When the proportion of one herbivore class is high, the remaining classes are typically lower, reflecting the varying ecological roles each class plays in different environments or their differing habitat requirements. A similar trend is found among the herbivorous dinosaurs of the Morrison Formation [22], although this pattern occurs across different depositional environments instead of biomes.

Arid climates typically host sparse, open environments more suited for the largest animals. Small herbivores may suffer not only from a lack of food, but suitable cover from predators. Increasing moisture levels lead to more abundant growth, which inhibits the largest herbivores through changes to habitat structure and/or resource distribution, while becoming more favorable for smaller herbivores. The overall proportion of carnivores appears largely independent from environmental conditions although the absence of smaller predators in arid environments may be related to the lack of appropriately-sized prey and/or successful competitive exclusion by larger predators.

Limitations of the Current Study

Despite the encouraging results, it is important to note some caveats. First, not all formations and regions have been equally explored. Historically, excavations have been conducted in North America and Europe more extensively than elsewhere. Spectacular discoveries from China in recent decades show that major strides in the number of vertebrate localities from a region can be accomplished rapidly. The attention now afforded to Africa and South America will hopefully yield similar results in the future.

The role of taphonomic factors presents the greatest uncertainty because its ultimate effect may depend on the scale of observation. Environment-specific taphonomic filtering may drive taxonomic variation between individual localities; aggregating the localities into assemblages minimizes this effect. However, at larger spatial scales climatic processes, especially the onset of arid conditions, affect vertebrate preservation [25], [31], [91], [105]. Under such conditions small size classes should be underrepresented in the fossil sample, potentially biasing the ecological profiles towards larger size classes with increasing aridity. The addition of vertebrate microfossil localities to an assemblage may help overcome this problem because these sites better reflect the community structure of the surrounding landscape, including both large and small taxa [106]. Dinosaur microfossil sites are found in the Morrison (M1–M4) and Camadas de Guimarota (E2) Formations, but are absent elsewhere [22], [97], [100]. Dinosaur microfossil remains usually consist of small theropod teeth, only some of which are assignable to useable taxa (see Materials and Methods and Table S1). The relationship between some assemblages change with removal of tooth taxa, but the climatic associations noted above remain more or less the same, indicating such localities are not necessary to assign climatic conditions but are useful is resolving ecological relationships among assemblages. Nevertheless, the role of microfossil localities and taphonomic filtering requires further scrutiny.

In addition to addressing taphonomic biases, it was necessary to group localities in order to achieve a minimum sample size for analysis. The relationships of assemblages with fewer than 10 taxa were found to be unresolvable, as in the case of S1. A small number of taxa skews the content of the categories since class data are calculated as a proportion of the total number of taxa and 10 is suggested here as a minimum sample for this type of study. While necessary, grouping spatially and stratigraphically distinct localities in this way increases the likelihood of including taxa and environments that never coexisted in life. Each assemblage therefore represents a coarse average of ecological conditions. Multiple studies have found overall taxonomic stability of the Morrison fauna through time [22], [99], [100]. The Tendaguru fauna of Africa was also similarly stable through the Late Jurassic [93]. The majority of formations included here lack such detailed biostratigraphic study. Paleoecological studies across a broad spectrum of scales, environments, and taxonomic groups have found that many past ecosystems maintained a stable structure over timescales of 100 ky to 2 my or more, despite major taxonomic turnovers or climatic events [107]–[113]. In this case, resolution of small-scale ecological differences are lost that may lead to interpretations of individual fossil localities disagreeing with the general results presented here.

A continuing challenge that requires more attention regards the assignment of ecological classes to dinosaurian taxa. In many cases the ecological niche of a dinosaur is still assessed using qualitative comparisons with living forms. Even with complex biomechanical models, very different interpretations of ecologically relevant morphological and behavioral reconstructions continue to arouse debate [114], [115]. This problem becomes more acute in taxa based on incomplete and fragmentary remains. Altering one or more classes for a single taxon has little effect, though more than these can alter the results. As always, new fossil finds may lead us to radically revise our view on the ecology of certain taxa. If multiple interpretations do exist separate analyses should be run using all combinations of interpretations to assess their effect on the results.

CONCLUSIONS

Our study demonstrates that ESA is a useful tool for quantifying ecological differences between Late Jurassic dinosaur assemblages. The grouping of climatically similar assemblages supports the preservation of ecological structure at large scales and helped assess the accuracy of two different paleoclimate models. Ecological similarities are most likely related to differences in habitat structure due to variation in moisture availability; most important among these are the relative proportions of herbivore and body size classes in an assemblage. Not only can these proportions be used as additional climate indicators, but also may provide evidence of ecosystem convergence when comparing taxonomically distinct dinosaur localities. Although most Late Jurassic fossil localities come from relatively arid or semi-arid environments [31], there is a surprising amount of ecological variation that warrants further study.

A more comprehensive understanding of broad-scale ecological patterns is important in understanding the effect of climate patterns on dinosaur ecology and evolution. Perhaps more importantly, this study provides a framework for studying the long-term evolutionary dynamics of terrestrial communities related to climate change, major adaptive radiations, or evolutionary events (e.g., angiosperm evolution) and whether these events had a significant impact on the subsequent structure of vertebrate communities. These data can then be used to test hypotheses related to community formation and ecosystem function. A great deal of effort has been spent quantifying global taxonomic diversity levels through the Phanerozoic [116], [117], however these estimates tell us little about the ecological factors responsible for producing that diversity, nor how it was distributed on the surface [118].

SUPPORTING INFORMATION

List of Late Jurassic dinosaur taxa and their assigned ecological categories used in this analysis, organized by assemblage. Diet: C = carnivore, HH = high browser, HI = intermediate browser, HL = low browser, HG = ground foraging. Locomotion: TQ = quadruped, TB = biped, TF = facultative biped, TA = arboreal biped. Body Mass: VS = very small, S = small, M = medium, L = large, VL = very large. See text and Table 1 for explanation. "PBDB Collection #" refers to the collection number of the specimen in the Paleobiology Database. Formations given in the table do not necessarily reflect all units in which the taxon is found. Data come from [51] and [52].

Table S1.: List of Late Jurassic dinosaur taxa and their assigned ecological categories used in this analysis, organized by assemblage. Diet: C=carnivore, HH=high browser, HI=intermediate browser, HL=low browser, HG=ground foraging. Locomotion: TQ=quadruped, TB=biped, TF=facultative biped, TA=arboreal biped. Body Mass: VS=very small, S=small, M=medium, L=large, VL=very large. See text and Table 1 for explanation. "PBDB Collection #" refers to the collection number of the specimen in the Paleobiology Database. Formations given in the table do not necessarily reflect all units in which the taxon is found. Data come from [51] and [52].

PBDB Collection#	Region	Country	Formation	Taxon Name	Trophic Mode	Locomotor Mode	Body Size
83734	South America	Chile	Toqui	Large theropod	C	TB	L
83734	South America	Chile	Toqui	Small theropod	C	TB	S
51938	South America	Argentina	Cañadón Cálcereo	Tehuelchesaurus	HI	TQ	VL
51287	South America	Argentina	Cañadón Cálcereo	Brachytrachelopan	HL	TQ	L
67425	South America	Argentina	Cañadón Cálcereo	Camarasauridae indet.	HH	TQ	VL
67425	South America	Argentina	Cañadón Cálcereo	Brachiosauridae indet.	HH	TQ	VL
47159	Africa	Tanzania	Tendaguru	Allosauridae indet. (includes "Allosaurus"? tendagurensis)	C	TB	L
47151	Africa	Tanzania	Tendaguru	Tetanurae indet. (=Coelurosaurier A)	C	TB	S
47151	Africa	Tanzania	Tendaguru	Abelisauroid indet. (=Coelurosaurier B, C)	C	TB	S
47151	Africa	Tanzania	Tendaguru	Ceratosaurus	C	TB	M
47153	Africa	Tanzania	Tendaguru	Elaphrosaurus	C	TB	M
47151	Africa	Tanzania	Tendaguru	Giraffatitan (=Brachiosaurus)	HH	TQ	VL
47151	Africa	Tanzania	Tendaguru	Dicraeosaurus	HI	TQ	L
48349	Africa	Tanzania	Tendaguru	Australodocus	HI	TQ	VL
47181	Africa	Tanzania	Tendaguru	Tornieria	HI	TQ	VL
47146	Africa	Tanzania	Tendaguru	Dryosaurus	HL	TB	M
47151	Africa	Tanzania	Tendaguru	Kentrosaurus	HL	TQ	M
47056	China	Xinjiang	Shishugou	Ankylosauria indet.	HL	TQ	L
93194	China	Xinjiang	Shishugou	Eugongbusaurus	HG	TB	VS
58489	China	Xinjiang	Shishugou	Guanlong	C	TB	M

in PBDB but no number	China	Xinjiang	Shishugou	Haplocheirus	C	TB	S
45172	China	Xinjiang	Kalazha	Hudiesaurus	HI	TQ	VL
91319	China	Xinjiang	Shishugou	Jiangjunosaurus	HL	TQ	L
90081	China	Xinjiang	Shishugou	Limusaurus	HG	TB	S
36569	China	Xinjiang	Shishugou	Mamenchisaurus	HI	TQ	VL
45192	China	Xinjiang	Shishugou	Sinraptor	C	TB	L
72107	China	Xinjiang	Shishugou	Yinlong	HG	TB	S
60067	China	Sichuan	Shangshaximiao	Chungkingosaurus	HL	TQ	M
89050	China	Sichuan	Shangshaximiao	Daanosaurus	HH	TQ	VL
67236	China	Sichuan	Shangshaximiao	Gigantspinosaurus	HL	TQ	M
60066	China	Sichuan	Shangshaximiao	Hypsilophodontidae	HG	TB	VS
92950	China	Sichuan	Shangshaximiao	Leshansaurus	C	TB	M
53662	China	Sichuan	Shangshaximiao	Mamenchisaurus	HI	TQ	VL
57771	China	Sichuan	Shangshaximiao	Omeisaurus	HI	TQ	VL
28234	China	Sichuan	Shangshaximiao	Shunosaurus	HI	TQ	VL
89156	China	Sichuan	Shangshaximiao	Sinraptor	C	TB	L
51914	China	Sichuan	Shangshaximiao	Tuojiangosaurus	HL	TQ	L
49790	China	Sichuan	Shangshaximiao	Yangchuanosaurus	C	TB	L
74916	Europe	France	Sables de Glos	Camarasauridae	HH	TQ	VL
26790	Europe	United Kingdom	Kimmeridge Clay	Camptosaurus	HL	TF	M
53966	Europe	France	Unnamed	Cetiosaurus	HI	TQ	VL
47420	Europe	France	Argiles d'Octeville	Dacentrurus	HL	TQ	L
35968	Europe	France	Marnes de Bléville	Dryosaurus	HL	TB	M
75743	Europe	United Kingdom	Kimmeridge Clay	Heterodontosauridae indet. (=Echinodon)	HG	TB	VS
65033	Europe	France	Unnamed	Iguanodon	HL	TF	M
47363	Europe	United Kingdom	Calcareous Grit	Ankylosauria indet. (=Priodontognathus)	HL	TQ	M
91991	Europe	United Kingdom	Corallian Oolite	Megalosaurus	C	TB	L
52556	Europe	United Kingdom	Oxford Clay	Metriacanthosaurus	C	TB	L
26795	Europe	United Kingdom	Kimmeridge Clay	Ornithopsis	HH	TQ	VL
48688	Europe	United Kingdom	Kimmeridge Clay	Stokesosaurus	C	TB	S
38680	Europe	Portugal	Camadas de Guimarota	Allosaurus	C	TB	L
24563	Europe	Portugal	Camadas de Guimarota	Archaeopterygiformes indet.	C	TA	T
24563	Europe	Portugal	Camadas de Guimarota	Aviatyrannis	C	TB	VS
38680	Europe	Portugal	Camadas de Guimarota	Camarasaurus	HH	TQ	VL
38680	Europe	Portugal	Camadas de Guimarota	Camptosauridae	HL	TF	M
49264	Europe	Portugal	Camadas de Guimarota	Ceratosaurus	C	TB	M

24563	Europe	Portugal	Camadas de Guimarota	Compsognathus	C	TB	VS
75968	Europe	Portugal	Camadas de Guimarota	Dacentrurus	HL	TQ	L
49304	Europe	Portugal	Camadas de Guimarota	Dinheirosaurus	HI	TQ	VL
38680	Europe	Portugal	Camadas de Guimarota	Diplodocoidea indet.	HI	TQ	VL
49262	Europe	Portugal	Unidade Bombarral	Draconyx	HL	TF	M
47364	Europe	Portugal	Freixial	Dracopelta	HL	TQ	M
38680	Europe	Portugal	Camadas de Guimarota	Dryosauridae indet.	HL	TB	M
53890	Europe	Spain	Villar del Arzobispo	Galveosaurus	HI	TQ	L
N/A	Europe	Portugal	Camadas de Guimarota	?Hypsilophodon	HG	TB	VS
39423	Europe	Spain	Unnamed	Losillasaurus	HI	TQ	VL
49261	Europe	Portugal	Unidade Sobral	Lourinhanosaurus	C	TB	L
49229	Europe	Portugal	Unidade Sobral	Lourinhasaurus	HI	TQ	VL
49256	Europe	Portugal	Unidade Sobral	Lusotitan	HH	TQ	VL
87089	Europe	Portugal	Unidade Sobral	Miragaia	HI	TQ	L
49229	Europe	Portugal	Unidade Sobral	Ornithopsis	HH	TQ	VL
73610	Europe	Portugal	Camadas de Guimarota	Stegosaurus	HL	TQ	L
68904	Europe	Portugal	Unidade Bombarral	Torvosaurus	C	TB	L
24563	Europe	Portugal	Camadas de Guimarota	Troodontidae indet.	C	TB	S
24563	Europe	Portugal	Camadas de Guimarota	Velociraptorinae indet.	C	TB	S
58509	United States	Wyoming	Morrison	Allosaurus	C	TB	L
58510	United States	Wyoming	Morrison	Apatosaurus	HI	TQ	VL
22678	United States	Wyoming	Morrison	Barosaurus	HI	TQ	VL
58507	United States	Wyoming	Morrison	Brachiosauridae	HH	TQ	VL
22650	United States	Wyoming	Morrison	Camarasaurus	HH	TQ	VL
22678	United States	Wyoming	Morrison	Camptosaurus	HL	TF	M
58509	United States	Wyoming	Morrison	Diplodocus	HI	TQ	VL
46465	United States	Wyoming	Morrison	Dryosaurus	HL	TB	M
58510	United States	Wyoming	Morrison	Ceratosauria indet. (= ?Elaphrosaurus)	C	TB	M
58510	United States	Wyoming	Morrison	Haplocanthosaurus	HI	TQ	VL
22678	United States	Wyoming	Morrison	Stegosaurus	HL	TQ	L
38866	United States	Montana	Morrison	Suuwassea	HI	TQ	L
62561	United States	Wyoming	Morrison	Torvosaurus	C	TB	L

49631	United States	South Dakota	Morrison	Tyrannosauroidea (=Stokesosaurus)	C	TB	S
57903	United States	Wyoming	Morrison	Allosaurus	C	TB	L
57903	United States	Wyoming	Morrison	Apatosaurus	HI	TQ	VL
39342	United States	Wyoming	Morrison	Barosaurus	HI	TQ	VL
22648	United States	Wyoming	Morrison	Brachiosaurus	HH	TQ	VL
57903	United States	Wyoming	Morrison	Camarasaurus	HH	TQ	VL
22648	United States	Wyoming	Morrison	Camptosaurus	HL	TF	M
49577	United States	Wyoming	Morrison	Ceratosaurus	C	TB	M
22648	United States	Wyoming	Morrison	Cetiosauridae indet.	HI	TQ	VL
39253	United States	Wyoming	Morrison	Coelurus	C	TB	S
57903	United States	Wyoming	Morrison	Diplodocus	HI	TQ	VL
56654	United States	Wyoming	Morrison	Drinker	HG	TB	VS
49577	United States	Wyoming	Morrison	Dryosaurus	HL	TB	M
49577	United States	Wyoming	Morrison	Gargoyleosaurus	HL	TQ	M
48837	United States	Wyoming	Morrison	Haplocanthosaurus	HI	TQ	VL
22648	United States	Wyoming	Morrison	Ornitholestes	C	TB	S
12816	United States	Wyoming	Morrison	Othnielosaurus	HG	TB	VS
38884	United States	Wyoming	Morrison	Stegosaurus	HL	TQ	L
12816	United States	Wyoming	Morrison	Stokesosaurus	C	TB	S
74664	United States	Wyoming	Morrison	Supersaurus	HI	TQ	VL
49577	United States	Wyoming	Morrison	Tanycolagreus	C	TB	S
49628	United States	Wyoming	Morrison	Torvosaurus	C	TB	L
74665	United States	Wyoming	Morrison	Troodontidae	C	TB	S
47196	United States	Utah	Morrison	Allosaurus	C	TB	L
58514	United States	Utah	Morrison	Apatosaurus	HI	TQ	VL
71476	United States	Colorado	Morrison	Brachiosaurus	HH	TQ	VL
47196	United States	Utah	Morrison	Camarasaurus	HH	TQ	VL
47196	United States	Utah	Morrison	Camptosaurus	HL	TF	M
64053	United States	Utah	Morrison	Ceratosaurus	C	TB	M

46439	United States	Utah	Morrison	Coelurus	C	TB	S
46437	United States	Utah	Morrison	Diplodocus	HI	TQ	VL
92360	United States	Colorado	Morrison	Fruitadens	HG	TB	T
46439	United States	Utah	Morrison	Marshosaurus	C	TB	M
47196	United States	Utah	Morrison	Othnielosaurus	HG	TB	VS
49626	United States	Utah	Morrison	Stegosaurus	HL	TQ	L
46439	United States	Utah	Morrison	Stokesosaurus	C	TB	S
46439	United States	Utah	Morrison	Tanycolagreus	C	TB	S
49614	United States	Utah	Morrison	Allosaurus	C	TB	L
45975	United States	Colorado	Morrison	Ankylosauria indet.	HL	TQ	L
11611	United States	Colorado	Morrison	Apatosaurus	HI	TQ	VL
11611	United States	Colorado	Morrison	Barosaurus	HI	TQ	VL
56683	United States	Colorado	Morrison	Brachiosaurus	HH	TQ	VL
11611	United States	Colorado	Morrison	Camarasaurus	HH	TQ	VL
21852	United States	Utah	Morrison	Camptosaurus	HL	TF	M
11611	United States	Colorado	Morrison	Ceratosaurus	C	TB	M
45975	United States	Colorado	Morrison	Coelurus	C	TB	S
68097	United States	Colorado	Morrison	Diplodocus	HI	TQ	VL
21852	United States	Utah	Morrison	Dryosaurus	HL	TB	M
45975	United States	Colorado	Morrison	Maniraptora	C	TB	T
N/A	United States	Colorado	Morrison	cf. Echinodon	HG	TB	VS
55350	United States	Utah	Morrison	Marshosaurus	C	TB	M
11611	United States	Colorado	Morrison	Mymoorapelta	HL	TQ	M
45975	United States	Colorado	Morrison	Ornitholestes	C	TB	S
21852	United States	Utah	Morrison	Stegosaurus	HL	TQ	L
45975	United States	Colorado	Morrison	Stokesosaurus	C	TB	S
45975	United States	Colorado	Morrison	Supersaurus	HI	TQ	VL
21852	United States	Utah	Morrison	Torvosaurus	C	TB	L
70027	United States	Colorado	Morrison	Allosaurus	C	TB	L

48161	United States	Colorado	Morrison	Ankylosauria indet.	HL	TQ	L
13341	United States	Colorado	Morrison	Apatosaurus	HI	TQ	VL
28366	United States	Colorado	Morrison	Brachiosaurus	HH	TQ	VL
48129	United States	Colorado	Morrison	Camarasaurus	HH	TQ	VL
28366	United States	Colorado	Morrison	Camptosaurus	HL	TF	M
28366	United States	Colorado	Morrison	Ceratosaurus	C	TB	M
28366	United States	Colorado	Morrison	Coelurus	C	TB	S
39376	United States	Colorado	Morrison	Diplodocus	HI	TQ	VL
48161	United States	Colorado	Morrison	Dryosaurus	HL	TB	M
48161	United States	Colorado	Morrison	Ceratosauria indet (= Elaphrosaurus sp.)	C	TB	M
48155	United States	Colorado	Morrison	Haplocanthosaurus	HI	TQ	VL
28366	United States	Colorado	Morrison	Stegosaurus	HL	TQ	L
48171	United States	Colorado	Morrison	Torvosaurus	C	TB	L
49596	United States	Okla-homa	Morrison	Allosaurus	C	TB	L
38867	United States	New Mexico	Morrison	Apatosaurus	HI	TQ	VL
49338	United States	Okla-homa	Morrison	Barosaurus	HI	TQ	VL
49338	United States	Okla-homa	Morrison	Brachiosaurus	HH	TQ	VL
70030	United States	Colorado	Morrison	Camarasaurus	HH	TQ	VL
49597	United States	Okla-homa	Morrison	Camptosaurus	HL	TF	M
49338	United States	Okla-homa	Morrison	Ceratosaurus	C	TB	M
49338	United States	Okla-homa	Morrison	Diplodocus	HI	TQ	VL
49338	United States	Okla-homa	Morrison	Saurophaganax	C	TB	L
49338	United States	Okla-homa	Morrison	Stegosaurus	HL	TQ	L

ACKNOWLEDGMENTS

Paleocoordinates for each locality were made available through the Paleobiology Database (http://paleodb.org) utilizing the paleogeographic reconstructions of Chris Scotese (University of Texas, Arlington). Oliver Rauhut (Bayerische Staatssammlung für Paläontologie und Geologie) provided important information on the Cañadón Calcáreo Formation. Thanks go to John Foster (Museum of Western Colorado) and John Fleagle, (Stony Brook University), whose insightful comments helped significantly improve earlier versions of this manuscript. Catherine Forster (George Washington University), Mike Bell (Stony Brook University), Louis Jacobs (Southern Methodist University), Chris Scotese, and John Foster provided important discussion and encouragement. The final version of this manuscript was significantly improved by the feedback of three anonymous reviewers.

AUTHOR CONTRIBUTIONS

Conceived and designed the experiments: CRN. Performed the experiments: CRN. Analyzed the data: CRN AG. Wrote the paper: CRN AG.

REFERENCES

1. Schiermeier Q (2003) Palaeobiology: Setting the record straight. Nature 424: 482–483.
2. Rees PM, Noto CR (2005) A new online database of dinosaur distributions. Journal of Vertebrate Paleontology 25: 103A.
3. Alroy J (2003) Global databases will yield reliable measures of global biodiversity. Paleobiology 29: 26–29.
4. Russell DA (1993) The role of Central Asia in dinosaurian biogeography. Canadian Journal of Earth Sciences 30: 2002–2012.
5. Holtz TR, Chapman RE, Lamanna MC (2004) Mesozoic biogeography of Dinosauria. In: Weishampel DB, Dodson P, Osmolska H, editors. The Dinosauria, second edition. Berkeley: University of California Press. pp. 627–642.
6. Forster CA (1999) Gondwanan dinosaur evolution and biogeographic analysis. Journal of African Earth Sciences 28: 169–185.
7. Upchurch P, Hunn CA, Norman DB (2002) An analysis of dinosaurian biogeography: evidence for the existence of vicariance and dispersal patterns caused by geological events. Proceedings of the Royal Society of London Series B-Biological Sciences 269: 613–621.
8. Sereno PC (1999) Dinosaurian biogeography: Vicariance, dispersal, and regional extinction. In: Tomida Y, Rich TH, Vickers-Rich P, editors. Proceedings of the Second Gondwanan Dinosaur Symposium. Tokyo: National Science Museum Monographs 15. pp. 249–257.
9. McCarthy D (2005) Biogeographical and geological evidence for a smaller, completely-enclosed Pacific Basin in the Late Cretaceous. Journal of Biogeography 32: 2161–2177.
10. Barrett PM, Hasegawa Y, Manabe M, Isaji S, Matsuoka H (2002) Sauropod dinosaurs from the Lower Cretaceous of Eastern Asia: Taxonomic and biogeographical implications. Palaeontology 45: 1197–1217.
11. Van Valkenburgh B, Molnar RE (2002) Dinosaurian and mammalian predators compared. Paleobiology 28: 527–543.
12. Carrano MT, Janis CM, Sepkoski JJ (1999) Hadrosaurs as ungulate parallels: Lost lifestyles and deficient data. Acta Palaeontologica Polonica 44: 237–261.
13. Currie PJ (1989) Long distance dinosaurs. Natural History 6: 60–65.
14. Bakker RT (1972) Anatomical and ecological evidence of endothermy in dinosaurs. Nature 238: 81–&.
15. Hopson JA (1975) The evolution of cranial display structures in hadrosaurian dinosaurs. Paleobiology 1: 21–43.
16. Seebacher F, Grigg GC, Beard LA (1999) Crocodiles as dinosaurs: Behavioural thermoregulation in very large ectotherms leads to high and stable body temperatures. Journal of Experimental Biology 202: 77–86.
17. Bailey JB (1997) Neural spine elongation in dinosaurs: Sailbacks or buffalo-backs? Journal of Paleontology 71: 1124–1146.
18. Geist NR, Jones TD (1996) Juvenile skeletal structure and the reproductive habits of dinosaurs. Science 272: 712–714.
19. Abler WL (1992) The serrated teeth of tyrannosaurid dinosaurs, and biting structures in other animals. Paleobiology 18: 161–183.
20. Farlow JO, Dodson P (1975) Behavioral significance of frill and horn morphology in ceratopsian dinosaurs. Evolution 29: 353–361.
21. Farlow JO, Dodson P, Chinsamy A (1995) Dinosaur biology. Annual Review of Ecology and Systematics 26: 445–471.
22. Foster JR (2003) Paleoecological analysis of the vertebrate fauna of the Morrison Formation (Upper Jurassic), Rocky Mountain Region, U.S.A. Bulletin of the New Mexico Museum of Natural History and Science 23: 1–95.
23. Barnosky AD, Hadly EA, Maurer BA, Christie MI (2001) Temperate terrestrial vertebrate faunas in North and South America: Interplay of ecology, evolution, and geography with biodiversity. Conservation Biology 15: 658–674.
24. Graham RW, Lundelius EL, Graham MA, Schroeder EK, Toomey RS, et al. (1996) Spatial response of mammals to late Quaternary environmental fluctuations. Science 272: 1601–1606.
25. Smith RMH, Swart R (2002) Changing fluvial environments and vertebrate taphonomy in response to climatic drying in a mid-Triassic rift valley fill: The Omingonde Formation (Karoo Supergroup) of Central Namibia. Palaios 17: 249–267.
26. Wendler J, Graefe K-U, Willems H (2002) Reconstruction of mid-Cenomanian orbitally forced palaeoenvironmental changes based on calcareous dinoflagellate cysts. Palaeogeography, Palaeoclimatology, Palaeoecology 179: 19–41.
27. Rees PM, Ziegler AM, Valdes PJ (2000) Jurassic phytogeography and climates: new data and model comparisons. In: Huber BT, MacLeod KG, Wing SL, editors. Warm Climates in Earth History. Cambridge, England: University of Cambridge Press. pp. 297–318.
28. Hallam A (1993) Jurassic climates as inferred from the sedimentary and fossil record. Philosophical Transactions of the Royal Society of London Series B: Biological Sciences 341: 287–296.
29. Hallam A (1998) The determination of Jurassic environments using palaeoecological methods. Bulletin De La Societe Geologique De France 169: 681–687.
30. Sellwood BW, Valdes PJ (2008) Jurassic climates. Proceedings of the Geologists' Association 119: 5–17.
31. Rees PM, Noto CR, Parrish JM, Parrish JT (2004) Late Jurassic climates, vegetation, and dinosaur distributions. Journal of Geology 112: 643–654.
32. Fastovsky DE, Huang Y, Hsu J, Martin-McNaughton J, Sheehan PM, et al. (2004) Shape of Mesozoic dinosaur richness. Geology 32: 877–880.

33. Ricklefs RE (2006) Evolutionary diversification and the origin of the diversity-environment relationship. Ecology 87: 3–13.
34. Harmon LJ, Kolbe JJ, Cheverud JM, Losos JB (2005) Convergence and the multidimensional niche. Evolution 59: 409–421.
35. Melville J, Harmon LJ, Losos JB (2006) Intercontinental community convergence of ecology and morphology in desert lizards. Proceedings of the Royal Society B: Biological Sciences 273: 557–563.
36. Lamouroux N, Poff NL, Angermeier PL (2002) Intercontinental convergence of stream fish community traits along geomorphic and hydraulic gradients. Ecology 83: 1792–1807.
37. Plotnick RE, Baumiller TK (2000) Invention by evolution: Functional analysis in paleobiology. Paleobiology 26: 305–323.
38. Taylor M, Wedel M, Naish D (2009) Head and neck posture in sauropod dinosaurs inferred from extant animals. Acta Palaeontologica Polonica 54: 213–220.
39. Rayfield EJ (2005) Aspects of comparative cranial mechanics in the theropod dinosaurs Coelophysis, Allosaurus and Tyrannosaurus. Zoological Journal of the Linnean Society 144: 309–316.
40. Manning PL, Payne D, Pennicott J, Barrett PM, Ennos RA (2006) Dinosaur killer claws or climbing crampons? Biology Letters 2: 110–112.
41. Hutchinson JR, Gatesy SM (2006) Dinosaur locomotion: Beyond the bones. Nature 440: 292–294.
42. Stevens KA, Parrish JM (2005) Neck posture, dentition, and feeding strategies in Jurassic sauropod dinosaurs. In: Tidwell V, Carpenter K, editors. Thunder-Lizards: The Sauropodomorph Dinosaurs. Bloomington: Indiana University Press. pp. 212–232.
43. Bonnan MF (2004) Morphometric analysis of humerus and femur shape in Morrison sauropods: Implications for functional morphology and paleobiology. Paleobiology 30: 444–470.
44. Magurran AE (1988) Ecological diversity and its measurement. Princeton: Princeton University Press. 179 p.
45. Miller ER (1999) Faunal correlation of Wadi Moghara, Egypt: Implications for the age of Prohylobates tandyi. Journal of Human Evolution 36: 519–533.
46. Nakaya H (1994) Faunal change of Late Miocene Africa and Eurasia: Mammalian fauna from the Namurungule Formation, Samburu Hills, Northern Kenya. African Study Monographs Supplementary Issue 1–112.
47. Andrews P (1996) Palaeoecology and hominoid paleoenvironments. Biological Reviews of the Cambridge Philosophical Society 71: 257–300.
48. Reed KE (1996) The paleoecology of Makapansgat and other African Plio-Pleistocene hominid localities [PhD]. Stony Brook: Stony Brook University. 624 p.
49. Reed KE (1998) Using large mammal communities to examine ecological and taxonomic structure and predict vegetation in extant and extinct assemblages. Paleobiology 24: 384–408.
50. Reed KE (2002) The use of paleocommunity and taphonomic studies in reconstructing primate behavior. In: Plavcan JM, editor. Reconstructing Behavior in the Primate Fossil Record. New York: Kluwer Academic/Plenum Publishers. pp. 217–259.
51. The Paleobiology Database (2010) http://paleodb.org. Accessed March 15, 2010.
52. Weishampel DB, Le Loeuff J, Coria RA, et al. (2004) Dinosaur distribution. In: Weishampel DB, Dodson P, Osmolska H, editors. The Dinosauria, second edition. Berkeley: University of California Press. pp. 517–606.
53. White CR, Phillips NF, Seymour RS (2006) The scaling and temperature dependence of vertebrate metabolism. Biology Letters 2: 125–127.
54. Loeuille N, Loreau M (2006) Evolution of body size in food webs: Does the energetic equivalence rule hold? Ecology Letters 9: 171–178.
55. Marquet PA, Quinones RA, Abades S, Labra F, Tognelli M, et al. (2005) Scaling and power-laws in ecological systems. Journal of Experimental Biology 208: 1749–1769.
56. Brown JH, Gillooly JF, Allen AP, Savage VM, West GB (2004) Toward a metabolic theory of ecology. Ecology 85: 1771–1789.
57. Greenwood JJD, Gregory RD, Harris S, Morris PA, Yalden DW (1996) Relations between abundance, body size and species number in British birds and mammals. Philosophical Transactions of the Royal Society of London Series B-Biological Sciences 351: 265–278.
58. Fleagle JG, Mittermeier RA (1980) Locomotor behavior, body size, and comparative ecology of seven Surinam monkeys. American Journal of Physical Anthropology 52: 301–314.
59. Milton K, May ML (1976) Body weight, diet, and home range area in primates. Nature 259: 459–462.
60. Christiansen P (2002) Mass allometry of the appendicular skeleton in terrestrial mammals. Journal of Morphology 251: 195–209.
61. Iossa G, Soulsbury CD, Baker PJ, Harris S (2008) Body mass, territory size, and life-history tactics in a socially monogamous canid, the red fox Vulpes vulpes. Journal of Mammalogy 89: 1481–1490.
62. Fa JE, Purvis AP (1997) Body size, diet and population density in Afrotropical forest mammals: A comparison with Neotropical species. The Journal of Animal Ecology 66: 98–112.
63. Seebacher F (2001) A new method to calculate allometric length-mass relationships of dinosaurs. Journal of Vertebrate Paleontology 21: 51–60.
64. Upchurch P, Barrett PM (2000) The evolution of sauropod feeding mechanisms. In: Sues H-D, editor. Evolution of herbivory in terrestrial vertebrates: Perspectives from the fossil record. Cambridge: Cambridge University Press. pp. 79–122.
65. Fiorillo AR (1998) Dental microwear patterns from the sauropod dinosaurs Camarasaurus and Diplodocus: Evidence for resource partitioning in the Late Jurassic of North America. Historical Biology 13: 1–16.

66. Wright JL (1999) Ichnological evidence for the use of the forelimb in iguanodontid locomotion. pp. 209–219. Cretaceous Fossil Vertebrates.

67. Christian A, Preuschoft H (1996) Deducing the body posture of extinct large vertebrates from the shape of the vertebral column. Palaeontology 39: 801–812.

68. Heinrich RE, Ruff CB, Weishampel DB (1993) Femoral ontogeny and locomotor biomechanics of Dryosaurus lettowvorbecki (Dinosauria, Iguanodontia). Zoological Journal of the Linnean Society 108: 179–196.

69. Sanz JL, Chiappe LM, Perez-Moreno BP, Buscalioni AD, Moratalla JJ, et al. (1996) An Early Cretaceous bird from Spain and its implications for the evolution of avian flight. Nature 382: 442–445.

70. Alonso PD, Milner AC, Ketcham RA, Cookson MJ, Rowe TB (2004) The avian nature of the brain and inner ear of Archaeopteryx. Nature 430: 666–669.

71. Long CA, Zhang GP, George TF, Long CF (2003) Physical theory, origin of flight, and a synthesis proposed for birds. Journal of Theoretical Biology 224: 9–26.

72. Thulborn RA (2003) Wind-assisted flight of Archaeopteryx. Neues Jahrbuch für Geologie und Palaontologie-Abhandlungen 229: 61–74.

73. Speakman JR (1993) Flight Capabilities in Archaeopteryx. Evolution 47: 336–340.

74. Carrano MT (1999) What, if anything, is a cursor? Categories versus continua for determining locomotor habit in mammals and dinosaurs. Journal of Zoology 247: 29–42.

75. Bonnan MF (2005) Pes anatomy in sauropod dinosaurs: Implications for functional morphology, evolution, and phylogeny. In: Tidwell V, Carpenter K, editors. Thunder-Lizards: The Sauropodomorph Dinosaurs. Bloomington: Indiana University Press. pp. 346–380.

76. Alexander RM (2006) Dinosaur biomechanics. Proceedings of the Royal Society B-Biological Sciences 273: 1849–1855.

77. Hutchinson JR (2006) The evolution of locomotion in archosaurs. Comptes Rendus Palevol 5: 519–530.

78. Sellers WI, Manning PL (2007) Estimating dinosaur maximum running speeds using evolutionary robotics. Proceedings of the Royal Society B: Biological Sciences 274: 2711–2716.

79. Willis KJ, McElwain JC (2002) The Evolution of Plants. Oxford: Oxford University Press. 392 p.

80. Sokal RR, Rohlf FJ (1995) Biometry. New York: W.H. Freeman and Company. 887 p.

81. Peres-Neto PR, Jackson DA, Somers KM (2003) Giving meaningful interpretation to ordination axes: Assessing loading significance in principal component analysis. Ecology 84: 2347–2363.

82. Hammer Ø, Harper D, Ryan P (2001) PAST: Paleontological statistics software package for education and data analysis. Palaeontologia Electronica 4: 9.

83. Abbink O, Targarona J, Brinkhuis H, Visscher H (2001) Late Jurassic to earliest Cretaceous palaeoclimatic evolution of the southern North Sea. Global and Planetary Change 30: 231–256.

84. Hallam A (1984) Continental humid and arid zones during the Jurassic and Cretaceous. Palaeogeography, Palaeoclimatology, Palaeoecology 47: 195–223.

85. Ziegler AM, Parrish JM, Jiping Y, Gyllenhaal ED, Rowley DB, et al. (1993) Early Mesozoic phytogeography and climate. Philosophical Transactions: Biological Sciences 341: 297–305.

86. Hallam A (1985) A review of Mesozoic climates. Journal of the Geological Society 142: 433–445.

87. Demko TM, Currie BS, Nicoll KA (2004) Regional paleoclimatic and stratigraphic implications of paleosols and fluvial/overbank architecture in the Morrison Formation (Upper Jurassic), Western Interior, USA. Sedimentary Geology 167: 115–135.

88. Turner CE, Peterson F (2004) Reconstruction of the Upper Jurassic Morrison Formation extinct ecosystem–a synthesis. Sedimentary Geology 167: 309–355.

89. Dunagan SP, Turner CE (2004) Regional paleohydrologic and paleoclimatic settings of wetland/lacustrine depositional systems in the Morrison Formation (Upper Jurassic), Western Interior, USA. Sedimentary Geology 167: 269–296.

90. Lucas SG (2001) Chinese fossil vertebrates. New York: Columbia University Press. 375 p.

91. Eberth DA, Brinkman DB, Chen P-J, Yuan F-T, Wu S-Z, et al. (2001) Sequence stratigraphy, paleoclimate patterns, and vertebrate fossil preservation in Jurassic–Cretaceous strata of the Junggar Basin, Xinjiang Autonomous Region, People's Republic of China. Canadian Journal of Earth Sciences 38: 1627–1644.

92. Tütken T, Pfretzschner HU, Vennemann TW, Sun G, Wang YD (2004) Paleobiology and skeletochronology of Jurassic dinosaurs: Implications from the histology and oxygen isotope compositions of bones. Palaeogeography, Palaeoclimatology, Palaeoecology 206: 217–238.

93. Aberhan M, Bussert R, Heinrich W-D, Schrank E, Schultka S, et al. (2002) Palaeoecology and depositional environments of the Tendaguru Beds (Late Jurassic to Early Cretaceous, Tanzania). Mitteilungen aus dem Museum für Naturkunde in Berlin Geowissenschaftliche Reihe 5: 19–44.

94. Salgado L, De La Cruz R, Suárez M, Fernández M, Gasparini Z, et al. (2008) First Late Jurassic dinosaur bones from Chile. Journal of Vertebrate Paleontology 28: 529–534.

95. Soto M, Perea D (2008) A Ceratosaurid (Dinosauria, Theropoda) from the Late Jurassic–Early Cretaceous of Uruguay. Journal of Vertebrate Paleontology 28: 439–444.

96. Hadly EA, Spaeth PA, Li C (2009) Niche conservatism above the species level. Proceedings of the National Academy of Sciences 106: 19707–19714.

97. Antunes MT, Mateus O (2003) Dinosaurs of Portugal. Comptes Rendus Palevol 2: 77–95.

98. Mateus O (2006) Late Jurassic dinosaurs from the Morrison Formation (USA), the Lourinha and Alcobaça Formations (Portugal), and the Tendaguru Beds

(Tanzania): A Comparison. New Mexico Museum of Natural History and Science Bulletin 36.

99. Turner CE, Peterson F (1999) Biostratigraphy of dinosaurs in the Upper Jurassic Morrison Formation of the Western Interior, U.S.A. In: Gillette DD, editor. Vertebrate Paleontology in Utah. Salt Lake City: Miscellaneous Publication - Utah Geological Survey. pp. 77–114.

100. Dodson P, Behrensmeyer AK, Bakker RT, McIntosh JS (1980) Taphonomy and paleoecology of the dinosaur beds of the Morrison Formation. Paleobiology 6: 208–232.

101. Engelmann GF, Chure DJ, Fiorillo AR (2004) The implications of a dry climate for the paleoecology of the fauna of the Upper Jurassic Morrison Formation. Sedimentary Geology 167: 297–308.

102. Farlow JO (1987) Speculations about the diet and digestive physiology of herbivorous dinosaurs. Paleobiology 13: 60–72.

103. Midgley JJ, Midgley G, Bond WJ (2002) Why were dinosaurs so large? A food quality hypothesis. Evolutionary Ecology Research 4: 1093–1095.

104. Hummel J, Gee CT, Südekum K-H, Sander PM, Nogge G, et al. (2008) In vitro digestibility of fern and gymnosperm foliage: Implications for sauropod feeding ecology and diet selection. Proceedings of the Royal Society B: Biological Sciences 275: 1015–1021.

105. Garcia AJV, da Rosa AAS, Goldberg K (2005) Paleoenvironmental and paleoclimatic control on early diagenetic processes and fossil record in Cretaceous continental sandstones of Brazil. Journal of South American Earth Sciences 19: 243–258.

106. Rogers RR, Brady ME (2010) Origins of microfossil bonebeds: Insights from the Upper Cretaceous Judith River Formation of north-central Montana. Paleobiology 36: 80–112.

107. Kidwell SM, Flessa KW (1995) The quality of the fossil record: Populations, species, and communities. Annual Review of Ecology and Systematics 26: 269–299.

108. Jablonski D, Sepkoski JJ (1996) Paleobiology, community ecology, and scales of ecological pattern. Ecology 77: 1367–1378.

109. Morris PJ, Ivany LC, Schopf KM, Brett CE (1995) The challenge of paleoecological stasis: Reassessing sources of evolutionary stability. Proceedings of the National Academy of Sciences 92: 11269–11273.

110. Van Valkenburgh B (1995) Tracking ecology over geological time—Evolution within guilds of vertebrates. Trends in Ecology & Evolution 10: 71–76.

111. Jackson JBC, Erwin DH (2006) What can we learn about ecology and evolution from the fossil record? Trends in Ecology & Evolution 21: 322–328.

112. McGill BJ, Hadly EA, Maurer BA (2005) Community inertia of Quaternary small mammal assemblages in North America. Proceedings of the National Academy of Sciences 102: 16701–16706.

113. DiMichele WA, Behrensmeyer AK, Olszewski TD, Labandeira CC, Pandolfi JM, et al. (2004) Long-term stasis in ecological assemblages: Evidence from the fossil record. Annual Review of Ecology, Evolution and Systematics 35: 285–322.

114. Antón M, Sánchez IM, Salesa MJ, et al. (2003) The muscle-powered bite of Allosaurus (Dinosauria; Theropoda): An interpretation of cranio-dental morphology. Estudios Geológicos 59:

115. Rayfield EJ, Norman DB, Horner CC, Horner JR, Smith PM, et al. (2001) Cranial design and function in a large theropod dinosaur. Nature 409: 1033–1037.

116. Raup DM (1972) Taxonomic diversity during Phanerozoic. Science 177: 1065–1068.

117. Benton MJ (1985) Patterns in the diversification of Mesozoic nonmarine tetrapods and problems in historical diversity analysis. Special Papers in Palaeontology 33: 185–202.

118. Vermeij GJ, Leighton LR (2003) Does global diversity mean anything? Paleobiology 29: 3–7.

CHAPTER 13

INTEGRATING PALEOECOLOGY AND GENETICS OF BIRD POPULATIONS IN TWO SKY ISLAND ARCHIPELAGOS

John E McCormack, Bonnie S Bowen and Thomas B Smith

BACKGROUND

Current patterns of genetic diversity in North American species were shaped to a large degree by quaternary glacial cycles [1, 2, 3, 4], but effects of the last (Wisconsin) glaciation and current interglacial have been difficult to assess because very recent genetic divergence is easily overlooked by limited sampling, and molecular clock estimates are vitiated by coalescent stochasticity and retained ancestral polymorphism [5, 6]. For this reason, fossil paleoecological data, which usually document relatively recent climate-induced habitat change, have been difficult to incorporate into genetic studies. Integrating paleoecology and genetics has, therefore, remained an elusive goal [7].

In the southwestern USA and northern Mexico, a detailed history of the last 40,000 years of habitat change has been reconstructed from fossilized plant material preserved in packrat (*Neotoma* spp.) middens [8]. According to these data, most of the southwestern USA and northern Mexico was connected by woodland at the last glacial maximum (LGM), approximately 18,000 years ago and probably until 8000 to 9000 years ago (Figure 1). Since then, warming has driven woodlands to higher elevations, culminating sometime in the Middle Holocene when grassland and desert displaced woodland in lowland basins. Resulting ‹sky islands› provide a natural laboratory of habitat replicates for the study of recent genetic divergence [9, 10, 11] and a geographical setting in which to test the paleoecological hypothesis of postglacial habitat fragmentation.

Citation: John E McCormack, Bonnie S Bowen and Thomas B Smith, Integrating paleoecology and genetics of bird populations in two sky island archipelagos, DOI: 10.1186/1741-7007-6-28.

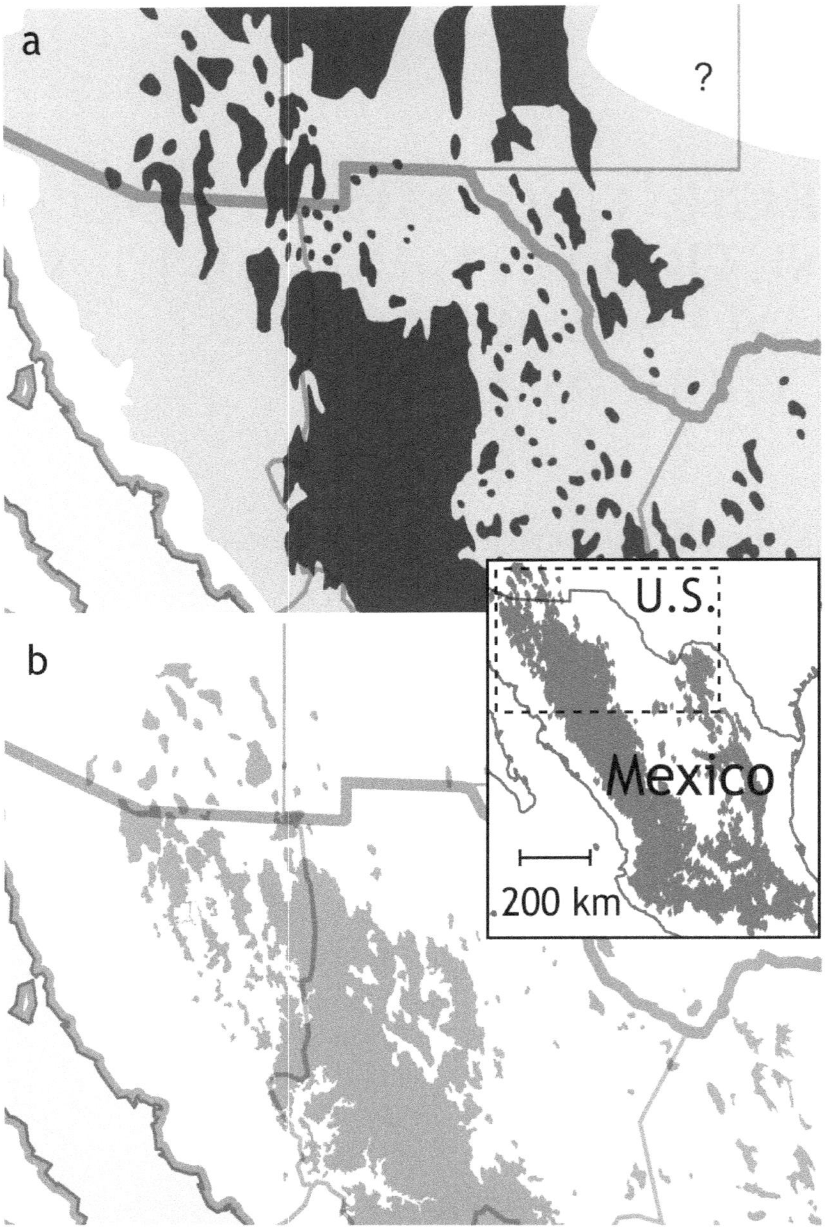

Figure 1 Habitat change from the last glacial maximum to the present. Since the last glacial maximum (a), patches of montane conifer forest and boreal forest (dark gray) within a matrix of mixed woodland with pinyon, oak and juniper (light gray) have given way to the present-day distribution of sky islands (b), patches of mixed pine-oak woodland (light gray) within a matrix of desertscrub, grassland or, further south in Mexico, thornscrub (white). Inset shows the current distribution of Mexican jays (shaded) and the region under study. Habitat at the last glacial maximum has been simplified for applicability to habitat requirements of Mexican jays and has been redrawn with permission from [8].

Despite their obvious appeal as a study system, little is known about genetic differentiation among sky islands. A few studies have suggested an effect of postglacial habitat fragmentation on current genetic structure [10, 12, 13], but other studies have failed to detect evidence for recent divergence [14, 15]. Furthermore, some genetic estimates of divergence times between sky islands point to ancient dates more than 1 million years ago (Ma), orders of magnitude older than those predicted by the midden data [11, 16].

With this study, we attempt to reconcile paleoecological and genetic estimates of the timing of postglacial habitat fragmentation for this region using as our model the Mexican jay (*Aphelocoma ultramarina*), a sedentary bird species found in pine-oak woodlands in the sky islands. Replication is rare in large-scale biogeographic studies, leading to the widespread use of comparative phylogeography of multiple codistributed species [16, 17, 18]. An alternative approach, which we use here, is to compare natural biogeographical replicates, in this case two independent sky-island archipelagos, that, according to the midden data, have experienced similar postglacial histories. This approach has the benefit of controlling for interspecific sources of variation due to climate and habitat change.

We predicted that populations from two sky-island archipelagos, one in Arizona and the other in the Trans-Pecos region of western Texas and northern Coahuila, Mexico, would show concordant evidence of postglacial genetic divergence (less than 18,000 years ago). To test this prediction, we assessed genetic diversity in mitochondrial DNA (mtDNA) and nuclear (microsatellite) DNA. We constructed minimum-spanning networks of mtDNA haplotypes to help clarify recent genealogies where ancestral and derived haplotypes often coexist, making inferences based on traditional bifurcating trees difficult [19]. Using observed patterns of shared and private haplotypes from the networks, we tested the postglacial fragmentation hypothesis against other plausible divergence scenarios such as ancient fragmentation and demographic expansion during the last glaciation (Figure 2). In addition, we estimated divergence times between sky islands with corrected genetic distance and a non-equilibrium coalescence model [20], using a joint data set, including both mtDNA and microsatellites. We show that patterns of unique and shared genes from mtDNA and nuclear microsatellites as well as divergence times corroborate the paleoecological hypothesis of postglacial fragmentation for the sky-island region, providing explicit genetic validation for one of the world's best fossil paleoecological records.

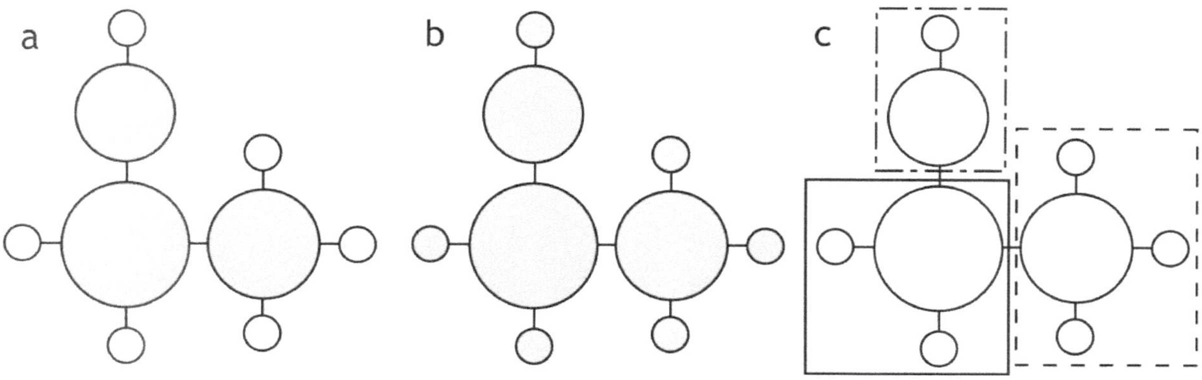

Figure 2 Predicted haplotype networks under different divergence scenarios. Haplotypes are represented as circles at the nodes and tips of trees, with the size of each circle proportional to the haplotype's overall frequency. In these networks, older haplotypes are often common and interior, whereas new haplotypes tend to be rare and peripheral [19]. (a) Expected pattern for postglacial sky-island divergence. Shared haplotypes (gray) resulting from population mixing at the last glacial maximum are common and interior, whereas new mutations subsequent to postglacial fragmentation produce unique haplotypes (white) on each sky island that are rare and peripheral. (b) Expected pattern for demographic expansion in a panmictic population with no postglacial divergence shows same star-shaped network, but recently derived haplotypes resulting from expansion are not necessarily partitioned among sky islands. (c) Expected pattern for ancient divergence among sky islands. Monophyletic complements of haplotypes occur on each sky island (enclosed by dotted lines) with no mixing. Note that intermediate patterns are possible.

RESULTS AND DISCUSSION

Evidence for postglacial genetic divergence

Our results validate the idea that the two archipelagos can be considered independent replicates for the purpose of investigating postglacial divergence. Sequences from 525 base pairs (bp) of the mtDNA control region of 265 Mexican jays from 10 different sky islands contained no missing data (for example, ambiguous base pairs) and revealed 32 variable sites, identifying 21 unique haplotypes in the Arizona archipelago and 18 in the Trans-Pecos archipelago (Figure 3). All sky islands contained some proportion of haplotypes shared with at least one other sky island, and eight out of nine contained unique haplotypes (Table 1). The two archipelagos shared no haplotypes and averaged 1.2% sequence divergence indicating a lack of recent gene flow, a pattern confirmed by a larger-scale study using both mtDNA and nuclear markers [21]. Paleoecological data suggest that continuous woodland connected the two archipelagos at the LGM (Figure 1a); yet, the amount of sequence divergence and lack of haplotype sharing shown here and in another study [21] suggest a historical barrier to gene flow, perhaps indicating that populations in the two archipelagos have acquired sufficient reproductive isolation to keep them distinct despite the potential for mixing at glacial maxima.

Figure 3 Minimum-spanning networks and haplotype frequencies in two sky-island archipelagos. Two sky-island archipelagos showing haplotype networks (insets) and haplotype frequencies for each sky island. Postglacial divergence is supported by the star-like pattern of shared (colored) and private (white or patterned) haplotypes. This pattern is nested within a more extensive genetic structure in the Arizona archipelago and haplotypes from the major genetic groups (I, II and III) are shared among sky islands, suggesting population mixing during the last glaciation. Other equally parsimonious links connect rare, peripheral haplotypes to each other or to common, interior haplotypes in genetic group III, but these links do not change the overall pattern and have been omitted for clarity of presentation. Sampling sites are: 1, Pajarito Mountains; 2, Huachuca Mountains; 3, Sierra San Jose; 4, Pinaleño Mountains; 5, Chiricahua Mountains; 6, Chisos Mountains; 7, Sierra del Carmen; 8, Serranías del Burros; 9, Sierra Santa Rosa; 10, Sierra Madre Oriental.

Table 1 Genetic diversity indices for sky islands

Sky island	n	H	H_p	A_p
Pajaritos (1)	19	6	4	7 (9.00)
Huachucas (2)	14	7	4	4 (4.12)
San Jose (3)	5	2	0	0
Pinaleños (4)	15	5	2	8 (10.0)
Chiricahuas (5)	41	9	5	7 (16.7)
Chisos (6)	56	4	1	2 (2.22)
Del Carmen (7)	155	10	3	8 (10.7)
Burros (8)	24	10	5	2 (3.22)
Santa Rosa (9)	15	7	2	1 (1.72)
Sierra Madre Oriental (10)	15	3	1	4 (7.27)

Sample size (n), number of control region haplotypes (H), number of private control region haplotypes (H_p), and number of private microsatellite alleles (A_p, summarized over 14 loci for sites 1 to 5 and over 13 loci for sites 6 to 10; percentage of total alleles in parentheses) for each sky island. Numbers after locations correspond to sampling sites in Figure 3.

Results from within each archipelago provide multiple lines of evidence to support the paleoecological hypothesis of postglacial sky-island fragmentation set forth by the fossil midden data. We first discuss patterns found in mtDNA haplotype networks and then divergence times estimated from corrected genetic distance and a coalescence model. The predicted mtDNA pattern for postglacial divergence, a 'star-shaped' haplotype network with rare, peripheral haplotypes unique to individual sky islands (Figure 2), was evident in both archipelagos (Figure 3), although in the Arizona archipelago this pattern was nested within a more extensive genetic structure (in particular, see group III, which has the largest sample size). High-frequency, widespread haplotypes often occur in the interior of haplotype networks, which can indicate that they are ancestral to haplotypes found on the network's periphery [20, 22]. In our networks, these interior haplotypes suggest the persisting genetic signature of a panmictic gene pool that existed during the last glaciation. By the same token, rare, private haplotypes located peripherally suggest *in situ* postglacial mutation within sky islands and limited gene flow among sky islands, in agreement with the natural history of the Mexican jay [23]. Alternatively, these rare haplotypes could have arrived in the region due to gene flow or from postglacial recolonization. However, this is unlikely because potential source populations in the northern Sierra Madre Occidental (for example, Durango) and Sierra Madre Oriental (for example, northern Nuevo León) do not share rare haplotypes with populations in our study [21]. Furthermore, Pleistocene niche models predict that Mexican jays remained in their same latitudinal distribution throughout the last glaciation [24]. Stability of geographic range through the last glaciation also makes our 'star-shaped' patterns unlikely to be caused by a recent demographic expansion resulting from recolonization, as has been shown for other passerine species [25, 26, 27, 28]. It should be noted that unlike many species, the habitable area for Mexican jays actually decreased following the LGM as woodland habitat fragmented and retreated to high elevations.

Private haplotypes could also represent haplotypes that went undetected in other sky islands due to limited sampling, but we believe this is unlikely because in most cases sample sizes were high relative to the number of haplotypes. For example, one sky island with a large sample size (site 7 where n = 155; Figure 3) did not contain even relatively high-frequency, private haplotypes from nearby sky islands (sites 6 and 8). Another possible explanation for our pattern is that genetic drift in small, shrinking populations resulted in the loss of existing rare haplotypes during postglacial forest fragmentation. However, this scenario, involving multiple, independent instances of haplotype loss in all sky islands but one, is less parsimonious than a scenario involving new mutation; thus we believe it is less likely to account for our pattern.

Stochastic processes, such as mutational variance and gene-sorting, can play a relatively large role in the generation and structuring of genetic diversity, making inferences based on networks or trees of a single locus inadvisable, especially for recent timescales [29]. To address this, we also assessed variation in another mtDNA marker (*ND2*) and nuclear microsatellites. The large-scale genetic structure seen in our data is largely corroborated by a smaller data set from *ND2* (Figure 4), suggesting that mutational variance is not driving the pattern. Furthermore, we found that, similar to results from mtDNA, most microsatellite alleles were shared among two or more sky islands, but nearly all sky islands (eight out of nine) contained a small percentage of private alleles (1.72% to 16.7%; Table 1). Bayesian analysis of population structure from microsatellite data also validates large-scale patterns seen in mtDNA, with east-west separation in the Arizona archipelago and assignment of some sky islands in the Trans-Pecos archipelago to separate genetic clusters (JEM, unpublished data). Taken as a whole, these results indicate that patterns of genetic diversity found in this study reflect the actual population history of Mexican jays and not stochastic factors peculiar to a single locus.

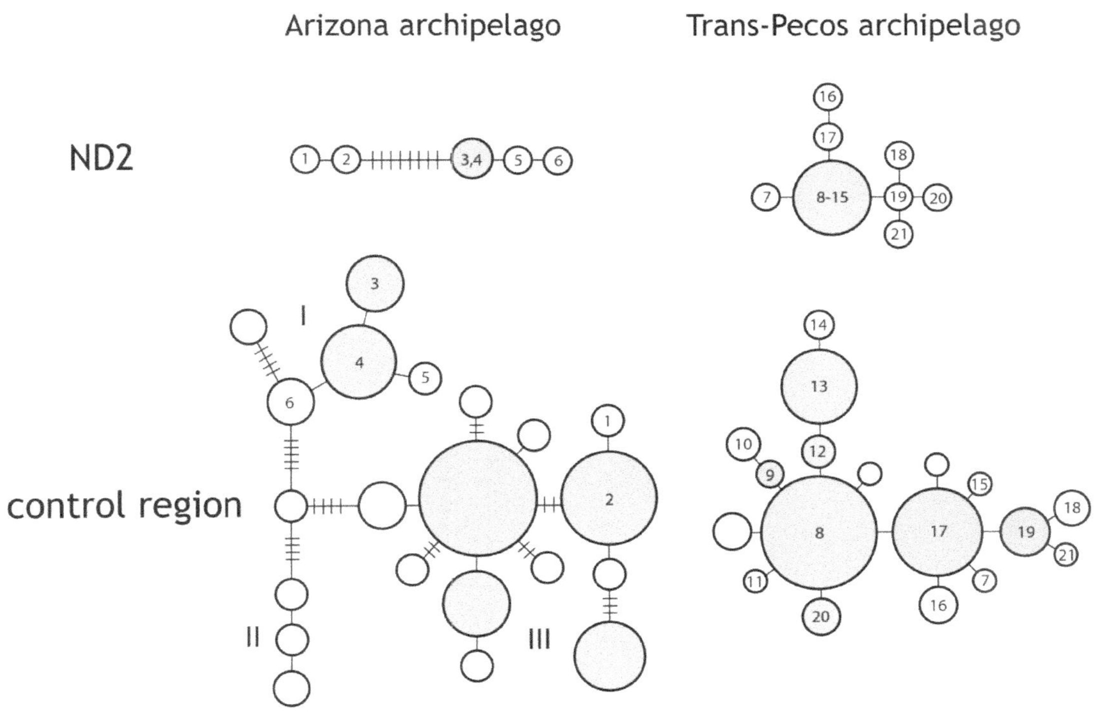

Figure 4 Comparison of mitochondrial DNA control region and *ND2* networks. A subset of individuals (identified by unique numbers) sequenced for both mitochondrial DNA control region and *ND2* shows similarities in large-scale genetic patterns between genes, suggesting that these patterns reflect the population history of Mexican jays and not stochastic genetic processes. In the Arizona archipelago, the same individuals fall into the same divergent genetic groups for both genes. *ND2* sample size for this archipelago is probably not large enough to detect the signature of postglacial divergence. In the Trans-Pecos archipelago, the pattern of postglacial divergence is corroborated. Note that many different haplotypes for the control region are grouped into one haplotype for *ND2*, indicating a slower mutation rate for *ND2*.

Divergence times based on corrected genetic distance, which used a conservative range of mutation rates, generally post-date the LGM (Tables 2 and 3; 200 to 12,000 years ago), except for comparisons between eastern and western sky islands in the Arizona archipelago, for which divergence estimates were older (see below). A coalescence-based approach [20] provides an independent and potentially powerful way to corroborate these dates because divergence times can be estimated from one or multiple loci in a coalescent (gene tree) framework that considers many possible gene trees consistent with the data and model, weights them according to their likelihood and, therefore, incorporates stochastic factors that limit the applicability of summary-statistic approaches [30]. Divergence times under an isolation model (see Methods) were highly consistent among sky-island pairs and across multiple runs (Figure 5), indicating postglacial divergence to the exclusion of later dates (Table 4). The inclusion of microsatellite data in the isolation model resulted in tighter likelihood curves and slightly earlier peak divergence times, with confidence intervals from several sky-island pairs overlapping values that were effectively zero.

Table 2 Corrected genetic distance and divergence times in Arizona archipelago

Sky island	1	2	4	5
Pajaritos (1)		0.02	0.29	0.27
Huachucas (2)	1.1–4.3		0.40	0.39
Pinaleños (4)	14 58	2–81		0.01
Chiricahuas (5)	14–55	20–79	0.3–1.1	

Corrected genetic distance (as a percentage) shown above the diagonal. Range of time since divergence (in thousands of years) using varying estimates for divergence rate (see Methods) appears below diagonal. Numbers of sky islands correspond to Figure 3.

Table 3 Corrected genetic distance and divergence times in Trans-Pecos archipelago

Sky island	6	7	8	9	10
Chisos (6)		0.01	0.02	0.01	0.01
Del Carmen (7)	0.8–3.4		<0.01	0.01	0.05
Burros (8)	1.0–3.9	0.2–0.8		0.01	0.06
Santa Rosa (9)	0.7–3.0	0.4–1.6	0.7–3.0		0.03
Sierra Madre Oriental (10)	0.6–2.6	2.7–11	3.1–12	1.6–6.2	

Corrected genetic distance (as a percentage) shown above the diagonal. Range of time since divergence (in thousands of years) using varying estimates for divergence rate (see Methods) appears below diagonal. Numbers after sky islands correspond to Figure 3.

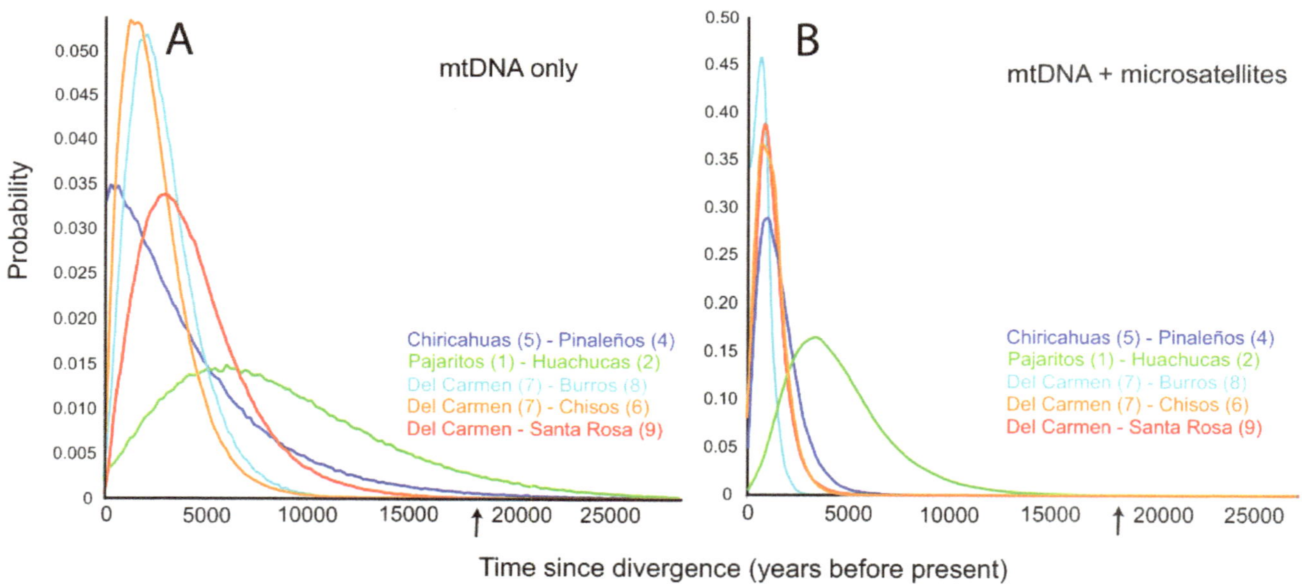

Figure 5 Divergence times between sky islands from coalescence model. Posterior probability densities for divergence times between five pairs of sky islands from coalescence model with mtDNA data only (a) and combined with microsatellite data (b). Divergence time estimates (t) have been converted to years by IM (see Methods). The last glacial maximum is indicated with an arrow.

Table 4 Divergence times between sky islands from coalescence model

Sky-island pair	t	T_{div} **(years)**
Mitochondrial DNA only		
Chiricahuas (5) – Pinaleños (4)	0.0075 (0.0025–0.2575)	300 (100–9900)
Pajaritos (1) – Huachucas (2)	0.1675 (0.0175–0.4675)	6500 (700–18000)
Sierra del Carmen (7) – Santa Rosa (9)	0.0325 (0.0075–0.1325)	1300 (300–5100)
Sierra del Carmen (7) – Burros (8)	0.0575 (0.0125–0.1525)	2200 (500–5900)
Sierra del Carmen (7) – Chisos (6)	0.0775 (0.0125–0.2175)	3000 (500–8400)
Mitochondrial DNA + microsatellites		
Chiricahuas (5) – Pinaleños (4)	0.0090 (0.0030–0.0310)	1100 (400–3800)
Pajaritos (1) – Huachucas (2)	0.0275 (0.0075–0.0725)	3700 (1000–9900)
Sierra del Carmen (7) – Santa Rosa (9)	0.0075 (0.0025–0.0375)	900 (300–4500)
Sierra del Carmen (7) – Burros (8)	0.0075 (0.0025–0.0425)	1000 (300–5600)
Sierra del Carmen (7) – Chisos (6)	0.0075 (0.0025–0.0325)	800 (300–3600)

Estimates of the t parameter from IM and conversion to years since divergence (T_{div}) using a range of mutation rates (see Methods). Peak values (90% highest probability density) are given. Parameters were not sampled at zero; therefore, the highest probability densities overlapping 0.0025 are effectively zero. Numbers after sky islands correspond to Figure 3.

Until now, evidence for postglacial fragmentation of sky islands has eluded intraspecific studies of poorly dispersing sky-island organisms. In a study of sky-island jumping spiders (*Habronattus pugillis*), Masta [11] found evidence for genetic differentiation among sky islands, but timing estimates were relatively old (some dates more than 1 Ma). No evidence was found for differentiation among sky-island populations of the insect herbivore grape phylloxera (*Daktulosphaira vitifoliae*) [15]. Of the studies reporting similar results to ours, one is from a different sky-island system in the Rocky Mountains, where populations of a flightless grasshopper (*Melanoplus oregonensis*) show evidence for recent divergence and *in situ* generation of unique haplotypes [10]. Quite recently, a study on rattlesnakes from three sky islands that form part of our Arizona archipelago (although not sampled in our study) indicated Holocene divergence [13]. Our results provide corroboration for this conclusion with multiple timing estimates throughout the region.

Importance of sampling and mutation rate in studies of recent genetic divergence

There are various reasons why previous studies might have failed to detect genetic divergence among sky islands. Reconstructing genetic histories for recently diverged lineages is hindered by stochasticity in the coalescent process and the confounding influences of migration and incomplete lineage sorting [5]. Thorough sampling from throughout the region of interest [31] and large sample sizes [6] are often necessary to overcome these difficulties (Figure 6). For example, if only one individual was sampled from each sky island in the Arizona archipelago, we might have uncovered only haplotypes from the three genetic groups (I, II and III in Figure 3). In doing so, we would have overlooked recent divergence and come to the spurious conclusion that divergence among sky islands was older (Figure 6b).

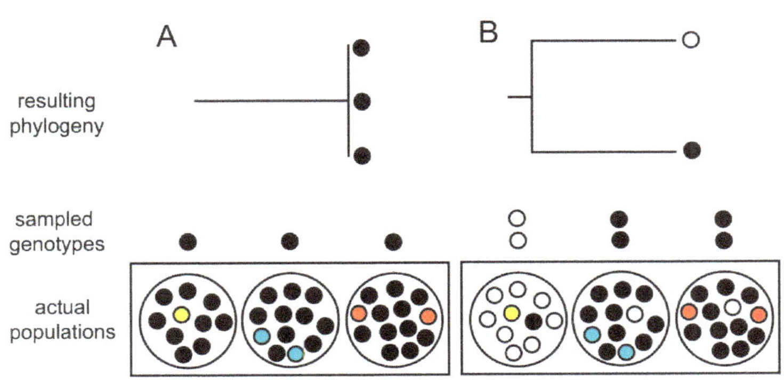

Figure 6 Reasons for failing to detect recent genetic divergence. (a) Limited sampling fails to uncover postglacial haplotypes (colored dots) resulting in a false conclusion of no divergence between sky islands. (b) Limited sampling fails to uncover postglacial haplotypes, but does detect shared haplotypes from historical divergence (black and white dots), leading to a false conclusion that the divergence of sky islands is older.

Uncovering recent divergence also depends critically on the use of genetic markers with mutation rates fast enough to record the history of population splitting. Substitution rates for the non-coding mtDNA control region are usually higher than those for protein-coding regions like *ND2*, especially over short timescales where saturation is not an important factor [32]. A qualitative assessment of the effect of mutation rate can be seen by comparing control region and *ND2* haplotype networks for the Trans-Pecos archipelago (Figure 4). In multiple instances, several unique control region haplotypes are collapsed into a single *ND2* haplotype, suggesting a slower substitution rate for *ND2* over recent timescales and reduced capacity to record information on recent population splitting events in the genetic code.

The ‹multilocus approach› to phylogeographic studies is gaining traction as more studies demonstrate that results from one gene often do not reflect the population history of organisms [29]. Our results, however, caution that to test hypotheses concerning recent genetic divergence, choosing a locus with a high mutation rate is also important, and thus, for many organisms, the mtDNA control region will still offer the best tool for this purpose. Multiple loci, if they do not mutate fast enough to record evidence of recent vicariance, will not provide informative data, no matter how many are marshaled to the task.

Evidence for differing glacial histories between archipelagos

Despite similarities in their postglacial genetic histories, differences in the broader genetic structure between the two archipelagos suggest contrasting glacial or pre-Wisconsin histories. The Arizona archipelago has much more genetic structure with three main genetic groups divergent from one another by four or five mutational steps (Figure 3). Haplotype frequencies shift dramatically from east to west within this archipelago. Estimates of divergence time across this divide range from 14,000 to 80,000 years ago (Table 2) and thus, generally date back to the last glaciation. This timing is corroborated by the fact that haplotypes from the different genetic groups mix in some sky islands, an event much more likely to have occurred when gene flow was facilitated by continuous woodland (for example, during the Wisconsin glaciation). Still, population mixing was not complete, indicating a hindrance to gene flow in the Arizona archipelago during the last glaciation. Jays in the Trans-Pecos archipelago, on the other hand, were probably panmictic during the LGM as evidenced by the relatively homogenous pool of shared haplotypes found in each sky island.

Studies on other organisms have shown large genetic [11, 15, 33], and in at least one case, phenotypic [34] breaks across an east-west gradient in the sky islands of Arizona, leading to speculation that woodland might have been discontinuous during the last glaciation. Another possibility that has not been previously considered is that montane conifer forest, currently associated with Steller's jays (*Cyanocitta stelleri*), which replace Mexican jays at high elevation throughout most of their distribution, provided a barrier to east-west dispersal of Mexican jays (and possibly other organisms) in Arizona during the last glaciation. Hypothetical maps of habitat at the LGM inferred from the midden data support this idea and suggest two glacial refugia, one in southwestern Arizona and another in southwestern New Mexico and northern Chihuahua, with connections between the two interrupted by large expanses of montane conifer forest (Figure 1a). This habitat arrangement over a period of about 80,000 years during the last glaciation might have resulted in two largely independent populations connected by limited gene flow. This possibility, and comparisons of the glacial histories of the two archipelagos, could be investigated further by combining fossil paleoecological data with niche-modeling [35], to produce finer-scale reconstructions of glacial habitat.

CONCLUSION

This study provides the most robust genetic support to date for the paleoecological hypothesis of postglacial fragmentation of sky islands. Previous studies might have been unable to detect postglacial divergence due either to small sample sizes or limited geographic coverage. This study and others demonstrate that sky islands can act as generators of genetic diversity over both recent and historical timescales [9, 10, 11, 12]. In some cases, sky islands have also facilitated rapid phenotypic differentiation [36, 37] and incipient speciation, for example, in jumping spiders [38]. Further studies incorporating genetic and phenotypic diversity of sky islands with their paleohistory promise to shed light on the evolutionary processes responsible for diversification in these systems. Human-caused climate change is radically altering the sky islands of the southwestern USA and northern Mexico by exacerbating droughts, fire and outbreaks of invasive insects [39]. Models and empirical data show that temperature increases of as little as a few degrees could lead to widespread extirpation of high-elevation species [40, 41]. It is especially urgent to understand the evolutionary processes at work in sky islands and incorporate them into the conservation agenda for this dynamic region where their role in generating diversity is so evident.

METHODS

Study species

The Mexican jay is a year-round resident of pine-oak woodlands in the highlands of Mexico (inset Figure 1) and the Madrean sky-island archipelagos of the southwestern USA. They have low dispersal capabilities and breed cooperatively in flocks of 5 to 25 individuals that defend permanent territories [23]. Records of wandering Mexican jays from desert basins between sky islands are extremely rare, thus, dispersal among sky islands appears to be infrequent, if it occurs at all. Populations from the two regions addressed in this study are currently classified as different subspecies: *A. u. arizonae* in southeastern Arizona and northern Sonora and *A. u. couchii* in southwestern Texas and northern Coahuila. These subspecies are genetically divergent with no evidence for recent gene flow [21].

Study areas and field sampling

Sky islands in southeastern Arizona and northern Sonora (‹Arizona archipelago›) and sky islands in southwestern Texas and northern Coahuila, Mexico (‹Trans-Pecos archipelago›) are composed of mountain ranges of roughly similar size and distance (Figure 3). For the Trans-Pecos archipelago, we included a ‹mainland› site in the Sierra Madre Oriental. From 2002 to 2006, we captured 226 jays in the Trans-Pecos archipelago (sites 6 to 10 in Figure 3), using mist-nets and baited ground traps. Blood samples from an additional 39 jays from the Chisos Mountains (site 6 in Figure 3) were provided by Nirmal Bhagabati. In the Arizona archipelago, 70 jays were collected in 1986 from three sky islands (sites 2, 4 and 5 in Figure 3) by Rolf Koford. Tissue samples from an additional 24 jays from two sky islands (sites 1 and 3 in Figure 3) were collected by A Townsend Peterson in 1987 and 1989 and werebprovided by the Field Museum in Chicago. Three to 10 flocks per sky island were sampled to ensure that the data set included many unrelated individuals. Blood was taken in the field by puncture of the subbrachial vein and kept in lysis buffer [42] at ambient temperature for transport to laboratory facilities where samples were stored at -80°C. Feathers collected in the field were kept in paper envelopes and later stored at -20°C. Tissue samples from animals collected in the field were stored in liquid nitrogen before being transferred to -80°C.

Laboratory techniques and analyses

We extracted genomic DNA from blood and tissue using a QIAGEN® DNeasy™ Tissue Kit (Qiagen, Inc., Valencia, CA, USA). We amplified a 525-base-pair portion of the mtDNA control region using primers JCR03 (5›-CCCCCCCATGTTTTTACR-3›) [43] and H1248 (5›-CATCTTCAGTGTCATGCT-3›) [44]. Reactions were 25 μl by volume and consisted of an initial denaturation at 94°C for 3 minutes, followed by 40 iterations of 45 seconds at 94°C, 45 seconds at 50°C and 30 seconds at 72°C, with a final extension period of 5 minutes at 72°C. We ran negative controls that contained no DNA template with each reaction to check for contamination.

Polymerase chain reaction (PCR) products were purified with an UltraClean™ GelSpin kit (Mo Bio Laboratories, Solana Beach, CA, USA). We obtained sequences from PCR products using a 10-μl total volume, dydeoxy-terminator cycle-sequencing reaction using a CEQ DTCS kit (Beckman Coulter, High Wycombe, UK). We purified products from this reaction with an ethanol precipitation and sequenced them in a Beckman-Coulter CEQ 2000 automated sequencer. We aligned sequences automatically using SEQUENCHER 4.1 (GeneCodes) and checked variable sites visually for accuracy. All sequences used in this study have been deposited in GenBank under accession numbers EU121284 to EU121375. We encountered no problems suggesting nuclear inserts (for example, double peaks); nevertheless, a subset of the data, including all unique haplotypes from blood extractions, was re-extracted from feather tissue and resequenced with identical results.

As differences in substitution patterns and stochasticity in the lineage-sorting process can result in disparity between gene history and population history [29], particularly for recent splitting events, we sought to corroborate control region patterns with those from another mtDNA gene and microsatellites. We amplified and sequenced a 636-bp portion of the NADH dehydrogenase, subunit 2 (*ND2*) gene for subset of our samples using primers L-5216 (5'-GGCCCATACCCCGRAAATG-3') and H-6313 (5'-CTCTTATTTAAGGCTTTGAAGGC-3') [45], using the same protocol as for the control region. Additionally, we screened individuals for genetic variation at 14 microsatellite loci: MJG1 and MJG8 [46], ApCo2, ApCo15, ApCo18, ApCo19, ApCo22, ApCo29, ApCo30, ApCo37, ApCo40, ApCo41, ApCo68 and ApCo97 [47]. For the Trans-Pecos archipelago, ApCo37 was excluded because it was not variable. Forward primers were tagged with a fluorescent dye (universal M13 forward primer) and used in a multiplex PCR kit (Qiagen Inc.) with groups of two to three primer combinations, differentiating between loci by prior knowledge of allele size ranges obtained from test runs. We ran PCR products on an ABI3730 capillary sequencer (Applied Biosystems) and scored alleles using GeneMapper software (Applied Biosystems). We counted the number of private alleles for each sky island (A_p) and calculated a percentage compared with total alleles.

Haplotype networks and divergence times

We constructed minimum-spanning networks of absolute distances between mtDNA haplotypes using the molecular variance parsimony algorithm [48] implemented in ARLEQUIN 2.0 [49]. We used the number of nucleotide changes between haplotypes to calculate corrected genetic distance between sky islands (D_{xy}), taking into account intra-population polymorphism [50], so that $D_{xy} = d_{ixy} - 0.5(d_{ix} + d_{iy})$, where x and y are the groups being compared and d_i is the uncorrected average genetic distance [51]. We calculated divergence times from these distances by applying a molecular clock. To account for varying estimates of mutation rate for the control region, we used a conservative range of divergence rates from 0.05 to 0.20 substitutions per site per million years, which were derived from several different bird species [32, 52] and have been applied to jays [53]. The Sierra San Jose population was omitted from these calculations because of a low sample size. We restricted our analysis of divergence time to control region data because sample sizes were much larger than those for *ND2*.

We also estimated divergence times using IM, a program that implements the isolation-with-migration coalescence model. IM uses a Markov chain Monte Carlo method to estimate jointly several demographic parameters of two diverging populations [20]. For this study, we focused on divergence time (t). Two types of data sets were used, one with only mtDNA sequence data under an Hasegawa-Kishino-Yano model of sequence evolution [54], and one using both mtDNA and microsatellite data, the latter under a step-wise mutation model [55]. Microsatellite loci appeared to conform to the step-wise mutation model, with no large gaps between repeats and roughly normal distribution of allele frequencies. One of the most challenging aspects of IM, particularly for data sets like ours that use microsatellite data involving many individuals and loci is slow convergence on the true posterior density of the parameters [20]. Running many Markov chains (Metropolis-coupling) speeds the search considerably, but achieving satisfactory mixing among the chains is critical, as it allows for full exploration of parameter space and prevents parameter estimates from reflecting local instead of global optima. We spent considerable time finding suitable conditions for runs, using initial runs to constrain t to certain intervals and to alter the heating scheme to achieve sufficient mixing among chains, as advised in IM documentation. Mixing was monitored by observing effective sample size (ESS) values and inspecting parameter plots for trends.

Although conditions varied among runs depending on the needs of the data, we ran each pairwise analysis for at least 10,000,000 steps after a long burn-in of 3,000,000 steps using 10 to 30 chains for the mtDNA data set and 40 to 70 chains for the joint data set. We used a geometric heating scheme, generally using high values for the heating parameters (for example, $g_1 = 0.9$, $g_2 = 0.8$), particularly for the joint data sets. MtDNA-only data sets generally achieved convergence quickly and the five different pairwise runs took between 3 and 7 days on a personal laptop with a dual-core Intel 1.83-GHz processor. Runs with the joint data sets took considerably longer, up to 30 days to exceed 10,000,000 steps and approach convergence. ESS values for these data sets were never particularly high (usually the final ESS value for *t* was about 50), and here we relied primarily on the lack of trends in the parameter plot for *t* to determine if convergence had been reached. Importantly, results for the joint data sets were consistent among independent runs and are very similar to results for mtDNA data alone, in which convergence was confidently achieved. We present results from the longest run for any particular data set and conditions.

IM allows a range of mutation rates to be input prior to the analysis for scaling parameter estimates in demographic units. A range of per locus mutation rates from 1.2975×10^{-5} to 5.19×10^{-5} was used for our 519-bp segment of the control region, reflecting the range of divergence rates described above. Using this mutation rate, IM calculates mutation rate scalars for the other loci and uses a geometric mean of all mutation rates to convert divergence times to demographic scales. Therefore, no mutation rate was specified for the microsatellite loci. An important assumption of IM is that the populations in question have most recently split from one another; thus, we limited IM analyses to adjacent sky islands. To reduce run times for the joint data set, we removed hatch-year and second-year individuals from the Sierra del Carmen population, resulting in a sample size of 93 individuals.

Initially, we estimated divergence times under an isolation-with-migration model that allowed the possibility of migration between sky islands. Previous results using MDIV[56], the predecessor to IM, indicated low migration rates and postglacial divergence between most sky-island pairs under an isolation-with-migration model (data not shown). However, population parameters, particularly migration rates, proved difficult to estimate in IM under this model. Similar difficulty in estimating parameters with high confidence in IMcompared with MDIV has been reported elsewhere [57], which could be due to the increased number of parameters in IM. When migration was unconstrained in IM, likelihood curves for this parameter were relatively flat and included zero as well as values that seemed unrealistically high (more than 20 migrants per generation). When migration was capped at more realistic values ($m_1 = m_2 \leq 4 \approx 2$ migrants per generation), divergence times were often multimodal, but had peak likelihood post-dating the LGM. However, estimates of migration rate under this model never peaked, with likelihood steadily increasing across the range of values. This could indicate insufficient information in the data to estimate migration and it is advisable in this situation to reduce the number of parameters (J Hey, personal communication). Importantly, postglacial divergence was not ruled out under realistic isolation-with-migration scenarios (data not shown). However, because for this study we were concerned with obtaining precise estimates of divergence time, and as our haplotype networks and knowledge of the natural history of the Mexican jay suggested little or no current gene flow among sky islands, we present results from an isolation model where migration rate was set to zero ($m_1 = m_2 = 0$). We acknowledge that some migration among sky islands likely occurred during the initial stages of habitat fragmentation, and this has probably contributed to the difficulty of finding a clear signature of migration in the results.

ACKNOWLEDGEMENTS

We thank T Peterson, the Field Museum (D Willard) and R Koford for making their samples available to us. R Skiles of Big Bend National Park, C Muzquiz, the Michaelis family, the Sellers family, Museo de las Aves de México and the Mexican cement company CEMEX provided land access and facilitated permits. The McKinneys, D Roe and Del Carmen Project staff provided logistical support. J Carrera, C Sifuentes and SEMARNAT permitted the work in Mexico. B Milá and G Castañeda gave critical assistance for all aspects of the project. A Alvarado, O Ballasteros, E Berg, M Buck, A Byrd, T Hanks, E Landay, B Larison, G Levandoski, E López-Medrano, V Rodríguez, L Rubio, M Starling and R Tinajero provided field assistance. G Castañeda, J Pollinger, B Starford and M Starling helped with the laboratory work. The manuscript was improved with comments from A Freedman, B Milá, F Hertel, T Price, T Van Devender, R Wayne and two anonymous reviewers. G Cabanne and J Hey helped with the IM analysis and P Klimov kindly loaned processor time. This research was funded by an American Ornithologists' Union research award, a University of California Los Angeles Latin American Center Doctoral Student Small Grant, a University of California Mexus Program Dissertation Research Grant, a National Science Foundation Graduate Research Fellowship to JEM, and National Science Foundation grants DEB-9726425 and IRCEB9977072 to TBS.

Authors' contributions

JEM collected samples, carried out the molecular genetic studies and analyses and drafted the manuscript. BSB collected samples and helped to draft the manuscript. TBS participated in the design of the project and helped to draft the manuscript. All authors read and approved the final manuscript.

REFERENCES

1. Avise JC, Walker D, Johns GC: Speciation durations and Pleistocene effects on vertebrate phylogeography. Proc R Soc Lond B Biol Sci. 1998, 265: 1707-1712. 10.1098/rspb.1998.0492.
2. Hewitt G: Some genetic consequences of ice ages, and their role in divergence and speciation. Biol J Linn Soc Lond. 1996, 58: 247-276.
3. Hugall A, Moritz C, Moussalli A, Stanisic J: Reconciling paleodistribution models and comparative phylogeography in the Wet Tropics rainforest land snail *Gnarosophia bellendenkerensis* (Brazier 1875). Proc Natl Acad Sci USA. 2002, 99: 6112-6117. 10.1073/pnas.092538699.

4. Lessa EP, Cook JA, Patton JL: Genetic footprints of demographic expansion in North America, but not Amazonia, during the Late Quaternary. Proc Natl Acad Sci USA. 2003, 100: 10331-10334. 10.1073/pnas.1730921100.

5. Arbogast BS, Edwards SV, Wakeley J, Beerli P, Slowinski JB: Estimating divergence times from molecular data on phylogenetic and population genetic timescales. Annu Rev Ecol Syst. 2002, 33: 707-740. 10.1146/annurev.ecolsys.33.010802.150500.

6. Maddison WP, Knowles LL: Inferring phylogeny despite incomplete lineage sorting. Syst Biol. 2006, 55: 21-30. 10.1080/10635150500354928.

7. Cruzan MB, Templeton AR: Paleoecology and coalescence: phylogeographic analysis of hypotheses from the fossil record. Trends Ecol Evol. 2000, 15: 491-496. 10.1016/S0169-5347(00)01998-4.

8. Betancourt JL, Van Devender TR, Martin PS: Packrat Middens: The Last 40,000 Years of Biotic Change. 1990, Tucson, AZ: University of Arizona Press

9. Knowles LL: Tests of Pleistocene speciation in montane grasshoppers (genus *Melanoplus*) from the sky islands of western North America. Evolution. 2000, 54: 1337-1348.

10. Knowles LL: Did the Pleistocene glaciations promote divergence? Tests of explicit refugial models in montane grasshoppers. Mol Ecol. 2001, 10: 691-701. 10.1046/j.1365-294x.2001.01206.x.

11. Masta SE: Phylogeography of the jumping spider *Habronattus pugillis* (Araneae: Salticidae): recent vicariance of sky island populations?. Evolution. 2000, 54: 1699-1711.

12. Sullivan RM: Micro-evolutionary differentiation and biogeographic structure among coniferous forest populations of the Mexican woodrat (*Neotoma mexicana*) in the American southwest: a test of the vicariance hypothesis. J Biogeogr. 1994, 21: 369-389. 10.2307/2845756.

13. Holycross AT, Douglas ME: Geographic isolation, genetic divergence, and ecological non-exchangeability define ESUs in a threatened sky-island rattlesnake. Biol Conserv. 2007, 134: 142-154. 10.1016/j.biocon.2006.07.020.

14. Lamb T, Jones TR, Wettstein PJ: Evolutionary genetics and phylogeography of tassel-eared squirrels (*Sciurus aberti*). J Mammal. 1997, 78: 117-133. 10.2307/1382645.

15. Downie DA: Phylogeography in a galling insect, Grape phylloxera, *Daktulosphaira vitifoliae* (Phylloxeridae) in the fragmented habitat of the Southwest USA. J Biogeogr. 2004, 31: 1759-1768. 10.1111/j.1365-2699.2004.01075.x.

16. Smith CI, Farrell BD: Range expansions in the flightless longhorn cactus beetles, *Moneilema gigas* and *Moneilema armatum*, in response to Pleistocene climate changes. Mol Ecol. 2005, 14: 1025-1044. 10.1111/j.1365-294X.2005.02472.x.

17. Weir J, Schluter D: Ice sheets promote speciation in boreal birds. Proc R Soc Lond B Biol Sci. 2004, 271: 1881-1887. 10.1098/rspb.2004.2803.

18. Zink RM: Comparative phylogeography in North American birds. Evolution. 1996, 50: 308-317. 10.2307/2410802.

19. Posada D, Crandall KA: Trees grafting into networks. Trends Ecol Evol. 2001, 16: 37-45. 10.1016/S0169-5347(00)02026-7.

20. Hey J, Nielsen R: Multilocus methods for estimating population sizes, migration rates and divergence time, with applications to the divergence of *Drosophila pseudoobscura* and *D. persimilis*. Genetics. 2004, 167: 747-760. 10.1534/genetics.103.024182.

21. McCormack J, Peterson AT, Bonaccorso E, Smith TB: Speciation in the highlands of Mexico: genetic and phenotypic divergence in the Mexican jay (*Aphelocoma ultramarina*). Mol Ecol. 2008, 17: 2505-2521. 10.1111/j.1365-294X.2008.03776.x.

22. Castelloe J, Templeton A: Root probabilities for intraspecific gene trees under neutral coalescent theory. Mol Phylogenet Evol. 1994, 3: 102-113. 10.1006/mpev.1994.1013.

23. McCormack JE, Brown JL: Mexican jay (*Aphelocoma ultramarina*). The Birds of North America Online. Edited by: Poole A. 2008, Ithaca, NY: Cornell Lab of Ornithology, [http://bna.birds.cornell.edu/review/species/118]

24. Peterson AT, Martínez-Meyer E, González-Salazar C: Reconstructing the Pleistocene geography of the *Aphelocoma* jays (Corvidae). Divers Distrib. 2004, 10: 237-246. 10.1111/j.1366-9516.2004.00097.x.

25. Milá B, Girman DJ, Kimura M, Smith TB: Genetic evidence for the effect of a postglacial population expansion on the phylogeography of a North American songbird. Proc R Soc Lond B Biol Sci. 2000, 267: 1033-1040. 10.1098/rspb.2000.1107.

26. Milá B, McCormack J, Castañeda G, Wayne R, Smith T: Recent postglacial range expansion drives the rapid diversification of a songbird lineage in the genus *Junco*. Proc R Soc Lond B Biol Sci. 2007, 274: 2653-2660. 10.1098/rspb.2007.0852.

27. Milá B, Smith TB, Wayne RK: Postglacial population expansion drives the evolution of long-distance migration in a songbird. Evolution. 2006, 60: 2403-2409.

28. Milá B, Smith TB, Wayne RK: Speciation and rapid phenotypic differentiation in the yellow-rumped warbler *Dendroica coronata* complex. Mol Ecol. 2007, 16: 159-173. 10.1111/j.1365-294X.2006.03119.x.

29. Edwards SV, Beerli P: Perspective: gene divergence, population divergence, and the variance in coalescence time in phylogeographic studies. Evolution. 2000, 54: 1839-1854.

30. Hey J, Machado CA: The study of structured populations – new hope for a difficult and divided science. Nat Rev Genet. 2003, 4: 535-543. 10.1038/nrg1112.

31. Funk DJ, Omland KE: Species-level paraphyly and polyphyly: Frequency, causes, and consequences, with insights from animal mitochondrial DNA. Annu Rev Ecol Syst. 2003, 34: 397-423. 10.1146/annurev.ecolsys.34.011802.132421.

32. Ruokonen M, Kvist L: Structure and evolution of the avian mitochondrial control region. Mol Phylogenet Evol. 2002, 23: 422-432. 10.1016/S1055-7903(02)00021-0.

33. Barber PH: Phylogeography of the canyon treefrog, *Hyla arenicolor* (Cope) based on mitochondrial DNA sequence data. Mol Ecol. 1999, 8: 547-562. 10.1046/j.1365-294x.1999.00593.x.

34. Maddison W, McMahon M: Divergence and reticulation among montane populations of a jumping spider (*Habronattus pugillis* Griswold). Syst Biol. 2000, 49: 400-421. 10.1080/10635159950127312.

35. Richards CL, Carstens BC, Lacey Knowles L: Distribution modelling and statistical phylogeography: an integrative framework for generating and testing alternative biogeographical hypotheses. J Biogeogr. 2007, 34: 1833-1845. 10.1111/j.1365-2699.2007.01814.x.

36. Boyd A: Morphological analysis of sky island populations of *Macromeria viridiflora* (Boraginaceae). Syst Bot. 2002, 27: 116-126.

37. Roberts JL, Brown JL, Schulte R, Arizabal W, Summers K: Rapid diversification of colouration among populations of a poison frog isolated on sky peninsulas in the central cordilleras of Peru. J Biogeogr. 2007, 34: 417-426. 10.1111/j.1365-2699.2006.01621.x.

38. Masta SE, Maddison WP: Sexual selection driving diversification in jumping spiders. Proc Natl Acad Sci USA. 2002, 99: 4442-4447. 10.1073/pnas.072493099.

39. Egan T: Heat invades cool heights over Arizona desert. New York Times, 27 March 2007, [http://www.nytimes.com/2007/03/27/us/27warming.html]

40. McDonald K, Brown J: Using montane mammals to model extinctions due to global change. Conserv Biol. 1992, 6: 409-415. 10.1046/j.1523-1739.1992.06030409.x.

41. Sekercioglu CH, Schneider SH, Fay JP, Loarie SR: Climate change, elevation range shifts, and bird extinctions. Conserv Biol. 2008, 22: 140-150. 10.1111/j.1523-1739.2007.00852.x.

42. Seutin G, White BN, Boag PT: Preservation of avian blood and tissue samples for DNA analyses. Can J Zool. 1991, 69: 82-90. 10.1139/z91-013.

43. Saunders MA, Edwards SV: Dynamics and phylogenetic implications of mtDNA control region sequences in New World jays (Aves: Corvidae). J Mol Evol. 2000, 51: 97-109.

44. Tarr CL: Primers for amplification and determination of mitochondrial control-region sequences in oscine passerines. Mol Ecol. 1995, 4: 527-529. 10.1111/j.1365-294X.1995.tb00251.x.

45. Sorenson MD, Ast J, Dimcheff D, Yuri T, Mindell DP: Primers for a PCR-based approach to mitochondrial genome sequencing in birds and other vertebrates. Mol Phylogenet Evol. 1999, 12: 105-114. 10.1006/mpev.1998.0602.

46. Li S-H, Huang Y-J, Brown JL: Isolation of tetranucleotide microsatellites from the Mexican jay *Aphelocoma ultramarina*. Mol Ecol. 1997, 6: 499-501. 10.1046/j.1365-294X.1997.00215.x.

47. Stenzler LM, Fitzpatrick JW: Isolation of microsatellite loci in the Florida Scrub-Jay *Aphelocoma coerulescens*. Mol Ecol Notes. 2002, 2: 547-550. 10.1046/j.1471-8286.2002.00312.x.

48. Excoffier L, Smouse PE: Using allele frequencies and geographic subdivision to reconstruct gene trees within a species: molecular variance parsimony. Genetics. 1994, 136: 343-359.

49. Schneider S, Roessli D, Excoffier L: ARLEQUIN version 2.000: A Software for Population Genetics Data Analysis. 2000, Geneva: University of Geneva, Genetics and Biometry Laboratory

50. Nei M: Molecular Evolutionary Genetics. 1987, New York, NY: Columbia University Press

51. Wilson AC, Cann RL, Carr SM, Geroge M, Gyllenstein UB, Helm-Bychowski KM, Higuchi RG, Palumbi SR, Prager EM, Sage RD, Stoneking M: Mitochondrial DNA and two perspectives on evolutionary genetics. Biol J Linn Soc Lond. 1985, 26: 375-400. 10.1111/j.1095-8312.1985.tb02048.x.

52. Baker AJ, Marshall HD: Mitochondrial control region sequences as tools for understanding evolution. Avian Molecular Evolution and Systematics. Edited by: Mindell DP. 1997, San Diego, CA: Academic Press, 49-80.

53. Delaney KS, Wayne RK: Adaptive units for conservation: population distinction and historic extinctions in the island scrub-jay. Conserv Biol. 2005, 19: 523-533. 10.1111/j.1523-1739.2005.00424.x.

54. Hasegawa M, Kishino H, Yano T-A: Dating of the human-ape splitting by a molecular clock of mitochondrial DNA. J Mol Evol. 1985, 22: 160-174. 10.1007/BF02101694.

55. Hey J, Won Y-J, Sivasundar A, Nielsen R, Markert JA: Using nuclear haplotypes with microsatellites to study gene flow between recently separated cichlid species. Mol Ecol. 2004, 13: 909-919. 10.1046/j.1365-294X.2003.02031.x.

56. Nielsen R, Wakeley J: Distinguishing migration from isolation: a Markov chain Monte Carlo approach. Genetics. 2001, 158: 885-896. Central

57. Johnson JA: Recent range expansion and divergence among North American prairie grouse. J Hered. 2008, 99: 165-173. 10.1093/jhered/esn002.

CHAPTER 14

GEOLOGY AND VASCULAR PLANTS EVOLUTION

Aldo Piombino

NATURAL DRIVERS OF CLIMATE CHANGES AND THEIR EFFECTS ON EVOLUTION

Climate is regulated firstly by the amount of solar radiation reaching the surface; this value is mainly a function of the latitude. But there are other astronomical factors regulating it. Earth orbit variations are globally felt: the effects of orbital forcing, known as Milankovitch cycles, have been well recognized in many sedimentary series in plenty of geochemical and stratigraphic parameters; there are also changes in solar activity rate at a pluricentennial scale: for example, in Holocene colder and warmer period (such as the Medieval climatic optimum and the Little Ice Age) are linked to these variations that can produce shifts of the limits of climatic belts. All these changes are poorly reflected in life evolution because they are too brief to promote evolution. From upper Tertiary the onset of the Antarctic glaciation amplified the climatic effects of orbital forcing.

Geological events are able to modify climate and environment at all the different scales, from a global perspective to strictly delimited areas, thus affecting and driving evolution on Earth. A main process that changes climate is Plate Tectonics: the wandering continents change their latitude, thus modifying the strength of solar radiation reaching their surface; the consequent climate and environment changes constitute major evolutionary forcing.

Other geological factors are orography, sea level, ocean, and landmasses distribution and their variations. Also marine currents and atmospheric winds regime are strongly influenced by landmass and ocean distribution.

The effects of the variation in landmasses distribution can be felt over long distances if they affect ocean currents. For example, in late Tertiary the northward drift of the Australian continent closed the seaway across which the warm waters from Central Pacific went to the Indian Ocean. In consequence the Indian Ocean cooled, and the East African region experienced a second drop in rain rate, after the one occurring when the mountains of the East African Rift rose [1].

The arrival of landmasses in polar area has strong consequences in their climate but also carries on global consequences: the development of ice caps drops sea level, while ice and permafrost retain variable amount of greenhouse gas like CO_2 and CH_4, triggering global temperatures falls.

After the break-up of a continent, new clades can rise from the division of formerly single populations, while the union of two continents results in a complex pattern of biotic interchanges, extinctions, and fast and spectacular radiations and differentiations.

Landmass union and intracontinental rifts have raised mountain belts that changed climate and created barriers for rains and biotic movements between the two sides of the uplifted belt.

Citation: Aldo Piombino, "Geology and Vascular Plants Evolution" International Journal of Evolutionary Biology, vol. 2016, Article ID 9264357, 9 pages, 2016. doi:10.1155/2016/9264357.

ATMOSPHERIC CO_2 AMOUNT AND LARGE IGNEOUS PROVINCES

The heavy links between life and continental drift have been well established in the second half of XX century (mainly in the 60s) while the effects of greenhouse changes (particularly CH_4 and, more, CO_2) have been poorly underscored until today, but global climate (and above all temperatures and what happens when they change) is a direct consequence of the amount of atmospheric CO_2, this value being influenced by many factors.

Among atmospheric factors, a main role is played by the greenhouse gas amount. Early Earth atmosphere was composed of more than 95% of CO_2, as the current ones of Venus and Mars. Young Sun radiation was weaker than in Phanerozoic times: only thanks to these high atmospheric pressure and greenhouse conditions Earth could retain liquid waters in Archean and Paleoproterozoic times, while Earth with the current atmosphere would be globally iced.

Volcanoes emit several hundreds of thousands of tons of CO_2 yearly: this "ordinary" gas flow from solid Earth to atmosphere is lower than the carbon dioxide demand of the Earth system; thus, currently, the residence time in the atmosphere of a carbon dioxide molecule is shorter than 4 years.

The best-known CO_2 users are plants and photosynthesis changed the world: more than 2 billion years ago, soon after the onset of this process, atmospheric carbon dioxide content dropped while the rise of oxygen occurred (the Great Oxygenation Event) because of photosynthetic unicellular algae; so these tiny organisms changed the environment from reducing to oxidizing conditions and also global temperature got down and the first global glaciation event occurred.

Sea water adsorbs CO_2 (the more its amount, the more the oceanic acidity); land surface and marine bottom use CO_2 in silicates minerals weathering and carbonates formations; another strong mineral sequestration process is the formation of hydrocarbons and coals reserves. Other CO_2 is trapped in marine sediments, returning in the mantle when their plate arrives at a subduction zone.

An important remover of atmospheric CO_2 is represented by polar areas: the arrival of landmasses in high latitude has strong consequences in their climate but also carries on global consequences: not only does the development of ice caps drop sea level, but, more, ice and permafrost retain huge amount of greenhouse gas like CO_2 and CH_4, triggering global temperatures falls. Atmospheric CO_2 dramatically dropped at the onset of Antarctic glaciation in Neogene and the current occurrence of ice caps is the main reason for the current very low CO_2 atmospheric content.

Only strong emissions from inner Earth can increase CO_2 atmospheric amount: this happens during the Large Igneous Provinces activity.

Large Igneous Province (from here LIP) is a concept introduced by Coffin and Eldhom in 1991 [4], joining into single definition continental flood basalts and oceanic plateaus. A LIP is an emplacement of hundreds of thousands of cubic kilometers (also millions of cubic kilometers) of mainly basaltic magmas in a very short period of time: flow-by-flow analysis of paleomagnetic directions along 10 traverses of the Western Ghats in India suggests that at the final Cretaceous a large fraction of the main sequence of the Deccan Traps was formed in a succession of less than 30 short eruptive events in a time interval lasting for less than 10 ka [5]. Similar considerations come from studies performed on Jurassic Karoo Basalts [6] and latest Permian Siberian Traps [7].

The emplacement of a LIP often results in a lot of environmental consequences at global scale, affecting atmosphere, soils, and hydrosphere because of the massive magma and volatiles production. This has been well established in Deccan Traps, the LIP that occurred at the end of Mesozoic, clearly related to dinosaur extinction (for a good examination of the topic see a recent volume of the Geological Society of America Special Publications [8]): single eruptive events of a LIP activity produce a lava volume ranging from 1,000 to 20,000 km^3 in few tens of years; during each of these short-lasting paroxysmal phases the releases of volatiles exceed the sequestration potential of Earth system: in consequence the accumulation in atmosphere of CO_2, SO_2, heavy metals, and other substances promotes continuous and strong environmental variations; in this way many parameters increase, like terrestrial weathering patterns, ocean and atmospheric temperatures, and water acidity in the seas and rains; many extinction events are also accompanied by severe sea level oscillations (particularly, a fast transgression preceded by a drop); in the worst cases also anoxic events with a worldwide distribution can occur. Figure 1 summarizes the climatic effects of LIP emissions in atmosphere.

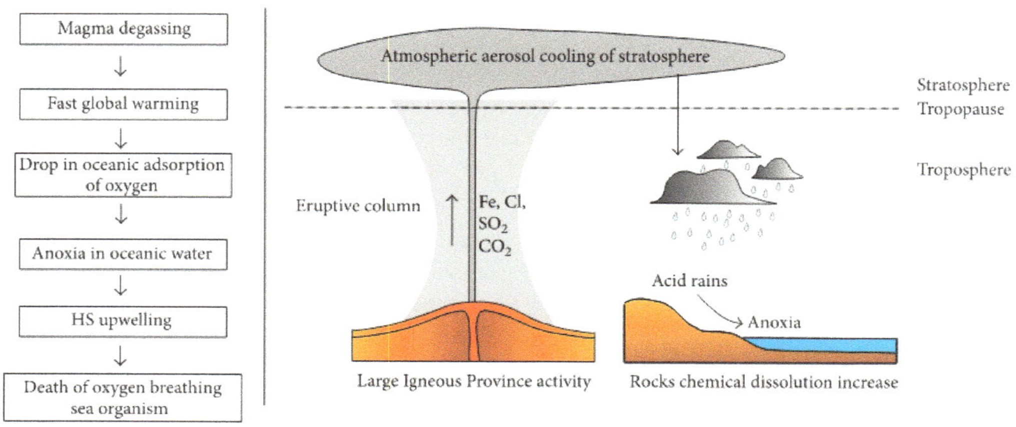

Figure 1: The effects on soil, atmosphere, oceanic waters, and rains of volatiles emitted by a Large Igneous Province.

Laki eruption in 1783 in Iceland can be a pale example of what happens during a LIP flow: in the island volatiles killed most of livestock and a strong famine occurred; in all Europe a sulphur cloud damaged crops and the winter 1784/84 recorded the worst mortality level in England and northern Europe of the century, especially for lung and heart diseases. And Laki produced only 16 cubic kilometers of lava in 8 months, a little quantity when compared with a single event of a LIP activity.

Environmental perturbations related to a LIP emplacement begin well before the spike of volcanic activity, probably because in the early stages the lava produces a huge amount of pyroclastics and volatiles coming from metasomatism and other reactions of the sediments in which they intrude; the higher the thickness and/or the evaporitic content of the sediments, the higher the production [9].

Figure 2, from [2], depicts the major LIPs events that occurred from early upper Permian (260 million years ago) until today. There is a clear coupling between LIP emplacements and biosphere changes, the strongest changes being the main mass extinction events. The very high speciation rate immediately before and after an extinction event reflects the continuous adaptation of the biosphere to new conditions: this is demonstrated by the biozones, which follow one another at a faster pace before and after the peak of the crisis, even if diversity generally falls because the extinction rate exceeds by far origination rate.

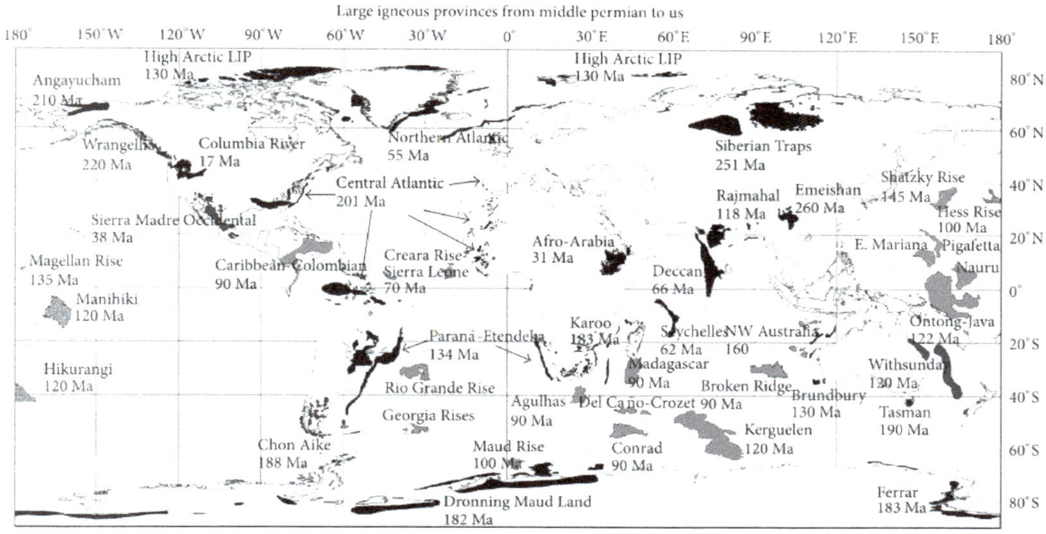

Figure 2: From [2], map of the Large Igneous Provinces emplaced from the Middle Permian (260 million years ago).

It is also noticeable that from the end of the Devonian to the beginning of the Eocene all the major chronological divisions in geological times are coeval with the occurrence of a LIP.

Figure 3, modified after [3], depicts the medium atmospheric CO_2 content from the Devonian to today, calculated analyzing carbon isotope composition from stomatal anatomy of fossil leaves: this method is based on the inverse relationship observed in many plants between stomatal number and CO_2. CO_2 was low (<500 ppm) during periods of long-lived and widespread continental glaciations and high (>1000 ppm) during warmer periods [10].

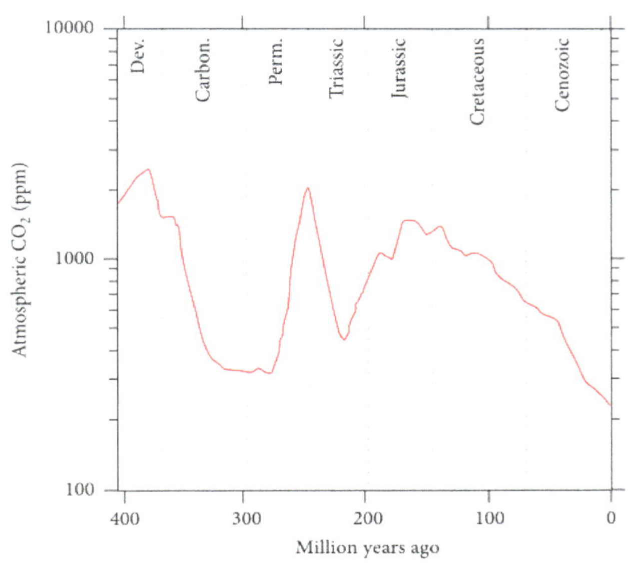

Figure 3: Atmospheric CO_2 in the last 400 Ma from [3].

The continuous drop in CO_2 atmospheric content in the Devonian can be related to plant colonization of the interior of continents [11]: if, during the Early to Middle Devonian, land plants were clustered in lower floodplain and coastal delta habitats, from the Middle to Late Devonian, they began to occupy more upland regions, thus increasing CO_2 consumption.

This trend was briefly interrupted in the Late Devonian because of the occurrence of some LIPs that injected in the atmosphere a great amount of this gas. A second process concurring with the atmospheric CO_2 drop started at the beginning of the Permo-Carboniferous glaciation: the arrival of landmasses in both the high latitude areas, Gondwana in the southern hemisphere and Siberia in the northern.

CO_2 atmospheric content grew at the end of the Permian abruptly, because of the final melting of the Permo-Carboniferous ice caps and because of the very high emissions produced by the Emeishan and Western Siberian Traps: for these reasons and for the occurrence of a large supercontinent in which moisture transport was difficult, the Early Triassic has been very hot and dry [12].

The high temperatures and CO_2 amount of Mesozoic atmosphere were a direct consequence of the widespread LIP magmatism of that time, related to the onset of the Gondwana break-up. So, LIP magmatism is the main dispenser of this gas and when the frequency of this activity dropped, also atmospheric CO_2 content did so. In Tertiary LIP activity has been weaker than in the Cretaceous (there have been only 3 smaller events on land and no one oceanic plateau has formed); contemporarily the Antarctic ice cap formation added a new, strong CO_2sequestration process, witnessed by the increase in the CO_2 drop rate from the end of the Oligocene to today. In consequence, CO_2 amount collapsed, as temperatures did, with plenty of environmental changes worldwide.

MASS EXTINCTIONS AND PLANTS EVOLUTION IN THE FOSSIL RECORD

Life on Earth has been largely affected by various mass extinction episodes, defined by Jack Sepkoski as short-lasting phases showing significant extinction rate in various nonrelated clades at a global scale [13]. Before and after these events life is very different since former dominant groups vanished and new groups, with a rapid origination, assume the leadership.

In 1983 Niklas et al. described four main phases of plant biodiversity growth [14]:(1)The first origination of early terrestrial plants and the rise of the first forest in Silurian times.(2)Ferns and lycophytes of the lower Carboniferous.(3) Gymnosperm diffusion from Upper Carboniferous.(4)The rise and diffusion of the angiosperms from the lower Cretaceous.A new method, PyRate, uses a Bayesian analytical framework to origination and extinction dynamics using fossil occurrence data of plant fossil occurrences from the Paleobiology Database (PBDB; https://paleobiodb.org/cgi-bin/bridge.pl) [15]. Compared to other approaches, this method has the advantage of using all fossil occurrences in the analysis (e.g., not only first and last appearances) and of jointly estimating (1) the preservation rate, (2) times of origination and extinction of all lineages, and (3) rates of origination and extinction through time. The inference is carried out in a sound statistical framework, which has been shown to be robust under many evolutionary and sampling scenarios through extensive simulations [16]. By exploiting the properties of Bayesian modeling, PyRate generates parameter estimates provided with credible intervals while accounting for different sources of data and parameter uncertainties.

PyRate tracks vascular plant origin and diversification from the Silurian to today, dividing them into four main clades: vascular plants, spore-bearing plants, nonflowering seed plants, and flowering seed plants. The obtained data are depicted in Figure 4, from [17].

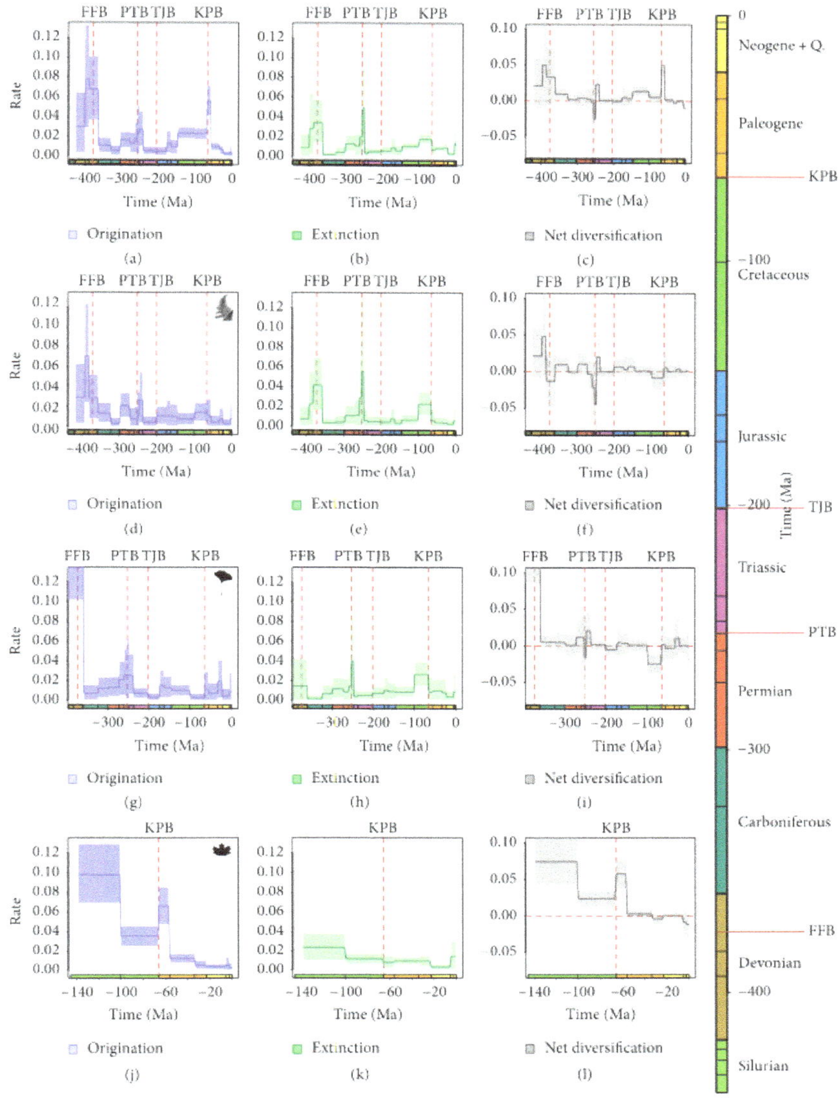

Figure 4: From [17], generic level diversification analysis of all vascular plants. Origination (blue) and extinction (green) rates were estimated using the Bayesian approach implemented in PyRate, within time bins defined as epochs of the geologic timescale (shown at the bottom). The timescale (-axis) is given in million years ago (Ma). Solid lines indicate mean posterior rates, whereas the shaded areas show 95% HPD intervals. The diversification dynamics were estimated for vascular plants as a whole (a–c), spore-bearing plants (d–f), nonflowering seed plants (g–i), and flowering seed plants (angiosperms: j–l). Net diversification rates (gray) are defined as origination minus extinction. The dashed lines indicate the major mass extinction events widely recognized: at the Frasnian-Famennian (FFB), Permian-Triassic (PTB), Triassic-Jurassic (TJB), and Cretaceous-Paleogene (KPB) boundaries. Plant silhouettes were obtained from http://phylopic.org/.

The oldest uncontroversial plant fossils are cryptospores of mid-Ordovician age [18]. At these times CO_2atmospheric content was very high, some 8–20 times the preindustrial atmospheric level [19]. Even if climate was not so hot as expected, high CO_2 levels are undoubtedly a good chance for plants growing, especially with concurrent humid conditions.

LOWER DEVONIAN EXPANSION OF VASCULAR PLANTS AND THE UPPER DEVONIAN BIOTIC CRISIS

The first vascular plants lived as early as the Early Silurian. PyRate shows a very high vascular plants origination rate in the Devonian, reaching the higher level ever in the first half of the Upper Devonian. This high level is an effect of the radiation of terrestrial plants during a huge increase in biomass amount that strongly changed continental grounds and modified hydrological cycle [20].

Nonflowering seeds plants originated in the Upper Devonian, exhibiting a high origination rate: the development of heterospory in progymnosperms with the spores transformed in seeds allowed them to reproduce also in harsher areas.

This novel reproduction system has been very useful in Devonian continents where arid or semiarid environment was widespread, raising the ability of plant to colonize larger areas than before: the fast radiation triggered a fast evolution and both origination and extinction rates have been constant up to the beginning of the Late Devonian.

In latest Devonian origination rate dropped two times, slightly before the Frasnian-Famennian border (from here FFB) and at the end of the Devonian, this second drop being huger than the first: PyRate shows an extinction rate spike some before the FFB, remaining constant throughout the border at slightly lesser levels and exhibiting a dramatic increase in the Famennian.

The Upper Devonian has been a very hard time: plant diversity modifications mirror two main faunal mass extinction events that occurred at the Frasnian-Famennian border (the Kellwasser event) and, soon after, at the Devonian-Carboniferous border (Hangenberg event). The end of Frasnian and the beginning of Famennian are characterized by fast climate changes, sea level oscillation, excursions in carbon and oxygen isotopic ratios, global anoxia, spike of atmospheric sulfide, and other phenomena.

This biotic event is one of the main Phanerozoic extinctions affecting animals, with a huge drop in biodiversity. PyRate demonstrates that also plants suffered a lot that time. Like most of mass extinctions, these events have been driven by the emplacement of Large Igneous Provinces: the Frasnian-Famennian border is clearly linked with the Viluy Traps in Yakutia [21]. The end Frasnian biotic crisis was followed by a second one at the very end of the period with widespread anoxic sediments and perturbation of the carbon isotopic ratio. Geological and geochemical observations witness that also the Hangenberg event is closely related to the emplacement of a LIP, but the position of the volcanic center has not been currently determined, Ogcheon Belt in Korea being the most candidate site today [22]. It is noticeable that the LIP activity briefly interrupted the dropping trend of atmospheric CO_2 content. PyRate demonstrates that the geological events of the Upper Devonian have played a major role in plants evolution, since their diversity at the beginning of the Carboniferous was well lower than at the onset of the Upper Devonian storm.

CARBONIFEROUS PLANTS: A TWO-FACED HISTORY

In the lower Carboniferous climate has been warm, thanks to the high CO_2 emissions from Late Devonian LIPs, and it was also very wet in the equatorial belt, where the main plant biomass was hosted. The climate was also stable that time, so, despite the floral abundance, either origination or extinction rate was slow.

We must also note a dramatic fall of atmospheric CO_2 throughout all the lower Carboniferous [23]. These dryer conditions have been harder for spore-bearing and favored nonflowering seed plants. This extinction event is not related to a LIP occurrence, as occurred well later, at the Eocene-Oligocene border: it is interesting that both these biotic crises are placed near the beginning of the two last Phanerozoic major glaciation phases.

So, we can say that mid-Carboniferous rainforest collapse witnesses another main link between geology and plant evolution, even if its exact causal mechanism is still poorly known. Climate becomes colder and dryer than before throughout the Upper Carboniferous, with very low atmospheric content of CO_2 and methane, also because they have been partly trapped in the large areas covered by thick ice caps and permafrost and in high latitude seafloor.

Whatever the geological origin of the problems affecting vascular plant in the mid-Carboniferous, in good agreement with the drying climatic trend, PyRate detects opposite behaviors in spore-bearing and nonflowering seed plants diversity between the lower and Upper Carboniferous: in the Upper Carboniferous spore-bearing plants reduced their diversity, with a huge drop in speciation and a very little extinction rate compared to the first half of the period, while nonflowering seeds plant performed a diversity growth even if they have been affected by a stronger extinction rate than before.

It is interesting that in Upper Carboniferous reptiles performed a wide radiation that, like the Devonian nonflowering seed plants diversification, has been driven by a reproductive step: the new eggs allowed the birth of an animal quite similar to an adult with very fewer needs of water than an aquatic tadpole.

PERMIAN AND THE END-PERMIAN GLOBAL CRISIS

Also in the Permian the lower and the upper part of the period show different trends in plant diversity, related to different climate conditions: the Early Permian marks the end of the major high latitude Gondwana glaciation [24]. The dropping CO_2 trend stopped. Thanks to a global temperature and moisture rise a huge increase in spore-bearing plants diversity occurred in the lower Permian, well detected by PyRate.

The warming trend can be partly ascribed to the earliest Permian Skagerrak basalt LIP emissions. The end of the Permian as well as the beginning of the Triassic has been one of the most difficult times for life on Earth: two main faunal extinction events occurred, one near the end of the Capitanian and the second at the end of the period, less than 8 Ma later. The faunal replacement has been so heavy that the end of the Permian has been the worst extinction event ever, establishing the limit between Paleozoic and Mesozoic era since the half of XIX century.

These events have been caused by the emplacement of two Large Igneous Provinces: the Emeishan Traps, now Southern China, and the Siberian Traps, east of the Ural Mountains.

Plants have been severally affected by the extinction: greenhouse gas and volatiles raised temperature and rain acidity in a drying world [25], triggering massive environmental changes that drove, either in spore-bearing plants or in nonflowering seed plants, the most dramatic diversity drop ever. But the diversity behavior shows a complex trend.

Spore-bearing plants slowed their origination rate in early upper Permian times: the onset of volcanic activity in Southern China coincided with their fall in origination rate and increase in extinction rate that began well before the faunal extinction spike of the upper Capitanian. In the latest Permian, between the two faunal extinction spikes, a lower origination rate is accompanied by a dramatic increase in extinction, prolonged toward the earliest Triassic, when it has reached its maximum level. Nonflowering seed plants show a similar pattern.

So, PyRate suggests heavy links between plant history and the environmental storm triggered by the volatiles released during the emplacement of Emeishan and Siberian basalts. It is noticeable that a spike in origination rate occurred after the second LIP activity in the earliest Triassic, but it could not compensate the even major extinction rate: thanks to the very high level of CO_2, partly of deglaciation origin, partly emitted during both LIPs' activity, these times have been the hottest and the driest in the Phanerozoic Eon [11], a very hostile period for life. The high plants origination and extinction rates demonstrate that environmental changes have been strong, frequent, and continuous during all the latest Permian and earliest Triassic.

MESOZOIC ERA AND THE RISE OF FLOWERING SEED PLANTS

After the first 6 Ma of the Triassic, temperatures went down, thanks to the various CO_2 sequestration processes, allowing plants and animals to reconquer the equatorial belt in the early Anisian. In this stage diversity fast increased either in spore-bearing or in nonflowering seed plants. During the following part of the Triassic there are no major changes in plant biodiversity patterns.

The Triassic ends with the emplacement of the Central Atlantic Magmatic Province, the initial stage of Pangea break-up; magmatic activity triggered a severe mass extinction event in animals: many clades of Archosaurs got extinct, as most of mammalian reptiles, and a fast dinosaurs radiation and differentiation took place. This event did not affect vascular plants.

Early Jurassic warm and wet environment promoted a high differentiation in spore-bearing plants. A peak in speciation and extinction is placed in the middle Jurassic, with a low drop in diversity of spore-bearing plants while diversity in nonflowering seed plants little increased. This scenario can be related to a minor extinction event at the end of the Toarcian, corresponding to a global anoxic stage, clearly related to the environmental effects of the Karoo-Ferrar LIP, immediately before the beginning of the Gondwana break-up: a little later the blocks formed by Africa-Arabia and South America split out from the block formed by India, Antarctica, and Australia.

In Cretaceous times the main event for vascular plants has been the strong angiosperms diversification.

The origin of this group is debated because molecular data point to an older, Triassic age for this clade, when the ancient most fossil has been found in lower Cretaceous sediments, even if there are controversial claims about some Triassic fossils.

Thus, there is a gap of several tens of millions of years between the molecular and the paleontological timing (no less than 60). Many explanations have been proposed for this bias: it could be simply of paleontological origin (the lack of fossils) but also methodological issues in the analysis of genetic data must be considered: for a good and recent examination of the topic see Beaulieu et al. (2015) and references therein [26].

Could the rise of angiosperms be well older than their huge radiation?

For Doyle Triassic molecular dates may be reconciled with the fossil record if the first angiosperms were restricted to wet forest understory habitats and did not radiate outside this environment until the Cretaceous [27].

If this last reconstruction is true, something has occurred promoting the nonflowering seed plants replacement with angiosperms.

Also, in this case, a geological trigger can be considered. Climate was hot thanks to the high, even if lower than in the early Triassic, CO_2 content that grew (up to four times higher than today), because of the widespread LIPs activity of the lower Cretaceous, during which the Caribbean, Ontong Java-Manihiki-Hikurangi, Madagascar, Kerguelen, and other more Large Igneous Provinces had been emplaced. Climate was wetter than the early Triassic, because of the completely different arrangement of the continents: Pangaea break-up resulted in newborn oceans interposed between smaller landmasses compared to the former supercontinent; these oceanic arms allowed humidity distribution in larger areas than before, diminishing the extension of arid and semiarid lands.

Moreover, most of the boreal continent Laurasia was placed at low northern latitudes, in the tropical area.

Thanks to both of these concurrent factors, higher CO_2 levels and wider humidity distribution, plant biomass was higher than before, resulting in an increase of photosynthetic activity: in this way also atmospheric oxygen increased. The higher atmospheric oxygen content raised the occurrence of wildfires that became very common later, in the upper Cretaceous, witnessed by the widespread occurrence of fullerenes in the sediments. Because of their resins content, wildfires can have represented a major challenge for conifers more than for angiosperms that replaced nonflowering seed plants in many environments [28].

In the upper Cretaceous, PyRate shows a drop in all vascular plants, also in angiosperms.

This is apparently strange, because the Gondwana break-up divided into some blocks the former greater continent and this event theoretically would have triggered a diversity increase.

Why is this decline in a commonly reputed happy period for life and of splitting continents? Climate becomes newly harsher, with pronounced seasonality also at low latitudes, and an aridification of waters, rains, and soils occurred, while the high atmospheric O_2 and the prolonging of dry seasons increased wildfires frequency, dramatically affecting plants diversity. Also various episodes of sea level oscillation can be blamed for this diversity drop which occurred also in animals.

The Cretaceous ended with the most famous of all mass extinctions, in which not only dinosaurs have been wiped off. As written before, for this event the Yucatan meteorite impact has been blamed but this hypothesis has been challenged, and now another LIP, the Deccan Traps, is currently the best candidate for the killer role [29], because the chain of events recorded in latest Cretaceous sediments is the typical sequence of the geological, geochemical, and environmental disturbance related to a LIP emplacement.

FROM THE PALEOCENE TO TODAY: VASCULAR PLANTS DIVERSITY IN THE GLOBAL COOLING TREND

As what happened in the other biotic restorations before mass extinctions, the Paleocene is characterized by a huge origination trend in all vascular plants, in the aftermath of K/T extinctions and environmental disturbance. From the Eocene to today, origination rates in nonflowering seeds plants and fern-bearing plants show trend opposite to that of the flowering plants: being higher in the Oligocene and Quaternary for the former, while the latter shows higher trends in the Eocene and Pliocene: crossing climatological data withPyRate-inferred plant diversity, in the Tertiary the warmer the period, the higher the origination rate in flowering plants, while colder periods show an increase in nonflowering and spore-bearing plants.

After the climax in the middle of the late Cretaceous, global temperatures began to fall: the Mesozoic greenhouse climate has been driven by the widespread occurrence of LIPs; but since their frequency lowered, Earth system has been able to soak up much more CO_2 than the ordinary volcanic emission amount: in consequence also temperatures trended downward after the highest point reached during the Turonian, 30 Ma before the end of the Cretaceous (90 Ma).

However Eocene climate has been still warm, accompanied by two thermal highs, at its beginning (PETM (Paleocene Eocene Thermal Maximum), synchronous with the main phase of the North Atlantic Igneous Province) and in the middle (MECO (middle Eocene climatic optimum)). Rainforest was widespread also in the areas now covered by deserts or savannas of Africa, Arabia, and SW Asia at time. But at the Eocene-Oligocene border the ongoing cooling trend triggered also a little faunal extinction event (Mongolian Remodelling orGrand Coupure).

There is probably a link between Eocene-Oligocene border cooling, Mongolian Remodelling, and the onset of Antarctic glaciation, an event for which many forcings acted:(i)The global scale atmospheric CO_2 continuous fall.(ii) The arrival of Antarctica near south pole.(iii)The split between Antarctica and Australia and the Drake Strait opening that allowed the circumpolar current formation, preventing since then the mixing of polar and warmer waters in southern ocean [30].From that moment permafrost and ice caps began their development in southern hemisphere (in northern hemisphere their onset is later, since North America and Eurasia arrived later than Antarctica at their current high latitudes), hence becoming a new, main, greenhouse gas sequestration process. From the beginning of the Oligocene, rainforest began to fall in higher latitudes, replaced by an open woodland environment also in Southern Africa.

It is noticeable also that these geological changes promoted in many areas C3 plants replacements with C4 plants.

CONCLUSIONS

Vascular plants diversity dropped during various mass extinction events, especially at the two Upper Devonian events, Permian-Triassic border, Toarcian and Cretaceous-Paleogene border, while they passed the Triassic-Jurassic border event without problems. Diversity pattern shows a growth before extinction peaks, a common behavior in the frame of the biosphere recovery after a mass extinction.

Plant recovery has been slightly delayed only after the End-Permian extinction, because of the extreme climatic conditions created by the double LIP event (Emeishan Traps, Permian Traps) and by the final sudden vanishing of the last Permo-Carboniferous ice caps.

The data inferred with PyRate are in excellent agreement with geological events, thus suggesting a heavy link between plant evolution and geological events, especially the occurrence of Large Igneous Provinces.

So, the four stages detected by Niklas et al. in 1983 [14] can be interpreted with environmental changes related to geological events:(1)The high CO_2 level allowed the origination of early terrestrial plants and the rise of the first forest in Silurian times.(2)The wide occurrence in the lower Carboniferous of lands along the equatorial belt allowed the rise of the tropical rainforest and the diffusion of ferns and lycophytes.(3)The arrival of large landmasses in polar positions and lowering in LIP occurrence rate originated the cooling and aridification trends of the Upper Carboniferous culminated with the onset of the Permo-Carboniferous glaciation, thus promoting the nonflowering seed plants diffusion.(4)The diffusion of the angiosperms from the lower Cretaceous has been a consequence of the wide wildfires occurrence, triggered by the climatic and environmental consequences of the massive volcanic activity of that time and by the Pangaea break-up in a world in which moisture was more widely distributed than before.

Conflict of Interests

The author declares there is no conflict of interests regarding the publication of this paper.

ACKNOWLEDGMENTS

The author thanks Nicola Casagli and Daniele Silvestro for the useful comments and discussion about the text.

REFERENCES

1. M. A. Cane and P. Molnar, "Closing of the Indonesian seaway as a precursor to east African aridification around 3-4 million years ago," Nature, vol. 411, no. 6834, pp. 157–162, 2001.
2. S. E. Bryan and R. E. Ernst, "Revised definition of Large Igneous Provinces (LIPs)," Earth-Science Reviews, vol. 86, no. 1–4, pp. 175–202, 2008.
3. P. J. Franks, D. L. Royer, D. J. Beerling et al., "New constraints on atmospheric CO_2 concentration for the Phanerozoic," Geophysical Research Letters, vol. 41, no. 13, pp. 4685–4694, 2014.
4. M. F. Coffin and O. Eldholm, "Large igneous provinces: JOI7USSAC workshop report," Tech. Rep. 114, The University of Texas at Austin Institute for Geophysics, 1991.
5. A.-L. Chenet, V. Courtillot, F. Fluteau, et al., "Determination of rapid Deccan eruptions across the Cretaceous-Tertiary boundary using paleomagnetic secular variation: 2. Constraints from analysis of eight new sections and synthesis for a 3500-m-thick composite section," Journal of Geophysical Research B: Solid Earth, vol. 114, no. 6, pp. 1–38, 2009.
6. M. Moulin, F. Fluteau, V. Courtillot et al., "An attempt to constrain the age, duration, and eruptive history of the Karoo flood basalt: Naude›s Nek section (South

Africa)," Journal of Geophysical Research, vol. 116, no. 7, Article ID B07403, 2011.

7. V. Courtillot and F. Fluteau, "A review of the embedded time scales of flood basalt volcanism with special emphasis on dramatically short magmatic pulses," Geological Society of America Special Papers, vol. 505, Article ID 301317, pp. 301–317, 2014.
8. G. Keller and A. C. Kerr, Volcanism, Impacts, and Mass Extinctions: Causes and Effects, Geological Society of America Special Publication no. 505, 2014.
9. M. T. Jones, D. A. Jerram, H. H. Svensen, and C. Grove, "The effects of large igneous provinces on the global carbon and sulphur cycles," Palaeogeography, Palaeoclimatology, Palaeoecology, vol. 441, part 1, pp. 4–21, 2016.
10. D. L. Royer, R. A. Berner, I. P. Montañez, N. J. Tabor, and D. J. Beerling, "CO_2 as a primary driver of Phanerozoic climate," GSA Today, vol. 14, no. 3, pp. 4–10, 2004.
11. G. Le Hir, Y. Donnadieu, Y. Goddéris, B. Meyer-Berthaud, G. Ramstein, and R. C. Blakey, "The climate change caused by the land plant invasion in the Devonian," Earth and Planetary Science Letters, vol. 310, no. 3-4, pp. 203–212, 2011.
12. Y. Sun, M. M. Joachimski, P. B. Wignall et al., "Lethally hot temperatures during the early triassic greenhouse," Science, vol. 338, no. 6105, pp. 366–370, 2012.
13. J. Jack Sepkoski, "Patterns of phanerozoic extinction: a perspective from global data bases," in Global Events and Event Stratigraphy in the Phanerozoic, O. H. Walliser, Ed., pp. 35–51, Springer, Berlin, Germany, 1986.
14. K. J. Niklas, B. H. Tiffney, and A. H. Knoll, "Patterns in vascular land plant diversification," Nature, vol. 303, no. 5918, pp. 614–616, 1983.
15. D. Silvestro, N. Salamin, and J. Schnitzler, "PyRate: a new program to estimate speciation and extinction rates from incomplete fossil data," Methods in Ecology and Evolution, vol. 5, no. 10, pp. 1126–1131, 2014. ·
16. D. Silvestro, J. Schnitzler, L. H. Liow, A. Antonelli, and N. Salamin, "Bayesian estimation of speciation and extinction from incomplete fossil occurrence data," Systematic Biology, vol. 63, no. 3, pp. 349–367, 2014.
17. D. Silvestro, B. Cascales-Miňana, C. D. Bacon, and A. Antonelli, "Revisiting the origin and diversification of vascular plants through a comprehensive Bayesian analysis of the fossil record," New Phytologist, vol. 207, no. 2, pp. 425–436, 2015.
18. P. Steemans, A. Le Hérissé, J. Melvin et al., "Origin and radiation of the earliest vascular land plants," Science, vol. 324, no. 5925, p. 353, 2009.
19. C. J. Yapp and H. Poths, "Ancient atmospheric CO_2 pressures interred from natural goethites," Nature, vol. 355, no. 6358, pp. 342–344, 1992.
20. T. J. Algeo and S. E. Scheckler, "Terrestrial-marine teleconnections in the devonian: links between the evolution of land plants, weathering processes, and marine anoxic events," Philosophical Transactions of the Royal Society B, vol. 353, no. 1365, pp. 113–130, 1998.
21. J. Ricci, X. Quidelleur, V. Pavlov, S. Orlov, A. Shatsillo, and V. Courtillot, "New $^{40}Ar/^{39}Ar$ and K-Ar ages of the Viluy traps (Eastern Siberia): further evidence for a relationship with the Frasnian-Famennian mass extinction," Palaeogeography, Palaeoclimatology, Palaeoecology, vol. 386, pp. 531–540, 2013.
22. V. A. Kravchinsky, "Paleozoic large igneous provinces of Northern Eurasia: correlation with mass extinction events," Global and Planetary Change, vol. 86-87, pp. 31–36, 2012.
23. R. A. Berner and Z. Kothavala, "Geocarb III: a revised model of atmospheric CO_2 over phanerozoic time," American Journal of Science, vol. 301, no. 2, pp. 182–204, 2001.
24. C. R. Fielding, T. D. Frank, J. L. Isbell, and C. R. Fielding, "The late Paleozoic ice age—a review of current understanding and synthesis of global climate patterns," Geological Society of America Special Paper, vol. 441, pp. 343–354, 2008.
25. M. J. Benton and A. J. Newell, "Impacts of global warming on Permo-Triassic terrestrial ecosystems," Gondwana Research, vol. 25, no. 4, pp. 1308–1337, 2014.
26. J. M. Beaulieu, B. C. O›Meara, P. Crane, and M. J. Donoghue, "Heterogeneous rates of molecular evolution and diversification could explain the Triassic age estimate for angiosperms," Systematic Biology, vol. 64, no. 5, pp. 869–878, 2015.
27. J. A. Doyle, "Molecular and fossil evidence on the origin of angiosperms," Annual Review of Earth and Planetary Sciences, vol. 40, pp. 301–326, 2012.
28. S. A. E. Brown, A. C. Scott, I. J. Glasspool, and M. E. Collinson, "Cretaceous wildfires and their impact on the Earth system," Cretaceous Research, vol. 36, pp. 162–190, 2012.
29. G. Keller and A. C. Kerr, "Foreword," in Volcanism, Impacts, and Mass Extinctions: Causes and Effects, Geological Society of America Special Paper 505, pp. 5–9, Geological Society of America, Boulder, Colo, USA, 2015.
30. M. E. Katz, B. S. Cramer, J. R. Toggweiler et al., "Impact of antarctic circumpolar current development on late paleogene ocean structure," Science, vol. 332, no. 6033, pp. 1076–1079, 2011

CHAPTER 15

FOSSIL RECORD OF MESOZOIC BIRDS: IMPLICATIONS FOR EARLY AVIAN EVOLUTION

Neil Brocklehurst, Paul Upchurch, Philip D. Mannion, Jingmai O'Connor

INTRODUCTION

Birds (Aves) represent one of the most diverse and abundant vertebrate groups, with over 10,000 species [1] and an estimated 300 billion individuals alive today [2]. The avian fossil record extends back to the Late Jurassic, or possibly further [2] (although see [3] and [4] and Materials and Methods below). This fossil record, in particular that of the Mesozoic, has recently undergone a revolution as a result of an explosion of newly discovered taxa during the last three decades. At present, over 120 avian species are known from the Mesozoic, from all continents except mainland Africa. Despite this new information, controversy surrounds several aspects of avian evolution, including the timing of the origin and diversification of modern birds (Neornithes).

Much of the debate surrounding neornithine evolution focuses on the apparent discrepancy between the time of their origins according to molecular data and their earliest appearance in the fossil record. The 'traditional' view of neornithine origins envisaged the evolution of modern groups in the Cretaceous [5]–[7]. This was based on the assignment of numerous species of Mesozoic bird fossils to extant orders [8], such as the placement of the Hesperornithiformes (toothed aquatic birds with reduced forelimbs from the Cretaceous) in a clade containing loons and grebes [6], even though this requires an evolutionary reversal to the plesiomorphic toothed condition [9]. Several studies based on molecular clocks support the traditional view of gradual neornithine diversification starting in the Early Cretaceous [10]–[13]. The exact timing of these events varies with each analysis; Kumar & Hedges [10] and Paton et al. [11] suggested that Neornithes originated during the Aptian (125–112 million years ago [mya]), whereas Cooper & Penny [12] and Brown et al. [13] proposed an origin as early as the Valanginian (140–133 mya) (Figure 1). Biogeographic analysis has also supported the idea of a Cretaceous origin: for example, Cracraft [14] found neornithine evolution to have been heavily influenced by vicariance (the isolation of lineages by the splitting of continents), and suggested that they diversified with the breakup of Gondwana during the Cretaceous [15]. The divergence of the majority of neornithine clades during the Cretaceous would suggest that the Cretaceous/Palaeogene (K/Pg) mass extinction had relatively little effect on this group, although Feduccia [16] considered this unlikely, since birds are often extremely sensitive to environmental perturbations.

Citation: Brocklehurst N, Upchurch P, Mannion PD, O'Connor J (2012) Fossil Record of Mesozoic Birds: Implications for Early Avian Evolution. PLoS ONE 7(6): e39056. doi:10.1371/journal.pone.0039056.

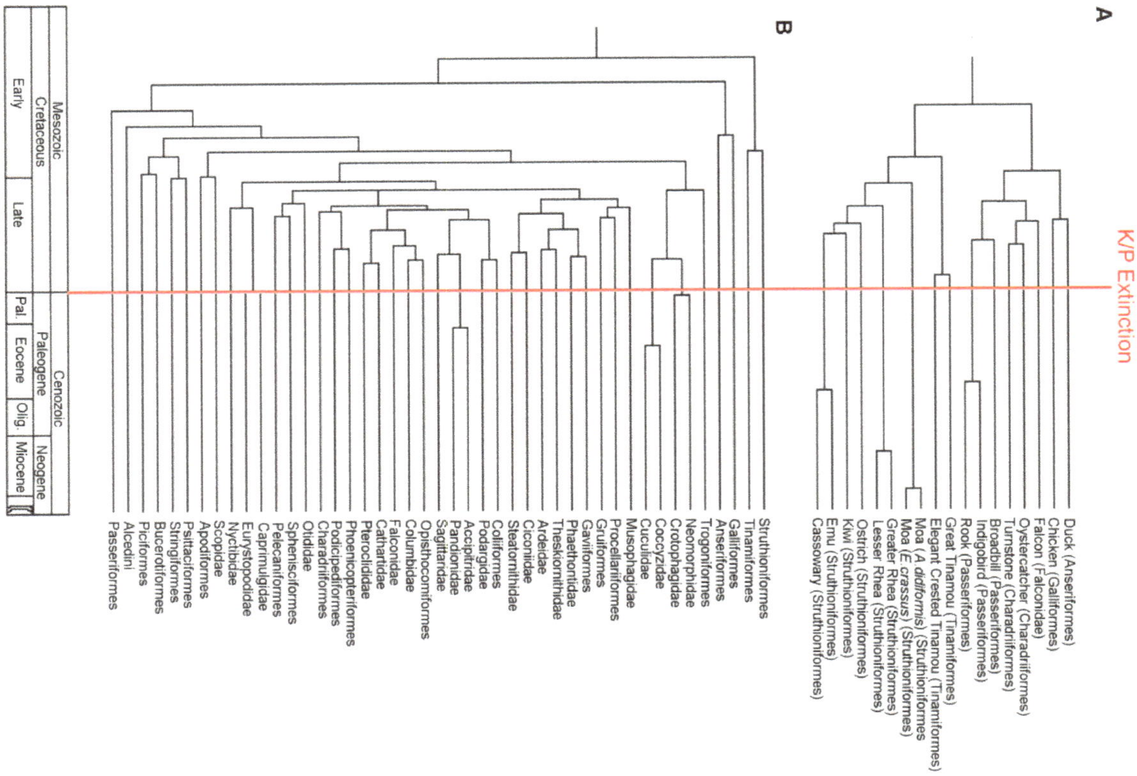

Figure 1. The two opinions on the timing of modern bird origins based on molecular clocks.

A) an origin 123 million years ago during the Aptian, modified from reference [11]; B) an origin 135 million years ago during the Valanginian modified from reference [13].

In contrast, many recent palaeobiological, and some molecular, studies have disputed such an early origin of Neornithes. Hope [17] documented 50 putative Cretaceous neornithine specimens from as early as the Coniacian, but these are all extremely fragmentary and none were subjected to formal phylogenetic analysis. The characters used to assign them to neornithine clades are often dubious or incorrect (see [18] for a summary), and attempts to place these fossils in modern groups have been hindered by poor knowledge of neornithine relationships [19]. There has been relatively little morphological analysis of modern clades incorporating their fossil representatives, and as such the relationships between the taxa and the characters that unite them are not well understood. Benton [20] argued that all records of Neornithes prior to the K/Pg boundary are either misdiagnosed or are from incorrectly dated localities; for example the Hornerstown Formation in the USA, originally dated as latest Maastrichtian and containing several putative neornithine species, has been reinterpreted by some as being from the earliest Palaeocene [21], although this remains uncertain [17]. A phylogenetically based study of the avifauna of the Hell Creek, Lance and Frenchman Formations of North America (the only formations containing bird specimens that can be reliably dated to the end of the Maastrichtian) [22] found no compelling evidence for a neornithine radiation prior to the K/Pg boundary. Instead, the majority of the birds were found to be more basal ornithurines, with three enantiornithines, none of which extend into the Palaeogene [22] (see also [23]).

Since the summary presented by Hope, more Mesozoic specimens have been assigned to the Neornithes. A coracoid, found in beds of Turonian–Coniacian age in Patagonia, was described as a galliform [24], while a quadrate originally assigned to *Cimolopteryx rara* from the Lance Formation [25], of late Maastrichtian age, was re-described as an anseriform [26]. A left carpometacarpus from the Campanian–Maastrichtian Allen Formation of Argentina has also been described as cf. Neornithes [27]. However all three of these specimens each consist of only a single bone, and their tentative assignments to neornithine clades were based on general comparisons rather than cladistic analysis. *Teviornis gobiensis*, another putative anseriform from the Maastrichtian Nemegt Formation of Mongolia [28], is better known, being represented by a complete forelimb, but again no formal cladistic analysis has been carried out, and its assignment to the Neornithes has been questioned [29]. In contrast, *Vegavis iaai* from the latest Cretaceous of Antarctica [30] has been subjected to phylogenetic analysis which supported a position within the Anseriformes. While this discovery pushes the neornithine record as far back as the Maastrichtian, this is still considerably later than is suggested by most molecular clock studies (see above).

Thus, the current palaeobiological perspective on these events is that most basal bird lineages suffered a catastrophic extinction at the K/Pg boundary, whereas the Neornithes originated in the latest Cretaceous and radiated in the early Cenozoic [14], [20], [30], [31] (Figure 2). This is supported by the molecular clock study of Ericson et al. [32], based on nuclear genes, which found that most extant avian lineages diverged after the K/Pg boundary, with only a few basal lineages appearing a short time before the end-Cretaceous mass extinction (although Ericson et al.›s results have been criticised because of poor choices of fossil calibrations and a lack of error bars around divergence time estimates [33]).

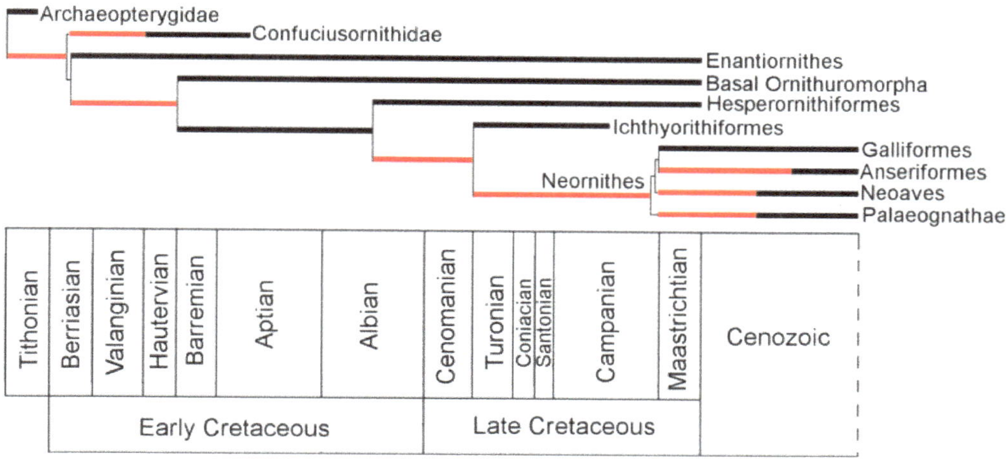

Figure 2. Neornithine evolution based on a 'literal' interpretation of the fossil record.

According to this scenario of avian evolution, the Neornithes did not appear until the latest Cretaceous, and then diversified rapidly in the Cenozoic, whereas all basal bird groups died out at or before the K/Pg boundary. Black lines represent lineages present in the fossil record, red lines represent ghost lineages inferred from phylogenetic analysis.

One way to reconcile the current fossil record of Mesozoic birds with the molecular clock-based Cretaceous divergence times for Neornithes is to propose that members of this clade were present earlier in the Cretaceous but that their fossil record is currently too incomplete for them to be unequivocally recognised by palaeobiologists. For example, Kumar & Hedges [10] implied that the reason the fossil record has not produced reliable Cretaceous neornithines is that birds are small animals with delicate skeletons, and therefore not easily preserved. However, palaeobiologists have not found this to be a compelling argument. In particular, Benton [20]noted that other animals with delicate skeletons, including basal birds, lizards, amphibians and small mammals, have been found throughout Cretaceous sediments, so their preservation is certainly possible. However, it is conceivable that a sampling bias or other taphonomic factor might have selectively preserved basal birds relative to Neornithes. For example, such a selective mechanism might have operated if basal bird species were much more abundant than early Neornithes: this is because population size might correlate with the probability of individuals making it into the fossil record. Another such bias might occur if neornithines originated and diversified in a part of the world that has a particularly poor avian fossil record, such as Africa. Finally, it is possible that neornithines showed a preference for different environments to basal birds and other small bodied taxa, and that these habitats varied in terms of their potential to preserve fossils [34].

Issues relating to the sampling and quality of the Cretaceous avian fossil record are clearly central to resolving the current discrepancies between molecular clock and palaeobiological estimates for the timing of the neornithine radiation. Several previous studies have examined the quality of the fossil record of Cretaceous birds, in order to estimate how much of the record might be missing. Fara & Benton [35] applied the simple completeness metric, which is a measure of the proportion of Lazarus taxa relative to observed taxa in the fossil record. These authors found a simple completeness metric value of 77.4% for Mesozoic birds, a high level of completeness relative to other vertebrate groups such as amphibians and squamates [35], implying a large proportion of the early avian fossil record is known. Bleiweiss [36] used gap analysis to estimate the extent to which three neornithine lineages can be extrapolated back in time beyond their first fossil occurrences, based on the number and length of 'gaps' in their fossil record. No support for a missing fossil record stretching far back into the Cretaceous was found. However, analysis of three clades represents only a small proportion of extant neornithine diversity. The accuracy of this technique also depends on the earliest known representative of a particular clade being identified and dated correctly, and this is clearly problematic for the earliest Neornithes (see above).

Fountaine et al. [37] used a different approach to assess the quality of the fossil record of Mesozoic birds. They quantified the completeness of each individual avian species, grading each on a scale of 1–4, whereby a species given a grade of 1 was represented by a single bone, 2 by more than one bone, 3 by a single nearly complete specimen and 4 by more than one nearly complete specimen. These completeness scores were then used to summarise the overall quality of the avian fossil record for time bins throughout the Mesozoic. It was found that most Cretaceous time bins had a 'good' fossil record (i.e. species graded 3 and 4 either out-numbered or were of similar frequency to those graded 1 and 2), except for the Maastrichtian, when fragmentary species were more common [37]. It was also argued that basal bird phylogeny was largely congruent with the sequence of appearance of clades in the fossil record, suggesting that there was little missing data [37]. Finally, the idea that the quality of the avian fossil record decreases with increasing stratigraphic age was disputed based on the observation that the number of bird-bearing localities and the number of described species are random with respect to geological age [37]. As such Fountaine et al. [37] deemed the avian fossil record of high enough quality to determine genuine biological signals. Consequently the poor completeness of putative Cretaceous neornithines was judged to represent their rarity at this time rather than any geological bias.

There are some drawbacks with the method used by Fountaine et al. to quantify specimen and species completeness (see 'Methods and Materials' below), and a re-evaluation of the Mesozoic avian fossil record is timely given the recent influx of new data (28 new species since Fountaine et al.›s study). In the current work, therefore, we present an updated and highly revised data set of Mesozoic birds (124 valid species), and assess fossil record quality using a recently developed character completeness metric [38]. We assess the impact of fossil record quality on the taxic diversity of Mesozoic birds by comparing completeness scores with observed diversity and a sampling-corrected diversity estimate. The impact of new discoveries is explored by comparing our updated dataset with that of Fountaine et al. [37]. Factors that might control or bias the quality of the avian fossil record are assessed by comparing completeness scores with a sea level curve, and by evaluating how fossil record quality varies with depositional environment and latitude. We also test whether the fossil record of small delicate organisms (birds) is better or worse than that of large robust forms (sauropodomorph dinosaurs). To conclude, we examine the implications of these analyses for claims concerning the presence/absence of neornithine fossils prior to the K/Pg boundary.

MATERIALS AND METHODS

Dataset

Data on the occurrences of all Mesozoic bird species were compiled from the published literature as well as the Paleobiology Database (PBDB: www.paleodb.org) and were then scrutinized for synonyms and *nomina dubia*. *Archaeopteryx* is here considered to be a bird, despite one recent phylogenetic study [39] that placed it closer to dromaeosaurs than to Aves (a further analysis, using the same dataset but applying a maximum likelihood analysis method, returned *Archaeopteryx* to Aves [40]). Since no published description of *Proornis coreae* exists, this species is a *nomen nudum* and excluded from the analysis. *Rahonavis ostromi* [41] was excluded following phylogenetic analyses [42]–[45] which recovered it as a non-avian theropod. The Late Triassic taxon *Protoavis texensis*, of dubious avian affinity [3], [4], was also not included, so the time period under study stretches from the Tithonian to the end of the Maastrichtian (150.8 to 65.5 Mya). The final dataset consists of the stratigraphic ranges, geographic distributions and character completeness metric scores (see below) for the 124 valid avian species in 82 genera. This dataset can be regarded as up-to-date as of May, 2011.

We have also used data on six other parameters: (1) the number of specimens named to species level in the literature per geological substage; (2) the environment in which the birds were preserved (data from the published literature and the PBDB); (3) the number of fluviolacustrine bird-bearing localities (data from the published literature and the PBDB); (4) the number of theropod-bearing collections per geological substage (data from the PBDB) (5) the number of dinosaur bearing formations per geological substage (6) the completeness metric values of Sauropodomorpha (data from reference [38]) and (7) sea level (data from reference [46]).

Completeness Metrics

The specimen completeness scoring systems proposed by Fountaine et al. [37] for Mesozoic birds and the similar method used by Benton [47] for dinosaurs are problematic because they are somewhat subjective and provide only coarse-grained quantifications of specimen quality. For example, where exactly is the boundary between a collection of associated skeletal elements (scored as '2' in Fountaine et al.'s scheme) and a nearly complete skeleton (scored as '3')? Different workers could assign different completeness scores to the same specimens, making it difficult to reproduce the results of analyses of specimen completeness. Also, the coarse-grained nature of completeness metrics based on just four or five categories means that important fluctuations in fossil record quality might be obscured. For these and other reasons, Mannion and Upchurch [38] introduced a fine-grained and more objective basis for assessing specimen and taxon completeness. These completeness metrics include: (1) the 'Skeletal Completeness Metric' (SCM) which attempts to capture the completeness of a specimen or taxon on the basis of the relative bulk and number of the elements that are preserved; and (2) the 'Character Completeness Metric' (CCM) which quantifies the amount of phylogenetically relevant information preserved in a specimen or taxon. In this study, we only apply the CCM because this is regarded as being more appropriate than the SCM for studies of the relationship between fossil record quality and taxic diversity [38].

In order to estimate the CCM for each species, it is necessary to obtain a list of phylogenetic characters for the group under study. Here, a list of such characters was compiled based on three phylogenetic analyses of basal birds [48]–[50], one of neornithine birds [51] (only osteological characters used), and one of coelurosaurian theropods [45]. These separate character sets were combined and duplicate characters were removed, leaving a list of 655 characters. The number of characters pertaining to different parts of the skeleton differs for coelurosaurs+birds relative to those of sauropodomorphs (see Table 1), but in both groups the complex cranial anatomy means that this region of the skeleton is responsible for a disproportionately large number of phylogenetic characters.

Table 1. The percentage of characters relating to each skeletal region in Aves and Sauropodomorpha (data on Sauropodomorpha from reference [33]).

Skeletal Region	Aves	Sauropodomorpha
Skull	32.37%	33.49%
Vertebral Column and Ribs	11.76%	25.78%
Pectoral Girdle	11.38%	3.19%
Forelimbs	19.39%	11.63%
Pelvic Girdle	10.76%	7.89%
Hindlimbs	16.18%	18.15%
Integument	0.46%	N/A

Mannion & Upchurch [38] proposed two ways of implementing the CCM. The first, CCM1, estimates the completeness of the most complete specimen of each species. The second, CCM2, assesses completeness based on the combined information from all specimens assigned to a species [38]. CCM2 was considered to be a more meaningful measure than CCM1, because it estimates the total information available from all known specimens rather than simply from the best preserved individual [38]. For example, if one specimen preserves the skull and neck region, and another preserves the neck and forelimbs, then CCM2 estimates the completeness of the taxon based on the characters that can be scored for the skull, neck and forelimbs. Here, CCM2 is preferred over CCM1 because the latter requires some species to be omitted from an analysis in cases where associations of disarticulated bones make it difficult to recognise 'the most complete individual' (e.g., as occurs in many bone bed deposits [38]). In any case, the choice of completeness metric might not be critical: Mannion & Upchurch [38] found strong positive correlations between all of the various metrics (SCM1, SCM2, CCM1, CCM2) for sauropodomorph dinosaurs.

The CCM2 score for a given Mesozoic avian species has been calculated as follows. Each element or portion of element preserved in a specimen can be scored for a given number of the total characters available. For example, a complete, well-preserved maxilla can be scored for seven characters (i.e., 1.07% of the 655 skeletal characters that can be scored for coelurosaurs+birds). Thus, if an extinct avian species is known solely from a complete maxilla, then the CCM2 for that species is 1.07%. If, however, a second specimen of this species is known, and if that specimen preserves a maxilla and a femur (the latter being scorable for 2.6% of the characters), then the CCM2 for the species is 1.07%+2.6%=3.67%. In other words, the CCM2 score for a species is the percentage of characters that can be scored for that taxon in the character list.

The characters referring to the elongated bones of the pectoral and pelvic girdles and the fore and hind limbs were divided into sets that pertain to the 'proximal end', 'distal end', and 'shaft'. Unlike Mannion & Upchurch's analysis of sauropodomorphs [38], the contribution of each skull element and each manual and pedal digit was scored individually. For each section of the vertebral column (cervical, thoracic, sacral and caudal), the characters were divided into four sets depending on whether they can only be scored when a single vertebra, an anterior vertebra, a posterior vertebra, or the entire series, is preserved. Because the neural spines are missing from the vertebrae in many specimens, characters pertaining to the neural spines were coded separately in each section of the vertebral column.

Data bins and average CCM2 scores

The quality of the avian fossil record is represented by taking the mean CCM2 score for all species occurring within a given time bin. These averages can then be plotted through time, providing a depiction of how avian fossil record quality fluctuated during the Mesozoic (see below). The standard deviation around the mean was also calculated. The time bins used were the geological stages of the Mesozoic. Each stage (the timescale provided by reference [52]) was divided into 'early' and 'late' substages with the boundary at the midpoint of the stage. When different specimens of one species occur in different substages, they were treated separately. If a specimen was of uncertain age, and could not be resolved to a particular substage, it was included as present in the entire range of substages to which it might have belonged. The same time bins have been used in the plots of raw taxic and residual diversity for Mesozoic birds, the CCM2 scores for sauropodomorphs, and the sea level curve (an interpolated version of the curve from reference [46] presented by Butler et al [53], calculating a mean average for each substage). Additionally, Mesozoic avian species were sorted by the modern latitude of the locality at which they were found. The mean completeness score for all the birds in each 5° latitudinal bin was calculated.

One problem with assessments of fossil record quality based on mean CCM2 scores is that these values might be strongly affected by sample size. A time or latitudinal bin that has yielded only a small number of specimens might have a very low mean CCM2 or very high mean CCM2 by chance, merely because the first few specimens to be found happen to be relatively incomplete or complete respectively. Such variation in mean CCM2 would not provide a very meaningful way of assessing differences in the general level of specimen completeness between data bins. Ideally, this issue should be addressed via, for example, a subsampling approach that would randomly select a specified number of specimens or species from each data bin. Unfortunately, this is not practical with regard to the current Mesozoic avian fossil record because of the relatively low numbers of species and specimens in many data bins. Such low sampling means that, either many data bins would have to be omitted from the subsampling analysis, or extremely low subsample sizes would have to be imposed on all data bins (the latter tending to result in artefactual dampening of fluctuations in completeness scores across data bins). Here, therefore, we show numbers of avian specimens appearing in the literature representing valid species in each temporal and latitudinal bin, alongside mean CCM2 scores, so that the reader can see which data bins are relatively well or poorly sampled. Conclusions based on data bins with particularly low sample sizes should be treated with caution, and such issues are highlighted in the relevant sections of the discussion.

Residual diversity estimate (RDE)

The taxic diversity counts for Mesozoic birds have been 'corrected' for potential sampling biases using the residuals method of Smith and McGowan [54]. This approach first organises the taxic diversity counts and a sampling metric into two data series so that each has its values ranked from low to high. The number of theropod (both avian and non-avian)-bearing collections (any collection [e.g. a quarry] containing theropod material) is used as the sampling metric in this case. This proxy is used since it shows a significant positive correlation with the taxic diversity, and also to address two criticisms raised against the use of proxies such as fossil-bearing collections and formations. The first is that they are redundant with diversity: if the diversity of birds decreased, one would expect a lower number of bird-bearing formations or collections [55], [56]. This problem may be mitigated by using a sampling proxy based on a group more inclusive than, but containing, the group under study [57], [58]. If the diversity of birds decreased, the diversity of other theropods would not necessarily show the same decrease, so one would not expect the number of theropod-bearing collections to decrease. The second criticism is that such proxies do not take into account non-occurrences i.e. instances where workers have looked for fossils in formations, but not found them [56]. Again, using a proxy based on a larger group that includes the group under study mitigates this concern. The number of theropod-bearing collections includes instances when searches have been made in rocks containing species closely related to birds, but no birds have been found.

All data were log transformed prior to the calculation of regression equations and statistical testing (to allow values of 0 to be log transformed, 1 was added to every value). A regression equation which expresses the relationship between these data series is then calculated: this equation represents a 'model' of the relationship between sampling and observed diversity in the fossil record. Residual diversity values are then calculated by subtracting the predicted diversity from the observed diversity (i.e., residuals represent the amount of diversity that cannot be explained by sampling) [53], [54], [58]–[63]. Confidence intervals were placed around the residual diversity using the standard deviations of the model, following the method of Lloyd [64].

Statistical tests

Two statistical tests were used to compare the time series of mean CCM2 scores to various other parameters. The Spearman's rank correlation coefficient is a non-parametric measure of how two variables are ordered [65]. Kendall's tau rank correlation coefficient is also a non-parametric statistic, which measures whether two curves change synchronously [65]. Generalised differencing was implemented to correct for trend and autocorrelation [65], [66]. To compare the completeness of bird specimens found in different environments, and the completeness of birds to that of sauropodomorphs, the Mann Whitney U-test was used. This is a non-parametric test comparing the medians and standard deviations of two data sets [65]. Statistics were executed using the computer program PAST [67].

RESULTS

Avian fossil record completeness

The mean CCM2 scores for each substage are plotted against time in Figure 3, along with their standard deviation and the number of specimens. Completeness levels are at their highest (76.84%) during the earliest substage (early Tithonian). Completeness remains high during the late Tithonian and early Berriasian, with only one taxon present in these bins (*Shenqiornis mengi* [68]; see 'Discussion' for further comments on how the uncertain date of this specimen may have affected the results). Following this, there is a decrease in completeness to 46.95% in the late Berriasian. Values start to rise again during the early Hauterivian, and continue to rise to a second peak of 66.50% in the early Aptian. The values decline slightly in the late Aptian, and at the Aptian/Albian boundary fall to 9.16%. The values remain between 0 and 20% for the rest of the Late Cretaceous, apart from a brief peak in the Coniacian and Santonian (the highest completeness score in the Late Cretaceous is 43.72% in the late Santonian, whereas the lowest is 1.53% in the late Turonian).

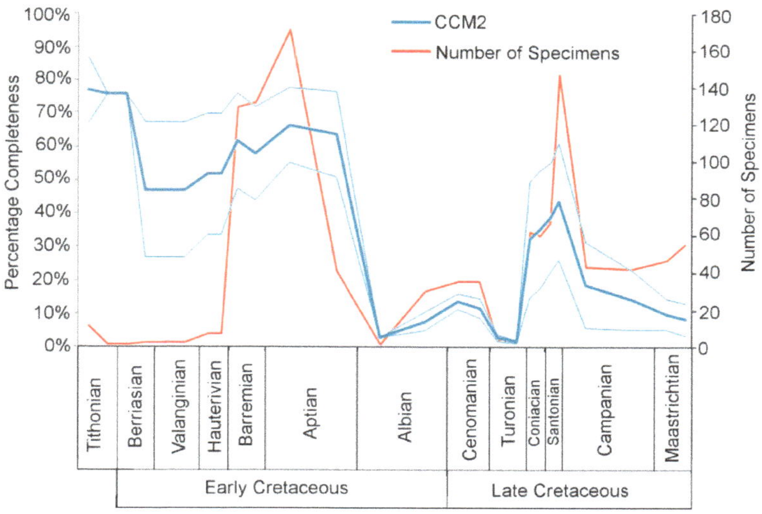

Figure 3. The mean character completeness metric scores for all Mesozoic birds in each substage.

The bold blue curve represents the mean CCM2 scores, while the thin blue lines above and below it represent one standard deviation above and below the mean. Numbers of specimens are shown by the red curve to indicate sample size.

Comparisons with Fountaine et al. [37]

The study of the quality of the avian fossil record by Fountain et al. [37] employed a different method for quantifying completeness (see above) and (inevitably given its earlier date of publication) a smaller dataset than that used herein. These differences provide an opportunity to investigate: (1) the impact of the influx of new data on Mesozoic birds during the past eight years (N.B. the dataset of Fountaine et al. was finalised in 2003); and (2) the extent to which conclusions regarding fossil record quality might vary depending on the method used to estimate completeness. The first issue is addressed by measuring the correlation between CCM2 values based on all currently known Mesozoic bird species and the CCM2 values for only the species available to Fountaine et al. (i.e. a pruned version of our dataset). The result (Table 2) demonstrates the presence of a strong positive correlation between these two CCM2 time series. The only periods in which the two curves (figure 4) differ greatly are the late Tithonian and early Berriasian. Since the only bird included in the dataset from these substages is the well preserved *Shenqiornis mengi* (although again the uncertain date of this taxon should be noted; see 'Discussion'), the completeness score for this substage in the complete dataset is 75.72%. However, this species was described after the study of Fountaine et al, so the completeness score for the late Tithonian and early Berriasian in the pruned dataset is 0%. Between the late Berriasian and the Aptian, the two curves show the same upward trend, although the complete dataset maintains a higher average completeness than the pruned dataset. Elsewhere the curves show the same peaks and troughs, although in some cases the height of these differs: the pruned dataset shows a higher completeness of species from the Coniacian, Santonian and Campanian than the complete dataset.

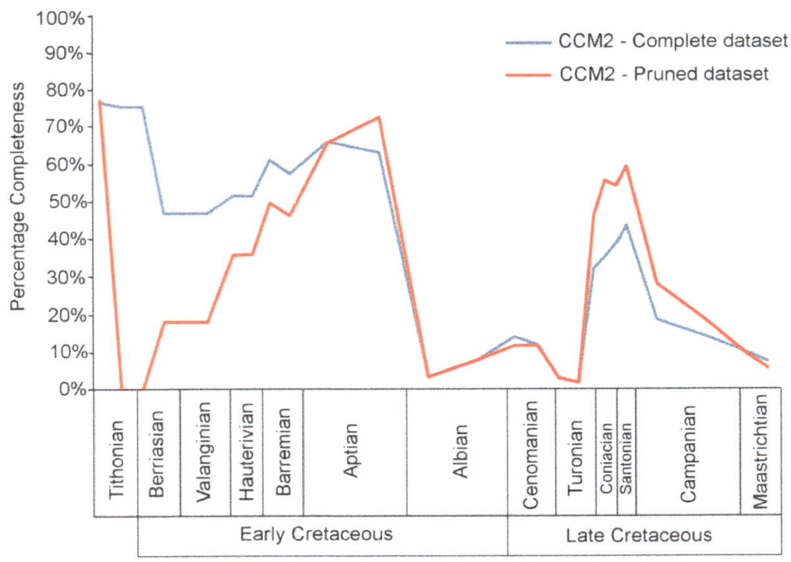

Figure 4. The CCM2 scores for Mesozoic birds, using both the complete dataset and the pruned dataset.

The complete dataset (blue curve) includes all 124 species, the pruned dataset (red curve) includes only species known to Fountaine et al [37].

Table 2. Kendall's tau and Spearman's rho correlation coefficients and uncorrected probability values (p) of the statistical comparisons of time series after generalised differencing.

Statistical Test	Kendall's tau	Spearman's rho
Theropod-bearing collections vs taxic diversity	0.38 ($p = 0.007757$)	0.52538 ($p = 0.0069971$)
Dinosaur-bearing formations vs taxic diversity	0.32923 ($p = 0.018352$)	0.44889 ($p = 0.021432$)
Mean CCM2 (Current dataset) vs mean CCM2 (taxa used by Fountaine et al. [32])	0.45231 ($p = 0.0011948$)	0.58154 ($p = 0.0018333$)
Mean CCM2 vs mean grades used by Fountaine et al. [32]	0.32308 ($p = 0.020648$)	0.47966 ($p = 0.013152$)
Mean CCM2 vs taxic diversity	0.36 ($p = 0.0099127$)	0.45505 ($p = 0.019504$)
Mean CCM2 vs residual diversity	0.39077 ($p = 0.0051217$)	0.55145 ($p = 0.0039007$)
Mean CCM2 (Aves) vs mean CCM2 (Sauropodomorpha)	−0.21231 ($p = 0.12829$)	−0.333265 ($p = 0.096826$)
Mean CCM2 (Aves) vs number of bird-containing fluviolacustrine localities	0.24308 ($p = 0.081634$)	0.33128 ($p = 0.09829$)
Mean CCM2 (Aves) vs sea level	−0.021538 ($p = 0.87738$)	0.0092308 ($p = 0.9643$)

The impact of the choice to use the 1–4 completeness grading system of Fountaine et al. [37], versus the CCM2, has been examined by calculating the mean completeness grade for all species in each substage (Figure 5). Comparison of these averaged completeness grades with CCM2 recovered a significant positive correlation (Table 2). However this correlation is not as strong as that between the CCM2 including the complete dataset and the pruned CCM2 including only species available to Fountaine et al. (Table 2). Thus it is clear that any differences in interpretation between the results presented here and those presented in Fountaine et al. [37] probably reflect the effects of choice of methodology rather than the addition of new data. The process of grading the specimens from 1 to 4 does produce noticeable differences in estimated fossil record quality compared to that based on mean CCM2 (see 'Discussion').

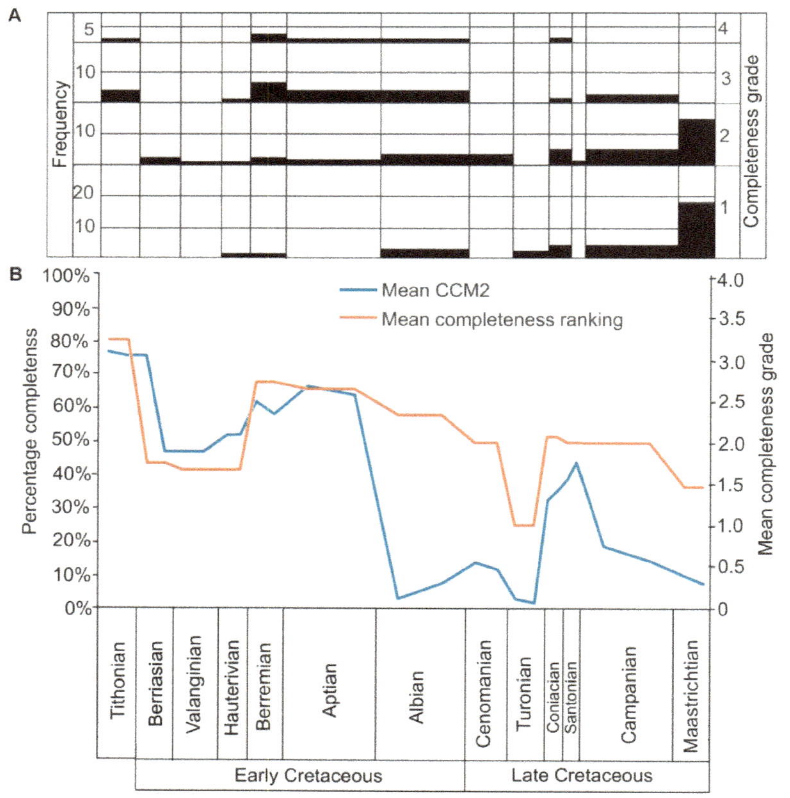

Figure 5. Comparison of the data collected by Fountaine et al. [37] **with the CCM2.**

(A) Modified from Fountaine et al.›s [37] assessment of the completeness of the fossil record of Mesozoic birds (bird species with a completeness grade of 1 are represented by one bone, those with 2 by more than one bone, those with 3 by a nearly complete specimen, and those with 4 by more than one specimen); (B) Comparison of the mean completeness grade of bird species using the method of Fountaine et al. (red curve) and the mean CCM2 scores determined in this study (blue curve).

Avian diversity and fossil record completeness

Raw (uncorrected) taxic diversity and our 'sampling-corrected' residual diversity estimates for Mesozoic birds are shown in Figure 6. The sampling corrected residual diversity curve was based on the number of theropod-bearing collections (see 'Materials and Methods'). There is a statistically significant positive correlation between the taxic diversity of birds and our completeness scores and an even stronger correlation between the latter and the residual diversity estimate (Table 2).

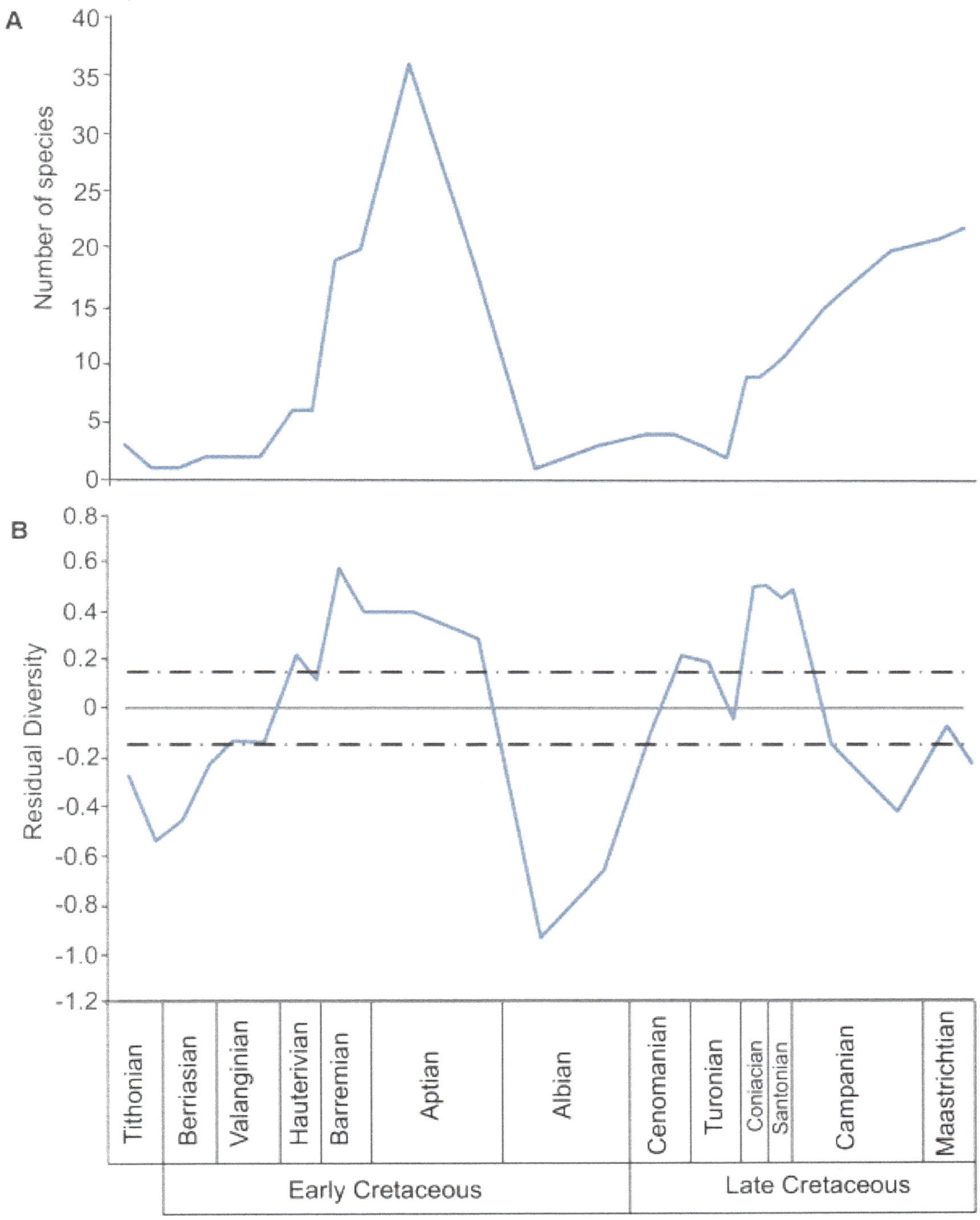

Figure 6. Two estimates of Mesozoic avian diversity.

(A) the raw taxic diversity estimate; (B) The residual diversity curve corrected for the number of theropod-bearing collections (dashed-dotted lines indicate standard deviation from the model).

Controls on avian fossil record completeness

Sea Level

The correlation between the mean CCM2 values and the sea level curve of Miller et al. [46](Figure 7) is weak and statistically non-significant (see Table 2).

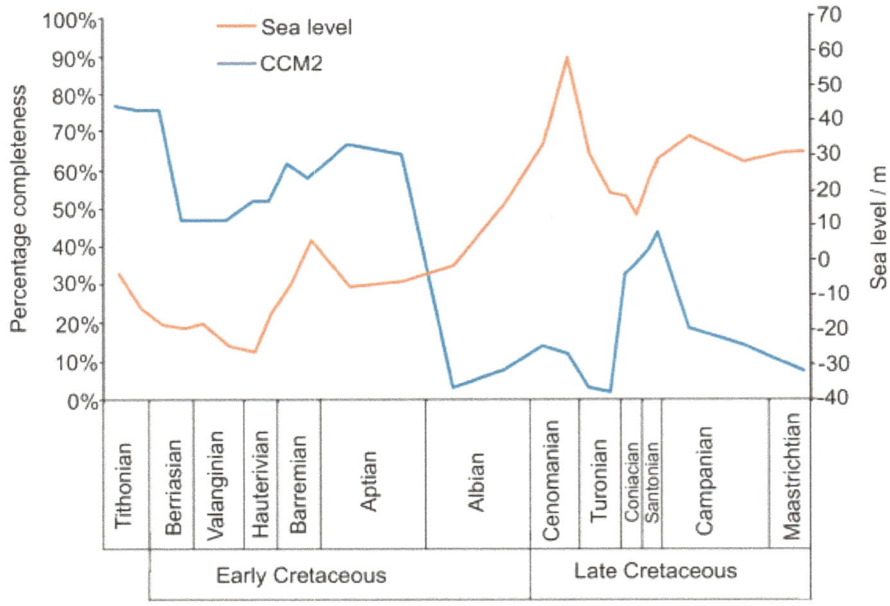

Figure 7. A comparison of sea level [53] **and the mean CCM2 scores of Mesozoic bird specimens.**

Depositional Environment

The localities that have yielded Mesozoic bird specimens were divided into three environmental categories: marine, fluviolacustrine, and non-fluviolacustrine terrestrial environments. The CCM2 scores for taxa from each type of environment were then compared using the Mann-Whitney U-test. The mean completeness scores of birds from fluviolacustrine localities were much higher than those found in the other environments (Figure 8). The Mann-Whitney test suggests that the completeness of birds from fluviolacustrine environments is significantly greater than that of species from the other two categories (Table 3). The CCM2 of birds from marine and non-fluviolacustrine terrestrial localities are more similar to each other, with no significant difference according to the Mann-Whitney test (Table 3). The mean CCM2 scores were then compared to the number of fluviolacustrine localities in each substage time bin (Figure 9) but there is no significant correlation between these two parameters (Table 2).

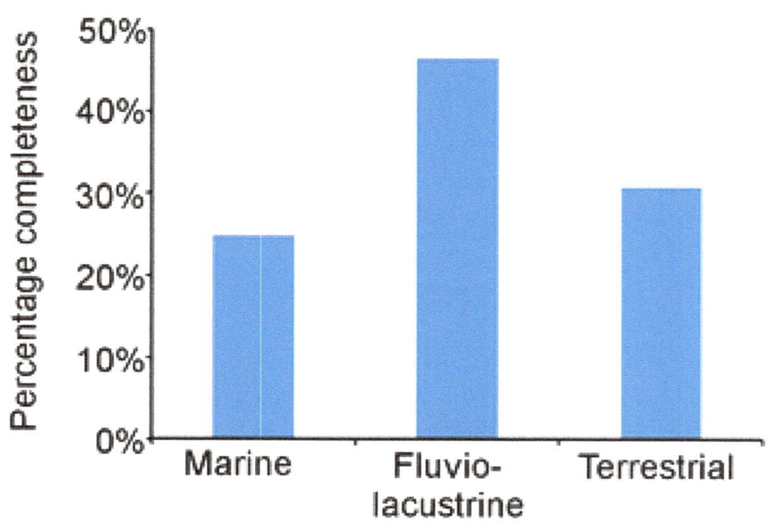

Figure 8. The mean CCM2 scores for birds found in three different palaeoenvironments: marine, fluviolacustrine, and non-fluviolactustrine terrestrial evnvironments.

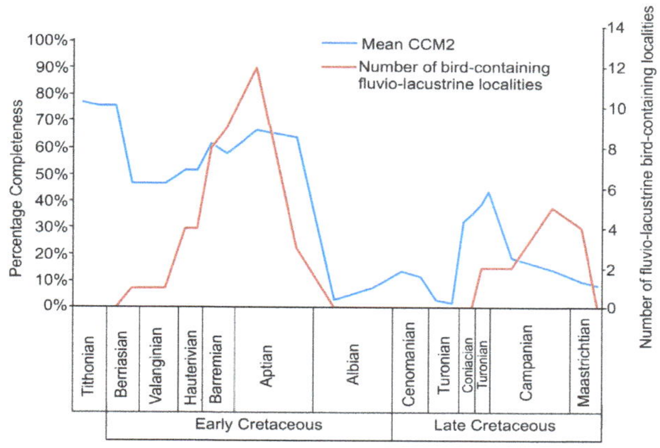

Figure 9. The number of fluvialolacustrine bird-bearing localities in each substage compared to mean CCM2.

Table 3. Mann-Whitney U-Test Scores and for the probability values of equality of medians (p), for comparisons of CCM2 scores of birds and sauropods, and the CCM2 scores of birds from different environments.

Statistical Test	Mann-Whitney U Score
Mean CCM2 of birds from fluvialolacustrine localities vs mean CCM2 of birds from other terrestrial localities	$U = 587$ ($p = 0.01273$)
Mean CCM2 of birds from fluvialolacustrine localities vs mean CCM2 of birds from marine localities	$U = 505$ ($p = 0.004826$)
Mean CCM2 of birds from marine vs mean CCM2 of birds from non-fluviolacustrine terrestrial localities	$U = 739$ ($p = 0.6983$)
Mean CCM2 of all Mesozoic birds vs mean CCM2 of Sauropodomorphs known from the Tithonian-Maastrichtian	$U = 9070$ ($p = 0.01912$)
Mean CCM2 of Late Cretaceous birds vs Mean CCM2 of Late Cretaceous sauropodomorphs	$U = 678$ ($p = 1 \times 10^{-7}$)

Avian Versus Sauropodomorph CCM2 Values

On average, the CCM2 scores for Mesozoic avian species are significantly lower than those of sauropodomorphs [38] when the entire dataset is considered, according to the Mann-Whitney test (Figure 10, Table 3). When we examine Late Cretaceous species alone, this difference between small-bodied birds with delicate skeletons and large-bodied robust sauropods becomes even more marked. When time series of mean CCM2 scores for Mesozoic birds and sauropodomorphs are compared (Figure 11), it is clear that the former series displays a much wider range of values (0–80%) whereas the latter has values that are much more restricted (20–50%). There is no significant correlation between these two time series of mean CCM2 scores (Table 2).

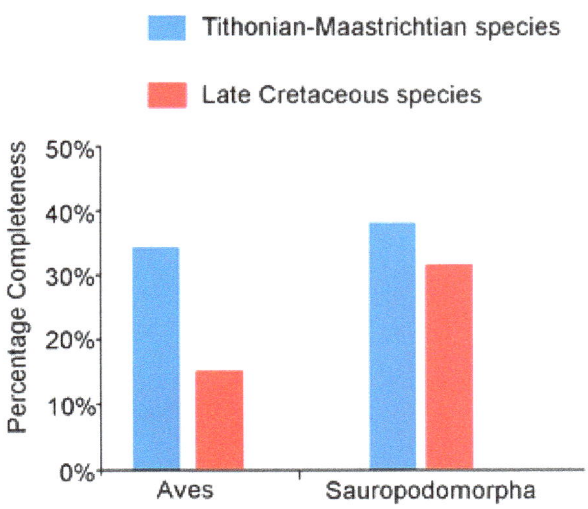

Figure 10. A comparison of the CCM2 scores of Mesozoic bird specimens and that of sauropodomorph specimens [38].

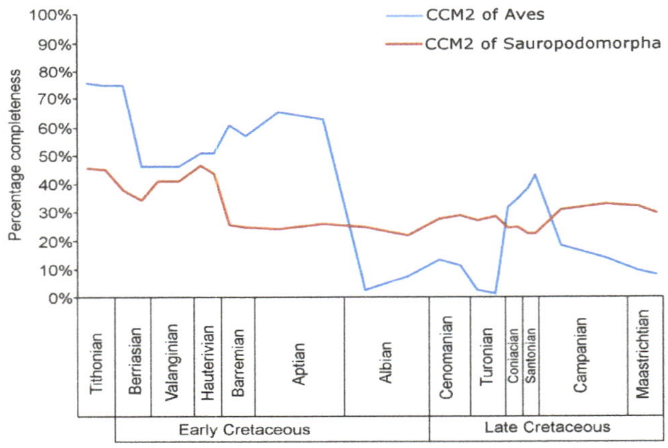

Figure 11. A comparison of the completeness of avian and sauropodomorph specimens.

The mean CCM2 scores of all birds (blue curve) and sauropodomorphs (red curve) in each substage from the Tithonian until the late Maastrichtian (data for Sauropodomorpha from reference [38]).

Latitude and Geographic Region

Mesozoic avian species are most diverse and abundant in the Northern Hemisphere between 30 and 60°N (present day coordinates), with the most complete specimens occurring between 40 and 45°N (Figure 12). There is one species (*Canadaga*) known from one locality in the 70–75°N latitudinal bin and another locality in the 75–80°N bin [69], [70], which has a much lower CCM2 score than taxa from the rest of the Northern Hemisphere. To date, no Mesozoic bird species have been found between 30°N and 15°S. The Southern Hemisphere record contains considerable gaps, with no birds found between 20 and 25°, 30 and 35°, 40 and 60°S, and none known from further south than 65°S. The latitudes from which the most Southern Hemisphere species are known are between 25 and 30°S, but the most complete specimens are found between 60 and 65°S. Similar latitudinal biases in the present day distribution of Mesozoic dinosaur fossils, including birds, were noted by Mannion et al. [71].

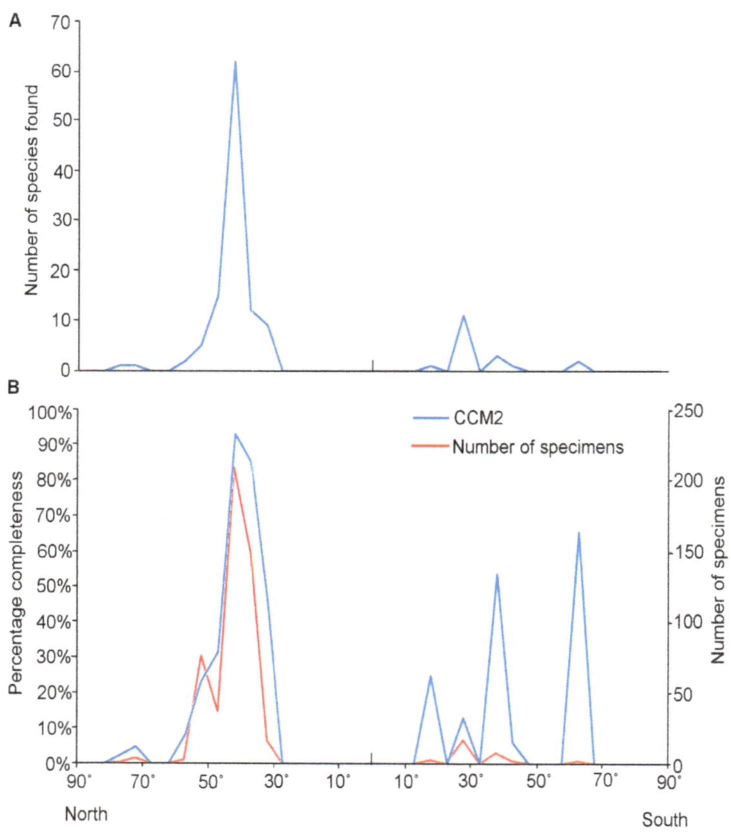

Figure 12. The number and mean CCM2 score of Mesozoic bird species found within modern latitudinal bins.

(A) The number of species found in each 5° latitudinal bin; (B) – the mean CCM2 score of all birds (blue curve) found in each latitudinal bin, with the number of specimens (red curve) to indicate sample size.

Mesozoic avian species are most diverse in Asia (Figure 13). The mean CCM2 value for specimens found in each landmass is also highest in Asia. However, the correspondence between higher species diversity and specimen completeness does not hold for other continents. For example, more species have been found in Europe and North America than in any of the Gondwanan landmasses, yet the mean CCM2 of species known from Antarctica is higher than that of Europe, and the mean CCM2 of species from Antarctica and completeness of the one species from the Arabian Peninsula are higher than that for North America (Figure 13). In Gondwana, most species are known from South America (Australasia, Madagascar and the Arabian Peninsula have each produced only one species, and Antarctica two, whereas South America has produced 15). However, the mean completeness of avian species from Antarctica and the completeness of the one species from the Arabian Peninsula are higher than that of South America (Figure 13). No Mesozoic bird species have been named from Africa or India; although one specimen has been found in Tanzania, it has not been described or named [72].

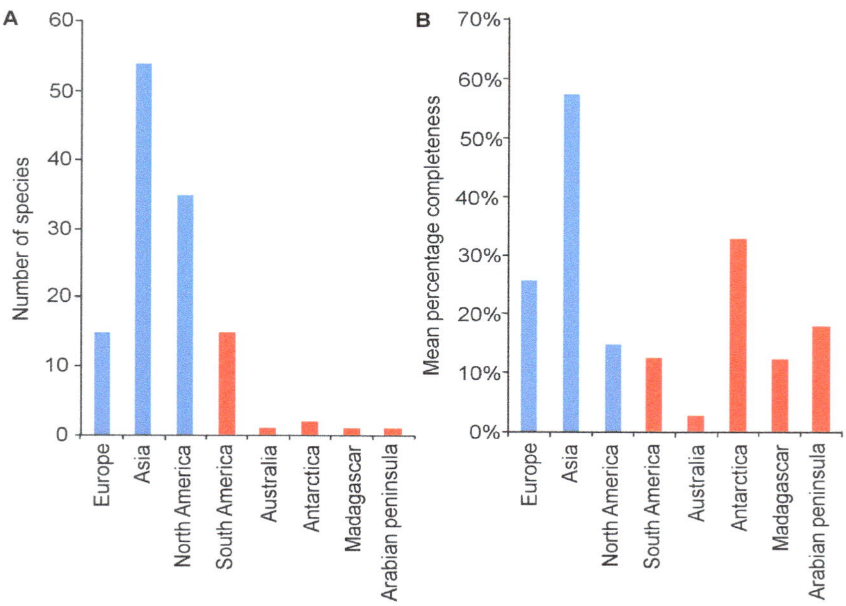

Figure 13. The geographical distribution and completeness (CCM2) of Mesozoic bird species.

(A) The number of avian species found in each landmass; (B) the mean CCM2 score of the avian species found on each landmass. Landmasses formerly part of Laurasia are represented by red bars, while those formerly part of Gondwana are represented by blue bars.

DISCUSSION

Mesozoic bird diversity

The current study presents the most complete and up-to-date analysis of Mesozoic avian diversity, including the first attempt to address potential sampling biases in the avian fossil record via the application of the residual diversity approach. Thus, although the main aim of this paper is to examine aspects of fossil record quality, a reassessment of how avian diversity fluctuated during the Mesozoic is warranted here.

The taxic diversity of Aves in the Mesozoic (Figure 6A) shows a significant positive correlation with the number of theropod-bearing collections from each substage (see Table 2). This suggests that the taxic diversity estimate is probably strongly influenced by collecting effort. Thus, the taxic diversity estimate is probably not an accurate representation of actual Mesozoic diversity. It should be noted that avian taxic diversity does not correlate as closely with the number of dinosaur-bearing formations known from each substage of the Mesozoic, a sampling metric shown to correlate with other dinosaur clades [49], [51], [56], [64]. It seems that in the case of Mesozoic birds, the effects of anthropogenic sampling biases are greater than those of the amount of sedimentary rock known from each time period.

The 'sampling corrected' residual diversity curve (Figure 6b) indicates a fall in diversity between the early and late Tithonian, a decline also apparent in the raw taxic diversity curve. In the early Tithonian time bin, three species are known: *Archaeopteryx lithographica* [74] and *Wellnhoferia grandis* [75], [76], both from the Solnhofen Limestone of Bavaria, Germany, and *Shenqiornis mengi*, from the Qiatou Formation of China. In the late Tithonian and early Berriasian, only *Shenquiornis* is present. It should be noted that the Qiatou formation has yet to be reliably dated beyond the fact that it underlies the Yixian Formation [77]. As such, in this dataset (where taxa of uncertain age were included as present in the entire range of substages to which it may have belonged) *Shenqiornis* is present from the Tithonian until the early Barremian.

Although both the raw taxic diversity curve and the residual diversity curve support a drop in avian diversity between the early and late Tithonian this is most likely a result of the early Tithonian Solnhofen Lagerstätte. This area of exceptional preservation will obviously lead to an increase in the diversity inferred from the raw species count, but can also affect the 'sampling corrected' diversity, since a relatively small amount of collecting effort can produce high numbers of specimens and new species. Thus, the decline in observed diversity seen during the Tithonian may merely reflect the presence of a Largerstätten in the early Tithonian and the absence of one in the late Tithonian.

During the Early Cretaceous, the residual curve shows a steady increase in avian diversity, which rises to a significant peak in the early Aptian. This radiation includes the appearance and diversification of many of the Mesozoic clades (see Figure 2). The earliest member of the Confuciusornithidae, *Eoconfuciusornis zhengi* [78], appears in the Hauterivian-aged Dabeigou Formation, although a ghost lineage for this family can be inferred at least as far back as the earliest member of the Enantiornithes, in this case *Shenqiornis* (as discussed above, we use the full possible stratigraphic range for this species, extending its range into the Tithonian as a result of the uncertain age of the Qiatou Formation). Otherwise, the enantiornithine *Noguerornis gonzalezi* [79], found in the late Berriasian-Valanginian Montsec Limestone of Spain [80], [81], would be considered the earliest member of this clade, although one study has provided a Hauterivian-early Barremian age for this formation [82]. The oldest members of the Ornithuromorpha appear in the Yixian Formation (Barremian–early Aptian [83]) e.g. *Archaeorhynchus spathula* [84], *Hongshanornis longicresta* [85], *Liaoningornis longidigitris* [86]and *Longicrusavis houi* [87].

At the Aptian/Albian boundary there is a significant drop in avian diversity in both the raw taxic curve and the RDE. Although this apparent extinction affected species numbers, most of the higher avian taxa survived. For example, the Confuciusornithidae had disappeared from the fossil record before this Aptian/Albian event, while the Enantiornithes and the Ornithuromorpha both survived. An alternative explanation for this pattern is that this decrease in diversity is an artefact of uneven sampling of the fossil record. The Aptian record is dominated by the Lagerstätten of the Chinese Yixian and Jiufotang Formations; as discussed above in reference to the Solnhofen Limestone this might have resulted in an increase in both raw taxic diversity and RDE. Thus, this decrease in diversity might be an artefact of, or have been exaggerated by, an overestimation of diversity in the Aptian rather than a true extinction in the Albian. A second possibility is that the perceived extinction is a result of very patchy geographical sampling; of the 23 species known from across this boundary, only two occur outside China. Such a localised record cannot be seen as representative of a worldwide evolutionary event (e.g. [58]).

Avian diversity recovered throughout the Albian and Cenomanian, before a plateau was reached. During the Coniacian and Santonian there is a second significant peak in diversity. This increase is associated with the radiation of the Cretaceous marine birds: the Hesperornithiformes and Ichthyornithiformes. Although Hesperornithiformes and Ichthyornithiformes first appear in the fossil record in the late Albian and the early Turonian respectively [88], [89], they did not become speciose until the Coniacian. This diversity peak was ephemeral, as at the Santonian/Campanian boundary diversity fell to levels lower than that of the late Turonian. As at the Aptian/Albian boundary, this decline does not seem to be accompanied by the extinction of any higher-level taxa; the Enantiornithes and Hesperornithiformes survived until the K/Pg boundary, while the Ichthyornithiformes survived into the Campanian (see figure 2).

During the early Maastrichtian, both the taxic and residual diversity estimates support an increase in diversity. However, many of the species in the Maastrichtian are too fragmentary to assign them to a particular clade, and so it is difficult to assess whether these end-Cretaceous radiations are related to the origination and diversification of new higher taxa. It should be noted that the earliest known unambiguous neornithine bird, the anatoid *Vegavis*, appears in Maastrichtian sediments [30]. This increase in diversity may reflect the diversification of this clade and other neornithine clades for which Maastrichtian ghost lineages may be inferred (see Figure 2). Longrich et al. [22] argued for an extensive radiation of more basal Ornithurae preceding the Cretaceous/Palaeogene extinction. However they noted that the only reliably dated record of Mesozoic birds immediately preceding the extinction is found in North America, and so our interpretations of the fossil record at this time should be treated with caution [22]. The residual diversity curve actually indicates a decline in diversity between the early and late Maastrichtian.

Comparisons with Fountaine et al.'s study

Fountaine et al. [37] found that no Mesozoic geological stage, with the exception of the Maastrichtian, preserves an overabundance of fragmentary avian material (Figure 5A). In contrast, mean CCM2 values presented here suggest that there is considerably more variation in fossil record quality between the stages, with extremely fragmentary material occurring in the late Berriasian, early Valanginian and several stages of the Late Cretaceous (Figure 3). This difference does not seem to reflect changes caused by the discovery of new species during the past eight years: when the CCM approach is applied to only species used by Fountaine et al., there is very little difference between the two CCM2 time series and they are strongly and positively correlated (Table 2, Figure 4). It seems that, although the 28 new species described since 2003 have improved the average completeness of taxa between the late Tithonian and the Barremian, the overall trend of completeness through time has not been affected. One must also note that the difference in the completeness scores between the complete and pruned datasets in the late Tithonian and early Berriasian is entirely due to the discovery of *Shenquiornis*. In the pruned dataset, no specimens are present in these substages, while in the complete dataset, one very complete specimen is present. Not only should the uncertain date of this specimen be emphasised (see above), but also the effect of the small sample size on our results. Other stages have shown no improvements in the completeness of specimens with the influx of new discoveries, indicating that one or more geological and/or geographical factors might be limiting the quality of specimens from such time periods. Similarly, this result indicates that recently discovered, high quality Mesozoic bird fossils have come from time bins that have previously yielded highly complete taxa.

This effect is presumably related to the continuing exploitation of Lagerstätten, such as the Liaoning avian fauna of China, which has produced 15 of the 28 new species discovered since 2003.

If the differences between Fountaine et al.'s study and that presented here do not stem from new discoveries, then they must instead reflect differences in methodology, as is demonstrated by the weaker correlation between the mean CCM2 and the mean completeness grades assigned by Fountaine et al (Figure 5b). The first distinction to note is the difference in temporal resolution: Fountaine et al. [37] calculated the completeness of the species in each stage of the Mesozoic, whereas the current study utilizes substages. Fountaine et al. also did not produce a time series with their data, but instead compared the ratios of the completeness grades in each time period. Assessment of the completeness of species based on either a 1–4 grading scheme or CCM2 results in important differences in interpretation. For example, consider how avian species completeness for the Cenomanian and Santonian are estimated using the two approaches. Fountaine et al. gave all Cenomanian and Santonian bird species a completeness grade of 2 (i.e. each species is represented by an association of a few disarticulated elements). This implies similar fossil record quality in both stages. In contrast, the mean CCM2 scores for the Cenomanian substages are 13.62% and 13.59%, whereas the Santonian substages have mean CCM2 values of 38.89 and 43.72%. Thus, CCM2 indicates a major difference in quality between the two stages. This disparity between estimates of fossil record quality stems from the relatively coarse-grained nature of Fountaine et al.'s grading system, compared to the fine-grained nature of the CCM2. For example, the Cenomanian bird species *Pasquiaornis hardiei*and *P. tankei* are represented by several bones [90], meaning they score a grade of 2 using Fountaine et al.›s method. The CCM2 produces low completeness estimates for these species (8.70% and 14.75% respectively) because almost all available bones are from the same skeletal region (the hindlimb) and several specimens duplicate the same element. Santonian bird species are also represented by collections of disarticulated bones; however, several of these species, such as *Hesperornis regailis*, *H. crassipes*, *Parahesperornis alexi*, *Baptornis advenus* and *Ichthyornis dispar*, are represented by very large collections of bones, covering a much wider range of body regions [89], [91]–[93], and thus many more phylogenetic characters can be scored for each species. This is reflected in a much higher mean CCM2 score for the Santonian. As the method of Fountaine et al. does not take into account differences in the anatomical positions of the known elements, this variation in taxon completeness is not observable in their dataset. Thus, we suggest that the CCM approach is preferable to completeness grading schemes such as those proposed by Fountaine et al. [37] and Benton [47] because the relatively coarse-grained nature of the latter results in failure to detect significant differences in fossil record quality.

Despite the differences noted above, there are some important points of agreement between our results and those of Fountaine et al. [37]. First, both studies support high completeness in the Tithonian (the earliest stage from which birds are known); the mean CCM2 value for this stage is higher than any other in the Mesozoic, and Fountaine et al. were able to give all species a grade of 3 or 4 (i.e. taxa are represented by one or more nearly complete skeletons). This is surprising: it might be expected that the quality of the fossil record would decrease with greater stratigraphic age [94] because older fossils have more time to be eroded, damaged or subducted. The high completeness score of this time period reflects the effect of Lagerstätten: two of the three early Tithonian birds are from Solnhofen in Germany, an area of exceptional preservation [95].

Second, both studies show low completeness scores for the Maastrichtian, the last stage of the Cretaceous. Fountaine et al. gave all species in this time period a grade of 1 or 2, reflecting the lack of nearly complete specimens. Comparably, the mean CCM2 values for the Maastrichtian range from only 7.60% to 9.82%. Again this contradicts the expectation that the youngest time bin should have a better fossil record. However both these studies support the notion that the fossil record›s completeness is in fact random with respect to geological age. In order to explain these low mean CCM2 values, we instead need to examine biotic and abiotic factors that might influence completeness.

Completeness Metrics and Diversity

During the past few decades, there has been a great deal of discussion concerning how uneven sampling of the fossil record might affect the accuracy of palaeobiological studies of taxic diversity [54]. Potential sampling biases include both geological factors (such as temporal fluctuations in the availability of fossiliferous rocks) and anthropogenic factors (such as variation in collecting effort with respect to stratigraphy and/or geographic region). Attempts have been made to measure and 'correct' these sampling biases using techniques based on ghost range estimation (e.g. [96]), subsampling (e.g. [97]–[99]), and sampling metrics such as quantification of rock volume or outcrop area, and counts of the numbers of geological formations, localities or collections, per time period (e.g. [54], [73], [100]). Mannion & Upchurch [38] suggested that their completeness metrics might provide an additional sampling metric that captures an aspect of sampling that is ignored by other approaches. Theoretically, time bins with low mean completeness values will produce low numbers of diagnosable species because most specimens are of poor quality and can only be confidently assigned to higher taxa. In contrast, time periods with good preservation will yield many specimens that are rich in diagnostic features, allowing taxonomists to identify numerous new species. If such a mechanism operated in the fossil record, then we would expect completeness metrics such as mean CCM2 to be positively correlated with observed taxic diversity but display little correlation with sampling-corrected diversity estimates. Benton et al. [55] found evidence for this phenomenon; using a previously published [101] assessment of the completeness of Permo-Triassic tetrapods from the South Urals with a four-level grading system, the 'quality measure' (number of 'good' specimens/total number of specimens) was found to correlate with genus-level diversity. There are, however, factors that might complicate the relationship between completeness metric values and observed taxic diversity in the fossil record. For example, Mannion & Upchurch [38] presented a hypothetical situation in which a time bin with low genuine diversity might have its raw taxic diversity count artificially inflated as a result of poor preservation of specimens. Essentially, the occurrence of fragmentary and largely non-overlapping specimens increases the likelihood that a taxonomist will recognise several diagnosable species based on isolated elements that actually belong to a single species. The extent to which poor preservation results in artificially inflated or artificially lowered estimates of taxic diversity will depend on the attitude of the taxonomists studying the fossil material. Those workers inclined towards taxonomic 'lumping' are more likely to underestimate true diversity, whereas those inclined towards taxonomic 'splitting' are more likely to overestimate it. This issue will be less problematic when taxonomists work with material from time periods that have yielded generally more complete specimens: such specimens are more likely to display anatomical overlap, making it easier to accurately refer specimens to existing taxa or distinguish them as new taxa.

It is also conceivable that genuine evolutionary events, such as changes in the abundance and/or diversity of a group could influence the completeness of fossils in each time bin. Time periods when species are particularly diverse, abundant and geographically widespread could have an increased probability of preserving highly complete specimens. If such factors operated in the Mesozoic avian fossil record, then we might expect a positive correlation between mean CCM2 scores and the 'sampling-corrected' residual diversity estimate.

In this context, it is interesting to note that the current study found a significant positive correlation between mean CCM2 and both raw taxic diversity and the residual diversity curve, but with the latter correlation the stronger of the two. The strong positive correlation with residual diversity, which should more closely represent the genuine diversity of Aves in the Mesozoic, suggests that their abundance and diversity has affected the probability of more complete specimens entering the fossil record and surviving to the present day. The correlation with the raw taxic diversity curve does suggest that there is also an influence of specimen completeness on the ability of taxonomists to recognise new taxa, as suggested by Mannion and Upchurch [38]. However, the effect of completeness on observed diversity seems to be less marked than the impact of genuine changes in diversity and/or abundance on completeness.

Controls on the Completeness of the Avian Fossil Record

Sea level

The impact of sea level change on fossil record quality and taxic diversity is complex and controversial. Fluctuations in sea level clearly have the potential to change preservation rates in particular environments. Some studies have found positive correlations between sea level and the raw taxic diversity of marine organisms, suggesting that increased formation and preservation of coastal deposits have resulted in a higher quality fossil record. A similar argument has been made for some terrestrial groups on the basis that higher sea level increases the preservation of terrestrial taxa whose remains are washed into deltas, estuaries, lagoons etc. [102]. Conversely, a negative correlation between sea level and the quality of the terrestrial fossil record is also supported (e.g. [38]), because higher sea level reduces the available land area and so decreases the amount of terrestrial sedimentary rock [54], [103]. The relationship between sea level and fossil record quality is further complicated by 'common cause' hypotheses which argue that rises in sea level simultaneously promote increased preservation of fossils and increases in diversity (the latter mediated by factors such as the radiation of groups living in the expanded near shore environments) [60], [104], [105]. On land, common cause could take the form of either a positive or a negative correlation between sea level and taxic diversity. One possibility is that rises in sea level result in fragmentation of terrestrial land areas and habitats, promoting an increase in diversity [57], [106], [107]. Alternatively, higher sea level reduces available land area and, according to the species-area relationship, this should result in decreases in the diversity of terrestrial taxa [104], [108], [109]. However, Butler et al. [53] demonstrated that sea level does not correlate with either the raw taxic diversity of dinosaurs or sampling metrics once time series data are detrended and the effects of autocorrelation are taken into account (more detailed discussion of the role of sea level on observed diversity in the fossil record, evolutionary radiations and extinctions, and sampling biases, can be found in references [53], [58], [62], [104]).

Mannion & Upchurch [38] noted that sea level varies inversely with the completeness of sauropodomorph specimens during the Cretaceous. This suggests that a high sea level, while not affecting dinosaur diversity, might affect the preservation potential of terrestrial organisms. Here, our results show no correlation between the mean CCM2 scores of birds and the sea level curve [53]. This could be because Cretaceous birds were not restricted to terrestrial environments, unlike sauropodomorphs. Mannion & Upchurch [110] demonstrated that only 0.01% of the sauropodomorph fossil record comes from marine deposits, and such fossils almost certainly represent rare instances where sauropod carcasses were washed out to sea. In contrast, several groups of Cretaceous bird flourished in marine environments, including the Hesperornithiformes (flightless aquatic birds) and Ichthyornithiformes (thought to be the ecological equivalents of modern gulls and terns) [89]. As such, fluctuations in sea level would not necessarily change the total area available for the preservation of birds: instead, they might merely shift preservation rates in favour of birds from particular habitats. Thus, at times of high sea level marine birds might have a higher preservation potential, whereas at times of low sea level terrestrial birds might be better preserved. If correct, the wider ecological range of Cretaceous birds might explain why their mean CCM2 values do not correlate with sea level.

Habitat and depositional environment

Another factor that might affect the quality of avian fossils is the environment of the locality in which they are preserved. Preservation should be best in low energy environments, where carcasses are less likely to be damaged by post-mortem transportation and/or erosion. Such environments include lakes, river floodplains, deltas and lagoons. The results of the analysis of the completeness of birds found in localities representing different environments (Figure 8) indicate that birds from fluviolacustrine localities are more completely preserved than those from marine and other terrestrial localities (mean CCM2 46.51%). The birds from marine and non-fluviolacustrine terrestrial localities had similar mean CCM2 values (24.97% and 30.87% respectively), significantly lower according to the

Mann-Whitney U test (Table 3), presumably reflecting the fact that fossils in both are at greater risk of erosion. Carcasses in fluvial environments are also at risk of erosion, but lacustrine, floodplain and deltaic environments, with sluggish water, are expected to yield more complete specimens.

Given the differences between the mean CCM2 values of the three depositional categories, it is possible that the number of localities in a particular environment controls the completeness of avian fossils in each time period. For example, it is interesting to note that during the Albian, Cenomanian, Turonian and the late Maastrichtian (times where the mean CCM2 is particularly low) none of the bird specimens come from fluviolacustrine environments. To investigate this possibility, the mean CCM2 values were compared to the number of fluviolacustrine bird-bearing localities per time bin (Figure 9). The Spearman's rank and Kendall's tau values indicate that there is no significant correlation between these two time series (Table 2). Therefore, although depositional environment does seem to affect the quality of the avian fossil record, it is not the dominant control.

Geographical Controls

Sampling of the fossil record varies not only across stratigraphic intervals, but also geographically. These spatial sampling biases result from both geological factors (such as how much fossiliferous rock of a given age occurs on each continent or within each latitudinal zone) and anthropogenic factors (such as the number of active palaeobiological researchers supported by different countries). For example, dinosaur diversity is dominated by species from North America and Asia [111], not necessarily because dinosaurs were more speciose in these regions, but because of greater collecting effort combined with exposure of larger tracts of fossiliferous rock. Mesozoic avian diversity follows a similar pattern, with the northern landmasses which made up the Laurasian supercontinent (Asia, North America and Europe) yielding considerably more species than the southern continents that formed Gondwana (South America, Australia, Madagascar, the Arabian Peninsula and Antarctica) (Figure 11). For example, Africa and the Indian subcontinent have produced no avian species at all, whereas Asia is responsible for 44.35% of the 124 valid Mesozoic bird species in our dataset. Laurasian landmasses as a whole account for 85.48% of Mesozoic bird species and 71.63% of species from the Late Cretaceous.

In contrast, geographical variation in mean CCM2 values does not display the same regional skews. Although Asian birds are both more diverse and more complete than those from other continents (Figure 13), this probably reflects the exceptional preservation of the Chinese specimens (i.e., a Lagerstätten effect produced by exceptional deposits such as the Yixian Formation). Aside from this skew caused by Chinese fossils, there is no correlation between the number of avian species known from a continent and the mean CCM2 score for those species. For example, the birds of the Arabian Peninsula and Antarctica (very under-sampled landmasses where a few avian fossils have been found only recently) have a higher mean CCM2 score than those found in North America, where Mesozoic bird fossils are abundant and have been known since the 1800s (Figure 13). The high completeness of the birds found in these as-yet unproductive areas could be an indication that the low number of species and specimens found is not caused by problems with preservation in these areas, but instead reflects the lack of sampling. The low number of specimens from these areas should be noted; it is possible that the high completeness of the bird species found here results from the random possibility of the first few specimens being found there being of high completeness. However, this does still indicate that the preservation of high-quality specimens is possible in these areas.

The number of taxa found within modern day 5°latitudinal bins is plotted in Figure 11A, while mean CCM2 values are plotted against latitude in Figure 11B. Sampling of the Southern Hemisphere is clearly poorer than that of the Northern Hemisphere. There are more empty latitudinal bins in the Southern Hemisphere (13 bins, compared to 10 in the Northern Hemisphere). No latitudinal bin in the Southern Hemisphere contains more than seven species. In contrast, in the Northern Hemisphere, four latitudinal bins have produced more than this number of taxa. Between 40 and 45°N, a total of 62 species have been recovered. However, the mean CCM2 values for those Southern Hemisphere latitudinal bins that have yielded Mesozoic birds are not substantially lower than the mean CCM2 values for equivalent bins in the Northern Hemisphere (typically between 30–60° north of the Equator) (Figure 11). These latitudinal analyses indicate regions which have potential to yield high quality specimens in the future. Although only two taxa have been recovered from high latitudes in the Southern Hemisphere (those further south than 50°S), the specimens are well preserved: *Vegavis iaai*and *Polarornis gregorii* have CCM2 scores of 40.00% and 25.65% respectively. Despite the logistical difficulties associated with fieldwork in Antarctica, our results suggest that this continent could provide a great deal of further information on avian evolution. In contrast, the tropical regions between 25°N and 20°S have produced no Mesozoic bird taxa. Finally, although the Northern Hemisphere has been well sampled in mid-latitudes, there is very little avian material of high quality from within the Arctic Circle, again presumably resulting from the difficulties of working in that environment. However, *Canadaga arctica*, known from two localities in northern Canada (between 70 and 80°N) [69], [70], indicates that there is future potential to sample specimens from a far-north avian fauna.

Taphonomic Effects: Body Size and Skeletal Robustness

Analysis of the mean CCM 2 data for Sauropodomorpha in each substage of the Mesozoic (figure 14) shows that there is no statistically significant correlation between the completeness of their record and that of contemporaneous Aves. This suggests that the completeness of large, robust animals is controlled by different factors to small animals with fragile skeletons (see also studies of the disarticulation of extant animals e.g. [112]). This is not surprising considering the different ways in which members of these two clades are preserved and discovered. Many of the most complete avian fossils are discovered as part and counterpart, whereby a block is split, revealing a flattened skeleton on one plate and an imprint on the other. Although this produces exceptional preservation of small animals, including soft tissues such as feathers, it cannot preserve complete skeletons of animals as large as sauropodomorphs. Therefore, a time period containing geological formations suitable for this preservation mode would lead to a large increase in the completeness of bird fossils, but have little effect on the completeness of sauropodomorphs. Conversely, time periods which include a preponderance of geological formations representing high energy environments, would generally yield better quality sauropodomorph material than avian material. Finally, as noted above, birds are often preserved in marine environments, whereas very few sauropodomorph fossils occur in such deposits. Time periods preserving numerous formations composed of marine deposits might therefore result in an increase in the completeness of bird specimens, but would make very little difference to the sauropodomorph record. Thus, there is little reason to expect that the avian and sauropodomorph mean CCM2 scores should correlate.

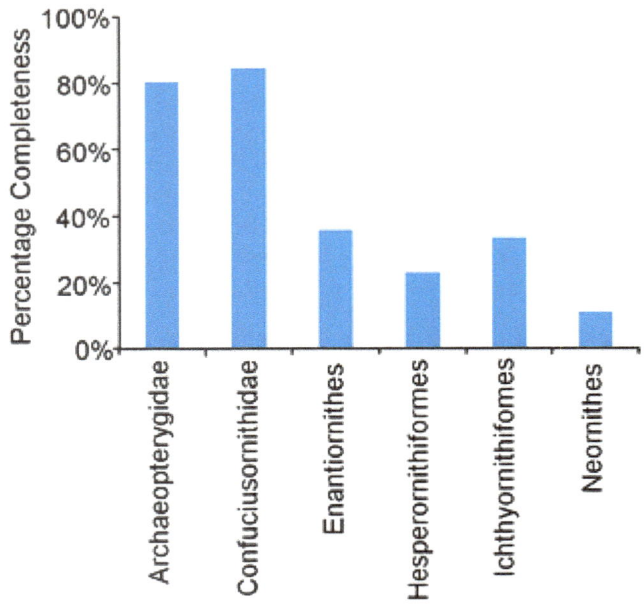

Figure 14. The mean CCM2 scores for birds assigned to the major Mesozoic avian clades.

The mean CCM2 scores of sauropodomorphs during the Cretaceous show less variation than those of birds. Apart from a peak in the Hauterivian of 47.33%, the completeness of sauropod species varies between a narrow range of 22.71% and 33.50% [38]. In contrast, the mean CCM2 scores for birds range from 1.53% to 75.72% over the same time period. This observation probably reflects differences in the factors that control the completeness of avian and sauropod specimens. In particular, the differences in body size and robustness of birds and sauropodomorphs might explain the greater variation in mean CCM2 values for the former and the lower variation for the latter. Depending on the precise geological setting, it is perhaps easier to utterly destroy an avian carcass during transportation or via erosion than a sauropodomorph carcass: however it is also easier to rapidly and completely bury a small skeleton (resulting in high completeness scores), whereas it is much less likely that a 20–30 m long sauropod skeleton will be preserved intact. This suggests that Largerstätten effects on observed diversity are likely to be far more severe for small delicate organisms such as birds than for large robust ones such as sauropods.

As noted in the 'Introduction' above, some molecular clock studies [10]–[13] have explained the absence of well-preserved Cretaceous neornithines in terms of the low preservation potential of birds. Avian species today, and in the past, are typically small-bodied and lightly built because of the constraints imposed by powered flight. We might expect, therefore, that bird carcasses would be particularly susceptible to damage and destruction during post-mortem transportation and erosion and consequently should have a poorer fossil record than larger and/or more robust contemporaneous taxa. The simple completeness metric for tetrapods [35], however, suggests that the record of small-bodied animals is no less complete than that of larger ones. This prediction can be tested by comparing the mean CCM2 values for birds and sauropodomorphs. In fact, when taking into account the entire period of the Mesozoic from which birds are known (from the Tithonian– Maastrichtian), despite the mean CCM2 of birds being only 4% less than that of sauropodomorphs (Figure 10), the difference between the medians is significant, as shown by the Mann-Whitney U-test (Table 3). When Late Cretaceous species alone are evaluated, the mean CCM2 score for birds is 16% less than that of sauropodomorphs (Figure 10, Table 3). Thus, contrary to the simple completeness metric results for tetrapods [35], CCM2 values suggest that larger and more robust organisms are better preserved than small delicate ones, at least in Dinosauria. Moreover, whatever factors are driving the differential preservation of small, delicate animals and large, robust ones, these factors do not appear to be constant through time.

The Origin of Neornithes

Claims that Neornithes are genuinely absent prior to the Maastrichtian (because they originated just before and radiated after the K/Pg mass extinction [16], [20], [31]) can only be supported if the Cretaceous avian fossil record is well sampled temporally, spatially and in terms of specimen/ taxon completeness. Fountaine et al. [37] argued that (with the exception of taxa of Maastrichtian age) Mesozoic birds are represented by high enough quality material to infer a genuine absence of Neornithes during the Cretaceous. However, this uniformity of relatively high quality preservation throughout much of the Cretaceous is not supported by our analyses. Mean CCM2 values display considerably more variation between time bins than do the completeness grades of Fountaine et al. We suggest that the relatively low variation in completeness scores recovered by Fountaine et al is an artefact generated by the application of a very coarse-grained completeness metric: the more sensitive mean CCM2 scores support the view that species completeness varied considerably during the Mesozoic, with peaks reflecting the occurrence of Lagerstätten such as Solnhofen and the Yixian Formation. Such Lagerstätten are absent from the Late Cretaceous, resulting in relatively low mean CCM2 values for most of this period. All stages of the Late Cretaceous have a mean CCM2 score of less than 19%, with the exception of the Santonian and Coniacian (Figure 3). The avian mean CCM2 scores are much lower than those produced for sauropodomorphs [38] for all stages of the Late Cretaceous, again with the exception of the Coniacian and the Santonian (Figure 14). The mean completeness of birds from the Late Cretaceous is considerably lower than that of all Mesozoic birds, and less than that of Late Cretaceous sauropodomorphs (Figure 10), indicating worse preservation of small delicate animals at this time.

The Cretaceous avian fossil record is also very patchy in terms of spatial sampling. This is particularly noticeable with regard to the tropics and high latitudes, with the majority of Mesozoic avian species (84%) being found between 30 and 60°North of the Equator (Figure 11). There are also entire regions, such as Africa and India, which have produced no diagnostic avian material. The relatively poor sampling of Gondwanan continents is particularly noteworthy given that some biogeographic analyses (e.g. [14]) have suggested that Neornithes originated in the Southern Hemisphere. Clearly, it is unrealistic to claim that the Cretaceous fossil record is adequate for determining the genuine absence of Neornithes if it transpires that this clade originated and initially radiated at high latitudes or in the tropics, or in regions such as Africa.

It should be noted that it is in the poorly sampled region of Antarctica that the two most complete pre-Palaeogene putative neornithines have been found. *Polarornis gregorii* from the Lopez de Bertodano Formation of Antarctica (late Campanian-Maastrichtian) [113] is well preserved, including a nearly complete skull, several vertebrae, a sternum and portions of the hindlimb [2] (CCM2 score =25.65%). Chatterjee [114] assigned this species to the neornithine family Gaviidae (loons). If both the age of the site and the affinities of the specimen are correct, this places the origin of the Gaviidae at a time which correlates well with the molecular clock study of Cooper & Penny [12]. However, the stratigraphic age of the Lopez de Bertodano Formation has yet to be verified [18], and the assignment of *Polarornis* to the loons was based on overall similarity rather than cladistic analysis. Of the six characters used to support the relationship with Gaviidae, five have been found in more basal birds and even non-avian theropod dinosaurs [18]. Thus, until cladistic analysis is applied, placement of *Polarornis* in Neornithes remains questionable. In contrast, *Vegavis iaai*, from the Maastrichtian of Antarctica, is known from a nearly complete postcranial skeleton (CCM2 score =40.00%), and has been subjected to phylogenetic analysis which placed it within the Anatoidea [30]. If this identification is correct, then a derived neornithine lineage was present in the latest Cretaceous, implying an earlier origin for Neornithes as a whole. This discovery indicates the possibility that Antarctica has more information to offer on this particular debate. The recent discovery of a possible neornithine carpometacarpus in Argentina [27] highlights the potential for future discoveries elsewhere in the Southern Hemisphere.

Comparisons of the CCM2 values for putative Cretaceous Neornithes versus other avian clades indicate that the former group are particularly poorly preserved (Figure 14). There are two possible explanations for the differences in completeness of neornithine and non-neornithine birds:

1. Fragmentary preservation leads to ambiguity of identification. It is conceivable that more complete Cretaceous bird specimens provide enough anatomical detail for them to be confidently assigned to non-neornithine lineages. The more fragmentary specimens assigned to Neornithes might actually belong to non-neornithines, but their poor preservation generates more ambiguity in their identification. For example, if some Cretaceous non-neornithines convergently acquired some derived character states that also occur in true Neornithes, then these convergences might only be detected when specimens are well preserved (thus providing additional character data that contradicts interpretation of these features as synapomorphies of the Neornithes). Conversely, convergence might go undetected when only fragmentary specimens are available.
2. The effects of abundance and habitat. A second possibility is that, during the Cretaceous, Neornithes were generally less abundant than other bird groups [37]. Lower numbers of individual birds might decrease the probability that some well-preserved specimens will successfully survive post-mortem transportation and fossilization, as well as increasing the chances of discovery. Similarly, it is also possible that the first Neornithes lived in habitats that are less likely to preserve highly complete specimens. For example, our results demonstrate that Mesozoic birds living in fluviolacustrine environments tend to be more complete than those living in marine or other terrestrial ones. If Neornithes radiated initially in other niches, their fossil record could be substantially poorer than that of non-neornithines occupying fluvial and lacustrine habitats.

In short, fossil evidence for the occurrence of Early or early Late Cretaceous neornithines, as predicted by molecular clocks, remains elusive. The Late Cretaceous avian fossil record is particularly fragmentary because of an absence of suitable Lagerstätten, and there are several large geographical regions, wide latitudinal zones and portions of the stratigraphic record that are poorly sampled or have yielded no avian fossils of any kind. Thus, it is premature for palaeobiologists to claim that the Cretaceous avian fossil record is sufficiently well sampled to determine that Neornithes were genuinely absent until after or shortly before the K/Pg mass extinction.

CONCLUSIONS

Although specimen/taxon completeness represents only one aspect of fossil record quality, it is a significant one because of its potential relationship to constraints on the accurate recognition of valid taxonomic units. To date, analyses of fossil record quality based on estimates of specimen/taxon completeness have been restricted to studies of dinosaurs (including birds) [37], [38], [42], Permo-Triassic tetrapods [55], [101] and echinoids [115]. Pioneering studies in this field utilised simple grading schemes, but these are problematic because of the arbitrary boundaries between grades and the coarse-grained picture of completeness they generate. A completeness metric based on scorable morphological characters, such as our character completeness metric (CCM2), circumvents both of these problems. Such completeness metrics can be used to generate estimates of fossil record quality, which in turn can be compared to various aspects of sampling, geological and environmental factors, diversity and other evolutionary events. There is growing interest and concern regarding the influence of Lagerstätten on diversity patterns observed directly in the fossil record and on supposedly 'sampling-corrected' estimates of palaeodiversity [59]: completeness metrics should make an important contribution to this field in the future given their ability to identify time periods where specimen/taxon completeness is unusually high or low.

The Mesozoic fossil record of birds has clearly been strongly influenced by uneven sampling. This is manifested in the positive correlations between sampling metrics and raw taxic diversity, the significant fluctuations in mean CCM2 values through time, and the very patchy spatial and temporal distribution of taxa. Mesozoic bird specimens are best preserved in conditions such as low energy lacustrine environments, where post-mortem transport and erosion are minimal. These conditions characterise deposits such as those of Solnhofen and the Yixian Formation that represent Lagerstätten. The avian fossil record is not noticeably poorer than that of large bodied robust sauropods except in the Late Cretaceous. The completeness of birds in different time bins is significantly more variable than that of sauropods, reflecting the possibility that bird carcasses are easier to destroy entirely but also easier to bury whole, than those of sauropods. It will be interesting to see if future studies of the completeness of other small and large bodied vertebrates conform to the same patterns. If these patterns do hold, it suggests that Lagerstätten effects on reconstructions of palaeodiversity are likely to be more significant for small delicate organisms than for large robust ones.

Avian diversity, indicated by the 'corrected' residual diversity estimate, increased steadily between the Berriasian and the Aptian. Between the Aptian and the Albian there appears to have been a large extinction, although this may be an artefact of sampling, in particular the effect of Lagerstätten. Diversity recovered during the Albian and Cenomanian, and then plateaued during the Turonian. The diversification of the Hesperornithiformes and Ichthyornithiformes during the Coniacian and Santonian led to a peak in the number of species in the Mesozoic. Diversity fell during the Campanian, before rising again in the Maastrichtian.

Both the residual diversity and the taxic diversity curves correlate significantly and positively with the CCM2; however, the correlation with the residual diversity curve is stronger. This suggests that biotic factors such as fluctuations in diversity, abundance and geographic range, have affected the frequency with which bird carcasses enter depositional environments where high quality preservation is possible. The significant, albeit weaker, correlation with the raw taxic diversity curve indicates that the completeness of specimens may place constraints on the ability of taxonomists to recognise new species and/or identify specimens as members of already known species.

The debate concerning the origin of Neornithes before or after the end-Cretaceous mass extinction cannot be settled by an analysis of fossil record quality alone. However, such an analysis can provide a valuable perspective on claims and counter-claims regarding the probability that Neornithes were genuinely absent in the pre-Maastrichtian Cretaceous or were present but have not been found yet. Our results indicate that most of the Late Cretaceous (not just the Maastrichtian as proposed by Fountaine et al. [37]) fossil record of birds is characterised by numerous highly fragmentary specimens. If, as has been proposed by several molecular and biogeographic studies, Neornithes originated in the Late Cretaceous in the Southern Hemisphere (especially at high latitudes), then it is quite plausible that we would not see any unambiguous Cretaceous neornithines in the currently available fossil record. A compelling case for the absence of pre-Maastrichtian Neornithes can be made only after significant gaps in the record (e.g. Africa) have been filled via the discovery of well-preserved non-neornithine birds.

The discrepancy between divergence time estimates based on molecular clocks and direct examination of the fossil record is not unique to the debate over neornithine origins. A very similar discussion has occurred in recent years concerning the origin of placental mammals before or after the K/Pg boundary (e.g. [20], [116]). It is hoped that the current study will stimulate further interest in the application of completeness metrics and other measures of fossil record quality in order to evaluate the likelihood that 'absence of evidence' might, or might not, also be 'evidence of absence'.

ACKNOWLEDGMENTS

This study is *Paleobiology Database* official publication #157. We are grateful for the efforts of Matthew T. Carrano and John Alroy in entering data on Mesozoic bird occurrences into *The Paleobiology Database*. We would also like to thank Marc E.H. Jones and Roger B.J. Benson for helpful discussion, as well as Gareth Dyke and an anonymous referee whose comments greatly improved this manuscript.

Author Contributions

Analyzed the data: NB. Wrote the paper: NB PU PDM. Scrutinised the dataset and removed taxa which were Nomina Dubia and Nomina Nuda, and suggested further taxa for inclusion: JOC.

REFERENCES

1. Gill FB (1990) Ornithology. New York: W. H. Freeman and Company. 660 p.
2. Chatterjee S (1997) The Rise of Birds. Baltimore and London: The John Hopkins University Press. 312 p.
3. Chiappe LM (1995) The first 85 million years of avian evolution. Nature 391: 147–152.
4. Witmer L (2002) The debate on avian ancestry: phylogeny, function and fossils. In: Chiappe LM, Witmer LM, editors. pp. 3–30. Berkeley, California: University of California Press.
5. Cracraft J (1971) Phylogeny and evolution of the ratite birds. Ibis 116: 494–521.
6. Cracraft J (1982) Phylogenetic relationships and monophyly of loons, grebes and hesperornithiform birds, with comments on the early history of birds. Sits Zoo 31: 35–56.
7. Sibley CG, Ahlquist JE (1990) Phylogeny and Classification of Birds: a Study in Molecular Evolution. New Haven: Yale University Press. 976 p.
8. Chiappe LM, Dyke GJ (2002) The Mesozoic radiation of birds. Annu Rev Ecol Syst 33: 91–124.
9. Olson SL (1985) The fossil record of birds. In: Farner DS, King J, Parkes KC, editors. pp. 79–238. New York: New York Academic Press.

10. Kumar S, Hedges SB (1998) A molecular timescale for vertebrate evolution. Nature 392: 917–920.
11. Paton T, Haddrath O, Baker AJ (2002) Complete mitochondiral DNA genome sequences show that modern birds are not descended from transitional shorebirds. P Roy Soc B-Biol Sci 269: 839–846.
12. Cooper A, Penny D (1997) Mass survival of birds across the Cretaceous-Tertiary boundary: molecular evidence. Science 275: 1109–1113.
13. Brown JW, Rest JS, Carcia Moreno J, Sorenson MD, Mindall DP (2008) Strong mitochondrial DNA support for a Cretaceous origin of modern avian lineages. BMC Biol 6: 1–18.
14. Cracraft J (2001) Avian evolution, Gondwana biogeography and the Cretaceous-Tertiary mass extinction event. P Roy Soc B-Biol Sci 268: 459–469.
15. Upchurch P (2008) Gondwanan breakup; legacies of a lost world? Trends Ecol Evol 23: 229–236.
16. Feduccia A (2003) 'Big bang' for tertiary birds? Trends Ecol Evol 18: 172–176.
17. Hope S (2002) The Mesozoic radiation of Neornithes. In: Chiappe LM, Witmer LM, editors. pp. 339–388. Berkeley, California: University of California Press.
18. Dyke GJ, Van Tuinen M (2004) The evolutionary radiation of modern birds (Neornithes): reconciling molecules, morphology and the fossil record. Zool J Linn Soc-Lond 141: 153–177.
19. Dyke GJ (2001) The evolutionary radiation of modern birds: systematics and patterns of diversification. Geol J 36: 305–315.
20. Benton MJ (1999) Early origins of modern birds and mammals: molecules vs morphology. BioEssays 21: 1043–1051.
21. Richards HG, Gallagher WB (1974) The problem of the Cretaceous-Tertiary boundary in New Jersey. Notul Nat Acad Nat Sci Philadelphia 449: 1–6.
22. Longrich NR, Tokaryk T, Field DJ (2011) Mass extinction of birds at the Cretaceous-Paleogene (K-Pg) boundary. Proc Natl Acad Sci USA 108: 15253–15257.
23. Mayr G (2009) Paleogene Fossil Birds. Berlin Heidelberg: Springer-Verlag. 262 p.
24. Agnolin FL, Novas FE, Lio G (2006) Neornithine bird coracoids from the Upper Cretaceous of Patagonia. Ameghiniana 43: 245–248.
25. Brodkorb P (1963) Birds from the Upper Cretaceous of Wyoming. In: Sibley CG, editor. pp. 55–70. Washington DC: American Ornithologists Union.
26. Elzanowski A, Stidham T (2011) Galloanserine quadrate from the Late Cretaceous Lance Formation of Wyoming. Auk 128: 138–145.
27. Agnolin FL, Novas FE (2012) A carpometacarpus from the upper cretaceous of Patagonia sheds light on the Ornithurine bird radiation. Palaeont Z 86: 85–89.
28. Kurochkin EN, Dyke GJ, Karhu AA (2002) A new presbyornithid bird (Aves, Anseriformes) from the Late Cretaceous of southern Mongolia. Am Mus Novit 3386: 1–11.
29. Clarke JA, Norrell MA (2002) New avialan remains and a review of the known avifauna from the Late Cretaceous Nemegt Formation of Mongolia. Am Mus Novit 3447: 1–12.
30. Clarke JA, Tambussi CP, Noriega JI, Erickson GM, Ketcham RM (2005) Definitive fossil evidence for the extant avian radiation in the Cretaceous. Nature 433: 305–308.
31. Feduccia A (1995) Explosive evolution in Tertiary birds and mammals. Science 267: 637–638.
32. Ericson PGP, Anderson CL, Britton T, Elzanowski A, Johansson US, et al. (2006) Diversification of Neoaves: integration of molecular sequence data and fossils. Biol Lett 2: 543–547.
33. Brown JW, Payne RB, Mindell DP (2007) Nuclear DNA does not reconcile 'rocks' and 'clocks' in Neoaves: a comment on Ericson et al. Biol Lett 3: 257–260.
34. Davis PG, Briggs DEG (1998) The impact of decay and disarticulation on the preservation of fossil birds. Palaios 13: 3–13.
35. Fara E, Benton MJ (2000) The fossil record of Cretaceous tetrapods. Palaios 15: 161–165.
36. Bleiweiss R (1998) Fossil gap analysis supports early Tertiary origin of trophically diverse avian orders. Geology 26: 323–326.
37. Fountaine T, Benton MJ, Dyke GJ, Nudds R (2005) The quality of the fossil record of Mesozoic birds. P Roy Soc B-Biol Sci 272: 289–294.
38. Mannion PD, Upchurch P (2010) Completeness metrics and the quality of the sauropodomorph fossil record through geological and historical time. Palaeobiology 36: 283–302.
39. Xu X, You H, Du K, Han F (2011) An Archaeopteryx-like theropod from China and the origin of Avialae. Nature 475: 465–470.
40. Lee MSY, Worthy TH (2012) Likelihood reinstates Archaeopteryx as a primitive bird. Biology Lett: In Press. DOI:10.1098/rsbl.2011.0884.
41. Forster CA, Sampson SD, Chiappe LM, Krause DW (1998) The theropod ancestry of birds: new evidence from the Late Cretaceous of Madagascar. Science 279: 1915–1919.
42. Makovicky PJ, Apesteguía S, Agnolín FL (2005) The earliest dromaeosaurid theropod from South America. Nature 437: 1007–1011.
43. Norell MA, Clark JM, Turner AH, Makovicky PJ (2006) A new dromaeosaurid theropod from Ukhaa Tolgod (Omnogov, Mongolia). Am Mus Novit 3545: 1–51.
44. Turner AH, Pol D, Clarke JA, Erickson GM, Norell MA (2007) A basal dromaeosaurid and size evolution preceding avian flight. Science 317: 1378–1381.
45. Senter P (2007) A new look at the phylogeny of Coelurosauria (Dinosauria: Theropoda). J Syst Palaeontol 5: 429–463.
46. Miller KG, Kominz MA, Browning JV, Wright JD, Mountain GS, et al. (2005) The Phanerozoic record of global sea-level change. Science 310: 1293–1298.

47. Benton MJ (2008) How to find a dinosaur, and the role of synonymy in biodiversity studies. Palaeobiology 34: 516–533.

48. Chiappe LM (1995) The phylogenetic position of the Cretaceous birds of Argentina: Enantiornithes and Patagopteryx deferrariisi. In: Peters DS, editor. pp. 55–63. Seckenberg, Germany: Courier Forschungsinstitut.

49. Zhou Z, Clarke J, Zhang F (2008) Insight into diversity, body size and morphological evolution from the largest Early Cretaceous enantiornithine bird. J Anat 212: 565–577.

50. O'Connor JK, Wang X, Chiappe L, Gao C, Meng Q, et al. (2009) Phylogenetic support for a specialized clade of Cretaceous Enantiornithine birds, with information from a new species. J Vertebr Palaeontol 29: 188–204.

51. Mayr G (2006) New specimens of the Eocene Messelirrisoridae (Aves: Bucerotes), with comments on the preservation of uropygial gland waxes in fossil birds from Messel and the phylogenetic affinities of Bucerotes. Paläeont Z 80: 390–405.

52. Gradstein FM, Ogg JG, Smith AG (2004) A Geologic Time Scale 2004. Cambridge: Cambridge University Press. 455 p.

53. Butler RJ, Benson RBJ, Carrano MT, Mannion PD, Upchurch P (2011) Sea level, dinosaur diversity, and sampling biases: investigating the 'common cause' hypothesis in the terrestrial realm. P Roy Soc B-Biol Sci 278: 1165–1170.

54. Smith AB, McGowan AJ (2007) The shape of the Phanerozoic palaeodiversity curve: how much can be predicted from the sedimentary rock record of western Europe. Palaeontology 50: 765–774.

55. Benton MJ, Dunhill AM, Lloyd GT, Marx FG (2011) Assessing the quality of the fossil record: insights from vertebrates. In: McGowan AJ, Smith AB, editors. pp. 63–94. London: Geological Society of London.

56. Benton MJ (2012) No gap in the Middle Permian record of terrestrial vertebrates. Geology: In Press. DOI:10.1130/G32669.1.

57. Upchurch P, Barrett PM (2005) Phylogenetic and taxic perspectives on sauropod diversity. In: Rogers KC, Wilson JA, editors. pp. 104–124. Berkley and Los Angeles, California: University of California Press.

58. Upchurch P, Mannion PD, Benson RBJ, Butler RJ, Carrano MT (2011) Geological and anthropogenic controls on the sampling of the terrestrial fossil record: a case study from the Dinosauria. In: McGowan A, Smith AB, editors. pp. 209–240. London: The Geological Society of London.

59. McGowan AJ, Smith AB (2008) Are global Phanerozoic marine diversity curves truly global? A study of the relationship between regional rock records and global Phanerozoic marine diversity. Paleobiology 34: 80–103.

60. Butler RJ, Barrett PM, Nowbath S, Upchurch P (2009) Estimating the effects of sampling biases on pterosaur diversity patterns: implications for hypotheses of bird/pterosaur competitive replacement. Paleobiology 35: 432–446.

61. Benson RB J, Butler RJ, Lindgren J, Smith AS (2010) Mesozoic marine tetrapod diversity: mass extinctions and temporal heterogeneity in geological megabiases affecting vertebrates. P Roy Soc B-Biol Sci 277: 829–834.

62. Benson RBJ, Butler RJ (2011) Uncovering the diversification history of marine tetrapods: ecology influences the effect of geological sampling biases. In: McGowan A, Smith AB, editors. pp. 191–208. London: The Geological Society of London.

63. Mannion P, Upchurch P, Carrano MT, Barrett PM (2011) Testing the effect of the rock record on diversity: a multidisciplinary approach to elucidating the generic richness of sauropodomorph dinosaurs through time. Biol Rev 86: 157–181.

64. Lloyd GT (2012) A refined modelling appreoach to assess the influence of sampling on palaeobiodiversity curves: new support for declining Cretaceous dinosaur richness. Biol Lett: In Press. DOI:10.1098/rsbl.2011.0210.

65. Hammer O, Harper DAT (2006) Palaeontological Data Analysis. Oxford: Blackwell Publishing. 351 p.

66. McKinney ML (1990) Classifying and analyzing evolutionary trends. In: McNamara KJ, editor. Tucson: University of Arizona Press. pp. 28–58. Evolutionary trends.

67. Hammer O, Harper DAT, Ryan PD (2001) PAST: palaeontological statistics software package for education and data analysis. Palaeontol Electron 4: 9.

68. Wang X, O'Connor JK, Zhao B, Chiappe LM, Gao C, et al. (2010) New species of Enantiornithes (Aves: Ornithothoraces) from the Qiaotou Formation in Northern Hebei, China. Acta Geol Sin 84: 247–256.

69. Hou L (1999) New hesperornithid (Aves) from the Canadian Arctic. Vertebrat PalAsiatic 37: 228–233.

70. Wilson LE, Chin K, Cumbaa S, Dyke G (2011) A high latitude hesperornithiform (Aves) from Devon Island: palaeobiogeography and size distribution of North American hesperornithiformes. J Syst Palaeontol 9: 9–23.

71. Mannion PD, Benson RJB, Upchurch P, Butler RJ, Carrano MT, et al. (2011) A temperate palaeodiversity peak in Mesozoic dinosaurs and evidence for Late Cretaceous geographical partitioning. Global Ecol Biogeogr In press. DOI: 10.1111/j1466–8238.2011.00735.

72. Krause DW, Gottfried MD, O'Connor PM, Roberts EM (2003) A Cretaceous mammal from Tanzania. Acta Palaeontol Pol 48: 321–330.

73. Barrett PM, McGowan AJ, Page V (2009) Dinosaur diversity and the rock record. P Roy Soc B-Biol Sci 276: 2667–2674.

74. Meyer H von (1861) Archaeopteryx lithographica (Vögel-Feder) und Pterodactylus von Solnhofen. Neues Jahrb Minerol. Geol. Palaontol. 1861: 678–679.

75. Wellnhofer P (1988) A new specimen of Archaeopteryx. Science 240: 1790–1792.

76. Elzanowski A (2001) A new genus and species for the largest specimen of Archaeopteryx. Acta Palaeontol Pol 46: 519–532.

77. Ji Q, Ji D, Lu J, You H, Chen W, et al. (2005) First avilian bird from China (Jingfengopteryx elegans gen. et sp. nov.). Geological Bulletin of China 24: 197–205.
78. Zhang Z, Zhou Z, Benton MJ (2008) A primitive confuciusornithid bird from China and its implications for early avian flight. Sci China Ser D 51: 625–639.
79. Chiappe LM, Lacasa-Ruiz A (2002) Noguerornis gonzalezi (Aves: Ornithothoraces) from the Early Cretaceous of Spain. In: Chiappe LM, Whitmer , LM , editors. pp. 230–239. Berkley, California: University of California Press.
80. Peybernés B, Oertli H (1972) La série de passage du Jurassique au Crétacé dans le Bassin sud-pyrénéen (Espagne). C R Acad Sci 276: 2501–2504.
81. Barale G (1973) Présence du genre Frenelopsios Schenk dans les calcaires lithographiques du Montsec, province de Lérida (Espagne). C R Acad Sci 227: 1340–1349.
82. Martin-Closas C, López-Morón (1995) The charophyte flora. In: Martinez Delclos X, editor. pp. 29–31. Lleida, Spain: Institut Estudis Ilerdencs.
83. Chang S, Zhung H, Renne PR, Fang Y (2009) High-precision 40Ar/39Ar age for the Jehol Biota. Palaeogeogr Palaeoclimatol Palaeocl 280: 94–104.
84. Zhou Z, Zhang F (2006) A beaked basal ornithurine bird (Aves, Ornithurae) from the Lower Cretaceous of China. Zool Scr 35: 363–373.
85. Zhou Z, Zhang F (2005) Discovery of an ornithurine bird and its implication for Early Cretaceous avian radiation. Proc Natl Acad Sci USA 102: 18998–19002.
86. Hou L (1997) Mesozoic Birds of China. Lagu Hsiang, Taiwan: Phoenix Valley Provincial Aviary. 153 p.
87. O'Connor JK, Gao KQ, Chiappe LM (2010) A new Ornithuromorph (Aves: Ornithothoraces) bird from the Jehol Group indicative of higher level diversity. J Vertebr Paleontol 30: 311–312.
88. Galton PM, Martin LD (2002) Enaliornis, an early Cretaceous Hesperornithiform bird from England, with comments on other Hesperornithiformes. In: Chiappe LM, Whitmer LM, editors. pp. 317–338. Berkley, California: University of California Press.
89. Clarke JA (2004) Morphology, phylogenetic taxonomy, and systematics of Ichthyornis and Apatornis (Avialae: Ornithurae). B Am Mus Nat Hist 286: 1–179.
90. Tokaryk TM, Cumbaa SL, Storer JE (1997) Early Late Cretaceous birds from Saskatchewan, Canada: the oldest diverse avifauna known from North America. J Vertebr Paleontol 17: 172–176.
91. Marsh OC (1880) Odontornithes: a monograph on the extinct toothed birds of North America. Washington DC: Government Printing Office.
92. Martin L (1984) A new Hesperornithid and the relationships of Mesozoic birds. Trans Kans Acad Sci 87: 141–150.
93. Martin LD, Tate J (1976) The skeleton of Baptornis advenus (Aves: Hesperornithiformes). Sm C Paleob 27: 35–66.
94. Raup DM (1972) Taxonomic diversity during the Phanerozoic. Science 177: 1065–1071.
95. Allison PA, Briggs DEG (1993) Exceptional fossil record: distribution of soft-tissue preservation through the Phanerozoic. Geology 21: 527–530.
96. Norrell MA (1992) Taxic origin and temporal diversity: The effect of phylogeny. In: Novacek MJ, Wheeler QD, editors. pp. 89–118. New York: Columbia University Press.
97. Bush AM, Markey MJ, Marshall CR (2004) Removing bias from diversity curves: the effects of spatially organized biodiversity on sampling standardization. Paleobiology 30: 666–686.
98. Alroy J, Aberhan M, Bottjer DJ, Foote M, Fürsich FT, et al. (2008) Phanerozoic trends in the global diversity of marine invertebrates. Science 321: 97–100.
99. Alroy J (2010) The shifting balance of diversity among major marine clades. Science 329: 1191–1194.
100. Crampton JS, Beu AG, Cooper RA, Jones CM, Marshall B, Maxwell PA (2003) Estimating the rock volume bias in paleobiodiversity studies. Science 301: 358–360.
101. Benton MJ, Tverdokhlebov VP, Surkov MV (2004) Ecosystem remodelling among vertebrates at the Permian-Triassic boundary. Nature 432: 97–100.
102. Haubold H (1990) Dinosaurs and fluctuating sea levels during the Mesozoic. Historical Biol 4: 75–106.
103. Smith AB (2001) Large scale heterogeneity of the fossil record: implications for Phanerozoic biodiversity studies. P Roy Soc B-Biol Sci 356: 351–367.
104. Benton MJ, Emerson BJ (2007) How did life become so diverse? The dynamics of diversification according to the fossil record and molecular phylogenetics. Palaeontology 50: 23–40.
105. Hannisdal B, Peters SE (2011) Phanerozoic Earth system evolution and marine biodiversity. Science 334: 1121–1124.
106. Bakker RT (1977) Tetrapod mass extinctions – a model of the regulation of speciation rates and immigration by cycles of topographic diversity. In: Hallam A, editor. pp. 439–468. Amsterdam: Elsevier.
107. Benton MJ (2009) The Red Queen and the Court Jester: species diversity and the role of biotic and abiotic factors through time. Science 323: 728–732.
108. Weishampel DB, Horner JR (1987) Dinosaurs, habitat bottlenecks, and the St. Mary River Formation. In: Currie PJ, Koster EH, editors. pp. 224–229. Drumheller, Canada: Royal Tyrell Museum.
109. Dodson P (1990) Counting dinosaurs: how many kinds were there? P Natl Acad Sci USA 87: 7608–7612.
110. Mannion PD, Upchurch P (2010) A quantitative analysis of environmental associations in sauropod dinosaurs. Paleobiology 36: 253–282.
111. Fastovsky DE, Huang Y, Hsu J, Martin-McNaughton J, Sheehan PM, et al. (2004) Shape of Mesozoic dinosaur richness. Geology 32: 877–880.

112. Brand LR, Hussey M, Taylor J (2003) Decay and disarticulation of small vertebrates in controlled experiments. J Taph 1: 69–95.

113. Chatterjee S (1989) The oldest Antarctic bird. J Vertebr Paleontol 8: 11A.

114. Chatterjee S (2002) The morphology and systematics of Polarornis, a Cretaceous loon (Aves: Gaviidae) from Antarctica. In: Zhou Z, Zhang F, editors Proceedings of the 5th Symposium of the Society of Avian Paleontology and Evolution, Beijing, 1–4 June 2000. Beijing: Science Press. pp. 125–155.

115. Smith AB (2007) Intrinsic versus extrinsic biases in the fossil record: contrasting the fossil record of echinoids in the Triassic and early Jurassic using sampling data, phylogenetic analysis and molecular clocks. Paleobiology 33: 310–323.

116. Goswami A, Prasad GVR, Upchurch P, Boyer DM, Seiffert ER, et al. 2011. A radiation of arboreal basal eutherian mammals beginning in the Late Cretaceous of India. Procf Natl Acad Sci USA (108): 16333–16338

CHAPTER 16

EVOLUTION OF NEW WORLD MONKEYS: AN ANALYSIS OF FOSSIL AND MOLECULAR DATA

S. Ivan Perez, Marcelo F. Tejedor, Nelson M. Novo, Leandro Aristide

INTRODUCTION

The estimation of phylogenetic relationships and divergence times among a group of organisms is a fundamental first step toward understanding its biological diversification [1], [2]. Because of the importance of macroevolutionary and macroecological studies for explaining current diversity and the recent development of statistics for evolutionary inference based on time calibrated phylogenies [2], [3], interest in estimating robust phylogenies and divergence times of different clades has grown significantly. Consequently, divergence times have been widely investigated among several key clades. The Order Primates is one of the most widely studied groups [4]–[6].

Among primates, the temporal divergence of the platyrrhine clade is one of the most controversial among scholars. Platyrrhines are a monophyletic group that migrated into South America and evolved in isolation from the Old World primates. Their current biodiversity stands at 100 to 125 extant species and at least 16 genera [7]–[9]. Within South America and the Caribbean, they experienced a broad radiation occupying a large range of ecological niches, resulting in a great variation in morphology and body size [7], [10]. Although the most recent estimations of platyrrhine phylogeny generated topologies that are generally similar – even considering differences in interpreting the position of *Aotus* [11] – the divergence times are a cause of considerable debate [12]–[17]. Moreover, these divergence time estimations have been used to support or contradict different higher order hypotheses which attempt to explain the shape of platyrrhine evolution. For example, Hodgson et al. [15] used mtDNA data and fossil calibrations to support the idea that platyrrhine diversification is characterized by two successive, sister-group radiations [16], the most recent of which is crown Platyrrhini, and to contradict the so called *long lineages hypothesis* of Rosenberger and co-workers [17]–[20], which interprets possibly all platyrrhines, living and extinct, as belonging to a single holophyletic group, and stresses the role of morphological stasis as a deep evolutionary phenomenon. The latter hypothesis considers the oldest records of platyrrhines (certainly those from the early to middle Miocene of Patagonia and Chile and possibly those from the late Oligocene of Bolivia) as part of the crown Platyrrhini, thus phylogenetically related to the lineages of anatomically modern forms. These also include the most diverse collection of platyrrhine fossils from La Venta, Colombia, deposits of middle Miocene age.

The recent studies of divergence times have used two sources of evidence to discuss estimations, the fossil record and molecular sequences [6], [21], [22]. The fossil record is the only direct source of evidence about the existence of an extant lineage during a time period in the past [4], [5] and it relies on establishing the phylogenetic relationships among fossil and extant forms based on morphological similarities. Particularly, the age of the geological formation containing the fossil provides an unobjectionable minimum boundary for the divergence time of the lineage it represents [5]. Molecular distances between DNA or protein sequences obtained from extant species provide indirect estimations

Citation: Perez SI, Tejedor MF, Novo NM, Aristide L (2013), Evolution of New World Monkeys: An Analysis of Fossil and Molecular, Data. PLoS ONE 8(6): e68029. doi:10.1371/journal.pone.0068029.

of divergence times, based on the substitutions accumulated along the phylogenetic branches during the divergence process. However, the molecular based method also use external sources of information to calibrate the substitution rates within lineages and derive estimates of divergence times of clades in years, generally fossil record and estimates of the substitution rate per generation [5], [6], [21], [22]. Both approaches have intrinsic sources of uncertainty for divergence time estimations. For the approach that uses the fossil record directly to calibrate the molecular divergence, the uncertainty is related to problems of misclassification or dating error of the fossils [23]. In the approach that uses substitution rate per generation there is uncertainty in the estimations of generation time and substitution rates [22].

These two complementary approaches have been not widely explored in studies of platyrrhine evolution, and thus the debate concerning the divergence schedule of this clade persists unabated. Here we used this arsenal of molecular based approaches to examine the problem, employing the whole mtDNA genome and large-scale nuclear sequence data that are now available for several platyrrhine species [15], [24]–[26]. For estimating divergence times we specifically use (1) a Bayesian Markov Chain Monte Carlo (MCMC) method that co-estimate phylogeny and divergence times under a relaxed-clock model [21] employing multiple fossil constraints and several topological hypotheses, and (2) an alternative approach that is independent of the fossil constraints and employs body size, substitution rate per generation and generation time estimates [22], [27]. We also explore the robustness of our estimations to change in several prior parameters and discuss the implications of our findings for understanding the platyrrhine evolutionary radiation. Specifically, we asked whether the molecular data, fossil constraints and substitution rate information are enough to confidently reject the hypothesis that crown Platyrrhini and/or the main platyrrhine lineages could have diverged at or before 20 Ma (megannum or million years ago). This would constitute a rejection of the long lineage hypothesis, which was supported by Hogdson et al. [15] in a recent influential molecular study.

South American Land Mammal Ages and the Fossils of the Basal Platyrrhine Radiation

As above mentioned, since the fossil record is the only direct source of evidence about the existence of an extant lineage during a time period in the past, we employ the South American fossil record of mammals, including primates, to establish minimum ages and calibrate phylogenetic trees. That record of mammals is rich in Patagonia, and especially in the primate containing formations, thus allowing correlations with other fossiliferous exposures in South America [20], [28]–[30]. To estimate the divergence times of the New World monkeys, we compiled information (presence/absence of platyrrhine fossils and body size estimations) of eight South American Land Mammal Ages (SALMAs), from the base of the Barrancan subage of the Casamayoran SALMA (41.6 Ma) to the Laventan SALMA (13.8 to 11.8 Ma). We summarized all these biochronological units starting with the oldest records of caviomorph rodents (Figure 1). From the eight SALMAs, primates are absent in the record from the Barrancan (middle Eocene) through the Tinguirirican (early Oligocene; see below), but rodents are relatively well represented by several extinct species.

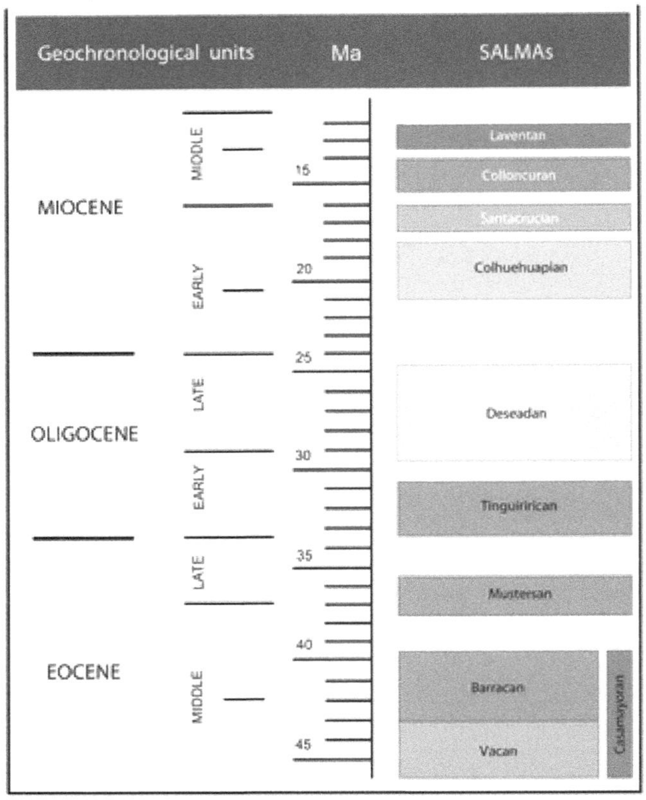

Figure 1. Geologic time scale.

Geochronological units and South American Land Mammal Ages (SALMAs) used in the text.

Except for the Mustersan, three of the oldest SALMAs discussed in this report preserved fossil caviomorph rodents, but no evidence of platyrrhine primates have yet appeared [31]–[37]. By Deseadan times (Figure 1), the oldest records of South American primates, *Branisella* and *Szalatavus*, were present in Salla, Bolivia, dating to the *Branisella*-zone fossiliferous level of about 26 Ma, although the generic status of *Szalatavus* is still debated (see [38]). These primates are known from dental and gnathic anatomical parts that exhibit a mix of unusual characteristics while other traits suggest a close relationship to cebids. Like callitrichines, they have subtriangular upper molars, relatively "waisted" upper premolars, and a conical, *Callimico*-like p2, and a V-shaped mandible. However, they also share unexpected traits such as high-crowned lower molars, and heavy wear on the occlusal surfaces, as well as very small canines. These primates have characteristics that may anticipate the Cebidae, but further studies and more material are needed for a more confident assessment. The estimated body size for these genera was established between 550 and 1000 grams [7]. After these scarce and isolated fossil primates, there is a gap of at least 6 million years with no records of fossil primates. In contrast, and following the previous radiations, Deseadan caviomorphs are abundant and diverse [39].

It is not until the Colhuehuapian SALMA (early Miocene; Figure 1), that primates reappear in Patagonia, Argentina. With 20 Ma *Dolichocebus gaimanensis* [16], [40], [41], from the locality of Gaiman, in Chubut Province, is represented by an edentulous skull and isolated teeth, and is recognized as possibly the earliest known cebine (see [28], [42]; but see [16]) on the basis of characters such as a relatively narrow interorbital septum, a relatively vaulted braincase, as well as oval, vertically oriented orbits and a narrow face. Also, dental traits show similarities to the Laventan *Neosamiri* and *Laventiana*, and by extension to *Saimiri*. But perhaps its most emblematic character, although controversial, is the presence of an interorbital fenestra that appears elsewhere only in the living *Saimiri* [42], thus strengthening their possible phylogenetic relationships. The estimated body size for this genus was established in 2700 grams [7]. Also Colhuehuapian in age, with 20 Ma *Tremacebus harringtoni*, from Chubut Province, Argentina, is recognized as the earliest known aotine based on cranial characters, especially its relatively large orbits, strong postorbital constriction, and short and abbreviated face [28], [43], with a body size estimated in 1800 grams [7]. The morphological debate for the assignment of *Dolichocebus* and *Tremacebus* to the Cebinae and Aotinae clades, respectively, will be dealt with in the Results and Discussion section. Also for the Colhuehuapian the recent description of *Mazzonicebus almendrae*, from Gran Barranca, in south-central Chubut Province [44], expands the record of pitheciines back to the early Miocene (see Results and Discussion) and reinforces the hypothesis of an ancient divergence of living clades. Pitheciines, living and extinct, are characterized mainly by their novel anterior dentition, which is distinctive in their adaptations toward the sclerocarpic harvesting [10], [45]. Another Colhuehuapian genus is *Chilecebus* [28], an unusual primate recovered in central Chile and known from a skull preserving the upper dentition. It shows some characters not found in other platyrrhines, including proportionally large teeth compared to the limited size of the palate. However, the transversely elongated, oval upper premolars are strongly reminiscent of a cebine.

The Santacrucian SALMA (Figure 1) presents a relatively abundant and diverse primate assemblage in Patagonia, and the slightly older Pinturas Formation has yielded four species of primates included in two genera. A third Pinturan genus is being described as well by two of us (MFT and NMN). At about 17 Ma [46], [47], these primates are members of the Pitheciidae, with at least two species of pitheciines allocated to *Soriacebus*. Both exhibit a derived anterior dentition resembling *Mazzonicebus*. There is another group of pitheciids in the Santacrucian as well. From Pinturas, *Carlocebus carmenensis* and *C. intermedius* represent generalized homunculines having close phylogenetic relationships with the younger *Homunculus*, from the Santa Cruz Formation in the southeastern coast of Santa Cruz province (late-early Miocene, 16.5 Ma). These pitheciid genera have body sizes estimated between 2000 and 2700 grams [7]. Finally, also from the Santa Cruz Formation, *Killikaike blakei* was recognized as a cebine closely related to the *Saimiri* lineage [48], mainly based on the morphology of the face (oval orbits vertically oriented, narrow interorbital septum, vaulted frontal bone with a relative anterior brain volume closer to the mean for *Saimiri*).

Thus far, the youngest primate record in Patagonia is *Proteropithecia*, from the Colloncuran SALMA (15.8 Ma; middle Miocene) of Neuquen Province [49]. This primate represents the only uncontroversial pitheciine from the southern regions. All agree it is a member of the crown Platyrrhini.

Far to the northern Neotropics, the middle Miocene primate fauna from La Venta, Colombia, represents the most diverse platyrrhine assemblage in South America and the Caribbean yet discovered [50], [51], as part of the Laventan SALMA (13.8 to 11.8 Ma; middle Miocene; [52]; Figure 1). The Laventan is not represented outside of Colombia, but some possible correlations have been suggested in Bolivia and Patagonia that are still under debate (Croft Quebrada Honda; Chubut Cerro Zeballos). There are 11 described primate species and most of them can be related to living clades, thus representing an uncontroversial cross section of the crown Platyrrhini at about 12.5 Ma. *Neosaimiri fieldsi* and *Laventiana annectens* are certainly cebines pertaining to the *Saimiri* lineage, based on their absolute dental and mandibular similarities to the living squirrel monkey [53]–[55]. *Aotus dindensis* is recognized as an extinct species of *Aotus* for the almost identical mandibular and dental morphology [56]. *Mohanamico hershkovitzi* [57] is probably most closely related to the callitrichines, possibly to the *Callimico*clade [58], based especially on its taller incisors and canines, large p2, and broader and longer trigonid in proportion to the talonid, all callitrichine characters [28]. *Stirtonia tatacoensis* and *S. victoriae* are known by several teeth, a mandible and a maxilla that closely resemble, and are almost indistinguishable from, the living *Alouatta* [53], [59]–[61]. These genera are characterized by a large body size reaching an estimated 10,000 grams [7]. Also with close affinities to a living group, this time with pitheciines, *Cebupithecia sarmientoi* [53] and *Nuciruptor rubricae* [62] exhibit a pitheciine-like molar relief, with low cusps and poorly develped crests, procumbent incisors and projecting canines (excepting *Nuciruptor*), and posteriorly deep mandibles. *Miocallicebus villaviejai* is poorly represented by a piece of maxilla with eroded molars [63], but it seems morphologically close to the living *Callicebus*. However, more remains are needed to strengthen this hypothesis. At least three other genera with uncertain affinities have also been described from the Laventan.

Other fossil platyrrhines have been recovered in younger sediments of South America and in the Greater Antilles (see [20], [28], and references therein). The recognition of phenetic similarities shared between some Caribbean primates and those from Patagonia [64] led us to suspect they are representatives of an old phylogenetic lineage within the crown group.

MATERIALS AND METHODS

Molecular Datasets

To estimate the divergence times of the New World monkeys, we analyzed 13 species of platyrrhines, 14 species of catarrhines, and one outgroup (*Tarsius bancanus*; Table 1). These species were selected (a) in order to provide nodes temporally constrained by well-supported fossil dates and molecular rates [4]–[6], [22]; (b) to take advantage of existing molecular rate estimations for catarrhine primates; and (c) because there are molecular genomic data for all 28 species. Two different molecular datasets were obtained. The first dataset was downloaded from GenBank and is composed of protein mtDNA sequences comprising a 12,996 bp matrix (Table 1). These mtDNA sequences were aligned using ClustalW. The alignment was in the reading frame and examined for ambiguous regions with BioEdit 7.0.0 software [65]. The dataset did not have important ambiguous regions. The second dataset was obtained from the supporting information in the Perelman *et al.* [26] study. This dataset is a post-GBLOCK editing alignment including 54 coding and non-coding nuclear sequences and comprises a 34,941 bp matrix (Table 1), constituting a representative stratified sample of the whole genome [26].

Table 1. Molecular data.

Species	mtDNA sequences	Nuclear sequences
Tarsius bancanus	NC_002811	Perelman et al. [26]
Homo sapiens (Cambridge)	NC_012920	Perelman et al. [26]
Pan paniscus	GU189661	Perelman et al. [26]
Pan troglodytes	NC_001643	Perelman et al. [26]
Pan troglodytes verus	X93335	Perelman et al. [26]
Gorilla gorilla	NC_001645	Perelman et al. [26]
Pongo pygmaeus	NC_001646	Perelman et al. [26]
Hylobates lar	NC_002082	Perelman et al. [26]
Macaca mulatta	NC_005943	Perelman et al. [26]
Macaca sylvanus	NC_002764	Perelman et al. [26]
Papio hamadryas	NC_001992	Perelman et al. [26]
Theropithecus gelada	FJ785426	Perelman et al. [26]
Cercopithecus aethiops	AY863426	Perelman et al. [26]
Chlorocebus sabaeus	EF597503	Perelman et al. [26]
Colobus guereza	AY863427	Perelman et al. [26]
Callithrix jacchus	AB572419	Perelman et al. [26]
Saguinus oedipus	FJ785424	Perelman et al. [26]
Cebus apella	JN380205	Perelman et al. [26]
Cebus albifrons	AJ309866	Perelman et al. [26]
Saimiri boliviensis boliviensis	HQ644339	Perelman et al. [26]
Saimiri oerstedii oerstedii	HQ644337	Perelman et al. [26]
Saimiri sciureus	FJ785425	Perelman et al. [26]
Aotus azarai azarai	JN161099	Perelman et al. [26]
Aotus lemurinus	FJ785421	Perelman et al. [26]
Aotus nancymaae	JN161101	Perelman et al. [26]
Aotus trivirgatus	AY250707	Perelman et al. [26]
Ateles belzebuth	FJ785422	Perelman et al. [26]
Callicebus donacophilus	FJ785423	Perelman et al. [26]

List of species used in the study and Genbank accession numbers.
doi:10.1371/journal.pone.0068029.t001

Phylogenetic Tree and Divergence Time Estimations

The best-fitting model of evolution for each sequence studied was estimated employing the Akaike information criterion with correction for sample size (AICc) implemented in jModelTest 0.1.1 [66]. The results are shown in table 2. Models of sequence evolution identified as optimal by jModelTest for both coding and non-coding sequences were implemented in the phylogenetic analyses.

Table 2. Substitution models.

Sequence	nst	rates	model	Sequence size
ABCA1	2	gamma	HKY+G	560
ADORA3	2	gamma	HKY+G	414
AFF2	6	gamma	GTR+G	500
AFF2.2	6	gamma	GTR+G	579
APP	6	gamma	GTR+G	672
AXIN1	6	gamma	HKY+I	949
BCOR	6	gamma	GTR+G	771
BDNF	2	gamma	HKY+G	561
BRCA2	6	gamma	GTR+G	1252
CFTR	2	gamma	HKY+G	791
CHRNA1	2	gamma	GTR+G	381
CNR1	2	gamma	HKY+I+G	998
CREM	2	gamma	HKY+G	428
DACH1	2	gamma	HKY+I+G	627
DMRT1	2	gamma	HKY+G	537
EDG1	2	gamma	HKY+G	967
FBN1	2	gamma	HKY+G	720
FES	2	gamma	HKY+G	469
FOXP1	6	gamma	GTR+G	564
GHR	2	gamma	HKY+G	646
KCNMA1	6	gamma	GTR+G	614
LRPPRC_169	2	gamma	HKY+G	792
LRPPRC_171	6	gamma	GTR+G	761
LUC7L	6	gamma	GTR+G	694
MAPKAP1	6	gamma	GTR+G	655
MBD5	2	gamma	HKY+I+G	558
NEGR1	6	gamma	GTR+G	540
NPAS3	6	gamma	GTR+G	605
NPAS3.2	6	gamma	GTR+G	650
PLCB4	6	gamma	GTR+G	338
RAG1	2	gamma	HKY+I+G	1071
RAG2	6	gamma	GTR+G	690
RPGRIP1	2	gamma	HKY+I+G	683
SGMS1	1	gamma	HKY+G	598
SIM1	2	gamma	HKY+G	646
SMCX	6	gamma	GTR+G	330
SMCY	2	gamma	HKY+G	940
SRY	2	gamma	HKY+G	467
TEX2	1	equal	HKY	156
TTR	6	gamma	GTR+G	877
TYR	2	gamma	HKY+G	475
USH2A	6	gamma	GTR+G	605
UTY	2	equal	GTR	371
ZFX	6	gamma	GTR+G	811
ZFY	6	gamma	GTR+G	853
ZIC3	2	equal	HKY	549
ATXN7	2	equal	HKY+I+G	523
BCHE	6	gamma	GTR+G	984
DCTN2	6	gamma	GTR+G	605
FAM123B	2	gamma	HKY+I+G	730
PNOC	2	gamma	HKY+G	313
POLA1	6	gamma	GTR+G	604
RAB6IP1	2	gamma	HKY+I+G	717
ERC2	6	gamma	GTR+G	750
Total nuclear	6	gamma	GTR+I+G	34941
mtDNA	6	gamma	GTR+I+G	12996

Coding and non-coding sequences used in the current study, sequence size (bp) and substitution models.

Two divergence time estimation approaches were used [21], [22], [27]. Firstly, the phylogenetic tree topology and divergence times were estimated jointly using the BEAST v1.6.1 package [21], [67]. We used the BEAUti program to unlink the substitution models of the data partitions and to implement the models of sequence evolution identified as optimal by jModelTest. We analyzed the sequences under a relaxed molecular clock model, which allows substitution rates to vary across branches according to an uncorrelated lognormal distribution [21], and set the species tree priors as a Yule Process. Two simultaneous analyses were performed using Markov Chain Monte Carlo (MCMC) simulations for 200,000,000 generations with a sampling frequency of 20,000. The convergence was determined with Tracer v1.5 [68] and the first 2,500 trees sampled were excluded using TreeAnnotator v1.4.8 [67]. FigTree v1.3.1 was used to plot all phylogenetic trees.

Uncertainty in divergence time estimation using BEAST could be mainly related to uncertainty in tree topology and fossil calibrations [21]. Because there are different hypotheses of topological relationships among the main extant lineages (families and subfamilies) of platyrrhines, as discussed above, we changed the best inferred topology for each dataset by enforcing monophyly constraints on several clades (Table 3). This procedure made the resulting trees consistent with previous studies of platyrrhine phylogeny [10], [11], [13], [26], [69]. We generated 4 alternative tree topologies: (1) Atelidae sister to Cebidae, with Aotinae as a branch external to the Cebidae family [69]; (2) Atelidae sister to Cebidae, with Aotinae as a branch external to the Callitrichinae subfamily [26]; (3) Atelidae sister to Cebidae, with Aotinae being related to Cebinae [13]; and (4) Pitheciidae sister to Atelidae, with Aotinae related to *Callicebus* [10], [18].

Table 3. Alternative topologies.

Topologies	Wildman et al [69]	Opazo et al. [13]*	Perelman et al. [26]**	Rosenberger [10,18]
Monophyly constrained clades	Cebinae (*Saimiri-Cebus*)	Cebinae (*Saimiri-Cebus*)	Cebinae (*Saimiri-Cebus*)	Cebinae (*Saimiri-Cebus*)
	Cebidae - Aotinae	Cebinae - Aotinae	Aotinae (*Aotus*) - Callitrichinae	Aotinae (*Aotus*) - *Callicebus*
	Cebinae - Callitrichinae	Cebidae - Aotinae	Cebidae - Aotinae	Pitheciidae (Aotinae-*Callicebus*) - Atelidae
	Cebidae-Aotinae-Atelidae	Cebidae-Aotinae-Atelidae	Cebidae-Aotinae-Atelidae	Cebinae - Callitrichinae

Monophyly constraints on platyrrhine clades.
*Best inferred topology for mtDNA. Topology inferred without monophyly constraints.
**Best inferred topology for nuclear data. Topology inferred without monophyly constraints.

For the four topologies, five fossil calibrations were selected based on criteria for choosing appropriate points [4], [6], two nodes each for the platyrrhine and catarrhine clades and one for the outgroup. Both minimum and maximum calibration bounds were set to the probability that the true divergence time is outside the bounds to be small, but non-zero (i.e., 'soft' for [6]). Fossil calibration for catarrhines was obtained from Benton et al. [4] and Steiper and Seiffert ([27]; Figure 2; Table 4). Because phylogenetic interpretations of the fossil record of platyrrhines is still debated [16], [17], [28], [70], we also set three fossil calibration hypotheses for the two nodes (Figure 2; Table 4). Our approach was designed to estimate the time of origin of the crown Platyrrhini without using any specific fossil-based calibration constraint for this particular node of the molecular tree; however, in the first and second hypotheses, *Dolichocebus* and *Tremacebus,* whose membership to the platyrrhine crown is contentious, are used as calibration points for extant families (see next and Results and Discussion). The first hypothesis is based on the most traditional phylogenetic interpretation for Patagonian fossils [17], [28]: (1) minimum divergence time of Cebinae was set at 20 Ma, based on *Dolichocebus gaimanensis*, a fossil from the valley of the Chubut river in Argentina, attributable to Cebinae; maximum divergence time of Cebinae was set at 26 Ma, based on the absence of Cebinae fossils previous to the Deseadan fauna of Salla, Bolivia; (2) minimum divergence time between Aotinae and Cebinae, Callitrichinae or *Callicebus* (depending on the topology) was set at 20 Ma, based on *Tremacebus harringtoni*, a fossil from Sacanana, Chubut Province in Argentina, attributable to Aotinae; maximum divergence time between Aotinae and Cebinae, Callitrichinae or *Callicebus* was set at 26 Ma, based on the absence of aotine fossils in the Deseadan fauna of Salla, Bolivia, and other South American formations of the same age. The second hypothesis is a modification of the first one based on an alternative phylogenetic interpretation for Patagonian fossils [15], [16], [70]. We modified the minimum divergence time of Cebinae and Aotinae using a calibration at 12.5 Ma, based on *Neosaimiri fieldsi* and *Aotus dindensis*, respectively, two fossil species from La Venta, in Colombia, which we recognize (and is now apparently the consensus view among active researchers) as cebines and aotines, respectively. The third hypothesis is a modification of the first one based on an alternative soft maximum divergence time of Cebinae and Aotinae using a calibration at 41 Ma, based on the absence of Cebinae and Aotinae fossils in the Contamana fauna, Peru, and other ancient South American formations of the same age [31]. All calibration points were implemented as log-normal distributions with an offset, mean, and standard deviation such that 95% of the prior distribution falls between the boundaries specified in figure 2 and table 4. This procedure allows molecular data to correct for conflicting fossil information and uncertainty in the in fossil evidence [6], [21].

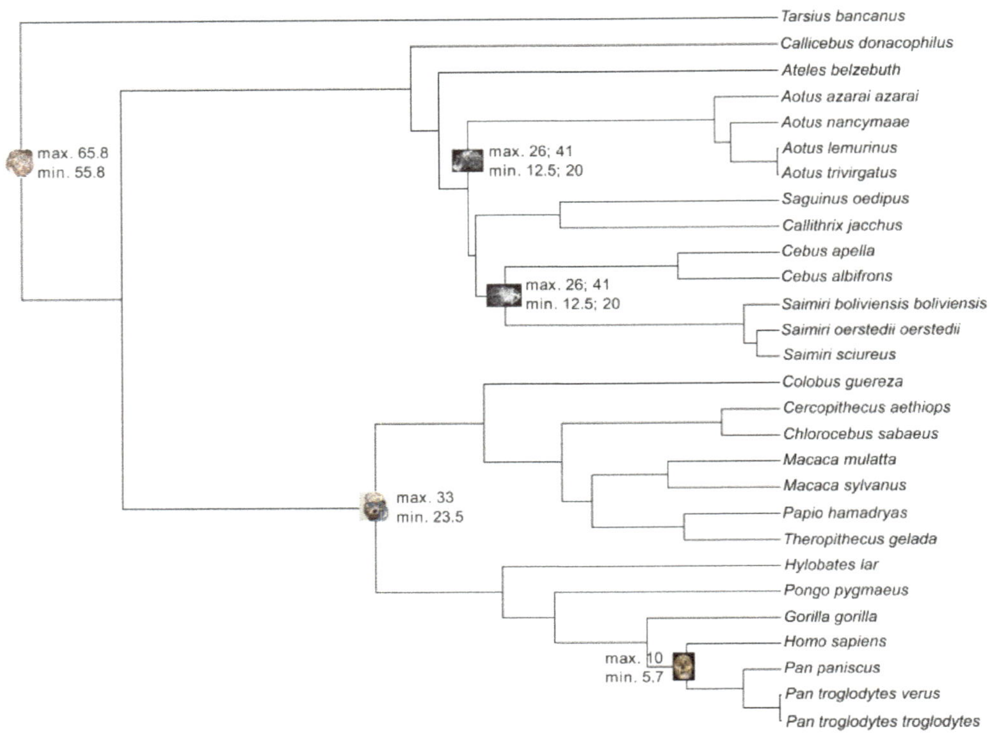

Figure 2. Fossil calibrations.

Phylogenetic tree of 28 primate species showing fossil calibrations. Calibration bounds are soft; i.e., the probability that the true divergence time is outside the bounds is small but non-zero [6]. The phylogenetic tree follows the Wildman et al. [69] hypothesis.

Table 4. Calibrations.

		Node	Minimum soft bound	Fossil minimum	Maximum soft bound	Evidence maximum	Parameters of the LogNormal distribution
Old World Primates		Root Haplorhini	55.80	*Teilhardina*	65.80	absence of euprimates in Paleocene	Offset (55.8), mean (1.3), and standard deviation (0.5), Initial Value = 60
		Catarrhini	23.50	*Proconsul*	33.00	absence of hominoids in late Eocene	Offset (23.5), mean (1.5), and standard deviation (0.5)
		Hominini	5.70	*Orrorin*	10.00	absence of hominines in mid-Miocene	Offset (5.7), mean (0.5), and standard deviation (0.5)
New World Primates	**First hypothesis***	Cebinae (*Saimiri-Cebus*)	20.00	*Dolichocebus*	26.00	absence of cebids after the Salla formation	Offset (20), mean (0.9), and standard deviation (0.4)
		Aotinae/–	20.00	*Tremacebus*	26.00	absence of aotines after the Salla formation	Offset (20), mean (0.9), and standard deviation (0.4)
	Second hypothesis**	Cebinae (*Saimiri-Cebus*)	12.50	*Neosaimiri*	26.00	absence of cebids after the Salla formation	Offset (12.5), mean (1.8), and standard deviation (0.4)
		Aotinae/–	12.50	*Aotus dindensis*	26.00	absence of aotines after the Salla formation	Offset (12.5), mean (1.8), and standard deviation (0.4)
	Third hypothesis***	Cebinae (*Saimiri-Cebus*)	20.00	*Dolichocebus*	41.00	absence of cebids after Contamana	Offset (20), mean (2), and standard deviation (0.5)
		Aotinae/–	20.00	*Tremacebus*	41.00	absence of aotines after Contamana formation	Offset (20), mean (2), and standard deviation (0.5)

Secondly, for comparative purposes we combined and applied a slight modification of the methods recently proposed by Steiper and Seiffert [27] and Langergraber et al. [22]. This method is based on the argument that by estimating an external molecular rate, or the rate at which the DNA sequence diverged in the genome, DNA differences can be converted into divergence times independently of fossil calibration constraints [22], [71]. The only available direct estimation of a molecular rate among primates is for the human lineage [22]. Since this rate is not necessarily the same for platyrrhines, a procedure to correct for possible differences is needed. The rationale behind the method proposed here is based on the Steiper and Seiffert work [27], which showed that there is a strong, inverse relationship between molecular substitution rates and body size for primates. It is also known that body size is correlated with primate life history (e.g. generation time) [72]. The relationship between generation time and substitution rate is based on the hypothesis that most germ-line mutations occur during DNA replication [73]. The obtained correlation coefficient between body mass and generation time is 0.89 (P>0.001) for extant platyrrhines and 0.907 (P>0.001) for all primates used in this study (see Material S1 for details). Using this, we employed body size and generation time estimates for extant and extinct platyrrhines to obtain a corrected substitution rate that is applicable to the different platyrrhine lineages, based on an estimation independent from the fossil constraint. This alternative method was only used with the nuclear DNA dataset (following Langergraber et al. [22]), since mitochondrial substitution rates are known to differ from nuclear ones and a time-dependent rate-curve effect is observed in mitochondrial DNA.

Uncertainty in divergence time estimations using an external molecular rate could be mainly related to uncertainty in substitution rate and generation time estimations. As in Langergraber et al. [22] we used the broadest available interval of substitution rates estimations –based on human mutation rates– to incorporate the first source of uncertainty in our analyses (9.70E-09 to 1.36E-08/site/generation [22]). Because a substitution rate independent from the fossil record is only available for a single species (*Homo sapiens*), we applied three imputation procedures to infer generation time and correct the substitution rate: a linear regression, a quadratic curve and the EM algorithm ([74]; see Material S1 for details). After careful inspection of the resulting imputation, we used the quadratic curve results in the following analyses (see Material S1; Table 5). We used the mean generation time inferred for each clade as the best estimation of generation times along their whole evolutionary history. This is a different approach to that of Steiper and Seiffert [27] (see Results and Discussion). In this way, the changes in substitution rates along the tree are a function of changes in body size and generation time of the studied primate species, as would be predicted by the hypothesis that most mutations occur during DNA replication [73].

Table 5. Size and generation time.

Clade	Genus	Body size in grams	Generation time in years
Atelidae	*Alouatta*	6404.2	12.0
Atelidae	*Ateles*	8276.3	15.0
Atelidae	*Brachyteles*	8840.0	20.0
Atelidae	*Lagothrix*	7150.0	15.0
Aotinae	*Aotus*	1018.7	8.0
Cebinae	*Cebus*	2475.1	15.0
Cebinae	*Saimiri*	786.9	8.0
Callitrichinae	*Saguinus*	444.4	6.0
Callitrichinae	*Leontopithecus*	471.4	7.0
Callitrichinae	*Callithrix*	351.2	6.0
Callitrichinae	*Callimico*	505.0	6.0
Pitheciidae	*Callicebus*	997.3	8.0
Pitheciidae	*Pithecia*	2003.5	9.0
Pitheciidae	*Cacajao*	2893.8	10.0
Pitheciidae	*Chiropotes*	2632.5	10.0
Pitheciidae	†*Soriacebus*	2000.0	9.0*
Pitheciidae	†*Carlocebus*	2000.0	9.0*
Pitheciidae	†*Homunculus*	2700.0	10.0*
Pitheciidae	†*Cebupithecia*	2200.0	9.0*
Pitheciidae	†*Nuciruptor*	2000.0	9.0*
Pitheciidae	†*Proteropithecia*	1600.0	9.0*
Aotinae	†*Tremacebus*	1800.0	9.0*
Aotinae	†*Aotus (dindensis)*	1000.0	8.0*
Cebinae	†*Dolichocebus*	2700.0	10.0*
Cebinae	†*Chilecebus*	1000.0	8.0*
Cebinae	†*Neosaimiri*	840.0	8.0*
Cebinae	†*Laventiana*	800.0	8.0*
Atelidae	†*Stirtonia*	5800.0	12.0*
Atelidae	†*Stirtonia*	10000.0	20.0*
Atelidae	†*Protopithecus*	23500.0	22.0*
Atelidae	†*Caipora*	24000.0	22.0*
Callitrichinae	†*Patasola*	1000.0	8.0*
Callitrichinae	†*Lagonimico*	1300.0	8.0*
Incertae sedis	†*Branisella*	1000.0	8.0*
Incertae sedis	†*Szalatavus*	550.0	7.0*
Hominidae	*Homo*	45000.0	29.0
Hominidae	*Pan*	33000.0	25.0
Hominidae	*Gorilla*	71000.0	19.0
Cercopithecinae	*Macaca*	9000.0	10.0

After correcting the substitution rates for each studied hominid and platyrrhine lineage, we estimated divergence time for each node of interest. For this, we estimated a Maximum Likelihood tree with a general time reversible substitution model and gamma distribution and then constructed a linearized tree using Mega 5.05 [75], [76]. For the different branches of this tree we specified the previously estimated different substitution rates. Prior to each calculation, we conducted Tajima's relative rate test [77], or molecular clock hypothesis test, for the molecular divergence between the two species compared using Mega 5.05 [76]. The test was only significant for comparisons that involved *Aotus*, and therefore this genus was excluded from the analysis. Generation time for extant platyrrhine species was obtained from IUCN [78]and average body mass for wild adults was obtained from Smith and Jungers [79] and for fossils from Fleagle [7].

RESULTS AND DISCUSSION

Phylogenetic Tree

We used the complete set of the protein-encoding genes from the mitochondrial genome and a large-scale stratified sampling of coding and non-coding nuclear sequences from several species taken from GenBank to represent the major platyrrhine lineages. We also used DNA data from several other haplorhine species with a well known fossil record to provide nodes that are temporally constrained (Table

4). Our chronophylogenetic trees based on these mtDNA and nuclear data with their maximum likelihood values are in agreement with other recent molecular trees (Figure 3), which support the division of platyrrhines into three monophyletic families (Atelidae, Cebidae, and Pitheciidae) and suggest a sister-group phylogenetic relationship between Atelidae and Cebidae [11], [13], [26], [69]. Within the family Cebidae, these trees display a branch for Cebinae, which includes *Cebus* and *Saimiri*, as well as a branch for Callitrichinae formed by *Saguinus* and *Callithrix*. The relationships among Old World monkeys are also in agreement with recent phylogenies [26]. These trees only differ in the *Aotus* position; it is phylogenetically related to Callitrichinae for the nuclear dataset, as per the Perelman et al. [26] tree, and to Cebinae for the mtDNA dataset, as in the Opazo et al. [13]tree, but both positions occur with low node support.

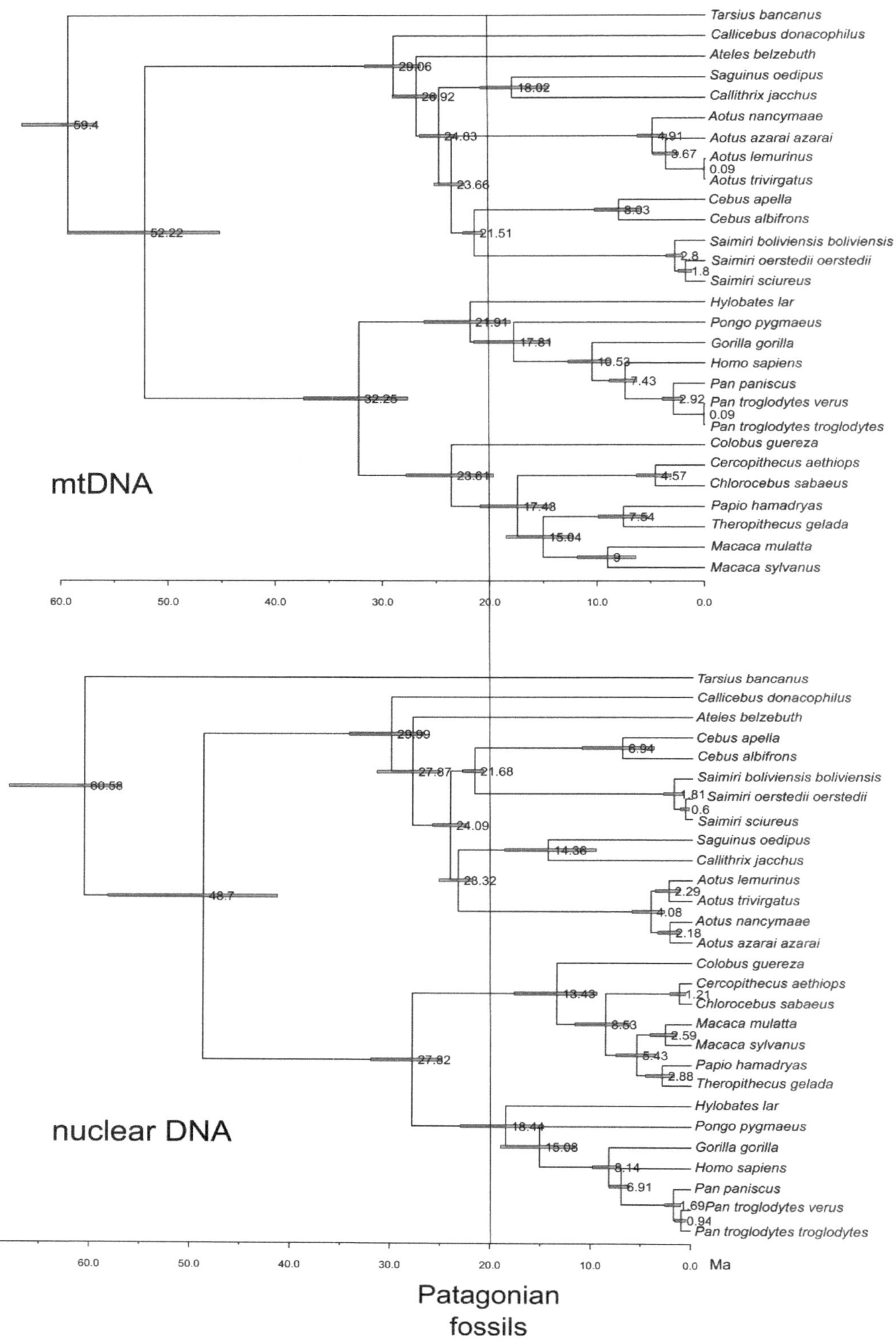

Figure 3. BEAST chronophylogenetic trees.

More likely chronophylogenetic tree from the BEAST analysis for 28 species of Primates using mtDNA and nuclear sequences. Mean node ages are depicted in each node. Blue horizontal bars represent the posterior 95% CI for the node ages. The vertical line shows the estimated earliest age of Patagonian lineages.

The alternative chronophylogenetic trees with constrained topologies based on previous molecular tree hypotheses display maximum likelihood values that are not significantly different from Perelman et al. [26] and Opazo et al. [13] for nuclear and mtDNA datasets, respectively (Table 6). This is not surprising given the existence of short branch lengths connecting Cebinae, Callitrichinae and Aotinae lineages (Figure 3) and the previous problems in estimating a robust platyrrhine species tree [11], [80]. On the other hand, the chronophylogenetic trees with a constrained topology based on the Rosenberger [10] morphological hypothesis display maximum likelihood values that are significantly lower than the Perelman et al. [26] and the Opazo et al. [13] trees (Table 6). We do not know whether one of these inferred phylogenetic trees is representative of the true branching process or history of platyrrhine species divergence. Although previous studies suggest that this is not problematic for divergence time estimation, our results suggest that the topology has great importance for inferring the divergence time of the main platyrrhine lineages (Table 7). For this reason we considered the alternative topologies in divergence time estimates that are discussed in the following section.

Table 6. Likelihood values.

Dataset	Topology	Likelihood mean	Likelihood median	95% HPD lower	95% HPD upper
mtDNA	Opazo	**−129818.64**	**−129818.30**	**−129829.13**	**−129807.70**
	Perelman	*−129824.11*	*−129823.68*	*−129835.01*	*−129813.66*
	Wildman	*−129831.34*	*−129831.05*	*−129842.25*	*−129820.89*
	Rosenberger	−129883.51	−129883.15	−129894.68	−129872.73
Nuclear DNA	Perelman	**−100064.87**	**−100064.51**	**−100076.67**	**−100053.04**
	Wildman	*−100068.80*	*−100068.46*	*−100080.92*	*−100057.59*
	Opazo	*−100073.10*	*−100072.69*	*−100085.33*	*−100061.60*
	Rosenberger	−100292.04	−100291.69	−100303.89	−100279.85

Table 7. Bayesian divergence time estimations.

Dataset	Approach	Node	Topology Wildman	Topology Perelman	Topology Opazo	Topology Rosenberger
mtDNA	**First hypothesis***	Crown Platyrrhini	**28.78 (26.37–31.50)**	**28.52 (26.12–31.22)**	**29.06 (26.52–31.75)**	**24.99 (23.58–26.68)**
		Atelidae branching	26.61 (24.76–28.73)	26.36 (24.58–28.52)	26.92 (25.02–29.14)	24.19 (22.72–25.99)
		Crown Cebidae	24.35 (22.95–25.94)	24.17 (22.78–25.70)	24.82 (23.39–26.62)	24.34 (22.90–25.95)
		Crown Anthropoidea	51.58 (44.10–58.81)	51.76 (44.57–58.65)	52.22 (45.25–59.44)	50.31 (43.28–57.95)
		Crown Catarrhini	31.74 (27.33–36.78)	32.03 (27.14–36.85)	32.25 (27.66–37.41)	31.43 (27.06–36.28)
		Homo/Pan	7.37 (6.33–8.87)	7.31 (6.30–8.62)	7.43 (6.33–8.87)	7.38 (6.36–8.77)
	Second hypothesis**	Crown Platyrrhini	**24.28 (21.22–27.93)**	**23.99 (20.89–27.37)**	**24.37 (21.25–27.70)**	**20.85 (18.45–23.78)**
		Atelidae branching	22.26 (19.55–25.53)	22.02 (19.47–25.15)	22.45 (19.61–25.32)	20.18 (17.68–23.07)
		Crown Cebidae	20.19 (17.79–23.10)	20.02 (17.52–22.64)	20.56 (18.05–23.16)	20.24 (17.79–23.02)
		Crown Anthropoidea	46.95 (39.29–55.49)	46.56 (39.71–55.21)	46.67 (39.47–54.46)	46.29 (39.44–54.27)
		Crown Catarrhini	29.12 (25.62–33.38)	29.07 (25.55–33.08)	29.07 (25.58–33.00)	29.03 (25.69–33.26)
		Homo/Pan	7.04 (6.20–8.25)	7.03 (6.19–8.33)	7.07 (6.19–8.29)	7.10 (6.19–8.36)
	Third hypothesis***	Crown Platyrrhini	**31.07 (27.93–34.57)**	**30.74 (27.78–34.49)**	**31.31 (27.89–35.36)**	**27.26 (24.76–30.12)**
		Atelidae branching	28.79 (26.25–31.93)	28.50 (25.98–31.66)	29.11 (26.27–32.48)	26.48 (24.05–29.45)
		Crown Cebidae	26.44 (24.18–29.10)	26.25 (24.06–28.91)	26.92 (24.55–25.75)	26.47 (24.17–29.34)
		Crown Anthropoidea	53.85 (46.14–60.52)	53.87 (45.75–60.53)	54.25 (46.91–60.38)	52.61 (45.08–59.54)
		Crown Catarrhini	33.01 (27.99–38.39)	33.11 (27.64–38.35)	33.52 (28.58–38.88)	32.69 (27.74–38.35)
		Homo/Pan	7.49 (6.30–9.14)	7.51 (6.34–8.97)	7.47 (6.30–8.99)	7.45 (6.25–8.88)
Nuclear	**First hypothesis***	Crown Platyrrhini	**30.06 (26.77–33.88)**	**29.99 (26.79–34.22)**	**30.04 (26.91–34.02)**	**25.65 (23.80–27.93)**
		Atelidae branching	27.91 (25.32–31.46)	27.87 (25.16–31.42)	27.87 (25.14–31.28)	25.52 (23.56–27.74)
		Crown Cebidae	24.27 (22.71–26.18)	24.09 (22.55–25.37)	24.04 (22.64–26.03)	23.88 (22.33–25.71)
		Crown Anthropoidea	48.80 (41.82–58.46)	48.70 (41.32–58.19)	48.65 (41.38–57.68)	46.16 (38.66–56.02)
		Crown Catarrhini	27.85 (24.94–32.04)	27.82 (24.82–31.92)	27.77 (25.02–32.82)	27.53 (24.90–31.22)
		Homo/Pan	6.91 (6.11–8.04)	6.91 (6.11–8.10)	6.91 (6.11–8.07)	6.91 (6.12–8.09)
	Second hypothesis**	Crown Platyrrhini	**25.57 (21.91–30.06)**	**25.61 (21.72–30.35)**	**25.65 (21.89–30.22)**	**21.89 (18.94–25.31)**
		Atelidae branching	23.58 (20.02–27.33)	23.61 (20.31–27.98)	23.68 (20.12–27.66)	21.79 (18.95–25.31)
		Crown Cebidae	19.97 (17.43–22.92)	19.93 (17.24–22.83)	19.95 (17.51–23.08)	19.79 (17.14–22.76)
		Crown Anthropoidea	44.17 (36.69–53.75)	44.55 (36.56–53.83)	44.49 (36.93–53.67)	43.45 (36.08–53.34)
		Crown Catarrhini	27.12 (24.8–30.41)	27.21 (24.69–30.80)	27.07 (24.58–30.59)	26.98 (24.65–30.36)
		Homo/Pan	6.83 (6.13–7.95)	6.81 (6.05–7.88)	6.82 (6.09–7.90)	6.81 (6.09–7.84)
	Third hypothesis***	Crown Platyrrhini	**32.92 (28.60–37.91)**	**32.94 (28.84–38.36)**	**33.06 (28.83–38.41)**	**28.67 (25.81–32.74)**
		Atelidae branching	30.72 (27.06–35.31)	30.69 (27.06–35.61)	30.82 (26.97–35.52)	28.55 (25.60–32.54)
		Crown Cebidae	26.89 (24.22–30.43)	26.78 (24.05–30.13)	26.72 (24.04–30.22)	26.34 (23.73–29.76)
		Crown Anthropoidea	51.04 (42.47–59.59)	51.61 (43.08–60.36)	51.42 (43.41–60.33)	49.04 (41.61–59.15)
		Crown Catarrhini	28.30 (25.06–33.23)	28.14 (25.09–33.00)	28.11 (25.00–32.87)	28.05 (24.94–32.46)
		Homo/Pan	6.96 (6.11–8.24)	7.00 (6.10–8.37)	6.98 (6.15–8.35)	6.96 (6.12–8.27)

Bayesian Divergence Times Estimations under a Relaxed-clock Model

The Bayesian phylogenetic method used in our analyses provide a framework to co-estimate phylogenetic relationships and divergence times under a relaxed molecular clock model [21]. Estimating divergence times using DNA data and fossil calibrations is a complex process as it accounts for fossil age constraints, tree topology and models of molecular evolution. In particular, the fossil calibration points and tree topology must be carefully considered [6], [21]because these parameters can generate very different divergence estimations. Our estimates for the Old World monkeys are in agreement with those obtained in recent studies (Table 7; [5], [6]). Divergence time estimates for platyrrhines based on the different molecular topologies display similar values, but the estimates based on the Rosenberger [10], [18] topology shows values *ca.* 4 Ma younger (Table 7; –S4). This result contrasts with previous studies that suggested that differences in tree topology among platyrrhine trees are not problematic for divergence time estimation [6], [15]. Therefore, we confirm previous suggestions that the Bayesian estimation of phylogeny and divergence time from DNA sequences may be biased when the tree topology is not adequately considered in the model [21].

Our divergence time estimates using the more likely topologies and the most conservative fossil evidence (Second Hypothesis) suggest that the last common ancestor (LCA) of extant platyrrhine primates existed at *ca.* 25 Ma, with the 95% confidence limit for the node ranging from *ca.* 21–30 Ma (Table 7; –S4). However, the fossil constraints also have an important influence over our divergence time estimates. The results of the divergence estimates using the alternative First and Third Hypotheses show older time values for the LCA than the estimates using the Second Hypothesis. The LCA value was *ca.* 29 Ma for the First Hypothesis (Figure 3) and *ca.* 32 Ma for the Third Hypothesis, with the 95% confidence limit for the node ranged from *ca.* 27–31 Ma and *ca.* 27–34 Ma, respectively (Table 7). The divergence time estimates based on the mtDNA dataset are always *ca.* 1 Ma younger than the ones based on the nuclear dataset (Table 7; –S4). For all the hypotheses, our results suggest that the extant platyrrhine families diverged before *ca.* 20 Ma (Figure 3; Table 7).

The use of prior lognormal distributions for calibration of fossil ages plus soft maximum ages allows the relaxed clock method to correct for conflicting fossil-based time constraints. Particularly, the estimated range age of crown Platyrrhini, 21–30 to 27–34 Ma for the different hypotheses, differs significantly from the minimum fossil age of 12.5 Ma based on La Venta fauna. This result also differs significantly from a recent study that used similar fossil constraints [15], but with different lognormal distribution parameters and different maximum fossil constraints. Hodgson and co-workers [15] pointed out that "the lack of lower bounds (the maximum bound in the present work) within the platyrrhines fully allows for the data to support the MSH" (*morphological stasis hypothesis* or *long lineages hypothesis* of Rosenberger and co-workers [17]–[20]). However, the parameters of the prior distribution for calibrating fossil ages used by Hodgson and co-workers [15] generate a very narrow distribution that does not allow the support of the *long lineages hypothesis*. These results suggest that modeling the parameters of the prior lognormal distributions for calibration of fossil ages is very important in divergence time estimation. Therefore, our results support previous studies suggesting that a comprehensive divergence time estimation should account for uncertainty in – among other sources – fossil calibrations, parameters of the prior lognormal distribution and tree topology [21].

Particularly, the uncertainty in fossil calibrations should be carefully considered because it generates serious questions about the credibility of divergence time estimations [81]. Using fossils that are phylogenetically misplaced or that have incorrect ages can introduce serious error into molecular dating. Therefore, we need to use an explicit protocol to justify phylogenetic position and chronological age for fossil specimens [81]. Here, we provide a discussion about phylogenetic position and geochronological age of the most controversial platyrrhine fossils used as constraints in this and previous works [13]–[15], [25]. The above mentioned extinct *Dolichocebus* has indeed an age of 20 Ma, and in our view is linked to the cebids on the basis of cranial characters such as relatively narrow interorbital septum, relatively vaulted braincase, presence of an interorbital fenestra, as well as oval and vertically oriented orbits and a narrow face, and dental traits showing similarities to the Laventan *Neosamiri* and *Laventiana*. However, the natural status of some of these traits, such as the interorbital fenestra, is a matter of discussion [16]. In recent works, Kay and co-workers [16], [44], [70] argued that *Dolichocebus*, like all the other Patagonian platyrrhines except *Proteropithecia*, is part of the stem Platyrrhini. However, there is reason to believe their analysis and interpretation is negatively influenced by the difficulty of establishing legitimate anatomical similarities among specimens suffering from poor preservation of the edentulous type skulls of two crucial Patagonian taxa, *Dolichocebus*and *Tremacebus* [17].

The aotine status of *Tremacebus* is justified for us, especially for its relatively large orbits, strong postorbital constriction, and short and abbreviated face [28], [43]. However, Kay et al. [16], [82] suggest that the orbits are not significantly enlarged, as in *Aotus*, and they maintain that the olfactory bulb of *Tremacebus* (judging by the dimensions of its olfactory fossa) was also not enlarged as in *Aotus,* leading these authors to question the nocturnal status of *Tremacebus*and its phylogenetic link with *Aotus.* However, this view could be difficult to sustain, and considering that nocturnal habits are a secondary acquisition in *Aotus*, it is expected that this adaptation once presented a more primitive state, and *Tremacebus* has all the basic morphological patterns predicted to represent the ancestral pattern of the nocturnal adaptations. *Tremacebus* may have been not strictly nocturnal but cathemeral, like most species of the living *Aotus* [83], [84].

Divergence Times Estimations Employing Body Size and Generation Time

By using generation times of extant New World monkey species, body size estimation for extant and extinct species and molecular rates directly observed in human families, we estimate rates of substitution per generation for the main platyrrhine lineages and the divergence times among these lineages without relying on external fossil calibration points. Our approach combines the Langergraber et al. [22] method which estimates divergence time from the estimated molecular rate per generation based on extant species, with the method proposed by Steiper and Seiffert [27] that corrects molecular rates in nuclear genomes of extant species using life-history variables, like body mass, inferred from the fossil record. We confirm the Steiper and Seiffert [27] and Langergraber et al. [22] observations about a relationship between body mass and generation times in primates using only platyrrhine data for extant species (r=0.89). This corroborates the idea that body mass is correlated with generation time, and then substitution rate [27], assuming that most germ-line mutations occur during DNA replication.

Steiper and Seiffert [27] suggest that molecular rates slowed down over the course of primate evolution because they find an inverse relationship between body mass and molecular rate. However, our results show a different picture concerning the evolution of platyrrhine body size than the Steiper and Seiffert [27] obtained by averaging across all primates. Fossil platyrrhines display body size values similar to the living species, suggesting that body mass estimates of fossil platyrrhine species fall within the range expected for each of the extant main lineages. Therefore, the first platyrrhines of each lineage were approximately the size of the extant species of the same clade, supporting an ancient shift in body size for each clade, such as was shown recently in Aristide et al. [85]. Moreover, because of the strong correlation between body mass and generation time among platyrrhines, it is likely that generation times were approximately the same along the lineages evolution. Using this correlation we were able to predict the generation time of different platyrrhine lineages from ancestral reconstructions of body mass (Table 5), and then correct the estimation of molecular rates. In this sense our work does not assume that the generation times calculated for present-day primates are valid proxies for their ancestors, as in Langergraber et al. [22]. However, we assume that the molecular rates estimated from present-day human families can be used (plus body size and generation time) as a starting line to estimate and correct substitution rates among all extinct primates.

The divergence time estimates based on the molecular rate for Old World primates are also in agreement with those obtained in a recent report (Table 8; [22]), suggesting that the method is also useful for the analyzed Perelman et al. [26] dataset, a stratified sample of the nuclear genome. Similarly, divergence time estimates for the platyrrhine LCA show ranges that include the estimates based on the relaxed-clock model and fossil constraints (Table 8), being closer in age to the earliest undoubted fossil platyrrhine, *Branisella* (*ca.* 26 Ma), and to the direct interpretations of the fossil record [19], [28]. Particularly, we estimate that the LCA of extant platyrrhine monkeys existed between *ca.* 21–29 Ma. Additionally, we estimate the branching of Atelidae between *ca.* 19–27 Ma and of the Cebidae LCA between *ca.* 16–22 Ma (Table 8). These results suggest that the previous differences observed in the length of platyrrhine branches [15], [60] compared with catarrhines is related to differences in generation time and not to time of divergence. It is important to highlight that by using a very wide range of molecular rates for extinct platyrrhine taxa – estimations based on conservative values of body size and generation time – we generated a wide range of divergence time uncertainty, similar to the BEAST estimations (see [22]).

Table 8. Generation based divergence time estimations.

Node	Lower and higher substitution rate	Lower and higher divergence time estimation
Crown Platyrrhini	8.5E-010–6.06E-010	20.31–28.49
Atelidae branching	8.5E-010–6.06E-010	19.05–26.72
Crown Cebidae	9.07E-010–6.47E-010	15.56–21.81
Crown Anthropoidea	8.5E-010–6.06E-010	36.88–51.73
Crown Catarrhini	6.97E-010–4.97E-010	24.29–34.06
Homo/Pan	5.04E-10–3.59E-010	7.31–10.26

Divergence Time and the Platyrrhine Radiation

As pointed out in the Introduction section, dating the basal crown platyrrhine has implications for understanding the platyrrhine evolutionary radiation. Our results suggest that molecular divergence times generated using fossil constraints and molecular rate information are not enough to confidently reject the hypothesis that crown Platyrrhini and the main platyrrhine lineages could have diverged at or before 20 Ma. This result is in marked contrast with the results of Hogdson et al. [15] molecular study. Therefore, molecular divergence time estimations cannot be used to support the idea that platyrrhine diversification is characterized by two successive, sister-group radiations [16] and to contradict the *long lineages hypothesis* of Rosenberger and co-workers [17]–[20].

In a recent submitted work [85] we explore other dimensions of platyrrhine diversification, such as the tempo and mode of species origination and the dynamics of body size evolution. In it, evidence is presented that suggests that platyrrhine evolution conforms to an adaptive radiation model, in which lineages are accumulated at a high rate during the early stages of a clade's evolutionary history. Moreover, body size variation is shown to have been partitioned among subclades early in the phylogenetic history of the platyrrhines, a pattern that is also in agreement with an adaptive radiation scenario and with body size estimations for fossil specimens. Taken together, the results of Aristide et al. [85] and the results of the present work, where we show that extant lineages probably have an ancient origin, are complementary to extend our understanding of the platyrrhine diversification history and stress the role of morphological stasis as a deep evolutionary phenomenon, providing new evidence that contribute to the long standing debate between contrasting hypotheses (*long lineages* vs. *successive radiations*).

CONCLUSION

In this work we used two largely independent molecular approaches (calibration bounds using BEAST and external molecular rates) to estimate the initial divergence time of platyrrhines. Both approaches have advantages and questions [86]. The approach based on calibration bounds using BEAST has the advantage of being relatively sequence independent, but indisputable and reliable calibration bounds are rarely available [21], [86]. The approach based on external molecular rates has the advantage of not requiring such calibration bounds [22], [27], [86]. However, dating methods based on external molecular rate estimations are in their initial stages of development and therefore not free of questions; particularly because they would yield younger divergence estimates than given for other methods (see [87], [88]). Nevertheless, our findings show that these methods are promising.

Our results suggest that several interpretations of the relationships between extant species and the ancient Patagonian fossil record are probably correct [17], [19], [20], [28], [58]. We also conclude that although the current platyrrhine fossil record is relatively scarce, it is not necessarily poorly sampled [7], [28], [89].The estimations based on the two approaches used in this study recalibrate the ages of the platyrrhine clades and make it possible to reconcile several points concerning the affinities of key fossils that have been contested. Contrary to the work of Hodgson et al. [15], our present work includes *Branisella boliviana* (*ca.* 26 Ma), which may fall within the platyrrhine crown group; *Dolichocebus gaimanensis* (*ca.* 20 Ma), which may represent the cebine lineage; and *Tremacebus harringtoni* (*ca.* 20 Ma), which may be an aotine. However, these estimates cannot resolve the controversy of whether these fossil species truly belong to the extant lineages or to closely related lineages [22], [81]. While that question can only be resolved by morphology, our study provides additional evidence that makes likely the broader evolutionary hypothesis that platyrrhine differentiation unfolded as a series of long-lived lineages with morphological stasis [19]. More generally, we show that the use of different approaches, considering molecular rate, fossil record and generation time, gives a more robust divergence time estimation for a clade and allows a more detailed discussion of its biological diversification.

ACKNOWLEDGMENTS

We are sincerely grateful to Don Colgan (Academic Editor, PLOS ONE), A. L. Rosenberger and one anonymous reviewer for their comments on the manuscript. We also thank to Marcela Moyano for helping us with the Fig. 1.

AUTHOR CONTRIBUTIONS

Conceived and designed the experiments: SIP. Performed the experiments: SIP. Analyzed the data: SIP. Contributed reagents/materials/analysis tools: SIP MFT NMN LA. Wrote the paper: SIP MFT NMN LA.

REFERENCES

1. Wiens JJ (2009) Estimating rates and patterns of morphological evolution from phylogenies: lessons in limb lability from Australian Lerista lizards. J Biol 8: 19. doi: 10.1186/jbiol112
2. Losos JB (2011) The limitations of phylogenies in comparative biology. Am Nat 177: 709–727. doi: 10.1086/660020
3. O'Meara BC (2012) Evolutionary inferences from phylogenies: A review of methods. Ann Rev Ecol Evol Syst 43: 267–285. doi: 10.1146/annurev-ecolsys-110411-160331
4. Benton MJ, Donoghue PCJ, Asher RJ (2009) Calibrating and constraining molecular clocks. In Hedges SB, Kumar S. eds. The Timetree of Life. Oxford: Oxford University Press. 35–86.
5. Wilkinson RD, Steiper ME, Soligo C, Martin RD, Yang Z, et al. (2011) Dating Primate divergences through an integrated analysis of palaeontological and molecular data. Syst Biol 60: 16–31.
6. dos Reis M, Inoue J, Hasegawa M, Asher RJ, Donoghue PCJ, et al. (2012) Phylogenomic datasets provide both precision and accuracy in estimating the timescale of placental mammal phylogeny. Proc R Soc B Biol Sci 279(1742): 3491–3500. doi: 10.1098/rspb.2012.0683
7. Fleagle JG (1999) Primate adaptation and evolution, 2nd edn. New York: Academic Press.
8. Groves CP (2001) Primate taxonomy. Washington: Smithsonian Institution Press.
9. Rosenberger AL (2012) New World Monkey nightmares: Science, art, use, and abuse (?) in platyrrhine taxonomic nomenclature. Am J Primatol 74: 692–695. doi: 10.1002/ajp.22037
10. Rosenberger AL (1992) The evolution of feeding niches in New World Monkeys. Am J Phys Anthropol 88: 525–562. doi: 10.1002/ajpa.1330880408
11. Perez SI, Klaczko J, dos Reis SF (2012) Species tree estimation for a deep phylogenetic divergence in the New World monkeys (Primates: Platyrrhini). Mol Phylogenet Evol 65: 621–630. doi: 10.1016/j.ympev.2012.07.014
12. Barroso CML, Schneider H, Schneider MPC, Sampaio I, Harada ML, et al. (1997) Update on the phylogenetic systematics of NewWorld monkeys: Further DNA evidence for placing pygmy marmoset (Cebuella) within the genus Callithrix. Int J Primatol 18: 651–674.
13. Opazo JC, Wildman DE, Prychitko T, Johnson RM, Goodman M (2006) Phylogenetic relationships and divergence times among New World monkeys (Platyrrhini, Primates). Mol Phylogenet Evol 40: 274–280. doi: 10.1016/j.ympev.2005.11.015
14. Schargo CG (2007) On the time scale of New World Primate diversification. Am J Phys Anthropol 132: 344–354. doi: 10.1002/ajpa.20459
15. Hodgson JA, Sterner KN, Matthews LJ, Burrell A, Jani RA, et al. (2009) Successive radiations, not stasis, in the South American primate fauna. Proc Natl Acad Sci U S A 106: 5534–5539. doi: 10.1073/pnas.0810346106
16. Kay RF, Fleagle JG, Mitchell TRT, Colbert M, Bown T, et al. (2008) The anatomy of Dolichocebus gaimanensis, a stem platyrrhine monkey from Argentina. J Hum Evol 54: 323–382. doi: 10.1016/j.jhevol.2007.09.002

17. Rosenberger AL (2010) Platyrrhines, PAUP, parallelism, and the Long Lineage Hypothesis: a reply to Kay et al. (2008). J Hum Evol 59: 214–217. doi: 10.1016/j.jhevol.2010.03.003
18. Rosenberger AL (2002) Platyrrhine paleontology and systematics: the paradigm shifts. In: Hartwig WC ed. The Primate fossil record. Cambridge: Cambridge University Press. 151–160.
19. Rosenberger AL, Tejedor MF, Cooke S, Halenar L, Pekkar S (2009) Platyrrhine ecophylogenetics, past and present. In: Garber PA, Estrada A, Bicca-Marques JC, Heymann EW, Strier KB, eds. South American primates: Comparative perspectives in the study of behavior, ecology and conservation. New York: Springer. 69–113.
20. Tejedor MF (2013) Sistemática, evolución y biogeografía de los primates Platyrrhini. Revista del Museo de La Plata. In press.
21. Drummond AJ, Ho SYW, Phillips MJ, Rambaut A (2006) Relaxed phylogenetics and dating with confidence. PLoS Biol 4: e88. doi: 10.1371/journal.pbio.0040088
22. Langergraber KE, Prüfer K, Rowney C, Boesch C, Crockford C, et al. (2012) Generation times in wild chimpanzees and gorillas suggest earlier divergence times in great ape and human evolution. Proc Nat Acad Sci U S A 109: 15716–15721. doi: 10.1073/pnas.1211740109
23. Yang Z, Rannala B (2006) Bayesian estimation of species divergence times under a molecular clock using multiple fossil calibrations with soft bounds. Mol Biol Evol 23: 212–226. doi: 10.1093/molbev/msj024
24. Babb PL, Fernandez-Duque E, Baiduc CA, Gagneux P, Evans S, et al. (2011) mtDNA Diversity in Azara's Owl Monkeys (Aotus azarai azarai) of the Argentinean Chaco. Am J Phys Anthropol 146: 209–224. doi: 10.1002/ajpa.21567
25. Chiou KL, Pozzi L, Lynch Alfaro JW, Di Fiore A (2011) Pleistocene diversification of living squirrel monkeys (Saimiri spp.) inferred from complete mitochondrial genome sequences. Mol Phylogenet Evol 59: 736–745. doi: 10.1016/j.ympev.2011.03.025
26. Perelman P, Johnson WE, Roos C, Seuánez HN, Horvath JE, et al. (2011) A molecular phylogeny of living Primates. PLoS Genetics 7: e1001342. doi: 10.1371/journal.pgen.1001342
27. Steiper ME, Seiffert ER (2012) Evidence for a convergent slowdown in primate molecular rates and its implications for the timing of early primate evolution. Proc Nat Acad Sci U S A 109: 6006–6011. doi: 10.1073/pnas.1119506109
28. Tejedor MF (2008) The origin and evolution of Neotropical Primates. Arq Mus Nac 66: 251–269.
29. Goin FJ, Gelfo JN, Chornogubsky L, Woodburne MO, Martin T (2010) Origins, radiations, and distribution of South American mammals. From Greenhouse to Icehouse Worlds. In: Patterson BD, Costa LP, ed. Bones, clones and biomes: The history and geography of recent Neotropical mammals. New York: University of Chicago.
30. Vucetich MG, Kramarz AG, Candela AM (2010) Colhuehuapian rodents from Gran Barranca and other Patagonian localities: the state of the art. In: Madden RH, Carlini AA, Vucetich MG, Kay RF, eds. The paleontology of Gran Barranca: Evolution and environmental change through the Middle Cenozoic of Patagonia. Cambridge: Cambridge University Press. 202–215.
31. Antoine PO, Marivaux L, Croft DA, Billet G, Ganerød M, et al. (2012) Middle Eocene rodents from Peruvian Amazonia reveal the pattern and timing of caviomorph origins and biogeography. Proc R Soc B Biol Sci 279: 1319–1326. doi: 10.1098/rspb.2011.1732
32. Flynn JJ, Wyss AR (1998) Recent advances in South American mammalian paleontology. Trends Ecol Evol 13: 449–454. doi: 10.1016/s0169-5347(98)01457-8
33. Poux C, Chevret P, Huchon D, de Jong WW, Douzery EJP (2006) Arrival and diversification of caviomorph rodents and platyrrhine primates in South America. Syst Biol 55: 228–244.
34. Frailey CD, Campbell KE Jr (2004) Paleogene rodents from Amazonian Peru: The Santa Rosa Local Fauna. In: Campbell KE Jr ed. The Paleogene mammalian fauna of Santa Rosa, Amazonian Peru. Los Angeles, CA: Natural History Museum of Los Angeles County. 71–130.
35. Wyss AR, Flynn JJ, Norell MA, Swisher CC III, Charrier R, et al. (1993) South America's earliest rodent and recognition of a new interval of mammalian evolution. Nature 365: 434–437. doi: 10.1038/365434a0
36. Bertrand OC, Flynn JJ, Croft DA, Wyss AR (2012) Two new taxa (Caviomorpha, Rodentia) from the Early Oligocene Tinguiririca fauna (Chile). Am Mus Novit 3750.
37. Vucetich MG, Vieytes EC, Pérez ME, Carlini AA (2010) The rodents from La Cantera and the early evolution of caviomorphs in South America. In: Madden RH, Carlini AA, Vucetich MG, Kay RF eds. The paleontology of Gran Barranca: Evolution and environmental change through the Middle Cenozoic of Patagonia. Cambridge: Cambridge University Press. 189–201.
38. Takai M, Anaya F, Shigehara N, Setoguchi T (2000) New fossil materials of the earliest New World monkey, Branisella boliviana, and the problem of platyrrhine origins. Am J Phys Anthropol 111: 263–281. doi: 10.1002/(sici)1096-8644(200002)111:2<263::aid-ajpa10>3.0.co;2-6
39. Vucetich MG (1991) Los roedores de Salla y Lacayani (Bolivia) y su correlación con los de otras faunas de Edad Deseadense. Rev Tec Yacim Pet Fisc Boliv 12: 625–629.
40. Bordas A (1942) Anotaciones sobre un "Cebidae" fósil de Patagonia. Physis 19: 265–269.
41. Kraglievich JL (1951) Contribuciones al conocimiento de los primates fósiles de la Patagonia. I. Diagnosis previa de un nuevo primate fósil del Oligoceno superior (Colhuehuapiano) de Gaiman, Chubut. Rev Mus Argent Cienc Nat 2: 57–82.
42. Rosenberger AL (1979) Cranial anatomy and implications of Dolichocebus a late Oligocene ceboid primate. Nature 279: 416–418. doi: 10.1038/279416a0

43. Fleagle JG, Rosenberger AL (1983) Cranial morphology of the earliest anthropoids. In: Sakka M ed. Morphologie, evolutive, morphogenese du crane et anthropogenese. Paris: CNRS. 141–153.

44. Kay RF (2010) A new primate from the early Miocene of Gran Barranca, Chubut Province, Argentina: Paleoecological implications. In: Madden RH, Carlini AA, Vucetich MG, Kay RF, eds. The paleontology of Gran Barranca: Evolution and environmental change through the Middle Cenozoic of Patagonia. Cambridge: Cambridge University Press. 220–240.

45. Kinzey WG (1992) Dietary and dental adaptations in the Pitheciinae. Am J Phys Anthropol 88: 499–514. doi: 10.1002/ajpa.1330880406

46. Fleagle JG, Perkins ME, Heizer T, Nash B, Bown M, et al.. (2012) Absolute and relative ages of fossil localities in the Santa Cruz and Pinturas Formations. In: Vizcaino SF, Kay RF, Bargo MS, eds. Early Miocene paleobiology in Patagonia: High-Latitude Paleocommunities of the Santa Cruz Formation. Cambridge: Cambridge University Press. 41–58.

47. Perkins ME, Fleagle JG, Heizer MT, Nash B, Bown TM, et al.. (2012) Tephrochronology of the Miocene Santa Cruz and Pinturas Formations, Argentina. In: Vizcaino SF, Kay RF, Bargo MS, eds. Early Miocene paleobiology in Patagonia: High-Latitude Paleocommunities of the Santa Cruz Formation. Cambridge: Cambridge University Press. 23–40.

48. Tejedor MF, Tauber AA, Rosenberger AL, Swisher CC III, Palacios ME (2006) New primate genus from the Miocene of Argentina. Proc Nat Acad Sci U S A 103: 5437–5441. doi: 10.1073/pnas.0506126103

49. Kay RF, Johnson D, Meldrum DJ (1998) A new pitheciine primate from the Middle Miocene of Argentina. Am J Primatol 45: 317–336. doi: 10.1002/(sici)1098-2345(1999)47:4<347::aid-ajp6>3.3.co;2-f

50. Kay RF, Ross CF, Williams BA (1997) Anthropoid origins. Science 275: 797–804. doi: 10.1126/science.275.5301.797

51. Hartwig WC, Meldrum DJ (2002) Miocene platyrrhines of the Northern Neotropics. In: Hartwig WC, ed. The primate fossil record. Cambridge: Cambridge University Press. 175–188.

52. Madden RH, Guerrero J, Kay RF, Flynn JJ, Swisher CC III, et al.. (1997) The Laventan stage and age. In: Kay RF, Madden RH, Cifelli RL, Flynn JJ, eds. Vertebrate paleontology in the Neotropics: The Miocene fauna of La Venta, Colombia. Washington: Smithsonian Institution Press. 499–519.

53. Stirton RA (1951) Ceboid monkeys from the Miocene of Colombia. Bull Dept Geol Univ Calif 28: 315–356.

54. Rosenberger AL, Hartwig W, Takai M, Setoguchi T, Shigehara N (1991) Dental variability in Saimiri and the taxonomic status of Neosaimiri fieldsi, an early squirrel monkey from La Venta, Colombia. Int J Primatol 12: 291–301. doi: 10.1007/bf02547590

55. Takai M (1994) New specimens of Neosaimiri fieldsi from La Venta, Colombia: a middle Miocene ancestor of the living squirrel monkeys. J Hum Evol 27: 329–360. doi: 10.1006/jhev.1994.1049

56. Setoguchi T, Rosenberger AL (1987) A fossil owl monkey from La Venta, Colombia. Nature 326: 692–694. doi: 10.1038/326692a0

57. Luchterhand K, Kay RF, Madden RH (1986) Mohanamico hershkovitzi, gen. et sp. nov., un primate du Miocène moyen d'Amérique du Sud. C R Acad Sci Paris, sér. II 303: 1753–1758.

58. Rosenberger AL, Setoguchi T, Shigehara N (1990) The fossil record of callitrichine primates. J Hum Evol 19: 209–236. doi: 10.1016/b978-0-12-260345-7.50012-6

59. Hershkovitz P (1970) Notes on Tertiary platyrrhine monkeys and description of a new genus from the late Miocene of Colombia. Folia Primatol 12: 1–37. doi: 10.1159/000155278

60. Delson E, Rosenberger AL (1984) Are there any anthropoid primate "living fossils"? In: Eldredge N, Stanley S, eds. Casebook on living fossils. New York: Fischer Publishers. 50–61.

61. Kay RF, Madden RH, Plavcan JM, Cifelli RL, Guerrero Díaz J (1987) Stirtonia victoriae, a new species of Miocene Colombian primate. J Hum Evol 16: 73–196. doi: 10.1016/0047-2484(87)90075-3

62. Meldrum DJ, Kay RF (1997) Nuciruptor rubricae, a new pitheciin seed predator from the Miocene of Colombia. Am J Phys Anthropol 102: 407–427. doi: 10.1002/(sici)1096-8644(199703)102:3<407::aid-ajpa8>3.0.co;2-r

63. Takai M, Anaya F, Suzuki H, Shigehara N, Setoguchi T (2001) A new platyrrhine from the Middle Miocene of La Venta, Colombia, and the phyletic position of Callicebinae. Anthropol Sciences 109: 289–307. doi: 10.1537/ase.109.289

64. Rosenberger AL, Cooke S, Rímoli R, Xijun Ni, Cardoso L (2011) First skull of Antillothrix bernensis, an extinct relict monkey from the Dominican Republic. Proc R Soc B Biol Sci 278: 67–74. doi: 10.1098/rspb.2010.1249

65. Hall T (2004) BioEdit 7.0.0. Available: http://www.mbio.ncsu.edu/bioedit/bioedit.html. Accessed 2011 Jan 8.

66. Posada D (2008) jModelTest: phylogenetic model averaging. Mol Biol Evol 25: 1253–1256. doi: 10.1093/molbev/msn083

67. Drummond AJ, Rambaut A (2007) BEAST: Bayesian evolutionary analysis by sampling trees. BMC Evol Biol 7: 214. doi: 10.1186/1471-2148-7-214

68. Rambaut A, Drummond AJ (2007) Tracer v1.4. Available: http://beast.bio.ed.ac.uk/Tracer. Accessed 2012 Jan 3.

69. Wildman DE, Jameson NM, Opazo JC, Yi SV (2009) A fully resolved genus level phylogeny of neotropical primates (Platyrrhini). Mol Phylogenet Evol 53: 694–702. doi: 10.1016/j.ympev.2009.07.019

70. Kay RF, Fleagle JG (2010) Stem taxa, homoplasy, long lineages, and the phylogenetic position of Dolichocebus. J Hum Evol 59: 218–222. doi: 10.1016/j.jhevol.2010.03.002

71. Lemey M, Salemi M, Vandamme AM (2009) The phylogenetic handbook: A practical approach to

phylogenetic analysis and hypothesis testing. Cambridge: Cambridge University Press.

72. Harvey PH, Clutton-Brock TH (1985) Life history variation in primates. Evolution 39: 559–581. doi: 10.2307/2408653

73. Wu CI, Li WH (1985) Evidence for higher rates of nucleotide substitution in rodents than in man. Proc Natl Acad Sci USA 82: 1741–1745. doi: 10.1073/pnas.82.6.1741

74. Schafer JL, Olsen MK (1998) Multiple imputation for multivariate missing-data problems: a data analyst's perspective. Multivariate Behav Res 33: 545–571. doi: 10.1207/s15327906mbr3304_5

75. Takezaki N, Rzhetsky A, Nei M (1995) Phylogenetic test of the molecular clock and linearized trees. Mol Phylogenet Evol 12: 823–833.

76. Tamura K, Peterson D, Peterson N, Stecher G, Nei M, et al. (2011) MEGA5: molecular evolutionary genetics analysis using maximum likelihood, evolutionary distance, and maximum parsimony methods. Mol Biol Evol 28: 2731–2739. doi: 10.1093/molbev/msr121

77. Tajima F (1993) Simple methods for testing molecular clock hypothesis. Genetics 135: 599–607.

78. IUCN (2012) IUCN Red List of Threatened Species. Version 2012.2. Available: http://www.iucnredlist.org. Accessed 2012 Jan 10.

79. Smith RJ, Jungers WL (1997) Body mass in comparative primatology. J Hum Evol 32: 523–559. doi: 10.1006/jhev.1996.0122

80. Rosenberger AL, Tejedor MF (2013) The misbegotten: long lineages, long branches and the interrelationships of Aotus, Callicebus and the saki-uakaris. In: Barnett AL, Veiga S, Ferrari S, Norconk MN, eds. Evolutionary biology and conservation of Titis, Sakis and Uacaris. Cambridge: Cambridge University Press.

81. Parham JF, Donoghue PC, Bell CJ, Calway TD, Head JJ, et al. (2012) Best practices for justifying fossil calibrations. Syst Biol 61: 346–59.

82. Kay RF, Campbell VM, Rossie JB, Colbert MW, Rowe TB (2004) Olfactory fossa of Tremacebus harringtoni (Platyrrhini, early Miocene, Sacanana, Argentina): Implications for activity pattern. Anat Rec 281A: 1157–1172. doi: 10.1002/ar.a.20121

83. Tattersall I (1987) Cathemeral activity in primates: a definition. Folia Primatol 49: 200–202. doi: 10.1159/000156323

84. Fernandez-Duque E (2003) Influences of moonlight, ambient temperature, and food availability on the diurnal and nocturnal activity of owl monkeys (Aotus azarai). Behav Ecol Sociobiol 54: 431–440. doi: 10.1007/s00265-003-0637-9

85. Aristide L, Rosenberger AL, Tejedor MF, Perez SI (2013) Modeling the timing of lineage and phenotypic diversification in the New World monkeys (Platyrrhini, Primates) radiation. Mol Phylogenet Evol.

86. Wilke T, Schultheiß R, Albrecht C (2009) As time goes by: A simple fool's guide to molecular clock approaches in invertebrates. Amer Malac Bull 27: 25–45. doi: 10.4003/006.027.0203

87. Gibb GC, Hills SFK (2013) Intergenerational mutation rate does not equal long-term evolutionary substitution rate. Proc Natl Acad Sci USA, 10.1073/pnas.1217134110.

88. Prüfer K, Langergraber KE, Pääbo S, Vigilant L (2013) Reply to Gibb and Hills: Divergence times, generation lengths and mutation rates in great apes and humans Proc Natl Acad Sci USA, 10.1073/pnas.1222027110.

89. Fleagle JG, Tejedor MF (2002) Early platyrrhines of southern South America. In: Hartwig WC, ed. The primate fossil record. Cambridge: Cambridge University Press. 161–174.

CHAPTER 17

THE ORIGIN AND EARLY EVOLUTION OF SAURIA

Martin D. Ezcurra, Torsten M. Scheyer, Richard J. Butler

INTRODUCTION

Saurians, or crown group diapsids, are highly taxonomically and morphologically diverse in extant ecosystems, with around 9,400 lepidosaur (snakes, lizards and rhynchocephalians) and 10,000 archosaur (birds and crocodilians) species, including cursorial, semi-aquatic, marine, fossorial and volant forms [1], [2]. The stem-groups of Lepidosauria (non-lepidosaurian Lepidosauromorpha) and Archosauria (non-archosaurian Archosauromorpha) also include several morphologically disparate saurian lineages that were mostly restricted in time to the Triassic. These lineages formed important components of Triassic continental assemblages, and include kuehneosaurids, rhynchosaurs, proterosuchids, erythrosuchids, euparkeriids, doswelliids and proterochampsids [3]–[8]. However, the earliest (i.e. pre-Mesozoic) evolutionary history of Sauria is poorly known and there has been substantial debate regarding the late Paleozoic (i.e. Permian) record of the group (e.g. [3], [9]–[18]).

The best source of information on the early history of Sauria comes from the numerous fossils of the well-known basal archosauromorph Protorosaurus speneri from the Late Permian of Germany and England [19]–[21]. Multiple less completely known specimens have been also argued to be Permian members of Sauria (e.g. Parrington's "problematic reptile" from Tanzania, UMZC T836 [9]). A better understanding of the Permian saurian record is fundamental for providing more accurate fossil constraints on the calibration of the crocodile-lizard (=bird-lizard) divergence, a major split within vertebrates that is of keen interest to molecular and evolutionary biologists and vertebrate paleontologists alike [22]–[25]. A better knowledge of Permian saurians is also necessary to improve understanding of phylogenetic relationships within early members of Diapsida, an area of key interest because of the controversial systematic affinities of several possible saurian lineages including turtles, choristoderans and sauropterygians (e.g. [26]–[39]). New information on the Permian saurian record may also yield fresh insights into survivorship of this clade across the Permian-Triassic mass extinction and the dynamics of the dramatic saurian radiation in post-extinction ecosystems.

Here, we revisit and reexamine the Permian record of Sauria to provide new information on the diversity, phylogeny, morphology, geographic distribution and physiology of Permian members of the clade, and the timing of the crocodile-lizard (or bird-lizard) split. We fully or partially redescribe some Permian saurian specimens (e.g. UMZC T836; BP/1/4220; Eorasaurus olsoni) and we erect a new genus and species of archosauromorph, Aenigmastropheus parringtoni, for a specimen from the middle Late Permian of Tanzania. Our new data provides an improved understanding of early saurian and early archosauromorph evolutionary history, including calibration dates for molecular biology.

MATERIALS AND METHODS

Access to specimens

The type and referred specimens of Eorasaurus olsoni (PIN 156/108–111), the new genus and species Aenigmastropheus parringtoni (UMZC T836), and Archosaurus rossicus (PIN 1100/55), as well as BP/1/4220 and all specimens that are used

Citation: Ezcurra MD, Scheyer TM, Butler RJ (2014), The Origin and Early Evolution of Sauria. PLoS ONE 9(2): e89165. doi:10.1371/journal.pone.0089165.

here for comparative purposes (indicated by the citation of their taxonomic name and respective collection accession numbers at relevant points in the manuscript) were studied at first-hand, with the explicit permission of appropriate curators and/or collection managers (see Acknowledgments), in recognized, scientifically accessible collections. Repository locations and abbreviations for all specimens discussed in the text and abbreviations listed in the Acknowledgments are listed below. No specimens were purchased or donated for the purpose of this study. The holotype of the new taxon Aenigmastropheus parringtoni was loaned from the collection of the UMZC with permission of the collection manager and returned before submission of the manuscript.

Terminology

We follow here the nomenclature for vertebral laminae and fossae of Wilson [40] and Wilson et al. [41]. We also follow the terminology of Rewcastle [42] for limb orientation in sprawling animals.

Histological analysis

The paleohistological sections of the holotype of the new taxon Aenigmastropheus parringtoni were prepared by one of us (TS) in the facilities of the PIMUZ using standard techniques. Paleohistological slices are reposited with the holotype specimen in the UMZC collections.

Phylogenetic analysis

In order to test quantitatively the phylogenetic relationships of early saurians, including the new taxon Aenigmastropheus parringtoni, we modified the data matrix of Reisz et al. [43]. This data matrix was employed by Reisz et al. [43] to test the position of the taxon Apsisaurus witteri among basal diapsids and early synapsids ("pelycosaurs"), demonstrating that Apsisaurus is a varanopid synapsid rather than a diapsid as originally identified. Similarly, we wished to test whether the new taxon Aenigmastropheus is a saurian diapsid, or belongs to some other group of early amniotes; thus, the scope and aims of the Reisz et al. [43] analysis broadly match those of this study.

The taxonomic sample of the data matrix of Reisz et al. [43] was enlarged with the addition of the new taxon Aenigmastropheus parringtoni and 13 early saurian species (Archosaurus rossicus, Eorasaurus olsoni, Erythrosuchus africanus, Euparkeria capensis, Gephyrosaurus bridensis, Howesia browni, Macrocnemus bassanii, Mesosuchus browni, Noteosuchus colletti, Paliguana whitei, Proterosuchus fergusi, Protorosaurus speneri, Tanystropheus longobardicus). Eorasaurus olsoni and Noteosuchus colletti are included for the first time in a quantitative phylogenetic analysis. The hypothesis of Dilkes [4] that Noteosuchus colletti is a junior synonym of Mesosuchus browni was not followed here because this proposal was based on generalized plesiomorphic similarities rather than autapomorphies and the temporal gap between the two species spans most of the Early Triassic [11]. As a result, Noteosuchus colletti was scored as an independent operational taxonomic unit (OTU) because of its potential to shed light on the minimal divergence time of Rhynchosauria. The derived rhynchosaur Hyperodapedon was pruned a priori from the data matrix because its advanced morphology is not congruent with that observed in basal members of the group (Mesosuchus browni, Howesia browni, Noteosuchus colletti) and the absence of species linking these basal forms and Hyperodapedon could cause artifacts in the optimization of characters within Archosauromorpha. Trilophosaurus was replaced with Trilophosaurus buettneri in order to avoid ambiguities in the scorings for the genus from the more poorly known Trilophosaurus jacobsi [44]. Several Triassic basal lepidosauromorphs were not included in this phylogenetic analysis because they come from a multi-taxic assemblage of amniote material [45], [46] and their hypodigms may represent more than one species (e.g. Marmoretta, Sophineta, Pamelina).

All the added OTUs were scored based on first hand observations, with the exception of Gephyrosaurus bridensis, which was scored following Evans [47], [48] (Table 1). Several scores for the non-saurian neodiapsids Coelurosauravus spp., Youngina capensis, and Acerosodontosaurus piveteaui, and the archosauromorphs Prolacerta broomi and Trilophosaurus buettneri were modified based on first hand observation of specimens or new published information.

Table 1. Specimens or references employed here to score the 14 species added to the taxonomic sample of Reisz et al. [43].

Archosaurus rossicus	PIN 1100/55 (only the holotype premaxilla was considered); [88]
Eorasaurus olsoni	PIN 156/108, 109, 110, 111; [89]
Erythrosuchus africanus	AMNH 5596, BP/1/2096, 2529, 3893, 4526, 4645, 4680, 5207; GHG AK-82-22; NHMUK R525, R533, R2790, R3592, R3762 (scapula), R3764, NHMUK unnumbered, NM QR1473, SAM-PK-905, 912, 978, 1315, 3028, 3612, 7684, 11330, K1098, K1118, K10025, UMZC T666, T700; [113]
Euparkeria capensis	GPIT 1681, SAM-PK-5867, 5883, 6047, 6048, 6049, 6050, 7696, 7700, 7712, 7713, 10671, 13664, 13665, 13666, K8050, K8309, UMZC T692, T921; [192]
Gephyrosaurus bridensis	[47,48]
Howesia browni	SAM-PK-5884, 5885, 5886; [118]
Macrocnemus bassanii	PIMUZ T2472, 4355, 4822; [225,226]
Mesosuchus browni	SAM-PK-5882, 6046, 6536, 7416; [4]
Noteosuchus colletti	AM 3591 (holotype); [11]
Paliguana whitei	AM 3585 (holotype); [10]
Proterosuchus fergusi	BP/1/3993, 4016, 4224; BSPG 1934-VIII-514; GHG 231; RC 59; SAM-PK-591 (holotype), 11208, K140, K10603, TM 201; [90,227]
Protorosaurus speneri	BSPG 1995 I 5 (cast of WMSN P47361), AS VII 1207; NHMW 1943I4 (lectotype); SMNS 55387 (cast of Simon/Bartholomäus specimen); USNM 442453 (cast of NMK S 180); ZMR MB R2171-3; [21]
Tanystropheus longobardicus	PIMUZ T2189, 2793, 2817, 2818, 3901; SMNS 54147, 54626, 54628, 54630–54632, 54634, 55341, 56289, 59380, 84821: [111,228]

The character list was enlarged with the addition of 107 characters from Dilkes [4], Müller [32] and Senter [49], plus some new characters. As a result, the new data matrix includes 40 taxa and 219 characters.

The data matrix was analyzed under equally-weighted parsimony using TNT 1.1 [50]. A heuristic search of 1,000 replications of Wagner trees (with random addition sequence) followed by the tree bisection and reconnection (TBR) branch-swapping algorithm (holding 10 trees per replicate) was performed. The best tree(s) obtained at the end of the replicates were subjected to a final round of TBR branch swapping. Zero length branches among any of the recovered most parsimonious trees (MPTs) were collapsed (rule 1 of Coddington and Scharff [51]). Characters 2, 6, 8, 10, 12, 16–18, 20, 21, 24, 26, 30, 31, 36–38, 40, 45, 49, 51–55, 57, 58, 62, 70, 73, 79, 91, 92, 94, 103, 143, 160, 198, 214, 216 and 219 represent nested sets of homologies and/or entail presence and absence information and as a result they were treated as additive (ordered).

As measures of tree support, decay indices (=Bremer support) were calculated and a bootstrap resampling analysis, with 10,000 pseudoreplicates, was performed. We report both absolute and GC (i.e. difference between the frequency that the original group and the most frequent contradictory group are recovered in the pseudoreplicates) frequencies. Taxa with high amounts of missing data may reduce node support values not as a result of a real low robustness of the node but because of ambiguous optimizations generated by unknown character states. Accordingly, a second round of decay indices and bootstrap resampling was conducted following the a posteriori pruning of saurian OTUs with large amount of missing data (i.e. Aenigmastropheus parringtoni, Archosaurus rossicus, Eorasaurus olsoni, Noteosuchus colletti, Paliguana whitei).

Nomenclatural Acts

The electronic edition of this article conforms to the requirements of the amended International Code of Zoological Nomenclature, and hence the new names contained herein are available under that Code from the electronic edition of this article. This published work and the nomenclatural acts it contains have been registered in ZooBank, the online registration system for the ICZN. The ZooBank LSIDs (Life Science Identifiers) can be resolved and the associated information viewed through any standard web browser by appending the LSID to the prefix "http://zoobank.org/". The LSID for this publication is: urn:lsid:zoobank.org:pub:A899B6F4-7362-44FB-88BD-188E8C642E0D. The electronic edition of this work was published in a journal with an ISSN, and has been archived and is available from the following digital repositories: PubMed Central and LOCKSS.

Institutional abbreviations

AM, Albany Museum, Grahamstown, South Africa; AMNH, American Museum of Natural History, New York, USA; BP, Evolutionary Studies Institute (formerly Bernard Price Institute for Palaeontological Research), University of the Witswatersrand, Johannesburg, South Africa; BSPG, Bayerische Staatssammlung für Paläontologie und Geologie, Munich, Germany; FC-DPV, Colección de Vertebrados Fósiles, Departamento de Paleontología, Facultad de Ciencias, Universidad de la República, Montevideo, Uruguay; FMNH, Field Museum of Natural History, Chicago, USA; GHG, Geological Survey, Pretoria, South Africa; GPIT, Paläontologische Sammlung der Universität Tübingen, Tübingen, Germany; IVPP, Institute of Vertebrate Paleontology and Paleoanthropology, Beijing, China; MB, Museum für Naturkunde – Leibniz-Institut für Evolutions- und Biodiversitätsforschung, Berlin, Germany; MCNAM, Museo de Ciencias Naturales y Antropológicas de Mendoza (J. C. Moyano), Mendoza, Argentina; MCP, Museu de Ciências e Tecnologia da Pontifícia Universidade Católica do Rio Grande do Sul, Porto Alegre, Brazil; MCZ, Museum of Comparative Zoology, Cambridge, USA; MNHN, Muséum national d'Histoire naturelle, Paris, France; NHMUK, The Natural History Museum, London, UK; NHMW, Naturhistorisches Museum Wien, Vienna, Austria; NM, National Museum, Bloemfontein, South Africa; PIMUZ, Paläontologisches Institut und Museum der Universität Zürich, Zurich, Switzerland; PIN, Paleontological Institute of the Russian Academy of Sciences, Moscow, Russia; PVL, Paleontología de Vertebrados, Instituto 'Miguel Lillo', San Miguel de Tucumán, Argentina; PVSJ, División de Paleontología de Vertebrados del Museo de Ciencias Naturales y Universidad Nacional de San Juan, San Juan, Argentina; SAM, Iziko South African Museum, Cape Town, South Africa; SMNS, Staatliches Museum für Naturkunde Stuttgart, Stuttgart, Germany; SSWG, Sektion Geologie, Ernst-Moritz-Arndt Universität, Greifswald, Germany; TM, Ditsong National Museum of Natural History (formerly Transvaal Museum), Pretoria, South Africa; TMM, Texas Memorial Museum, Austin, USA; UMZC, University Museum of Zoology, Cambridge, UK; USNM, National Museum of Natural History (formerly United States National Museum), Smithsonian Institution, Washington, D.C., USA; UTGD, School of Earth Sciences, University of Tasmania, Hobart, Australia; WMSN, Westfälisches Museum für Naturkunde, Münster, Germany; ZAR, Muséum national d'Histoire naturelle (Zarzaitine collection), Paris, France.

RESULTS

Phylogenetic analysis

The phylogenetic analysis recovered one MPT of 861 steps, with a consistency index (CI) of 0.337, a retention index (RI) of 0.645, and the best score hit 320 times out of the 1000 replications. The topology of the tree is completely congruent with that obtained by Reisz et al. [43] and only differs in the resolution of the polytomies recovered by Reisz et al. [43] within Squamata and Varanopidae. As a result, only the relationships within Sauria, which includes the species that we have added to the phylogenetic dataset, will be described here.

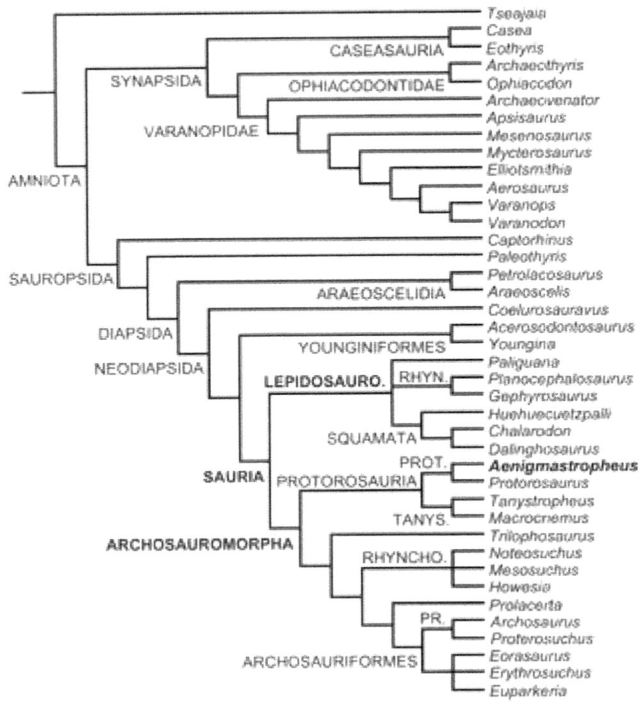

Figure 1. Phylogenetic relationships of Aenigmastropheus parringtoni **and other basal archosauromorphs.**

Single most parsimonious tree recovered here with zero length branches collapsed. Abbreviations: Lepidosauro., Lepidosauromorpha; Pr., Proterosuchidae; Prot., Protorosauridae; Rhyn., Rhynchocephalia; Rhyncho., Rhynchosauria; Tanys., Tanystropheidae.

The new taxon Aenigmastropheus parringtoni was recovered as a basal archosauromorph, being nested within Protorosauria as the sister-taxon of Protorosaurus speneri from the Late Permian of Europe. The decay indices for Archosauromorpha, Protorosauria and Protorosauridae (i.e. the Protorosaurus+Aenigmastropheus clade) were minimal (i.e. 1) and the bootstrap resampling frequencies below 50% (although the absolute bootstrap support of Archosauromorpha is 49%) (Fig. 2). When Aenigmastropheus and other saurians with high amounts of missing data (i.e. Paliguana, Noteosuchus, Eorasaurus, Archosaurus) were a posteriori pruned, the new decay index for Archosauromorpha was 11 and the absolute and GC bootstrap resampling frequencies rose to 97% and 96%, respectively.

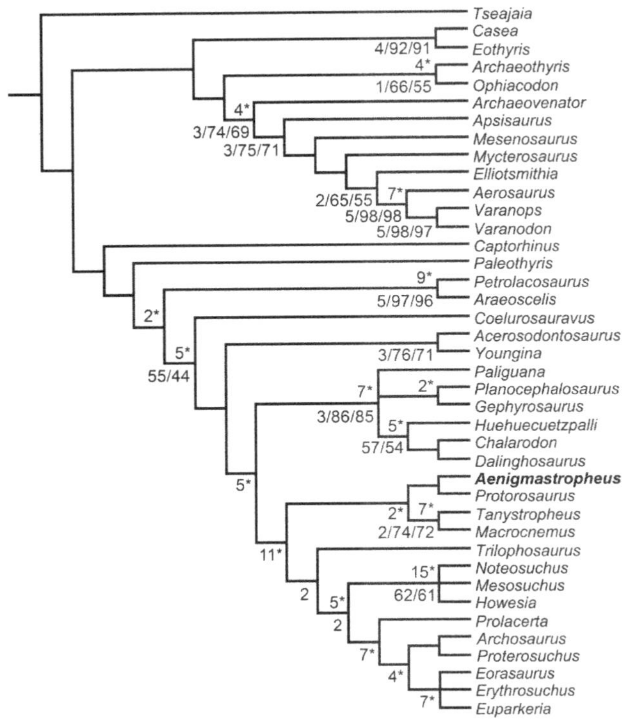

Figure 2. Bremer support and bootstrap resampling frequencies for the phylogenetic hypothesis presented here.

Numbers below the nodes are Bremer support, absolute bootstrap and GC bootstrap resampling frequencies, respectively, for the single most parsimonious tree recovered here. Numbers above the nodes and with an asterisk are Bremer support values after a posteriori pruning of fragmentary saurian terminals (see text).

The Middle Triassic taxa Tanystropheus longobardicus and Macrocnemus bassanii were also found as members of Protorosauria, being sister-taxa and comprising the clade Tanystropheidae (sensu Dilkes [4]). Following the pruning of fragmentary saurians, the decay index for Proterosauria increased to 2 but the bootstrap resampling frequencies remained lower than 50%. Support values were moderately high for Tanystropheidae, with a decay index of 2 and absolute and GC bootstrap resampling frequencies of 74% and 72%, respectively. Following the pruning of fragmentary saurians the decay index for Tanystropheidae rose to 7 and the resampling frequencies rose to 93% and 91%, respectively.

Protorosauria was recovered as the most basal clade within Archosauromorpha, being the sister group of a clade composed of Trilophosaurus buettneri and more derived archosauromorphs (Rhynchosauria, Prolacerta broomi and Archosauriformes). The decay index for the latter clade was 2 and absolute and GC bootstrap resampling frequencies remained below 50% even when fragmentary saurians were pruned.

The recovery of Trilophosaurus buettneri as the sister-taxon of the Rhynchosauria+Prolacerta+Archosauriformes clade agrees with several previous hypotheses (e.g. [4], [52]–[54]) but contrasts with some studies that recovered a Trilophosaurus+Rhynchosauria clade (e.g. [21], [55]) or a position of Trilophosaurus as closer to archosauriforms than to rhynchosaurs (e.g. [56], [57]). The decay index of the Rhynchosauria+Prolacerta+Archosauriformes clade was 2 and the absolute and GC bootstrap resampling frequencies were below 50%. Following the a posteriori pruning of poorly known saurians the decay index rose to 5 and the bootstrap frequencies rose to 72% and 64%, respectively.

Howesia browni, Mesosuchus browni and Noteosuchus colletti were found within a monophyletic Rhynchosauria (sensu Dilkes [4]), but relationships among these three taxa were unresolved. The decay index was minimal, but absolute and GC bootstrap resampling frequencies were 72% and 64%. This was the first quantitative analysis to include Noteosuchus colletti and the results support its identification as the oldest known rhynchosaur [4], [11], [58]. Following the pruning of fragmentary saurian (including Noteosuchus) the decay index of Rhynchosauria rose to 15 and the resampling frequencies both rose to 99%.

Prolacerta broomi was found as the sister-taxon of Archosauriformes. As a result, Prolacertiformes sensu Benton [52] was polyphyletic, as also recovered by several other recent analyses [21], [53], [54]. The decay index for the Prolacerta+Archosauriformes clade was minimal and the absolute and GC bootstrap resampling frequencies were below 50%, but following the pruning of fragmentary saurians the decay index rose to 7 and the resampling frequencies rose to 80% and 76%, respectively.

Two clades were found within Archosauriformes, one composed of the proterosuchids Proterosuchus fergusi and Archosaurus rossicus and the other consisting of a trichotomy including the Late Permian Eorasaurus olsoni and the early Middle Triassic Euparkeria capensis and Erythrosuchus africanus. This topology is consistent with the interrelationships recently recovered for the clade by other analyses [59] (see below for a discussion of the phylogenetic position of Eorasaurus olsoni). Decay indices for Archosauriformes and clades nested within Archosauriformes were minimal and absolute and GC bootstrap resampling frequencies were below 50%, although the Proterosuchus+Archosaurus clade had bootstrap resampling frequencies of 47% and 43%, respectively. However, following the pruning of fragmentary saurians the decay index of Archosauriformes rose to 4 and the bootstrap resampling frequencies rose to 82% and 73%, respectively, and the decay index of the Erythrosuchus+Euparkeria clade rose to 7 and the bootstrap resampling frequencies to 93% and 90%, respectively.

Lepidosauromorpha was composed by three lineages: Paliguana whitei, Rhynchocephalia (Gephyrosaurus bridensis and Planocephalosaurus robinsonae) and Squamata (Dalinghosaurus longidigitus, Huehuecuetzpalli mixtecus and Chalarodon madagascariensis), a result congruent with that of Evans and Borsuk-Białynicka [45]. The relationships between these three lineages were unresolved. Support values for Lepidosauromorpha were relatively high, with a decay index of 3 and absolute and GC bootstrap resampling frequencies of 86% and 85%, respectively. The decay index of Lepidosauromorpha rose to 7 and the bootstrap resampling frequencies increased to 94% and 90%, respectively, following the pruning of fragmentary saurians. Support values were low for Rhynchocephalia (absolute and GC bootstrap resampling frequencies of 48% and 35%). Following the pruning of fragmentary saurians, the decay index of Rhynchocephalia increased to 2 and the absolute and GC bootstrap resampling frequencies increased to 76% and 62%. The decay index of Squamata was minimal and the absolute and GC bootstrap resampling frequencies were 57% and 54%. Following the pruning of fragmentary taxa the decay index rose to 5 and the bootstrap resampling frequencies increased to 90% and 85%, respectively.

Review of the Permian saurian record

Sauria comprises Lepidosauromorpha, Archosauromorpha, their most recent common ancestor, and all their extinct descendants [57]. Several footprints and ichnotaxa potentially attributable to both lepidosauromorphs (e.g. Ganasauripus ladinus, Paradoxichnium radeinensis) and archosauromorphs (e.g. Protochirotherium isp., Synaptichnium isp.) have been described from Upper Permian beds of southern Europe [60], [61] and northern Africa [62]. However, here we focus solely on the body fossil record of Permian saurians.

"Younginiformes".

Several Late Permian diapsid species (all referred to at various points as "eosuchians") from South Africa (Youngina capensis [63], [64]), Tanzania (Tangasaurus mennelli [65], [66]) and Madagascar (Hovasaurus boulei [67], [68]; Acerosodontosaurus piveteaui [69], [70]; Thadeosaurus colcanapi [71]) have been considered by some authors to form a monophyletic Younginiformes [52], [56], [66] within Lepidosauromorpha [52], [56], [57], [72]. However, subsequent work has suggested that "Younginiformes" form a paraphyletic assemblage [70], and these species are now widely accepted as basal non-saurian neodiapsids and therefore not lepidosauromorphs (e.g. [4], [13], [32], [43], [49], [70], [73]–[75]) (Fig. 3A). "Younginiformes" do not, therefore, represent Permian saurians.

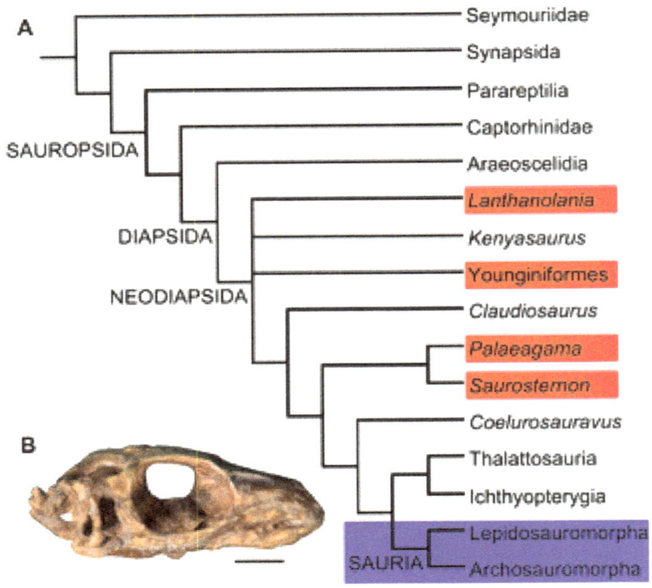

Figure 3. Simplified phylogenetic relationships of Diapsida.

(A) Phylogenetic positions of Sauria (blue box) and several species previously considered as Permo-Triassic lepidosauromorphs (red boxes), based upon the phylogenetic analysis of Bickelmann et al. [70] (illustration simplified) (B) Holotype (AM 3585) of Paliguana whitei, the oldest known lepidosauromorph from the Early Triassic of South Africa, in right lateral view. Scale bar equals 5 mm.

Other putative Permian lepidosauromorph records

Five basal diapsid species (all referred to at various points as "eolacertilians") have been proposed as possible Permian lepidosauromorphs, and have often been identified as lizards: Saurosternon bainii, Palaeagama vielhaueri, Paliguana whitei and Lacertulus bipesfrom South Africa [10], [76]–[80], the former three species forming the "Paliguanidae" of Carroll [10], and Lanthanolania ivakhnenkoi from the Middle Permian of Russia [15].

Among the South African specimens, the exact stratigraphic position of the type and only known specimen of Saurosternon bainii is poorly constrained [76], but it is of definite Late Permian age [10]. By contrast, the type and only specimen of Palaeagama vielhaueri [79]cannot be stratigraphically constrained with certainty beyond a Late Permian–Early Triassic age, although an Early Triassic age may be more likely [10]. The stratigraphic position of the type and only specimen of Lacertulus bipes is not constrained beyond Permian–Triassic [80]. Whereas Carroll [10] considered Paliguana whitei [77] to be of uncertain Late Permian–Triassic age, Kitching [81] and Groenwald & Kitching [82] listed this species as derived from the Lystrosaurus Assemblage Zone of Early Triassic age (Induan–?Olenekian [83]–[85]).

Several of these species are of uncertain phylogenetic position. Saurosternon bainii and Palaeagama vielhaueri were assigned to Lepidosauromorpha by Gauthier et al. [57], whereas Evans [3] considered Saurosternon bainii as a possible basal lepidosauromorph and Palaeagama vielhaueri as an indeterminate diapsid. However, more recent quantitative phylogenetic analyses identified both species as non-saurian basal diapsids [18], possibly forming a monophyletic clade [32], [70] (Fig. 3A). These results do not therefore support the positions of Saurosternon bainii and Palaeagama vielhaueri within Sauria. Lacertulus bipesis not a squamate, but its phylogenetic relationships cannot be further determined because of the poor preservation of the specimen [3], [86].

Paliguana whitei (Fig. 3B) possesses a quadrate conch ([17], [45], [87]; AM 201, MDE pers. obs.), a character widely accepted as a diagnostic feature of Lepidosauromorpha [57]. As a result, Evans and Borsuk-Białynicka [45] and Evans and Jones [17] considered Paliguana whitei as referable to Lepidosauromorpha. In agreement with this hypothesis, our phylogenetic results recovered Paliguana whitei as a basal lepidosauromorph. However, as discussed above, Paliguana whitei is currently considered Early Triassic in age. Therefore, although Paliguana whitei is accepted as one of the oldest known lepidosauromorphs [17] it does not represent a Permian record of the group.

Finally, in the original description of the species Lanthanolania ivakhnenkoi from the Middle Permian of Russia, Modesto and Reisz [15] recovered this species as a lepidosauromorph in some of the most parsimonious trees and as the sister-taxon of Sauria in others. These results suggested possible but uncertain saurian affinities. However, a more recent phylogenetic analysis recovered Lanthanolania ivakhnenkoi close to the base of Neodiapsida [75] and outside of Sauria, thus contradicting the possible inclusion of this species within Lepidosauromorpha or Sauria.

In summary, there is currently no Permian specimen that can unambiguously assigned to Lepidosauromorpha. The earliest known member of Lepidosauromorpha (Paliguana whitei) comes from lowermost Triassic (Induan–?Olenekian) rocks that were deposited in the aftermath of the Permo-Triassic mass extinction.

Permian records of Archosauromorpha

Only three Permian species are currently considered as unambiguous members of Archosauromorpha: Protorosaurus speneri [19] from Germany and England, and Archosaurus rossicus [88] and Eorasaurus olsoni [89], both from Russia. In addition to these species, several incomplete specimens of Permian or possible Permian age have been considered as possible members of Archosauromorpha. This material includes a "problematic reptile" from Tanzania [9], an isolated cervical vertebra from South Africa [90], and some vertebral material from Uruguay [91]. We discuss these three unambiguous Permian archosauromorphs and the additional possible records of the clade in more detail below.

"Acanthotoposaurus bremneri", based upon a single specimen (SAM-PK-K6888) from the Late Permian of South Africa, has been also considered an early member of Archosauromorpha [92]. However, Reisz et al. [14] provided a strong rebuttal to this interpretation and considered "Acanthotoposaurus bremneri" to be a subjective junior synonym of the "younginiform" Youngina capensis and, as a result, a non-saurian diapsid (see above). Another South African Late Permian taxon, Heleosaurus scholtzi, was suggested as a possible archosaur ancestor by Carroll [93]. However, Heleosaurus has never been formally referred to Archosauromorpha, and Reisz & Modesto [16] reinterpreted Heleosaurus as a varanopid synapsid, and thus non-diapsid. Mesenosaurus romeri from the Permian of Russia was reinterpreted by Ivachnenko [12] as the oldest known archosaur, rather than a "pelycosaurian" synapsid as described by previous authors [94], [95]. However, additional specimens and new anatomical work have demonstrated that Mesenosaurus romeri is a varanopid synapsid [43], [96].

Protorosaurus Speneri

Protorosaurus speneri [19] was a quadrupedal archosauromorph reaching a body length of 1.5–2 meters [21], known from numerous specimens from the Upper Permian Kupferschiefer Formation of Germany and England (Fig. 4). The first fossil specimen of Protorosaurus speneri was discovered in Germany in 1706, and Spener [97] published a description of this specimen (identifying it as the remains of a Nile crocodile), making the taxon one of the first fossil reptiles ever described. Meyer [19], [98], [99] identified the remains as of a previously unknown extinct reptile, erected the new species Protorosaurus speneri, and published a monographic description. Subsequently, Protorosaurus remains were also recovered from England [20], and Gottman-Quesada & Sander [21] recently published a full monographic redescription of Protorosaurus speneri, based on the abundant German material.

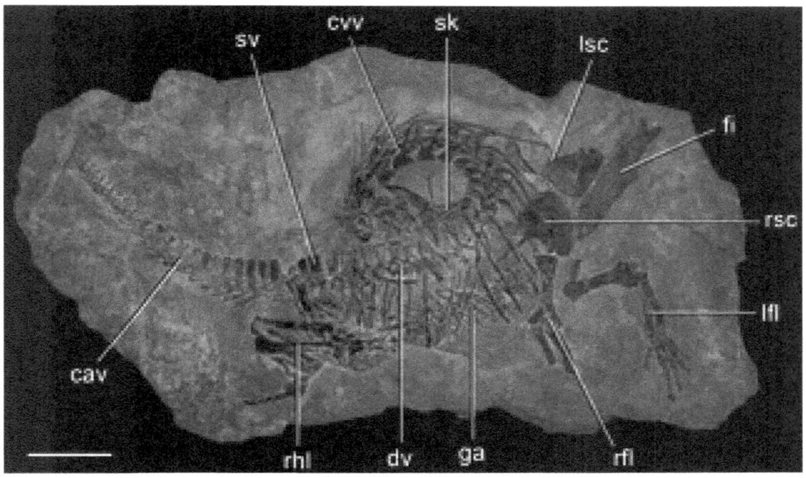

Figure 4. Protorosaurus speneri, a protorosaurian archosauromorph from the middle Late Permian of Western Europe.

Axial skeleton primarily exposed in right lateral view (BSPG 1995 I 5, cast of WMSN P47361) collected near Münster, Germany. Abbreviations: cav, caudal vertebrae; cvv, cervical vertebrae; dv, dorsal vertebrae; fi, fisch; ga, gastralia; lfl, left forelimb; lsc, left scapula and coracoid; rfl, right forelimb; rhl, right hindlimb; rsc, right scapula and coracoid; sk, skull; sv, sacral vertebrae. Scale bar equals 10 cm.

At least 28 Protorosaurus speneri specimens are known from the states of Thuringia and Hesse in central Germany. All of these German specimens come from the Kupferschiefer, part of the classic Permian Zechstein Group, which is divided into six cycles (Z1–Z6; e.g. [100]). The Kupferschiefer forms part of the basal cycle of the Zechstein (Z1) and is a dark bituminous and calcareous shale deposited in a marine environment. The Kupferschiefer is often given as Tatarian (equivalent to the Wordian–Wuchiapingian) (e.g. [21]) or Capitanian (e.g. [24]) in age. Brauns et al. [101] reported a date of 257.3±2.6 Ma (Late Permian/Lopingian: Wuchiapingian) for the Kupferschiefer based on a Re-Os geochronological study. The presence of the conodont Mesogondolella britannica in Kupferschiefer equivalents supports a middle Wuchiapingian age for the Protorosaurus-bearing levels ([102], [103]; Schneider pers. comm. 2012). Protorosaurus specimens from northwest England have been discovered in the Marl Slate [20], considered a lateral equivalent of the Kupferschiefer on the basis of independent geological data. A putative second species of Protorosaurus from England, P. huxleyi [104], was referred to the genus Adelosaurus by Evans [56] and considered as a probable diapsid of uncertain affinities.

The phylogenetic position of Protorosaurus speneri within Archosauromorpha has been widely accepted and is uncontroversial [4], [21], [52], [56], [72]–[74], [105], [106] and supported by our phylogenetic results. Protorosaurus speneri has generally been considered to belong to a clade of otherwise Triassic archosauromorphs referred to either as Prolacertiformes or Protorosauria, although the composition and monophyly of this grouping is debated (see summary in Gottman-Quesada & Sander [21]). Our phylogenetic results support referral of this species to Protorosauria, and suggest that "Prolacertiformes" is polyphyletic.

Sues & Munk [107] briefly mentioned archosauromorph cranial and postcranial remains from fissure fill deposits at Korbach, in Hesse, central Germany, including "a Protorosaurus-like form and tooth-bearing jaw fragments of a large, as yet unidentifiable taxon". The formation and infilling of this fissure was inferred to have taken place during the Z2 cycle of the Zechstein, indicating that these archosauromorph remains are slightly younger than Protorosaurus speneri. Unfortunately, more detailed descriptions of this material have not yet been published.

Eorasaurus Olsoni

Sennikov [89] erected Eorasaurus olsoni based on a sequence of cervico-dorsal vertebrae with one dorsal rib in articulation and some additional bone fragments (Figs. 5, 6). This material was collected from the bank of the Volga River in Tatarstan in European Russia during the 1930s. Eorasaurus olsoni comes from the upper substage of the Severodvinian regional stage [89], [108]. Recent magnetostratigraphic evidence suggests that the base of the Severodvinian stage is within the Capitanian (approximately middle Capitanian), but there exists uncertainty regarding the age of the upper boundary of the Severodvinian, which may be close to the Wuchiapingian–Changhsingian boundary ([109]: fig. 8) or to the Capitanian–Wuchiapingian boundary ([109]: p. 46). Accordingly, Eorasaurus olsoni is late Capitanian–Wuchiapingian in age and, as a result, roughly contemporaneous with (or possibly slightly older than) the middle Wuchiapingian Protorosaurus speneri.

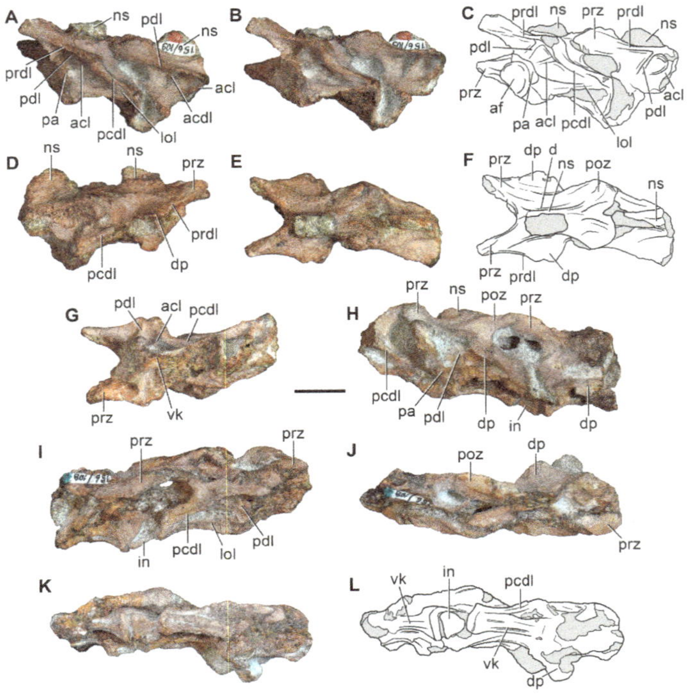

Figure 5. Eorasaurus olsoni, a possible early archosauriform from the late Middle–early Late Permian of Russia.

Middle (PIN 156/109: A–G) and posterior (PIN 156/108, holotype: H–L) cervical vertebrae in left lateral (A, H), left ventrolateral (B, C), right lateral (D, I), dorsal (E, F, J), and ventral (G, K, L) views. Abbreviations: acl, accessory lamina; af, anterior articular facet; d, depression; dp, diapophysis; in, intercentrum; lol, longitudinal lamina; ns, neural spine; pa, parapophysis; pcdl, posterior centrodiapophyseal lamina; pdl, paradiapophyseal lamina; poz, postzygapophysis; prdl, prezygodiapophyseal lamina; prz, prezygapophysis; vk, ventral keel. Scale bar equals 1 cm.

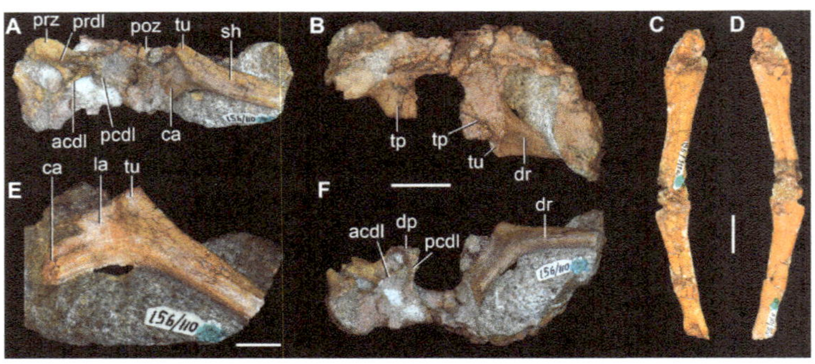

Figure 6. Eorasaurus olsoni, a possible early archosauriform from the late Middle–early Late Permian of Russia.

Anterior dorsal vertebrae and rib in articulation (PIN 156/110: A, B, F), close-up of the anterior dorsal rib (E), and probable long bones (PIN 156/111a, b: C, D) in left lateral (A, E), dorsal (B), and ventral (F) views. Abbreviations: acdl, anterior centrodiapophyseal lamina; ca, capitulum; dp, diapophysis; dr, anterior dorsal rib; la, lamina; pcdl, posterior centrodiapophyseal lamina; prdl, prezygodiapophyseal lamina; poz, postzygapophysis; prz, prezygapophysis; sh, shaft; tp, transverse process; tu, tuberculum. Scale bars equal 1 cm in (A–D, F) and 5 mm in (E).

Sennikov [89] considered Eorasaurus olsoni to be closely related to Protorosaurus speneriand considered both taxa to be members of Protorosauridae. Eorasaurus olsoni was diagnosed by Sennikov on the basis of a combination of characters of the cervico-dorsal axial skeleton, such as moderately elongated and strongly parallelogram-shaped vertebral centra, well-developed ridges situated below the diapophyses, absence of intercentra and three-headed anterior dorsal ribs ([89]: p. 95). Despite the importance of Eorasaurus olsonias one of the oldest known archosauromorphs, this taxon has been largely ignored by subsequent authors. Sennikov [89] provided a detailed description and drawings of Eorasaurus olsoni, and, as a result, a full redescription is not necessary here. However, we complement the original description of the species with some additional observations and provide a few reinterpretations based upon first hand examination of the specimens.

We agree with Sennikov [89] in considering the holotype (PIN 156/109) and referred specimens (PIN 156/108, 110, 111) of Eorasaurus olsoni to belong to a single individual, because they possess the same mode of preservation and are congruent in size and morphology. The preserved bones of Eorasaurus olsoni are generally well preserved, but there are several broken areas and damaged surfaces. The vertebrae of Eorasaurus olsoniprobably represent a continuous series of nine postaxial vertebrae, including middle (PIN 156/109; Fig. 5A–G) and posterior (PIN 156/108; Fig. 5H–L) cervical vertebrae and anterior dorsal vertebrae (PIN 156/110; Fig. 6A, B, E, F) (Table 2). The vertebrae of PIN 156/108 are interpreted as belonging to the posterior cervical series because the parapophyses are situated on the dorsal halves of the centra and the vertebrae of PIN 156/109 are identified as middle cervicals because the parapophyses are situated at mid-height on the anterior margins of the centra (Fig. 5: pa). By contrast, Sennikov [89] originally interpreted the holotype vertebrae of PIN 156/109 as posterior cervical vertebrae and those of PIN 156/108 as more anterior cervicals. No traces of neurocentral sutures were observed in the vertebrae of Eorasaurus olsoni, suggesting that the specimen was not a juvenile when it died [110].

Table 2. Measurements of the preserved bones of Eorasaurus olsoni **(PIN 156/108, 109, 110) in millimeters.**

PIN 156/108 (most anterior almost complete vertebra)			
	Centrum length	18.3	
	Anterior articular facet height	(8.4)	
	Anterior articular facet width	ca. 6.4	
	Posterior articular facet width	10.2	
	Maximum height of the vertebra	(21.8)	
	Length along zygapophyses	28.8	
	Length of base of neural spine	(11.4)	
	Intercentrum length	4.7	
	Intercentrum width	5.7	
PIN 156/109 (most anterior almost complete vertebra)			
	Centrum length	16.9	
	Anterior articular facet height	10.0	
	Anterior articular facet width	9.2	
	Posterior articular facet height	10.4	
	Posterior articular facet width	ca. 9.8	
	Maximum height of the vertebra	(18.8)	
	Neural canal height	2.9	
	Neural canal width	6.7	
	Length along zygapophyses	28.2	
	Length of base of neural spine	11.0	
PIN 156/110	**Vertebra**	**A**	**B**
	Centrum length	16.5	-
	Anterior articular facet height	(8.4)	-
	Anterior articular facet width	(6.8)	-
	Maximum height of the vertebra	(17.4)	(11.0)
	Length along zygapophyses	(23.3)	ca. 21.2
	Transverse process width	(5.8)	11.5
	Transverse process length at distal end	-	7.6
	Neural spine length	(8.7)	(9.6)
Anterior dorsal rib			
	Length	(23.9)	
	Anteroposterior proximal depth	(15.8)	
	Length tubercular facet	3.6	
PIN 156/111	**Long bone**	**A**	**B**
	Length	(46.4)	(38.7)

Values between brackets indicate incomplete measurements and the value given is the maximum measurable. The length along the zygapophyses is the maximum anteroposterior length between the anterior tips of the prezygapophyses and the posterior tips of the postzygapophyses. In PIN 156/110 the vertebrae were labeled as A or B, where A corresponds to the most anterior element in the specimen. Maximum deviation of the digital caliper equals 0.02 mm but measurements were rounded to the nearest 0.1 millimeter.

The centra of the middle cervical vertebrae (PIN 156/109; Fig. 5A–G) possess low and well defined, median longitudinal ventral keels (Fig. 5: vk). The lateral surface of the centrum possesses a horizontal lamina that extends from the base of the parapophysis to the posterior margin of the centrum (Fig. 5: lol), resembling the condition of Macrocnemus bassanii (PIMUZ T4822) and Tanystropheus longobardicus (PIMUZ T2818). The lateral surface of the centrum immediately dorsal to the horizontal lamina is strongly concave, but the degree of concavity seems to be exaggerated due to the collapse of the cortical bone. The neural canal is considerably wider than tall in anterior view. Well-developed paradiapophyseal, posterior centrodiapophyseal and prezygodiapophyseal laminae extend away from the base of the diapophysis (Fig. 5: acdl, pcdl, pdl, prdl), as also occurs in the posterior cervical and/or dorsal vertebrae of the enigmatic neodiapsid Helveticosaurus zollingeri (PIMUZ T4352), numerous basal archosauromorphs (e.g. Tanystropheus longobardicus [111]: figs. 52–54; Protorosaurus speneri, BSPG 1995 I 5, cast of WMSN P47361; Spinosuchus caseanus [112]), and several basal archosauriforms and crown group archosaurs (e.g. Erythrosuchus africanus, NHMUK R3592, [113]; Euparkeria capensis, UMZC T921; Bromsgroveia walkeri [114]; Hypselorhachis mirabilis [115]; Silesaurus opolensis [116]; Herrerasaurus ischigualastensis, PVSJ 373, [117]). The posterior centrodiapophyseal lamina extends to the posterodorsal corner of the centrum and contacts in this area the horizontal lamina of the centrum.

The neural arch laminae delimit centrodiapophyseal and prezygapophyseal centrodiapophyseal fossae, but the postzygapophyseal centrodiapophyseal fossa is absent. The centrodiapophyseal fossa is subdivided by an accessory lamina that extends anteriorly from the posterior centrodiapophyseal lamina and contacts the base of the parapophysis and the paradiapophyseal lamina (Fig. 5: acl). This accessory lamina is not present in other basal diapsids and represents an autapomorphy of Eorasaurus olsoni. A ridge extends anteriorly from the base of the postzygapophysis onto the lateral surface of the neural arch, and curves ventrally, being positioned between the prezygapophysis and diapophysis but without reaching either of these structures. This ridge delimits the lateral margin of a shallow depression positioned next to the base of the neural spine (Fig. 5: d). A similar depression is also found in the cervical vertebrae of several basal archosauromorphs (e.g. Protorosaurus speneri, BSPG 1995 I 5; Prolacerta broomi, BP/1/2675; Proterosuchus fergusi, GHG 231). The base of the neural spine is transversely thin and not as wide as it appears in the drawing in the original description ([89]: fig. 1d). There is no evidence of intercentra in PIN 156/109 [74].

The posterior cervical vertebrae (PIN 156/108) (Fig. 5H–L) possess a morphology that is very similar to that found in the middle cervical vertebrae, including the presence of a ventral longitudinal keel and the same suite of laminae on the centrum and neural arch. The accessory lamina that divides the centrodiapophyseal fossa is even more extensively developed laterally in the posterior cervical vertebrae than the middle cervicals. The neural arches of the posterior cervicals of PIN 156/108 each possess an incipient postzygapophyseal centrodiapophyseal fossa, consistent with their more posterior position in the axial series than the middle cervical vertebrae of PIN 156/109. There is no depression lateral to the base of the neural spine. Two intercentra are present in PIN 156/108 but were overlooked in the original description of the specimen. The presence of intercentra resembles the condition observed in the postaxial cervical vertebrae of several basal archosauromorphs (e.g. Macrocnemus bassanii, PIMUZ T4822; Trilophosaurus buettneri [44]: fig. 30; Proterosuchus fergusi, NM QR 1484). By contrast, postaxial cervical intercentra are absent in Tanystropheus longobardicus (PIMUZ T2817), Protorosaurus speneri [21], Mesosuchus browni [4] and Howesia browni [118]. The intercentra are situated anterior to the most complete vertebrae of PIN 156/108 (Fig. 5: in). The intercentra are proportionally large and subtriangular in ventral view, with a transversely broad posterior margin and a tapering anterior end.

The general morphology of the anterior dorsal vertebrae (PIN 156/110) is congruent with that of the cervical vertebrae, but in the anterior dorsals the centrum is subrectangular in lateral view (Fig. 6A, B, E, F). The neural arches of the anterior dorsal vertebrae possess prezygodiapophyseal, posterior centrodiapophyseal and anterior centrodiapophyseal/paradiapophyseal laminae (Fig. 6: acdl, pcdl, prdl). It is not possible to determine whether or not the latter lamina reached the parapophysis because the relevant area is damaged. The centrodiapophyseal, prezygapophyseal centrodiapophyseal, and postzygapophyseal centrodiapophyseal fossae are present and the latter fossa is better developed than in the posterior cervical vertebrae (PIN 156/108). There is no accessory lamina subdividing the centrodiapophyseal fossa, contrasting with the condition in the cervical vertebrae (PIN 156/108, 109). The left transverse process of the third anterior dorsal vertebra of PIN 156/110 is complete and is very strongly developed laterally, with a transverse length to centrum length ratio of 0.70 (Fig. 6B: tp). This ratio resembles that observed in the anterior dorsal vertebrae of Trilophosaurus buettneri (0.84 [44]: fig. 37), Proterosuchus fergusi (0.95, NM QR 1484) and Erythrosuchus africanus (0.85, NHMUK R3592). By contrast, proportionally shorter transverse processes are present in the anterior dorsal vertebrae of Youngina capensis (0.46, BP/1/3859), early lepidosaurs (e.g. Gephyrosaurus bridensis [48]: figs. 5, 6; Planocephalosaurus robinsonae, 0.18–0.25 [119]: figs. 5, 6), protorosaurs (Protorosaurus speneri, 0.38–0.45, BSPG 1995 I 5; Tanystropheus longobardicus, 0.46, SMNS 54628), Macrocnemus bassanii (0.56, PIMUZ T2472), Mesosuchus browni (approximately 0.5 [4]: p. 513) and Prolacerta broomi (0.55, BP/1/2675). The transverse process of Eorasaurus olsoni is slightly anteroposteriorly compressed close to its base, but possesses an overall subrectangular outline in dorsal view. There is no depression on the neural arch lateral to the base of the neural spine, similar to the condition in the posterior cervical vertebrae (PIN 156/108) but contrasting with condition in the middle cervical vertebrae (PIN 156/109).

The proximal half of the left dorsal rib is preserved in near articulation with the third vertebra of PIN 156/110 (Fig. 6A, B, E, F). The capitulum of this anterior dorsal rib lacks its distal end, but the process is relatively long. The tuberculum is complete and is very short. The articular facet of the tuberculum is flat and oval, with an acute posterior margin. The tuberculum is not well differentiated from the rest of the rib due to the presence of a thin lamina of bone that connects it with the capitulum (Fig. 6E: la). The lamina extends up to the same level as the articular facet of the tuberculum. An apparent notch between the capitulum and the lamina is the result of breakage.

There is no conclusive evidence for the presence of a third articular facet on the anterior dorsal rib (contra Sennikov [89]). Although the lamina between the capitulum and tuberculum resembles a similar lamina that houses the third articular facet in Prolacerta broomi (BP/1/2675), Proterosuchus fergusi (NM QR 1484) and Erythrosuchus africanus [113], the preserved portion of the lamina in PIN 156/110 lacks the transverse thickening that bears the facet in the those taxa.

PIN 156/111 is represented by two long bones (PIN 156/111a, 111b; Fig. 6C, D) and a small block of matrix with some unidentified partial bones (PIN 156/111). PIN 156/111a and 111b are interpreted as two limb bones in articulation. They do not seem to be rib shafts because they lack the curvature that would be expected for a rib and the proximal and distal ends of the bones are subequally expanded (contra Sennikov [89]). Neither bones appears to be a femur and, as a result, they may represent a humerus and an ulna or radius. The long bones are strongly flattened, resembling the condition of the forelimb bones of protorosaurian archosauromorphs (e.g. Tanystropheus longobardicus [111]). However, there are no clear features that would allow a confident identification of these bones; as a consequence, they are not very informative.

The morphology of Eorasaurus olsoni is congruent with that observed in basal archosauromorphs (e.g. presence of well developed prezygodiapophyseal and anterior and posterior centrodiapohyseal laminae, anterior dorsal zygapophyses close to the sagittal plane of the axial skeleton; cf. Sennikov [89]). Indeed, our phylogenetic results recovered Eorasaurus olsoni as a derived archosauromorph within Archosauriformes.

Our reexamination of the anatomy of Eorasaurus olsoni allowed us to reinterpret some characters that were included in the original diagnosis of the species, including the supposed absence of intercentra and the presence of a dorsal rib with three articular facets. Accordingly, we provide here an emended diagnosis for the species. Eorasaurus olsoni is a small archosauromorph that differs from other diapsids in possessing the following combination of characters: presence of prezygodiapophyseal and anterior and posterior centrodiapophyseal laminae; an accessory lamina that extends anteroventrally from the posterior centrodiapophyseal lamina and subdivides the centrodiapophyseal fossa (autapomorphic); strongly parallelogram-shaped middle and posterior cervical centra; and presence of postaxial cervical intercentra.

Archosaurus rossicus

Tatarinov [88] described Archosaurus rossicus on the basis of fragmentary cranial and postcranial remains from the upper substage of the Vyatskian regional stage at the Vyazniki locality in Vladamir region, Russia. Archosaurus rossicus is considered to belong to the uppermost part of the Tatarian series (e.g. [120], [121]), and part of the "Vyazniki Biotic Assemblage" which also includes pareiasaurs, chroniosuchians, therocephalians, and dicynodonts [122]. The late Tatarian and Vyazniki Biotic Assemblage have in recent years been generally correlated to the late Changhsingian, or terminal Permian [84], [122]–[125].

Sennikov [120] subsequently referred to Archosaurus rossicus an additional dentary from the same general locality as the holotype and paratypes, as well as two isolated elements from a second locality. These referrals were based on the presence of congruent proterosuchid morphology. Sennikov [120] also revised the taxonomic status of Archosaurus rossicus (see also [121], [126]). However, we note that some caution is warranted with regard to the assignment of the paratype and referred specimens from the type locality to Archosaurus rossicus given that they come from three different geographical points and different stratigraphic levels within a geographically large locality with a stratigraphic thickness of around 25 meters (Sennikov pers. comm., 2013). In addition to the holotype premaxilla (PIN 1100/55; Fig. 7; Table 3), we consider that the only previously referred specimens of Archosaurus rossicus that can be confidently identified as referable to Archosauriformes are the left dentary (PIN 1100/78), skull roof (PIN 1100/48) and possibly a tooth crown (PIN 1100/85). In addition, the cervical vertebrae (PIN 1100/66, 66a, 66b) referred by Tatarinov [88] to Archosaurus rossicus possess a morphology that is very similar to and congruent with that of the cervical vertebrae of Proterosuchus fergusi (NM QR 1484, GHG 236), and therefore they also possibly belong to an archosauriform.

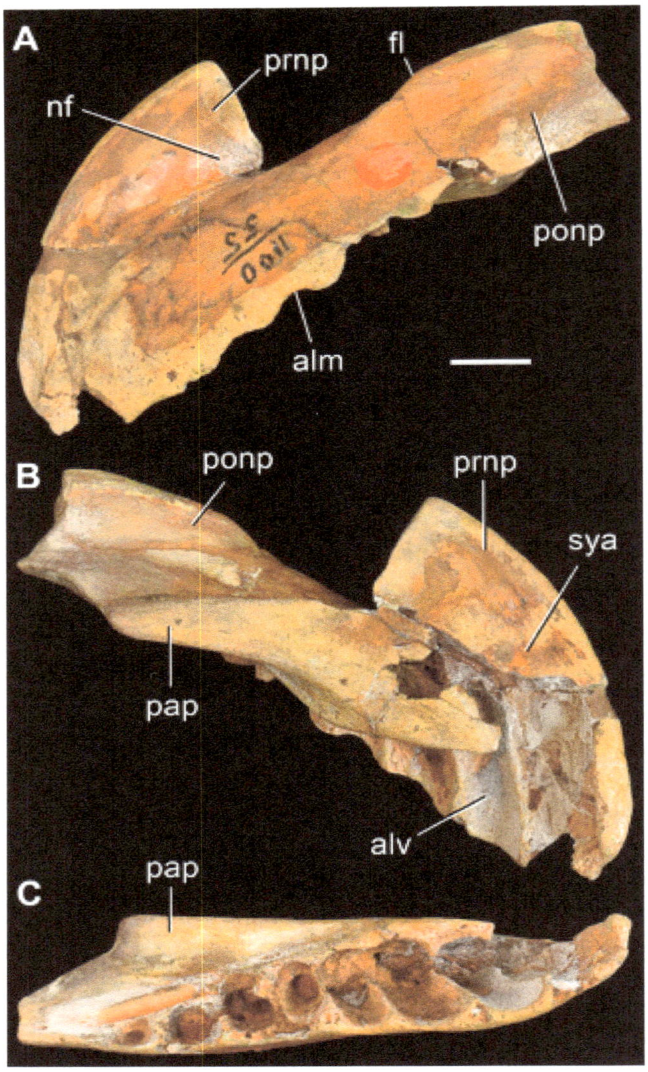

Figure 7. Archosaurus rossicus, a proterosuchid archosauriform from the latest Permian of Russia.

Left premaxilla (PIN 1100/55, holotype) in lateral (A), medial (B) and ventral (C) views. Abbreviations: alm, alveolar margin; alv, alveolus; fl, lateral flange; nf, narial fossa; pap, palatal process; ponp, postnarial process; prnp, prenarial process; sya, symphyseal area. Scale bar equals 1 cm.

Table 3. Measurements of the holotype of Archosaurus rossicus **(PIN 1100/55) in millimeters.**

Premaxilla		
	Length	(83.3)
	Height of the premaxillary body	19.8
	Length of the premaxillary body	73.6
	Maximum height	(34.6)
	Length first alveolus	11.1
	Length second alveolus	7.8
	Length third alveolus	8.3
	Length fourth alveolus	6.7
	Length fifth alveolus	5.6
	Length sixth alveolus	6.3
	Length seventh alveolus	7.1
	Length eighth alveolus	5.4

Values between brackets indicate incomplete measurements and the value given is the maximum measurable. Maximum deviation of the digital caliper equals 0.02 mm but measurements were rounded to the nearest 0.1 millimeter.
doi:10.1371/journal.pone.0089165.t003

The bone identified as a squamosal (PIN 1100/84a) and referred to Archosaurus rossicusby Tatarinov [88] does not possess a morphology congruent with that of a squamosal. For example, it lacks a facet for articulation with the quadrate head and possesses a tuberosity on the posterodorsal border of the supposed supratemporal fenestra. Moreover, the anterior process is unusually transversely thick. Accordingly, we doubt the identification of this bone and, as a result, its archosauriform affinities.

The holotype premaxilla of Archosaurus rossicus (Fig. 7) differs from most basal archosauromorphs in that the first four premaxillary alveoli open lateroventrally (Fig. 7C), contrasting with the mostly ventrally opening anterior alveoli of Prolacerta broomi(BP/1/471), Sarmatosuchus otschevi (PIN 2865/68-9), Tasmaniosaurus triassicus (UTGD 54655) and Erythrosuchus africanus (NHMUK R3592). In addition, the angle formed between the anterior margin of the premaxillary body and the alveolar margin is more acute in Archosaurus rossicus than in Proterosuchus fergusi (RC 59, SAM-PK-11208) and "Chasmatosaurus" yuani (IVPP V90002, V4067). Accordingly, the holotype specimen of Archosaurus rossicus is diagnostic and, as a result, the genus and species can be considered valid.

Archosaurus rossicus has been widely accepted as a proterosuchid archosauriform [88], [120], [121], [127], and quantitative phylogenetic support for this position has been recovered by Nesbitt [59] and by our phylogenetic analysis.

Putative proterosuchian from the Late Permian of South Africa.

Cruickshank [90] reported that all known specimens of the early archosauriform Proterosuchus came from the lowermost Triassic Lystrosaurus Assemblage Zone of South Africa, with one possible exception: a cervical vertebra (BP/1/4220; Fig. 8; Table 4) collected from the Upper Permian upper Cistecephalus Assemblage Zone. BP/1/4220 was collected in May 1969 at the farm Gegund 532 near Harrismith in Free State. Cruickshank ([90]: table 1) identified BP/1/4220 as ?Proterosuchus sp. and figured the specimen ([90]: fig. 4a). Subsequently, Reisz et al. [14] briefly noted that they could not find any evidence of archosauromorph features in BP/1/4220. Additionally, Reisz et al. ([14]: 443) cast doubts on the exact providence of the specimen and concluded that it could be Triassic rather than Permian in age, but without providing supporting evidence.

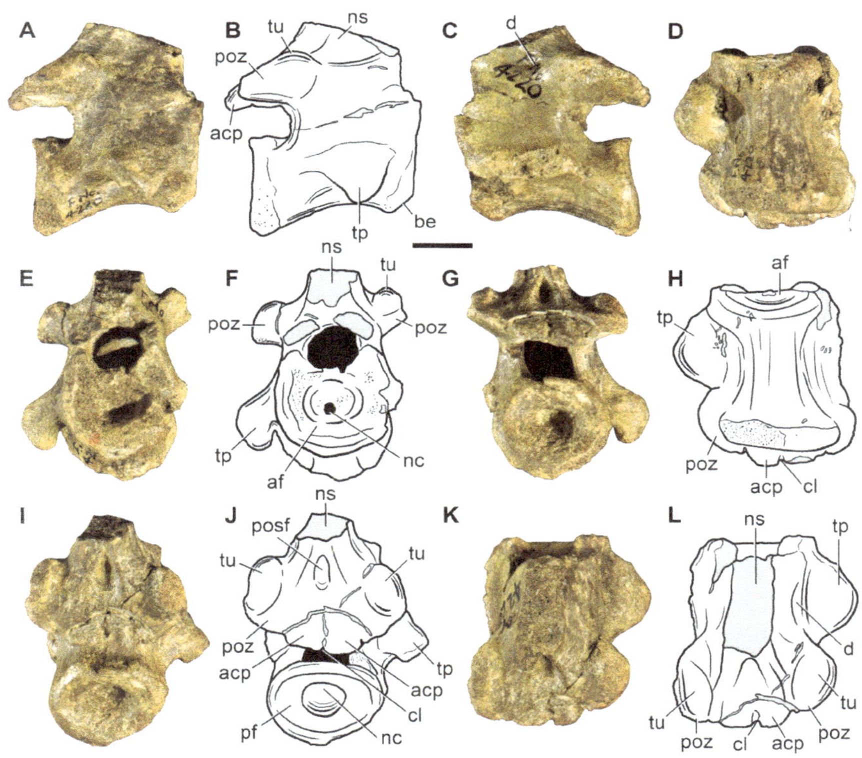

Figure 8. Indeterminate reptiliomorph from the Late Permian–Early Triassic of South Africa.

Cervical vertebra (BP/1/4220) in right lateral (A), left lateral (C), ventral (D, H), anterior (E, F), posterior (G), posterodorsal (I, J) and dorsal (K, L) views. Abbreviations: acp, accessory (interpostzygapophyseal) process; af, anterior facet; be, ventral beveling; cl, median cleft; d, depression; nc, notochordal canal; ns, neural spine; tu, tuberosity; pf, posterior facet; poz, postzygapophysis; posf, postspinal fossa; tp, transverse process. Scale bar equals 1 cm.

Table 4. Measurements of the preserved bones of BP/1/4220 in millimeters.

Cervical vertebra		
	Centrum length	28.5
	Anterior articular facet height	20.6
	Anterior articular facet width	20.3
	Posterior articular facet height	17.7
	Posterior articular facet width	20.1
	Maximum height of the vertebra	(38.5)
	Length along zygapophyses	(28.8)
	Length of base of neural spine	18.6
	Width of neural spine	9.4
	Width along postzygapophyses	26.4
	Width of accessory processes	13.2
rib		
	Length	(109.5)

Values between brackets indicate incomplete measurements and the value given is the maximum measurable. The length along the zygapophyses is the maximum anteroposterior length between the anterior tips of the prezygapophyses and the posterior tips of the postzygapophyses. Maximum deviation of the digital caliper equals 0.02 mm but measurements were rounded to the nearest 0.1 millimeter.
doi:10.1371/journal.pone.0089165.t004

Re-examination of BP/1/4220 revealed a morphology that does not conform to that expected for a basal archosauriform (cf. Reisz et al. [14]). However, we did observe some unusual features not reported in any other tetrapod that we are aware of. Because no detailed description of the specimen has ever been published, we here describe BP/1/4220 in detail for the first time and reassess its possible phylogenetic position.

BP/1/4220 (Fig. 8) is an almost complete vertebra that is not an axis because the transverse processes are well-developed, but may have belonged to the middle cervical series based on the presence of a parallelogram-shaped centrum in lateral view (the anterior articular surface is positioned distinctly dorsal to the posterior surface) and a diapophysis that is placed well below the neural arch (ventral to the level of the dorsal margin of the centrum). The vertebra is notochordal, with an open notochordal canal that is wider than tall and which completely pierces the centrum (Fig. 8: nc). Series of concentric bony laminae surround the notochordal canal, indicating the partial resorption of the notochord during life. In addition, the neurocentral suture is obliterated, suggesting that the animal was not a juvenile at the time of its death [110], and that the presence of an open notochordal canal is therefore not a result of an early ontogenetic stage. The persistence of an open notochordal canal in a non-juvenile individual resembles the condition in multiple lineages of basal reptiliomorphs, parareptiles, basal synapsids, basal sauropsids, basal lepidosauromorphs, and the new archosauromorph species Aenigmastropheus parringtoni from the Late Permian of Tanzania ([48], [70], [119], [128]–[134], see below).

The anterior and posterior articular surfaces of the centrum are wider than tall. The anterior and posterior borders of the centrum are strongly beveled on their ventral margin in lateral view (Fig. 8: be), indicating the probable presence of small intercentra. The centrum is slightly transversely compressed at mid-length, and has a spool-shape in ventral view. The ventral surface of the centrum is mostly planar and well differentiated from the lateral surfaces. The lateral surfaces of the centrum are concave in ventral view and possess shallow and poorly defined fossae. The centrum lacks parapophyses and it is likely that the parapophyses would have been placed on the intercentrum, as occurs in several basal amniotes (e.g. Procolophon trigoniceps: [135]). Subcentral foramina are absent in BP/1/4220.

In the neural arch, the transverse processes are robust and directed posterolaterally and slightly ventrally. BP/1/4220 completely lacks any development of centrodiapophyseal or paradiapophyseal laminae, contrasting with the condition observed in some caudatans, "pelycosaurian" synapsids, basal diapsids and archosauromorphs (see below). The postzygapophyses are separated from the posterior end of the centrum by a tall and deep notch in lateral view. Only the bases of the prezygapophyses are preserved and they are well separated transversely from one another, as is also the case for the postzygapophyses. The presence of broadly separated zygapophyses contrasts with the condition observed in the cervical vertebrae of most archosauromorphs and some "pelycosaurian" synapsids, in which the zygapophyses are placed close to each other in dorsal view (e.g. Ophiacodonsp., MCZ 1426; Prolacerta broomi, BP/1/2675; Protorosaurus speneri, ZMR MB R2173). The neural arch lacks a prezygodiapophyseal lamina, contrasting with the condition observed in the varanopid Apsisaurus witteri ([73]: fig. 6; sensu Reisz et al. [43]) and several archosauromorphs (e.g. Tanystropheus longobardicus, SMNS 54628; Protorosaurus speneri, BSPG 1995 I 5; Erythrosuchus africanus, NHMUK R3592; Garjainia prima, PIN 2394/5-16).

The articular facets of the postzygapophyses of BP/1/4220 are oval, anteroposteriorly long and transversely wide, and face ventrally and slightly laterally. The dorsal surfaces of the postzygapophyses possess thick, rounded tuberosities (Fig. 8: tu) that resemble the epipophyses present in the cervical vertebrae of dinosaurs [136], the tanystropheids Macrocnemus bassanii (PIMUZ T4822) and Tanystropheus longobardicus (SMNS 54630, 54654), and derived rhynchosaurs [137]. However, in contrast to the latter taxa, in BP/1/4220 the tuberosity is situated not on the posterior half of the postzygapophysis, but at the level of the anterior margin of its articular facet. The tuberosity possesses a rugose surface, which may suggest a tendinous attachment.

The neural arch possesses two posteriorly directed median or interpostzygapophyseal processes between the postzygapophyses that lack their most posterior ends (Fig. 8: acp). Despite being damaged posteriorly, the interpostzygapophyseal processes project further posteriorly than do the postzygapophyses. The position of these interpostzygapophyseal processes in BP/1/4420 is similar to the transpostzygapophyseal lamina of trilophosaurids [98] and the accessory intervertebral articular processes of some saurians (i.e. the non-homologous hyposphene of archosauriforms and the zygosphene of squamate lepidosauromorphs and sauropterygians) [138]. The interpostzygapophyseal processes of BP/1/4220 are oval, posteriorly and slightly ventrally oriented, and separated from one another by a deep but transversely narrow median cleft (Fig. 8: cl). The presence of a cleft between the interpostzygapophyseal processes and the posterior extension of the processes beyond the level of the postzygapophyses differs from the morphology of the archosauromorph hyposphene and is in complete contrast with the depressed morphology of the lepidosauromorph zygosphene. No articular facet is discernable on the preserved portions of the interpostzygapophyseal processes of BP/1/4220. The interpostzygapophyseal processes of BP/1/4220 also differ from the accessory articular processes of tangasaurids (e.g. Hovasaurus [68]), which are vertically oriented and placed dorsal to the postzygapophyses at the base of the neural spine, and from those of diadectomorphs and seymouriamorphs [139], in which the accessory processes are medioventral projections of the postzygapophyses. The presence of a median cleft and the possible absence of articular facets in the interpostzygapophyseal processes of BP/1/4220 resemble the condition present in the transpostzygapophyseal lamina of trilophosaurids [112], but in the latter taxa the lamina does not extend posteriorly beyond the level of the posterior margin of the postzygapophysis. Accordingly, the condition observed in BP/1/4220 does not match with the morphology of any other tetrapod of which we are aware.

The neural spine is transversely thick at its base and moderately expanded anteroposteriorly (Fig. 8: ns). The neural arch possesses a shallow depression lateral to the base of the neural spine on its left side (Fig. 8: d). This condition resembles that observed in some "pelycosaurian" synapsids (e.g. Apsisaurus witteri [73]) and archosauromorphs (e.g. Protorosaurus speneri, BSPG 1995 I 5; Prolacerta broomi, BP/1/2675), but contrasts with the deeper and better-defined depressions of the araeoscelidians Araeoscelis gracilis [128] and Petrolacosaurus kansensis [131]. Nevertheless, this depression is absent on the right side of the neural arch of BP/1/4220. The neural spine possesses a very deep and transversely wide postspinal fossa that is well defined laterally by sharp edges forming the posterolateral corners of the neural spine (Fig. 8: posf). The postspinal fossa is not completely preserved dorsally, but it is shallow at its most dorsal preserved portion suggesting that it would have extended only along the ventral portion of the neural spine. The postspinal fossa is subtriangular in posterior view.

Three indeterminate bone fragments and a possible fragment of rib shaft are also preserved in BP/1/4220. The possible rib shaft possesses a plate-like end that becomes rod-like, with an elliptical cross-section, towards the other end of the bone. No articular facet is preserved on this fragment of bone.

We are unable to recognize any synapomorphies that would allow assignment of BP/1/4220 to Archosauromorpha (see also Reisz et al. [14]), Lepidosauromorpha or Sauria. Indeed, BP/1/4220 differs from several archosauromorphs (e.g. the new archosauromorph species Aenigmastropheus parringtoni, see below: UMZC T836; Prolacerta broomi, BP/1/2675; Proterosuchus fergusi, GHG 231) in possessing postzygapophyses that are strongly divergent posteriorly (although this condition is present in Trilophosaurus and rhynchosaurs; see below) and the absence of laminae on the neural arch (laminae are also absent in rhynchosaurs; see below). BP/1/4220 further differs from saurians in the absence of parapophyses on the centrum and the extreme transverse thickness of the neural spine at its base. As a result, the assignment of BP/1/4220 to Archosauromorpha by Cruickshank [90] is not followed here.

BP/1/4220 possesses a striking combination of features unknown in any amniote that we are familiar with (e.g. notochordal centrum, thick and anterodorsally oriented neural spine, large tubercle on the dorsal surface of the postzygapophysis, interpostzygapophyseal processes). Although BP/1/4220 appears to represent a distinct amniote taxon we do not erect a new species for it due to the highly incomplete nature of the specimen. BP/1/4220 can be unambiguously assigned to Reptiliomorpha (diadectomorphs+amniotes) based on the presence of a large pleurocentrum (with a reduced intercentrum, if present; cf. Romer [138]). However, we could not identify any feature that would allow the specimen to be assigned to a less inclusive reptiliomorph clade. BP/1/4220 was not included in the phylogenetic analysis conducted here because of its highly incomplete condition, and because of the absence of some major amniote clades in the taxonomic sample of the analysis (e.g. parareptiles). In summary, we interpret BP/1/4220 as belonging to Reptiliomorpha, and it may represent a previously unrecognized reptiliomorph lineage within the Late Permian of South Africa.

Specimens identified as either varanopid "pelycosaurs" or basal archosauromorphs from the Permo-Triassic of Uruguay.

Piñeiro et al. [140] described multiple isolated dorsal and caudal vertebrae from the Buena Vista Formation of northwestern Uruguay. This sedimentary unit was deposited during the Late Permian and probably also during the Early Triassic as part of the infill of the Paraná Basin [141]. Piñeiro et al. [140] assigned the vertebrae to varanopid "pelycosaurs", noting strong resemblances to the Permian species Mycterosaurus longiceps and Mesenosaurus romeri.

Subsequently, Dias-da-Silva et al. [91] stated that the identification of "pelycosaurian" synapsids in the Buena Vista Formation was unwarranted and that the isolated vertebrae described by Piñeiro et al. [140] closely resembled those of the basal archosauromorph Prolacerta broomi. Dias-da-Silva et al. [91] concluded that the vertebrae reported from the Buena Vista Formation may belong to a basal archosauromorph or to another kind of diapsid. At the same time, Dias-da-Silva et al. [91] pointed out that the other tetrapods (i.e. temnospondyl and procolophonoid remains) collected from the Buena Vista Formation [142], [143] are not strongly indicative of a Late Permian age.

Our re-examination of the isolated vertebrae described by Piñeiro et al. [140] (FC-DPV 1182, 1183, 1189, 1194, 1199, 1200 and 1333) does not reveal the presence of any archosauromorph synapomorphies in these specimens (e.g. there are no anterior or posterior centrodiapophyseal or prezygodiapophyseal laminae). The overall morphology of these vertebrae is congruent with the vertebrae of basal archosauromorphs (e.g. Prolacerta broomi; see also Dias-da-Silva et al. [91]), but also with those of some varanopid "pelycosaurs" [140]. As a result, we do not support an unambiguous assignment of these vertebrae to Archosauromorpha. Nevertheless, some unpublished specimens also collected from the Buena Vista Formation can be assigned to archosauromorphs that are probably closely related to protorosaurs, Prolacerta and proterosuchids (MDE pers. obs.). These specimens will be described elsewhere.

In sum, although the Buena Vista Formation yields (currently unpublished) saurian remains, the current poor stratigraphic constraints on its age mean that the putative Permian age of specimens from this unit is ambiguous.

"Problematic reptile" from the Late Permian of Tanzania.

Parrington [9] described the remains (several vertebrae and some fragmentary forelimb elements) of an enigmatic Permian specimen (UMZC T836) collected in the Ruhuhu Valley of Tanzania. He highlighted the apparent contrast between the primitive appearance of the forelimb bones and the more derived appearance of the vertebrae, with neural arch laminae ("buttresses") and articular rib facets resembling those of archosaurs. Parrington [9]concluded that the bones of UMZC T836 did not bear close resemblances to any known synapsid, and suggested instead that the specimen might have close affinities with archosaurs because of the vertebral morphology and the presence of hollow limb bones and an ectepicondylar groove on the humerus.

Subsequently, Hughes [144] noted that the vertebrae of UMZC T836 were not as archosaurian in appearance as Parrington originally thought and that laminae on the neural arch also occur in "pelycosaurian" synapsids. Hughes [144] further noted that a notochordal centrum is present in "pelycosaurs", but is unknown among archosaurs. However, Hughes [144] concluded that the combination of a derived vertebral column and a primitive limb structure occurs in proterosuchian archosauromorphs, and suggested that UMZC T836 might possibly be an "incipient proterosuchian" (i.e. a proterosuchian ancestor). Reig [145] noted that the vertebrae of UMZC T836 were transitional between those of "pelycosaurs" and archosaurs. Charig and Sues [127] listed this specimen as a possible member of Proterosuchidae in their review of "Proterosuchia", but also highlighted the skepticism raised by Hughes [144] as to the archosaurian affinities of UMZC T836. Gower and Sennikov [121] noted that UMZC T836 is probably indeterminate, but could possibly be archosaurian. Most recently, Ezcurra et al. [5] indicated that UMZC T836 is likely not referable to Archosauriformes (the archosauromorph clade that includes proterosuchians).

Parrington [9] reported that he collected UMZC T836 in the Ruhuhu Valley of Tanzania in 1933. These fossil-bearing levels correspond to locality B35 of Stockley [146], which is located near the town of Ruanda in the Songea District of southern Tanzania ([146]: plate 38; [147]: fig. 1) (Fig. 9). Stockley [146] considered locality B35 to be part of the "Lower Bone Bed", corresponding to his K6 horizon of the Songea Series. The K6 horizon is currently assigned to the Usili Formation (formerly the Kawinga Formation) of the Songea Group of the Ruhuhu Basin. Wopfner et al. [148] and Sidor et al. [149] described the Usili Formation as a 260 meters thick fluviolacustrine succession made up of a lowermost conglomeratic interval that is approximately 5 meters thick, grading up into a trough cross-bedded, coarse-grained, sandstone-dominated interval that is 25–40 meters thick, overlain by massive nodular siltstone and laminated mudstone beds with minor ribbon sandstones forming the bulk of the succession.

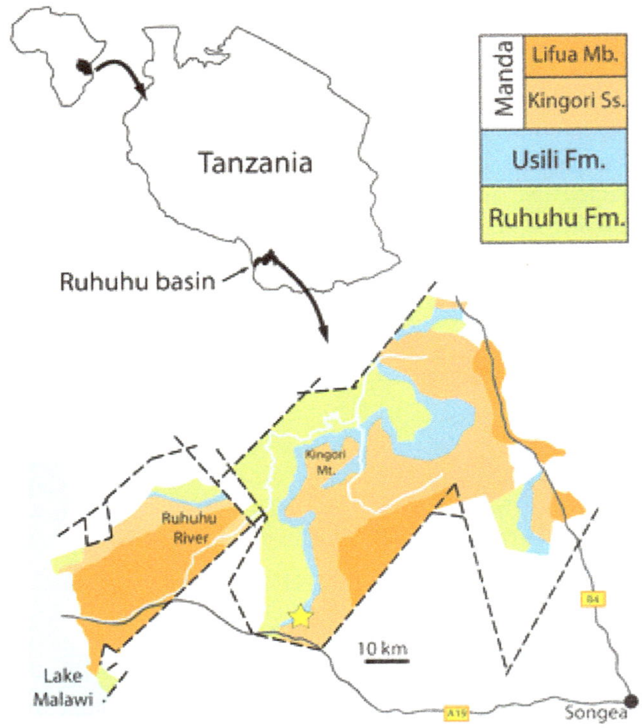

Figure 9. Type locality of Aenigmastropheus parringtoni **in the Ruhuhu Basin, southwestern Tanzania, Africa.**

Star indicates the approximate geographic and stratigraphic occurrence of Aenigmastropheus parringtoni (locality B35). Abbreviations: Fm, formation; Mb, member; Mt, mountain; Ss, sandstone. Figure modified from Nesbitt et al. [224].

Sidor et al. [149] recognized a single tetrapod faunal assemblage in the Usili Formation, which includes, in addition to UMZC T836, temnospondyls, pareiasaurs, gorgonopsians, therocephalians, cynodonts, and dicynodonts [147], [149]–[151]. In particular, the locality from which UMZC T836 was collected also yielded an isolated maxilla of a dicynodont listed by Parrington [9] as cf. "Esoterodon" uniseries (UMZC T969), as well as other dicynodont (UMZC T779, T1170) and gorgonopsid (UMZC T882, T883) remains ([9], [152]; UMZC catalogue and unpublished field notes of Parrington in UMZC collections). Parrington [9]proposed that locality B35 is equivalent in age to the South African horizons that yield Endothiodon (currently known in the late Pristerognathus, Tropidostoma, and early Cistecephalus assemblage zones of South Africa: [84], [149], [153]–[155]) because of the presence of cf. "Esoterodon" uniseries ("Esoterodon" is currently considered to be a junior synonym of Endothiodon [156]). More recently, Angielczyk et al. [155] considered that the common presence of the dicynodonts Dicynodon huenei and possibly Katumbia parringtoniallow a direct correlation between the faunistic associations of the Usili Formation and the Zambian Upper Madumabisa Mudstone. As a result, the well-supported correlation of the Upper Madumabisa Mudstone with the rocks of the Cistecephalus Assemblage Zone in the South African Karoo Basin implies that the Usili Formation can be considered a lateral equivalent of the Cistecephalus Assemblage Zone [155], constrained to the middle–late Wuchiapingian (ca. middle Late Permian, 260–255 Ma [157]).

Several authors commented on the phylogenetic relationships of UMZC T836 following the original description of Parrington [9]. However, a detailed redescription, illustrations and comparisons of the specimen are currently lacking. The unusual combination of archosauromorph-like features and amniote plesiomorphies recognized in UMZC T836 by Parrington [9] led us to revisit its anatomy and phylogenetic relationships, and allowed us to recognize this specimen as a new taxon, Aenigmastropheus parringtoni gen. et sp. nov.

Systematic Paleontology

AMNIOTA Haeckel, 1866 [158]

DIAPSIDA Osborn, 1903 [159] sensu Laurin (1991) [73]

SAURIA Gauthier, 1984 [72] sensu Gauthier et al. (1988) [57]

ARCHOSAUROMORPHA Huene, 1946 [160] sensu Dilkes (1998) [4]

?PROTOROSAURIA Huxley, 1871 [161] (new explicit definition)

Phylogenetic definition.

Protorosauria Huxley, 1871 [161] is a stem-based clade that includes all taxa more closely related to Protorosaurus speneri Meyer 1830 [98] than to Varanus komodoensis Ouwens 1912 [162] or Crocodylus niloticus Laurenti 1768 [163] (new definition).

?PROTOROSAURIDAE Lydekker, 1888 [164] (new explicit definition)

Phylogenetic definition.

Protorosauridae Lydekker, 1888 [164] is a stem-based clade that includes all taxa more closely related to Protorosaurus speneri Meyer 1830 [98] than to Tanystropheus longobardicus Bassani 1886 [165], Prolacerta broomi Parrington 1935 [166], Sharovipteryx mirabilis (Sharov, 1971) [167], Drepanosaurus unguicaudatus Pinna 1980 [168] or Varanus komodoensis Ouwens 1912 [162] (new definition).

Aenigmastropheus gen. nov.

urn:lsid:zoobank.org:act:354E966B-CDA9-4509-84F5-2F130E23B2B5

Aenigmastropheus parringtoni sp. nov.

urn:lsid:zoobank.org:act: 78DF791F-C4F4-4592-8C3E-D333C8C91E58

(Figures 10–15, 16B, 17A, D, 18A)

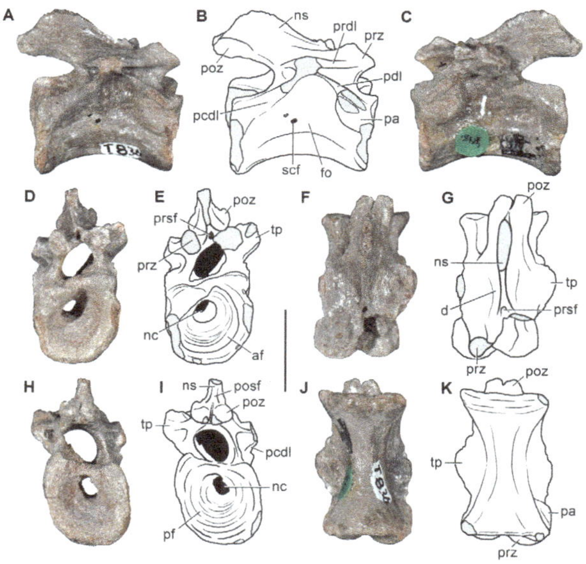

Figure 10. Aenigmastropheus parringtoni, an early archosauromorph from the middle Late Permian of Tanzania.

Cervical vertebra (vertebra 1 sensu Parrington [9]) (UMZC T836, holotype) in right lateral (A, B), left lateral (C), anterior (D, E), dorsal (F, G), posterior (H, I) and ventral (J, K) views. Abbreviations: af, anterior facet; d, depression; fo, lateral fossa; nc, notochordal canal; ns, neural spine; pa, parapophysis; pcdl, posterior centrodiapophyseal lamina; pdl, paradiapophyseal lamina; pf, posterior facet; posf, postspinal fossa; poz, postzygapophysis; prdl, prezygodiapophyseal lamina; prz, prezygapophysis; prsf, prespinal fossa; scf, subcentral foramen; tp, transverse process. Scale bar equals 1 cm.

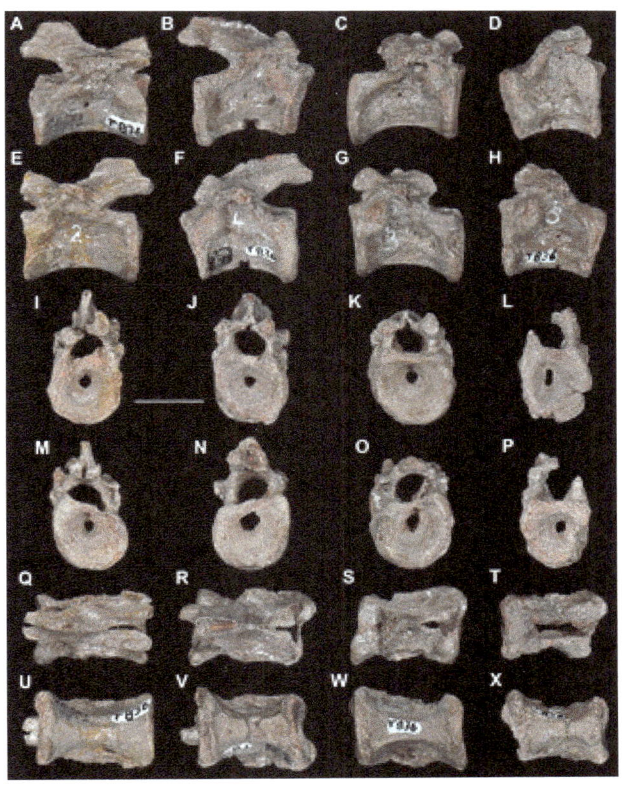

Figure 11. Aenigmastropheus parringtoni, an early archosauromorph from the middle Late Permian of Tanzania.

Cervico-dorsal vertebrae (UMZC T836, holotype) in right lateral (A–D), left lateral (E–H), anterior (I–L), posterior (M–P), dorsal (Q–T) and ventral (U–X) views. Vertebrae 2 (A, E, I, M, Q, U), 3 (C, G, K, O, S, W), 4 (B, F, J, N, R, V) and 5 (D, H, L, P, T, X) sensu Parrington [8]. Scale bar equals 1 cm.

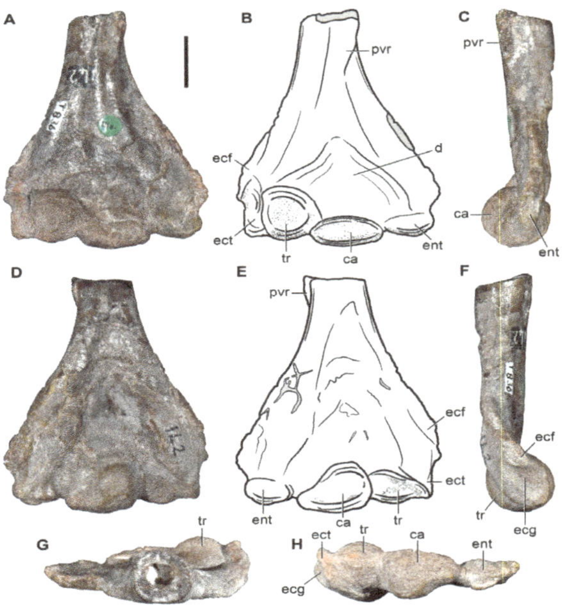

Figure 12. Aenigmastropheus parringtoni, an early archosauromorph from the middle Late Permian of Tanzania.

Distal half of the right humerus (UMZC T836, holotype) in ventral (A, B), posterior (C), dorsal (D, E), anterior (F), proximal (G) and distal (H) views. Abbreviations: ca, capitellum (radial condyle); d, depression; ecf, ectepicondylar flange; ecg, ectepicondylar groove; ect, ectepicondyle; ent, entepicondyle; pvr, posteroventral ridge; tr, trochlea (ulnar condyle). Scale bar equals 1 cm.

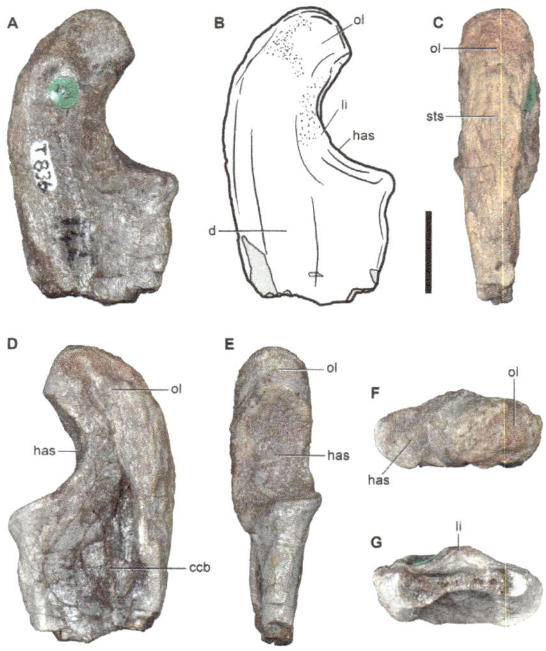

Figure 13. Aenigmastropheus parringtoni, an early archosauromorph from the middle Late Permian of Tanzania.

Proximal end of the right ulna (UMZC T836, holotype) in anterior (A, B), dorsal (C), posterior (D), ventral (E), proximal (F) and distal (G). Abbreviations: ccb, collapsed cortical bone; d, depression; has, humeral articular surface; li, lip; ol, olecranon; sts, striated surface. Scale bar equals 1 cm.

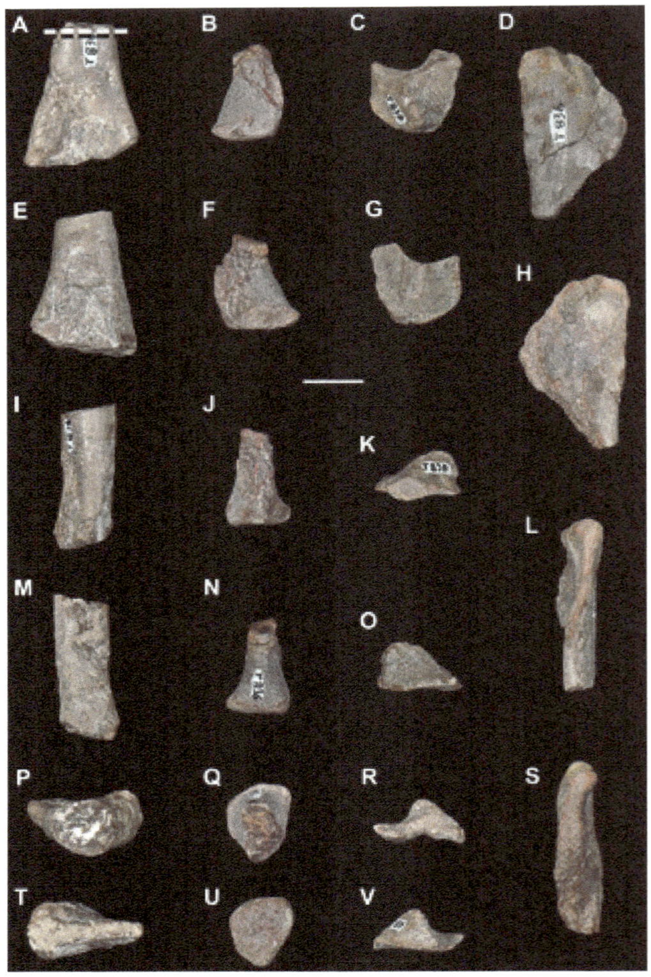

Figure 14. Aenigmastropheus parringtoni, an early archosauromorph from the middle Late Permian of Tanzania.

Possible left humeral shaft (A, E, I, M, P, T), possible distal end of radius (B, F, J, N, Q, U) and two indeterminate bones (C, D, G, H, K, L, O, R, S, V) (UMZC T836, holotype) in several views. The dashed line indicates the area sampled for the paleohistological study. Scale bar equals 1 cm.

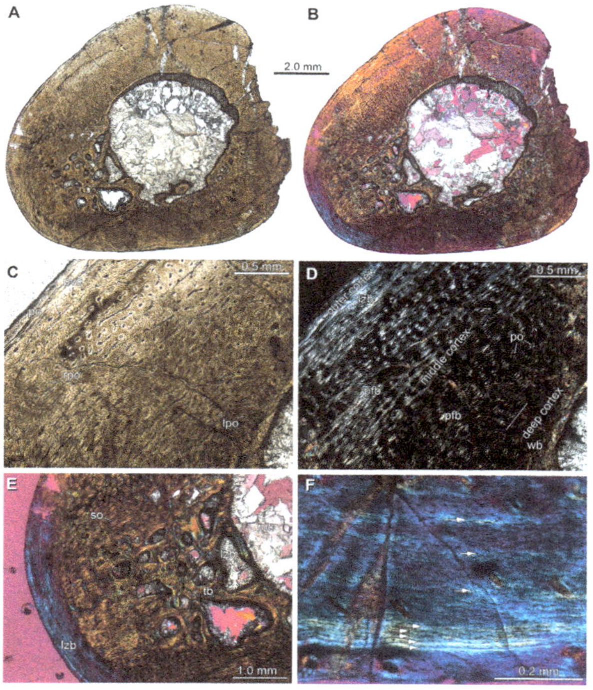

Figure 15. Aenigmastropheus parringtoni, an early archosauromorph from the middle Late Permian of Tanzania.

Paleohistological slices of possible left humerus (UMZC T836, holotype). Overview of the bone section in normal light (A) and cross-polarized light with lambda compensator (B); upper left quadrant showing the complete section of the cortex, in which the medullary cavity is to the lower right of the image, in normal (C) and cross-polarized light (D); lower left quadrant showing the complete section of the cortex in cross-polarized light with lambda compensator (E); and close-up of the outer part of the periosteal cortex in cross-polarized light with lambda compensator (F). White arrows in (F) indicate lines of arrested growth (LAGs). Abbreviations: lpo, longitudinal primary osteons; lzb, lamellar-zonal bone; mc, medullary cavity; pfb, parallel-fibred bone; po, primary osteons; pvc, primary vascular canal; rpo, reticular primary osteons; so, secondary osteons; tb, trabecular bone; wb, woven bone.

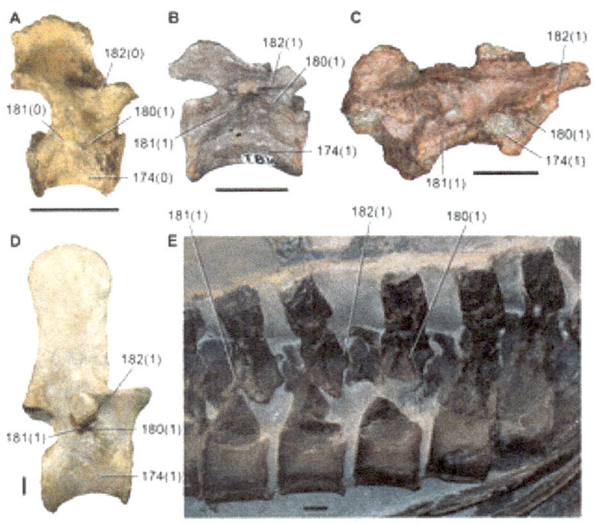

Figure 16. Character states supporting the phylogenetic positions of Aenigmastropheus parringtoni **and** Eorasaurus olsoni.

Anterior and middle dorsal vertebrae of Youngina capensis (BP/1/3859) (A), Tanystropheus longobardicus (SMNS 55341, reversed) (D) and Helveticosaurus(PIMUZ T4352, holotype, reversed) (E) and middle-posterior cervical vertebrae of Aenigmastropheus parringtoni (UMZC T836, vertebra 1, holotype, reversed) (B) and Eorasaurus olsoni (PIN 156/109) (C) in right lateral views. Scale bar equals 5 mm in (A) and 1 cm in (B–E).

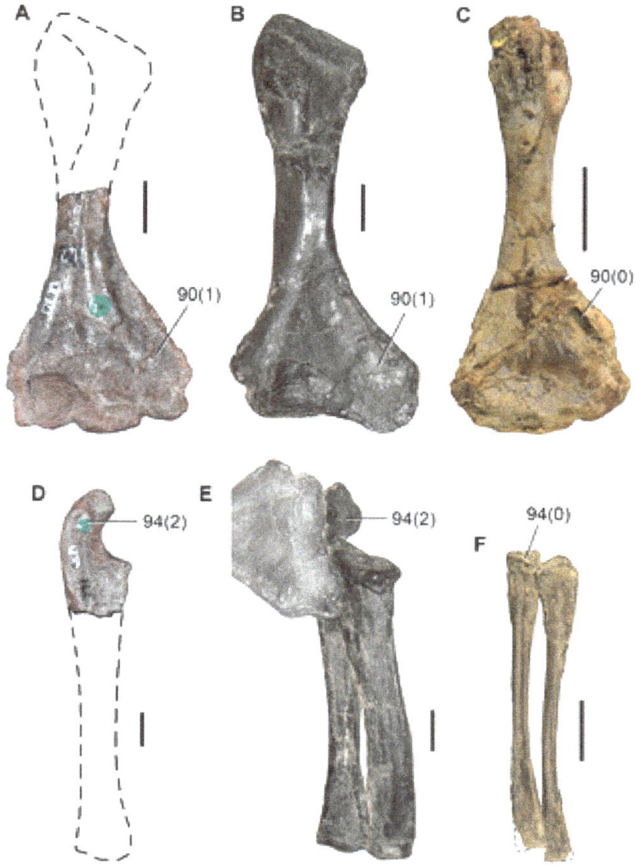

Figure 17. Character states supporting the phylogenetic position of Aenigmastropheus parringtoni.

Humeri of Aenigmastropheus parringtoni (UMZC T836, holotype) (A), Protorosaurus speneri (BSPG 1995 I 5, cast of WMSN P47361, reversed) (B) and Youngina capensis (BP/1/3859) (C) in ventral views; ulna of Aenigmastropheus parringtoni(UMZC T836, holotype) (D) and ulnae and radii of Protorosaurus speneri (BSPG 1995 I 5, cast of WMSN P47361) (E) and Macrocnemus bassanii (PIMUZ T4355) (F) in anterior views. Scale bar equal 1 cm in (A, B, D–F) and 5 mm in (C).

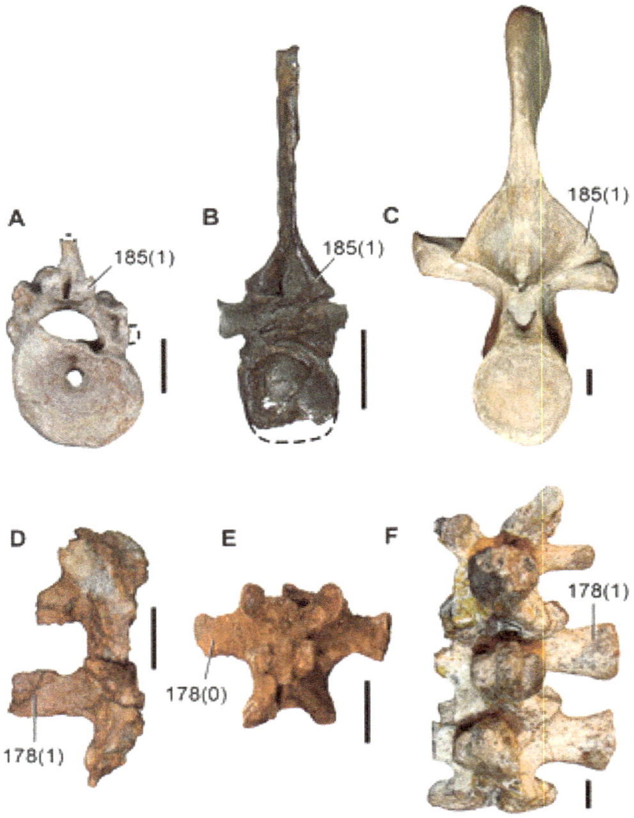

Figure 18. Character states supporting the phylogenetic positions of Aenigmastropheus parringtoni and Eorasaurus olsoni.

Cervical and dorsal vertebrae of Aenigmastropheus parringtoni (UMZC T836, vertebra 2, holotype) (A), Protorosaurus speneri (ZMR MB R2173) (B) and Tanystropheus longobardicus (SMNS 55341) (C) in posterior views, and cervico-dorsal vertebrae of Eorasaurus olsoni (PIN 156/110) (D), Prolacerta broomi(BP/1/2675) (E) and Proterosuchus fergusi (GHG 363) (F) in dorsal views. Scale bars equal 5 mm in (A, E) and 1 cm in (B, C, D, F).

Etymology

The generic name ("enigmatic vertebra") is derived from the Latin word aenigma(enigmatic) and the Greek word stropheus (vertebra) in allusion to the problematic taxonomic history of the holotype specimen. The specific name honors the British paleontologist Dr. F. R. Parrington for his contribution to the understanding of Permo-Triassic amniotes and his discovery and initial description of the holotype specimen.

Holotype

UMZC T836, partial postcranial skeleton including five posterior cervical–anterior dorsal vertebrae, distal half of the right humerus, fragment of probable left humeral shaft, proximal end of the right ulna, and three indeterminate fragments of bone (one of which may represent part of a radius) (Figs. 10–15).

Diagnosis

Aenigmastropheus parringtoni is a medium-sized archosauromorph saurian distinguished from other amniotes by the following combination of features: posterior cervical and anterior dorsal vertebrae notochordal, with well-developed anterior and posterior centrodiapophyseal and prezygodiapophyseal laminae, and sub-triangular neural spines in lateral view; humerus with a strong diagonal ridge on the anterior surface of the shaft (autapomorphy); humerus with strongly developed capitellum (radial condyle) and trochlea (ulnar condyle) and without entepicondylar and ectepicondylar foramina; ulna with strongly developed olecranon process forming a single ossification with the rest of the bone.

Locality

Locality B35 of Stockley, close to the road near Ruanda, Songea District, Ruhuhu Valley, southern Tanzania [9], [146] (Fig. 9).

Horizon and Age

Usili Formation (formerly the Kawinga Formation), deposited during the middle–late Wuchiapingian (middle Late Permian; [155]), Songea Group, Ruhuhu Basin.

Comments

Aenigmastropheus possesses a striking combination of features that complicates assessment of its phylogenetic relationships (see Discussion). However, these features also support its distinctiveness from other amniotes, including diapsids and "pelycosaur" synapsids. The combination of posterior centrodiapophyseal and prezygodiapophyseal laminae, zygapophyses that are positioned close to each other medially, and the absence of an entepicondylar foramen in the distal end of the humerus distinguish Aenigmastropheus from non-archosauromorph sauropsids. Aenigmastropheus differs from the enigmatic neodiapsid Helveticosaurus zollingeri due to the presence of low neural arches with subtriangular neural spines in the cervico-dorsal transition region, well-developed distal condyles of the humerus and a well-developed olecranon process on the proximal end of the ulna. Within Archosauromorpha, Aenigmastropheus differs from other basal members of the group due to the presence of notochordal vertebrae, a strongly developed olecranon process as part of a single ossification with the rest of the ulna (convergently acquired in some crown archosaurs) and a thick posteroventral ridge along the humeral shaft. The latter character is an autapopmorphy of Aenigmastropheus parringtoni among basal diapsids.

Anatomical description

Parrington [9] apparently considered all the bones catalogued as UMZC T836 to belong to a single individual. The five preserved vertebrae do indeed possess a congruent morphology and similar size, consistent with belonging to a single individual (Figs. 10, 11; Table 5). The right humerus and ulna are also consistent in size with belonging to a single individual, and the trochanter (ulnar condyle) of the humerus fits quite well when articulated with the proximal articular facet of the ulna (Figs. 12, 13; Table 6). However, assessing the assignment of the axial and appendicular elements to a single individual is less straightforward. The ratio between the posterior widths of the centra and the maximum distal width of the humerus ranges between 0.29–0.33 in Aenigmastropheus parringtoni. This range is very similar to or overlaps the ratio observed in basal reptiliomorphs (e.g. 0.31 in the diadectomorph Diadectes [169]), synapsids (e.g. 0.35–0.37 in the varanopid Varanops, TMM 43628-1 in Reisz et al. [43]; 0.27–0.29 in Ophiacodon mirus, FMNH (WM) 671 in Romer and Price [95]; 0.27–0.32 in Dimetrodon loomsi, FMNH (WM) 114 in Romer and Price [95]) and diapsids (e.g. 0.38–0.41 in the archosauriform Erythrosuchus africanus, SAM-PK-905). Accordingly, this ratio supports the interpretation of Parrington [9] that all the bones of UMZC T836 belong to a single individual.

Table 5. Measurements of the preserved axial bones of Aenigmastropheus parringtoni **nov. gen. et nov. sp. (UMZC T836) in millimeters.**

Vertebrae (sensu Parrington)	1	2	3	4	5
Centrum length	16.4	16.5	15.17	16.6	(15.7)
Anterior articular facet height	10.2	11.3	11.2	(11.3)	11.5
Anterior articular facet width	(9.9)	10.8	11.4	(11.0)	10.4
Posterior articular facet height	11.6	(10.9)	10.9	10.8	(10.4)
Posterior articular facet width	10.2	11.0	11.3	11.3	(9.8)
Maximum height of the vertebra	(19.8)	(18.6)	(17.4)	(19.5)	(16.9)
Length along zygapophyses	(19.5)	(19.1)	(12.7)	(17.7)	(10.3)

Values between brackets indicate incomplete measurements (due to post-mortem damage) and the value given is the maximum measurable. The length along the zygapophyses is the maximum anteroposterior length between the anterior tips of the prezygapophyses and the posterior tips of the postzygapophyses. Maximum deviation of the digital caliper is 0.02 mm but measurements were rounded to the nearest 0.1 millimeter.

Table 6. Measurements of the preserved forelimb bones of Aenigmastropheus parringtoni **nov. gen. et nov. sp. (UMZC T836) in millimeters.**

Humerus		
	Length	(46.4)
	Width close to mid-shaft	9.5
	Depth close to mid-shaft	10.0
	Perimeter close to mid-shaft	32
	Distal width	34.1
	Distal depth	10.5
	Width of entepicondyle	7.6
	Depth of entepicondyle	5.9
	Width of trochlea	12.0
	Depth of trochlea	9.2
	Width of capitellum	8.9
	Depth of capitellum	8.7
	Width of ectepicondyle	3.3
	Depth of ectepicondyle	8.2
Ulna		
	Length	(37.5)
	Length of articular facet for humerus	15.5
	Width of articular facet for humerus	10.0
	Depth at distal broken surface	15.9
	Width at distal broken surface	4.8

Values between brackets indicate incomplete measurements (due to post-mortem damage) and the value given is the maximum measurable. For the humerus the width is measured in the anteroposterior plane and the depth in the dorsoventral plane. Maximum deviation of the digital caliper is 0.02 mm but measurements were rounded to the nearest 0.1 millimeter. The perimeter close to mid-shaft was rounded to the nearest millimeter because the measurement cannot be made directly with the caliper.
doi:10.1371/journal.pone.0089165.t006

The bones are generally well preserved, but possess some damaged surfaces, some of which possibly result from preparation with acetic acid. In a few areas the cortical bone has collapsed or has broken away, and some degree of post-mortem distortion is evident in some elements (most notably in the vertebrae).

Five presacral **vertebrae** are preserved (Figs. 10, 11), and exhibit slight post-mortem distortion, with the left sides (e.g. zygapophyses and transverse processes) of some of the vertebrae having being displaced dorsally relative to the right side. Unfortunately, the prezygapophyses, transverse processes and neural spines are incomplete in all of the vertebrae. Parrington ([9]: fig. 1) designated the vertebrae of UMZC T836 as specimens "1"–"5" for descriptive purposes (hereafter referred as vertebrae 1–5, retaining Parrington's original numbering), and these numbers are written on the left lateral surfaces of the elements. Although we cannot assess with confidence whether or not the preserved vertebrae represent a continuous series, vertebra 4 at least articulates well at its posterior end with vertebra 3 and vertebra 3 at its posterior end with vertebra 5. Thus, it is likely that these three were continuous elements. Our interpretation of their order therefore differs from the numbering used by Parrington. In addition, all five of the vertebrae were likely close to one another within the axial series based on the similar positions of the parapophyses and the similar morphologies of the laminae of the neural arches. The parapophyses are placed primarily on the anterodorsal corners of the centra and extend only a short distance onto the base of the neural arch. Thus, as Parrington [9] noted, these vertebrae likely correspond to the region of the cervico–dorsal transition. Vertebra 1 is here interpreted as the most anterior preserved element because of the relatively ventral position of the parapophysis. Vertebrae 2 and 4 most likely successively followed vertebra 1, because they possess slightly more dorsally placed parapophyses and similarly elongated centra. The most posterior preserved elements seem to be vertebrae 3 and 5, with vertebra 5 the most posterior of the two. These elements exhibit more dorsally positioned parapophyses and likely correspond to anterior dorsal vertebrae.

All the preserved vertebrae possess a completely open notochordal canal (Fig. 10: nc), but the notochordal canal is still connected to the neural canal through an hourglass-shaped opening in at least some elements [9]. This is best observed in vertebra 5, in which most of the neural arch is broken away. The neurocentral sutures are closed, without any trace of the suture remaining visible on the external surface. Thus, UMZC T836 was likely not a juvenile individual on the basis of the closed neurocentral sutures, but possibly also not fully-grown based upon the connected notochordal and neural canals [9].

The centra are anteroposteriorly elongated, with a length to anterior height ratio of 1.61 in vertebra 1, 1.46 in vertebra 2, and 1.40 in vertebra 3 (Table 5), resembling the elongated cervical vertebrae of some basal synapsids (e.g. Apsisaurus witteri [73]) and several diapsids (e.g. Protorosaurus speneri, BSPG 1995 I 5; Araeoscelis gracilis [128]; Endennasaurus acutirostris [170]; Boreopricea funerea [105]; Trilophosaurus buettneri[44]). Vertebrae 1 and 2 have sub-rectangular centra in lateral view. By contrast, in vertebrae 3 and 4 the posterior articular surface terminates in a more ventral position than the anterior one in lateral view. This condition would likely have resulted in a dorsally curved cervico–dorsal transition in lateral view, when the vertebrae were articulated with one another. The condition in vertebra 5 cannot be determined due to extensive damage to the posterior end of the centrum. In ventral view, the centra are hourglass-shaped, being transversely constricted at mid-length, as commonly occurs in many amniote lineages [171]. In lateral view, the ventral margin of the centrum is arched upwards. None of the preserved vertebrae possess a transversely thin median ventral keel, contrasting with the condition commonly found among the cervico–dorsal vertebrae of some basal parareptiles (e.g. Millerosaurus pricei [129]; Procolophon trigoniceps [135]), some "pelycosaurian" synapsids (e.g. Apsisaurus witteri [73]; Varanops brevirostris [172]), araeoscelidians (e.g. Petrolacosaurus kansensis [131]; Araeoscelis gracilis [128]) and some basal archosauromorphs (e.g. Trilophosaurus buettneri [44]; Euparkeria capensis, SAM-PK-5867). Instead, the ventral surface of the centrum is flattened in vertebrae 3 to 5, with subtle longitudinal ridges laterally delimiting these planar ventral surfaces. In vertebrae 1 and 2 this flattening of the ventral surface is less well developed.

The ventral half of the lateral surface of the centrum is planar to very gently concave in all the vertebrae; more dorsally, the centrum possesses a deeper longitudinal fossa without well-defined margins, directly below the inferred position of the neurocentral suture (Fig. 10: fo). The presence of lateral fossae in the vertebral centrum has been also described for the dorsal vertebrae of probable choristoderan basal diapsids (e.g. Pachystropheus rhaeticus[173]) and the presacral vertebrae of numerous archosauriforms (e.g. Koilamasuchus gonzalezdiazi [174]; Tarjadia ruthae [175]; Erythrosuchus africanus, NHMUK R3592; Euparkeria capensis, UMZC T692j; Cuyosuchus huenei, MCNAM 2669; Pseudopalatus mccauleyi [110]; Arizonasaurus babbitti [176]; Aetosauroides scagliai [177]; Marasuchus lilloensis, PVL 3870; Pantydraco caducus [178]).

Parrington [9] described foramina that pierce the lateral faces of the centra (Fig. 10: scf). These foramina lie within the lateral fossae in UMZC T836. A pair of foramina seems to be present on the right side of vertebra 1, but only the ventral margin of the most ventral foramen can be confidently identified as a natural border. Furthermore, it should be noted that Parrington ([9]: fig. $1A_1$) only figured one foramen for this vertebra, which corresponds to the more ventral of the two openings currently present. Accordingly, the second "foramen" currently observed on vertebra 1 seems to be a break in the lateral surface of the centrum. The foramen of vertebra 1 is positioned slightly posterior to the mid-length of the centrum. No foramen is observed within the lateral fossa on the left side of the same vertebra. In vertebra 2 a well-defined circular foramen with well-preserved natural borders is present on the right side of the element, being of similar size and placed in the same position to that of vertebra 1. A circular foramen with well-defined natural borders is present on both right and left sides of vertebra 3, and is identical in position to the foramina of the more anterior elements, but the foramen on the left side is considerably smaller than that on the right side. In vertebra 4, a foramen is also present in the right side in an identical position to the foramina of the other vertebrae, but the pair of foramina figured by Parrington [9] on the left side of this vertebra cannot be identified and the surface of the left side seems instead to be devoid of foramina. Finally, a foramen is observed on the left side of vertebra 5, but the condition on the right side cannot be assessed due to damage. Accordingly, the foramina are irregular in their occurrence on each vertebra [9], but usually at least one foramen is present on at least one side of the centrum. The positions, sizes and shapes of the foramina are similar through the preserved vertebrae.

Parrington [9] considered that these foramina would have had a nutritive function and were related to the persistence of the notochord. A foramen of similar shape and position to that of Aenigmastropheus parringtoni is present and termed a subcentral foramen in several tetrapods, such as amphiumid caudatans (e.g. Amphiuma means [179]), "younginiforms" (e.g. Acerosodontosaurus piveteaui [70]), sauropterygians (e.g. Bobosaurus forojuliensis[180]), and basal lepidosauromorphs (e.g. Gephyrosaurus bridensis [48]). Nevertheless, subcentral foramina occur in amniotes with both notochordal and non-notochordal vertebrae (e.g. sauropterygians [180]). Similar circular vascular foramina are also present in the non-notochordal vertebrae of extant lepidosaurs and crocodiles [114], [180]–[182], which sometimes occur within a lateral fossa that is associated with fat deposits (e.g. Crocodylus acutus [182]). Accordingly, the nutritive function for the subcentral foramina in Aenigmastropheus parringtoni proposed by Parrington [9] seems reasonable but their proposed association with the notochordal canal is ambiguous.

The anterior and posterior articular surfaces of the vertebrae are roughly sub-circular in outline, being slightly higher than wide in vertebra 2 and 5, and slightly wider than tall in vertebra 3. In all the vertebrae both anterior and posterior articular surfaces are strongly amphicoelous and the centrum is completely pierced by a notochordal canal positioned slightly dorsal to the center of the centrum in anterior or posterior view (Fig. 10: nc). The presence of a notochordal canal in Aenigmastropheus parringtoni resembles the condition observed in several basal reptiliomorphs (e.g. Tseajaia campi [130]), parareptiles (e.g. Millerosaurus pricei [129]; Procolophon trigoniceps [135]), basal synapsids (e.g. Casea broilii, Archaeothyris florensis, Ophiacodon sp., Mycterosaurus longiceps, Mesenosaurus romeri, Aerosaurus wellesi, Varanops brevirostris, Varanodon agilis, Archaeovenator hamiltonensis, Apsisaurus witteri [73], [43]: Appendix S2, [95], [172]), basal sauropsids (e.g. Captorhinus aguti [183]; Coelurosauravus jaekeli [132]; Petrolacosaurus kensensis [131]; Araeoscelis gracilis [128]; Acerosodontosaurus piveteaui [70]; Youngina capensis, BP/1/3859), the enigmatic neodiapsids Helveticosaurus zollingeri [133] and Hypuronector limnaios [134], and basal lepidosauromorphs (e.g. Gephyrosaurus bridensis [48]; Planocephalosaurus robinsonae [119]). The presence of a notochordal canal was described in the basal archosauromorph Jesairosaurus lehmani [74], but unequivocal evidence for the presence of this feature could not be identified during direct restudy of the specimens of this taxon (in the holotype, ZAR 06, the center of an anterior dorsal centrum that is exposed in cross-section is filled by quartz crystals, whereas in ZAR 13 the exposed section of the dorsal vertebra possesses several trabeculae, but no notochordal canal).

The dorsal borders of the anterior articular surfaces of vertebrae 1, 3 and 4 possess a subtle notch at the midline, whereas in vertebrae 2 and 5 this border is not preserved due to damage. The centra do not possess beveled surfaces or facets for articulation with intercentra at the ventral margins of the articular surfaces, but it cannot be assessed with certainty whether intracentra were present or not. The parapophyses are not raised on peduncles, and they are oval, with the long axis being orientated posterodorsally in lateral view. The articular surfaces of the parapophyses are slightly concave and bounded by thick lips.

The most complete neural arches are preserved in vertebrae 1, 2 and 4. Vertebra 3 preserves the pedicles of the neural arch and the base of the prezygapophyses, whereas in vertebra 5 only parts of the neural arch pedicles are available. In all of the available vertebrae of Aenigmastropheus parringtoni, the dorsoventral height of the neural arch, from its base up to the base of the neural spine, is lower than that of the centrum. A similar condition is usually found in the posterior cervical and anterior dorsal vertebrae of basal amniotes, such as the synapsids Apsisaurus witteri [73] and Varanops brevirostris [172], and the diapsids Araeoscelis gracilis [128], Petrolacosaurus kensensis [131], Youngina capensis (BP/1/3859), Prolacerta broomi (BP/1/2675), Tanystropheus longobardicus(PIMUZ T2817), Macrocnemus bassanii (PIMUZ T2472) and Protorosaurus speneri (BSPG 1995 I 5). By contrast, in the basal lepidosaurs Gephyrosaurus bridensis [48] and Planocephalosaurus robinsonae [119] the neural arch is considerably higher than the centrum.

The neural canal of Aenigmastropheus parringtoni is oval in outline in the less-deformed vertebrae, being wider than tall. The base of the transverse process is anteroposteriorly short and dorsoventrally compressed, and did not extend onto the lateral margin of the prezygapophysis. A series of thin and well-developed laminae connect the transverse processes with other structures in all the preserved vertebrae [9]. A paradiapophyseal lamina connects the transverse process with the parapophysis (Fig. 10: pdl), whereas a posterior centrodiapophyseal lamina extends posteroventrally from the transverse process towards the posterodorsal corner of the centrum (Fig. 10: pcdl). The posterior centrodiapophyseal lamina becomes thicker and lower posteriorly. The paradiapophyseal and posterior centrodiapophyseal laminae bound a shallow concave depression that is separated from the lateral fossa of the centrum by a gently convex surface, and which can be recognized as a centrodiapophyseal fossa [41]. A prezygodiapophyseal lamina connects the transverse process with the prezygapophysis (Fig. 10: prdl), and bounds together with the paradiapophyseal lamina a deep and sub-triangular prezygapophyseal centrodiapophyseal fossa that opens anterolaterally. The transverse process extends posteriorly as a thin ridge. However, this ridge cannot be identified as a postzygodiapophyseal lamina in a strict sense, because it does not reach the postzygapophysis (see Wilson [40]), but it is topologically equivalent with such a feature. The ridge extending from the posterior end of the transverse process and the posterior centrodiapophyseal lamina bound a deep and sub-triangular depression, which is located in the same position as the postzygapophyseal centrodiapophyseal fossa of archosaurs (see Wilson et al. [41]). This depression is considerably shorter anteroposteriorly than the prezygapophyseal centrodiapophyseal fossa.

The presence of well-developed laminae in the neural arch is also observed in some caudatans, basal synapsids, basal diapsids and several archosauropomorphs. For example, the ophiacodontid synapsid Ophiacodon sp. (MCZ 1426), the basal diapsids Acerosodontosaurus piveteaui (MNHN 1908-32-57) and Youngina capensis (BP/1/3859), and the basal archosauriforms Proterosuchus fergusi (SAM-PK-K140, GHG 363) and "Chasmatosaurus" yuani (IVPP V2719) have paradiapophyseal or anterior centrodiapophyseal laminae. The amphiumid caudatan Amphiuma means [179] has anterior and posterior centrodiapophyseal laminae (in Amphiuma the structures have been termed alar processes [sensu Gardner [179]] but they are considered here as topologically equivalent to amniote laminae, although not homologous). The varanopid synapsid Apsisaurus witteri ([73]: fig. 6; sensu Reisz et al. [43]) and the basal archosauromorphs Prolacerta broomi (BP/1/2675), Tasmaniosaurus triassicus (UTGD 54655) and Sarmatosuchus otschevi (PIN 2865/68) possess both prezygodiapophyseal and anterior centrodiapophyseal (or paradiapophyseal) laminae. The basal archosauromorph Trilophosaurus buettneri possesses prezygodiapophyseal and postzygodiapophyseal laminae (S. Nesbitt pers. comm. 2013; USNM mounted skeleton, MDE pers. obs.). Finally, the paradiapophyseal or anterior centrodiapophyseal, posterior centrodiapophyseal and prezygodiapophyseal laminae are all present, as in Aenigmastropheus parringtoni, in the enigmatic neodiapsid Helveticosaurus zollingeri (PIMUZ T4352), the basal archosauromorphs Tanystropheus longobardicus (SMNS 54628, [111]: fig. 52–54), Protorosaurus speneri (BSPG 1995 I 5), Spinosuchus caseanus [112] and Macrocnemus bassanii (PIMUZ T2472, 4822; although the presence of a posterior centrodiapophyseal lamina cannot be determined in this species), the basal archosauriforms Erythrosuchus africanus (NHMUK R3592 [99]), Garjainia prima (PIN 2394/5-16), Shansisuchus shansisuchus ([184]: fig. 21) and Euparkeria capensis (UMZC T921), and several archosaurs (e.g. Bromsgroveia walkeri [114]; Hypselorhachis mirabilis [115]; Silesaurus opolensis [116]; Herrerasaurus ischigualastensis, PVSJ 373).

Only the bases of the prezygapophyses are preserved in vertebrae 1–4, and the prezygapophyses are missing completely in vertebra 5. The prezygapophyses are upturned in lateral view, being anterodorsally directed in all the vertebrae (Fig. 10: prz), which is in agreement with the wide notch formed by the postzygapophyses and the posterodorsal corner of the centra in lateral view. The prezygapophyses possess a thin, medially developed ridge that runs along the ventromedial edge of the articular surface in vertebrae 1 and 2, and along the mid-height of the structure in vertebrae 3 and 4. As a result, the prezygapophyses of vertebrae 1 and 2 are L-shaped in cross-section and those of vertebrae 3 and 4 are T-shaped. These ridges constrain transversely the space between the prezygapophyses at the midline to an extremely narrow longitudinal notch. The most complete prezygapophysis, on the left side of vertebra 2, indicates that, at least in this vertebra, the prezygapophyses extended anteriorly beyond the anterior margin of the centrum.

The postzygapophyses are short in all the preserved vertebrae and poorly laterally distinguished from the base of the neural spine (Fig. 10: poz), resembling the condition observed in most archosauromorphs (e.g. Protorosaurus speneri, ZMR MB R2173; Prolacerta broomi, BP/1/2675; Proterosuchus fergusi, GHG 231) and some "pelycosaurian" synapsids (e.g. Varanops brevirostris [172]; Ophiacodon sp., MCZ 1426). By contrast, the zygapophyses are laterally deflected from the rest of the neural arch in most basal sauropsids and lepidosaurs, such as Araeoscelis gracilis [128], Petrolacosaurus kansensis[131], Youngina capensis (BP/1/3859), Gephyrosaurus bridensis [48] and Planocephalosaurus robinsonae [119]. In vertebrae 1 and 2 the postzygapophyseal articular facets are mainly laterally facing, with only a slight ventral orientation [9]. This orientation of the postzygapophyseal articular facets would have mostly prevented lateral movement between the vertebrae in that region of the axial skeleton, which would probably correspond to the posterior end of the neck. By contrast, in vertebra 4 the articular facet of the preserved left postzygapophysis faces ventrolaterally, which would have allowed both lateral and dorsoventral movements among the vertebrae of this region, probably corresponding to the anterior end of the trunk. In vertebrae 1, 2 and 4 the postzygapophyses extend posteriorly beyond the posterior end of the centrum, but the condition cannot be assessed in vertebrae 3 and 5. The postzygapophyses are separated from one another along the posterior midline by a deep vertical furrow that does not extend dorsally along the neural spine.

The neural arches of vertebrae 1, 2 and 4 do not possess the depressions on both sides at the base of the neural spine (Fig. 10: d) that are usually observed in varanopid "pelycosaurs" [172], [185], [186], araeoscelidians [128], [131] and some archosauromorphs (e.g. Protorosaurus speneri, BSPG 1995 I 5; Mesosuchus browni, SAM-PK-6046; Prolacerta broomi, BP/1/2675; Proterosuchus fergusi, GHG 231). The bases of the neural spines are anteroposteriorly long, extending from the base of the prezygapophyses up to a point just anterior to the posterior ends of the postzygapophyses. The preserved regions of the neural spines of vertebrae 1 and 2 suggest that the anterior portion of the anterior margin possessed a low angle to the long axis of the vertebra up to a point approximately level with the mid-length of the centrum. Beyond this point, the anterior margin became sharply upturned to form the neural spine (Fig. 10: ns). Unfortunately, the total height of the neural spine and the morphology of its distal end cannot be assessed in any of the preserved vertebrae.

The microstructure of the vertebrae can be observed due to breakages and damaged surfaces and its pattern varies in different regions of the axial elements. The articular face of the centrum possesses multiple layers of bone laminae distributed in a concentric pattern that follows the outline of the notochordal canal. These bone layers may represent successive sequences of ossification of the vertebral centrum along the internal surface of the notochordal canal, implying at least partial reabsorption of the notochord during ontogeny. Within the external borders of the centrum and along the neural arch a typical trabecular bone microstructure is visible.

The distal half of a right **humerus** is preserved, and has a well-preserved external bone surface (Fig. 12). The preserved portion of the shaft is roughly oval in cross-section, being slightly dorsoventrally deeper than anteroposteriorly wide (Table 6). The outline of the shaft in cross-section is asymmetric, with convex ventral, dorsal and anterior margins. The posterior margin is slightly sigmoid due to the presence of a posterior depression extending proximodistally along the shaft that is ventrally bounded by a thick ridge (Fig. 12: pvr). This ridge rises from the central portion of the ventral surface of the shaft and extends more posteriorly towards its proximal end. It becomes thicker towards the proximal end of the bone; distally, the ridge does not reach the distal end of the humerus but fades out on the ventral surface. The posteroventral ridge observed in the humerus of Aenigmastropheus parringtoni does not appear to be homologous with that of basal synapsids (e.g. dicynodonts [151]), which connects the deltopectoral crest with the entepicondyle, because in basal synapsids the ridge is directed anteriorly towards its proximal end. A similar ridge is not present in basal diapsids (e.g. Petrolacosaurus kansensis [131]; Araeoscelis gracilis, MCZ 4383; Youngina capensis, BP/1/3859; Protorosaurus speneri, BSPG 1995 I 514; Prolacerta broomi, BP/1/2675). This ridge is identified here as autapomorphic for Aenigmastropheus parringtoni.

The distal end of the humerus of Aenigmastropheus parringtoni is strongly anteroposteriorly expanded, being around 3.6 times wider than the proximal end of the shaft at the point at which it is broken, resembling the condition in several amniotes (e.g. Barasaurus besairiei[187]: fig. 3d; Millerosaurus pricei [129]; Varanosaurus acutirostris, Dimetrodon kempae[95]; Dicynodontoides spp. [151]; Captorhinus aguti [188]; Araeoscelis gracilis, MCZ 4383; Boreopricea funerea [105]; Trilophosaurus buettneri [44]). In addition, the most proximal portion of the preserved shaft indicates that the bone was still tapering proximally and the ratio of the anteroposterior width of the distal end to the minimum shaft width would have been even greater than currently preserved. By contrast, the dorsoventral thickness of the distal end of the bone is less than that of the shaft.

The distal end of the humerus possesses four well-developed and distinct distal articular condyles, which represent the entepicondyle, ectepicondyle, capitellum (radial condyle) and trochlea (ulnar condyle) (Fig. 12: ca, ect, ent, tr). The presence of four well-developed distal articular condyles resembles the condition in several basal amniotes (e.g. parareptiles [129], [187]; basal synapsids [95], [151], [189]; captorhinids [188]; basal diapsids [190]) and the basal archosauromorph Trilophosaurus buettneri [44]. Protorosaurus speneri possesses at least three distinct distal condyles, which as preserved are considerably less well developed those of Aenigmastropheus parringtoni. Nevertheless, it is likely that the degree of development of the distal humeral condyles of Protorosaurus speneri is underestimated due to the strong compression that specimens suffered during fossilization (e.g. BSPG 1995 I 5, BSPG AS VII 1207). In other archosauromorphs only the ulnar and radial condyles are distinctly developed (e.g. Boreopricea funerea [105]; Malutinisuchus gratus [191]; Tanystropheus longobardicus [111]; Mesosuchus browni [4]; Proterosuchus fergusi [90]; Erythrosuchus africanus [113]; Euparkeria capensis [192]).

The surfaces of the distal articular condyles are porous and covered by low striations, indicating that they were probably covered by hyaline cartilage that participated in a synovial elbow joint. The ventral surface of the humerus, proximal to the distal condyles, possesses a complex topography. Two ridge-like convexities extend from the shaft in an inverted Y-shaped pattern. The thinner convexity is anterodistally directed and contacts the base of the ectepicondyle, whereas the thicker convexity is posterodistally directed and almost reaches the base of the entepicondyle. Both convexities and the capitellum and trochlea define a sub-triangular depressed area (Fig. 12: d), the depth of which appears to be exaggerated by damage and partial collapse of the cortical bone. This ventral depression would have housed the attachment area for the antebrachial ligaments (see Angielczyk et al. [151]). Both ectepicondylar and entepicondylar foramina are absent from the distal end of the humerus of Aenigmastropheus parringtoni. Either both foramina, or the entepicondylar foramen alone, occur widely among amniotes (e.g. parareptiles [129], [187]; basal synapsids [95], [151], [189]; captorhinids [188]; basal diapsids [138]). By contrast, the absence of both foramina from the humerus of Aenigmastropheus parringtoni is a condition shared with the enigmatic neodiapsid Helveticosaurus zollingeri [133] and almost all archosauromorphs (e.g. Mesosuchus browni [4]; Hyperodapedon gordoni [193]; Prolacerta broomi, BP/1/2675; Tanystropheus longobardicus [111]; Macrocnemus bassanii, PIMUZ T4355; Protorosaurus speneri [21]; Boreopricea funerea [105]; Trilophosaurus buettneri[44]; Proterosuchus fergusi [90]; Erythrosuchus africanus [113]; Euparkeria capensis, SAM-PK-5867), with the exception of the putative protorosaurs Czatkowiella harae, which possesses an entepicondylar foramen [46], and Jesairosaurus lehmani, which possesses an ectepicondylar foramen (ZAR 09; [74]).

The anterior margin of the distal end of the humerus, above the ectepicondyle, has a prominent supinator ridge or ectepicondylar flange (Fig. 12: ecf), which was likely the area of origin of the M. supinator (e.g. [193]). The posterior margin of the distal end of the bone is not completely preserved, but the available portion is very thin and suggests that only a small portion of bone has been lost. This posterior margin is folded ventrally, and delimits together with the thicker of the ventral convexities a proximodistally-extending concave depression that terminates a substantial distance from the base of the entepicondyle. The dorsal surface of the distal end of the humerus is mostly occupied by a large and sub-triangular depression that would have borne the origin of the M. triceps humeralis medialis(see Angielczyk et al. [151]). The depth of the central and proximal areas of this depression is exaggerated by damage with collapse of cortical bone. A raised shelf of bone that is continuous with the ectepicondyle delimits the anterior border of this dorsal depression and possesses a slightly rugose dorsal surface that would have housed the area for origin of the antebrachial extensor muscles (see Angielczyk et al. [151]). The posterior border of the depression is bounded by a faint posterodistally extending convexity that is continuous with the entepicondyle. The surface of the bone posterior to this convexity is slightly rugose, representing the area of origin of the antebrachial flexor muscles (see Angielczyk et al. [151]).

The ectepicondyle is the smallest of the distal condyles and is restricted to a rounded structure with a dorsoventrally oriented main axis. The anterior surface of the ectepicondyle possesses a shallow teardrop-shaped depression that corresponds to the ectepicondylar groove [9] for the passage of the radial nerve [95] (Fig. 12: ecg). This feature is widely distributed among amniotes (e.g. Apsisaurus witteri [73]; Varanops brevirostris [172]; Dimetrodon sp., SAM-PK-K8670; Youngina capensis, BP/1/3859; Prolacerta broomi, BP/1/2675; Erythrosuchus africanus [113]) and the ectepicondylar groove of Aenigmastropheus parringtoni is considerably less well-developed than in some basal archosauromorphs (e.g. Tanystropheus conspicuous [111]; Trilophosaurus buettneri [44]: figs. 66–68; Otischalkia elderae [194]), in which the passage is represented by a deep notch.

The capitellum (Fig. 12: ca), which articulated with the radius, is a ball-shaped, proximoventrally-projecting condyle.

The trochlea is the widest and deepest of the condyles of the distal end (Fig. 12: tr). It possesses a slightly convex ventral surface and a strongly convex dorsal surface that articulated with the U-shaped proximal articular surface of the ulna. In distal view, the trochlea has a comma-shaped outline, with a concavity on its posterodorsal margin. The entepicondyle (Fig. 12: ent) is smaller than the capitellum and trochlea, but larger than the ectepicondyle. It has an overall morphology that is very similar to that of the trochlea and its convex surface is dorsally oriented. The posterior surface of the entepicondyle possesses a small and deep oval concavity.

A fragment from the shaft of a long bone is interpreted as part of a probable left humerus (Fig. 14A, E, I, M, P, T). One of its ends is roughly sub-circular in cross-section and the opposite end is strongly compressed. The compressed end appears to have collapsed cortical bone on both of its main surfaces; it can thus be inferred to have had an originally sub-triangular cross-section. The size and shape of the end with a sub-circular cross-section is very similar to that of the shaft of the partial right humerus. Although this fragment of bone could belong to part of the left humeral shaft, it might also represent a fragment of tibial or femoral shaft.

The proximal end of the right **ulna** is preserved (Fig. 13). It has a well-preserved external bone surface, but the cortical bone on the posterior surface of the element distal to the olecranon process has collapsed, probably exaggerating the anteroposterior compression of the shaft (Fig. 13: ccb). This collapse is also evident in the constriction of the medullary space of the bone on the broken cross-section of the shaft (Fig. 13G). The ulna has a very large olecranon process (Fig. 13: ol), resembling the condition of some basal synapsids (e.g. Ophiacodon navajovicus, Edaphosaurus boanerges [95]; Dinodontosaurus turpior[189]), basal sauropsids (e.g. Captorhinus aguti [183]; Thuringothyris mahlendorffae [195]), the basal archosauromorphs Protorosaurus speneri [21] and Trilophosaurus buettneri [44], and several archosaurs (e.g. Typothorax coccinarum [196]; Fasolasuchus tenax, PVL 3850; Saturnalia tupiniqium, MCP 3845-PV, [197]; Eodromaeus murphi, PVSJ 560). By contrast, the olecranon process is poorly developed or absent in varanopid synapsids [172], “younginiforms” (e.g. Youngina capensis [64]; Acerosodontosaurus piveteaui [70]), the neodiapsid Helveticosaurus zollingeri (PIMUZ T4352), and most basal archosauromorphs (e.g. Prolacerta broomi [64]; Tanystropheus longobardicus, PIMUZ T2817; Macrocnemus bassanii, PIMUZ T4822; Mesosuchus browni [4]).

The olecranon process of Aenigmastropheus parringtoni tapers gradually towards its apex, contrasting with that of at least some “pelycosaurian” synapsids, in which the process possesses an anteroposterior expansion towards its apex (e.g. Dimetrodon sp., SAM-PK-K8670; Ophiacodon sp., MCZ 1426). There are no traces of sutures in the olecranon of Aenigmastropheus parringtoni; thus, it seems that this structure formed a single ossification with the ulna. By contrast, in the basal archosauromorph Protorosaurus speneri the olecranon is developed as a separate ossification from the rest of the ulna [21].

The proximal bone surface of the olecranon process is damaged. The olecranon process curves ventrally in order to form a U-shaped, large, oval and deeply concave ventral articular facet for the reception of the humeral trochlea (Fig. 13: has). The surface of this facet is porous and covered by faint striations, indicating that it was covered by hyaline cartilage in life. A thick lip bounds the anterior margin of the articular surface (Fig. 13: li). Dorsal to this lip, the olecranon has a concave longitudinal depression that extends distally along the anterior surface of the ulnar shaft (Fig. 13: d). The dorsal surface of the olecranon process is strongly convex and possesses a series of roughly longitudinal striations that represent the scars of the area of insertion of the Mm. triceps (Fig. 13: sts). The posterior surface of the olecranon possesses a longitudinal furrow, but this seems to be an artifact resulting from the collapse of the cortical bone. A lip also bounds the posterodistal border of the articular surface, but this lip is considerably lower than the anterior one. The ulnar shaft is strongly anteroposteriorly compressed, and the anterior and posterior longitudinal depressions (although the posterior depression has been exaggerated by damage) give the shaft a figure-of-eight-shaped cross-section at the point at which it is broken.

Parrington [9] included the end of a bone with a planar articular surface among the indeterminate elements of UMZC T836, and this is here identified as a possible **radius** (Fig. 14B, F, J, N, Q, U). Although one of the borders of the bone has broken away, the preserved portion of the shaft suggests that it was oval in cross section when complete. The bone clearly does not belong to a humerus, femur or tibia because it is proportionally too small. It does not appear to represent the distal end of an ulna because distal ulnae are usually considerably more compressed in amniotes. However, the morphology of this bone matches to that of a distal radius, particularly in possessing an expanded medial border, an approximately planar distal articular surface, and well-developed scars for probable ligament attachments. Indeed, the morphology of the distal articular surface in distal view is very similar to and matches that of the right radius of some amniotes (e.g. Captorhinus aguti[188]), with both tapering and more rounded borders. Nevertheless, the identity of this fragmentary bone cannot be established with certainty, nor can it be determined to which side it belonged.

Parrington [9] also considered two other fragments of bone within UMZC T836 as indeterminate elements. One of the elements is a flattened and plate-like element (Fig. 14D, H, L, S). Four distinct borders can be recognized when the element is viewed perpendicular to its main plane. Two of these borders are broken margins, whereas the other two are damaged but maintain their overall shape. A short convex margin possesses a sub-triangular concavity that likely formed an articular area. A straight margin extends away from this short convex margin, and gradually increases in thickness, forming a low rounded tuberosity. One of the main surfaces of the bone is convex and the other one concave. This bone might represent part of a thickened cranial bone or a pectoral/pelvic girdle element.

A small, approximately pyramidal bone is also included among the indeterminate bones listed by Parrington [9] (Fig. 14C, G, K, O, R, V). The bone is comma-shaped when it is viewed perpendicular to its main plane, but one of the tapering ends of the element in this view is broken away. Thus, the bone would likely have had an “L”-shaped morphology when it was complete. One of the main surfaces of the element is planar, whereas the opposite surface bears a high and asymmetrically placed keel formed by two concave surfaces, one deeper than the other. The textures of the planar and the deepest concave surfaces suggest that they were articular surfaces. The remaining shallow concavity appears to be a non-articular surface. The morphology of this bone does not agree with that of a proximal end of a metacarpal or metatarsal because it does not have any process with a circular or oval cross section that could represent a shaft and the positions of the articular surfaces do not match the morphology expected for a metapodial. The size of the bone is very large relative to that expected for a carpal or distal tarsal bone. As such, we are unable to provide an identification for this element.

Histological description

Two thin section slices were made of the fragment of shaft of the possible left humerus (Fig. 14A). If this bone does not actually represent a humerus, it would instead represent another limb bone with the same kind of preservation as the remaining elements of UMZC T836. As described above, the cross-section of the bone is ovoid to triangular-shaped, but part of the outer cortex is damaged in one quadrant (diameter ranging from 9.1 to 10.2 mm, the latter value is based on reconstructed cortex) (Fig. 15A, B). The thickness of the periosteal cortex ranges between 1.8 and 4.0 mm. The section reveals a circular medullary cavity (4.3 to 4.7 mm in diameter), which is placed slightly off center. Few bone trabeculae are present deep within the cortex, and are present mainly in the quadrant where the cortical bone is thickest. Otherwise the interior of the medulla is devoid of trabeculae.

The cortex consists of fibrolamellar bone in the deepest parts, with a woven bone matrix that is vascularized by circumferentially arranged longitudinal primary osteons (Fig. 15C, D). In the middle part of the cortex, the woven-fibered matrix is substituted to a large degree by parallel-fibered bone in the middle parts of the cortex, although a few interspaced, thin layers of woven bone do occur. Although longitudinally arranged primary osteons constitute the majority of the vascular system, reticular patterns are locally developed. The outer part of the cortex is composed of lamellar-zonal bone, which is weakly vascularized by scattered primary osteons and more radially oriented short simple vascular canals (Fig. 15C–F). In this part of the cortex, eight growth cycles are visible with the fourth to eighth growth mark (i.e. lines of arrested growth, LAG) being closely spaced just below the external bone surface (Fig. 15F: white arrows). The first (inner) three growth marks are less conspicuous and represent annuli instead of LAGs. Locally, the outermost cortical layer does not show growth marks.

In the areas of the bone where no trabecular bone has been developed, a secondary endosteal layer of lamellar bone lines the marrow cavity. The bone trabeculae are also mostly secondary in nature, consisting of lamellar bone, with only a few interstitial areas of primary bone matrix being preserved deep within some of the thicker trabeculae (Fig. 15E). Towards the more external parts of the cortex, the cavities decrease in size and only few scattered secondary osteons are present in the surrounding primary bone matrix. Sharpey›s fibers are rare in the section and restricted to few locations in the outermost parts of the cortex, indicating few localized areas of tendinous muscle attachment.

DISCUSSION

Phylogenetic position of Aenigmastropheus parringtoni

The position of Aenigmastropheus parringtoni within Archosauromorpha is supported by three unambiguous synapomorphies in our analysis, the presence of parallelogram-shaped centra, in which the anterior articular surface is placed dorsal to the posterior surface, in the cervico-dorsal vertebrae in lateral view (character 174, state 1, CI=0.5000; RI=0.8750; Fig. 16), prezygodiapophyseal lamina present in posterior cervical and anterior–middle dorsal vertebrae (character 182, state 1, CI=0.2500; RI=0.6667; Fig. 16), and the absence of an entepicondylar foramen in the distal end of the humerus (character 90, state 1, CI=1.0000; RI=1.0000; Fig. 17A–C).

An entepicondylar foramen is absent in almost all archosauromorphs, with exception of Czatkowiella harae [46]. However, the nature of the multitaxonomic bonebed that yielded Czatkowiella harae suggests that caution may be necessary in referring humeri with this morphology to Czatkowiella harae (the holotype of which is based on a right maxilla) rather than to a non-archosauromorph diapsid.

The presence of parallelogram-shaped centra in the cervico-dorsal vertebrae in lateral view is present in all the archosauromorphs sampled in this analysis (i.e. Aenigmastropheus parringtoni; Trilophosaurus buettneri [44]; Prolacerta broomi, BP/1/2675; Proterosuchus fergusi, NM QR 1484; Erythrosuchus africanus, NHMUK R3592; Tanystropheus longobardicus, PIMUZ T2817T, SMNS 54628, 54630; Euparkeria capensis, SAM-PK-5867; Protorosaurus speneri, BSPG 1995 I 5; Macrocnemus bassanii, PIMUZ T4355; Eorasaurus olsoni, PIN 156/108, 109) and was convergently acquired by the varanopid synapsid Apsisaurus witteri ([73]: fig. 6a).

A prezygodiapophyseal lamina in posterior cervical and anterior–middle dorsal vertebrae is present in Aenigmastropheus parringtoni and several archosauromorphs, including Trilophosaurus buettneri (S. Nesbitt pers. comm. 2013; MDE pers. obs. USNM mounted skeleton), Prolacerta broomi (BP/1/2675), Erythrosuchus africanus (NHMUK R3592), Tanystropheus longobardicus (PIMUZ T2817T), Euparkeria capensis (UMZC T921), Protorosaurus speneri (BSPG 1995 I 5), Macrocnemus bassanii (PIMUZ T4822) and Eorasaurus olsoni (PIN 156/108–110). This feature is convergently present in the varanopid synapsid Apsisaurus witteri ([73]: fig. 6a) and within Archosauromorpha it is interpreted as secondarily lost in rhynchosaurs (Noteosuchus colletti, AM 3591; Mesosuchus browni, SAM-PK-6046; Howesia browni, SAM-PK-5886) and Proterosuchus fergusi (NM QR 1484, SAM-PK-K140).

The position of Aenigmastropheus parringtoni within Archosauromorpha is better supported by two character states that optimize as synapomorphies of Protorosauria: dorsal vertebrae with anterior centrodiapophyseal lamina or paradiapophyseal lamina (character 180, state 1, CI=0.1667; RI=0.6154; Fig. 16), and with posterior centrodiapophyseal lamina (character 181, state 1, CI=0.5000; RI=0.8000; Fig. 16). The presence of this suite of laminae in the dorsal vertebrae is reconstructed as independently acquired in Protorosauria and in the Prolacerta+Archosauriformes clade because these laminae are absent in Trilophosaurusand rhynchosaurs. However, this optimization should be considered as a preliminary result because several archosauromorphs (e.g. Pamelaria dolichotrachela [198]; Spinosuchus caseanus [112]; Augustaburiania vatagini [108]) and the enigmatic neodiapsid Helveticosaurus zollingeri also possess laminae in the neural arches of their dorsal vertebrae and were not included in this analysis (Fig. 16E). For example, the position of Helveticosaurus zollingeri among neodiapsids may affect the optimization of the presence of laminae in the dorsal vertebra and the absence of an ectepicondylar foramen in the distal end of the humerus. Some authors favored a position of Helveticosaurus zollingeri as closely related to or included within sauropterygians, but outside Sauria [39], [199], [200], or nested within lepidosauromorphs as a sauropterygian [32]. If either of these positions is correct, and depending on the phylogenetic interrelationships within Lepidosauromorpha, the optimization of these character states could turn out to be an ambiguous synapomorphy of Sauria or an independent acquisition in archosauromorphs and Helveticosaurus zollingeri. By contrast, Rieppel [133] concluded that Helveticosaurus zollingeri should be considered a basal member of Archosauromorpha. In this hypothesis, these character states will likely be optimized as synapomorphies of Archosauromorpha and Aenigmastropheus parringtoni would probably be recovered as the most basal archosauromorph. In any case, the combination of the presence of strongly developed laminae in the dorsal vertebrae and the absence of an ectepicondylar foramen in the distal end of humerus is congruent with archosauromorph affinities for Aenigmastropheus parringtoni. The exact optimization of these characters will remain an open question until the development of a more robust consensus for diapsid interrelationships, a topic beyond the scope of this paper.

Among archosauromorphs, the position of Aenigmastropheus parringtoni as more closely related to Protorosaurus speneri than to any other member of the group is supported by the following synapomorphies: anterior and mid-dorsal vertebrae with zygapophyses placed close to one another medially (character 185, state 1, CI=0.2500; RI=0.6250; Fig. 18A–C), and ulna with a strongly developed olecranon process, being higher than its transverse depth at its base (character 94, state 2, CI=0.2500; RI=0.7143; Fig. 17D–F). The absence of zygapophyses that are mainly sagittally oriented in the dorsal vertebrae of Tanystropheus longobardicus (SMNS 54630), Macrocnemus bassanii (PIMUZ T4822), Trilophosaurus buettneri [44] and Mesosuchus browni (SAM-PK-6046) results in the optimization of this character state as independently acquired in the Aenigmastropheus+Protorosaurus and Prolacerta+Archosauriformes clades and in some ophiacodontid and varanopid "pelycosaurs". A strongly developed olecranon process occurs rarely among "younginiforms" and basal archosauromorphs (e.g. tanystropheids, rhynchosaurs, Prolacerta and basal archosauriforms). In Trilophosaurus the olecranon process is relatively well developed but to a lesser degree than in Aenigmastropheus parringtoni and Protorosaurus speneri.

Under constrained topologies only one additional step is necessary to position Aenigmastropheus parringtoni outside Archosauromorpha and as the sister-taxon of the "pelycosaurian" synapsid Ophiacodon (Templeton test, p-value=1.0000, non-significant [NS]). This result is not unexpected because of the fragmentary condition of the type specimen of Aenigmastropheus parringtoni, which can be scored for less than 10% of the characters of our data set, and previous authors have already highlighted similarities between the vertebrae of this taxon and those of "pelycosaurian" synapsids [144]. However, we still consider the position of Aenigmastropheus parringtoni within "pelycosaurian" synapsids to be unlikely because it possesses several features that are not congruent with the typical morphology of the group, such as the presence of posterior centrodiapophyseal and prezygodiapophyseal laminae and low and sub-triangular neural spines in the cervico-dorsal vertebrae, and the absence of both entepicondylar and ectepicondylar foramina on the distal end of the humerus.

Three additional steps are necessary to recover Aenigmastropheus parringtoni as a non-diapsid sauropsid or a very basal diapsid (Templeton test for the most basal neodiapsid hypothesis, p-value=0.5078 [NS]). Two additional steps are necessary to position Aenigmastropheus parringtoni as the most basal archosauromorph (Templeton test, p-value=0.6875 [NS]) and five for it to be positioned as the sister-taxon of Sauria (Templeton test, p-value=0.1797 [NS]). Accordingly, although the support for the archosauromorph position of Aenigmastropheus parringtoni is relatively low, based on the currently available information for the species and the MPTs obtained in our phylogenetic analysis we favor the hypothesis that Aenigmastropheus parringtoni is a Late Permian archosauromorph.

Phylogenetic position of Eorasaurus olsoni

Eorasaurus olsoni was included for the first time in a quantitative phylogenetic analysis and our recovery of this species within Archosauromorpha partially supports the original interpretation of Sennikov [89]. However, the position of Eorasaurus olsoni within Archosauriformes within our phylogenetic results differs from the phylogenetic hypothesis of Sennikov [89] that Eorasaurus olsoni was closely related to Protorosaurus speneri and, as a result, a member of Protorosauridae. The inclusion of Eorasaurus olsoni within Archosauriformes is supported by the presence of diapophyses in the anterior dorsal vertebrae that are wider than 65% of the length of the centrum (character 178, state 1, CI=0.2000; RI=0.2000; Fig. 18D–F).

The position of Eorasaurus olsoni as an archosauriform closer to Erythrosuchus africanusand Euparkeria capensis than to Proterosuchidae is supported by the presence of a posterior centrodiapophyseal lamina in the posterior cervical and dorsal vertebrae (character 181, state 1; Fig. 16). This character state is also present in protorosaurs (i.e. Aenigmastropheus parringtoni, Protorosaurus speneri, Tanystropheus longobardicus, Macrocnemus bassanii), but is absent in Trilophosaurus buettneri, rhynchosaurs, Prolacerta broomi and Proterosuchus fergusi. As a result, the present analysis optimizes independent acquisitions of a posterior centrodiapophyseal lamina for protorosaurs and the Eorasaurus+Erythrosuchus+Euparkeria clade (see above). Eorasaurus olsoni differs from protorosaurs in the presence of proportionally shorter cervical vertebrae and the presence of a very well developed, laterally projected diapophysis in anterior dorsal vertebrae. As a result, two extra steps are necessary to recover Eorasaurus olsoni within Protorosauria

(Templeton test, p-value=0.6250 [NS]). Only one additional step is required to position Eorasaurus olsoni as the sister-taxon of Archosauriformes (Templeton test, p-value=1.000 [NS]), three additional steps to be placed as the sister-taxon of Prolacerta broomi(Templeton test, p-value=0.2500 [NS]), four additional steps to be positioned at the base of Archosauromorpha (Templeton test, p-value=0.2187 [NS]), and five additional steps to be found outside Sauria, being nested within Varanopidae (Templeton test, p-value=0.1250 [NS]). As a result, the nesting of Eorasaurus within Sauria and Archosauromorpha seems to be quite well supported (but is not statistically significant).

The position of Eorasaurus olsoni as a member of Archosauriformes would imply that it is the oldest known member of the group, but the phylogenetic position recovered here for the species should be considered tentative given that the known material is highly incomplete and its phylogenetic position labile in sub-optimal trees. The position of Eorasaurus olsonishould be further tested in phylogenetic analyses that include wider taxonomic samples of archosauromorphs.

Timing of the crocodile-lizard (or bird-lizard) divergence and recommendations for molecular calibrations

Our revision of the Permian saurian record indicates that only four species can be considered as well-supported pre-Mesozoic members of the group: the approximately contemporaneous early–middle Late Permian Protorosaurus speneri, Aenigmastropheus parringtoni and Eorasaurus olsoni and the latest Permian Archosaurus rossicus (Table 7). Other supposed saurian taxa are identified as more basal diapsid forms (e.g. Saurosternon, Youngina) or as indeterminate reptiliomorphs (e.g. BP/1/4220), or as of Early Triassic or indeterminate Late Permian–Early Triassic age. All the Permian saurian taxa recognized here can be assigned to Archosauromorpha and, as a result, there is no fossil record of Lepidosauromorpha prior to the Early Triassic (Paliguana whitei) (Fig. 19). The radioisotopic estimate for the age of Protorosaurus establishes a minimum divergence time for the crocodile-lizard split (origin of Sauria; equivalent to the bird-lizard split).

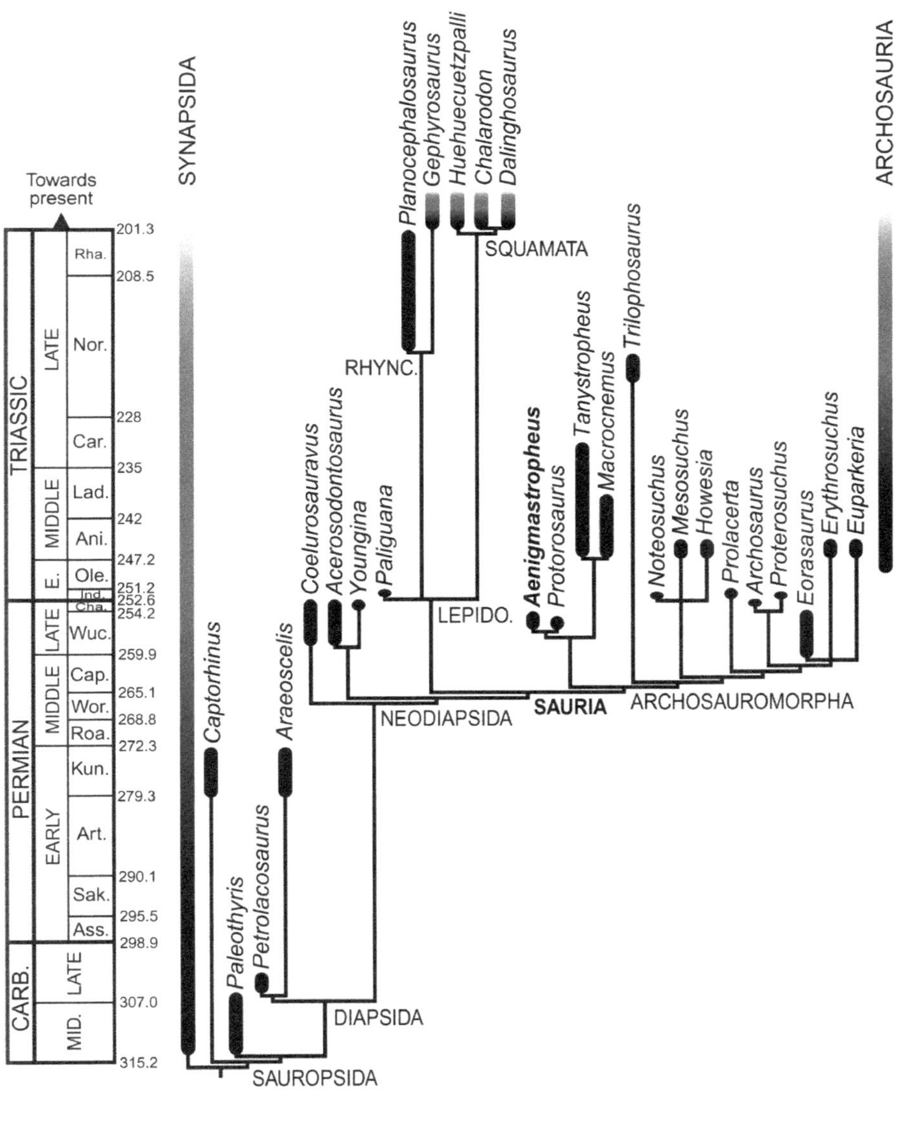

Figure 19. Time-calibrated cladogram showing the basal saurian interrelationships recovered here.

Geological timescale after Gradstein et al. [157].

Table 7. List of specimens/species previously considered as pre-Mesozoic saurians and their age and current taxonomic assignment.

Specimen/Taxon	Age	Occurrence	Reported taxonomic assignment	Revised taxonomic assignment
Aenigmastropheus parringtoni	**middle–late Wuchiapingian, middle Late Permian**	**Tanzania**	**"Proterosuchia" (Charig & Sues [127])**	**Archosauromorpha, ?Protorosauria**
Archosaurus rossicus	**Changhsingian, latest Permian**	**Russia**	**Proterosuchidae (Tatarinov [88]; Nesbitt [59])**	**Archosauriformes, Proterosuchidae**
BP/1/4220	Late Permian–Early Triassic	South Africa	*?Proterosuchus* sp. (Cruickshank [90])	Reptiliomorpha
Buena Vista Fm specimens	Late Permian–Early Triassic	Uruguay	Synapsida (Piñeiro et al. [140])	Amniota
Eorasaurus olsoni	**late Capitanian–Wuchiapingian, late Middle Permian–early Late Permian**	**Russia**	**Archosauromorpha, Protorosauria (Sennikov [89])**	**Archosauromorpha, ?Archosauriformes**
Lanthanolania ivakhnenkoi	late Wordian, middle Middle Permian	Russia	Lepidosauromorpha/sister-taxon of Sauria (Modesto & Reisz [15]); non-saurian Neodiapsida (Reisz et al. [75])	Non-saurian Neodiapsida
Lacertulus bipes	Late Permian–Early Triassic	South Africa	Lepidosauromorpha (Carroll & Thompson [80]); non-squamate Diapsida (Estes [86]; Evans [3])	Non-squamate Diapsida
Palaeagama vielhaueri	Late Permian–Early Triassic	South Africa	Lepidosauromorpha (Carroll [10]; Gauthier et al. [57]); non-saurian Neodiapsida (Müller [32])	Non-saurian Neodiapsida
Paliguana whitei	Early Triassic	South Africa	Lepidosauromorpha (Carroll [10,93]; Evans [56])	Lepidosauromorpha
Protorosaurus speneri	**middle Wuchiapingian, middle Late Permian**	**Germany and England**	**Archosauromorpha Protorosauria (Benton [52]; Gottman-Quesada & Sander [21])**	**Archosauromorpha, Protorosauria**
Saurosternon bainii	Late Permian	South Africa	Lepidosauromorpha (Carroll [10]; Gauthier et al. [57]); non-saurian Neodiapsida (Müller [32])	Non-saurian Neodiapsida

Following the explicit five-point protocol for fossil calibrations proposed by Parham et al. [201], we make the following recommendations for molecular biologists wishing to use fossil data to calibrate the crocodile-lizard (or bird-lizard) split: (1) the split can be calibrated on the basis of the voucher specimen NHMW 1943 I 4, an almost complete skeleton missing the skull, lectotype of Protorosaurus speneri Meyer 1830 [98] (see Gottmann-Quesada & Sander [21] for designation of lectotype material); (2) Protorosaurus speneri is universally considered and strongly supported as a non-archosaurian member of Archosauromorpha upon the basis of the explicit morphological phylogenetic analyses conducted here, and many other studies [4], [21], [52], [56], [72], [73], [74], [105], [106]; (3) results of morphological phylogenetic analyses for major saurian relationships are generally consistent with molecular analyses, with the exception of the highly controversial phylogenetic position of turtles (e.g. [26]–[39]). However, because the earliest fossil turtle remains are Late Triassic in age [202] the problematic position of turtles has no impact on the calibration proposed here; (4) NHMW 1943I4 comes from the locality Glücksbrunn in Thuringia, central Germany, and is from the Kupferschiefer (cycle Z1) of the Zechstein Group; (5) as discussed above, the Kupferschiefer is dated as 257.3±2.6 Ma (Late Permian/Lopingian: Wuchiapingian) based on a Re-Os geochronological study [101]. This age is consistent with biostratigraphic data from the conodont Mesogondolella britannica supporting a middle Wuchiapingian age [102], [103] (Schneider pers. comm. 2012).

These data suggest a minimum or soft calibration date for the crocodile-lizard split of 254.7 Ma (the youngest date for the Kupferschiefer suggested by geochronology). This is younger than the minimum calibration date proposed by Benton & Donoghue [24], who proposed a calibration of 259.7 Ma, based on their interpretation of the Kupferschiefer as older (Capitanian) in age, and slightly older than the 252 Ma calibration recommended by Reisz & Müller [22].

A maximum or hard calibration date for the crocodile-lizard split is more difficult, and Benton & Donoghue [24] proposed that Apsisaurus from the Archer City Formation of Texas, dated as Asselian (298.9±0.2 Ma to 295.5±0.4 Ma; [157]) can be used to constrain the maximum date of divergence, and therefore recommended a maximum or hard calibration date of 299.1 Ma. However, Apsisaurus was recently re-interpreted as a varanopid synapsid [43]. Accordingly, we propose here that the basal neodiapsid Lanthanolania ivakhnenkoi be used to constrain the maximum divergence time of Sauria because it is the closest of the sister taxa to Sauria [75] that is unambiguously older than the earliest known saurian fossils. Lanthanolania comes from the Wordian of Russia [15], [75] (268.8±0.5 Ma to 265.1±0.4 Ma; [157]) and implies a maximum calibration date of 269.3 Ma, which is around 30 Ma younger than the date previously recommended by Benton & Donoghue [24].

Ghost lineages and archosauriform divergence

As discussed above, Archosauromorpha and Lepidosauromorpha diverged prior to 254.7–259.9 Ma (middle–late Wuchiapingian; [157], [203]) (Fig. 19). This minimum divergence date matches that of the immediate successive outgroups of Sauria, such as "younginiforms" and tangasaurids [75], implying the absence of an unambiguous ghost lineage for Sauria (Fig. 19). A middle–late Wuchiapingian age for the origin of the group is consistent with dates estimated by some recent molecular clocks, such as those of Alfaro et al. [204] of 257–292 Ma and Sanders & Lee [25] of 249.5–269.1 Ma. Conversely, the age estimated here for the origin of Sauria is considerably younger than dates estimated by some other recent molecular clock analyses that have proposed an Early Permian divergence data (e.g. 285–289 Ma by Hugall et al. [205]; 276–295 Ma by Shen et al. [38]). As a result, the ghost lineage between the origin of Sauria (and several non-saurian neodiapsids) and the first appearance of the group in the fossil record would be of 21 to 40 million years based on the latter molecular estimates. This extensive ghost lineage could be explained by gaps in the fossil record or by errors in the estimations based on molecular evidence (such as a more rapid rate of molecular evolution than those expected by the molecular models). Accordingly, although the oldest known unambiguous members of Sauria are middle Late Permian in age, it would not be surprising to find early members of the group in Middle Permian rocks. Such discoveries would result in a considerably older minimum origin time for the group, as has also recently been documented for neodiapsid sauropsids [75].

The recovery of the Russian Eorasaurus olsoni within Archosauriformes in our phylogenetic analysis implies that it represents the oldest member of the group, also suggesting a minimum middle Wuchiapingian divergence time for rhynchosaurians, prolacertids, proterosuchids and the Erythrosuchus+Euparkeria clade (Fig. 19). As a result, this phylogenetic position would indicate that archosauriforms are not a group that appeared immediately before the Permo-Triassic mass extinction event and, conversely, had already undergone substantial taxonomic diversification by the Late Permian, with the presence of proterosuchids and the Eorasaurus lineage. However, the fragmentary nature of the known specimens of Eorasaurus olsoni and the weak support for its position within archosauriforms (e.g. only one extra step is necessary to recover it outside Archosauriformes) suggests that the phylogenetic position recovered here for this taxon should be considered as tentative.

Paleobiogeography

The presence of early archosauromorphs in the middle Upper Permian of Germany, England, Russia and Tanzania indicates a broad geographical distribution for the group during the late Paleozoic, spanning from close to the paleo-Equator (Germany) to a paleolatitude of 30°N (Russian localities) in the northern hemisphere to high paleolatitudes of 55°S (Tanzania) in southern Pangea (Fig. 20; paleomap generated using Fossilworksbased upon data from the Paleobiology Database: [206]). This broad paleobiogeographic distribution during the Permian undermines previous hypotheses that proposed a dispersal event for archosauromorphs from Eurasia to southern high latitudes (southern Africa) following the Permo-Triassic mass extinction [14]. The occurrences of Protorosaurus speneri and Aenigmastropheus parringtoni imply either a northern–southern dispersal event or a wider paleobiogeographic distribution for archosauromorphs during the Late Permian, which is not currently well documented in the fossil record. In addition, previous authors have discussed the presence of endemic taxa in the Usili Formation of Tanzania (e.g. [147], [149], [155], [207]–[210]), which yielded the holotype of Aenigmastropheus parringtoni. The occurrence of the basal archosauromorph Aenigmastropheus parringtoniin the Usilli Formation contrasts with the current absence of the group in the Late Permian of the Upper Madumabisa Mudstone of Zambia and the considerably better sampled Cistecephalus Assemblage Zone of the South African Karoo Basin. Thus, Aenigmastropheus parringtoni enlarges the list of endemic taxa for the Tanzanian unit (see Angielczyk et al. [155]) and contrasts with previous hypotheses of a more homogenous Pangean fauna during the Late Permian (e.g. [211]).

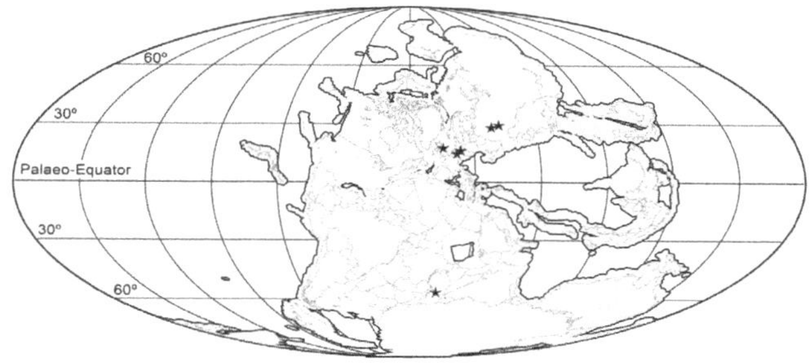

Figure 20. Paleobiogeographical distribution of Archosauromorpha across Pangea during the late Middle–Late Permian.

Black stars indicate archosauromorph records (paleomap for 260 Ma downloaded from Fossilworks using data from the Paleobiology Database: [206]).

Paleobiology

TheosteohistologicalevidencegatheredfromAenigmastropheus parringtoni has potential implications for the understanding of the paleobiology of early archosauromorphs. The analyzed bone of Aenigmastropheus parringtoni belonged to an animal that exhibited high growth rates during early development, as exemplified by the short period of fibrolamellar bone formation. The steady decrease in vascularization coupled with the increase in spatial organization of the bone matrix up until the formation of lamellar-zonal bone indicates, however, that the animal was growing more slowly for most of its life. The closely spaced growth marks found in the lamellar-zonal bone of the outer part of the cortex could be interpreted as an outer circumferential layer (OCL, see Ponton et al. [212]; often referred to also as the external fundamental system [EFS]), in which case the animal could be considered to have reached skeletal maturity (Fig. 15D, E). It remains unclear though why the outermost thin layer of bone directly below the external bone surface does not show growth marks. Growth marks in this layer could be simply obscured in some areas or they could be truly absent. If the latter condition is assumed to be correct, this would argue against the presence of an OCL, as the supposed OCL would not reach the external bone surface. Instead, the closely spaced LAGs would instead indicate that the animal had a prolonged period in life (e.g. several years) as an adult in which it experienced unfavorable growth conditions (e.g. nutritional shortage, drought, disease, etc.) leading to restricted growth (Fig. 15F). The growth pattern observed in Aenigmastropheus parringtoni resembles those of Prolacerta broomi and the basal archosauriforms Proterosuchus fergusi, Erythrosuchus africanus and Chanaresuchus bonapartei, with rapid growth during their early development [213]. By contrast, extant lepidosauromorphs [214], [215] and the basal archosauromorphs Trilophosaurus buettneri and hyperodapedontid rhynchosaurs [216], [217] exhibit a slow overall growth pattern. Accordingly, the currently available osteological evidence indicates the presence of diverse growth strategies among basal members of Archosauromorpha, but further sampling of early members of the group is required to establish the optimization of growth strategies and its implications for the early evolution of archosauromorphs.

The early archosauromorph Protorosaurus speneri was interpreted as a fully terrestrial animal found in marine paleoenvironments as an allochthonous element, in which its carcasses were transported from nearby islands [21], [218]. The presence of terrestrial plant remains (i.e. ovules of a conifer) in the gut content of one Protorosaurus speneri specimen provides support for this hypothesis [219]. The fragmentary condition of the holotype of Aenigmastropheus parringtoni complicates an assessment of the paleobiology of the species. However, the strongly developed distal humeral condyles and olecranon process of the ulna suggest a fully terrestrial mode of life (Fig. 21). The sectioned probable humerus of Aenigmastropheus parringtoni has a relatively thick cortex (k=0.44, represented by the ratio between the inner diameter and outer diameter of the bone; [220]) that closely resembles the condition of some bones of terrestrial basal archosauriforms, such as the humerus and tibia of Euparkeria capensis (k=0.44, 48) and the fibula of Proterosuchus fergusi (0.37) [213]. The k value observed in Aenigmastropheus parringtoni is higher than those observed in aquatic and semi-aquatic animals (k<0.30) [221]–[223], supporting a terrestrial mode of life for this species and thus also for the earliest known archosauromorphs.

Figure 21. Late Permian of Tanzania at the time of deposition of the Usili Formation.

Life restoration of Aenigmastropheus parringtoni (foreground), and herd of the dicynodont Endothiodon being pursued by a gorgonopsian. All these vertebrates were found together at the type locality of Aenigmastropheus parringtoni (locality B35). Drawing by Emilio López-Rolandi.

ACKNOWLEDGMENTS

We thank Mathew Lowe and Jennifer Clack (UMZC) for the loan of UMZC T836 and permission to carry out histological sampling, and Renate Liebreich (BSPG) for preparing a mold and cast of the sampled limb bone prior to sectioning. We thank Andrey Sennikov for discussion and providing useful information on the locality of Archosaurus rossicus. We thank Grzegorz Niedźwiedzki for pointing out the presence of Permian footprints assigned to saurian trackmakers. Discussions with Bruce Rubidge, Fernando Abdala, Juan Carlos Cisneros, David Gower, Sterling Nesbitt, Jörg Schneider and Michael Szurlies are appreciated. We thank Gabriela Sobral and Marc Jones for their suggestions and comments, which improved the quality of the final manuscript. We thank Kenneth Angielczyk for sharing unpublished information about the biostratigraphy and age of the Usili Formation. We thank the following curators, researchers and collection managers that provided access to specimens under their care for the purpose of this research: Billy de Klerk (AM); Carl Mehling and Steve Brusatte (AMNH); Bernhard Zipfel, Bruce Rubidge and Fernando Abdala (BP); Markus Moser (BSPG); Graciela Piñeiro, Alejandra Rojas and Andrea Corona (FC-DPV); Ellen de Kock (GHG); Jun Liu and Corwin Sullivan (IVPP); Daniela Schwarz-Wings (MB); Susana Devinvenzi González (MCNAM); Jessica Cundiff (MCZ); Ronan Allain (MNHN); Sandra Chapman and Lorna Steel (NHMUK); Elize Butler and Jennifer Botha-Brink (NM); Heinz Furrer (PIMUZ); Andrey Sennikov (PIN); Jaime Powell (PVL); Ricardo Martínez (PVSJ); Sheena Kaal and Roger Smith (SAM); Rainer Schoch (SMNS); Heidi Fourie (TM); Mathew Lowe (UMZC); Michael Brett-Surman and Hans-Dieter Sues (USNM); Max Langer (USP); and Isabella von Lichtan (UTGD). We thank Emilio López-Rolandi for the life reconstruction of Aenigmastropheus parringtoni (Fig. 21). Access to the free version of TNT 1.1 was possible due to the Willi Henning Society.

Author Contributions

Conceived and designed the experiments: MDE RJB TMS. Performed the experiments: MDE TMS. Analyzed the data: MDE TMS. Contributed reagents/materials/analysis tools: MDE TMS RJB. Wrote the paper: MDE RJB TMS.

REFERENCES

1. Pyron RA, Burbrink FT, Wiens JJ (2013) A phylogeny and revised classification of Squamata, including 4161 species of lizards and snakes. BMC Evol Biol 13: 93. doi: 10.1186/1471-2148-13-93
2. Clements JF (2007) The Clements checklist of the birds of the world, 6th edition. New York: Cornell University Press. 864 p.
3. Evans SE (2003) At the feet of the dinosaurs: the early history and radiation of lizards. Biol Rev 78: 513–551. doi: 10.1017/s1464793103006134
4. Dilkes DW (1998) The Early Triassic rhynchosaur Mesosuchus browni and the interrelationships of basal archosauromorph reptiles. Philos Trans R Soc Lond B Biol Sci 353: 501–541. doi: 10.1098/rstb.1998.0225
5. Ezcurra MD, Butler RJ, Gower DJ (2013) 'Proterosuchia': the origin and early history of Archosauriformes. In: Geol Soc, London, Spec Publ Nesbitt SJ, Desojo JB, Irmis RB, Anatomy, phylogeny and palaeobiology of early archosaurs and their kin 379: 9–33. doi: 10.1144/sp379.11
6. Sookias RB, Butler RJ (2013) Euparkeriidae. In: Geol Soc, London, Spec Publ Nesbitt SJ, Desojo JB, Irmis RB, Anatomy, phylogeny and palaeobiology of early archosaurs and their kin 379: 35–48. doi: 10.1144/sp379.6
7. Sues H-D, Desojo JB, Ezcurra MD (2013) Doswelliidae: a clade of unusual armoured archosauriforms from the Middle and Late Triassic. In: Geol Soc, London, Spec Publ Nesbitt SJ, Desojo JB, Irmis RB, Anatomy, phylogeny and palaeobiology of early archosaurs and their kin 379: 49–58. doi: 10.1144/sp379.13

8. Trotteyn MJ, Arcucci AB, Raugust T (2013) Proterochampsia: an endemic archosauriform clade from South America. In: Geol Soc, London, Spec Publ Nesbitt SJ, Desojo JB, Irmis RB, Anatomy, phylogeny and palaeobiology of early archosaurs and their kin 379: 59–90. doi: 10.1144/sp379.23
9. Parrington FR (1956) A problematic reptile from the Upper Permian. Ann Mag Nat Hist 12: 333–336. doi: 10.1080/00222935608655819
10. Carroll RL (1975) Permo-Triassic 'lizards' from the Karroo. Palaeontol Afr 18: 71–87.
11. Carroll RL (1976) Noteosuchus—the oldest known rhynchosaur. Ann S Afr Mus 72: 37–57.
12. Ivachnenko MF (1978) The Permian and Triassic procolophonians from the Russian platform. Akademiya Nauk SSSR, Paleontologicheskiy Institut, Trudy 164: 1–80.
13. Gaffney ES (1980) Phylogenetic relationships of the major groups of amniotes. In: Panchen AL, editor. The terrestrial environment and the origin of land vertebrates. Academic Press, London: 593–610.
14. Reisz RR, Modesto SP, Scott D (2000) Acanthotoposaurus bremneri and the origin of the Triassic archosauromorph reptile fauna of South Africa. S Afr J Sci 96: 443–445.
15. Modesto SP, Reisz RR (2002) An enigmatic new diapsid reptile from the Upper Permian of Eastern Europe. J Vert Paleontol 22: 851–855. doi: 10.1671/0272-4634(2002)022[0851:aendrf]2.0.co;2
16. Reisz RR, Modesto SP (2007) Heleosaurus scholtzi from the Permian of South Africa: a varanopid synapsid, not a diapsid reptile. J Vert Paleontol 27: 734–740. doi: 10.1671/0272-4634(2007)27[734:hsftpo]2.0.co;2
17. Evans SE, Jones MEHJ (2010) The origins, early history and diversification of lepidosauromorph reptiles. In: New aspects of Mesozoic biodiversity, lecture notes in Earth Sciences Bandyopadhyay S, editor. 132: 27–44. doi: 10.1007/978-3-642-10311-7_2
18. Jones MEH, Anderson CL, Hipsley CA, Müller J, Evans SE, et al. (2013) Integration of molecules and new fossils supports a Triassic origin for Lepidosauria (lizards, snakes, and tuatara). BMC Evol Biol 13: 208. doi: 10.1186/1471-2148-13-208
19. von Meyer H (1832) Palaeologica zur Geschichte der Erde und ihrer Geschöpfe. Frankfurt am Main: Verlag von Siegmund Schmerber. 560 p.
20. Evans SE, King MS (1993) A new specimen of Protorosaurus (Reptilia: Diapsida) from the Marl Slate (Late Permian) of Britain. Proc Yorkshire Geol Soc 49: 229–234. doi: 10.1144/pygs.49.3.229
21. Gottmann-Quesada A, Sander PM (2009) A redescription of the early archosauromorph Protorosaurus spenseri Meyer, 1832 and its phylogenetic relationships. Palaeontographica A 287: 123–220.
22. Reisz RR, Müller J (2004) Molecular timescales and the fossil record: a paleontological perspective. Trends Genet 20: 237–241. doi: 10.1016/j.tig.2004.03.007
23. Müller J, Reisz RR (2005) Four well-constrained calibration points from the vertebrate fossil record for molecular clock estimates. BioEssays 27: 1069–1075. doi: 10.1002/bies.20286
24. Benton MJ, Donoghue PCJ (2007) Paleontological evidence to date the tree of life. Mol Biol Evol 24: 26–53. doi: 10.1093/molbev/msl150
25. Sanders KL, Lee MSY (2007) Evaluating molecular clock calibrations using Bayesian analyses with soft and hard bounds. Biol Letters 3: 275–279. doi: 10.1098/rsbl.2007.0063
26. Gregory WK (1946) Pareiasaurs versus placodonts as near ancestors to turtles. Bull Am Mus Nat Hist 86: 275–326.
27. Rieppel O, deBragga M (1996) Turtles as diapsid reptiles. Nature 384: 453–455. doi: 10.1038/384453a0
28. deBraga M, Rieppel O (1997) Reptile phylogeny and the relationships of turtles. Zool J Linn Soc 120: 281–354. doi: 10.1111/j.1096-3642.1997.tb01280.x
29. Platz JE, Conlon JM (1997) Reptile relationships turn turtle… and turn back again. Nature 389: 245–46.
30. Zardoya R, Meyer A (1998) Complete mitochondrial genome suggests diapsid affinities of turtles. Proc Natl Acad Sci USA 95: 14226–14231. doi: 10.1073/pnas.95.24.14226
31. Rieppel O, Reisz R (1999) The origin and evolution of turtles. Annu Rev Ecol Syst 30: 1–22. doi: 10.1146/annurev.ecolsys.30.1.1
32. Müller J (2004) The relationships among diapsid reptiles and the influence of taxon selection. In: München: Verlag Dr Friedrich Pfeil Arratia G, Wilson MVH, Cloutier R, Recent advances in the origin and early radiation of vertebrates 379–408.
33. Hill RV (2005) Integration of morphological data sets for phylogenetic analysis of Amniota: the importance of integumentary characters and increased taxonomic sampling. Syst Biol 54: 530–547.
34. Bhullar B-AS, Bever GS (2009) An archosaur-like laterosphenoid in early turtles (Reptilia: Pantestudines). Breviora 518: 1–11. doi: 10.3099/0006-9698-518.1.1
35. Lyson TR, Bever GS, Bhullar B-AS, Joyce WG, Gauthier JA (2010) Transitional fossils and the origin of turtles. Biol Letters 2010: 830–833. doi: 10.1098/rsbl.2010.0371
36. Lee MSY (2011) Palaeontology: Turtles in Transition. Currently Biol 23: 513–515.
37. Lyson TR, Bever GS, Scheyer TM, Hsiang AY, Gauthier JA (2013) Evolutionary origin of the turtle shell. Current Biol 23: 1113–1119. doi: 10.1016/j.cub.2013.05.003
38. Shen X-X, Liang D, Wen J-Z, Zhang P (2011) Multiple genome alignments facilitate development of NPCL markers: a case study of tetrapod phylogeny focusing on the position of turtles. Mol Biol Evol 28: 3237–3252. doi: 10.1093/molbev/msr148
39. Neenan JM, Klein N, Scheyer TM (2013) European origin of placodont marine reptiles and the evolution of crushing dentition in Placodontia. Nat Comm 4: 1621 doi:10.1038/ncomms2633.
40. Wilson JA (1999) Vertebral laminae in sauropods and other saurischian dinosaurs. J Vert Paleontol 19: 639–653. doi: 10.1080/02724634.1999.10011178

41. Wilson JA, D'Emic MD, Ikejiri T, Moacdieh EM, Whitlock JA (2011) A nomenclature for vertebral fossae in sauropods and other saurischian dinosaurs. PLOS ONE 6: e17114 doi:10.1371/journal.pone.0017114.

42. Rewcastle SC (1980) Form and function in lacertilian knee and mesotarsal joints; a contribution to the analysis of sprawling locomotion. J Zool 191: 147–170. doi: 10.1111/j.1469-7998.1980.tb01454.x

43. Reisz RR, Laurin M, Marjanović D (2010) Apsisaurus witteri from the Lower Permian of Texas: yet another small varanopid synapsid, not a diapsid. J Vert Paleontol 30: 1628–1631. doi: 10.1080/02724634.2010.501441

44. Spielmann JA, Lucas SG, Rinehart LF, Heckert AB (2008) The Late Triassic archosauromorph Trilophosaurus. New Mexico Mus Nat Hist Sci Bull 43: 1–177.

45. Evans SE, Borsuk-Białynicka M (2009) A small lepidosauromorph reptile from the Early Triassic of Poland. Palaeontol Polonica 65: 179–202.

46. Borsuk–Białynicka M, Evans SE (2009) A long-necked archosauromorph from the Early Triassic of Poland. Palaeontol Polonica 65: 203–234.

47. Evans SE (1980) The skull of a new eosuchian reptile from the Lower Jurassic of South Wales. Zool J Linn Soc 70: 203–264. doi: 10.1111/j.1096-3642.1980.tb00852.x

48. Evans SE (1981) The postcranial skeleton of Gephyrosaurus bridensis (Eosuchia: Reptilia). Zool J Linn Soc 73: 81–116. doi: 10.1111/j.1096-3642.1981.tb01580.x

49. Senter P (2004) Phylogeny of Drepanosauridae (Reptilia: Diapsida). J Syst Palaeontol 2: 257–268. doi: 10.1017/s1477201904001427

50. Goloboff PA, Farris JS, Nixon KC (2008) TNT, a free program for phylogenetic analysis. Cladistics 24: 774–786. doi: 10.1111/j.1096-0031.2008.00217.x

51. Coddington JA, Scharff N (1994) Problems with zero-length branches. Cladistics 10: 415–423. doi: 10.1111/j.1096-0031.1994.tb00187.x

52. Benton MJ (1985) Classification and phylogeny of the diapsid reptiles. Zool J Linn Soc 84: 97–164. doi: 10.1111/j.1096-3642.1985.tb01796.x

53. Sues H-D (2003) An unusual new archosauromorph reptile from the Upper Triassic Wolfville Formation of Nova Scotia. Can J Earth Sci 40: 635–649. doi: 10.1139/e02-048

54. Modesto SP, Sues H-D (2004) The skull of the Early Triassic archosauromorph reptile Prolacerta broomi and its phylogenetic significance. Zool J Linn Soc 140: 335–351. doi: 10.1111/j.1096-3642.2003.00102.x

55. Chatterjee S (1986) Malerisaurus langstoni, a new diapsid reptile from the Triassic of Texas. J Vertebr Paleontol 6: 297–312. doi: 10.1080/02724634.1986.10011627

56. Evans SE (1988) The early history and relationships of the Diapsida. In: Benton MJ, editor. The phylogeny and classification of the tetrapods. Oxford: Oxford University Press. pp. 221–260.

57. Gauthier JA, Kluge AG, Rowe T (1988) Amniote phylogeny and the importance of fossils. Cladistics 4: 105–209. doi: 10.1111/j.1096-0031.1988.tb00514.x

58. Watson DMS (1912) Mesosuchus browni, gen. et spec. nov. Rec Albany Mus 2: 298–299.

59. Nesbitt SJ (2011) The early evolution of archosaurs: relationships and the origin of major clades. Bull Am Mus Nat Hist 352: 1–292. doi: 10.1206/352.1

60. Conti MA, Leonardi G, Mietto P, Nicosia U (2000) Orme di tetrapodi non dinosauriani del Paleozoico e Mesozoico in Italia. In: Leonardi G, Mietto P, editors. Dinosauri in Italia. Le orme giurassiche dei Lavini di Marco (Trentino) e gli altri resti fossili italiani. Pisa: Accademia Editoriale: 297–320.

61. Avanzini M, Bernardi M, Nicosia U (2011) The Permo-Triassic tetrapod faunal diversity in the Italian southern Alps. In: Ahmad Dar I, Ahmad Dar M, editors. Earth and environmental sciences. InTech Ed. pp. 591–608.

62. Hmich D, Schneider JW, Saber H, Voigt S, El Wartiti M (2006) New continental Carboniferous and Permian faunas of Morocco: implications for biostratigraphy, palaeobiogeography and palaeoclimate. Geol Soc London Spec Publ 265: 297–324. doi: 10.1144/gsl.sp.2006.265.01.14

63. Broom R (1914) A new thecodont reptile. Proc Zool Soc London 1914: 1072–1077.

64. Gow CE (1975) The morphology and relationships of Youngina capensis Broom and Prolacerta broomi Parrington. Palaeontol Afr 18: 89–131.

65. Haughton SH (1924) On a new type of thecodont from the Middle Beaufort Beds. Ann Transvaal Mus 11: 93–97.

66. Currie P (1982) The osteology and relationships of Tangasaurus mennelli Haughton (Reptilia, Eosuchia). Ann South Afr Mus 86: 247–265.

67. Piveteau J (1926) Paléontologie de Madagascar, XIII—Amphibiens et reptiles Permiens. Ann Paléontol 15: 55–178.

68. Currie P (1981) Hovasaurus boulei, an aquatic eosuchian from the Upper Permian of Madagascar. Palaeontol Afr 24: 99–168.

69. Currie P (1980) A new younginid (Reptilia: Eosuchia) from the Upper Permian of Madagascar. Can J Earth Sci 17: 500–511. doi: 10.1139/e80-046

70. Bickelmann C, Müller J, Reisz RR (2009) The enigmatic diapsid Acerosodontosaurus piveteaui (Reptilia: Neodiapsida) from the Upper Permian of Madagascar and the paraphyly of "younginiform" reptiles. Can J Earth Sci 46: 651–661. doi: 10.1139/e09-038

71. Carroll RL (1981) Plesiosaur ancestors from the Upper Permian of Madagascar. Philos Trans R Soc Lond 293: 315–383. doi: 10.1098/rstb.1981.0079

72. Gauthier JA (1984) A cladistic analysis of the higher categories of the Diapsida. Berkeley: PhD thesis, University of California. 564 p.

73. Laurin M (1991) The osteology of a Lower Permian eosuchian from Texas and a review of diapsid phylogeny. Zool J Linn Soc 101: 59–95. doi: 10.1111/j.1096-3642.1991.tb00886.x

74. Jalil N-E (1997) A new prolacertiform diapsid from the Triassic of North Africa and the interrelationships of

the Prolacertiformes. J Vert Paleontol 17: 506–525. doi: 10.1080/02724634.1997.10010998

75. Reisz RR, Modesto SP, Scott DM (2011) A new Early Permian reptile and its significance in early diapsid evolution. Proc R Soc B 278: 3731–3737. doi: 10.1098/rspb.2011.0439

76. Huxley TH (1868) On Saurosternon Bainii, and Pristerodon McKayi, two new fossil lacertillian reptiles from South Africa. Geol Mag 5: 201–205. doi: 10.1017/s0016756800207784

77. Broom R (1903) On a new reptile (Proterosuchus fergusi) from the Karroo beds of Tarkastad, South Africa. Ann S Afr Mus 4: 159–164.

78. Broom R (1925) On the origin of lizards. Proc Zool Soc Lond 1925: 1–16.

79. Broom R (1926) On a nearly complete skeleton of a new eosuchian reptile (Palaeagama vielhaueri, gen. et sp. nov.). Proc Zool Soc Lond 1926: 487–492. doi: 10.1111/j.1469-7998.1926.tb08109.x

80. Carroll RL, Thompson P (1982) A bipedal lizardlike reptile from the Karroo. J Paleontol 56: 1–10.

81. Kitching JW (1977) The distribution of the Karroo vertebrate fauna. Bernard Price Institute for Palaeontological Research Memoir 1: 1–131.

82. Groenewald GH, Kitching JW (1995) Biostratigraphy of the Lystrosaurus Assemblage Zone. In: South African Committee for Stratigraphy, Biostratigraphic Series Rubidge BS, Biostratigraphy of the Beaufort Group (Karoo Supergroup) 1: 35–39.

83. Damiani RJ, Neveling J, Hancox PJ, Rubidge B (2000) First trematosaurid temnospondyl from the Lystrosaurus Assemblage Zone of South Africa and its biostratigraphic implications. Geol Mag 137: 659–665. doi: 10.1017/s0016756800004660

84. Rubidge BS (2005) Re-uniting lost continents – fossil reptiles from the ancient Karoo and their wanderlust. South Afr J Geol 108: 135–172. doi: 10.2113/108.1.135

85. Lucas SG (2010) The Triassic timescale based on non-marine tetrapod biostratigraphy and biochronology. In: Geol Soc, London, Spec Publ Lucas SG, The Triassic timescale 334: 447–500. doi: 10.1144/sp334.15

86. Estes R (1983) Sauria terrestria, Amphisbaenia. Stuttgart: Handbuch der Paläoherpetologie 10A. 249 p.

87. Evans SE (1984) The classification of the Lepidosauria. Zool J Linn Soc 82: 87–100. doi: 10.1111/j.1096-3642.1984.tb00537.x

88. Tatarinov LP (1960) Discovery of pseudosuchians in the Upper Permian of SSSR. Paleontol J 1960: 74–80.

89. Sennikov AG (1997) An enigmatic reptile from the Upper Permian of the Volga River Basin. Paleontol J 31: 94–101.

90. Cruickshank ARI (1972) The proterosuchian thecodonts. In: Joysey KA, Kemp TS, editors. Studies in vertebrate evolution. Edinburgh: Oliver and Boyd. pp. 89–119.

91. Dias-da-Silva S, Modesto SP, Schultz CL (2006) New material of Procolophon (Parareptilia: Procolophonidae) from the Lower Triassic of Brazil, with remarks on the ages of the Sanga do Cabral and Buena Vista formations of South America. Can J Earth Sci 43: 1685–1693. doi: 10.1139/e06-043

92. Evans SE, van den Heever JA (1987) A new reptile (Reptilia: Diapsida) from the Upper Permian Daptocephalus zone of South Africa. S Afr J Sci 83: 724–730.

93. Carroll RL (1976) Eosuchians and the origin of archosaurs. In: Churcher CS, editor. Athlon: Essays on palaeontology, Misc. Publs R. Ont. Mus. Life Sci. pp. 58–79.

94. Efremov JA (1938) Some new Permian reptiles of the USSR. Academy of Sciences URSS CR 19: 121–126.

95. Romer AS, Price LI (1940) Review of the Pelycosauria. Geol Soc Am Spec Paper 28: 1–538. doi: 10.1130/spe28-p1

96. Reisz RR, Berman DS (2001) The skull of Mesenosaurus romeri, a small varanopsid (Synapsida: Eupelycosauria) from the Upper Permian of the Mezen River Basin, Northern Russia. Ann Carnegie Mus 70: 113–132.

97. Spener CM (1710) Disquistio de crocodile in lapide scissile expresso, aliisque Lithozois. Misc Berol ad increment sci, ex scr Soc Regiae Sci exhibits ed I: 92–110.

98. von Meyer H (1830) Protorosaurus. Isis von Oken 1830: 517–519.

99. von Meyer H (1856) Zur Fauna der Vorwelt. Saurier aus dem Kupferschiefer der Zechstein-Formation. Frankfurt: Verlag Heinrich Keller (vormals S. Schmerber'sche Buchhandlung). pp. 1–24.

100. Strohmenger C, Voigt E, Zimdars J (1996) Sequence stratigraphy and cyclic development of Basal Zechstein carbonate-evaporite deposits (Upper Permian, northwest Germany). Sed Geol 102: 33–54. doi: 10.1016/0037-0738(95)00058-5

101. Brauns CM, Pätzold T, Haack U (2003) A Re–Os study bearing on the age of the Kupferschiefer at Sangerhausen (Germany). Utrecht: Abstracts XVth Intern Congress on Carboniferous and Permian Stratigraphy. p. 66.

102. Legler B, Gebhardt U, Schneider JW (2005) Late Permian non-marine–marine transitional profiles in the central Southern Permian Basin. Int J Earth Sci 94: 851–862. doi: 10.1007/s00531-005-0002-5

103. Legler B, Schneider JW (2008) Marine ingressions into the Middle/Late Permian saline lake of the Southern Permian Basin (Rotliegend, northern Germany) possibly linked to sea-level highstands in the Arctic rift system. Palaeogeogr Palaeoclimatol Palaeoecol 267: 102–114. doi: 10.1016/j.palaeo.2008.06.009

104. Hancock A, Howse R (1870) On Proterosaurus speneri, von Meyer, and a new species, Proterosaurus huxleyi, from the Marl-Slate of Midderidge, Durham. Quarterly J Geol Soc 26: 565–572. doi: 10.1144/gsl.jgs.1870.026.01-02.53

105. Benton MJ, Allen JL (1997) Boreopricea from the Lower Triassic of Russia, and the relationships of the prolacertiform reptiles. Palaeontology 40: 931–953.

106. Rieppel O, Fraser NC, Nosotti S (2003) The monophyly of Protorosauria (Reptilia, Archosauromorpha): a preliminary analysis. Atti Soc ital sci nat, Mus civ stor nat Milano 144: 359–382.

107. Sues H-D, Munk W (1996) A remarkable assemblage of terrestrial tetrapods from the Zechtsein (Upper Permian: Tatarian) near Korbach (northwestern Hesse). Paläont Z 70: 213–223. doi: 10.1007/bf02988279

108. Sennikov AG (2011) New tanystropheids (Reptilia: Archosauromorpha) from the Triassic of Europe. Paleontol J 45: 90–104. doi: 10.1134/s0031030111010151

109. Taylor GK, Tucker C, Twitchett RJ, Kearsey T, Benton MJ, et al. (2009) Magnetostratigraphy of Permian/Triassic boundary sequences in the Cis-Urals, Russia: no evidence for a major temporal hiatus. Earth Planet Sci Lett 281: 36–47. doi: 10.1016/j.epsl.2009.02.002

110. Irmis RB (2007) Axial skeleton ontogeny in the Parasuchia (Archosauria: Pseudosuchia) and its implications for ontogenetic determination in archosaurs. J Vert Paleontol 27: 350–361. doi: 10.1671/0272-4634(2007)27[350:asoitp]2.0.co;2

111. Wild R (1973) Die Triasfauna dr Tessiner Kalkalpen. XXIII Tanystropheus longobardicus (Bassani) (Neue Ergebnisse). Abhandlungen Schweizerische Paläontologische Gesellschaft 95: 1–162.

112. Spielmann JA, Lucas SG, Heckert AB, Rinehart LF, Richards HR III (2009) Redescription of Spinosuchus caseanus (Archosauromorpha: Trilophosauridae) from the Upper Triassic of North America. Palaeodiversity 2: 283–313.

113. Gower DJ (2003) Osteology of the early archosaurian reptile Erythrosuchus africanus Broom. Ann S Afr Mus 110: 1–84. doi: 10.1111/j.1469-7998.1997.tb03855.x

114. Butler RJ, Barrett PM, Gower DJ (2012) Reassessment of the evidence for postcranial skeletal pneumaticity in Triassic archosaurs, and the early evolution of the avian respiratory system. PLOS ONE 7 (3) e34094 doi:10.1371/journal.pone.0034094.

115. Butler RJ, Barrett PM, Abel RL, Gower DJ (2009) A possible ctenosauriscid archosaur from the Middle Triassic Manda Beds of Tanzania. J Vert Paleontol 29: 1022–1031. doi: 10.1671/039.029.0404

116. Piechowski R, Dzik J (2010) The axial skeleton of Silesaurus opolensis. J Vert Paleontol 30: 1127–1141. doi: 10.1080/02724634.2010.483547

117. Novas FE (1993) New information on the systematics and postcranial skeleton of Herrerasaurus ischigualastensis (Theropoda: Herrerasauridae) from the Ischigualasto Formation (Upper Triassic) of Argentina. J Vert Paleontol 13: 400–423. doi: 10.1080/02724634.1994.10011523

118. Dilkes DW (1995) The rhynchosaur Howesia browni from the Lower Triassic of South Africa. Palaeontology 38: 665–685.

119. Fraser NC, Walkden GM (1984) The postcranial skeleton of the Upper Triassic sphenodontid Planocephalosaurus robinsonae. Palaeontology 27: 575–595.

120. Sennikov AG (1988) The role of the oldest thecodontians in the vertebrate assemblage of Eastern Europe. Paleontol J 1988: 78–87.

121. Gower DJ, Sennikov AG (2000) Early archosaurs from Russia. In: Benton MJ, Shishkin MA, Unwin DM, Kurochkin EN, editors. The age of dinosaurs in Russia and Mongolia. Cambridge: Cambridge University Press. pp. 140–159.

122. Sennikov AG, Golubev VK (2006) Vyazniki biotic assemblage of the terminal Permian. Paleontol J 40: S475–S481. doi: 10.1134/s0031030106100078

123. Benton MJ, Tverdokhlebov VP, Surkov MV (2004) Ecosystem remodelling among vertebrates at the Permo-Triassic boundary in Russia. Nature 432: 97–100. doi: 10.1038/nature02950

124. Krassilov V, Karasev E (2009) Paleofloristic evidence of climate change near and beyond the Permian–Triassic boundary. Palaeogeogr Palaeoclimatol Palaeoecol 284: 326–336. doi: 10.1016/j.palaeo.2009.10.012

125. Sennikov AG, Golubev VK (2012) On the faunal verification of the Permo–Triassic boundary in continental deposits of eastern Europe: 1. Gorokhovets–Zhukov Ravine. Paleontol J 46: 313–323. doi: 10.1134/s003103011203015x

126. Sennikov AG (2008) Archosauromorpha. In: Ivakhnenko MF, Kurochkin EN, editors. Fossil vertebrates from Russia and adjacent countries, Fossil Reptiles and Birds Part 1. Moscow: Russian Academy of Sciences Paleontological Institute. pp. 266–318.

127. Charig AJ, Sues H-D (1976) Proterosuchia. In: Kuhn O, editor. Handbuch der Paläoherpetologie 13. Stuttgart: Gustav Fischer. pp. 11–39.

128. Vaughn PP (1955) The Permian reptile Araeoscelis restudied. Bull Harvard Mus Comp Zool 113: 305–467.

129. Watson DMS (1957) On Millerosaurus and the early history of the sauropsid reptiles. Philos Trans R Soc Lond B Biol Sci 240: 325–400. doi: 10.1098/rstb.1957.0003

130. Moss JL (1972) The morphology and phylogenetic relationships of the Lower Permian tetrapod Tseajaia campi Vaughn (Amphibia: Seymouriamorpha). Uni Calif Publ Geol Sci 98: 1–72.

131. Reisz RR (1981) A diapsid reptile from the Pennsylvanian of Kansas. Uni Kansas Publ Mus Nat Hist 7: 1–74. doi: 10.5962/bhl.title.8440

132. Evans SE, Haubold H (1987) A review of the Upper Permian genera Coelurosauravus, Weigeltisaurus and Gracilisaurus (Reptilia: Diapsida). Zool J Linn Soc 90: 275–303. doi: 10.1111/j.1096-3642.1987.tb01356.x

133. Rieppel O (1989) Helveticosaurus zollingeri Peyer (Reptilia, Diapsida): skeletal paedomorphosis; functional anatomy and systematic affinities. Palaeontographica A 208: 123–152.

134. Colbert EH, Olsen PE (2001) A new and unusual aquatic reptile from the Lockatong Formation of New Jersey (Late Triassic, Newark Supergroup). Am Mus Novit 3334: 1–24. doi: 10.1206/0003-0082(2001)334<0001:anauar>2.0.co;2

135. deBraga M (2003) The postcranial skeleton, phylogenetic position and probable lifestyle of the Early Triassic reptile Procolophon trigoniceps. Can J Earth Sci 40: 527–556. doi: 10.1139/e02-082

136. Langer MC, Benton MJ (2006) Early dinosaurs: a phylogenetic study. J Syst Palaeontol 4: 309–358. doi: 10.1017/s1477201906001970

137. Montefeltro FC, Bittencourt JS, Langer MC, Schultz CL (2013) Postcranial anatomy of the hyperodapedontine rhynchosaur Teyumbaita sulcognathus (Azevedo and Schultz, 1987) from the Late Triassic of southern Brazil. J Vert Paleontol 33: 67–84. doi: 10.1080/02724634.2012.710285

138. Romer AS (1956) Osteology of the reptiles. Chicago: University of Chicago Press. 772 p.

139. Sumida SS (1991) Vertebral morphology, alternation of neural spine height, and structure in Permo-Carboniferous tetrapods, and a reappraisal of primitive modes of terrestrial locomotion. Univ Calif Publ Zool 122: 1–135.

140. Piñeiro G, Verde M, Ubilla M, Ferigolo J (2003) First basal synapsids ("pelycosaurs") from the Upper Permian-?Lower Triassic of Uruguay, South America. J Paleontol 77:389–392. doi: 10.1666/0022-3360(2003)077<0389:fbspft>2.0.co;2

141. Bossi J, Navarro R (1991) Geología del Uruguay. Montevideo: Departamento de Publicaciones de la Universidad de la República 1: 1–453.

142. Marsicano C, Perea D, Ubilla M (2000) A new temnospondyl amphibian from the Lower Triassic of South America. Alcheringa 24: 119–123. doi: 10.1080/03115510008619528

143. Piñeiro G, Rojas A, Ubilla M (2004) A new procolophonoid (Reptilia, Parareptilia) from the Upper Permian of Uruguay. J Vert Paleontol 24: 814–821. doi: 10.1671/0272-4634(2004)024[0814:anprpf]2.0.co;2

144. Hughes B (1963) The earliest archosaurian reptiles. S Afr J Sci 59: 221–241.

145. Reig OA (1970) The Proterosuchia and the early evolution of the archosaurs; an essay about the origin of a major taxon. Bull Mus Comp Zool 139: 229–292.

146. Stockley GM (1932) The geology of the Ruhuhu coalfields, Tanganyika Territory. Quart J Geol Soc Lond 88: 610–622. doi: 10.1144/gsl.jgs.1932.088.01-04.20

147. Gay SA, Cruickshank ARI (1999) Biostratigraphy of the Permian tetrapod faunas from the Ruhuhu Valley, Tanzania. J Afr Earth Sci 29: 195–210. doi: 10.1016/s0899-5362(99)00089-5

148. Wopfner H, Markwort S, Semkiwa PM (1991) Early diagenetic laumontite in the Lower Triassic Manda Beds of the Ruhuhu Basin, southern Tanzania. J Sed Petrol 61: 65–72. doi: 10.1306/d426768b-2b26-11d7-8648000102c1865d

149. Sidor CA, Angielczyk KD, Weide DM, Smith RMH, Nesbitt SJ, et al. (2010) Tetrapod fauna of the lowermost Usili Formation (Songea Group, Ruhuhu Basin) of southern Tanzania, with a new burnetiid record. J Vert Paleontol 30: 696–703. doi: 10.1080/02724631003758086

150. Schlüter T, Kohring R (1997) The fossil record of the Karoo in East Africa: history, palaeoecology and biostratigraphy. Neues Jahrbuch Geol Palaontol Abhand 204: 1–17.

151. Angielczyk KD, Sidor CA, Nesbitt SJ, Smith RMH, Tsuji LA (2009) Taxonomic revision and new observations on the postcranial skeleton, biogeography, and biostratigraphy of the dicynodont genus Dicynodontoides, the senior subjective synonym of Kingoria (Therapsida, Anomodontia). J Vert Paleontol 29: 1174–1187. doi: 10.1671/039.029.0427

152. Kemp TS (1969) On the functional morphology of the gorgonopsid skull. Philos Trans R Soc Lond B 256: 1–83. doi: 10.1098/rstb.1969.0036

153. Rubidge BS (1995) Biostratigraphy of the Beaufort Group (Karoo Supergroup). Pretoria: South African Committee for Stratigraphy, Biostratigraphic Series 1. 46 p.

154. Fröbisch J (2009) Composition and similarity of global anomodont-bearing tetrapod faunas. Earth Sci Rev 95: 119–175. doi: 10.1016/j.earscirev.2009.04.001

155. Angielczyk KD, Steyer J-S, Sidor CA, Smith RMH, Whatley RL, et al. (In press) Permian and Triassic dicynodont (Therapsida: Anomodontia) faunas of the Luangwa Basin, Zambia: taxonomic update and implications for dicynodont biogeography and biostratigraphy. In: Kammerer CF, Angielczyk KD, Fröbisch J, editors. The early evolutionary history of the Synapsida. Dordrecht: Springer.

156. Cox CB (1964) On the palate, dentition, and classification of the fossil reptile Endothiodon and related genera. Am Mus Novit 2171: 1–25.

157. Gradstein FM, Ogg JG, Schmitz MD, Ogg G (2012) The geologic time scale 2012. Boston: Elsevier. 1176 p.

158. Haeckel E (1866) Generelle Morphologie der Organismen. Allgemeine Grundzüge der organischen Formen-Wissenschaft, mechanisch begründet durch die von Charles Darwin reformirte Deszendenz Theorie. Zweiter Band: Allgemeine Entwicklungsgeschichte der Organismen. Berlin: Verlag von Georg Reimer. 462 p.

159. Osborn HF (1903) The reptilian subclasses Diapsida and Synapsida and the early history of the Diaptosauria. Mem Am Mus Nat Hist 1: 449–507.

160. von Huene F (1946) Die grossen Stämme der Tetrapoden in den geologischen Zeiten. Biol Zent Bl 65: 268–275.

161. Huxley TH (1871) A manual of the anatomy of the vertebrated animals. London: J & A Churchill. 432 p.

162. Ouwens PA (1912) On a large Varanus species from the island of Komodo. Bull Jardin Botanique Buitenzorg ser 2: 1–3.

163. Laurenti JN (1768) Specimen medicum, exhibens synopsin reptilium emendatam, cum experimentis circa venena et antidota reptilium Austriacorum. Vienna: Joan Thom Nob de Trattnem.

164. Lydekker R (1888) Catalogue of the fossil Reptilia and Amphibia in the British Museum (Natural History). Part I, the orders Ornithosauria, Crocodilia, Dinosauria, Squamata, Rhynchocephalia and Proterosauria. London: British Museum (Natural History). 309 p.

165. Bassani F (1886) Sui fossili e sull'eta degli schisti bituminosi triasici di Besano in Lombardia. Atti Societa Italiana di Scienze Naturali 29: 15–72.

166. Parrington FR (1935) On Prolacerta broomi, gen. et sp. n. and the origin of lizards. Ann Mag Nat Hist 16: 197–205. doi: 10.1080/00222933508655037

167. Sharov AG (1971) New flying reptiles from the Mesozoic of Kazakhstan and Kirghizia. Transactions of the Paleontological Institute, Akademia Nauk, USSR, Moscow 130: 104–113.

168. Pinna G (1980) Drepanosaurus unguicaudatus, nuovo genere e nuova specie di lepidosauro del Trias alpino (Reptilia). Atti Soc Ital Sci Nat Mus Civ Stor Nat Milano 121: 181–192.

169. Romer AS (1944) The Permian cotylosaur Diadectes tenuitectes. American J Sci 242: 139–144. doi: 10.2475/ajs.242.3.139

170. Renesto S (1992) The anatomy and relationships of Endennasaurus acutirostris (Reptilia: Neodiapsida) from the Norian (late Triassic) of Lombardy. Riv Ital Paleontol Stratigr 97: 409–430.

171. Wedel MJ (2007) What pneumaticity tells us about 'prosauropods', and vice versa. Spec Papers Palaeontol 77: 207–222.

172. Campione NE, Reisz RR (2010) Varanops brevirostris (Eupelycosauria: Varanopidae) from the Lower Permian of Texas, with discussion of varanopid morphology and interrelationships. J Vert Paleontol 30: 724–746. doi: 10.1080/02724631003762914

173. Storrs GW, Gower DJ (1993) The earliest possible choristodere (Diapsida) and gaps in the fossil record of semi-aquatic reptiles. J Geol Soc 150: 1103–1107. doi: 10.1144/gsjgs.150.6.1103

174. Ezcurra MD, Lecuona A, Martinelli A (2010) A new basal archosauriform diapsid from the Lower Triassic of Argentina. J Vert Paleontol 30: 1433–1450. doi: 10.1080/02724634.2010.501446

175. Arcucci A, Marsicano CA (1998) A distinctive new archosaur from the Middle Triassic (Los Chañares Formation) of Argentina. J Vert Paleontol 18: 228–232. doi: 10.1080/02724634.1998.10011046

176. Nesbitt SJ (2005) The osteology of the Middle Triassic pseudosuchian archosaur Arizonasaurus babbitti. Hist Biol 17: 19–47. doi: 10.1080/08912960500476499

177. Desojo JB, Ezcurra MD (2011) A reappraisal of the taxonomic status of Aetosauroides (Archosauria: Aetosauria) specimens from the Late Triassic of South America and their proposed synonymy with Stagonolepis. J Vert Paleontol 31: 596–609. doi: 10.1080/02724634.2011.572936

178. Yates AM (2003) A new species of the primitive dinosaur Thecodontosaurus (Saurischia: Sauropodomorpha) and its implications for the systematics of early dinosaurs. J Syst Palaeontol 1: 1–42. doi: 10.1017/s1477201903001007

179. Gardner JD (2003) The fossil salamander Proamphiuma cretacea Estes (Caudata; Amphiumidae) and relationships within the Amphiumidae. J Vert Paleontol 23: 769–782. doi: 10.1671/1828-4

180. Dalla Vecchia FB (2006) A new sauropterygian reptile with plesiosaurian affinity from the Late Triassic of Italy. Rivista Italiana di Paleontologia e Stratigrafia 112: 207–225.

181. Sood MS (1948) The anatomy of the vertebral column in serpentes. Proc Indian Aca Sci B 28: 1–26.

182. O'Connor PM (2006) Postcranial pneumaticity: an evaluation of soft-tissue influences on the postcranial skeleton and the reconstruction of pulmonary anatomy in archosaurs. J Morph 267: 1199–1226. doi: 10.1002/jmor.10470

183. Fox RC, Bowman MC (1966) Osteology and relationships of Captorhinus aguti (Cope) (Reptilia: Captorhinomorpha). Uni Kansas Paleontol Contr, Series Vertebrata 11: 1–79.

184. Young C-C (1964) The pseudosuchians in China. Palaeontol Sinica 151: 1–205.

185. Berman DS, Reisz RR (1982) Restudy of Mycterosaurus longiceps (Reptilia, Pelycosauria) from the Lower Permian of Texas. Ann Carnegie Mus 51: 423–453.

186. Reisz RR, Dilkes DW (2003) Archaeovenator hamiltonensis, a new varanopid (Synapsida: Eupelycosauria) from the upper Pennsylvanian of Kansas. Can J Earth Sci 40: 667–678. doi: 10.1139/e02-063

187. Cisneros JC (2008) Phylogenetic relationships of procolophonid parareptiles with remarks on their geological record. J Syst Palaeontol 6: 345–366. doi: 10.1017/s1477201907002350

188. Holmes RB (1977) The osteology and musculature of the pectoral limb of small captorhinids. J Morph 152: 101–140. doi: 10.1002/jmor.1051520107

189. Cox CB (1965) New Triassic dicynodonts from South America, their origins and relationships:. Philos Trans R Soc Lond B 248: 457–516. doi: 10.1098/rstb.1965.0005

190. Bulanov VV, Sennikov AG (2010) New data on the morphology of Permian gliding weigeltisaurid reptiles of Eastern Europe. Paleontol J 44: 682–694. doi: 10.1134/s0031030110060109

191. Sennikov AG (2005) A new specialized prolacertilian (Reptilia: Archosauromorpha) from the Lower Triassic of the Orenburg Region. Paleontol J 39: 199–209.

192. Ewer RF (1965) The anatomy of the thecodont reptile Euparkeria capensis Broom. Phil Trans R Soc London B 751: 379–435. doi: 10.1098/rstb.1965.0003

193. Benton MJ (1983) The Triassic reptile Hyperodapedon from Elgin: functional morphology and relationships. Philos Trans R Soc Lond B 302: 605–717. doi: 10.1098/rstb.1983.0079

194. Hunt AP, Lucas SG (1991) A new rhynchosaur from the Upper Triassic of West Texas, and the biochronology of Late Triassic rhynchosaurs. Palaeontology 34: 927–938.

195. Müller J, Berman DS, Henrici AC, Sumida SS, Martens T (2006) The basal eureptile Thuringothyris mahlendorffae from the Lower Permian of Germany. J Paleontol 80: 726–739. doi: 10.1666/0022-3360(2006)80[726:tbrtma]2.0.co;2

196. Heckert AB, Lucas SG, Rinehart LF, Celesky MD, Spielmann JA, et al. (2010) Articulated skeletons of the aetosaur Typothorax coccinarum Cope (Archosauria:

Stagonolepididae) from the Upper Triassic Bull Canyon Formation (Revueltian: early-mid Norian), eastern New Mexico, USA. J Vert Paleontol 30: 619–642. doi: 10.1080/02724631003763524

197. Langer MC, França MAG, Gabriel S (2007) The pectoral girdle and forelimb anatomy of the stem-sauropodomorph Saturnalia tupiniquim (Late Triassic, Brazil). In: Spec Papers Palaeontol Barrett PM, Batten DJ, Evolution and palaeobiology of early sauropodomorph dinosaurs. 77: 113–137.
198. Sen K (2003) Pamelaria dolichotrachela, a new prolacertid reptile from the Middle Triassic of India. J Asian Earth Sci 21: 663–681. doi: 10.1016/s1367-9120(02)00110-4
199. Peyer B (1955) Die Triasfauna der Tessiner Kalkalpen. XVIII. Helveticosaurus zollingeri, n.g. n.sp. Schweizerische Paläontol Abhand 72: 3–50.
200. Nosotti S, Rieppel O (2003) Eusaurosphargis dalsassoi n. gen n. sp., a new, unusual diapsid reptile from the Middle Triassic of Besano (Lombardy, N Italy). Memorie della Società Italiana di Scienze Naturali e del Museo Civico di Storia Naturale di Milano 31: 3–33.
201. Parham JF, Donoghue PCJ, Bell CJ, Calway TD, Head JJ, et al. (2012) Best practices for justifying fossil calibrations. Syst Biol 61: 346–359.
202. Li C, Wu X-C, Rieppel O, Wang L-T, Zhao L-J (2008) An ancestral turtle from the Late Triassic of southwestern China. Nature 456: 497–501. doi: 10.1038/nature07533
203. Roscher M, Schneider JW (2006) Permo-Carboniferous climate; Early Pennsylvanian to Late Permian climate development of Central Europe in a regional and global context. Non-marine Permian biostratigraphy and biochronology: In: Lucas SG, Cassinis G, Schneider JW, editors. Geol Soc, London, Spec Publ 265: 95–136. doi: 10.1144/gsl.sp.2006.265.01.05
204. Alfaro ME, Santinia F, Brock C, Alamillo H, Dornburg A, et al. (2009) Nine exceptional radiations plus high turnover explain species diversity in jawed vertebrates. Proc Nat Acad Sci 106: 13410–1341. doi: 10.1073/pnas.0811087106
205. Hugall AF, Foster R, Lee MS (2007) Calibration choice, rate smoothing, and the pattern of tetrapod diversification according to the long nuclear gene RAG-1. Syst Biol 56: 543–563.
206. Alroy J (2013) Online paleogeographic map generator. Available: http://fossilworks.org/?a=mapForm
207. Maisch MW (2002) Observations on Karoo and Gondwana vertebrates. Part 3: notes on the gorgonopsians from the Upper Permian of Tanzania. Neues Jahrb Geol Palaontol Monatsh 2002: 237–251.
208. Abdala F, Allinson M (2005) The taxonomic status of Parathrinaxodon proops (Therapsida: Cynodontia), with comments on the morphology of the palate in basal cynodonts. Palaeontol Afr 41: 45–52.
209. Angielczyk KD (2007) New specimens of the Tanzanian dicynodont "Cryptocynodon" parringtoni von Huene, 1942 (Therapsida, Anomodontia), with an expanded analysis of Permian dicynodont phylogeny. J Vertebr Paleontol 27: 116–131. doi: 10.1671/0272-4634(2007)27[116:nsottd]2.0.co;2
210. Weide DM, Sidor CA, Angielczyk KD, Smith RMH (2009) A new record of Procynosuchus delaharpeae (Therapsida: Cynodontia) from the Upper Permian Usili Formation, Tanzania. Palaeontol Afr 44: 21–26.
211. Colbert EH (1973) Continental drift and the distributions of fossil reptiles. In: Tailing DH, Runcorn SK, editors. Implications of continental drift for the Earth sciences. London: Academic Press. pp. 395–412.
212. Ponton F, Elzanowski A, Castanet J, Chinsamy A, de Margerie E, et al. (2004) Variation of the outer circumferential layer in the limb bones of birds. Acta Ornithol 39: 21–24. doi: 10.3161/068.039.0210
213. Botha-Brink J, Smith RMH (2011) Osteohistology of the Triassic archosauromorphs Prolacerta, Proterosuchus, Euparkeria, and Erythrosuchus from the Karoo Basin of South Africa. J Vert Paleontol 31: 1238–1254. doi: 10.1080/02724634.2011.621797
214. de Buffrénil V, Castanet J (2000) Age estimation by skeletochronology in the Nile Monitor (Varanus niloticus), a highly exploited species. J Herpetol 34: 414–424. doi: 10.2307/1565365
215. de Ricqlès AJ, Padian K, Horner JR (2003) On the bone histology of some Triassic pseudosuchian archosaurs and related taxa. Ann Paléontol 89: 67–101. doi: 10.1016/s0753-3969(03)00005-3
216. de Ricqlès AJ, Padian K, Knoll F, Horner JR (2008) On the origin of high growth rates in archosaurs and their ancient relatives: complementary histological studies on Triassic archosauriforms and the problems of a "phylogenetic signal" in bone histology. Ann Paléontol 94: 57–76. doi: 10.1016/j.annpal.2008.03.002
217. Werning S, Irmis RB (2010) Reconstructing the ontogeny of the Triassic basal archosauromorph Trilophosaurus using bone histology and limb bone morphometrics. J Vertebr Paleontol 30: 185A–186A.
218. Weigelt J (1989) Recent vertebrate carcasses and their paleobiological implications. Chicago: University Chicago Press. 204 p.
219. Munk W, Sues H-D (1993) Gut contents of Parasaurus (Pareiasauria) and Protorosaurus (Archosauromorpha) from the Kupferschiefer (Upper Permian) of Hessen, Germany. Paläont Z 67: 169–176. doi: 10.1007/bf02985876
220. Currey JD, Alexander RM (1985) The thickness of the walls of tubular bones. J Zool 206: 453–468. doi: 10.1111/j.1469-7998.1985.tb03551.x
221. Wall WP (1983) The correlation between limb-bone density and aquatic habits in recent mammals. J Vertebr Paleontol 57: 197–207.
222. Stein BR (1989) Bone density and adaptation in semi-aquatic mammals. J Mammal 70: 467–476. doi: 10.2307/1381418
223. Germain D, Laurin M (2005) Microanatomy of the radius and lifestyle in amniotes (Vertebrata, Tetrapoda). Zool Scripta 34: 335–350. doi: 10.1111/j.1463-6409.2005.00198.x

224. Nesbitt SJ, Butler RJ, Gower DJ (2013) A New Archosauriform (Reptilia: Diapsida) from the Manda Beds (Middle Triassic) of Southwestern Tanzania. PLoS ONE 8 (9) e72753 doi:10.1371/journal.pone.0072753.

225. Peyer B (1937) Die Triasfauna der Tessiner Kalkalpen XII. Macrocnemus bassanii Nopcsa. Abh Schweizerische Palaontol Geol Ges 1–140.

226. Rieppel O (1989) The hind limb of Macrocnemus bassanii (Nopcsa) (Reptilia, Diapsida): development and functional anatomy. J Vertebr Paleontol 9: 373–387. doi: 10.1080/02724634.1989.10011771

227. Welman J (1998) The taxonomy of the South African proterosuchids (Reptilia, Archosauromorpha). J Vert Paleontol 18: 340–347. doi: 10.1080/02724634.1998.10011062

228. Nosotti S (2007) Tanystropheus longobardicus (Reptilia, Protorosauria): re-interpretations of the anatomy based on new specimens from the Middle Triassic of Besano (Lombardy, Northern Italy). Mem Soc Ital Sci Nat Mus Civ Stor Nat Milano 35: 1–88

CHAPTER 18 INTEGRATING INSIGHTS ACROSS DIVERSE TAXA: CHALLENGES FOR UNDERSTANDING SOCIAL EVOLUTION

Mark A. Elgar

Keywords: Reproductive altruism, cooperation, polyandry, caste, facultative eusociality, cooperative breeding, social communication, pheromones

While it is broadly true that all sexually reproducing organisms have a "social life" (Trivers, 1985), social behavior outside courting and mating is not the modal animal lifestyle. Nevertheless, there is remarkable phylogenetic breadth and evolutionary convergence of the diverse forms of social behavior (e.g., Wilson, 1975), perhaps helping explain why it has captured the enduring attention of the curious. Social behavior is especially puzzling from an evolutionary perspective in at least two ways. The first reflects a Public Goodsor Tragedy of the Commons dilemma (e.g., Rankin et al., 2007): although social individuals can benefit from the underlying cooperation among group members, they are nonetheless susceptible to exploitation. The second perspective, arguably attracting far greater attention, concerns reproductive altruism—the question why individuals might forgo personal reproduction in order to assist the reproductive output of others. These two issues are not mutually exclusive: social systems that involve reproductive altruism are also vulnerable to commons tragedies. Nor are they particular to a given taxonomic group.

An enviably strong theoretical framework guides much of our broad understanding of the evolutionary significance of these and other issues in social behavior, so the devil is in the detail—to identify the nature of proximate and ultimate processes responsible for the evolution and maintenance of all manifestations of social behavior. Yet the empirical insights that derive from such investigations seem to be too frequently informed by taxonomically constrained traditions and, as a consequence, advances in the field are surprisingly patchy across different taxa. There are at least three sources of this unevenness: historical contingencies; logistical challenges with data collection; and different evolutionary processes. All are apparent in studies of the evolutionary significance of social behavior. My intention here is to highlight that unevenness, and to encourage a taxonomically more inclusive research agenda, which will guarantee that the conceptual models that attend to the detail are challenged for taxonomic generality. Embracing studies across diverse species, from the "model species" to the bizarre, and ensuring that these empirical insights cut across major taxonomic boundaries will help achieve this goal. For example, our understanding of cooperative breeding in vertebrates is overwhelmingly informed by studies of mammals and birds (despite only 8% of avian species being cooperative breeders, Jetz and Rubenstein, 2011), and of social behavior in invertebrates by studies of insects. This taxonomic imbalance potentially creates a distorted picture by overshadowing investigations of social behavior in other taxa, such as reptiles (e.g., While et al., 2009; Clark et al., 2012), marine invertebrates (e.g., Duffy et al., 2000) or spiders (e.g., Whitehousel and Lubin, 2005; Yip and Rayor, 2014). Such taxonomic unevenness is not restricted to studies of social evolution: past research into sexual selection through female choice was overwhelmingly dominated by studies of birds, while sexual selection through sperm competition was dominated by studies of insects. The taxonomic divide is less apparent in contemporary studies of sexual selection, although there is an emerging risk of theory being challenged by only a few "model species" (Zuk et al., 2014).

REPRODUCTIVE ALTRUISM

Darwin (1859) did not explicitly state the challenge of explaining reproductive altruism, but he certainly recognized

Citation: Mark A. Elgar, Integrating insights across diverse taxa: challenges for understanding social evolution, http://dx.doi.org/10.3389/fevo.2015.00124.

the problem that workers "often differ widely in instinct and in structure from both the males and fertile females, and yet, from being sterile, they cannot propagate their kind." For Darwin, the "difficulty, which at first appeared …fatal to my whole theory" was how the sterile workers "could have transmitted successively acquired modifications of structure or instinct to its progeny." Darwin's Difficulty of Sterile Castes is often interpreted as that of "reproductive altruism" (see Herbers, 2009), but it is clear that he was concerned how selection acts on any trait (including caring for others› offspring) in an organism that does not reproduce (see Ratnieks et al., 2011). The broad solution to his problem emerged some 100 years later.

Following Hamilton (1964a,b, 1972), the role of inclusive fitness, which includes the direct and indirect components of natural selection, has dominated our understanding of the evolution of reproductive altruism, and of sociality more generally (see the Special Issue in Biology Letters, Herbers, 2013). Historically, the genetic component of this conceptual framework attracted far greater interest in empirical studies of social insects than in other taxa (see also Herbers, 2013). Perhaps this is hardly surprising, given the apparently exquisite link between the sex-determining mechanism of hymenopterans, which predisposes a high degree of relatedness between sisters, and the frequency with which eusociality has evolved within that taxon. This meant that, initially, the primary focus of research in social insects, and especially the hymenoptera, was on the underlying genetic conditions, and in particular the degree of relatedness between cooperating individuals. In contrast, research into the costs and benefits component of Hamilton›s Rule was a far greater focus for studies of cooperative breeding in vertebrates (see Cockburn, 1998; but see Hatchwell et al., 2014). The taxonomic divide is less apparent now, especially since it became clear that haplodiploidy is not the primary driver of eusociality in the hymenoptera (Gadagkar, 1991), with maternal care and nest defense more likely candidates (see Ross et al., 2013). Nowadays, differences of perspective more obviously reflect the important taxonomic difference in the nature of cooperative behaviors: investigating direct direct fitness and cost/benefit issues are important where no permanent caste division occurs, while genetic considerations become more important in species in which the allocation to a particular caste is permanent. Thus, vertebrate "helpers," capable of breeding, typically characterize the former, while studies of social insects have tended to focus on the highly eusocial species that characterize the latter (e.g., Jetz and Rubenstein, 2011; Bourke, 2014).

Nevertheless, the similarities between cooperatively breeding vertebrates and the primitively social bees and wasps often remain overlooked, despite the successful application of reproductive skew models (see Nonacs and Hager, 2011) as a shared conceptual framework. Empirical research into the evolution of cooperative breeding in vertebrates and invertebrates still seems to proceed largely independently, with reviews of cooperative breeding often retaining a taxonomic focus (but see Russell and Lummaa, 2009; Bourke, 2014; English et al., 2015). Field experiments of insects are often logistically less challenging than for vertebrates, but the insights are no less transferable (e.g., Cockburn et al., 2008; Leadbeater et al., 2011; Schwarz et al., 2011). For example, field experiments on primitively social wasps successfully identified the role of key ecological selection pressures favoring extended care and additional careers (e.g., Field et al., 2000; Field and Brace, 2004). Species that adjust their social behavior facultatively are also ideal for helping us understand the ecological and genetic influences on the evolution of social behavior (e.g., Field et al., 2010; Hatchwell et al., 2013; Rehan et al., 2014; Kapheim et al., 2015b). Jeremy Field and colleagues revealed, through clever field transplant experiments, that eusocial behavior in the sweat bee Halictus rubicundus is more likely if there is more time available for offspring production (Field et al., 2010). These results lend support to the view that iteroparity, rather than semelparity, is the predominant pattern for insects with subsocial behavior (Trumbo, 2013), which is typically regarded as a precursor to eusociality (e.g., Queller and Strassmann, 1998). Maintaining a strong taxonomic focus may also obscure broader patterns. Interestingly, Ben Hatchwell and colleagues long-term study of the facultatively cooperative breeding long-tailed tit Aegithalos caudatus reveals a negative correlation between breeding season length and the number of helpers in the social group (Hatchwell et al., 2013). This apparent inconsistency may derive from the temporal scale of iteropary that reflects different life-cycles: Halictus "non-helpers" breed in the same season as the social bees, but long-tailed tits may be "non-helpers" in order to improve the likelihood of surviving to breed the following season.

The heuristic value of inclusive fitness theory in helping explain the evolution of social behavior, and a raft of other biological adaptations, is widely recognized (e.g., Abbot et al., 2011; Bourke, 2011a,b, 2014). In particular, support comes from many empirical studies of social insects that use genetic and demographic data to test directly Hamilton›s Rule (Bourke, 2014), especially when applied to conflicts of interest within social insect societies (Bourke, 2011a; see also Rubenstein, 2012). Perhaps the number of studies might be greater, if there was broader clarity of how inclusive fitness can be measured (see Nonacs and Richards, 2015). But arguably more importantly, Andrew Bourke documents the success of Hamilton›s Rule in providing a unifying, cross-taxa framework for identifying the ecological, demographic and physiological processes that are likely to favor various forms of social behavior, including when cooperative behavior is (e.g., Schneider and Bilde, 2008) or is not necessarily restricted to interactions between kin (e.g., Clutton-Brock, 2002; Riehl, 2013).

Mating Systems

Indeed, Hamilton's Rule should focus attention on the ecological and life-history traits that influence the elements of the inequality, $rb–c > 0$. For example, theoretical modeling has long highlighted how mating systems can have an impact on relatedness (e.g., Crozier and Pamilo, 1996), but subsequent empirical investigations seem to have focused more on the outcome for sex allocation, perhaps following Trivers and Hare (1976). Koos Boomsma (2007, 2009, 2013) noted that the evolution of sterile castes is more likely to occur under monandry, since this ensures high levels of within-group relatedness. This view seems to have galvanized a taxonomically integrated investigation, as support is provided by inter-specific comparative studies of insects, birds, and mammals. Hughes et al. (2008) used ancestral state reconstruction to show that monandry was the ancestral state in eight independent lineages leading to the evolution of eusociality in the hymenoptera. Similarly, the non-cooperative breeding ancestors of cooperative breeding birds typically had lower levels of polyandry than non-cooperative ancestors of non-cooperative breeders (Cornwallis et al., 2010). Lukas and Clutton-Brock (2012) also report that the evolution of cooperative breeding in mammals was confined to socially monogamous species, whose offspring are likely to be close kin since socially monogamous species are typically not polyandrous. The consistency of these patterns across these major taxonomic groups is impressive, especially because each study used slightly different methods, reflecting in part the different logistical challenges of data collection in different taxonomic groups. For example, Hughes et al. (2008) inferred the mating system of species from genetic data, while Lukas and Clutton-Brock (2012) inferred it primarily from behavioral data. Cornwallis et al. (2010) obtained a continuous measure of polyandry, inferred from genetic data, whereas Hughes et al. (2008) distinguished between monandry and polyandry according to whether the female mated with fewer or more than two males respectively. The patterns for vertebrates are interesting because the mating system is expected to specifically influence the evolution of eusociality (Boomsma, 2007), which Hughes et al. (2008) tested by comparing species in which workers are totipotent (can reproduce) or not. It is, perhaps, intriguing that such relaxed definitions of monogamy yield significant patterns, because even low levels of polyandry will substantially reduce the levels of within-brood relatedness (but see Boomsma and d›Ettorre, 2013).

Perhaps more intriguing, is the absence (or at least very low levels) of polyandry in the ancestors of species with reproductive altruism. The ubiquity of polyandry across multicellular animals suggests that it is the ancestral state, allowing females to exercise strategic, non-random mating decisions to improve their reproductive success (Kvarnemo and Simmons, 2013). Indeed, studies of eusocial insects were amongst the first to identify the benefits of polyandry (e.g., Schmid-Hempel and Crozier, 1999). However, polyandry reduces male fitness through shared paternity, and thus creates a sexual conflict (Arnqvist and Rowe, 2005; Parker and Birkhead, 2013), with male-imposed monandry a widespread, and probably more common, consequence. Monogamy in many parasitoid wasps appears to be male induced (e.g., Leonard and Boake, 2008; Ablard et al., 2011), and in other species is favored when local mating patchiness increases (e.g., Boulton and Shuker, 2015). If monandry is a significant pre-condition of cooperative and eusocial behavior under certain circumstances (see Leggett et al., 2012; Nonacs, 2014), then it would be interesting to know why it occurs, given there is likely to be strong opposing selection as a female mating strategy. Further, the relationship between polyandry (or multiple mating) and mixed parentage is not necessarily straightforward, especially if there is an element of cryptic female choice (e.g., Den Boer et al., 2010; Pryke et al., 2010; Boomsma, 2013; Kvarnemo and Simmons, 2013). Indeed, the term promiscuous, frequently used in the context of polyandry and social behavior, may be unhelpful as it conflates the nature (discriminating or not) with the frequency of mating (Elgar et al., 2013), thereby obscuring the importance of female and male mating strategies in the evolution of mating systems (e.g., Leggett et al., 2012). In any event, monandry is unlikely to have evolved as a mechanism to promote social behavior, and as Boomsma (2013) wisely counsels, we require a much richer understanding of the interplay between female mating and parental care strategies across the taxonomic spectrum.

These phylogenetic analyses also highlight the multiple losses of cooperative or social breeding in diverse species (e.g., Danforth et al., 2003, 2013; Hughes et al., 2008; Cornwallis et al., 2010). Investigations of the evolutionary loss of social behavior could provide significant insights, but remarkably this has not been a fashionable field of investigation (see Kranz et al., 2002; Beauchamp, 2004; Liebert et al., 2005; Jiricny et al., 2014). As already mentioned above, studies of facultative reproductive altruism are likely to be helpful, as it is possible to investigate simultaneously the factors favoring a loss or gain in social behavior.

Phylogenies, Genomics and Insights into Social Evolution

Comparative analyses have been a powerful analytic tool for studies of social evolution, and the current armory of molecular genetic techniques provide further exciting opportunities to address a range of issues (Evans and Wheeler, 2001; Page and Amdam, 2007; Danforth et al., 2013). For example, differences in gene expression and DNA methylation across different castes in social insects are providing remarkable insights into how the interplay between genes and environment cause both caste differentiation and subsequent behavioral plasticity (see Linksvayer et al., 2012; Libbrecht et al., 2013; Yan et al., 2015). Linking population-wide variation in social behavior (Kocher et al., 2015) with genomics may provide rich insights into the separate and combined influences of genetic and environmental factors on social behavior (Kocher et al., 2013), and their consequences for other life history attributes. Inter-specific comparisons of the genome sequences of bees with varying complexity of social behavior reveal that while independent evolutionary transitions to eusociality have different genetic underpinnings, they nonetheless increase in the complexity of gene networks (Kapheim et al., 2015a). A fascinating comparative transcriptome-wide analysis of single representatives of bees, ants and wasps reveals that the convergent evolution of caste determination across these lineages involves similar metabolic pathways and molecular functions, but not exactly the same genes (Berens et al., 2015). The importance of social communication is highlighted by Zhou et al. (2015), whose comparison of the chemoreceptor genomes of a solitary wasp, a facultatively eusocial halictid bee and several species of ants revealed positive selection on these chemoreceptor genes.

The recent publication of almost 50 avian genomes, representing at least one species of all avian orders (see Jarvis et al., 2014; Zhang et al., 2014), will likely ensure that we can similarly draw on a rich diversity of avian species, rather than focus on a few model systems.

Cooperation Beyond Reproduction

My focus thus far has been on the varying manifestations of reproductive altruism, perhaps reflecting the majority interest in this form of social behavior. Indeed, the term social insect seems to have a rather specific meaning—namely those species that exhibit varying levels of reproductive altruism—despite the diversity of social behaviors among insects and other invertebrates that fall outside the traditional classifications that emphasize parent-offspring interactions (e.g., Wilson, 1971, 1975; Choe and Crespi, 1997; Costa, 2006; Székely et al., 2010). While these aggregations can have defensive properties (Costa, 2006), they may also involve subtle levels of cooperation over foraging: for example, aggregations of many insect larvae comprise leaders and followers, with the former initiating movement of the group from their roosting to foraging sites. Leadership behavior occurs in diverse social organizations across the taxonomic spectrum, and is thought to improve the efficiency of group decision-making (Conradt and Roper, 2007; King and Cowlishaw, 2009). Drawing on insights from across these social organizations has proved helpful. Leadership in some species is despotic (e.g., King et al., 2008), suggesting the benefits of leadership may come at a cost to followers. However, consensual leadership in social insects benefits the following workers, through the collective success of the colony (e.g. Seeley, 2010). The latter highlights how our understanding of consensual leadership requires consideration of the benefits of leadership to followers. For example, leaders of foraging aggregations of the steel-blue sawfly Perga affinis do not have higher growth rates than followers, but individual sawflies in experimental groups comprising leaders and followers have higher growth rates than those in groups of either only leaders or only followers (Hodgkin et al., 2014).

Social behavior involves cooperation, which is inevitably vulnerable to exploitation. For example, the benefits that individuals in a group obtain from their collective foraging activities will be diminished if some individuals do not forage but rather depend on the searching skills of others (Barnard and Sibly, 1981). Similarly, the early detection of predators may depend on the corporate vigilance of individuals in the group, but an individual could benefit more by spending less time vigilant, and thus more time foraging, without substantially reducing the likelihood that the group detects the predator (see Elgar, 1989). More complex social systems are similarly vulnerable to exploitation, as Jon Seger so eloquently observed: "insect sociality ... portrays an almost melodramatically volatile politics, in which individuals attempt to gain advantage by taking part in profoundly cooperative relationships, but none the less find themselves constantly tempted to cheat" (Seger, 1991, p. 338).

An extensive body of game theory examines the cooperative resolution of public goods, prisoner›s dilemma, and other social games (extensively reviewed in Sigmund, 2010)—far exceeding the number of empirical tests beyond laboratory games involving university students of psychology or business studies. The temptation to cheat occurs in social systems with or without reproductive altruism, and there are both vertebrate and invertebrate examples of behaviors in the former social systems that have evolved apparently in response to these conflicts (e.g. Mulder and Langmore, 1993; Wenseleers and Ratnieks, 2006; Ratnieks and Wenseleers, 2008; Zanette et al., 2012; Fischer et al., 2014). The underlying mechanisms that allow individuals to recognize "cheating" events and respond appropriately have been well documented in social insects, but seem to be less well understood in vertebrate systems. Arguably, more complex mechanisms of recognizing and responding to cheating are expected among those inter-specific cooperative, or mutualistic, associations that involve social behavior (Raihani et al., 2012)—such as between ants and the larvae of lycaenid butterflies (e.g., Pierce et al., 2002) or associations involving cleaner fish (Vail et al., 2013). Particularly interesting are those inter-specific systems—from "slave-making ants" (e.g., Buschinger, 2009) to microbes (e.g., Lopez et al., 2011) in which the exploited species is unable to detect or respond to the exploiting species.

Interestingly, microorganisms are emerging as especially tractable model systems to test theoretical accounts of public goods games. The behavior of these organisms has strikingly similar social functions with that of invertebrates and vertebrates, including group hunting, social communication, creation of domiciles and specialized roles (see Crespi, 2001). These systems have provided remarkable accounts of the conflicts and confluences of interest in social organizations (e.g., Jiricny et al., 2010; Strassmann and Queller, 2011; Cordero et al., 2012; Popat et al., 2012; Ghoul et al., 2014; Oliveira et al., 2014), providing insights into the evolution of multicellularity (e.g., Michod and Roze, 2001; Grosberg and Strathmann, 2007; Biernaskie and West, 2015; Levin et al., 2015), and revealing intriguing practical implications (e.g., Leggett et al., 2014; Webster, 2014).

Communication, Signals and Control

Communication is crucial for cooperative relationships. Broadly, social communication can provide information about individual or group identity, and/or about fitness-relevant traits, such as fertility. At the very least, individuals must produce and perceive signals that identify themselves to appropriate partners. Our understanding of the role of these signals in social behavior varies across taxa. Chemical signals involved in the communication systems of social insects, especially involving individual or colony recognition, have been investigated very extensively (van Zweden and d'Ettorre, 2010; Sturgis and Gordon, 2012; Richard and Hunt, 2013), with varying degrees of experimental rigor (van Wilgenburg and Elgar, 2013). In contrast, fewer studies have investigated how cooperative behavior in vertebrates is similarly facilitated by either olfactory (e.g., Mateo, 2003; Charpentier et al., 2010; Le Vin et al., 2010; Krause et al., 2012; Leclaire et al., 2013) or vocal (e.g., Rendall et al., 1996; Sharp et al., 2005; McDonald and Wright, 2011; Keen et al., 2013) recognition cues.

For social insects, arguably the most crucial impact that pheromones can have on fitness-relevant traits is their role in the reproductive division of labor (see Oi et al., 2015a). It is widely understood that queen derived pheromones effectively suppress worker reproduction, and the identity of these pheromones has been identified in several species (e.g., Van Oystaeyen et al., 2014; Oi et al., 2015a). However, there is some lively debate whether these pheromones represent queen control of worker reproduction, against the interests of the workers, or a signal informing the workers of queen's reproductive state and thus colony health (Oi et al., 2015a). These "alternative" positions may, in fact, represent explanations at different levels of analysis (Peso et al., 2015): the strong association between the presence of queen derived pheromone and little, or no, worker reproduction is consistent with the view that, at the proximate level, the pheromone controls the behavior of the workers. However, at the ultimate level, a pheromone that represents a signal against the interests of the receiver will exert strong selection on the receiver to avoid that control. One outcome of the resulting co-evolutionary arms race is a proliferation of diverse "worker control" pheromones—as selection favors modified queen pheromone to counteract selection on worker "resistance" (Peso et al., 2015). Such a diversity of pheromones is not apparent (Holman et al., 2013; van Zweden et al., 2013; Van Oystaeyen et al., 2014; Oi et al., 2015b), suggesting that the pheromone is more likely an honest signal of queen fertility, and the workers respond accordingly.

One surprisingly neglected component of communication concerns the reception capacity of receivers (Gill et al., 2013). Typically, studies of signaling and communication evolution focus on the signal and how it might reflect attributes of the signaler. But communication involves both signalers and receivers, with theory predicting co-evolution between the two. Thus, features of the receiver exert selection on the design of the signal and features of the signal exert selection on the ability of the receiver to detect it, with the latter manifested in the structure of the receptors and/or receptor organs (e.g., Endler, 1993). Perhaps social evolution, with its greater reliance on communication, also requires the evolution of increasingly sensitive sensory receptors. For example, the sensory structures of facultatively social species may vary with the complexity of the social organization.

Driven by Theory, Unconstrained by Taxa

Our broad understanding of the evolution of social behavior has benefited from an enviably strong, taxonomically unconstrained, conceptual framework that guides empirical studies. New theoretical perspectives continue to emerge (e.g., Port et al., 2011; Van Dyken and Wade, 2012a,b; McGlothlin et al., 2014; Van Cleve and Akçay, 2014; Avila and Fromhage, 2015) as well as new methodological approaches (e.g., Kurvers et al., 2014), and both will undoubtedly guide future avenues for research. And yet, some quite fundamental questions remain—such as why is eusociality so much more prevalent in terrestrial than aquatic or marine ecosystems (Ruxton et al., 2014)? While important discoveries will continue to emerge from studies of model species, our understanding of social evolution will undoubtedly be enriched by studies of both "model" and less fashionable or quirky taxa, and by a far greater integration of these empirical insights across taxonomic divides. Conceptual models should be tested against diverse taxa (and even against different populations of the same species to ensure genuine replication—see Parker and Nakagawa, 2014), and synthetic reviews should be taxonomically inclusive. The challenge for Frontiers in Social Evolution is to facilitate this integration, by providing a taxonomically unconstrained focus on the ideas and empirical insights about the evolution of all manners of social behavior, thereby allowing a richer appreciation of among the most impressive and exquisite consequences of natural selection.

Conflict of Interest Statement

The author declares that the research was conducted in the absence of any commercial or financial relationships that could be construed as a potential conflict of interest.

ACKNOWLEDGMENTS

I thank Dustin Rubenstein, Sebastian Pohl, and the two reviewers for their insightful comments that vastly improved the breadth and quality of this piece, and the Australian Research Council for support (DP120100162).

REFERENCES

1. Abbot, P., Abe, J., Alcock, J., Alizon, S., Alpedrinha, J. A. C., Andersson, M., et al. (2011). Inclusive fitness theory and eusociality. Nature 471, E1–E2. doi: 10.1038/nature09831
2. Ablard, K., Fairhurst, S., Andersen, G., Schaefer, P., and Gries, G. (2011). Mechanisms, functions, and fitness consequences of pre- and post-copulatory rituals of the parasitoid wasp Ooencyrtus kuvanae. Ent. Exp. Appl. 140, 103–111. doi: 10.1111/j.1570-7458.2011.01137.x
3. Arnqvist, G., and Rowe, L. (2005). Sexual Conflict. Princeton, NJ: Princeton University Press.
4. Avila, P., and Fromhage, L. (2015). No synergy needed: ecological constraints favor the evolution of eusociality. Am. Nat. 186, 31–40. doi: 10.1086/681637
5. Barnard, C. J., and Sibly, R. M. (1981). Producers and scroungers - a general model and its application to captive flocks of house sparrows. Anim. Behav. 29, 543–550. doi: 10.1016/S0003-3472(81)80117-0
6. Beauchamp, G. (2004). Reduced flocking by birds on islands with relaxed predation. Proc. R. Soc. B271, 1039–1042. doi: 10.1098/rspb.2004.2703
7. Berens, A. J., Hunt, J. H., and Toth, A. L. (2015). Comparative transcriptomics of convergent evolution: different genes but conserved pathways underlie caste phenotypes across lineages of eusocial insects. Mol. Biol. Evol. 32, 690–703. doi: 10.1093/molbev/msu330
8. Biernaskie, J. M., and West, S. A. (2015). Cooperation, clumping and the evolution of multicellularity. Proc. R. Soc. B 282, 20151075. doi: 10.1098/rspb.2015.1075
9. Boomsma, J. J. (2007). Kin selection versus sexual selection: why the ends do not meet. Curr. Biol.17, R673–R683. doi: 10.1016/j.cub.2007.06.033
10. Boomsma, J. J. (2009). Lifetime monogamy and the evolution of eusociality. Phil. Trans. R. Soc. B364, 3191–3207. doi: 10.1098/rstb.2009.0101
11. Boomsma, J. J. (2013). Beyond promiscuity: mate-choice commitments in social breeding. Phil. Trans. R. Soc. B 368, 20120050. doi: 10.1098/rstb.2012.0050
12. Boomsma, J. J., and d›Ettorre, P. (2013). Nice to kin and nasty to non-kin: revisiting Hamilton›s early insights on eusociality. Biol. Lett. 9, 20130444. doi: 10.1098/rsbl.2013.0444
13. Boulton, R. A., and Shuker, D. M. (2015). The costs and benefits of multiple mating in a mostly monandrous wasp. Evolution 69, 939–949. doi: 10.1111/evo.12636
14. Bourke, A. F. G. (2011a). Principles of Social Evolution. (Oxford: Oxford University Press).
15. Bourke, A. F. G. (2011b). The validity and value of inclusive fitness theory. Proc. R. Soc. B 278, 3313–3320. doi: 10.1098/rspb.2011.1465

16. Bourke, A. F. G. (2014). Hamilton›s rule and the causes of social evolution. Phil. Trans. R. Soc. B369, 20130362. doi: 10.1098/rstb.2013.0362

17. Buschinger, A. (2009). Social parasitism among ants: a review (Hymenoptera: Formicidae). Myrmecol. News 12, 219–235.

18. Charpentier, M. J. E., Crawford, J. C., Boulet, M., and Drea, C. M. (2010). Message 'scent': lemurs detect the genetic relatedness and quality of conspecifics via olfactory cues. Anim. Behav. 80, 101–108. doi: 10.1016/j.anbehav.2010.04.005

19. Choe, J. C., and Crespi, B. J. (eds.). (1997). Social Behavior in Insects and Arachnids. (Cambridge: Cambridge University Press).

20. Clark, R. W., Brown, W. S., Stechert, R., and Greene, H. W. (2012). Cryptic sociality in rattlesnakes (Crotalus horridus) detected by kinship analysis. Biol. Lett. 8, 523–525. doi: 10.1098/rsbl.2011.1217

21. Clutton-Brock, T. (2002). Kin selection and mutualism in cooperative vertebrates. Science 296, 69–72. doi: 10.1126/science.296.5565.69

22. Cockburn, A. (1998). Evolution of helping behavior in cooperatively breeding birds. Ann. Rev. Ecol. Syst. 29, 141–177. doi: 10.1146/annurev.ecolsys.29.1.141

23. Cockburn, A., Osmond, H. L., Mulder, R. A., Double, M. C., and Green, D. J. (2008). Demography of male reproductive queues in cooperatively breeding superb fairy-wrens Malurus cyaneus. J. Anim. Ecol. 77, 297–304. doi: 10.1111/j.1365-2656.2007.01335.x

24. Conradt, L., and Roper, T. J. (2007). Democracy in animals: the evolution of shared group decisions. Proc. R. Soc. B, 274, 2317–2326. doi: 10.1098/rspb.2007.0186

25. Cordero, O. X., Wildschutte, H., Kirkup, B., Proehl, S., Ngo, L., Hussain, F., et al. (2012). Ecological populations of bacteria act as socially cohesive units of antibiotic production and resistance. Science 337, 1228–1231. doi: 10.1126/science.1219385

26. Cornwallis, C. K., West, S. A., Davis, K. E., and Griffin, A. S. (2010). Promiscuity and the evolutionary transition to complex societies. Nature 466, 969–972. doi: 10.1038/nature09335

27. Costa, J. T. (2006). The Other Insect Societies. (Cambridge, MA: Harvard University Press).

28. Crespi, B. J. (2001). The evolution of social behavior in microorganisms. Trends Ecol. Evol. 16, 178–183. doi: 10.1016/S0169-5347(01)02115-2

29. Crozier, R. H., and Pamilo, P. (1996). Evolution of Social Insect Colonies: Sex Allocation and Kin-selection. (Oxford: Oxford University Press).

30. Danforth, B. N., Cardinal, S., Praz, C., Almeida, E. A. B., and Michez, D. (2013). The impact of molecular data on our understanding of bee phylogeny and evolution. Annu. Rev. Entomol. 58, 57–78. doi: 10.1146/annurev-ento-120811-153633

31. Danforth, B. N., Conway, L., and Ji, S. Q. (2003). Phylogeny of eusocial Lasioglossum reveals multiple losses of eusociality within a primitively eusocial clade of bees (Hymenoptera: Halictidae). Syst. Biol. 52, 23–36. doi: 10.1080/10635150390132687

32. Darwin, C. (1859). On the origin of species. (London: Murray).

33. den Boer, S. P., Baer, B., and Boomsma, J. J. (2010). Seminal fluid mediates ejaculate competition in social insects. Science 327, 1506–1509. doi: 10.1126/science.1184709

34. uffy, J. E., Morrison, C. L., and Ríos, R. (2000). Multiple origins of eusociality among sponge-dwelling shrimps (Synalpheus). Evolution 54, 503–516. doi: 10.1111/j.0014-3820.2000.tb00053.x

35. Elgar, M. A. (1989). Predator vigilance and group size in mammals and birds: a critical review of the empirical evidence. Biol. Rev. 64, 13–33. doi: 10.1111/j.1469-185X.1989.tb00636.x

36. Elgar, M. A., Jones, T. M., and McNamara, K. B. (2013). Promiscuous words. Front. Zool. 10:66. doi: 10.1186/1742-9994-10-66

37. Endler, J. A. (1993). Some general comments on the evolution and design of animal communication systems. Phil. Trans. R. Soc. B 340, 215. doi: 10.1098/rstb.1993.0060

38. English, S., Browning, L. E., and Raihani, N. J. (2015). Developmental plasticity and social specialization in cooperative societies. Anim. Behav. 106, 37–42. doi: 10.1016/j.anbehav.2015.05.006

39. Evans, J. D., and Wheeler, D. E. (2001). Gene expression and the evolution of insect polyphenisms. Bioessays 23, 62–68. doi: 10.1002/1521-1878(200101)23:1<62::AID-BIES1008>3.3.CO;2-Z

40. Field, J., and Brace, S. (2004). Pre-social benefits of extended parental care. Nature 428, 650–652. doi: 10.1038/nature02427

41. Field, J., Paxton, R. J., Soro, A., and Bridge, C. (2010). Cryptic plasticity underlies a major evolutionary transition. Curr. Biol. 20, 2028–2031. doi: 10.1016/j.cub.2010.10.020

42. Field, J., Shreeves, G., Sumner, S., and Casiraghi, M. (2000). Insurance-based advantage to helpers in a tropicalhover wasp. Nature 404, 869–871. doi: 10.1038/35009097

43. Fischer, S., Zöttl, M., Groenewoud, F., and Taborsky, B. (2014). Group-size dependent punishment of idle subordinates in a cooperative breeder where helpers pay to stay. Proc. R. Soc. B 281, 20140184. doi: 10.1098/rspb.2014.0184

44. Gadagkar, R. (1991). On testing the role of genetic asymmetries created by haplodiploidy in the evolution of eusociality in the Hymenoptera. J. Genet. 70, 1–31. doi: 10.1007/BF02923575

45. Ghoul, M., West, S. A., Diggle, S. P., and Griffin, A. S. (2014). An experimental test of whether cheating is context dependent. J. Evol. Biol. 27, 551–556. doi: 10.1111/jeb.12319

46. Gill, K. P., van Wilgenburg, E., Macmillan, D., and Elgar, M. A. (2013). Density of antennal sensilla influences

efficacy of communication in a social insect. Am. Nat. 182, 834–840. doi: 10.1086/673712

47. Grosberg, R. K., and Strathmann, R. R. (2007). The evolution of multicellularity: a minor major transition? Ann. Rev. Ecol. Evol. Syst. 38, 621–654. doi: 10.1146/annurev.ecolsys.36.102403.114735
48. Hamilton, W. D. (1964a). The genetical evolution of social behaviour. I. J. Theor. Biol. 7, 1–16. doi: 10.1016/0022-5193(64)90038-4
49. Hamilton, W. D. (1964b). The genetical evolution of social behaviour. II. J. Theor. Biol. 7, 17–52. doi: 10.1016/0022-5193(64)90039-6
50. Hamilton, W. D. (1972). Altruism and related phenomena, mainly in social insects. Ann. Rev. Ecol. Syst. 3, 192–232. doi: 10.1146/annurev.es.03.110172.001205
51. Hatchwell, B. J., Gullett, P. R., and Adams, M. J. (2014). Helping in cooperatively breeding long-tailed tits: a test of Hamilton›s rule. Phil. Trans. R. Soc. B 369, 20130565. doi: 10.1098/rstb.2013.0565
52. Hatchwell, B. J., Sharp, S. P., Beckerman, A. P., and Meade, J. (2013). Ecological and demographic correlates of helping behaviour in a cooperatively breeding bird. J. Anim. Ecol. 82, 486–494. doi: 10.1111/1365-2656.12017
53. Herbers, J. M. (2009). Darwin›s 'one special difficulty': celebrating Darwin 200. Biol. Lett. 5, 214–217. doi: 10.1098/rsbl.2009.0014
54. Herbers, J. M. (2013). 50 years on: the legacy of William Donald Hamilton. Biol. Lett. 9, 214–217. doi: 10.1098/rsbl.2013.0792
55. Hodgkin, L. K., Symonds, M. R. E., and Elgar, M. A. (2014). Leaders benefit followers in the collective movement of a social sawfly. Proc. R. Soc. B 281, 20141700. doi: 10.1098/rspb.2014.1700
56. Holman, L., Lanfear, R., and d›Ettorre, P. (2013). The evolution of queen pheromones in the ant genus Lasius. J. Evol. Biol. 26, 1549–1558. doi: 10.1111/jeb.12162
57. Hughes, W. O. H., Oldroyd, B. P., Beekman, M., and Ratnieks, F. L. W. (2008). Ancestral monogamy shows kin selection is key to the evolution of eusociality. Science 320, 1213–1216. doi: 10.1126/science.1156108
58. Jarvis, E. D., Mirarab, S., Aberer, A. J., Li, B., House, P., Li, C., et al. (2014). Whole-genome analyses resolve early branches in the tree of life of modern birds. Science 346, 1320–1331. doi: 10.1126/science.1253451
59. Jetz, W., and Rubenstein, D. R. (2011). Environmental uncertainty and the global biogeography of cooperative breeding in birds. Curr. Biol. 21, 72–78. doi: 10.1016/j.cub.2010.11.075
60. Jiricny, N., Diggle, S. P., West, S. A., Evans, B. A., Ballantyne, G., Ross-Gillespie, A., et al. (2010). Fitness correlates with the extent of cheating in a bacterium. J. Evol. Biol. 23, 738–747. doi: 10.1111/j.1420-9101.2010.01939.x
61. Jiricny, N., Molin, S., Foster, K., Diggle, S. P., Scanlan, P. D., Ghoul, M., et al. (2014). Loss of social behaviours in populations of Pseudomonas aeruginosa infecting lungs of patients with cystic fibrosis. PLoS ONE, 9:e83124. doi: 10.1371/journal.pone.0083124
62. Kapheim, K. M., Nonacs, P., Smith, A. R., Wayne, R. K., and Wcislo, W. T. (2015b). Kinship, parental manipulation and evolutionary origins of eusociality. Proc. R. Soc. B 282, 20142886. doi: 10.1098/rspb.2014.2886
63. Kapheim, K. M., Pan, H., Li, C., Salzberg, S. L., Puiu, D., Magoc, T., et al. (2015a). Genomic signatures of evolutionary transitions from solitary to group living. Science 348, 1139–1143. doi: 10.1126/science.aaa4788
64. Keen, S. C., Meliza, C. D., and Rubenstein, D. R. (2013). Flight calls signal group and individual identity but not kinship in a cooperatively breeding bird. Behav. Ecol. 24, 1279–1285. doi: 10.1093/beheco/art062
65. King, A. J., and Cowlishaw, G. (2009). Leaders, followers and group decision-making. Commun. Integr. Biol. 2, 147–150 doi: 10.4161/cib.7562
66. King, A. J., Douglas, C. M. S., Huchard, E., Isaac, N. J. B., and Cowlishaw, G. (2008). Dominance and affiliation mediate despotism in a social primate. Curr. Biol. 18, 1833–1838. doi: 10.1016/j.cub.2008.10.048
67. Kocher, S. D., Li, C., Yang, W., Tan, H., Yi, S. V., Yang, X., et al. (2013). The genome of a socially polymorphic halictid bee, Lasioglossum albipes. Genome Biol. 14:R142. doi: 10.1186/gb-2013-14-12-r142
68. Kocher, S. D., Pellissier, L., Veller, C. M., Purcell, J., Nowak, M. A., Chapuisat, M., et al. (2015). Transitions in social complexity along altitudinal gradients reveal a combined impact of altitude and season length on social evolution. Proc. R. Soc. B. 281:20140627. doi: 10.1098/rspb.2014.0627
69. Kranz, B. D., Schwarz, M. P., Morris, D. C., and Crespi, B. J. (2002). Life history of Kladothrips ellobus and Oncothrips rodwayi: insight into the origin and loss of soldiers in gall-inducing thrips. Ecol. Entomol. 27, 49–57. doi: 10.1046/j.1365-2311.2002.0380a.x
70. Krause, E. T., Krüger, O., Kohlmeier, P., and Caspers, B. A. (2012). Olfactory kin recognition in a songbird. Biol. Lett. 8, 327–329. doi: 10.1098/rsbl.2011.1093
71. Kurvers, R. H. J. M., Krause, J., Croft, D. P., Wilson, A. D. M., and Wolf, M. (2014). The evolutionary and ecological consequences of animal social networks: emerging issues. Trends Ecol. Evol. 29, 326–335. doi: 10.1016/j.tree.2014.04.002
72. Kvarnemo, C., and Simmons, L. W. (2013). Polyandry as a mediator of sexual selection before and after mating. Phil. Trans. Royal Soc. B 368, 20120042. doi: 10.1098/rstb.2012.0042
73. Leadbeater, E., Carruthers, J. M., Green, J. P., Rosser, N. S., and Field, J. (2011). Nest inheritance is the missing source of direct fitness in a primitively eusocial insect. Science 333, 874–876. doi: 10.1126/science.1205140
74. Leclaire, S., Nielsen, J. F., Thavarajah, N. K., Manser, M., and Clutton-Brock, T. H. (2013). Odour-based kin discrimination in the cooperatively breeding

meerkat. Biol. Lett. 9, 20121054. doi: 10.1098/rsbl.2012.1054

75. Leggett, H. C., Brown, S. P., and Reece, S. E. (2014). War and peace: social interactions in infections. Phil. Trans. R. Soc. B 369, 20130365. doi: 10.1098/rstb.2013.0365

76. Leggett, H. C., El Mouden, C., Wild, G., and West, S. (2012). Promiscuity and the evolution of cooperative breeding. Proc. R. Soc. B 279, 1405–1411. doi: 10.1098/rspb.2011.1627

77. Leonard, J. E., and Boake, C. R. B. (2008). Associations between male courtship and female polyandry in three species of wasp, Nasonia (Hymenoptera: Pteromalidae). Anim. Behav. 76, 637–647. doi: 10.1016/j.anbehav.2008.04.013

78. Le Vin, A. L., Mable, B. K., and Arnold, K. E. (2010). Kin recognition via phenotype matching in a cooperatively breeding cichlid, Neolamprologus pulcher. Anim. Behav. 79, 1109–1114. doi: 10.1016/j.anbehav.2010.02.006

79. Levin, S. R., Brock, D. A., Queller, D. C., and Strassmann, J. E. (2015). Concurrent coevolution of intra-organismal cheaters and resisters. J. Evol. Biol. 28, 756–765. doi: 10.1111/jeb.12618

80. Libbrecht, R., Oxley, P. R., Kronauer, D. J. C., and Keller, L. (2013). Ant genomics sheds light on the molecular regulation of social organization. Genome Biol. 14:212. doi: 10.1186/gb-2013-14-7-212

81. Liebert, A. E., Nonacs, P., and Wayne, R. K. (2005). Solitary nesting and reproductive success in the paper wasp Polistes aurifer. Behav. Ecol. Sociobiol. 57, 445–456. doi: 10.1007/s00265-004-0875-5

82. Linksvayer, T. A., Fewell, J. H., Gadau, J., and Laubichler, M. D. (2012). Develop- mental evolution in social insects: regulatory networks from genes to societies. J. Exp. Zool. B Mol. Dev. Evol. 318, 159–169. doi: 10.1002/jez.b.22001

83. Lopez, M. A., Nguyen, H. T., Oberholzer, M., and Hill, K. L. (2011). Social parasites. Curr. Op. Microbiol. 14, 642–648. doi: 10.1016/j.mib.2011.09.012

84. Lukas, D., and Clutton-Brock, T. H. (2012). Cooperative breeding and monogamy in mammalian societies. Proc. R Soc. B 279, 2151–2156. doi: 10.1098/rspb.2011.2468

85. Mateo, J. M. (2003). Kin recognition in ground squirrels and other rodents. J. Mammal. 84, 1163–1181. doi: 10.1644/BLe-011

86. McDonald, P. G., and Wright, J. (2011). Bell miner provisioning calls are more similar among relatives and are used by helpers at the nest to bias their effort towards kin. Proc. R. Soc. B 278, 3403–3411. doi: 10.1098/rspb.2011.0307

87. McGlothlin, J. W., Wolf, J. B., Brodie, E. D. III, and Moore, A. J. (2014). Quantitative genetic versions of Hamilton›s rule with empirical applications. Phil. Trans. R. Soc. B 369, 20130358. doi: 10.1098/rstb.2013.0358

88. Michod, R. E., and Roze, D. (2001). Cooperation and conflict in the evolution of multicellularity. Heredity 86, 1–7. doi: 10.1046/j.1365-2540.2001.00808.x

89. Mulder, R. A., and Langmore, N. E. (1993). Dominant males punish helpers for temporary defection in superb fairy-wrens. Anim. Behav. 45, 830–833. doi: 10.1006/anbe.1993.1100

90. Nonacs, P. (2014). Resolving the evolution of sterile worker castes: a window on the advantages and disadvantages of monogamy. Biol. Lett. 10, 20140089. doi: 10.1098/rsbl.2014.0089

91. Nonacs, P., and Hager, R. (2011). The past, present and future of reproductive skew theory and experiments. Biol. Rev. 86, 271–298. doi: 10.1111/j.1469-185X.2010.00144.x

92. Nonacs, P., and Richards, M. H. (2015). How (not) to review papers on inclusive fitness. Trends Ecol. Evol. 30, 235–237. doi: 10.1016/j.tree.2015.02.007

93. Oi, C. A., Van Oystaeyen, A., Oliveira, R. C., Millar, J. G., Kevin, J., Verstrepen, K. J., et al. (2015b). Dual effect of wasp queen pheromone in regulating insect sociality. Curr. Biol. 25, 1638–1640. doi: 10.1016/j.cub.2015.04.040

94. Oi, C. A., van Zweden, J. S., Oliveira, R. C., Van Oystaeyen, A., Nascimento, F. S., and Wenseleers, T. (2015a). The origin and evolution of social insect queen pheromones: novel hypotheses and outstanding problems. Bioessays 37, 808–821. doi: 10.1002/bies.201400180

95. Oliveira, N. M., Niehus, R., and Foster, K. R. (2014). Evolutionary limits to cooperation in microbial communities. Proc. Natl. Acad. Sci. U.S.A. 111, 17941–17946. doi: 10.1073/pnas.1412673111

96. Page, R. E. Jr., and Amdam, G. V. (2007). The making of a social insect: developmental architectures of social design. Bioessays 29, 334–343. doi: 10.1002/bies.20549

97. Parker, G. A., and Birkhead, T. R. (2013). Polyandry: the history of a revolution. Phil. Trans. R. Soc. B 368, 20120335. doi: 10.1098/rstb.2012.0335

98. Parker, T. H., and Nakagawa, S. (2014). Mitigating the epidemic of type I error: ecology and evolution can learn from other disciplines. Front. Ecol. Evol. 2:76. doi: 10.3389/fevo.2014.00076

99. Peso, M., Elgar, M. A., and Barron, A. B. (2015). Pheromonal control: reconciling physiological mechanism with signalling theory. Biol. Rev. 90, 542–559. doi: 10.1111/brv.12123

100. Pierce, N. E., Braby, M. F., Heath, A., Lohman, D. J., Mathew, J., Rand, D. B., et al. (2002). The ecology and evolution of ant association in the Lycaenidae (Lepidoptera). Ann. Rev. Entomol. 47, 733–771. doi: 10.1146/annurev.ento.47.091201.145257

101. Popat, R., Crusz, S. A., Messina, M., Williams, P., West, S. A., and Diggle, S. P. (2012). Quorum-sensing and cheating in bacterial biofilms. Proc. R. Soc. B 279, 4765–4771. doi: 10.1098/rspb.2012.1976

102. Port, M., Kappeler, P. M., and Johnstone, R. A. (2011). Communal defense of territories and the evolution of sociality. Amer. Nat. 178, 787–800. doi: 10.1086/662672

103. Pryke, S. R., Rollins, L. A., and Griffith, S. C. (2010). Females use multiple mating and genetically loaded sperm

competition to target compatible genes. Science 329, 964–967. doi: 10.1126/science.1192407

104. Queller, D. C., and Strassmann, J. E. (1998). Kin selection and social insects. Bioscience 48, 165–175. doi: 10.2307/1313262

105. Raihani, N. J., Thornton, A., and Bshary, R. (2012). Punishment and cooperation in nature. Trends Ecol. Evol. 27, xxx. doi: 10.1016/j.tree.2011.12.004

106. Rankin, D. J., Bargum, K., and Kokko, H. (2007). The tragedy of the commons in evolutionary biology. Trends Ecol. Evol. 22, 643–651. doi: 10.1016/j.tree.2007.07.009

107. Ratnieks, F. L. W., Foster, K. R., and Wenseleers, T. (2011). Darwin›s special difficulty: the evolution of "neuter insects" and current theory. Behav. Ecol. Soc. Biol. 65, 481–492. doi: 10.1007/s00265-010-1124-8

108. Ratnieks, F. L. W., and Wenseleers, T. (2008). Altruism in insect societies and beyond: voluntary or enforced? Trends Ecol. Evol. 23, 45–52. doi: 10.1016/j.tree.2007.09.013

109. Rehan, S. M., Richards, M. H., Adams, M., and Schwarz, M. P. (2014). The costs and benefits of sociality in a facultatively social bee. Anim. Behav. 97, 77–85. doi: 10.1016/j.anbehav.2014.08.021

110. Rendall, D., Rodman, P. S., and Emond, R. E. (1996). Vocal recognition of individuals and kin in free-ranging rhesus monkeys. Anim. Behav. 51, 1007–1015. doi: 10.1006/anbe.1996.0103

111. Richard, F. –J., and Hunt, J. H. (2013). Intracolony chemical communication in social insects. Insect. Soc. 60, 275–291. doi: 10.1007/s00040-013-0306-6

112. Riehl, C. (2013). Evolutionary routes to non-kin cooperative breeding in birds. Proc. R. Soc. B 280, 20132245. doi: 10.1098/rspb.2013.2245

113. Ross, L., Gardner, A., Hardy, N., and West, S. A. (2013). Ecology, not the genetics of sex determination, determines who helps in eusocial populations. Curr. Biol. 23, 2383–2387. doi: 10.1016/j.cub.2013.10.013

114. Rubenstein, D. R. (2012). Family feuds: social competition and sexual conflict in complex societies. Phil. Trans. R. Soc. B 367, 2304–2313. doi: 10.1098/rstb.2011.0283

115. Russell, A. F., and Lummaa, V. (2009). Maternal effects in cooperative breeders: from hymenopterans to humans. Phil. Trans. R. Soc. B 364, 1143–1167. doi: 10.1098/rstb.2008.0298

116. Ruxton, G. D., Humphries, S., Morrell, L. J., and Wilkinson, D. M. (2014). Why is eusociality an almost exclusively terrestrial phenomenon? J. Anim. Ecol. 83, 1248–1255. doi: 10.1111/1365-2656.12251

117. Schmid-Hempel, P., and Crozier, R. H. (1999). Polyandry versus polygyny versus parasites. Phil. Trans. R. Soc. B, 354, 507–515. doi: 10.1098/rstb.1999.0401

118. Schneider, J. M., and Bilde, T. (2008). Benefits of cooperation with genetic kin in a subsocial spider. Proc. Natl. Acad. Sci. U.S.A. 105, 10843–10846. doi: 10.1073/pnas.0804126105

119. Schwarz, M. P., Tierney, S. M., Rehan, S. M., Chenoweth, L. B., and Cooper, S. J. B. (2011). The evolution of eusociality in allodapine bees: workers began by waiting. Biol. Lett. 7, 277–280. doi: 10.1098/rsbl.2010.0757

120. Seeley, T. D. (2010). Honeybee Democracy. (Princeton, NJ: Princeton University Press).

121. Seger, J. (1991). "Cooperation and conflict in social insects," in Behavioural Ecology: an Evolutionary Approach, 3rd Edn., eds J. R. Krebs and N. B. Davies (Oxford: Blackwells), 338–373.

122. Sharp, S. P., McGowan, A., Wood, M. J., and Hatchwell, B. J. (2005). Learned kin recognition cues in a social bird. Nature 434, 1127–1130. doi: 10.1038/nature03522

123. Sigmund, K. (2010). The Calculus of Selfishness. (Princeton, NJ: Princeton University Press).

124. Strassmann, J., and Queller, D. C. (2011). Evolution of cooperation and control of cheating in a social microbe. Proc. Natl. Acad. Sci. U.S.A. 108, 10855–10862. doi: 10.1073/pnas.1102451108

125. Sturgis, S. J., and Gordon, D. M. (2012). Nestmate recognition in ants (Hymenoptera: Formicidae): a review. Myrmecol. News 16, 101–110.

126. Székely, T., Moore, A. J., and Komdeur, J. (2010). Social Behaviour: Genes, Ecology and Evolution. (Cambridge: Cambridge University Press).

127. Trivers, R. L. (1985). Social Evolution. (San Francisco, CA: Benjamin/Cummings Co).

128. Google Scholar

129. Trivers, R. L., and Hare, H. (1976). Haplodiploidy and the evolution of the social insects. Science191, 249–263. doi: 10.1126/science.1108197

130. Trumbo, S. T. (2013). Maternal care, iteroparity and the evolution of social behavior: a critique of the semelparity hypothesis. Evol. Biol. 40, 613–626. doi: 10.1007/s11692-013-9237-4

131. Vail, A. L., Manica, A., and Bshary, R. (2013). Referential gestures in fish collaborative hunting. Nat. Commun. 4:1765. doi: 10.1038/ncomms2781

132. Van Cleve, J., and Akçay, E. (2014). Pathways to social evolution: reciprocity, relatedness, and synergy. Evolution 68, 2245–2258. doi: 10.1111/evo.12438

133. Van Dyken, J. D., and Wade, M. J. (2012a). Origins of altruism diversity I: the diverse ecological roles of altruistic strategies and their evolutionary responses to local competition. Evolution 66, 2484–2497. doi: 10.1111/j.1558-5646.2012.01630.x

134. Van Dyken, J. D., and Wade, M. J. (2012b). Origins of altruism diversity II: runaway coevolution of altruistic strategies via "reciprocal niche construction". Evolution 66, 2498–2513. doi: 10.1111/j.1558-5646.2012.01629.x

135. Van Oystaeyen, A., Oliveira, R. C., Holman, H., van Zweden, J. S., Romero, C., Oi, C. A., et al. (2014). Conserved class of queen pheromones stops social insect workers from reproducing. Science343, 287–290. doi: 10.1126/science.1244899

136. van Wilgenburg, E., and Elgar, M. A. (2013). Confirmation bias in studies of nestmate recognition:

a cautionary note for research into the behaviour of animals. PLoS ONE 8:e53548. doi: 10.1371/journal.pone.0053548

137. van Zweden, J. S., Bonckaert, W., Wenseleers, T., and d›Ettorre, P. (2013). Queen signaling in social wasps. Evolution 68, 976–986. doi: 10.1111/evo.12314

138. van Zweden, J. S., and d›Ettorre, P. (2010). "Nestmate recognition in social insects and the role of hydrocarbons," in Insect Hydrocarbons: Biology, Biochemistry, and Chemical Ecology, eds G. J. Blomquist, A.-G. Bagnéres (Cambridge: Cambridge University Press), 222–243.

139. Webster, N. S. (2014). Cooperation, communication, and co-evolution: grand challenges in microbial symbiosis research. Front. Microbiol. 5:164. doi: 10.3389/fmicb.2014.00164

140. Wenseleers, T., and Ratnieks, F. L. W. (2006). Comparative analysis of worker reproduction and policing in eusocial hymenoptera supports relatedness theory. Am. Nat. 168, E163–E179. doi: 10.1086/508619

141. While, G. M., Uller, T., and Wapstra, E. (2009). Family conflict and the evolution of sociality in reptiles. Behav. Ecol. 20, 245–250. doi: 10.1093/beheco/arp015

142. Whitehouse, M. E. A., and Lubin, Y. (2005). The functions of societies and the evolution of group living: spider societies as a test case. Biol. Rev. 80, 347–361. doi: 10.1017/S1464793104006694

143. Wilson, E. O. (1971). Insect Societies. Cambridge: Harvard University Press.

144. Wilson, E. O. (1975). Sociobiology: The New Synthesis. Cambridge: Harvard University Press.

145. Yan, H., Bonasio, R., Simola, D. F., Liebig, J., Berger, S. L., and Reinberg, D. (2015). DNA methylation in social insects: how epigenetics can control behavior and longevity. Annu. Rev. Entomol. 60, 435–452. doi: 10.1146/annurev-ento-010814-020803

146. Yip, E. C., and Rayor, L. S. (2014). Maternal care and subsocial behaviour in spiders. Biol. Rev. 89, 427–449. doi: 10.1111/brv.12060

147. Zanette, L. R. S., Miller, S. D. L., Faria, C. M. A., Almond, E. J., Huggins, T. J., Jordan, W. C., et al. (2012). Reproductive conflict in bumblebees and the evolution of worker policing. Evolution 66, 3765–3777. doi: 10.1111/j.1558-5646.2012.01709.x

148. Zhang, G., Li, C., Li, Q., Li, B., Larkin, D. M., Lee, C., et al. (2014). Comparative genomics reveals insights into avian genome evolution and adaptation. Science 346, 1311–1320. doi: 10.1126/science.1251385

149. Zhou, X., Rokas, A., Berger, S. L., Liebig, J., Ray, A., and Zwiebel, L. J. (2015). Chemoreceptor evolution in hymenoptera and its implications for the evolution of eusociality. Genome Biol. Evol. 7, 2407–2416. doi: 10.1093/gbe/evv149

150. Zuk, M., Garcia-Gonzalez, F., Herberstein, M. E., and Simmons, L. W. (2014). Model systems, taxonomic bias, and sexual selection: beyond Drosophila. Annu. Rev. Entomol. 59, 321–338. doi: 10.1146/annurev-ento-011613-162014

INDEX

O

P

Q

R

S

T

U

V